Pocket Companion to

Small Animal Clinical Nutrition

4th Edition

Editors

Editors for Pocket Companion to Small Animal Clinical Nutrition, 4th ed.
Michael S. Hand, DVM, PhD
Diplomate, American College of Veterinary Nutrition
Mesilla, New Mexico

Bruce J. Novotny, DVM
Helios Communications, LLC, Shawnee, Kansas

Editors for Small Animal Clinical Nutrition, 4th ed.
Michael S. Hand, DVM, PhD
Diplomate, American College of Veterinary Nutrition
Vice President of Research (Emeritus), Hill's Science and Technology Center,
Topeka, Kansas
Chair, Board of Directors, Mark Morris Institute, Topeka, Kansas

Craig D. Thatcher, DVM, PhD
Diplomate, American College of Veterinary Nutrition
Professor and Head, Department of Large Animal Clinical Sciences,
Virginia-Maryland Regional College of Veterinary Medicine,
Virginia Polytechnic Institute and State University, Blacksburg, Virginia

Rebecca L. Remillard, PhD, DVM
Diplomate, American College of Veterinary Nutrition
Staff Nutritionist, Angell Memorial Animal Hospital, Boston, Massachusetts
Clinical Assistant Professor, School of Veterinary Medicine, Tufts University,
North Grafton, Massachusetts

Philip Roudebush, DVM
Diplomate, American College of Veterinary Internal Medicine (Internal Medicine)
Veterinary Fellow, Hill's Science and Technology Center, Topeka, Kansas
Adjunct Professor, Department of Clinical Sciences, College of Veterinary Medicine,
Kansas State University, Manhattan
Associate, Mark Morris Institute, Topeka, Kansas

Consulting Editors for Small Animal Clinical Nutrition, 4th ed.
Mark L. Morris, Jr., DVM, PhD
Diplomate, American College of Veterinary Nutrition, Topeka, Kansas

Bruce J. Novotny, DVM
Helios Communications, LLC, Shawnee, Kansas

MARK MORRIS INSTITUTE

Pocket Companion
to
Small Animal Clinical Nutrition, 4th Edition

ISBN 0-945837-06-2
Library of Congress Catalog Card Number: 2002100212

Printed in the United States of America.

For more information about this book contact:
Mark Morris Institute
P.O. Box 2097
Topeka, Kansas 66601-2097
Phone 785-286-8101
Facsimile 785-286-8173

Last digit is the print number: 9 8 7 6 5 4 3 2 1

Preface

The fourth edition of *Small Animal Clinical Nutrition* was published in January 2000 with the goal of providing basic and applied information about small animal clinical nutrition to veterinary students and veterinarians worldwide. The book contained many features to promote learning and ease of use. Most chapters were organized according to a two-step iterative process (See Chapter 1.). Two years after its publication, more than 54,000 English, Spanish and Japanese copies have been distributed.

While it would be ideal to have a copy of *Small Animal Clinical Nutrition, 4th ed.* available at all times, the sheer bulk and weight of the book make that impossible. Thus, the editors condensed relevant content from *Small Animal Clinical Nutrition, 4th ed.* into the less cumbersome *Pocket Companion* that you now hold. The goal was to preserve the iterative process in a quick, pocket-sized reference for in-clinic use.

The *Pocket Companion* contains abbreviated information from 26 of the 30 chapters and four of the 22 appendices in *Small Animal Clinical Nutrition, 4th ed.* Key information is included to help readers rapidly select appropriate foods for pets in health and disease. Numerous links to the parent book are included to facilitate in-depth study. These connections include page numbers in major headings and at the end of chapters directing readers to where more information and cases can be found in *Small Animal Clinical Nutrition, 4th ed.* to reinforce the content included in the *Pocket Companion to Small Animal Clinical Nutrition, 4th ed.* Much information from *Small Animal Clinical Nutrition, 4th ed.* has not been included in the *Pocket Companion* for space considerations. This unduplicated information includes many tables, figures and sidebars; four chapters and 18 appendices have been omitted in their entirety. Cases to reinforce the clinical nutrition process can only be found in *Small Animal Clinical Nutrition, 4th ed.* Furthermore, references directing where readers can go for more information are only published in the parent book. Likewise, credits for tables and figures are only published in *Small Animal Clinical Nutrition, 4th ed.*

It is important to emphasize that this *Pocket Companion* is meant to complement not replace *Small Animal Clinical Nutrition, 4th ed.* The *Pocket Companion* is designed to be used in clinics and other environments when a brief introduction to a topic is needed and time doesn't permit consultation with *Small Animal Clinical Nutrition, 4th ed.* Thus, readers are advised to consult *Small Animal Clinical Nutrition, 4th ed.* for in-depth information about clinical nutrition. The editors consider *Small Animal Clinical Nutrition, 4th ed.* and the *Pocket Companion to Small Animal Clinical Nutrition, 4th ed.* a single, complementary education package.

THE EDITORS

Notice

Companion animal practice, clinical nutrition and the commercial pet food industry are ever changing fields. The editors of this *Pocket Companion to Small Animal Clinical Nutrition, 4th ed.* have carefully checked the trade names and nutrient levels of commercial pet foods and verified food and drug dosages to ensure that information is precise and in accordance with standards accepted at the time of publication. Readers are advised, however, to check the product information currently provided by the manufacturer of each food or drug to be administered to ensure that changes have not been made in the nutrient profile, recommended feeding guide or in the contraindications for administration. This caution is particularly important in regard to new or infrequently used foods or drugs. It is the responsibility of those recommending a food or administering a drug, relying on their professional skill and experience, to determine the best treatment for the patient and the appropriate food, food dosage, drug and drug dosage. The publisher and editors cannot be responsible for misuse or misapplication of the material in this book.

Names and nutrient information of commercial pet foods in this book are based on manufacturer's published information or product analysis. Readers are advised, however, to check product information currently provided by the manufacturer. Individual products listed in the chapters and appendices are based on overall market share and availability of product information.

THE EDITORS

Contents

(Continued on next page.)

Contents

A four-color anatomic guide to common clinical conditions is found between pages 408 and 409.

*For in-depth review, see the same chapter in *Small Animal Clinical Nutrition, 4th ed.*

CHAPTER 1

Small Animal Clinical Nutrition: An Iterative Process

For a review of the unabridged chapter, see Thatcher CD, Hand MS, Remillard RL. Small Animal Clinical Nutrition: An Iterative Process. In: Hand MS, Thatcher CD, Remillard RL, et al, eds. Small Animal Clinical Nutrition, 4th ed. Topeka, KS: Mark Morris Institute, 2000; 1-19.

THE TWO-STEP ITERATIVE PROCESS OF CLINICAL NUTRITION (Page 3)*

Figure 1-1 depicts the two-step clinical nutrition process. The American College of Veterinary Nutrition (ACVN) has recommended that nutrition problem solving include assessment of the patient, the food and the feeding method. This assessment process is step one. Development of a feeding plan is step two and includes recommendations for food and feeding methods. If the assessment process indicates that the current food and feeding method are appropriate, the current feeding plan can remain in place. However, if the assessment indicates otherwise, a new feeding plan should be formulated and implemented.

After a suitable period of time (the length of which depends on the patient's condition), the two-step process is repeated to determine the appropriateness or effectiveness of the new feeding plan. Thus, the patient is reassessed and, if its status has changed, a new feeding plan is implemented. This is the iterative or repetitive part of the process. Any number of iterations or repetitions of the two-step process can occur, depending on the needs of each patient. A critically ill patient may need to be reassessed every few hours, whereas a normal adult dog or cat may be reassessed annually. The subsequent reassessment of the patient at each cycle is also referred to as monitoring.

ASSESSMENT (Pages 3-14)

Assess the Animal (Pages 3-8)

The goal of animal assessment is to establish a dog's or cat's nutrient needs and feeding goals in light of its physiologic or disease condition. The animal's nutrient needs are the benchmark for assessing the ani-

---

*Page numbers in headings refer to Small Animal Clinical Nutrition, 4th ed., where additional information may be found.

mal's food. Assessment of dogs and cats to determine their nutritional status should be a structured process that includes: 1) review of the history and medical record, 2) physical examination, 3) laboratory tests and other diagnostic procedures and 4) estimation of the target levels for the key nutritional factors based on the patient's physiologic state and medical diagnosis.

Obtain an Accurate History and Review the Medical Record
(Pages 3-6)
Obtaining the animal's history and reviewing the medical record help determine the nutritional status of the patient. The signalment is part of the history and defines the patient's physiologic state and includes: 1) species, 2) breed, 3) age, 4) gender, 5) reproductive status, 6) activity level and 7) environment.

A complete history should also include questions about the pet's weight and therapies (medical, surgical, etc.) that may affect appetite, nutrient metabolism or both. An accurate description of the current feeding plan, including the animal's food, eating and drinking habits and feeding methods should be obtained from the client. Intakes of treats and nutritional supplements should be recorded.

Review of the medical record provides objective historical information and documents the pet's previous health status, health maintenance procedures that were performed and medications that were prescribed. Veterinarians should evaluate this information to determine if any of these factors are related to the animal's current nutritional status. This review permits early nutritional intervention in the treatment of established malnutrition (under- or overnutrition) and in the prevention of malnutrition in individuals at risk.

Conduct a Physical Examination (Page 6)
A thorough physical examination can help define an animal's nutritional status as well as identify diseases that may have a nutritional component. Physical findings should be recorded in the patient's medical record. Veterinarians should examine each body system for problems that are responsive to nutritional intervention. An animal's body condition will likely reflect abnormalities of major organ systems.

Body condition can be subjectively assessed by a process called body condition scoring. In general, this process assesses a patient's fat stores and, to a lesser extent, muscle mass. Fat cover is evaluated over the ribs, down the topline, around the tailbase and ventrally along the abdomen. Body condition score (BCS) descriptors have been developed with respect to the species (dogs and cats) and age of the patient (**Figures 1-2** and **1-3**). Score descriptors vary due to the structural differences

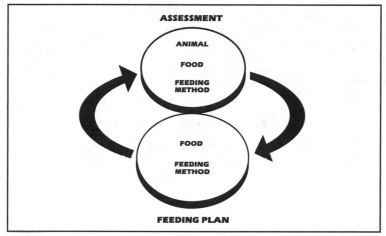

Figure 1-1. The two-step process of veterinary clinical nutrition.

between species and between young and adult pets. The scores range from 1 to 5 with 1 being very thin, 5 being grossly obese; 3 is ideal.

A history of rapid weight loss and a reduced BCS may indicate a catabolic condition with a marked loss of lean tissue, dehydration or both. A history of progressive weight gain and an increased BCS may indicate an anabolic condition with an excessive accumulation of fat, water or both.

Conduct Necessary Laboratory Tests and Other Diagnostics
(Pages 6-7)

No single laboratory test or other diagnostic procedure can accurately assess a patient's nutritional status. Routine complete blood counts, urinalyses and biochemistry profiles, however, can provide insight into the presence of metabolic disorders and other diseases. Albumin concentration, lymphocyte count, packed cell volume and serum total protein values may serve as general indicators of nutritional status. Other chapters in this pocket companion and the textbook from which it was derived will discuss specific laboratory tests and other diagnostic procedures that may help assess healthy and sick patients.

Determine the Key Nutritional Factors and Their Target Levels
(Pages 7-8)

The concept of key nutritional factors is fundamental to the practical application of clinical nutrition.

BCS 1. Very thin. The ribs are easily palpable with no fat cover. The tailbase has a prominent raised bony structure with no tissue between the skin and bone. The bony prominences are easily felt with no overlying fat. Dogs over six months of age have a severe abdominal tuck when viewed from the side and an accentuated hourglass shape when viewed from above.

BCS 2. Underweight. The ribs are easily palpable with minimal fat cover. The tailbase has a raised bony structure with little tissue between the skin and bone. The bony prominences are easily felt with minimal overlying fat. Dogs over six months of age have an abdominal tuck when viewed from the side and a marked hourglass shape when viewed from above.

BCS 3. Ideal. The ribs are palpable with a slight fat cover. The tailbase has a smooth contour or some thickening. The bony structures are palpable under a thin layer of fat between the skin and bone. The bony prominences are easily felt under minimal amounts of overlying fat. Dogs over six months of age have a slight abdominal tuck when viewed from the side and a well-proportioned lumbar waist when viewed from above.

BCS 4. Overweight. The ribs are difficult to feel with moderate fat cover. The tailbase has some thickening with moderate amounts of tissue between the skin and bone. The bony structures can still be palpated. A moderate layer of fat covers the bony prominences. Dogs over six months of age have little or no abdominal tuck or waist when viewed from the side. The back is slightly broadened when viewed from above.

BCS 5. Obese. The ribs are very difficult to feel under a thick fat cover. The tailbase appears thickened and is difficult to feel under a prominent layer of fat. A moderate to thick layer of fat covers the bony prominences. Dogs over six months of age have a pendulous ventral bulge and no waist when viewed from the side due to extensive fat deposits. The back is markedly broadened when viewed from above. A trough may form when epaxial areas bulge dorsally.

Figure 1-2. Body condition score (BCS) descriptors for dogs in a five-point system.

PROCESS

BCS 1. Very thin. The ribs are easily palpable with no fat cover. The bony prominences are easily felt with no overlying fat. Cats over six months of age have a severe abdominal tuck when viewed from the side and an accentuated hourglass shape when viewed from above.

BCS 2. Underweight. The ribs are easily palpable with minimal fat cover. The bony prominences are easily felt with minimal overlying fat. Cats over six months of age have an abdominal tuck when viewed from the side and a marked hourglass shape when viewed from above.

BCS 3. Ideal. The ribs are palpable with a slight fat cover. The bony prominences are easily felt under a slight amount of overlying fat. Cats over six months of age have an abdominal tuck when viewed from the side and a well-proportioned lumbar waist when viewed from above.

BCS 4. Overweight. The ribs are difficult to feel with moderate fat cover. The bony structures can still be palpated. A moderate layer of fat covers the bony prominences. Cats over six months of age have little or no abdominal tuck or waist when viewed from the side. The back is slightly broadened when viewed from above. A moderate abdominal fat pad is present.

BCS 5. Obese. The ribs are very difficult to feel under a thick fat cover. A moderate to thick layer of fat covers the bony prominences. Cats over six months of age have a pendulous ventral bulge and no waist when viewed from the side due to extensive fat deposits. The back is markedly broadened when viewed from above. A marked abdominal fat pad is present. Fat deposits may be found on the limbs and face.

Figure 1-3. Body condition score (BCS) descriptors for cats in a five-point system.

In 1992 and 1993, the Association of American Feed Control Officials (AAFCO) published recommended nutrient profiles for dog and cat foods, respectively. These nutrient profiles have been republished yearly and have replaced the National Research Council (NRC) bulletins as the official source for nutrient profiles for dog and cat foods in the United States.

The AAFCO nutrient profiles include safety factors similar to those in the recommended dietary allowances (RDAs) that have been established for people. These safety factors compensate for changes in a food's nutrient availability due to ingredient and processing variables and for individual differences in nutrient requirements within dog and cat populations. Because of these safety factors, the term "allowance" is better suited to describe AAFCO values than "requirements." AAFCO values are adequate to meet the known nutrient needs of almost all healthy dogs and cats and are a better source of feeding recommendations for most dogs and cats than are minimum requirements. Besides recommendations for lower limits, AAFCO prescribes upper limits for certain nutrients with the obvious implication that some nutrient excesses can be harmful. As with RDAs for people, AAFCO allowances for pet food nutrient profiles are not necessarily optimums. No nutrient profiles have been established by AAFCO for geriatric dogs or cats or those with specific disease processes.

Key nutritional factors encompass nutrients of concern and other food characteristics. The concept of nutrients of concern greatly simplifies the approach to clinical nutrition because most commercial pet foods sold in the United States provide at least AAFCO allowances of all nutrients. Thus, if a commercial food is fed, veterinarians and their health care teams need only to understand and focus on delivering the target levels for a few nutrients (nutrients of concern) rather than all 43 nutrients currently recognized for cats and all 36 nutrients currently recognized for dogs.

Nutrients of concern encompass nutritional risk factors for disease treatment and prevention as well as nutrients that are key to optimizing normal physiologic processes such as growth, gestation, lactation and physical work. The following elements must be considered in determining key nutritional factors and their target levels: 1) the patient's lifestage and physiologic state, 2) environmental conditions such as temperature, housing and pet-to-pet competition, 3) the nature of any disease or injury, 4) the known nutrient losses through skin, urine and intestinal tract, 5) the interactions of medications and nutrients, if applicable, 6) the known capacity of the body to store certain nutrients and 7) the interrelationships of various nutrients.

Depending on the patient's needs, food characteristics other than the nutrient content may be important to consider. These characteristics include the food's influence on systemic acid-base balance and urinary pH, the food's texture, digestibility, osmolality and whether or not it contains few/novel protein sources or unique ingredients.

Chapters 9 through 11 and Appendix J (See Small Animal Clinical Nutrition, 4th ed., pp 1048-1063.) list key nutritional factors and their target levels for healthy dogs and cats. The key nutritional factors and their target levels for dogs and cats with specific disease complexes can be found in Chapters 12 through 26. Regardless of which nutrients are considered, the reader must understand how nutrient needs are expressed.

The three methods for expressing an animal's nutrient needs are: 1) dry matter, energy density defined, 2) energy basis and 3) absolute basis.

Dry matter basis, energy density defined is the percentage or quantity of a nutrient in the food's dry matter that is needed by the animal. It is the most common method of expressing an animal's nutrient needs. This measure describes what is required in a food and indicates an animal's nutrient needs. Dry matter refers to that weight of the food remaining when the water content is subtracted (**Tables 1-1** and **1-2** demonstrate methods of calculating dry matter). Dry matter values are most meaningful if the energy density of the food's dry matter is specified because most animals eat, or are fed, to meet their energy requirements.

Energy basis refers to the quantities of nutrients per animal's energy requirement. Units of measure are typically nutrient amounts per 100 kcal or 1 MJ metabolizable energy (ME). Occasionally an animal's protein, fat and soluble carbohydrate needs are expressed as a percentage of the animal's total energy needs.

Absolute basis refers to the unit measure (usually weight) of a nutrient that is needed by an animal in a 24-hour period. These needs are expressed as quantities per kg of body weight per day.

Assess the Food (Pages 9-11)

After the nutritional status of the patient has been assessed and the key nutritional factors and their target levels determined, the adequacy of the food is assessed. The components to food assessment include: 1) physical evaluation of the food, 2) evaluation of the product label, including the ingredients used and whether or not feeding tests were conducted, 3) evaluation of the food's nutrient content relative to the animal's nutrient needs (key nutritional factors) and 4) determination of the presence or absence of specific food characteristics.

Physical Evaluation (Page 9)

Physical evaluation of the food can provide information about package quality (which may or may not reflect product quality), consistency and presence or absence of extraneous materials. Physical evaluation of the food is probably most useful for assessing whether or not the food has spoiled. (See Chapter 7, Small Animal Clinical Nutrition, 4th ed., pp 183-198.)

Table 1-1. How to convert from as fed basis to dry matter basis.

Step 1. Obtain the food's dry matter content by subtracting the water content from the as fed amount of the food.
Example A: If a moist food contains 75% water, 25% of the food is dry matter: 100% as fed − 75% water = 25% food dry matter
Example B: If a dry food contains 10% water, 90% of the food is dry matter: 100% as fed − 10% water = 90% food dry matter

Step 2. Convert the percentage as fed nutrient content of the food to a dry matter basis by dividing the percentage of the nutrient content on an as fed basis by the percentage dry matter.
Example A: If the moist food above contained 10% protein on an as fed basis, on a dry matter basis it would contain 40% protein:
10% protein as fed basis ÷ 25% dry matter = 40% protein dry matter basis
Example B: If the dry food above contained 18% protein on an as fed basis, on a dry matter basis it would contain 20% protein:
18% protein as fed basis ÷ 90% dry matter = 20% protein dry matter basis

Table 1-2. Shorthand method for converting from as fed basis to dry matter basis.

A less accurate, shorthand method for converting from an as fed basis to a dry matter basis (**Table 1-1**) is to simply multiply the percentage nutrient content on an as fed basis by four for moist foods or add 10% for dry foods. This method is based on the assumption that moist foods contain approximately 75% water and dry foods contain approximately 10% water. Check the guaranteed analysis on the product label.
Example A: If a moist food contains 10% protein on an as fed basis, on a dry matter basis it would contain 40% protein:
10% protein as fed basis x 4 (factor for moist foods) = 40% protein dry matter basis*
Example B: If a dry food contains 18% protein on an as fed basis, on a dry matter basis it would contain 20% protein:
18% protein as fed basis + 10% (factor for dry food) = approximately 20% protein dry matter basis*

*Compare these results with those obtained in **Table 1-1** for moist and dry foods with the same moisture content.

Label Evaluation *(Page 9)*

Evaluation of the product label is also of limited value. (See Chapter 5, Small Animal Clinical Nutrition, 4th ed., pp 147-161.) The ingredient panel of the pet food label provides general information about which ingredients were used and their relative amounts. The ingredients used in the product are listed in descending order by weight in many countries. The ingredient panel can be useful if specific ingredients are contraindicated for certain animals or an owner has an ingredient concern. However, the quality of the ingredients cannot be determined from the label and there is much misinformation about pet food ingredients. (See Chapter 4, Small Animal Clinical Nutrition, 4th ed., pp 127-146.)

The three most useful components when assessing food are to: 1) ensure that the food has been tested or fed to animals, 2) determine the food's nutrient content (especially for the key nutritional factors) and 3) compare the food's nutrient content with the animal's nutrient needs (again, paying particular attention to the key nutritional factors).

FEEDING TESTS Whether or not commercial foods for healthy pets have been animal tested can usually be determined from the nutritional adequacy statement on the product's label. (See Chapter 5, **Table 5-2**, for examples of such statements.) However, some brands of these products have passed regulatory agency (AAFCO) prescribed feeding tests although the product label may not include such information.

Commercial pet foods that have undergone AAFCO-prescribed or similar feeding tests provide reasonable assurance of nutrient availability and sufficient palatability to ensure acceptability (i.e., food intake sufficient to meet nutrient needs). Feeding tests also provide some assurance that a product will adequately support certain functions such as gestation, lactation and growth. However, even controlled animal testing is not infallible.

In the United States, the AAFCO testing protocol for adult maintenance lasts six months, requires only eight animals per group and monitors a limited number of parameters. (See Chapter 5 and Appendix J, Small Animal Clinical Nutrition, 4th ed., pp 147-161 and 1048-1063.) Passing such tests does not ensure the food will be effective in preventing long-term nutrition/health problems or detect problems with prevalence rates less than 15%. Likewise, these protocols are not intended to ensure optimal growth or maximize physical activity. Thus, in addition to having passed AAFCO tests, the food should be evaluated to ensure that key nutritional factors are at levels appropriate for treatment of disease, for promotion of long-term health or for optimal performance. Few, if any, homemade recipes have been animal tested according to prescribed feeding protocols.

Determine the Food's Nutrient Content (Pages 9-10)

Appendices 2 and 3 provide partial nutrient profiles for selected commercial foods and treats sold in the United States, Canada and Europe. The levels of the nutrients of concern for most of the commonly used commercial foods are listed in tables in the individual chapters of this pocket companion and the textbook from which it was derived. In most instances, these profiles will provide the necessary food nutrient content information. If the food in question cannot be found in this book, refer to the sidebar "Four Ways to Determine the Nutrient Content of a Food" (See Small Animal Clinical Nutrition, 4th ed., p 7.) for other ways to determine its nutrient content.

As fed basis simply refers to the quantity of nutrients in a food as it is fed. This method ignores moisture and energy content. The units of measure are percentages or quantities of nutrients per unit weight (kg) of food.

Dry matter is that weight of the food remaining after the water content has been subtracted from the as fed amount. Dry matter basis, therefore, is the amount of nutrients in the food's dry matter. It accounts for variability in water content but not variability in energy density. The units of measure are percentages or quantities of nutrients per unit weight (kg) of food dry matter. The usefulness of dry matter basis is limited because the energy density of individual foods can vary widely. This consideration will be further explained below (dry matter basis, energy density defined). **Tables 1-1** and **1-2** show the conversion from as fed basis to dry matter basis.

Dry matter basis, energy density defined is the same as dry matter but specifies a food's energy density, thus accounting for potential variability. The units of measure are the same as those used with dry matter basis but are further qualified by expressing the energy density of the food. For example, recommended nutrient values for canine and feline foods (See Appendix J, Small Animal Clinical Nutrition, 4th ed., pp 1048-1063.) are based on an energy density of 3.5 and 4.0 kcal ME/g of food dry matter, respectively (14.64 and 16.74 kJ ME/g). Dry matter basis, energy density defined is probably the most widely used method of expressing a food's nutrient content.

Energy basis refers simply to the amount of nutrients per 100 kcal or 1 megajoule ME of food.

Both dry matter basis, energy density defined and energy basis are reasonably accurate methods of expressing a food's nutrient content.

Compare the Food's Nutrient Content with the Animal's Nutrient Needs (Pages 10-11)

When comparing a food's nutrient content with an animal's nutrient needs, methods of expressing nutrient content and nutrient/requirements (same units) must be compatible. Also, when using dry matter basis, energy density defined to compare foods or to compare foods with animal requirements, the energy densities must be the same for the comparisons to be meaningful. **Table 1-3** shows how to convert to the same energy density. In some cases it will be desirable to convert food nutrient content on an as fed basis to dry matter basis, energy density defined.

Comparing a food's nutrient content with the animal's nutrient needs will help identify any significant nutritional imbalances in the food being

Table 1-3. How to convert to the same energy density.

Correcting energy densities in order to make valid nutrient comparisons, either between foods or between a food and an animal's requirement, is based on the assumption that the relationship between nutrient content and energy density is directly proportional. A simple ratio can be established to generate a multiplier that converts the units of the animal's requirements to those of the food; then the animal's requirement and the food's nutrient content can be compared. The multiplier is obtained by dividing the energy density of the food by the requirement energy density.

Example: Is a food that provides 0.72% potassium and 4 kcal (16.74 kJ)/g, on a dry matter basis, adequate for canine adult maintenance?
1) The requirement for potassium is 0.6% (dry matter basis) in an adult dog food that provides 3.5 kcal (14.64 kJ)/g.
2) Convert the requirement to the same energy density as the food by generating the multiplier.
   Multiplier
   = food energy density ÷ requirement energy density
   = 4.0 kcal (16.74 kJ)/g dry matter ÷ 3.5 kcal (14.64 kJ)/g dry matter
   = 1.14
3) To obtain the equivalent nutrient requirement for a food providing 4 kcal (14.74 kJ)/g, on a dry matter basis, multiply the requirement by the multiplier.
   Equivalent nutrient requirement
   = 1.14 x 0.06% potassium
   = 0.68% potassium, 4 kcal (14.74 kJ)/g, on a dry matter basis
4) The amount of potassium in the food (0.72%) is compared to the animal's equivalent nutrient requirement (0.68%) and is found to be adequate.
5) The multiplier obtained above (1.14) can be used to convert the other nutrient requirements to the same basis as the food to compare the adequacy of their levels, if desired.

Once the energy densities of the food and the animal's needs are converted to the same units, the comparison is simple as shown in the following example. If a cat food supplies 0.27% magnesium on a dry matter basis (4.0 kcal [16.74 kJ] ME/g food dry matter) and the cat requires 0.04% magnesium on a dry matter basis (4.0 kcal [16.74 kJ] ME/g food dry matter), the food contains excess magnesium.

fed. This comparison is fundamental to determining whether or not to feed a different food.

Compare Specific Food Characteristics with the Animal's Needs (Page 11)

Besides requiring specific levels of certain nutrients, some patients have other food-related needs. These needs might include management of acute or chronic acidosis, maintenance of a specific urinary pH range, certain kibble texture, a specific range of digestibility or osmolality, specific ingredients and avoidance of certain protein sources. As mentioned earlier, the presence or absence of specific protein sources or other ingredients in a food can be obtained from the product label. Other information about food characteristics should be available from product manufacturers. Pet food labels contain addresses and toll-free phone numbers of the manufacturer. When available and where applicable, this information is included in the individual chapters of this pocket companion and the textbook from which it was derived and Appendices 2 and 3.

Assess the Feeding Method (Pages 11-14)

Feeding methods relate directly to the physiologic or disease state of the animal and the food or foods being fed. Thus, the information obtained by assessing the animal and the food is fundamental to assessing the feeding method. There are at least three things to consider regarding feeding methods: 1) feeding route, 2) amount fed and 3) how the food is offered (when, where, by whom and how often).

Feeding Route (Pages 11-12)

Whether or not the feeding route is appropriate depends on the animal's condition. Although most animals are able to feed themselves, orphans and some critical care patients may require assistance. Assisted-feeding methods are described in detail in Chapter 12 and Appendix 4. Assisted-feeding methods include enteral feeding by syringe or tube (several approaches) and parenteral feeding.

Amount Fed (Pages 12-13)

The nutrient needs of an animal are met by a combination of the nutrient levels in the food and the amount of food fed. Even if a food has an appropriate nutrient profile, significant over- or undernutrition could result if too much or too little is consumed. Thus, it is important to know if the amount being consumed is appropriate.

The amount of food being fed should have been determined when the history of the animal was obtained. Although many animals are fed free choice, owners should still be able to provide a reasonable estimate of the actual amount being consumed. The owner may need to return home and measure the amount the pet consumes before reporting this. The amount actually being consumed can then be compared with the amount that should be fed. If the animal in question has a normal BCS (3/5) and no history of weight changes, the amount fed is probably appropriate. Exceptions to this generalization include growing animals, animals that are gestating or lactating and hunting dogs and other canine athletes early in the athletic event season.

The appropriate amount to feed can be difficult to determine precisely, but can be estimated. For most commercial pet foods, food dosage estimates can be found in the feeding guidelines on the product label. However, food dosages can be calculated if guidelines are not available. The precision of feeding guidelines or calculated food dosages is limited because the efficiency of food use varies among individuals because of differences in physical activity, metabolism, body condition, insulative characteristics of the coat and external environment. Even when environmental conditions and physical activity are similar, sizable individual differences can exist.

The total amount of energy needed by dogs and cats for maintenance, even under similar environmental conditions can vary two- to threefold. Therefore, a calculated food dosage should only be considered an estimate or a starting point that may have to be adjusted.

Calculations to estimate food dosage are based on the assumption that if a food contains the proper proportions of nutrients relative to its energy density, and is fed to meet an animal's energy requirement, then the animal's requirements for nonenergy nutrients will automatically be met. Food dosage estimation has four steps:
1) Estimate the energy requirement of the animal (**Table 1-4**).
2) Determine the appropriate category of food to be fed. Select a food that is suitable for the intended application (i.e., growth, gestation/lactation, maintenance, weight control [increasing or decreasing weight], old age, physical work or a specific disease). If you are unsure whether the food is balanced for the intended application, evaluate it yourself as described above under food evaluation.
3) Determine the energy density of the food (kcal or kJ ME/g food, as fed basis). Sources include product labels, product literature and company personnel. (See Tables 1-8 and 1-9, Small Animal Clinical Nutrition, 4th ed., p 14 for calculation methods using modified Atwater values.)
4) Divide the energy requirement of the animal by the energy density of the food to determine the daily amount to feed (food dosage).

How the Food is Offered (Pages 13-14)

The amount fed is usually offered one of three ways: 1) free-choice feeding (dogs and cats), 2) food-restricted meal feeding (dogs and cats) and 3) time-restricted meal feeding (dogs). The number of feedings per day must be considered when the last two methods are used.

Free-choice feeding (also referred to as ad libitum or self feeding) is a method in which more food than the dog or cat will consume is always available; therefore, the animal can eat as much as it wants, whenever it chooses. The major advantage of free-choice feeding is that it is quick and easy. All that is necessary is to ensure that reasonably fresh food is always

Table 1-4. Calculation of energy requirements.

Calculation of daily energy requirement (DER) is based on the resting energy requirement (RER) for the animal modified by a factor to account for normal activity or production (e.g., growth, gestation, lactation, work). RER is a function of metabolic body size. RER is calculated by raising the animal's body weight in kg to the 0.75 power. The average RER for mammals is about 70 kcal/day/kg metabolic body size: RER (kcal/day) = $70(BW_{kg})^{0.75}$ or $30(BW_{kg})$ + 70 (if the animal weighs between 2 and 45 kg). Expressed in kJ, the average RER for mammals is about $293(BW_{kg})^{0.75}$. These energy requirements should be used as guidelines, starting points or estimates of energy requirements for individual animals and not as absolute requirements.

Feline DER
Maintenance (0.8 to 1.6 x RER)
Neutered adult   = 1.2 x RER
Intact adult   = 1.4 x RER
Active adult   = 1.6 x RER
Obese prone   = 1.0 x RER
Weight loss   = 0.8 x RER
Critical care   = 1.0 x RER
Weight gain   = 1.2-1.4 x RER at ideal weight
Gestation
Energy requirement increases linearly during gestation in cats.
Energy intake should be increased to 1.6 x RER at breeding and gradually
  increased through gestation to 2 x RER at parturition.
Free-choice feeding of pregnant queens is also recommended.
Lactation
Lactation is nutritionally demanding and the physiologic and nutritional
  equivalent of heavy work.
Recommend 2 to 6 x RER (depending on number of kittens nursing) or free-
  choice feeding.
The following table may also be used to estimate the DER of lactating
  queens.

Weeks of lactation	DER
Weeks 1-2	RER + 30% per kitten
Week 3	RER + 45% per kitten
Week 4	RER + 55% per kitten
Week 5	RER + 65% per kitten
Week 6	RER + 90% per kitten

Growth
Daily energy intake for growing kittens should be about 2.5 x RER. Free-
  choice feeding is recommended. (Continued on next page.)

available. Free-choice feeding is the method of choice during lactation. Free-choice feeding also has a quieting effect in a kennel and timid dogs have a better chance of getting their share if dogs are fed in a group.

Table 1-4. Calculation of energy requirements (Continued.).

Canine DER
Maintenance (1.0 to 1.8 x RER)

Neutered adult	= 1.6 x RER
Intact adult	= 1.8 x RER
Obese prone	= 1.4 x RER
Weight loss	= 1.0 x RER
Critical care	= 1.0 x RER
Weight gain	= 1.2-1.4 x RER at ideal weight

Work

Light work	= 2 x RER
Moderate work	= 3 x RER
Heavy work	= 4-8 x RER

Gestation
First 42 days: feed as an intact adult.
Last 21 days: use 3 x RER. (This quantity may need to be increased to maintain normal body condition for some dogs, especially larger breeds.)

Lactation
Lactation is nutritionally demanding and the physiologic and nutritional equivalent of heavy work.
Recommend 4 to 8 x RER (depending on number of puppies nursing) or free-choice feeding.
The following table may also be used to estimate the DER of lactating bitches.

Number of puppies	DER
1	3.0 x RER
2	3.5 x RER
3-4	4.0 x RER
5-6	5.0 x RER
7-8	5.5 x RER
≥9	≥6.0 x RER

Growth
Daily energy intake for growing puppies should be 3 x RER from weaning until four months of age.
At four months of age energy intake should be reduced to 2 x RER until the puppy reaches adult size.

Disadvantages include: 1) anorectic animals may not be noticed for several days, especially if two or more animals are fed together, 2) if food is always available, some dogs and cats will continuously overeat and may become obese (such animals should be meal fed) and 3) moist foods and moistened dry foods left at room temperature for prolonged periods can spoil and are not appropriate for free-choice feeding. (See Chapter 7.)

When changing a dog from meal feeding to free-choice feeding, first feed it the amount of the food it is used to receiving at a meal. After this food has been consumed and the dog's appetite has been somewhat satisfied, set out the food to be fed free choice. This transitioning method helps prevent engorgement by dogs unaccustomed to free-choice feeding. Engorgement is generally not a problem when transitioning cats to free-choice feeding. Although dogs and cats unaccustomed to free-choice feeding may overeat initially, they generally stop doing so within a few days, once they learn that food is always available. Avoid taking the food away at any time during this transition period. Each time food is taken away increases the difficulty in changing the animal to a free-choice feeding regimen.

With food-restricted meal feeding, the dog or cat is given a specific, but lesser amount of food than it would eat if the amount fed were not restricted. Time-restricted meal feeding is a method in which the animal is given more food than it will consume within a specified period of time, generally five to 15 minutes. Time-restricted meal feeding is of limited usefulness with dogs and has little if any practical application in cats. Many dogs can eat an entire meal in less than two to three minutes. Both types of meal feeding are repeated at a specific frequency such as one or more times a day. Some people combine feeding methods such as free-choice feeding a dry or semi-moist food and meal feeding a moist food or other foods such as meat or table scraps.

Frequent food consumption resulting from frequent meal and free-choice feeding has several advantages. Feeding small meals frequently throughout the day results in a greater loss of energy as a result of an increase in daily meal-induced heat production. Also, it generally results in greater total food intake than does less frequent feeding. Frequent feeding of small meals benefits animals with dysfunctional ingestion, digestion, absorption or use of nutrients.

Frequent feeding is also desirable in normal animals that require a high food intake. Puppies less than six months old, some dogs engaged in heavy work (high levels of physical activity), dogs and cats experiencing ambient temperature extremes, bitches and queens during the last month of gestation and bitches and queens that are lactating should be fed at least three times per day to ensure that their nutritional needs are met. These animals may require one and one-half to four times as much food per unit of body weight than most normal adult dogs and cats. A reduced frequency might limit total food intake in these situations. Also, more frequent feeding during periods of variable appetite suppression, such as occurs with psychologic stress or high ambient temperatures, helps ensure adequate food intake.

Most clinically normal adult dogs that are not lactating, working or experiencing stress will have a sufficient appetite and physical capacity to consume all of the food required daily in a single 10-minute period (assuming food of typical nutrient density [about 3.5 kcal/g or 14.64 kJ/g dry matter]). Cats are less likely to eat their entire meal in one 10-minute sitting, but once-a-day feeding is adequate for most healthy adults. Although many dogs and cats are fed once daily without noticeable detrimental effects, at least twice daily feeding is generally recommended.

FEEDING PLAN (Pages 14-16)

A feeding plan should be formulated based on realistic and quantifiable nutritional objectives after the animal, food and feeding method have been assessed. The feeding plan will guide the veterinarian's selection of food or foods and the feeding method. The two steps to formulating the feeding plan are to determine: 1) what food or foods to feed and 2) the feeding method.

Determine What Food to Feed (Pages 14-15)

As explained above (See Assessment.), comparing the food's levels of key nutritional factors with the animal's needs allows the veterinarian to determine whether nutrient excesses or deficiencies exist. If the animal's current food is adequate (key nutritional factors in balance with the animal's needs) then the food currently being fed can continue to be fed. The animal's current food must be modified or "balanced" if significant nutrient excesses or deficiencies exist.

There are numerous approaches to balancing foods. Some are rather extensive. This section will review the most practical methods, including: 1) food replacement and 2) simple mathematical ration balancing (Pearson square). Alternatively, veterinarians can contact a person who specializes in veterinary nutrition.

Food Replacement (Pages 14-15)

If food assessment indicates that an animal's nutrient requirements are not being met, the most practical way to balance a food is to simply select a different food (i.e., one that does a better job of meeting the animal's requirements). The most likely application of this method occurs when one commercial food is substituted for another. One homemade food can be replaced by another if the nutrient profiles of both recipes are available. (See Chapter 6.)

The process is straightforward and simple. The nutrient contents of other foods are evaluated to see which food most closely meets the animal's requirements. Assuming acceptable palatability, the "winner" of this comparison replaces the previous food.

Changing foods for most healthy dogs and cats is of minor consequence. Some owners switch their pets from one food to another daily. Most dogs and cats tolerate these changes. However, vomiting, diarrhea,

Table 1-5. Recommended short- and long-term food transition schedules for dogs and cats.

Short schedule*	Long schedule**		Food percentages	
Dogs and cats (days)	Dogs (days)	Cats (weeks)	Previous food	New food
1,2	1-3	1	75	25
3,4	4-6	2	50	50
5,6	7-9	3	25	75
7	10	4	0	100

*Recommended for most healthy dogs and cats.
**Recommended for situations in which the food change is known to be significant, the dog or cat has demonstrated low tolerance to such changes in the past or food refusal is anticipated.

belching, flatulence or a combination of signs may occur with sudden, rapid switching of foods, probably because of ingredient differences. It is prudent, therefore, to recommend that owners change their pet's food over the course of at least three days. A seven-day period is even better, as owners increase the proportion of new food and decrease the proportion of old food (**Table 1-5**). Nearly all pets readily tolerate a seven-day transition period. A much longer transitional period is recommended in cases in which the food change is known to be significant, the pet has demonstrated a poor tolerance to such changes in the past or food refusal is expected (**Table 1-5**). For example, a long transition schedule is likely to be needed for an old cat recently diagnosed with kidney disease when the food must be switched from a highly palatable grocery "gourmet" food to an appropriate veterinary therapeutic food.

Simple Mathematical Ration Balancing (Pearson Square) (Page 15)

The Pearson square is another useful diet balancing tool. This handy method can be used to combine any two foods, supplements or ingredients to yield a mixture with a desired nutrient content. **Figure 1-4** shows how the Pearson square method is used to balance a diet. Here's how to use the Pearson square:

1) A small square is drawn and the desired nutrient concentration of the proposed mixture is written in the middle of the square.
2) The nutrient concentration of one component of the mixture is written at the upper left corner of the square.
3) The nutrient concentration of the other component of the mixture is written at the lower left corner of the square.
4) The nutrient values at the corners are subtracted from those in the center of the square. The smaller number is always subtracted from the larger and the differences written diagonally at the right corners of the square.

As an example, the Pearson square can be used to solve the following problem: How much calcium carbonate containing 36% calcium must be added to a meat-based food to increase its calcium content from 0.01% to 0.3% on an as fed basis?
Assume you are making 5 kg of the mixture. The problem is set up and worked as follows:

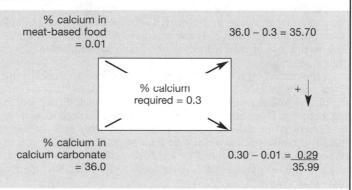

% calcium in
meat-based food
= 0.01

36.0 − 0.3 = 35.70

% calcium
required = 0.3

% calcium in
calcium carbonate
= 36.0

0.30 − 0.01 = 0.29 / 35.99

The final step converts fractions to percentages by dividing the numerator of the fractions by the denominator and multiplying by 100.
        Meat-based food: (35.70 ÷ 35.99) x 100 = 99.19%
        Calcium carbonate: (0.29 ÷ 35.99) x 100 = 0.81%
If the total mixture is 5 kg, then 99.19% (4.96 kg) should be a meat-based food and 0.81% (0.04 kg, or 40 g) should be calcium carbonate.

Figure 1-4. Example of how to use the Pearson square.

5) The differences are added together and the sum is written below each difference as the denominator of a fraction.
6) The fractions are converted to percentages. These percentages are the proportion of each component of the mixture in the corners directly to the left. When combined in those percentages, the constituent components will yield a mixture having the same concentration as the number in the center of the square.

Determine the Feeding Method (Pages 15-16)

As mentioned above, veterinarians and their health care teams should consider several aspects of feeding methods. These include the feeding route, the amount to be fed and how the amount fed is provided. In addition, feeding factors that affect compliance should be considered, such as whether or not the animal has access to other foods and who provides the food.

The feeding route will depend on the ability of the animal to self-feed. Animals unable to self-feed will need assistance such as syringe feeding, tube feeding or parenteral feeding.

Estimates of the amount of a new food to feed can be obtained from product information such as feeding guidelines on the product label. The food dosage estimate can also be calculated as shown above. (See Assess the Feeding Method.) Recall that individual variability is large and that either method is, at best, an estimate. The amount fed should be adjusted to ensure a normal BCS (3/5) and to avoid intolerance (vomiting or diarrhea in the case of enteral feeding or metabolic complications with parenteral feeding).

As mentioned earlier in the chapter, how the food is provided and how often it is provided depend on the animal's condition and in some cases the lifestyle of the owner. Each animal's situation will dictate which feeding method is most desirable (free choice, time-restricted meal feeding or food-restricted meal feeding). For many physiologic and disease conditions this consideration will not be important. For others it will be very important. Recommendations for the best method of providing the food and the number of times per day the food is offered are included in each individual chapter.

Owner compliance is necessary for effective clinical nutrition. Feeding methods should reinforce or enable compliance. Enabling compliance includes limiting access to other foods and knowing who provides the food. An animal from a multi-pet household may have access to the other pets' food. If so, such access needs to be denied or limited. Restriction can be difficult in some homes. In such cases the veterinary health care team and pet owner may need to compromise.

Compliance can be eroded if everyone in the family is not supportive of the feeding plan. Whoever feeds the pet must understand the consequences when the wrong foods are fed or even when the right foods are fed in the wrong amounts. Client education is essential for the successful outcome of any feeding plan. Specific client education must be provided for feeding healthy pets and for those with specific disease problems. Both oral and written instructions encourage compliance with feeding plans.

Veterinarians and their health care teams should actively involve clients in the formulation of the feeding plan to ensure commitment to the plan. The hospital staff should strive to uncover issues that clients may have about the feeding plan and negotiate mutually acceptable solutions. Open communication about the client's and the hospital's objectives, concerns and shared responsibilities is necessary for successful implementation of the feeding plan. Authoritarian approaches are unlikely to be effective because they discount the high degree of independent decision making that clients have based on their own perceptions of nutrition. Veterinarians and their health care teams can guide clients and enable them to make informed decisions.

REASSESSMENT (Page 16)

Finally, monitoring, or reassessment of the animal, should be performed at appropriate intervals to evaluate the effectiveness of the feeding plan. For patients undergoing intensive care, reassessment may need to be done every few hours, whereas pets in a health maintenance program could be reassessed annually. Reassessment signals the initiation of the iterative step of the clinical nutrition process. Involving the client in the action plan is an essential component of the veterinarian-client relationship. The reader is referred to the remaining chapters of this book for information about specific feeding plans and practices according to pet nutritional needs in health and in specific diseases.

Clinical cases that illustrate and reinforce the nutritional concepts presented in this chapter can be found in Small Animal Clinical Nutrition, 4th ed., pp 17-19.

CHAPTER 2

Nutrients

For a review of the unabridged chapter, see Gross KL, Wedekind KJ, Cowell CS, et al. Nutrients. In: Hand MS, Thatcher CD, Remillard RL, et al, eds. Small Animal Clinical Nutrition, 4th ed. Topeka, KS: Mark Morris Institute, 2000; 21-107.

INTRODUCTION (Page 21)*

Proper nutrition is among the more important considerations in maintenance of health and is key to the management of many diseases. A basic knowledge of nutrients, requirements, availability and consequences of deficiencies or excesses is important to feed animals correctly and give advice about feeding.

A nutrient is any food constituent that helps support life. Nutrients are essential in that they are involved in all the basic functions of the body including: 1) acting as structural components, 2) enhancing or being involved in chemical reactions of metabolism, 3) transporting substances into, throughout or out of the body, 4) maintaining temperature and 5) supplying energy.

Nutrients are divided into six basic categories: 1) water, 2) carbohydrates, 3) protein, 4) fat, 5) minerals and 6) vitamins. Some nutrients fulfill a number of functions. For example, water and several minerals are needed for all the functions described above except supplying energy. Carbohydrates, fats and proteins may be used for energy and serve as structural components. Vitamins are involved primarily with metabolic functions.

This chapter is organized into seven sections. The nutrients will be covered in the order discussed above. Energy, a non-nutrient, but nonetheless essential for life will be covered after water.

WATER (Pages 21-25)

Water is vital to life and is considered the most important nutrient. Water performs the following important functions in animals:
1) Water is the solvent in which substances are dissolved and transported around the body.
2) Water is necessary for the chemical reactions that involve hydrolysis (e.g., enzymatic digestion of carbohydrates, proteins and fats).

*Page numbers in headings refer to Small Animal Clinical Nutrition, 4th ed., where additional information may be found.

3) Water assists in the regulation of body temperature.
4) Water provides shape and resilience to the body.

Water is one of the largest constituents of the animal body, varying from 40 to greater than 80% of the total.

Water Requirements (Pages 24-25)
Dogs and cats meet most of their water requirement through water ingested as food or drink. Animals consuming commercial moist foods will drink less liquid than those fed dry foods because of the higher water content of moist foods (>75% water). As a general guideline, the daily water requirement of dogs and cats, expressed in ml/day, is roughly equivalent to the daily energy requirement (DER) in kcal/day (for dogs 1.6 x resting energy requirement [RER], for cats 1.2 x RER).

ENERGY (Pages 25-36)

Living organisms need energy to fuel all body functions. Although energy itself is not a nutrient, fats, carbohydrates and amino acids contain energy in the form of chemical bonds and are the energy-containing nutrients in food. Once eaten, these nutrients are digested, absorbed and transported to body cells where they are used to generate energy.

Function (Page 26)
Animals use energy for pumping ions, molecular synthesis and to activate contractile proteins. These three processes essentially describe the total use of energy by an animal. Without energy supplied by food, these reactions would rapidly cease and death would occur.

Importance of Energy (Page 26)
The energy content of a food ultimately determines the quantity of food that is eaten each day and therefore affects the amount of all other nutrients that an animal ingests. Animals should be fed enough food to meet their energy requirements and the non-energy nutrients in the food should be balanced relative to energy density to ensure adequate nutrient intake. Animals eating an energy-dense food consume less of the food to meet energy needs; therefore, the concentration of other critical nutrients must be higher to ensure sufficient intake. Conversely, animals must consume more of a low-energy food to meet energy needs. Therefore, the concentration of non-energy nutrients should be lower to avoid excessive intake and maintain nutrient balance.

Energy Storage (Page 29)

After the body has enough energy to meet demands, the pathways of glycogen and fat synthesis are simultaneously accelerated and excess dietary energy is stored as glycogen and body fat. These energy stores can then be used to generate ATP later when needed. Generally, in fasting animals, when the body needs energy, it uses glycogen first, fat stores second and finally, as a last resort, amino acids from body protein. (See Chapter 12.) In fed animals, food energy is primarily used for meeting body energy needs, thus preserving body tissues by preventing catabolism.

Energy Requirements (Pages 30-36)

Knowledge of energy requirements is needed to determine how much food to feed an animal.

RER represents the energy requirement for a normal animal at rest under thermoneutral but not fasted conditions. DER represents the average daily energy requirement of an animal. DER depends on lifestage and activity. DER equals RER plus energy needed for physical activity and production. **Table 1-4** (Pages 14 and 15) summarizes energy requirements for cats and dogs.

■ CARBOHYDRATES INCLUDING FIBER (Pages 36-48)

Simple Carbohydrates and Starches (Pages 36-42)
Definition (Page 36)

Carbohydrates are composed of carbon, hydrogen and oxygen in the general formula $(CH_2O)_n$. Carbohydrates encompass: 1) simple sugars such as monosaccharides (e.g., glucose) and disaccharides (e.g., sucrose), 2) oligosaccharides (three to nine sugar units; e.g., raffinose, stachyose) and 3) polysaccharides (more than nine sugar units). Examples of polysaccharides include starches (amylose, amylopectin, glycogen), hemicellulose, cellulose, pectins, gums, etc.

Complex carbohydrates that are digested by the animal's endogenous digestive enzymes are designated starches, whereas those polysaccharides that are resistant to enzymatic digestion and thus are fermented by intestinal microbes are labeled fibers.

Function (Page 38)

The body uses simple carbohydrates and starches in foods as a source of glucose. As such, they have several major functions. First, they provide energy (ATP) via glycolysis and the TCA cycle. Second, when metabolized for energy to carbon dioxide and water, they are a source of heat for the body. Third, as they proceed through metabolic pathways, certain products can be used as building blocks for other nutrients, such as nonessential amino acids, glycoproteins, glycolipids, lactose, vitamin C, etc. Finally, simple carbohydrates and starches in excess of the body's immediate energy needs are stored as glycogen or converted to fat.

IMPORTANCE OF CARBOHYDRATES The primary purpose for adding carbohydrates and starches to pet foods is to supply energy. Generally, assuming an average digestibility (84%), carbohydrates supply about 3.5 kcal/g. Although there is no minimum dietary requirement for simple carbohydrates or starches per se, certain organs and tissues (e.g., brain and red blood cells) require glucose for energy.

When energy needs are high and tissue accretion is occurring (e.g., during growth, gestation and lactation), adequate dietary carbohydrates or glucose precursors are necessary to maintain metabolic processes. In these situations, carbohydrates become conditionally essential; therefore, foods fed to growing animals and those with high-energy needs should contain at least 20% carbohydrates.

Analyses (Pages 40-41)
The total carbohydrate content of pet foods and ingredients is not typically determined directly by analysis but indirectly by difference. Nitrogen-free extract (NFE) is the carbohydrate fraction of a proximate analysis. NFE is determined by adding the percentages of water, crude protein, crude fat, ash and crude fiber and subtracting from 100%. NFE is primarily made up of readily digestible carbohydrates (e.g., sugars and starches).

Requirements/Deficiencies/Excesses (Pages 41-42)
Dogs and cats do not have an absolute dietary requirement for carbohydrates in the same way that essential amino acids or fatty acids must be provided. They do, however, have a requirement for adequate glucose or glucose precursors to provide essential fuel for the central nervous system. When energy needs are high and anabolic processes are proceeding at an active rate (e.g., during growth, gestation and lactation), it is best to supply a food containing readily digestible carbohydrates and starches. Without dietary carbohydrates, there is added strain on lipid and protein metabolic pathways to supply glucose precursors.

Most commercially prepared pet foods contain carbohydrates well in excess of glucose requirements. Grains such as corn, rice, wheat, barley and oats provide the bulk of starch in commercial pet foods and are well digested and absorbed due to the cooking and extrusion processes used to make pet foods.

CANINE CARBOHYDRATE REQUIREMENTS Overall, a minimum of 23% carbohydrate is recommended in foods for gestating and lactating bitches. Excess starch in the food typically does not cause health problems in dogs. Dry extruded dog foods typically contain 30 to 60% carbohydrate, mostly starch, and cause no adverse effects.

FELINE CARBOHYDRATE REQUIREMENTS Cats have some unique metabolic differences that limit their ability to efficiently use large amounts of absorbed dietary carbohydrate. For example, cats have low activities of the intestinal disaccharidases sucrase and lactase; furthermore, the sugar transportation system in the feline intestine does not adapt to various levels of dietary carbohydrates. Cats produce only 5% of the pancreatic amylase that dogs produce. Unlike dogs, cats lack hepatic glucokinase activity, which limits their ability to metabolize large amounts of simple carbohydrates.

The metabolic differences between cats and dogs support the classification of cats as strict carnivores, adapted to a low-carbohydrate diet, and dogs as omnivores. If large amounts of carbohydrates are fed to cats (e.g., more than 40% of the food on a dry matter basis [DMB]), signs of maldigestion occur (e.g., diarrhea, bloating and gas) and adverse metabolic effects can occur (e.g., hyperglycemia and excretion of significant amounts of glucose in urine). Despite the limitations of digestive capacity and metabolism, the starch levels found in commercial cat foods (up to 35% of the food DMB) are well tolerated.

Fiber (Pages 42-48)
Definition (*Page 42*)
Fibers differ from starches in that fibers resist enzymatic digestion in the small intestine. Microbes in the colon usually ferment fibers.

Fiber sources can be described as rapidly to slowly fermentable (**Figure 2-1**). Rapidly fermented fibers produce more short-chain fatty acids and gases in a shorter period of time vs. fiber sources that are fermented more slowly. Slowly fermentable fiber sources used in pet foods contain primarily celluloses and hemicelluloses including purified cellulose and peanut hulls. Citrus and apple pectin and most gums are rapidly fermented. Fiber sources that contain mixtures of pectins, hemicelluloses and celluloses (e.g., rice bran, oat bran, wheat bran, soy fibers, soy hulls and beet pulp) are moderately fermentable. The rate and extent of fiber fermentation are important distinguishing characteristics when discussing physiologic functions of fiber. As the fermentation rate of fiber increases, GI transit time decreases, fecal bulk decreases and fecal bile acid excretion increases.

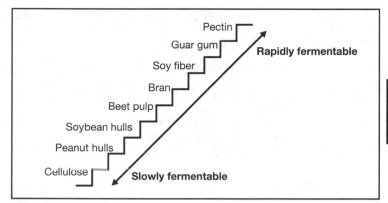

Figure 2-1. Relative degree of fermentation of various dietary fiber sources in the GI tract of dogs and cats. At the extremes are pectins and gums, which are rapidly fermented, and cellulose, which is slowly fermented. Other fiber sources such as soy hulls and beet pulp are intermediate and termed moderately fermentable.

The rate of fermentation of a fiber interacts with water-holding capacity and viscosity to affect the degree of fecal bulking (volume). For moderately and slowly fermentable fibers, the degree of fecal bulking is related to water-holding capacity. Slowly fermented fibers (e.g., cellulose) are the most effective stool bulking agents because they retain their structure longer and are thus able to bind water. For rapidly fermentable fibers, the increased bound water reduces fecal bulk. In fact, most fermentable fibers have laxative effects and may produce diarrhea if fed at high levels.

The primary function and benefit of adequate dietary fiber are to increase bulk and water in the intestinal contents. Fiber appears to shorten intestinal transit rate in dogs with normal or slow transit time and prolongs it in dogs with rapid transit rate. Together, these factors help to promote and regulate normal bowel function. In addition, the typical end products of microbial fermentation of fiber (acetate, propionate and butyrate) are important in maintaining the health of the colon.

Research results demonstrate the need for some fiber in foods to maintain health and optimal function of the entire GI tract, but especially for colonocytes.

OBESITY AND BODY WEIGHT MANAGEMENT A pet food containing slowly fermentable fiber can be very effective for controlling body weight and treating obesity. (See Chapter 13.) Slowly fermentable fibers, such as cellulose and peanut hulls, increase bulk in the stomach and intestines and help promote a feeling of satiety when fewer calories are consumed. Pets in weight-control programs can eat more total food when the calories are diluted by fiber; thus, the animal eats fewer calories and loses weight.

DIARRHEA AND CONSTIPATION Fiber normalizes intestinal water content; fiber absorbs water in cases of diarrhea and adds moisture in cases of constipation. The more fermentable fibers (e.g., gums and soy fibers) can help pets with diarrhea and constipation by moderating the water content of the stool, thereby making a watery stool drier and a dry stool moister.

The binding and gelling properties of fiber also assist in management of diarrhea. In constipated pets, the fermentable fibers increase stool weight and moisture content, making the stool softer.

DIABETES MELLITUS Management of diabetes mellitus in people, dogs and cats includes dietary changes. Both slowly and rapidly fermentable fiber types help control blood glucose levels in diabetic animals.

FIBER INTERACTIONS WITH NUTRIENT DIGESTIBILITY AND AVAILABILITY The amount and type of fiber in a pet food have the greatest overall effect on digestibility of all the nutrients. In general, foods containing slowly fermentable fiber sources will have lower overall dry matter digestibility than foods without fiber or those containing rapidly fermentable fiber sources. Also, as the level of fiber in the food increases, the dry matter digestibility of the food decreases. Apparent and true digestibility of fats, starch and energy are unaffected by the type and amount of fiber in the food.

Analyses *(Pages 45-46)*

In the United States, regulations require that the maximum amount of crude fiber be listed on the label of all pet foods. (See Chapter 5.) Determination of crude fiber adequately represents total fiber in a pet food when most of the fiber is slowly fermentable; however, the analysis excludes the more rapidly fermentable pectins and gums. Because the crude fiber analysis underestimates fermentable fiber, it does not accurately represent the total fiber in a pet food.

Requirements/Deficiencies/Excesses *(Pages 46-48)*

Fiber is not considered essential in the diets of normal cats and dogs, although it is often included in commercial foods. Overall, dogs and cats do not derive much energy from absorbing the typical end products of bacterial fermentation; however, short-chain fatty acids are important in maintaining the health of the colon. Therefore, a small amount of fiber (<5%) that contains both rapidly and slowly fermentable fibers is recommended in foods for healthy pets. However, as mentioned above, fiber also aids in the management of diseases such as obesity, diabetes mellitus, diarrhea, colitis and constipation. The types and amounts of dietary fiber required to assist in the management of these diseases can be found in Chapters 13, 22 and 24.

Total deficiency of fiber in typical pet foods is not a practical problem because many ingredients used contain some fiber.

Excess fiber may have undesirable effects. For instance, certain fiber types decrease the absorption of minerals. The effects on mineral absorption vary by fiber type and the mineral. More rapidly fermentable fibers (e.g., pectins and guar gum) apparently decrease availability of some minerals, whereas fibers that contain more cellulose have little effect on mineral absorption.

Excess fiber can dilute the energy and nutrient content of the food to such an extent that an animal may have difficulty eating enough of the food to meet its needs. As mentioned above, controlled levels of fiber are advantageous in weight-reducing foods for dogs and cats; however, such foods are fortified so that only total energy intake is low and other nutrients are present in adequate amounts to meet daily requirements.

PROTEIN/AMINO ACIDS (Pages 48-59)

Definition (Page 48)

Proteins are large, complex molecules composed of hundreds to thousands of amino acids. Amino acids are composed of carbon, hydrogen, oxygen, nitrogen and sometimes sulfur and phosphorus. Although hundreds of amino acids exist in nature, only 20 are commonly found as protein components.

Function (Page 51)

Proteins are the principal structural constituents of body organs and tissues including: 1) collagen and elastin found in cartilage, tendons and ligaments, 2) the contractile proteins actin and myosin in muscles, 3) keratin proteins in skin, hair and nails and 4) blood proteins including hemoglobin, transferrin, albumin and globulin. Proteins also function as enzymes, hormones (e.g., insulin) and antibodies. Amino acids can serve as a source of energy after the nitrogen-containing amino group is removed by deamination or transamination.

Importance of Amino Acids (Page 51)

Several amino acids are classified as essential or indispensable (10 for dogs and 11 for cats) (**Table 2-1**). These amino acids cannot be synthesized by the body in sufficient quantities and therefore must be supplied by food. Many of the remaining amino acids are nonessential or dispensable; they can be synthesized in the body from carbon and nitrogen building blocks and need not be present in the food if adequate nitrogen and energy are available. Some amino acids are conditionally essential. These amino acids ordinarily are not required in the food except during certain physiologic or pathologic conditions when they may not be synthesized in adequate quantities.

Unlike other animals, cats conjugate bile acids only to taurine and not to glycine. Furthermore, cats have an obligatory loss of taurine in the feces due to bacterial degradation in the intestine and intestinal losses of taurine through enterohepatic circulation. This obligatory loss coupled with a minimal capacity for cats to synthesize taurine makes taurine an essential amino acid for this species.

Documented signs of taurine deficiency include reproductive failure in queens, developmental abnormalities in kittens, central retinal degeneration and dilated cardiomyopathy. Taurine deficiency is more likely to occur in cats that are fed dog foods, homemade foods and vegetarian foods that are not supplemented with taurine.

Taurine requirements in cats are highly dependent on ingredient sources and processing. Current recommendations are to include at least 1 g of purified crystalline taurine/kg (0.1% DM) in dry foods and at least 2 g of purified crystalline taurine/kg for moist foods (0.2% DM).

Protein Usage (*Page 52*)

Absorbed amino acids and small di- and tripeptides are reassembled into "new" proteins by the liver and other tissues of the body. Amino acids from food are transported from the liver to other tissues by serum albumin or as free amino acids. The fate of amino acids after absorption falls into three general categories: 1) tissue protein synthesis, especially in muscles and liver, 2) synthesis of enzymes, albumin, hormones and other nitrogen-containing compounds and 3) deamination and use of the remaining carbon skeletons for energy.

The body is able to synthesize new proteins and enzymes provided all the necessary amino acids are available. The source of amino acids is not important. Cells use amino acids from a variety of sources including those derived from food proteins, single amino acids added to the food and amino acids synthesized by the body. In addition, cells synthesizing new protein cannot distinguish between amino acids from grains (e.g., corn and rice) and those from meats (e.g., chicken and beef). The only criterion is that all the amino acids needed to synthesize a particular protein be present in sufficient quantities when necessary. Protein synthesis will be limited when certain amino acids are not present or available in the quantities needed.

Table 2-1. Amino acids and essentiality.

Name	Essentiality
Alanine	Nonessential
Arginine	Essential
Asparagine	Nonessential
Aspartate	Nonessential
Cysteine	Nonessential, but can provide for up to 50% of methionine requirements
Glutamate	Nonessential
Glutamine	Conditionally essential
Glycine	Nonessential
Histidine	Essential
Isoleucine	Essential
Leucine	Essential
Lysine	Essential
Methionine	Essential
Phenylalanine	Essential
Proline	Nonessential
Serine	Nonessential
Taurine	Essential for cats, possibly conditionally essential for dogs
Threonine	Essential
Tryptophan	Essential
Tyrosine	Nonessential, but can provide for up to 50% of phenylalanine requirements
Valine	Essential

Dietary protein must be consumed each day to replace amino acids lost to catabolism. Trauma, infection, severe sepsis and burns increase protein turnover and nitrogen losses, whereas nitrogen losses are reduced during long-term fasting and starvation. Nitrogen is normally lost from the body in feces (nitrogen, proteins, cells), in urine and through skin desquamation and hair loss.

Protein Storage (Page 52)
Liver and muscle protein and serum albumin synthesis increase after consumption of a protein-containing meal. Once protein synthesis is maximized, excess amino acids are deaminated and transaminated to yield amino groups and carbon skeletons. The carbon skeletons can be used for many purposes including glucose precursors, which can be stored as glycogen or converted to fatty acids and acetyl-CoA, which can be used immediately for fuel.

Protein Excretion (Page 53)
Amino acid catabolism produces waste nitrogen, if it is not used for purine or pyrimidine synthesis. Catabolism of amino acids typically leads

to formation of ammonia, which is toxic to body cells. Ammonia must be converted to a less toxic form that can be transported in blood and excreted. Ammonia formation occurs in all tissues, but is especially prevalent in the liver and kidneys during gluconeogenesis. More than 90% of the nitrogen resulting from protein degradation is converted to urea in the liver and kidneys and then excreted in the urine.

Protein Quality (Pages 53-55)

Because protein-containing ingredients are added to pet foods to supply required amino acids, the amino acid composition and digestibility of the protein ingredients are important considerations. The quality of the protein affects the amount that must be provided to meet an animal's protein requirement. The amino acids of proteins that are easily digested are readily available for absorption. Such proteins have high nutritional value and less of them are needed to meet an animal's amino acid requirement. As protein quality increases, the amount of the protein required in the pet food decreases.

Another important factor contributing to protein value is the balance of essential amino acids. Even proteins that are easily digested can have low nutritional value if they have imbalances or deficiencies of essential amino acids. For example, gelatin is a protein source derived from animal collagen. Although gelatin is highly digestible it is a very poor quality protein because it is deficient in the essential amino acid tryptophan.

Requirements/Deficiencies/Excesses (Pages 55-56)

New proteins can be synthesized from dietary amino acids or nonessential amino acids that were previously synthesized by the body. By definition, essential amino acids used in protein synthesis must be provided by food. Therefore, animals do not have a requirement for protein per se but have an amino acid requirement. The amount of each amino acid that an animal requires varies based on factors such as growth, pregnancy, lactation and some disease states. In addition to requiring specific essential amino acids, dogs and cats have a requirement for building blocks (carbon skeletons and nitrogen) for nonessential amino acids. The building blocks for nonessential amino acids can either be derived from excess essential amino acids that are broken down and reassembled into nonessential amino acids or from other nonessential amino acids in food. Thus, a complete statement of amino acid requirements should include all the essential amino acids and an amount of amino nitrogen that can be used for synthesis of nonessential amino acids.

Practically speaking, because most foods contain whole proteins and not individual amino acids, and because exact individual amino acid requirements are not known for every situation, it is fair to consider a protein requirement instead of individual amino acid requirements in most instances when discussing companion animal nutrition.

The amount of protein that must be included in a pet food also depends on how much food the animal consumes. It is easy to understand why animals that are growing, pregnant or lactating require dietary protein to support new tissue growth and milk production. If an animal only consumes small quantities of food to meet its energy requirement then the food needs to have a greater protein concentration to meet the animal's protein requirement. For example, high-calorie pet foods should have more protein as a percentage of the food than low-calorie foods. The opposite is also true. Larger portions of low-calorie foods are typically consumed; therefore, animals can adequately meet their daily requirements with a food that has a lower percentage of protein.

Adult animals also need dietary protein to replace the amino acids that enter pathways of amino acid catabolism and are permanently lost. Healthy adults have a daily requirement for protein to replace nitrogen lost as urea, ammonia, creatinine, nitrate in urine and feces, sloughing of epithelial cells in skin and the GI tract, sweat, hair, nasal secretions, semen from males and secretions due to reproductive cycles in females. Dietary protein that must be consumed each day to replace the obligatory nitrogen loss is termed the maintenance protein requirement.

Adult and growing animals have maintenance requirements for protein, but only growing animals have the additional requirement of protein for growth. The additional protein required by pregnant and lactating animals really supports growth.

Nitrogen balance is the difference between the nitrogen consumed and the amount lost each day. Growing animals, pregnant females and any animals that are replenishing or rebuilding tissue are in positive nitrogen balance. Zero nitrogen balance occurs in normal healthy adults receiving minimally adequate or more than adequate dietary protein when nitrogen output equals nitrogen intake. Negative nitrogen balance can occur during lactation, starvation or fasting when there is inadequate or no protein intake. Excessive body protein catabolism due to burns, injury, fever, infections, hormonal imbalance or psychological causes can also cause negative nitrogen balance. Amino acid imbalances and antagonisms can cause negative nitrogen balance even when adequate amounts of protein are being consumed.

The absolute minimum dietary protein requirement can be estimated by feeding extremely high-quality protein or commonly used protein sources. If the estimate is based on feeding high-quality protein (e.g., lactalbumin), a growing dog requires approximately 9.5% DM protein and an adult dog about 6.0% DM protein. The Association of American Feed Control Officials (AAFCO) has established that foods for growth should contain at least 22% DM protein, and foods for adult maintenance should contain at least 18% DM protein, for dog foods containing commonly used protein ingredients. It is important to note that AAFCO recommendations should be interpreted as daily allowances, not as absolute minimum requirements.

Growing kittens and adult cats have higher protein requirements than most other domestic species. The minimum requirement has been estimated to be about 24% (DMB) for kittens and 14% (DMB) for adult cats, assuming use of extremely high-quality protein sources. For commercial foods made from commonly available protein sources, AAFCO has recommended that foods for kittens and adult cats contain at least 30 and 26% protein on a DMB, respectively. Again, AAFCO recommendations for protein should be interpreted as a daily allowance of protein, not as an absolute minimum requirement.

The requirement for taurine increases slightly with increased dietary protein. Certain proteins (e.g., isolated soya protein) and the canning process increase the dietary taurine allowance compared with freezing or using casein as the protein source. Processing neither destroys nor binds appreciable quantities of taurine; however, processing apparently alters food so that enhanced numbers of intestinal bacteria degrade taurine.

Clinical signs of protein deficiency include reduced growth rate, anorexia, anemia, infertility, reduced milk production, alopecia, brittle hair and a poor coat.

Commercial dog foods contain three to seven times the minimum protein requirement. Some commercial cat foods contain two to four times the minimum protein requirement. Excess dietary protein can be problematic for dogs and cats with specific disease conditions. For example, any disease that affects organs involved with conversion of ammonia to urea and waste nitrogen disposal can result in accumulation of toxic byproducts of protein metabolism. In particular, protein intake above requirement must be carefully monitored in animals with impaired renal or liver function. (See Chapters 19 and 23.) In other situations, such as struvite urolith dissolution in dogs and adverse food reactions in cats and dogs, minimizing excesses of dietary protein is a beneficial part of therapy. (See Chapters 14 and 20.) In these cases, excess dietary protein could be considered "conditionally toxic."

Feeding protein above requirements or recommendations for healthy dogs and cats does not result in a true toxicity because the excess amino acids from the protein are catabolized and the waste nitrogen is excreted.

Protein excesses are found in pet foods for several reasons. Cats are strict carnivores (See Chapter 11, Small Animal Clinical Nutrition, 4th ed., pp 291-347.) and have a higher protein requirement than dogs, which are omnivores. (See Chapter 9, Small Animal Clinical Nutrition, 4th ed., pp

213-260.) However, some pet food companies have perpetuated the myth that dogs are carnivores and that meat-based, high-protein foods are more natural and thus better than lower protein foods that contain both animal and plant sources of protein. Other fallacies, such as high levels of dietary protein build more muscle or a thicker coat, have contributed to pet owners' mistaken perception that higher protein is indicative of a higher quality pet food.

Excess protein adds unnecessary cost to foods. Excess protein is used for energy. As an energy source, protein is no better than soluble carbohydrate; however, protein is a more expensive energy source. The increased costs associated with increased dietary protein are invariably passed on to pet owners.

There are no nutritional reasons that support providing excessive amounts of dietary protein. After the protein/amino acid requirements are met, more protein provides no additional benefits. Thus, dog foods for adult maintenance should not exceed 30% DM protein. Cat foods for adult maintenance should not exceed 45% DM protein.

LIPIDS (Pages 59-66)

Function (Page 60)

Intake of lipids benefits the animal by supplying energy, essential fatty acids (EFA) and a positive environment for absorption of fat-soluble vitamins. Dietary lipid may be assimilated and stored as fat in adipocytes, incorporated into functional lipid or catabolized for fuel, depending on the energy status of the animal. On a per weight basis, the energy value of dietary fat is approximately 2.25 times that of protein or carbohydrate.

Some lipids required for adequate physiologic function, such as certain long-chain fatty acids, cannot be synthesized de novo and are thus required in food. These fatty acids are called EFA because a lack of them in foods results in classic signs of deficiency. A small amount of lipid (1 to 2% of total food) of no specific structure is also required in foods for proper absorption of fat-soluble vitamins (A, D, E and K).

Nonesterified fatty acids (NEFA) consist of hydrocarbon chains ranging from two to 24 carbons or more, with a carboxylic acid group on one terminus and a methyl group on the opposite terminus. NEFA that contain 14 to 24 carbons are classified as long chain, eight to 12 medium chain and two to six short chain.

Fatty acids with the first double bond between the third and fourth carbon are in the n-3 family, sixth and seventh carbon the n-6 family and ninth and tenth carbon the n-9 family. The n-3 and n-6 fatty acid families are EFA because they cannot be synthesized de novo in mammals; lack of

EFA in foods results in suboptimal physiologic activity. Mammals are capable of de novo synthesis of saturated fatty acids and fatty acids of the n-9 series up to 18 carbons. Subsequently, mammals may elongate and desaturate de novo or dietary fatty acids of all classes via enzymes specific for certain carbons in the hydrocarbon chain. However, mammals cannot desaturate fatty acids between the n-1 carbon and the n-3, n-6 or n-9 double bond. Because of the specificity of these enzyme systems, unsaturated fatty acids cannot be converted between families (e.g., n-6 or n-9 to n-3). Also, because of limitations and specificity in metabolism, monounsaturated and saturated fatty acids cannot be converted to EFA (e.g., n-9 or stearate to n-6).

Members of the n-6 family include linoleic acid (18:2n-6), γ-linolenic acid (18:3n-6) and arachidonic acid (20:4n-6). Dogs, but not cats, are able to elongate and desaturate linoleic acid to form arachidonic acid. Thus, linoleic acid is usually listed as an EFA for dogs, whereas both linoleic acid and arachidonic acid are listed as EFA for cats. The n-6 fatty acid family is required for growth, reproduction and precursors of eicosanoid and prostaglandin synthesis.

Essentiality *(Page 64)*

Fish oils have recently garnered the attention of nutritionists and veterinarians for their purported positive effects in management of a variety of disease processes. The oils most commonly associated with fish are those of the n-3 and n-6 families. The primary fish oil supplement is derived from menhaden and is very high in the n-3/n-6 families, compared with animal fat. The n-3 family usually predominates over the n-6 family in fish and shellfish, whereas polyunsaturated fatty acids (PUFA) from vegetable sources are usually higher in the n-6 family. Threefold differences in fatty acid composition may occur in fish depending on season and geographic locale of the catch. Fish oil compositional profiles depend on dietary intake, type of fish (carnivore vs. plankton eater), warm vs. fresh water and season of catch. Unfortunately, data about specific variations based on the above factors are unavailable.

The eicosanoids resulting from n-3 fatty acid metabolism are less immunologically stimulating than those resulting from n-6 fatty acids. Thus, feeding n-3 fatty acids has been recommended in situations in which a reduced inflammatory response is desired such as: 1) before and after surgery, 2) after trauma, injury, burns and some types of cancer and 3) to assist in control of dermatitis, arthritis, inflammatory bowel disease and colitis.

Although more research with n-3 fatty acids in companion animals needs to be done, it is prudent to conclude that n-3 fatty acids will be shown to be essential for normal function of the retina and brain as well as for physiologic homeostasis. At this time, there are no conclusive data proving the optimal level or relationship of n-3 fatty acids to n-6 fatty acids for any species at any specific lifestage. The optimal relationship will likely depend on many individual parameters and differ depending on individual physiologic function.

Requirements and Deficiencies (Page 65)

Dogs and cats have a "requirement" for lipid to enhance the absorption of the fat-soluble vitamins A, D, E and K. Dietary fat provides a physical environment in the gut that enhances the absorption of fat-soluble vitamins. This requirement is in the range of 1 to 2% of the food and is not specific for any type of fat. When the total bulk of the food is a limiting factor, increasing dietary fat allows for increased energy consumption. Increased aerobic capacity during exercise is supported by increased fat consumption because of the enhanced use of fat calories when compared with calories from carbohydrate. (See Chapter 10.)

A concentration of 1% of the food dry matter as linoleic acid is a safe and effective concentration for dogs. Foods for cats should contain at least 0.5% linoleic acid and at least 0.02% arachidonic acid (DMB).

Deficiencies of fatty acids impair wound healing, cause a dry lusterless coat and scaly skin and change the lipid film on the skin, which may predispose the animal to pyoderma.

High dietary fat concentration requires increased antioxidant protection, such as added vitamin E. In the absence of adequate antioxidants, dietary lipids will become rancid. (See Chapter 4, Small Animal Clinical Nutrition, 4th ed., pp 127-146.) Rancidity adversely affects the animal through reduced palatability, reduced vitamin activity and possibly subsequent oxidation of body fat.

MINERALS (Pages 66-80)

Definition (Page 66)

The term mineral is generally used to denote all inorganic elements in a food. More than 18 mineral elements are believed to be essential for mammals. By definition, macrominerals are required by the animal in the diet in percentage amounts, whereas microminerals or "trace" minerals are required at the parts per million (ppm) level. There are seven macrominerals: calcium, phosphorus, sodium, magnesium, potassium, chloride and sulfur. There are at least 11 trace elements or micronutrient minerals: iron, zinc, copper, iodine, selenium, manganese, cobalt, molybdenum, fluorine, boron and chromium. The last six are assumed to be essential for dogs and cats by analogy with other species. For a variety of reasons (See Chapter 2, Small Animal Clinical Nutrition, 4th ed., p 66.), only calcium, phosphorus, magnesium, potassium, sodium, chloride, iron, zinc, copper, manganese, iodine, selenium and chromium are discussed in the following text.

Availability (Pages 70-71)

Evaluation of feeds as sources of minerals depends not only on what the feed contains (i.e., the analyzed nutrient content), but also on how much of the mineral can be used by the animal. The adequacy of a food, as determined by its analytical mineral concentration, can be misleading because a number of factors can influence the mineral availability. These include: 1) the chemical form (which influences solubility), 2) the amounts and proportions of other dietary components with which it interacts metabolically, 3) the age, gender and species of the animal, 4) intake of the mineral and the need (body stores) and 5) environmental factors.

The availability of different forms of a mineral can vary widely even among inorganic mineral supplements. In general, different forms of trace minerals (iron, zinc, manganese and copper) differ in availability as follows: sulfate and chloride forms >carbonates >oxides. The oxides of iron and copper are poorly available and should not be used as mineral supplements in pet food.

In general, meat-derived foodstuffs are considered a more available source of certain minerals than plant-derived foodstuffs. The organic forms of minerals found in meats are often more available or as available as those from inorganic mineral supplements, whereas the minerals in plants are often less available. This finding is truer for iron, zinc and copper than for selenium.

In addition, meats, unlike plants, do not contain anti-nutritional factors, such as phytate, oxalate, goitrogens and fiber, which reduce mineral availability.

Research in chicks and puppies indicates marked differences about how fiber sources affect mineral availability. Beet pulp consistently reduces the availability of minerals (zinc, calcium, phosphorus and iron); however, cellulose, corn bran and sunflower hulls have negligible effects. Pea fiber, peanut hulls and soy hulls inhibit availability of some but not all minerals.

Calcium (Pages 71-73)

Calcium serves two important functions: 1) as a structural component in bones and teeth and 2) as an intracellular second messenger that enables cells to respond to stimuli such as hormones and neurotransmitters. Calcium's two major physiologic functions in bone are to serve as a structural material and as an ion reservoir.

The amount of true calcium absorption may range from 25 to 90%, depending upon calcium status, calcium form or intake. This exchangeable pool consists of the small amount of calcium in blood, lymph and other body fluids, and accounts for 1% of the total body calcium. The remaining 99% is located in bones and teeth. Active calcium absorption is affected by the physiology of the host; that is, calcium and vitamin D status, age, pregnancy and lactation.

Passive absorption is a nonsaturable, paracellular route that is independent of vitamin D regulation. The amount of calcium absorbed in this way depends primarily on quantity and availability of calcium in the food. Vitamin D is the most important regulator of calcium absorption.

Requirements/Deficiencies/Excesses (Pages 72-73)

Deficiencies and excesses of calcium, as well as calcium-phosphorus imbalances, should be avoided in dogs and cats. The "ideal" calcium-phosphorus ratio recommended for simple-stomached animals is generally between 1:1 and 2:1. A number of factors, however, influence the importance of this ratio. Increasing levels of vitamin D reduce the significance of adverse calcium-phosphorus ratios. Further, the ratio can differ markedly with the form and availability of the calcium and phosphorus supplied in the diet. For example, animals eating foods high in phytate phosphorus require greater phosphorus intake to meet their needs. Thus, the ideal calcium-phosphorus ratio would be lower when foods with these dietary characteristics are fed vs. foods composed of mostly meat ingredients. A food grossly deficient in calcium, but adequate in phosphorus can cause secondary hyperparathyroidism. An all-meat diet devoid of bones, for example, is a very poor source of calcium.

Hypocalcemia is a common problem in diseased states (chronic or acute renal failure, pancreatitis, eclampsia, etc.), and parenteral supplementation of calcium and/or calcitriol (1,25 dihydroxycholecalciferol) is sometimes warranted. Calcium excess is probably more detrimental in rapidly growing animals than in adults, especially large- and giant-breed puppies. (See Chapter 17.) **Table 2-2** describes signs of calcium deficiency and excess.

The current AAFCO canine and feline recommendations for calcium are 1.0% for growth/reproduction and 0.6% for adult maintenance (DMB for all values). For dogs, this calcium requirement is based on an energy density of 3.5 kcal/g metabolizable energy (ME), whereas an energy density of 4.0 kcal/g ME is assumed for cats.

The following meat meals are rich sources of calcium because of their bone content: poultry byproduct meal, lamb meal and fish meal. Grains (corn, rice, etc.), on the other hand, are generally poor sources of calcium. Soybean meal and flaxseed have calcium contents between those of meat meals and grains. The most common calcium supplements used in pet foods are limestone (calcium carbonate), calcium sulfate, calcium chloride, calcium phosphate and bone meal, ranging in calcium from 16 to 39%.

Phosphorus (Pages 73-74)

After calcium, phosphorus is the second largest constituent of bone and teeth. Phosphorus is a structural component of RNA and DNA, high-energy phosphate compounds such as ATP and cell membranes composed largely of phospholipids.

About 60 to 70% of phosphorus is absorbed from a typical diet. In general, phosphorus availability is greater from animal-based ingredients than from plant-based ingredients. Phosphorus in meat is found mainly in the organic form, whereas in plants, phosphorus is in the form of phytic acid. Phytate phosphorus is only about one-third available to monogastric animals but availability from different grains can vary markedly.

Under conditions of low dietary phosphorus intake, the intestine increases its absorptive efficiency to maximize phosphorus absorption and the kidneys increase renal phosphorus transport or minimize urinary phosphorus losses. Conversely, under conditions of dietary excess, the kidneys increase excretion of minerals. Avoiding excess dietary phosphorus slows progression of kidney disease. (See Chapter 19.)

Requirements (Page 74)

The AAFCO recommendation for phosphorus, for both dogs and cats, is 0.8% for growth and reproduction and 0.5% for adult maintenance (DMB). **Table 2-2** describes the effects of phosphorus deficiency and excess.

Magnesium (Pages 74-75)

Magnesium is the third largest mineral constituent of bone, after calcium and phosphorus. Magnesium is involved in the metabolism of carbohydrates and lipids and acts as a catalyst for a wide array of enzymes. In light of these functions, it is not surprising that magnesium deficiency in animals is manifested clinically in a wide range of disorders, including retarded growth, hyperirritability and tetany, peripheral vasodilatation, anorexia, muscle incoordination and convulsions.

From 20 to 70% of dietary magnesium is absorbed. A number of dietary and physiologic factors negatively influence magnesium absorption, including high levels of dietary phosphorus, calcium, potassium, fat and protein.

The kidneys play a critical role in magnesium homeostasis. Approximately 70% of serum magnesium is filtered by glomeruli; healthy kidneys reabsorb about 95% of the filtered magnesium. Several physiologic and metabolic factors, drugs and disease states influence magnesium reabsorption in nephrons. Certain drugs can cause increased renal wasting of magnesium, including diuretics, aminoglycosides, cisplatin, cyclosporin, amphotericin and methotrexate.

Table 2-2. Mineral functions and effects of deficiencies and excesses.

Mineral	Function	Deficiency	Excess
Calcium	Constituent of bone and teeth, blood clotting, muscle function, nerve transmission, membrane permeability	Decreased growth, decreased appetite, decreased bone mineralization, lameness, spontaneous fractures, loose teeth, tetany, convulsions, rickets (osteomalacia in adults)	Decreased feed efficiency and feed intake, nephrosis, lameness, enlarged costochondral junctions. Increased calcium intake is a risk factor for calcium-containing urinary precipitates; however, moderate- to high-calcium levels may be protective against calcium oxalate precipitates. Calcium in meals may bind with oxalate in the gut decreasing the risk.
Phosphorus	Constituent of bone and teeth, muscle formation, fat, carbohydrate and protein metabolism, phospholipids and energy production, reproduction	Depraved appetite, pica, decreased feed efficiency, decreased growth, dull coat, decreased fertility, spontaneous fractures, rickets	Bone loss, uroliths, decreased weight gain, decreased feed intake, calcification of soft tissues, secondary hyperparathyroidism
Potassium	Muscle contraction, transmission of nerve impulses, acid-base balance, osmotic balance, enzyme cofactor (energy transfer)	Anorexia, decreased growth, lethargy, locomotive problems, hypokalemia, heart and kidney lesions, emaciation	Rare. Paresis, bradycardia
Sodium and chloride	Osmotic pressure, acid-base balance, transmission of nerve impulses, nutrient uptake, waste excretion, water metabolism	Inability to maintain water balance, decreased growth, anorexia, fatigue, exhaustion, hair loss	Occurs only if there is inadequate good-quality water available. Thirst, pruritus, constipation, seizures and death

(Continued on next page.)

Table 2-2. Mineral functions and effects of deficiencies and excesses (Continued.).

Mineral	Function	Deficiency	Excess
Magnesium	Component of bone and intracellular fluids, neuromuscular transmission, active component of several enzymes, carbohydrate and lipid metabolism	Muscular weakness, hyperirritability, convulsions, anorexia, vomiting, decreased mineralization of bone, decreased body weight, calcification of aorta	Uroliths, flaccid paralysis
Iron	Enzyme constituent, activation of O_2 (oxidases and oxygenases), oxygen transport (hemoglobin, myoglobin)	Anemia, rough coat, listlessness, decreased growth	Anorexia, weight loss, decreased serum albumin concentrations, hepatic dysfunction, hemosiderosis
Zinc	Constituent or activator of 200 known enzymes (nucleic acid metabolism, protein synthesis, carbohydrate metabolism), skin and wound healing, immune response, fetal development, growth rate	Anorexia, decreased growth, alopecia, parakeratosis, impaired reproduction, vomiting, hair depigmentation, conjunctivitis	Relatively nontoxic. Reported cases of zinc toxicity from consumption of die-cast zinc nuts or pennies
Copper	Component of several enzymes (oxidases), catalyst in hemoglobin formation, cardiac function, cellular respiration, connective tissue development, pigmentation, bone formation, myelin formation, immune function	Anemia, decreased growth, hair depigmentation, bone lesions, neuromuscular disorders, reproductive failure	Hepatitis, increased liver enzyme activity

(Continued on next page.)

NUTRIENTS

Table 2-2. Mineral functions and effects of deficiencies and excesses (Continued.).

Mineral	Function	Deficiency	Excess
Manganese	Component and activator of enzymes (glycosyl transferases), lipid and carbohydrate metabolism, bone development (organic matrix), reproduction, cell membrane integrity (mitochondria)	Impaired reproduction, fatty liver, crooked legs, decreased growth	Relatively nontoxic
Selenium	Constituent of glutathione peroxidase and iodothyronine 5'-deiodinase, immune function, reproduction	Muscular dystrophy, reproductive failure, decreased feed intake, subcutaneous edema, renal mineralization	Vomiting, spasms, staggered gait, salivation, decreased appetite, dyspnea, oral malodor, nail loss
Iodine	Constituent of thyroxine and tri-iodothyronine	Goiter, fetal resorption, rough coat, enlarged thyroid glands, alopecia, apathy, myxedema, lethargy	Similar to those caused by deficiency. Decreased appetite, listlessness, rough coat, decreased immunity, decreased weight gain, goiter, fever
Chromium	Potentiates insulin action, therefore improves glucose tolerance	Impaired glucose tolerance, increased serum triacylglyceride and cholesterol concentrations	Trivalent form less toxic than hexavalent. Dermatitis, respiratory irritation, lung cancer

Requirements and Excesses *(Pages 74-75)*

AAFCO recommends 0.08% DM magnesium for growth and reproduction and 0.04% DM magnesium for adult maintenance for cats. The AAFCO magnesium recommendation for dogs is 0.04% (DMB) for both lifestages.

Avoiding excess dietary magnesium is recommended for the prevention of struvite urinary precipitates in cats and dogs; however, magnesium deficiency is reported to increase the risk of calcium oxalate urolithiasis in rats. **Table 2-2** describes signs of magnesium deficiency and excess.

Potassium *(Page 75)*

Potassium is the most abundant intracellular cation and the third most abundant mineral in the body. Potassium is involved in a number of functions, including: 1) maintaining acid-base balance, 2) maintaining osmotic balance, 3) transmitting nerve impulses, 4) facilitating muscle contractility and 5) serving as a cofactor in several enzyme systems (energy transfer and use, protein synthesis and carbohydrate metabolism).

Potassium is absorbed primarily by simple diffusion from the upper small intestine, although some absorption also occurs in the lower small intestine and large intestine. Potassium availability is relatively high (95% or higher) for most foodstuffs. Yet, in contrast to most minerals, potassium is not readily stored and must be supplied daily in the diet. Thus, it is important that foods for dogs and cats contain adequate potassium. Increased intake of potassium is unlikely to cause sustained hyperkalemia unless renal excretion of potassium is impaired.

Requirements *(Page 75)*

AAFCO recommends 0.6% potassium (DMB) for both dogs and cats for all lifestages. **Table 2-2** describes signs of potassium deficiency and excess.

Sodium and Chloride *(Page 75)*

Sodium and chloride, in addition to potassium, are important for maintaining osmotic pressure, regulating acid-base equilibrium and transmitting nerve impulses and muscle contractions via Na-K-ATPase (sodium pump).

Hormones, acting to maintain a constant sodium-potassium ratio in extracellular fluid, regulate sodium concentrations in the body. Aldosterone, secreted from the adrenal cortex, regulates reabsorption of sodium from the renal tubules. Antidiuretic hormone from the posterior pituitary responds to osmotic pressure changes in the extracellular fluid. Both hormones maintain a constant sodium-potassium ratio.

Requirements and Excesses (*Page 75*)

The sodium requirement is influenced by a number of factors. The requirement is increased during reproduction, lactation, rapid growth, heat stress and with high dietary potassium levels. In people, the average sodium intake exceeds the recommended requirement by 15-fold. Likewise, typical sodium content of pet foods exceeds the recommended level by four- to 15-fold.

The AAFCO recommendation for sodium in cats is 0.2% (DMB) for both lifestages, whereas in dogs, the recommendation is 0.3% for growth and reproduction (DMB) and 0.06% for adult maintenance (DMB).

In the absence of studies establishing chloride requirements for dogs or cats, the recommendation for chloride is 1.5 times that of sodium. This value is comparable to the Na:Cl requirement ratio for other species. Thus, the chloride recommendation for dogs is 0.45% for growth/reproduction and 0.09% for adult maintenance (DMB). The chloride recommendation for cats is 0.3% for all lifestages (DMB).

The effect of dietary sodium chloride on blood pressure has generally been attributed to the sodium ion. However, it is clear from a number of studies that both sodium and chloride are necessary to inhibit renin production. **Table 2-2** describes signs of sodium chloride deficiency and excess.

Iron (Pages 75-76)

Iron is present in several enzymes and other proteins responsible for oxygen activation (oxidases and oxygenases), for electron transport (cytochromes) and for oxygen transport (hemoglobin, myoglobin).

The amount of iron absorbed from food is determined by three factors: 1) iron status of the body, 2) availability of dietary iron (as affected by other ingredients/nutrients) and 3) amounts of heme and nonheme iron in food.

Requirements and Excesses (*Page 76*)

The iron requirement of kittens and puppies fed a phytate-free purified diet is 80 mg iron/kg of food (DMB). This requirement is the AAFCO recommendation for iron for dogs and cats, for both growth/reproduction and adult maintenance. Most pet foods are high in iron because of the high iron concentrations found in meat ingredients, especially organ meats.

Although iron levels may be high in pet foods (levels sometimes exceed the requirement by 15-fold without supplementation), AAFCO has set a maximum level of 3,000 mg iron/kg of food for dogs (no maximum is established for cats), which clearly exceeds dietary concentrations of iron in most typical pet foods. Iron excesses should be avoided because of potential antagonisms with other minerals (e.g., zinc and copper). **Table 2-2** describes signs of iron deficiency and excess.

Iron oxide is often added to pet foods to impart a "meaty red" color. A relatively high level of iron oxide is added (up to 0.04%) when iron oxide is used as a pigment in pet foods. Analytically, a pet food containing iron oxide will appear to be high in iron, but may not be high in available iron.

Zinc (Pages 76-77)

Zinc is a constituent or activator of more than 200 enzymes; therefore, it is involved in a number of diverse physiologic functions including: 1) nucleic acid metabolism, 2) protein synthesis, 3) carbohydrate metabolism, 4) immunocompetence, 5) skin and wound healing, 6) cell replication and differentiation, 7) growth and 8) reproduction.

Requirements and Deficiencies (*Page 77*)

Zinc deficiency is probably more of a practical concern with pet foods than is toxicity, because: 1) zinc is relatively nontoxic and 2) its availability is decreased by a number of factors (phytate, high dietary levels of calcium, phosphate, copper, iron, cadmium and chromium). The antagonistic effects of calcium are greatest when phytate is also present, resulting in formation of highly insoluble complexes of calcium, phytate and zinc. Signs of zinc deficiency have been reported to occur in dogs fed cereal-based dry foods (e.g., grains may contain significant concentrations of phytate), even when the zinc content of the food exceeded National Research Council (NRC) minimum requirements. **Table 2-2** describes signs of zinc deficiency and excess.

AAFCO recommends 120 ppm zinc (DMB) for dogs and 75 ppm zinc for cats (DMB). For trace minerals, AAFCO makes the same recommendations for adult maintenance and growth/reproduction foods.

Copper (Pages 77-78)

Of the many copper-containing proteins, four enzyme systems may play key roles in the clinical signs associated with copper deficiency: 1) the ferroxidase activity of ceruloplasmin explains in part the disturbances of hematopoiesis in copper deficiency, 2) the monoamine oxidase enzymes may account for the role of copper in pigmentation and control of neurotransmitters and neuropeptides, 3) lysyl oxidase is essential for maintaining the integrity of connective tissue, a function that explains disturbances in lungs, bones and the cardiovascular system and 4) the copper enzymes cytochrome C oxidase and superoxide dismutase (SOD) play a central role in the terminal steps of oxidative metabolism and the defense against the superoxide radicals, respectively. These functions have been postulated to account for the disturbances of the nervous system as seen in neonatal ataxia in several animal species with copper deficiency.

Requirements and Deficiencies (Page 77)

Dietary copper deficiency has been reported to occur in dogs and cats and thus is of practical concern. Availability of copper from different foods and supplements can vary greatly, thus the requirement for copper is difficult to define. The requirement can vary several-fold depending on the source of copper in the food and the level of other ingredients/nutrients/non-nutrients.

The AAFCO recommendation for copper in dogs is 7.3 ppm (DMB). Separate copper requirements are recommended for extruded cat foods (15 ppm) vs. moist cat foods (5 ppm) during growth/reproduction. The recommended AAFCO copper level for maintenance of adult cats is 5 ppm, regardless of the food form.

AAFCO has recommended that pet food companies discontinue the use of copper oxide as a copper source based on studies of swine, poultry and dogs and unpublished data for cats in which researchers have demonstrated the poor availability of copper from copper oxide.

Signs of copper deficiency in cats include poor reproductive performance, early fetal loss, fetal deformities, cannibalism, coat hypopigmentation, kinked tails and inverted carpi. Clinical signs in dogs include hair depigmentation and hyperextension of the distal forelimbs. Bedlington terriers are predisposed to hereditary copper hepatotoxicosis. **Table 2-2** lists signs of copper deficiency and excess.

Manganese (Page 78)

Manganese functions as an enzyme activator and as a constituent of metalloenzymes. Manganese is also essential in bone and cartilage development. In addition, manganese is involved in reproduction and lipid metabolism (e.g., biosynthesis of choline and cholesterol).

Requirements and Deficiencies (Page 78)

AAFCO recommends 5 ppm manganese for dogs and 7.5 ppm for cats (DMB). Manganese deficiency is of little practical relevance to dogs and cats, but is of practical relevance to birds. **Table 2-2** lists signs of manganese deficiency and excess.

Iodine (Pages 78-79)

Iodine is a constituent of the thyroid hormones 3,5,3',5'-tetraiodothyronine (thyroxine, T_4) and 3,5,3'-triiodothyronine (T_3). Thyroid hormones have an active role in thermoregulation, intermediary metabolism, reproduction, growth and development, circulation and muscle function. Thyroid hormones also: 1) influence physical and mental growth and differentiation and maturation of tissues, 2) affect other endocrine glands, especially the hypophysis and the gonads, 3) influence neuromuscular functioning and 4) have an effect on the integument and coat. The thyroid glands actively trap iodine daily to ensure an adequate supply of thyroid hormone.

Requirements/Excesses/Deficiencies (Pages 78-79)

AAFCO recommends an iodine level of 1.5 ppm (DMB) for dogs. This margin of safety is necessary in practical foods to overcome effects of goitrogens and negative mineral antagonisms. The iodine requirement for cats has been extrapolated from other species. AAFCO recommends 0.35 ppm iodine for cats. Recent research suggests that the iodine requirement for adult cats may be higher than the minimum iodine requirement currently recommended by NRC or AAFCO. Findings suggest that levels less than 2.4 ppm iodine may be deficient.

The iodine requirement is influenced by physiologic state and diet. Lactating animals require more dietary iodine because about 10% of the iodine intake is normally excreted in milk. Likewise, the presence of goitrogenic substances and nutrient excesses such as calcium and potassium increase the need for iodine. Potential sources of goitrogens in pet foods include peas, peanuts, soybeans and flaxseed.

Since the late 1970s, feline hyperthyroidism has become a more frequently diagnosed condition. However, much remains to be learned about this endocrinopathy (e.g., prevalence and cause). (See Chapter 24.) Hypothyroidism is a much more prevalent thyroid disorder in dogs. Both iodide excess and deficiency may result in subclinical or overt hypothyroidism. **Table 2-2** lists signs of iodine deficiency and excess.

Selenium (Page 79)

Selenium is an essential constituent of glutathione peroxidase, which helps protect cellular and subcellular membranes from oxidative damage. Glutathione peroxidase and vitamin E work synergistically to reduce the destructive effects of peroxidative reactions on living cells. Selenium spares vitamin E in at least three ways: 1) preserves the integrity of the pancreas, which allows normal fat digestion, and thus normal vitamin E absorption, 2) reduces the amount of vitamin E required to maintain integrity of lipid membranes via glutathione peroxidase and 3) aids retention of vitamin E in plasma by an unknown mechanism.

Vitamin E reduces the selenium requirement in at least two ways: 1) maintains body selenium in an active form, or prevents losses from the body and 2) prevents destruction of lipids within membranes, thereby inhibiting production of hydroperoxides and reducing the amount of the selenium-dependent enzyme needed to destroy peroxides formed in cells. Selenium also has a vital role in maintaining normal thyroid hormone and iodine metabolism, particularly through the control of deiodinase enzymes that regulate conversion of T_4 to T_3.

Requirements/Excesses/Deficiencies (Page 79)

AAFCO selenium recommendations are 0.11 mg/kg of food (DMB) for dogs and 0.1 mg/kg of food for cats (DMB). Recent research in puppies showed that the level of selenium needed to maximize sera glutathione peroxidase and selenium concentrations is 0.21 ppm, which is two times current AAFCO recommendations. Although selenium deficiency has been observed experimentally in dogs, the incidence of selenium deficiency has not been reported for dogs and cats. Likewise, selenium toxicity has not been noted in dogs and cats, despite high concentrations (>4 mg selenium/kg food) in seafood and fish-containing cat foods. **Table 2-2** lists signs of selenium deficiency and excess.

Selenium concentrations in some commercial pet foods may not be adequate due to low availability of selenium in pet food ingredients.

Chromium (Pages 79-80)

Several studies in people and other animals have demonstrated beneficial effects of chromium supplementation (in chromium deficiency or diabetics) including: 1) improved glucose tolerance, 2) reversed hyperglycemia and glycosuria, 3) decreased circulating insulin concentrations, 4) decreased plasma lipid concentrations, 5) decreased body fat, 6) increased protein accretion, 7) improved immune response and 8) reduced cortisol production in response to heat and transport stress. Not all studies have shown improvements in these variables. **Table 2-2** lists signs of chromium deficiency and excess.

VITAMINS (Pages 80-95)

Definition (Page 80)

In order for a substance to be classified as a vitamin, it must have five basic characteristics: 1) it must be an organic compound different from fat, protein and carbohydrate, 2) it must be a component of the diet, 3) it must be essential in minute amounts for normal physiologic function, 4) its absence must cause a deficiency syndrome and 5) it must not be synthesized in quantities sufficient to support normal physiologic function.

A vitamer is chemically the same compound as a vitamin, but may exert varying physiologic effects because it is an isomer. Vitamin E is a good example of vitamers, because of its many forms. α-tocopherol is the most biologically active form, whereas γ-tocopherol has little biologic function, but acts as an in vitro antioxidant. A provitamin is a compound that requires an activation step before it becomes biologically active. β-carotene, for example, is cleaved by enzymatic processes to release two molecules of retinol (vitamin A).

Function (Page 80)

The two main categories of vitamins are distinguished by their miscibility in either lipids (fat soluble) or water (water soluble). There are four fat-soluble vitamins (A, D, E and K) and 10 generally recognized water-soluble vitamins (thiamin [B_1], riboflavin [B_2], niacin, pyridoxine [B_6], pantothenic acid, folic acid, cobalamin [B_{12}], choline, biotin and vitamin C). AAFCO only recognizes three fat-soluble and eight water-soluble vitamins for dogs (vitamin K, biotin and vitamin C are not recognized), and four fat-soluble and nine water-soluble vitamins for cats (vitamin C is not included).

Vitamins have incredibly diverse physiologic functions. Vitamins act as potentiators or cofactors in enzymatic reactions. They also play a significant role in DNA synthesis, energy release from nutrient substrates, bone development, calcium homeostasis, normal eye function, cell membrane integrity, blood clotting, free radical scavenging, amino acid and protein metabolism and nerve impulse transduction (**Table 2-3**).

Because of the differences between fat and water solubility and in chemical structure, vitamins are absorbed in the body through a variety of means. Fat-soluble vitamins require bile salts and fat to form micelles for absorption. They are then passively absorbed, usually in the duodenum and ileum, and transported in conjunction with chylomicrons to the liver via the lymphatic system. In contrast, most of the water-soluble vitamins are absorbed via active transport. Some vitamins (e.g., cobalamin) require a carrier protein called "intrinsic factor" whereas others need a sodium-dependent, carrier-mediated absorption pump.

Deficiency/Adequacy/Toxicity (Pages 80-82)

In general, fat-soluble vitamins are stored in the lipid depots of all tissues, making them more resistant to deficiency, but also more likely to cause toxicity. Conversely, water-soluble vitamins are depleted at a faster rate because of limited storage; therefore, they are less likely to cause toxicity and more likely to be acutely deficient. **Table 2-3** lists signs of deficiency and toxicity.

Factors Affecting Requirements (Pages 82-83)

Different lifestages of animals affect vitamin requirements. Growing and reproducing animals accrete tissues and therefore require higher levels of vitamins, minerals, protein and energy for optimal performance.

Table 2-3. Summary of names, functions and clinical syndromes associated with deficiency and toxicity of vitamins.

Vitamin	Function	Deficiency	Toxicity
Vitamin A	Component of visual proteins (rhodopsin, iodopsin), differentiation of epithelial cells, spermatogenesis, immune function, bone resorption	Anorexia, retarded growth, poor coat, weakness, xerophthalmia, nyctalopia, increased CSF pressure, aspermatogenesis, fetal resorption	Cervical spondylosis (cats), tooth loss (cats), retarded growth, anorexia, erythema, long-bone fractures
Vitamin D	Calcium and phosphorus homeostasis, bone mineralization, bone resorption, insulin synthesis, immune function	Rickets, enlarged costochondral junctions, osteomalacia, osteoporosis	Hypercalcemia, calcinosis, anorexia, lameness
Vitamin E	Biologic antioxidant, membrane integrity through free radical scavenging	Sterility (males), steatitis, dermatosis, immunodeficiency, anorexia, myopathy	Minimally toxic, fat-soluble vitamin antagonism, increased clotting time (reversed with vitamin K)
Vitamin K	Carboxylation of clotting proteins II (prothrombin), VII, IX, X and other proteins, cofactor of the bone protein osteocalcin	Prolonged clotting time, hypoprothrombinemia, hemorrhage	Minimally toxic, anemia (dogs)
Thiamin (B$_1$)	Component of thiamin pyrophosphate (TPP), cofactor in decarboxylase enzyme reactions in the TCA cycle, nervous system	Anorexia, weight loss, ataxia, polyneuritis, ventriflexion (cats), paresis (dogs), cardiac hypertrophy (dogs), bradycardia	Decreased blood pressure, bradycardia, respiratory arrhythmia
Riboflavin (B$_2$)	Component of flavin adenine dinucleotide (FAD) and flavin mononucleotide (FMN) coenzymes, electron transport in oxidase and dehydrogenase enzymes	Retarded growth, ataxia, collapse syndrome (dogs), dermatitis, purulent ocular discharge, vomition, conjunctivitis, coma, corneal vascularization, bradycardia, fatty liver (cats)	Minimally toxic

(Continued on next page.)

NUTRIENTS

Table 2-3. Summary of names, functions and clinical syndromes associated with deficiency and toxicity of vitamins.

Vitamin	Function	Deficiency	Toxicity
Niacin (B₃)	Component of nicotinamide-adenine dinucleotide (NAD) and adenine dinucleotide phosphate (NADP) coenzymes, hydrogen donor/acceptor in energy-releasing dehydrogenase reactions	Anorexia, diarrhea, retarded growth, ulceration of soft palate and buccal mucosa, necrosis of the tongue (dogs), reddened ulcerated tongue (cats), cheilosis, uncontrolled drooling	Low toxicity, bloody feces, convulsions
Pyridoxine (B₆)	Coenzyme in amino acid reactions (transaminases and decarboxylases), neurotransmitter synthesis, niacin synthesis from tryptophan, heme synthesis, taurine synthesis, carnitine synthesis	Anorexia, retarded growth, weight loss, microcytic hypochromic anemia, convulsions, renal tubular atrophy, calcium oxalate crystalluria	Low toxicity, anorexia, ataxia (dogs)
Pantothenic acid	Precursor to coenzyme A (CoA), protein, fat and carbohydrate metabolism in the TCA cycle, cholesterol synthesis, triacylglyceride synthesis	Emaciation, fatty liver, depressed growth, decreased serum cholesterol and total lipids, tachycardia, coma, lowered antibody response	No toxicity established in dogs and cats
Folic acid	Methionine synthesis from homocysteine (vitamin B₁₂ dependent), purine synthesis, DNA synthesis	Anorexia, weight loss, glossitis, leukopenia, hypochromic anemia, increased clotting time, elevated plasma iron, megaloblastic anemia (cats), sulfa drugs interfere with gut synthesis, cancer drugs (methotrexate) are antagonistic	Nontoxic

(Continued on next page.)

Table 2-3. Summary of names, functions and clinical syndromes associated with deficiency and toxicity of vitamins.

Vitamin	Function	Deficiency	Toxicity
Biotin	Component of four carboxylase enzymes: pyruvate carboxylase, acetyl-CoA carboxylase, propionyl-CoA carboxylase and 3-methylcrotonyl CoA carboxylase	Hyperkeratosis, alopecia (cats), dry secretions around eyes, nose and mouth (cats), hypersalivation, anorexia, bloody diarrhea	No toxicity established in dogs and cats
Cobalamin (B_{12})	Coenzyme functions in propionate metabolism, aids tetrahydrofolate-containing enzymes in methionine synthesis, leucine synthesis/degradation	Cessation of growth (cats), methylmalonic aciduria, anemia	Altered reflexes (reduction in vascular conditioned reflexes and an exaggeration of unconditioned reflexes)
Vitamin C	Cofactor in hydroxylase enzyme reactions, synthesis of collagen proteins, synthesis of carnitine, enhances iron absorption, free radical scavenging, antioxidant/pro-oxidant functionality	Liver synthesis precludes dietary requirement, no signs of deficiency have been described in normal cats and dogs	No toxicity established in dogs and cats
Choline	Component of phosphatidyl-choline found in membranes, neurotransmitter acetylcholine, methyl group donor	Fatty liver (puppies), increased blood prothrombin times, thymic atrophy, decreased growth rate, anorexia, perilobular fat infiltration of the liver (cats)	None described for cats and dogs
Carnitine*	Transports long-chain fatty acids into the mitochondria for use in β-oxidation	Hyperlipidemia, cardiomyopathy, muscle asthenia	None described for cats and dogs

*Carnitine is a vitamin-like substance.

Various disease conditions also affect vitamin status. Prolonged anorexia deprives animals of vitamins and other nutrients and depletes vitamin stores. Polyuric diseases such as diabetes mellitus and chronic renal failure may increase the excretion of water-soluble vitamins. Renal failure can also lead to a secondary vitamin D deficiency by reducing the final hydroxylation step converting 25-hydroxyvitamin D_3 to 1,25-dihydroxyvitamin D_3, which occurs in the proximal tubules of the kidneys.

In addition, certain drugs (e.g., antibiotics) may decrease the intestinal microflora responsible for vitamin K synthesis. Also, diuretic therapy may increase excretion of water-soluble vitamins.

Some vitamin requirements depend on levels of other nutrients. The amount of cobalamin required is related to the amount of folic acid, choline and methionine present because these nutrients interact metabolically and are dependent on each other. In addition, the amount of tryptophan influences niacin requirements because tryptophan is the precursor for that vitamin.

Supplementation (Page 83)

All commercial pet foods contain added vitamins. The body uses synthetic and naturally formed vitamins in the same way, although they may have different availabilities. The effects of processing on vitamin stability, availability in conjunction with disputed requirement levels in complex foods make fortification necessary. (See Chapter 4.)

Commercial pet foods, therefore, are fortified to meet an animal's vitamin requirement for a given lifestage, overcome processing and storage losses and avoid toxicity. Because pet foods are fortified with vitamins, it is usually unnecessary, and perhaps contraindicated, to concurrently give multi-purpose vitamin supplements. Supplementation may be warranted in the management of diseases that affect vitamin metabolism, but should be monitored if long-term treatment is planned.

Fat-Soluble Vitamins (Pages 83-87)

Vitamin A (Pages 83-84)

Vitamin A is added almost universally to animal foods. Plants do not contain vitamin A per se, but instead contain provitamins in the form of carotenes (α, β, γ carotenes and cryptoxanthin are considered to be the most potent of the carotenoids). The vitamin A activity of β-carotene is markedly greater than that of other carotenoids, but it has only half the potency of pure vitamin A.

Nutritionally, vitamin A is also important because it is essential for a number of distinct biologic functions. It is necessary for normal vision, bone and muscle growth, reproduction and maintenance of healthy epithelial tissue.

REQUIREMENTS/DEFICIENCIES/EXCESSES In contrast with most other species, dogs and cats have a unique way of metabolizing vitamin A. Cats require preformed vitamin A because they lack the dioxygenase enzyme necessary for β-carotene cleavage. In addition, studies have shown that cats and dogs do not depend on retinol-binding protein to transport vitamin A in plasma.

Recognizing that dogs are more tolerant of vitamin A toxicosis than other animals are and prompted by recent research, the Canine Nutrition Expert/Feline Nutrition Expert (CNE/FNE) Panels of AAFCO increased the maximum vitamin A allowance for dogs from 50,000 to 250,000 IU/kg of food. For cats, the maximum dietary allowance for vitamin A is 750,000 IU/kg of food.

Unlike dogs, cats cannot meet any of their vitamin A requirements from carotenoids. The AAFCO recommendation for vitamin A is 5,000 IU/kg of food (DMB) for dogs (growth/reproduction and maintenance requirements are the same) and 9,000 and 5,000 IU/kg of food (DMB) for growth/reproduction and maintenance in cats, respectively. Little is known about the vitamin A requirements during gestation and lactation. However, a minimum vitamin A allowance has been established for cats at 6,000 IU/kg of food (DMB), which is approximately double the requirement for growth (3,333 IU/kg of food) (DMB).

The appreciable stores of vitamin A in the body are mobilized as needed to mitigate against the effects of low dietary intakes of the vitamin. The only unequivocal signs of vitamin A deficiency are the ocular lesions nyctalopia (night blindness) and xerophthalmia (extreme dryness of the conjunctiva). Other signs include anorexia, weight loss, ataxia, skin lesions, increased susceptibility to infection, retinal degeneration, poor coat, weakness, increased cerebrospinal fluid pressure, nephritis, skeletal defects (periosteal overgrowth and narrowing of foramina) and impaired reproduction.

Vitamin A toxicities have been encountered in numerous species. The most characteristic signs of hypervitaminosis A are skeletal malformation, spontaneous fractures and internal hemorrhage. Other signs include anorexia, slow growth, weight loss, skin thickening, suppressed keratinization, increased blood clotting time, reduced erythrocyte count, enteritis, congenital abnormalities, conjunctivitis, fatty infiltration of the liver and reduced function of the liver and kidneys. **Table 2-3** lists signs of vitamin A deficiency and toxicity.

Vitamin D *(Pages 84-86)*

Vitamin D exists in two forms: cholecalciferol (vitamin D_3, which occurs in animals) and ergocalciferol (vitamin D_2, which occurs predominantly in plants). Cholecalciferol can be produced in the skin of most mammals from the provitamin 7-dehydrocholesterol via activation with ultraviolet-B light.

The primary function of vitamin D is to enhance intestinal absorption and mobilization, as well as retention and bone deposition of calcium and phosphorus. This function is manifested through its activity as a hormone (i.e., 1,25-dihydroxyvitamin D_3).

Vitamin D is distributed relatively evenly among the various tissues where it resides in lipid depots. Vitamin D can be found in fatty tissues such as adipose, kidneys, liver, lungs, aorta and heart.

Several factors tightly regulate the vitamin D endocrine system: 1,25-dihydroxyvitamin D_3, PTH, calcitonin, several other hormones and circulating levels of calcium and phosphate.

REQUIREMENTS/DEFICIENCIES/EXCESSES Requirements for vitamin D depend on: 1) environmental conditions (e.g., exposure to ultraviolet-B light), 2) dietary concentrations (ratio and forms) of calcium and phosphorus, 3) physiologic stage of development and 4) perhaps gender and breed. The AAFCO allowance for vitamin D is 500 IU/kg of food (DMB) for dogs (growth/reproduction and maintenance allowances are the same), and 750 and 500 IU/kg of food (DMB) for cats for growth/reproduction and maintenance, respectively. Research has shown, however, that the vitamin D requirement for growing kittens is markedly lower (250 IU D_3/kg of food) than current AAFCO recommendations.

Signs of vitamin D deficiency are frequently confounded by a simultaneous deficiency or imbalance of calcium and phosphorus. Clinical signs generally include rickets (young animals), enlarged costochondral junctions, osteomalacia (adult animals), osteoporosis (adult animals) and decreased serum calcium and inorganic phosphorus concentrations. Experimental vitamin D deficiency has been produced in cats, resulting in neurologic abnormalities associated with degeneration of the cervical spinal cord. Other signs included hypocalcemia, elevated PTH concentrations, posterior paralysis, ataxia and eventual quadriparesis.

Excessive intake of vitamin D is associated with increases in 25-hydroxyvitamin D_3 levels, with the D_3 form being more toxic than the D_2 form. Excessive vitamin D concentrations may result in hypercalcemia, soft-tissue calcification and ultimately death. **Table 2-3** lists signs of vitamin D deficiency and toxicity.

Vitamin E (Pages 86-87)

Vitamin E functions as an antioxidant in vivo and in vitro. α-tocopherol is the most active biologic form. The γ isomer is the most active in vitro form and is widely used in pet food manufacturing to prevent lipid oxidation in products. It now appears that vitamin E in cellular and subcellular membranes is the first line of defense against peroxidation of vital phospholipids.

Selenium, as part of the enzyme glutathione peroxidase, is a second line of defense that destroys peroxides before they damage membranes. Therefore, selenium, vitamin E and sulfur-containing amino acids, through different biochemical mechanisms, are capable of preventing some of the same nutritional diseases. Vitamin E prevents fatty acid hydroperoxide formation, sulfur-containing amino acids are precursors of glutathione peroxidase and selenium is a component of glutathione peroxidase.

All tissues show linear increases in tocopherol concentrations with increases in tocopherol intake. This relationship differs from that of most other vitamins, which usually have distinct deposition thresholds in tissues other than the liver, and may provide an explanation for the pharmacologic effects of vitamin E. The vitamin is most concentrated in membrane-rich cell fractions such as mitochondria and microsomes.

The need for vitamin E in the diet is markedly influenced by dietary composition. The requirement increases with increasing levels of PUFA, oxidizing agents, vitamin A, carotenoids and trace minerals and decreases with increasing levels of fat-soluble antioxidants, sulfur-containing amino acids and selenium.

REQUIREMENTS/DEFICIENCIES/EXCESSES The AAFCO allowance for vitamin E is 50 IU/kg of food (DMB) for dogs and 30 IU/kg of food (DMB) for cats, for all lifestages. For foods containing fish oils, AAFCO recommends an addition (i.e., above the minimum level) of 10 IU vitamin E/g of fish oil/kg of food.

Signs of vitamin E deficiency are mostly attributed to membrane dysfunction as a result of the oxidative degradation of polyunsaturated membrane phospholipids and disruption of other critical cellular processes. Clinical findings of vitamin E deficiency in dogs include degenerative skeletal muscle disease associated with muscle weakness, degeneration of testicular germinal epithelium and impaired spermatogenesis, failure of gestation, brown pigmentation (lipofuscinosis) of intestinal smooth muscle and decreased plasma tocopherol concentrations. In cats, deficiency signs include steatitis, focal interstitial myocarditis, focal myositis of skeletal muscle and periportal mononuclear infiltration in the liver.

Vitamin E is one of the least toxic vitamins. Animals and people apparently tolerate high levels without adverse effects (i.e., at least two orders of magnitude above nutritional requirements [1,000 to 2,000 IU/kg of food]). However, at very high doses, antagonism with other fat-soluble vitamins may occur, resulting in impaired bone mineralization, reduced hepatic storage of vitamin A and coagulopathies as a result of decreasing absorption of vitamins D, A and K, respectively. A maximum of 1,000 IU/kg food was set by AAFCO for dogs; however, there is no evidence that levels above this are toxic to dogs and may even be beneficial. **Table 2-3** lists signs of vitamin E deficiency and toxicity.

Only plants synthesize vitamin E. The richest sources of vitamin E are vegetable oils and, to a lesser extent, seeds and cereal grains.

Vitamin K (Page 87)

Phylloquinone (K_1) and menaquinone (K_2) are the natural forms of vitamin K. Vitamin K plays a major role in the carboxylation of proteins (factors II, VII, IX, X and proteins C and S) to convert prothrombin to thrombin for normal blood clotting. Vitamin K is also involved in the synthesis of osteocalcin, a protein that regulates the incorporation of calcium phosphates in growing bone.

Any food can be expected to contain a mixture of menaquinones and phylloquinones. In general, such mixtures appear to be absorbed with 40 to 70% efficiency. Although phylloquinones and menaquinones are ingested, much of the vitamin K in tissues is from bacterial origin.

REQUIREMENTS/DEFICIENCIES/EXCESSES Because microbially synthesized vitamin K_2 is readily absorbed by passive diffusion in the colon in most mammalian species, dietary supplementation is unnecessary for most cats and dogs. Coprophagy increases vitamin K absorption in dogs. AAFCO lists no recommendations for vitamin K for dogs, but does recommend supplementation in feline foods containing more than 25% fish (0.1 ppm for all lifestages, DMB). This recommendation is warranted because vitamin K deficiency has been observed in cats fed certain commercial foods containing high levels of salmon or tuna.

Malabsorption diseases, ingestion of coagulant antagonists (e.g., coumarin, indanedione), destruction of gut microflora by antibiotic therapy (sulfonamides and broad-spectrum antibiotics) and congenital defects may influence vitamin K requirements. Vitamin K_3 (menadione) has lower lipid solubility and is the most effective form of vitamin K for cases of malabsorption. Vitamin K_1 is the only form of vitamin K effective in anticoagulant antagonism.

Phylloquinone produces no adverse effects when administered to animals in massive doses by any route. The menaquinones are similarly thought to have negligible toxicity. Menadione, however, can produce fatal anemia, hyperbilirubinemia and severe jaundice. The intoxicating doses appear to be at least three orders of magnitude above those levels required for normal physiologic function. The only reported case of toxicity in dogs occurred when warfarin was ingested, followed by intravenous treatment with 30 mg vitamin K_1 in 5% dextrose and lactated Ringer's solution. **Table 2-3** lists signs of vitamin K deficiency and toxicity.

Water-Soluble Vitamins (Pages 87-94)

Deficiency of B vitamins occurs in veterinary medicine but may be difficult to specifically diagnose because analytical tests are not readily available. Therefore, diagnosis relies almost entirely upon clinical signs and nutrient intake history.

B vitamins are relatively nontoxic and may be supplied to veterinarians in individual or combination forms. Because many of the B-vitamin deficiencies present with overlapping clinical signs, it may be prudent to treat deficiency with vitamin-B complex. If signs are specific for a particular B-vitamin deficiency, and if the single preparation form of the vitamin is available, individual targeted treatment may be initiated. However, individual preparations of B vitamins are often more expensive, and the relative nontoxic levels of B vitamins warrant treatment with the combination form.

Thiamin (Page 88)

Thiamin pyrophosphate (TPP) is the major coenzymatic form of thiamin and is required for only a small number of enzymatic reactions. TPP is involved in the following general scheme of reactions: 1) nonoxidative decarboxylation of alpha-ketoacids, 2) oxidative decarboxylation of alpha-ketoacids and 3) transketolation reactions.

REQUIREMENTS/DEFICIENCIES/EXCESSES Table 2-4 lists AAFCO thiamin allowances for dogs and cats.

Thiamin deficiency may result from inadequate intake of thiamin, attributable to foods with low-thiamin content or processing losses, or high intake of thiamin antagonists. The processing conditions used to prepare commercial pet foods are destructive to thiamin. However, this anticipated loss may be overcome by adding synthetic thiamin before processing.

Thiamin antagonists may be synthetic or natural compounds that modify the thiamin structure rendering it inactive. The natural antagonists include thiaminases (enzymes that degrade thiamin), and polyhydroxyphenols (caffeic acid, chlorogenic acid, tannins), which inactivate thiamin by an oxidoreductase process. Thiaminases are found in high concentrations in raw fish, shellfish, bacteria, yeast and fungi (**Table 2-5**). Cooking destroys thiaminases.

Clinical thiamin deficiency is rarely observed in dogs and cats because most commercial pet foods have adequate supplementation. Signs of defi-

Table 2-4. Blood levels, allowances and tests for B-complex vitamins.

Vitamin	Blood level	Cats Allowance*	Best test
Thiamin	20-90 ng/ml (WB)	5 mg/kg	Erythrocyte transketolase activity
Riboflavin	196-660 ng/ml (WB)	4 mg/kg	Erythrocyte glutathione reductase**
			Urine riboflavin
Niacin	1.8-5.8 µg/ml (WB)	60 mg/kg	Urine methyl nicotinamide or methylpyridones**
Pantothenic acid	104-270 ng/ml (WB)	5 mg/kg	Urinary excretion of pantothenate
Pyridoxine	86-350 ng/ml (P)	4 mg/kg	Blood levels of pyridoxine
			Urinary metabolites of pathway intermediates
Folate	3.2-34 ng/ml (P)	0.8 mg/kg	Serum folate
Cobalamin	120-1,200 pg/ml (WB)	20 µg/kg	Blood levels of cobalamin
			Serum and urine methylmalonic acid
Biotin	1,000-3,000 pg/ml (WB)	70 µg/kg	Urinary biotin
			Urinary organic acids
Choline	180-490 mg/ml (P)	2,400 mg/kg	Plasma choline and phosphatidylcholine

(Continued on next page.)

Table 2-4. Blood levels, allowances and tests for B-complex vitamins (Continued.).

Vitamin	Blood level	Dogs Allowance*	Best test
Thiamin	46-112 ng/ml (WB)	1 mg/kg	Erythrocyte transketolase activity
Riboflavin	185-420 ng/ml (WB)	2.2 mg/kg***	Erythrocyte glutathione reductase**
			Urine riboflavin
Niacin	2.7-12 µg/ml (WB)	11.4 mg/kg	Urine methyl nicotinamide or methylpyridones**
Pantothenic acid	120-380 ng/ml (WB)	10 mg/kg	Urinary excretion of pantothenate
Pyridoxine	40-270 ng/ml (P)	1 mg/kg	Blood levels of pyridoxine
			Urinary metabolites of pathway intermediates
Folate	4-26 ng/ml (P)	0.18 mg/kg	Serum folate
Cobalamin	135-950 pg/ml (WB)	22 µg/kg	Holotranscobalamin II**
Biotin	530-5,000 pg/ml (WB)	None established	Urinary biotin
			Urinary organic acids
Choline	235-800 µg/ml (P)	1,200 mg/kg	Plasma choline and phosphatidylcholine

Key: WB = whole blood, P = plasma.
*Allowances are similar for growth and adult maintenance and are expressed on dry matter basis (AAFCO Official Publication, 2001).
**Not currently available in veterinary medicine.
***Investigators have shown a riboflavin requirement approximately 20 to 33% higher than the AAFCO allowance listed here.

Table 2-5. Thiaminase activity in fish products.

Food	Thiaminase activity*
Marlin	0
Yellowfin tuna	265
Red snapper	265
Skipjack tuna	1,000
Dolphin (mahi mahi)	120
Ladyfish	35
Clam	2,640

*mg thiamin destroyed/100 g fish per hour.

ciency include anorexia, failure to grow, muscle weakness and neurologic dysfunction such as ventriflexion of the head in cats with paresis, and cardiac hypertrophy and ataxia in dogs. **Table 2-3** lists signs of deficiency and toxicity. **Table 2-4** lists thiamin blood values for dogs and cats.

Riboflavin (Pages 88-89)

Flavins are used as coenzymes in about 50 enzymes in mammals. Flavins participate in intermediary energy metabolism and function mainly in oxidoreductase reactions.

REQUIREMENTS/DEFICIENCIES/EXCESSES **Table 2-4** lists AAFCO riboflavin allowances for dogs and cats. Recently, the requirement of riboflavin for adult dogs at maintenance was suggested to be approximately 20 to 33% higher than values currently recommended by AAFCO (2.7 mg/kg vs. 2.2 mg/kg of food). Most commercial pet foods are supplemented with synthetic riboflavin. There appears to be little storage of riboflavin in the body, thus it is critical to supply the vitamin in the daily diet.

Deficiency of riboflavin in dogs and cats is uncommon but may manifest as dermatitis, erythema, weight loss, cataracts, impaired reproduction, neurologic changes and anorexia. Toxicity has not been reported in dogs and cats. **Table 2-3** lists signs of deficiency. **Table 2-4** lists riboflavin blood values for dogs and cats.

Niacin (Pages 89-90)

Niacin is the generic term used to describe compounds that exhibit vitamin B_3 activity (i.e., nicotinamide and nicotinic acid). Tryptophan metabolism is intrinsically linked to niacin requirements and may even affect niacin requirements when both tryptophan and niacin are limiting in food.

Niacin, in its cofactor form, is essential to several physiologic reactions: 1) oxidoreductase reactions, 2) nonredox reactions, 3) cleavage of β-N-glycosidic bonds with transfer of ADP-ribose to proteins (post-translational modification) and 4) formation of cyclic ADP-ribose (mobilizes intracellular calcium).

REQUIREMENTS/DEFICIENCIES/EXCESSES Table 2-4 lists AAFCO niacin allowances for dogs and cats. Cats have the enzymatic machinery to efficiently make niacin from tryptophan; however, one of the intermediates is rapidly siphoned away by another pathway. Thus, cats do not synthesize substantial amounts of niacin from tryptophan and have a strict dietary requirement for preformed niacin. Niacin is a very stable vitamin that is found in a variety of foodstuffs.

Deficiency of niacin results in pellagra with its classic 4D signs: dermatitis, diarrhea, dementia and death. Clinical deficiency is not common in dogs because most commercial foods are supplemented with niacin. Cats, however, are more likely to develop signs of deficiency because of their strict requirement. **Table 2-3** lists signs of deficiency and toxicity. **Table 2-4** lists niacin blood values for dogs and cats.

Pyridoxine (Pages 90-91)
    The biologically active forms of pyridoxine (vitamin B_6) are the coenzymes pyridoxal phosphate (PLP) and pyridoxamine phosphate (PMP). PLP is involved in most reactions of amino acid metabolism, including transamination, decarboxylation, desulfhydration and nonoxidative deamination. PLP is also involved in the catabolism of glycogen and metabolism of lipids. As a coenzyme for decarboxylase enzymes, PLP functions in the synthesis of serotonin, epinephrine, norepinephrine and γ-aminobutyric acid (GABA). Pyridoxine is involved in vasodilatation through the production of histamine and is required in the pathway in which niacin is produced from tryptophan. Pyridoxine helps catalyze the synthesis of taurine from cysteine and participates with ascorbic acid and NAD in the synthesis of carnitine from the amino acid lysine. Pyridoxine is also involved with the synthesis of the heme precursor porphyrin.

REQUIREMENTS/DEFICIENCIES/EXCESSES The dietary requirement for vitamin B_6 is influenced by a number of factors including lifestage, dietary composition and microbial synthesis. AAFCO recommends 4 mg vitamin B_6/kg of food (DMB) for cats and 1 mg vitamin B_6/kg of food (DMB) for dogs. The requirement is the same for growth/reproduction and maintenance for dogs and cats.

Reduced growth, muscle weakness, neurologic signs, (e.g., hyperirritability and seizures), mild microcytic anemia, irreversible kidney lesions and anorexia are signs of pyridoxine deficiency. Oxalate cystalluria is also a notable sign in pyridoxine-deficient cats. **Table 2-3** lists signs of deficiency. **Table 2-4** lists pyridoxine blood values for dogs and cats.

The prevalence of vitamin B_6 toxicity appears to be low. Earliest detectable signs include ataxia and loss of small motor control. Many of the signs of toxicity resemble those of vitamin B_6 deficiency: ataxia, muscle weakness and loss of balance (**Table 2-3**).

Pantothenic Acid *(Page 91)*

Although this vitamin is found in practically all foodstuffs, the quantity present is generally insufficient for most monogastric species. Pantothenic acid occurs mainly in bound form (i.e., acetyl-CoA and acyl carrier protein). Acetyl-CoA is found in all tissues and is one of the most important coenzymes for tissue metabolism. Acetyl-CoA is a major substrate in the TCA cycle for production of ATP from fat (glycerol and fatty acids), glucose and amino acids. Acetyl-CoA is also involved in the synthesis of fatty acids, steroid hormones and cholesterol. Acetyl-CoA is necessary for oxidation of fatty acids, pyruvate and ketoglutarate.

REQUIREMENTS/DEFICIENCIES/EXCESSES The requirement for pantothenic acid is affected by dietary composition and lifestage. Less pantothenic acid is apparently required to optimize growth when high-protein foods are fed, whereas high-fat diets may increase the requirement for pantothenic acid. AAFCO recommends 10 ppm and 5 ppm (DMB) per day for dogs and cats, respectively.

Dogs with pantothenic acid deficiency have erratic appetites, depressed growth, fatty livers, decreased antibody response, hypocholesterolemia and coma, in later stages. Pantothenic acid-deficient cats develop fatty livers and become emaciated. **Table 2-3** lists signs of deficiency. **Table 2-4** lists pantothenic acid blood values for dogs and cats.

Pantothenic acid is generally regarded as nontoxic. No adverse reactions or clinical signs other than gastric upset have been observed in any species following ingestion of large doses.

Folic Acid *(Pages 91-92)*

Folic acid is the name commonly used to designate a family of vitamers with related biologic activity. Other common designations are folate, folates and folacin. Folic acid functions as a one-carbon (methylene, methenyl, methyl) donor and acceptor molecule in intermediary metabolism. Specific pathways include nucleotide biosynthesis, phospholipid synthesis, amino acid metabolism, neurotransmitter production and creatinine formation. In addition, vitamin B_{12} is closely paired with folate in the production of methionine from homocysteine.

REQUIREMENTS/DEFICIENCIES/EXCESSES Table 2-4 lists AAFCO folic acid allowances for dogs and cats. Folate is required in the diet daily; no reserves are kept in the body.

Folate deficiency is characterized by poor weight gain, megaloblastic anemia, anorexia, leukopenia, glossitis and decreased immune function (**Table 2-3**). Folate levels in blood may be measured to confirm a deficiency suggested by clinical signs. **Table 2-4** lists folic acid blood values for dogs and cats. There have been no reported cases of folate toxicity.

Biotin *(Page 92)*

Biotin is an essential cofactor for four different carboxylase reactions in mammals. These carboxylases have important functions in the metabolism of lipids, glucose, some amino acids and energy.

REQUIREMENTS/DEFICIENCIES/EXCESSES Mammalian tissues are incapable of synthesizing biotin. Microbes in the intestine are thought to produce enough biotin to meet one-half of the daily requirement; the other half is supplied by the diet. **Table 2-4** lists AAFCO biotin allowances for dogs and cats.

Naturally occurring biotin deficiency is very rare in dogs and cats. Feeding raw egg whites or oral antimicrobials are probably the two most common causes. Raw egg whites contain the glycoprotein avidin, which binds biotin rendering it unavailable for absorption. Feeding avidin to cats may result in signs of biotin deficiency that include dermatitis, alopecia and a dull coat. Because gut microbial synthesis probably accounts for half of the biotin requirement, antimicrobials that decrease the population of the intestinal microflora may also result in signs of biotin deficiency. Clinical signs include poor growth, dermatitis, lethargy and neurologic abnormalities. **Table 2-3** lists signs of deficiency. Biotin toxicity has not been reported. **Table 2-4** lists biotin blood values for dogs and cats.

Cobalamin *(Pages 92-93)*

The cobalamins are important in one-carbon metabolism. Methylcobalamin, which contains cobalt in the 1+ state, is a coenzyme for methionine synthase. 5'deoxyadenosylmethionine, which contains cobalt in the 2+ state, is a coenzyme for methylmalonyl-CoA mutase. Vitamin B_{12} is required by the enzyme (methionine synthase) that removes a methyl group from methyl tetrahydrofolate (THF) to regenerate THF, which is needed for pyrimidine biosynthesis.

REQUIREMENTS/DEFICIENCIES/EXCESSES **Table 2-4** lists AAFCO cobalamin allowances for dogs and cats. Microbes and yeast can make vitamin B_{12} for absorption by animals. Plants generally contain very low amounts of vitamin B_{12}. Meat and, to some degree, milk products are good sources of vitamin B_{12}. Most commercial pet foods are supplemented with stable, pharmaceutical grade vitamin B_{12}.

Vitamin B_{12} deficiency is very rare but may result in poor growth and neuropathies. Because vitamin B_{12} is only made by microbes and found in animal tissue, a vegetarian diet may lead to deficiency. Toxicity of vitamin B_{12} has not been reported to occur in dogs and cats unless excessive amounts are given parenterally. **Table 2-3** lists signs of deficiency and toxicity. **Table 2-4** lists cobalamin blood values for dogs and cats.

Choline (Pages 93-94)

Choline is classified as one of the B-complex vitamins though it does not entirely satisfy the strict definition of a vitamin. Choline, unlike other B vitamins, can be synthesized by the liver, and is required in the body in greater amounts than the other B vitamins (i.e., 1,200 to 2,400 ppm vs. only 0.02 to 60 ppm). Thus, although choline is an essential nutrient for all animals, not all animals require it as a dietary supplement (e.g., no requirement is established for people and adult rats). Furthermore, choline does not function as a coenzyme or cofactor as do most vitamins.

Choline has four basic functions in metabolism: 1) as phosphatidylcholine, it is a structural element of biologic membranes, 2) as phosphatidylcholine, it promotes lipid transport (as a "lipotrope"), 3) as acetylcholine, it is a neurotransmitter and 4) after conversion to betaine, it is a source of labile methyl groups for transmethylation reactions (e.g., the formation of methionine from homocysteine, or creatine from guanidoacetic acid).

Most species can synthesize choline, as phosphatidylcholine, by the sequential methylation of phosphatidylethanolamine. The activity is greatest in the liver, but is also found in many other tissues. In most species, the choline requirement is greater for younger animals than for adults. Choline may not be required in some adult species.

The requirement for choline is affected greatly by dietary factors such as methionine, betaine, myoinositol, folate, and vitamin B_{12}, as well as the combination of different levels and composition of fat, carbohydrate and protein in the diet. In addition, age, gender, caloric intake and growth rate influence the lipotrophic action of choline and thereby its requirement.

REQUIREMENTS/DEFICIENCIES/EXCESSES AAFCO recommends 1,200 ppm choline (DMB) for dogs and 2,400 ppm for cats (DMB). Methionine can completely replace choline as a methyl donor. For example, in cat foods, if dietary methionine exceeds 0.62% DMB, then choline supplementation is not required.

Choline deficiency in most animal species is characterized by depressed growth, hepatic steatosis and hemorrhagic renal degeneration. Additional signs of choline deficiency in dogs include thymic atrophy and elevated plasma phosphatase values and increased blood prothrombin time. **Table 2-3** lists signs of deficiency. **Table 2-4** lists choline blood values for dogs and cats.

Vitamin C (Page 94)

Because of de novo synthesis, vitamin C is not technically a vitamin for healthy dogs and cats. (See vitamin definition.) However, it is included here because of its biochemical functions, including in vivo and in vitro antioxidant properties.

Vitamin C primarily functions in the body as an antioxidant and free radical scavenger. Ascorbic acid is best known for its role in collagen synthesis, where it is involved in hydroxylation of prolyl and lysyl residues of procollagen. It is also involved in drug, steroid and tyrosine metabolism, and electron transport in cells. Ascorbic acid is also necessary for synthesis of carnitine, an important carrier of acyl groups across mitochondrial membranes. Normal circulating plasma levels are 4 µg/ml in dogs and 3 µg/ml in cats.

More recently, research into the role of ascorbic acid has shifted from prevention of deficiency to the treatment and prevention of disease. Because ascorbic acid protects against free-radical damage induced by the "oxidative burst" of neutrophils, and stimulates the phagocytic effect of leukocytes, it plays a role in immune function. Larger doses may play a protective role against carcinogenesis.

REQUIREMENTS/DEFICIENCIES There are no AAFCO recommendations for vitamin C for dogs or cats. **Table 2-3** lists signs of deficiency.

Vitamin-Like Substances (Pages 94-95)

Vitamin-like substances are compounds that exhibit properties similar to those of vitamins, but do not fit the strict definition of a vitamin. They have physiologic functionality, but questionable essentiality. These compounds can be "conditionally essential" depending upon the metabolic capacity of the animal.

Carnitine (Page 94)

L-carnitine is a natural component of all animal cells. Its primary function is to transport long-chain fatty acids across the inner mitochondrial membrane into the mitochondrial matrix for β-oxidation. Skeletal and cardiac muscle contain 95 to 98% of the L-carnitine in the body.

Clinical signs of L-carnitine deficiency include chronic muscle weakness, fasting hypoglycemia, cardiomyopathy, hepatomegaly and dicarboxylic aciduria (**Table 2-3**). In many cases of L-carnitine deficiency, no clinical signs are apparent.

Carotenoids (Page 95)

A group of pigments called carotenoids also exhibit vitamin-like activity. More than 600 different compounds are classified as carotenoids, but fewer than 10% can be metabolized into vitamin A. (See vitamin A.)

Although carotenoids do not strictly fit the definition of a vitamin for mammalian species, they have biologic activity beyond their provitamin A role. Carotenoids with nine or more double bonds function as antioxidants by squelching singlet oxygen and other reactive oxygen species such as hydroxyl radicals, superoxide anion radicals and hydrogen peroxide, which are produced in normal metabolism. This function is accomplished by the carotenoids sacrificing highly reactive multiple double bonds to the free radicals via hydrogen donation, thereby stabilizing the reactive products. Carotenoids also protect cell membranes by stabilizing the oxygen radicals produced when phagocytic granulocytes undergo respiratory bursts that destroy intracellular pathogens.

Bioflavonoids (Page 95)

The bioflavonoids are another group of red, blue and yellow pigments (noncarotenoids) that have vitamin-like activity. Flavonoids have the ability to perform similarly to vitamin C: reduce capillary fragility and permeability and chelate the divalent metal ions copper and iron. (See vitamin C.) Flavonoid reactions are involved in the antioxidant system for lipid and aqueous environments. Flavonoids have been reported to influence several critical enzyme systems.

Clinical cases that illustrate and reinforce the nutritional concepts presented in this chapter can be found in Small Animal Clinical Nutrition, 4th ed., pp 101-107.

Commercial Pet Foods

For a review of the unabridged chapter, see Crane SW, Griffin RW, Messent PR. Introduction to Commercial Pet Foods and Cowell CS, Stout NP, Brinkmann MF, et al. Making Commercial Pet Foods. In: Hand MS, Thatcher CD, Remillard RL, et al, eds. Small Animal Clinical Nutrition, 4th ed. Topeka, KS: Mark Morris Institute, 2000; 111-126 and 127-146.

PET FOOD FORMS (Pages 111-117)*

COM. FOODS

Commercial pet foods are available in three basic forms: 1) dry, 2) semi-moist and 3) moist. As suggested by the category names, water content differs markedly among the three forms.

Moist Foods (Pages 111-113)

The moisture content of moist foods varies from 60 to more than 87%. The dry matter portion of the food contains all the nonwater nutrients: protein, fat, carbohydrate, vitamins and minerals. Small differences in moisture content greatly affect a moist food's dry matter content. For example, if the moisture contents of Food A, Food B and Food C are 72, 78 and 82%, respectively, the dry matter percentages differ markedly.

Dry matter in Food A = (100 − 72) = 28%
Dry matter in Food B = (100 − 78) = 22%
Dry matter in Food C = (100 − 82) = 18%

The % dry matter difference = (28 − 22) ÷ 22 = 6 ÷ 22 x 100 = 27% more nutrients in Food A than in Food B and (28 − 18) ÷ 18 = 10 ÷ 18 x 100 = 56% more nutrients in Food A than in Food C.

Many moist pet foods contain high levels of meat and meat byproducts. These foods typically contain higher levels of protein, phosphorus, sodium and fat than semi-moist or dry forms.

The high palatability of moist foods is a primary reason this form is fed. The high preference that pets express for moist foods requires portion-controlled feeding to prevent overconsumption.

Moist foods have a low caloric density as fed and typically yield 0.7 to 1.4 kcal (2.93 to 5.86 kJ) metabolizable energy/g food. The lower caloric density and higher packaging costs translate to a higher cost per calorie. Correspondingly, moist foods have the highest daily feeding cost.

---

*Page numbers in headings refer to Small Animal Clinical Nutrition, 4th ed., where additional information may be found.

Dry Foods (Page 113)

Dry pet foods contain 3 to 11% water. The average dry pet food is lower in protein, fat and most minerals on a dry matter basis than the average moist pet food. Dry foods cost about one-third as much as moist foods on a cost-per-calorie basis.

Dry foods are usually acceptable to most pets, but have reduced average preference when compared with moist or semi-moist foods. Dry foods are often perceived as providing dental hygiene benefits. However, the perception that dry foods are superior for dental hygiene is a generalization. (See Chapter 16 and Small Animal Clinical Nutrition, 4th ed., pp 475-504.)

Semi-Moist and Soft-Dry Foods (Pages 113-115)

Semi-moist and soft-dry pet foods have an intermediate water content (25 to 35%), falling in between moist and dry pet foods. These pet food forms are highly palatable and have an average intermediate preference between moist and dry pet foods.

Treats (Pages 115-116)

Pet owners and those taking and interpreting a dietary history can easily ignore the variable contribution of treats to the daily nutrient intake. As a generalization, dietary balance is maintained when less than 10% of the daily intake consists of table scraps or treats and the remainder is a prepared food that is complete and balanced. At low levels, treats can be considered nutritionally trivial except in certain medical conditions. However, excess treat intake interferes with normal appetite and dietary balance and can contribute to obesity. The nutritional content of selected commercial pet treats and human foods used as treats is found in Appendix 3. (Also see Appendix M, Small Animal Clinical Nutrition, 4th ed., pp 1084-1091.)

Supplements (Pages 116-117)

Supplements are different from treats, although the two are sometimes confused. Treats are often nutritionally trivial, but supplements are very concentrated nutrient modules. The proper role of a supplement is to correct a diagnosed nutrient deficiency. Unfortunately, many supplements are over-prescribed and present some risk for abuse and toxicosis.

The most common form of veterinary supplements is a wide variety of vitamin and vitamin-mineral combinations that are used by about 10% of animal owners. Mineral and electrolyte supplements include calcium, phosphorus, sodium, potassium, magnesium, iron and zinc.

Routine use of vitamin-mineral supplements is not needed when dogs and cats eat typical commercial pet food.

PET FOOD MARKETING CONCEPTS (Pages 117-119)

Specific-Purpose Foods (Page 117)

The objective of the specific-purpose concept is to provide a specialized nutrient profile for a particular feeding application. Specific-purpose foods can be divided into lifestage and special needs groups. Lifestage products are formulated to provide appropriate nutrition based on pet age or "lifestage." The primary lifestage types are: 1) growth or puppy/kitten foods, 2) adult or maintenance foods and 3) senior/geriatric foods.

Special needs products provide specialized nutrition for individual pet needs. For example, rapidly growing puppies of the large and giant breeds have increased risk for developmental orthopedic disorders. (See Chapter 17 and Small Animal Clinical Nutrition, 4th ed., pp 505-528.) Therefore, these puppies may benefit from a growth-type food specially modified to control nutritional risk factors such as excess calcium and energy intake. Other examples are light products for overweight animals and active products for animals with higher caloric requirements.

All-Purpose Foods (Page 117)

Many people assume all-purpose foods are manufactured for adult animals. This assumption is based on the fact they are not called puppy or kitten foods and have pictures of adult animals on the package. However, these foods must be balanced to support the nutritional requirements for growth and lactation, even if they are fed to adult or geriatric animals. Thus, all-purpose foods provide nutrients in excess of allowances for adult and geriatric pets.

MEASURING PET FOOD FEATURES AND BENEFITS (Pages 120-126)

Pet Food Preference and Acceptability (Pages 121-125)

The two primary assessment tools are the one-pan acceptance (monadic) test and the two-pan preference test. The one-pan test measures acceptability; that is, it simply determines if a given food is palatable enough to be eaten in sufficient quantity to maintain the subject's body weight in a neutral state. The two-pan preference test measures "choice" between a pair of test foods that are fed simultaneously side by side. In application, the results of a two-pan preference test may indicate that animals distinctly prefer one of the foods, thus the preferred food would be termed more "palatable." However, the "losing" food could still be quite palatable and could be

COM. FOODS

consumed in sufficient quantity to support body weight. For more information, see Small Animal Clinical Nutrition, 4th ed., pp 121-125.

Factors Affecting Food Preferences (Pages 123-125)

WATER CONTENT On average, pets prefer moist foods to semi-moist foods and semi-moist foods to dry foods. This effect is maintained but less clear-cut when intake is adjusted for caloric consumption. Adding water to dry food increases preference for some pets. On the other hand, some animals refuse to eat dry kibbles softened with water. Some dogs and cats strongly prefer or are "addicted" to one food form.

NUTRIENT CONTENT AND INGREDIENT SELECTION Increasing protein levels in typical pet foods has a positive effect on preference for dogs and cats. Savory characteristics may be especially strong for cats; cats seem to prefer the inclusion of some liver in their foods. Cats prefer liver to muscle meats and muscle meats to lung tissue. Dogs prefer beef, pork and lamb to chicken and liver. Horse meat is highly palatable to dogs. Both species prefer fish as a protein source in moist foods; however, the quality of fish is critically important to preference. The type of fish (white vs. "oily" species), the season of the catch and the cut and freshness of the fish are important variables. Other animal-source proteins some pets prefer include whey, cheese and egg.

Although dogs and cats prefer a high-meat content to a high-cereal content, cereal-based foods are acceptable to many animals. Dry foods must be cereal-based for manufacturing purposes. The specific cereal grain(s) and quality and processing parameters affect the olfactory and gustatory characteristics of dry pet foods.

COOKING EFFECTS AND FOOD TEMPERATURE Dogs and cats prefer cooked meats to fresh meats. However, overcooking decreases preference, which is especially important if a "burned" flavor permeates moist cat foods. The controlled cooking of cereal starches during extrusion of dry foods is important to starch digestibility and the pet's preference for the final food. The serving temperature of pet food modifies olfactory cues and mouth feel. Dogs and cats prefer food served at body temperature. Rewarming refrigerated moist foods in a microwave oven can produce "hot spots." Learned aversions to foods may occur following accidental mouth burning.

PALATABILITY ENHANCERS Most dry pet food particles are coated with flavor-enhancing agents such as "digests" of animal tissues. Digests are animal tissues enzymatically altered by proteolytic enzymes. When the tissue digestion process produces a desired amino acid and peptide content, sterilization and acidification of the proteolyzed slurry stop the enzymatic action. The digest is then applied to kibbles as a topical liquid or a coating powder.

Other palatability enhancers include salt, topically or internally applied fats, L-lysine, L-cysteine, monosodium glutamate, sugar and soy sauce. Blood and feather meals, nucleotides, yeasts, whey, cheese powder, fermented meats and yeasts, meat slurries injected at extrusion, hydrolyzed vegetable protein, egg and onion and garlic powders have all been used by various manufacturers to enhance palatability. Artificial flavor technology is becoming increasingly evident; people may detect the odor of bacon, cheese and liquid hickory smoke in some pet foods and treats.

EFFECTS OF PAST FEEDING PATTERNS ON CURRENT FOOD INTAKE
Aversion to new and unfamiliar foods and flavors occurs most commonly when animals receive a single food from an early age. "Novelty" is the behavior of enjoying new foods and flavors. In studies, dogs preferred novel foods and flavor changes when exposed to food rotation from weaning to two years of age. Experience-based ingestive imprinting, aversion and novelty behaviors may help wild animals survive by allowing them to adapt to foods they are unaccustomed to when their typical food becomes scarce.

The surroundings in which a pet eats may also influence conditioned ingestive behaviors. Cats prefer a novel food when fed in their normal housing, but became aversive when the same food is presented in an unfamiliar environment. Additionally, preference tests can differ between a laboratory setting and a home-feeding environment. One effect of presenting a new food is a measurable, transitory increase in food intake. This novelty response may occur even if the old flavor is preferred to the new one. The pet owner's anthropomorphic inference is that flavor boredom is a problem and the experience prompts more frequent flavor rotations. These events may set the stage for pets becoming "finicky."

FINICKY BEHAVIOR Finickiness is defined as excessively particular or fastidious behavior. This behavior is commonly described as a human-caused problem resulting from a pet's conditioned expectations for frequent changes in food variety or flavor. Supermarket shelves contain a proliferating number of varieties and flavors. Some pet owners take advantage of this phenomenon and rotate the flavor they feed daily. Clearly, the emphasis given by many owners to satisfying the food preferences of their pets is a strong indication of how much pets are viewed as human surrogates.

Finicky can also be an intermittent, slow or "picky" eating pattern. In these circumstances, pet health care providers should consider the possibility that the pet is simply being overfed or the owner is confusing an appropriate autoregulation of food consumption with food refusal or flavor boredom. In either case, pet owners may be concerned when their pet's consumption doesn't match the high consumption/gusto portrayed by television advertising.

The pet's body condition score (BCS) will need to be evaluated when helping clients deal with finicky pets. If the BCS is normal (3/5) or the pet is overweight (4/5 to 5/5), the finicky behavior was probably acquired from excessive flavor rotation. Behavior modification, or gradually weaning the pet from a high-frequency flavor rotation to a more stable platform of less frequent changes, may correct the problem. Ritualizing the feeding routine to the same time, place, quantity and brand of food may also help.

FOOD ADDICTIONS Single-ingredient food addictions almost always cause compositional incompleteness or an imbalance in the nutrient profile, leading to nutritional deficiency, or toxicity syndromes. Progressive counter-conditioning (adding dilute pepper sauce to the addicting ingredient) while concurrently offering a complete and balanced pet food of the same general flavor as the addicting substance can be successful.

Digestibility (Page 125)

Digestion is the various mechanical, chemical and bacteriologic degradation processes that reduce a complex food substance into absorbable entities such as amino acids, peptides, fatty acids and disaccharide and monosaccharide sugars. Digestibility is an important pet food feature.

Two measurable aspects are "apparent" and "true" digestibility. Apparent digestibility is quantified by measuring the difference between the dry matter content of an individual nutrient in the food and the quantity in the feces. As an example, the % apparent protein digestibility is calculated:

$$\frac{\text{Protein food} - \text{Protein feces}}{\text{Protein food}} \times 100$$

Nondietary features may influence some fecal nutrient levels. An example would be the protein or other sources of nitrogen in the feces contributed by sloughing intestinal cells, bacteria, mucus, blood, ammonia and urea. Nondietary factors that increase the fecal protein level reduce apparent digestibility.

True digestibility is a calculated value that must be established by first measuring the baseline value of endogenous output when a food devoid of a given nutrient is fed. As an example, the % true protein digestibility is calculated as follows:

$$\frac{\text{Protein food} - (\text{Protein feces} - \text{Endogenous fecal protein})}{\text{Protein food}} \times 100$$

High digestibility yields more available nutrients for passive or active transport in intestinal absorption. Another benefit of increased digestibility is less food is needed to meet a pet's energy and nutrient requirements. Accordingly, high digestibility reduces food costs such that a pet food that appears more expensive to purchase on a unit price basis may actually be a better value than less expensive foods with lower digestibility and caloric density.

The primary determinants of digestibility are differences in ingredient selection and processing. For example, undercooked carbohydrates markedly reduce digestibility. The undigested residue can also alter the pH of intestinal chyme and may produce osmotic effects expressed as decreased stool quality and diarrhea. Additionally, interbreed anatomic differences influence food digestibility in some dogs. In one study, Great Dane dogs had reduced relative gastrointestinal (GI) tract mass (weight) when compared with beagles. Giant-breed dogs also had more rapid oral-colon transit times, more voluminous feces and a higher content of fecal water and electrolytes. These effects were independent of food composition and form. These findings suggest that, compared with smaller dogs, some large dog breeds are more prone to loose stools and may benefit more from highly digestible foods.

Stool Quality (Page 125)

Fecal volume and consistency are of concern to many pet owners. In normal animals, fecal volume correlates with overall dry matter digestibility of the food, whereas the consistency of feces is affected by overall GI motility and colonic function. Higher digestibility influences the quantity and quality of feces. Reduced dry matter intake reduces stool volume and may also improve the form and texture attributes relating to easy "clean-up." Fecal volume, water content and firmness are especially important to owners of urban dogs who must pick up feces from gutters and curbs. These fecal attributes are also important to animal caretakers who care for dogs and cats in kennels and colonies where sanitation may be facilitated by washing elimination areas with high-pressure water sprayers. House-training puppies will also be easier if fecal volume is small and bowel movements are infrequent.

Energy Content and Feeding Costs (Page 126)

For many pet owners, low cost is an important criterion for selecting pet foods. The unit price (cost per weight) is the most obvious way for consumers to compare cost but may be a poor method of judging value. Value-for-money is best evaluated by actually measuring feeding costs (cost per calorie or cost per day or year). Actual feeding cost evaluation may reveal that there are only small price differences between pet foods perceived as "inexpensive" and those perceived as "expensive."

The energy content and digestibility of a pet food directly affect feeding costs. The methods by which energy content can be determined and stated are regulated to ensure standardized reporting, which supports fairness to consumers. In the United States, any label statement for energy content must be limited to kcal of metabolizable energy per kg food and familiar measuring units (per can or measuring cup). Feeding costs are directly related to the energy provided by a given volume of food and the cost of that food volume. True costs of feeding are best reflected by the cost of the food per day or year or the cost per calorie.

The following example demonstrates that veterinarians and their health care team members need to discuss the true cost of feeding with pet owners when clients are concerned about the price of a particular food. A 4.5-kg, three-year-old neutered male cat is diagnosed with lower urinary tract disease due to struvite urolithiasis. A moist veterinary therapeutic food (Food A) is recommended to help prevent further episodes of struvite urolithiasis. The cat's owner is concerned about the "high cost" of the veterinary therapeutic food, but would be willing to use Food A if it costs the same as what she now feeds her cat (Food B, a gourmet grocery brand). This calculation shows that the veterinary therapeutic food costs markedly less to feed than the cat's current food.

	Food A	Food B
Cost/can	$1.46	$0.50
Size of container	425 g	100 g
Cost/g	$0.003	$0.005
Feeding amount (300 kcal/1,255 kJ)	214 g	350 g
Cost/day	$0.74	$1.75
Cost/year	$270	$639

COMMON PET FOOD INGREDIENTS (Pages 138-143)**

Ingredients available for use in the pet food industry range from human nonedible pet food grade byproducts to human grade ingredients found in grocery stores. In the United States, ingredients are legally defined in the official handbook of the Association of American Feed Control Officials (AAFCO) and are listed on the label in order of predominance. (See Chapter 5 and Small Animal Clinical Nutrition, 4th ed.,

**The remainder of this chapter was taken from Chapter 4, Making Commercial Pet Foods, Small Animal Clinical Nutrition, 4th ed., pp 127-146. The content here covers only selected ingredients commonly used in commercial pet foods. See Chapter 4 for an overview of how pet foods are made.

Table 3-1. Common grain ingredients and their carbohydrate or nitrogen-free extract (NFE) concentrations (as fed).

Grains	NFE (%)*
Barley	76
Corn	81
Corn flour	85
Cornstarch	88
Grain sorghum	80
Oat groats	70
Rice	90
Rice bran	46
Rice flour	90
Wheat	78
Wheat flour	82
Wheat middlings	66

*NFE is the nonfiber carbohydrate fraction and is calculated as follows: % NFE = 100% – % moisture – % crude protein – % crude fat – % crude fiber – % ash.

pp 147-161.) This section will discuss the benefits and characteristics of ingredients available to pet food manufacturers.

Carbohydrate Ingredients (Pages 138-139)

Grains are typically classified as carbohydrate ingredients. These ingredients are composed primarily of starch (>60%) with protein, fat and fiber fractions making up most of the balance. The amount of starch, as a percentage of the whole ingredient, will vary depending on the degree of milling, as in the case of whole corn vs. cornstarch. Examples of carbohydrate ingredients used in pet foods are found in **Table 3-1**.

Nutritional Characteristics (Page 139)

Grains primarily provide energy to the food, but are also a source of many different nutrients (protein, fat, fiber, minerals, vitamins) at various proportions. For example, corn contains average levels of protein, but is a good source of linoleic acid. By comparison, rice is lower in fat, phosphorus and magnesium. Barley and rice bran are higher in protein, but also higher in phosphorus. Wheat middlings are higher in protein, but also higher in fiber, and higher in nutrient variability. Grains differ analytically and in nutrient availability. Although corn, rice, barley and oats have nearly identical ileal and total tract carbohydrate digestibility (all >98%), whole grains have differing dry matter digestibility (rice >corn >barley >oats). Varying fiber levels between the grains account for this difference.

Dry Protein Ingredients (Pages 139-140)

Protein ingredients contain more than 20% protein. Protein ingredients typically used in dry pet foods and their protein contents are listed in

Table 3-2. Dry protein sources commonly used in commercial pet foods and their protein concentrations (as fed).

Ingredients	Protein (%)
Chicken meal	63-67
Corn gluten meal	60-64
Dried egg product	43-48
Fish meal	60-65
Lamb meal	48-55
Meat and bone meal	50-55
Poultry byproduct meal	65-70
Rice gluten meal	40-50
Soybean meal	46-50

Table 3-2. Protein ingredients vary widely in the levels of protein and other nutrients they deliver to a formulation. For example, ash represents the total mineral element of the formula (the sum of calcium, phosphorus, magnesium, potassium, sodium, etc.). Ash is the material that remains after combustion or hydrolysis of the organic material. The protein-to-ash ratio is a good indicator of an ingredient's efficiency in providing protein and the ingredient's dry matter digestibility. The higher an ingredient's ash content, the lower its digestibility.

Nutritional Characteristics (Page 140)

Dogs are omnivorous and have lower protein requirements than cats. Therefore, formulations for dogs are more flexible and may include more vegetable proteins. Vegetable proteins typically have higher protein-to-ash ratios and contain some fiber. Soybean meal is an excellent source of the amino acids lysine and tryptophan. However, because dogs prefer animal tissues to vegetable meals, it is advantageous to add animal source proteins to the formulation. A blend of animal tissue meals and vegetable meals is appropriate and often optimal.

Wet Protein Ingredients (Pages 140-141)

Wet protein ingredients are classified as fresh or frozen meats and meat byproducts. These ingredients generally have moisture contents greater than 60%. **Table 3-3** lists the typical protein ingredients used in canned pet foods. Controlling excess minerals in moist foods is easier because the protein-to-ash ratio for fresh meat ingredients is higher overall (because they contain less bone) than that of rendered meat meals used in dry pet foods.

Fiber Ingredients (Page 141)

Fiber ingredients contain levels of crude fiber between 18 and 80%. **Table 3-4** lists fiber ingredients typically used in pet foods and their fiber content.

Table 3-3. Wet protein sources used in commercial pet foods and their protein concentrations (as fed).

Ingredients	Protein (%)
Beef (carcass)	18-22
Chicken (whole, backs, necks)	10-12
Fish (freshwater)	12-15
Fish (ocean)	20-27
Liver (pork, beef, turkey, sheep)	17-22
Meat byproducts (lungs, spleens, kidneys)	15-20

Table 3-4. Common fiber ingredients used in commercial pet foods and their crude fiber concentration (as fed).

Ingredients	Crude fiber (%)
Beet pulp	17-20
Cellulose	72-78
Peanut hulls	52-58
Soymill run	32-36
Wheat bran	13-16

COM. FOODS

Nutritional Characteristics *(Page 141)*

Fiber may be classified as soluble or insoluble based on solubility in water. Soluble fiber is easily fermented in the gut by intestinal flora and provides energy and substrates for colonocyte health. (See Chapter 2 and Small Animal Clinical Nutrition, 4th ed., pp 21-107.) Examples of soluble fiber include pectin, gum and hemicellulose. Beet pulp, citrus pulp and soymill run are good sources of soluble fiber. These types of fibers improve stool consistency without compromising total digestibility.

Insoluble fiber, which is found in cellulose and peanut hulls, improves stool quality (i.e., adds bulk and holds water) and modulates GI motility. Insoluble fiber is also useful in obesity management because it dilutes calories, maintains satiety and can be used at higher levels without causing flatulence. The efficiency of the fiber source is critical for formulation.

Fat Ingredients (Pages 141-142)

Fat ingredients contain more than 50% fat. Fat ingredients typically used in pet foods are animal fat (pork fat, beef tallow, poultry fat) and various types of vegetable oil (soybean, sunflower, corn). Each type of fat has several different grades of quality, as measured by peroxide value and free fatty acids, which indicate rancidity. Selection of high-quality fat ensures a low oxidative potential and increases the palatability of the finished pet food product.

Nutritional Characteristics *(Page 142)*
Fat ingredients are extremely efficient in delivering energy to a food. Fats contribute calories at 2.25 times the rate of carbohydrates or proteins. Use of fat ingredients is the most efficient method of increasing the energy density of a food to limit a pet's consumption of other nutrients. However, preventing or managing obesity in sedentary pets limits the broad application of this approach.

Ingredient Myths and Facts (Page 140)
Sometimes pet foods are marketed on ingredient stories that have consumer appeal. Ingredient stories are simple and believable but sometimes mislead consumers. Animals require nutrients, not ingredients. Ingredients are the means to achieve the nutritional and palatability goals of a product. What are some of the myths and facts surrounding ingredients commonly used in pet foods?

Myth No. 1: Corn is a filler, is poorly digested and causes allergies.
Fact: Fillers are ingredients that serve no nutritional purpose, and corn does not fit that description. Corn is a nutritionally superior grain compared with others used in pet foods because it contains a balance of nutrients not found in other grains. Corn provides a highly available source of complex carbohydrates and substantial quantities of linoleic acid, an essential fatty acid important for healthy skin. Corn also provides essential amino acids and fiber. See the response to Myth 3 below for information about corn as an allergen.

Myth No. 2: Soybean meal causes bloat in dogs.
Fact: Bloat, or gastric dilatation-volvulus, is a condition usually seen in large, deep-chested dogs. Research has shown that food ingredients (moist meat-based vs. dry cereal-based food) do not affect gastric motility and emptying. (See Chapter 22 and Small Animal Clinical Nutrition, 4th ed., pp 725-810.)

Myth No. 3: Corn is highly allergenic.
Fact: There have been only six confirmed cases of allergy to corn in dogs reported in the veterinary literature out of 253 total cases. (See Chapter 14 and Small Animal Clinical Nutrition, 4th ed., pp 431-453.) This equates to a 2.4% incidence rate. Foods most often cited as causing canine food allergy are beef, dairy products and wheat.

Myth No. 4: Chicken meal is superior to poultry byproduct meal.
Fact: Both chicken meal and poultry byproduct meal contain quality protein that is digestible and palatable. Chicken meal, however, contains most-

ly rendered chicken necks and backs, which means it provides more ash per unit protein than poultry byproduct meal. This characteristic may make it less desirable for use in formulations in which controlling the mineral content of the product is indicated. Poultry byproduct meal is a slightly more concentrated protein source (**Table 3-2**).

Myth No. 5: Byproducts are of lesser quality than meat.
Fact: Pet food ingredients including muscle meat are by nature byproducts. Some of the byproducts used in pet foods are ingredients that are considered human grade both domestically and internationally. Examples of these are pork and beef liver, tripe and spleen. Many byproducts such as liver offer superior palatability to muscle meats when used in dog and cat foods.

Myth No. 6: There is one best fiber source.
Fact: Various fiber types can be used to provide distinct functions in pet foods. Though fiber does not serve as a major energy source for dogs or cats, it can help promote normal bowel function, maintain the health of the intestinal tract and aid in the nutritional management of certain diseases. No single fiber source or type can optimally deliver all the benefits fiber can provide in pet nutrition. Insoluble fiber is preferred in weight-loss regimens. Soluble fiber is more appropriate in the maintenance of intestinal tract health. It is important to use the fiber source or sources that achieve the nutritional goals of the product. (See Chapter 2 and Small Animal Clinical Nutrition, 4th ed., pp 21-107.)

Myth No. 7: Cellulose fiber binds minerals and decreases the digestibility of other nutrients.
Fact: As with other fibers, dry matter digestibility decreases with increasing cellulose levels. However, research has shown that fiber type does not affect protein digestibility in dogs. In addition, purified cellulose does not decrease protein digestibility in cats. Purified cellulose is inert when it comes to mineral binding and has no effect on calcium or zinc availability in chicks or iron in dogs. More soluble fibers such as beet pulp bind calcium and zinc in chicks and iron in growing puppies. (See Chapter 2 and Small Animal Clinical Nutrition, 4th ed., pp 21-107.)

Additives (Pages 142-143)

Since 1920, legally sanctioned food additives have been used commonly in human and animal foods. Pet food manufacturers use various additives to generate products with visual appeal, prolonged nutritional quality, palatability and a long shelf life.

Because most commercial pet foods are designed as complete foods, nutrient enrichment with vitamins and minerals is the most important and beneficial use of pet food additives. Most ingredients with unfamiliar, chemical-sounding names are, in fact, nutrients. In general, additives other than vitamins and minerals are found least often or in smallest amounts in moist foods, and most commonly or in largest amounts in dry foods, semi-moist foods, treats and snacks.

The term "additive" is inclusive for anything imparting increased nutritional, gustatory or cosmetic appeal. Additives commonly used in prepared human and pet foods include colors, flavors, flavor enhancers, emulsifying agents, gelling substances, stabilizers, thickeners and processing aids. The terms preservative and additive are often used synonymously, but they are distinctly different. Preservatives are substances added to foods to protect or retard decay, discoloration or spoilage under normal use or storage conditions. Thus, all preservatives are additives, but not all additives serve a preservative function. Many additives have multiple purposes in pet foods as outlined in **Table 3-5**.

PRODUCT SHELF LIFE (Pages 145-146)

Shelf life is the amount of time a product maintains nutritional, microbial, physical and organoleptic (sensory) integrity. The main cause of diminished shelf life in dry products is oxidation. Fat, either bound in the ingredient matrix or applied to the surface of dry products is subject to the second law of thermodynamics, which states that a system follows an irreversible cascade toward entropy, or disorder. The double bonds of polyunsaturated fatty acids are particularly susceptible to attack by oxygen molecules to form fatty acid radicals and peroxide byproducts. This process is initiated by oxygen and catalyzed by iron, copper, light and warm temperatures to create a series of chemical reactions called auto-oxidation. Unless checked, auto-oxidation will decrease palatability and destroy fat and fat-soluble vitamins. Oxidation does not occur in an environment lacking oxygen (e.g., moist pet food); therefore, moist products have a longer shelf life.

There are no recognized industry standards for shelf life, but 12 to 18 months for dry pet foods, nine to 12 months for semi-moist pet foods and 24 months for moist pet foods are reasonable estimates. Dry pet foods preserved with natural antioxidants may have a shelf life markedly shorter than 12 to 18 months because these antioxidants are not as effective as synthetic antioxidants. Shelf-life information for products should be available from manufacturers.

Table 3-5. Common pet food additives.

Antioxidant preservatives
Butylated hydroxyanisole (BHA)
Butylated hydroxytoluene (BHT)
Ethoxyquin
Propyl gallate
Rosemaric acid/rosmarequinone
Tertiary butylhydroquinone (BHQT)
Tocopherols

Antimicrobial preservatives
Calcium propionate
Citric acid
Fumaric acid
Hydrochloric acid
Phosphoric acid
Potassium sorbate
Propionic acid
Pyroligneous acid
Sodium nitrite
Sodium propionate
Sorbic acid

Humectants
Cane molasses
Corn syrups
Propylene glycol
Sorbitol
Sucrose/dextrose

Coloring agents/preservatives
Aluminum potassium sulfate
Artificial color(s)
Azo dyes (tartrazine [FD&C yellow
    No. 5], sunset yellow [FD&C
    yellow No. 6], allura red
    [FD&C red No. 40])
Caramel color
Iron oxide
Natural color(s)
Nonazo dyes (brilliant blue [FD&C
    blue No. 1], indigotin [FD&C
    blue No. 2])
Sodium erythrobate
Sodium metabisulfite
Sodium nitrite
Titanium dioxide

Flavors/flavor enhancers
Artificial flavors
Citrus bioflavonoids

Flavors/flavor enhancers (cont.)
Dehydrated cheese/dried cheese
    powder
Digests
Liver meal
Monosodium glutamate
Natural flavors
Natural smoke flavor

Palatability enhancers
Acidified yeast
Digests
Garlic powder/oil
Hydrochloric acid
L-lysine
Meat extracts (beef, chicken, turkey)
Onion powder/oil
Phosphoric acid
Spices
Sucrose, dextrose cane molasses
Water (moist)
Whey

**Emulsifying agents, stabilizers
    and thickeners**
Diglycerides (of edible fats and oils)
Glycerin
Glyceryl monostearate
Gums (hydrocolloids)
    Chemically modified plant materials
        (sodium carboxymethylcellulose)
    Microbial gums (xanthan gum)
    Seaweed extracts (carrageenan,
        alginates)
    Seed gums (guar gum)
Modified starch
Monoglycerides
    (of edible fats and oils)

Miscellaneous
Charcoal
Mineral oil (reduces dust)
Polyphosphates
    Disodium phosphate
    Sodium tripolyphosphate
    Tetrasodium pyrophosphate
Yucca schidigera extract
    (flavor, odor control)

Antioxidants (Page 146)

Antioxidants are a class of compounds that function as one or more of the following: 1) electron donors, 2) oxygen scavengers, 3) free radical scavengers or 4) hydrogen donors. **Table 3-5** lists common antioxidants used in pet foods. Antioxidants can be synthetic or natural, used in combination with other antioxidants or alone. They also gain synergism with mineral chelators (e.g., citric and ascorbic acid), and emulsifiers (e.g., lecithin, propyl gallate) and have vastly different potencies depending on the matrix being modified and the antioxidant used. Antioxidants bind with free radicals breaking the cascade of auto-oxidation. Synthetic antioxidants (e.g., ethoxyquin and butylated hydroxyanisole [BHA]) are much more effective than the same quantities of natural antioxidants, such as mixed tocopherols or ascorbic acid. Synthetic antioxidants better resist processing losses and are effective longer, thereby extending shelf life.

Nutrient Stability (Page 146)

The oxidation cascade not only creates rancidity with its objectionable odors and flavors, but also destroys the functionality of nutrients. Pet foods contain fat, which provides essential fatty acids and the fat-soluble vitamins A, D, E and sometimes K. These compounds can be markedly reduced by oxidation, possibly leading to a food with vitamin deficiencies. A robust antioxidant system is required to protect these essential nutrients.

CHAPTER 5

Pet Food Labels

For a review of the unabridged chapter, see Roudebush P, Dzanis DA, Debraekeleer J, et al. Pet Food Labels. In: Hand MS, Thatcher CD, Remillard RL, et al, eds. Small Animal Clinical Nutrition, 4th ed. Topeka, KS: Mark Morris Institute, 2000; 147-161.

INTRODUCTION (Page 147)*

The pet food label is the primary means by which specific product information is communicated between a manufacturer or distributor and consumers, veterinarians and regulatory officials.

Reading and interpreting pet food labels is one way veterinarians and pet owners obtain information about a pet food; however, labels do not provide information about digestibility and biologic value. Contacting the manufacturer or nutrition experts for additional information is the best way to compare the quality of pet foods.

PET FOOD LABELS IN THE UNITED STATES
(Pages 147-157)

Regulation in the United States (Pages 147-150)

Several federal and state governmental agencies and organizations are responsible for regulating pet foods in the United States. For more information, see Small Animal Clinical Nutrition, 4th ed., pp 147-150.

Label Design (Pages 150-157)

A pet food label is divided into two main parts: 1) the principal display panel and 2) the information panel (**Figure 5-1**). The Food and Drug Administration (FDA) defines the principal display panel as "the part of a label that is most likely to be displayed, presented, shown or examined under customary conditions of display for retail sale." The principal display panel is the primary means of attracting the consumer's attention to a product and should immediately communicate the product identity. The information panel is defined as "that part of the label immediately contiguous and to the right of the principal display panel" and usually

*Page numbers in headings refer to Small Animal Clinical Nutrition, 4th ed., where additional information may be found.

Figure 5-1. Typical pet food label with all elements.

contains important information about the product. In the United States and some other countries, several items are required by law to be included on the principal display panel and information panel (**Table 5-1**). The following discussion will focus on the major features found on these two portions of the pet food label.

Principal Display Panel (Pages 151-153)

PRODUCT IDENTITY The product identity is the primary means by which a specific pet food is identified by consumers. In the United States, the product identity must legally include a product name but may also include a manufacturer's name, a brand name or both. The brand name is the name by which pet food products of a given company are identified and usually conveys the overall image of the product. The product name provides information about the individual identity of the particular product within the brand. The manufacturer or distributor is not required to include its name as part of the product identity on the principal display panel, but must include its name and address on the label.

Initial assessment of pet foods is best determined by looking at the product name on the principal display panel. The product name is usually descriptive of the food and in the United States is subject to AAFCO

Table 5-1. Key elements found on pet food labels in the United States and Canada.

Principal display panel	Information panel
Product identity	Ingredient statement*
Manufacturer's name	Guaranteed analysis*
Brand name	Nutritional adequacy statement*
Product name*	(Product description)*
Designator*	Feeding guidelines*
(Statement of intent)*	Statement of calorie content
Net weight*	Manufacturer or distributor*
Product vignette	Universal product code
Nutrition claim	Batch information
Bursts and flags	Freshness date

*Elements required on pet food labels in the United States, on labels certified by the Canadian Veterinary Medical Association (CVMA) Program and in some other countries.

(Association of American Feed Control Officials) regulations about composition of ingredients. Percentage rules are important; beef ingredients will be used as an example: 1) the term "Beef" in a product name requires that beef ingredients be at least 70% of the total product (stated another way, beef must be 95% or more of the total weight of all ingredients exclusive of water used in processing), 2) the term "Beef dinner," "Beef platter," "Beef entree" or any similar designation requires that beef ingredients be at least 10% of the total product (stated another way, beef must be at least 25% but not more than 95% of the total weight of all ingredients exclusive of water used in processing), 3) the term "with Beef" is intended to highlight minor ingredients and this example requires that beef ingredients be at least 3% of the total product and 4) the term "Beef flavor" only requires that beef is "recognizable by the pet." The beef flavor designation usually indicates that beef is less than 3% of the total product. An ingredient that gives the characterizing flavor (e.g., beef digest, beef byproducts) can be used instead of the actual named flavor, beef. In fact, some ingredients may be less than 1% of the total product and still appear in the product name as a flavor. This type of regulation is also found in human foods in which the product names cranberry juice, cranberry juice cocktail and cranberry drink indicate different levels of actual juice in the product.

Percentage rules also apply to product names and moisture content of foods. In the United States, the maximum moisture content in all pet foods should not exceed 78%. However, pet foods can have moisture contents higher than 78% if they consist of stew, gravy, broth, juice or a milk replacer that is so labeled. High-moisture pet foods in cans or tins will have a product name with the terms "in sauce," "in aspic," "in gravy" or some similar designation.

NUTRITION CLAIM Nutrition statements appearing on the principal display panel are usually brief. Examples include the terms "complete and nutritious," "100% nutritious," "100% complete nutrition" or some similar designation. A nutritional adequacy statement on the information panel must substantiate nutrition claims such as these on the principal display panel. Manufacturers can substantiate these nutrition claims by meeting the appropriate AAFCO nutrient profile or successfully completing a protocol feeding trial. Nutrition claims substantiation is discussed in more detail below.

Information Panel (Pages 153-157)
INGREDIENT STATEMENT Pet foods sold in the United States must list each ingredient of the food in the ingredient statement. Ingredients are listed in descending order by their predominance by weight according to the product's formula. AAFCO has established the name and definition of a wide variety of ingredients. The ingredient names must conform to the AAFCO name (e.g., poultry byproduct meal, corn gluten meal, powdered cellulose) or should be identified by the common or usual name (e.g., beef, lamb, chicken). Brand or trade names cannot be used in the ingredient statement and no reference to quality or grade of ingredients can be made. Collective terms (e.g., "animal protein products"), allowed for use on livestock and poultry feed labels, are not allowed on pet food labels in the United States.

The list of ingredients may be helpful, although it has some shortcomings that limit its usefulness for evaluating pet foods. The nutritive value of ingredients cannot be identified from the ingredient statement. A consumer must rely on the reputation or word of the manufacturer to assess the nutritive value and safety of the ingredients appearing on the list. A serious limitation of the ingredient statement is that terms such as "meat byproducts" are difficult to evaluate. The nutritive value of various meat byproducts varies widely. As an example, meat byproducts such as liver, kidney and lungs have excellent nutritive value whereas other meat byproducts such as udder, bone and connective tissue have poor nutrient availability.

Manufacturers can also misrepresent the ingredient content of pet foods. A pet food that lists several different forms of the same ingredient separately (e.g., wheat germ meal, wheat middlings, wheat bran, wheat flour) makes wheat-based ingredients appear to be a lower portion of the food than is fact. Because ingredients are listed in descending order by weight, this also allows dry ingredients to appear lower on the list than ingredients that are naturally high in moisture.

This basic principle is commonly used in moist meat-type dog foods in which textured vegetable protein (TVP) is a major portion of the product.

The ingredient list may look like this for food named a "beef dinner": Water sufficient for processing, chicken, meat byproducts, beef, soy flour, food starch. . . . In this type of food, water is typically combined with soy flour to produce TVP. The TVP makes up a predominant portion of the food, but soy flour appears lower on the ingredient statement because it is a "dry" ingredient whereas other components of the food are added "wet." The consumer thinks he or she is purchasing a meat-based moist food (e.g., beef dinner) when the predominant ingredient is soy flour.

This same principle is used in dry pet foods in which "fresh" meats are highlighted. The ingredient list may look like this for a lamb and rice dog food that claims to provide "real lamb meat": Lamb, brewers rice, ground yellow corn, corn gluten meal, oat groats, poultry byproduct meal, beef tallow. . . . Lamb appears first on the ingredient list because its moisture content is higher than that of the other dry ingredients. The predominant portion of the food contains a mixture of grains (rice, corn, oats) rather than "real meat." Pet food additives such as vitamins, minerals, antioxidant preservatives, antimicrobial preservatives, humectants, coloring agents, flavors, palatability enhancers and emulsifying agents that are added by the manufacturer must be listed in the ingredient statement.

GUARANTEED ANALYSIS In the United States, pet food manufacturers are required to include minimum percentages for crude protein and crude fat and maximum percentages for crude fiber and moisture. Guarantees for other nutrients may follow moisture, but a nutrient need not be listed unless its presence is highlighted elsewhere on the label (e.g., "contains taurine," "calcium enriched"). The sliding scale method of listing guarantees as percentage ranges (e.g., 15 to 18%) is not allowed. It is important to recognize that these percentages generally indicate the "worst case" levels for these nutrients in the food and do not reflect the exact or typical amounts of these nutrients. This differs from pet food labels in Europe where "typical" percentages are used.

Crude Protein Crude protein refers to a specific analytical procedure that estimates protein content by measuring nitrogen. Crude protein is an index of protein quantity but does not indicate protein quality (amino acid profile) or digestibility. (See Chapter 2 and Small Animal Clinical Nutrition, 4th ed., pp 21-107.)

Crude Fat Crude fat refers to a specific analytical procedure that estimates the lipid content of a food obtained through extraction of fat from the food with ether. In addition to lipids, this procedure also isolates certain organic acids, oils, pigments, alcohols and fat-soluble vitamins. Because fats have more than twice the energy density of protein and carbohydrates, crude fat can be used to estimate the energy density of the food. If the moisture and crude fiber content of two foods are somewhat similar, the food with the higher crude fat guarantee will usually have the higher energy density.

Crude Fiber Crude fiber represents the organic residue that remains after plant material has been treated with dilute acid and alkali solutions. It is determined by a specific analytical procedure that was originally developed for the wood pulp industry and then applied to animal foods. Although crude fiber is used to report the fiber content of commercial pet foods, it usually underestimates the true level of fiber in the product. Crude fiber is an estimate of the indigestible portion of the food for dogs and cats. (See Chapter 2 and Small Animal Clinical Nutrition, 4th ed., pp 21-107.) The crude fiber method typically recovers a large percentage of cellulose and lignin in a sample, and a variable percentage of hemicellulose and even ash.

Moisture Moisture is determined by drying a sample of the product to a constant weight. The drying procedure measures water in the product as a whole, but does not distinguish between added water and water in the ingredients. Subtle differences in moisture content of moist products can result in marked differences in dry matter content and therefore the economics of feeding a given pet food. Remember, the dry matter content of the food contains all of the nutrients except water. For example, compare the dry matter content of three different moist cat foods, 1) Food A contains 72% moisture, 2) Food B contains 78% moisture and 3) Food C contains 82% moisture.

Food A	100 − 72% water = 28% dry matter
Food B	100 − 78% water = 22% dry matter
Food C	100 − 82% water = 18% dry matter

28 − 22 ÷ 22 x 100 = 27% more dry matter in Food A (72% moisture) vs. Food B (78% moisture)

28 − 18 ÷ 18 x 100 = 55% more dry matter in Food A (72% moisture) vs. Food C (82% moisture)

Therefore, what appears to be a small difference in water content of a food produces a marked difference in dry matter content. Guarantees are usually expressed on an "as is" or "as fed" basis. It is important to remember to convert these guarantees to a dry matter basis when comparing foods with differing moisture content (e.g., moist vs. dry foods).

Ash Although a maximum ash guarantee is not required in the United States, many pet food manufacturers include one on the labels of their foods. In the United States, "low ash" claims are not allowed because "ash" per se is of no true significance. "Low magnesium" claims on cat food labels are allowed if the food meets certain FDA criteria. In such cases, a magnesium guarantee is required. To be labeled as a "low magnesium" food, the product must contain less than 0.12% magnesium, on a dry matter basis, and 25 mg per 100 kcal metabolizable energy.

Ash consists of all noncombustible materials in the food, usually salt and other minerals. A high-ash content in dry and semi-moist foods generally indicates a high magnesium content. However, the ash content of moist cat foods usually correlates poorly with the magnesium content. Excessive magnesium intake may be one risk factor for feline struvite urolithiasis. (See Chapter 21 and Small Animal Clinical Nutrition, 4th ed., pp 689-723.)

NUTRITIONAL ADEQUACY STATEMENT Since 1984, regulations in the United States have required that all pet food labels, with the exception of products clearly labeled as treats and snacks, contain a statement and validation of nutritional adequacy (product description). When a claim of "complete and balanced," "100% nutritious" or some similar designation is used, manufacturers must indicate the method and lifestage that was used to substantiate this claim (**Table 5-2**).

AAFCO regulations allow two basic methods to substantiate claims. The formulation method requires that the manufacturer formulate the food to meet the AAFCO Dog Food Nutrient Profiles or Cat Food Nutrient Profiles. The feeding trial (protocol) method requires that the manufacturer perform an AAFCO-protocol feeding trial using the food as the sole source of nutrition.

AAFCO nutrient profiles are published for two categories: 1) growth and reproduction and 2) adult maintenance. The formulation method allows the manufacturer to substantiate a "complete and balanced" claim by calculating the nutrient content of a food using standard nutrient information about ingredients. **Table 5-2** lists some of the wording that connotes this type of claim. The formulation method is less expensive and time-consuming, but has been criticized because it does not account for acceptability of the food or nutrient availability. A report in 1991 documented that some commercial pet foods that made "complete and balanced" claims by formulation methods alone did not provide adequate growth of normal animals because of poor availability of nutrients in the food.

The feeding trial (protocol) method is the preferred method for substantiating a claim. Feeding tests can result in a nutritional adequacy claim for one or more of the following categories: 1) gestation and lactation, 2) growth, 3) maintenance and 4) complete for all lifestages. AAFCO has published minimum testing protocols for adult maintenance, growth and gestation/lactation. (See Appendix J, Small Animal Clinical Nutrition, 4th ed., pp 1048-1063.) A food that successfully completes a sequential growth and gestation/lactation trial can make a claim for all lifestages. The required terminology for labels of pet foods that have passed these tests is as follows: "Animal feeding tests using AAFCO procedures substantiate that (brand) provides complete and balanced nutrition for (lifestage)." The inclusion of the term "feeding test," "AAFCO feeding studies" or "AAFCO feeding protocols" in a nutritional adequacy statement supports the idea that the food has successfully completed a minimum feeding protocol (**Table 5-2**). The same statement can also be used on product family members found to be "nutritionally similar" to the tested product.

AAFCO feeding trials are minimum protocols. As an example, the adult maintenance protocol uses eight animals that are fed the food as the sole source of nutrition for six months. A veterinarian examines the animals for signs of nutritional deficiency or excess at the beginning of the study and at the end of 26 weeks. Body weight is recorded weekly and minimal laboratory evaluations (total erythrocyte count, hemoglobin, packed cell volume, serum alkaline phosphatase, serum albumin and whole blood taurine in cats) are performed. This type of protocol will usually detect nutrient deficiencies but might not detect some nutrient excesses that may be harmful when fed over a longer period. In this respect, the AAFCO profiles are better because maximum levels of some nutrients are also established. Growth protocols include feeding the food for a minimum of 10 weeks.

Pet foods that are clearly labeled as snacks or treats may make a nutritional adequacy claim but are not required to do so. Pet foods that do not meet AAFCO requirements by either of the standard methods will have a nutritional statement as follows: "This product is intended for intermittent or supplemental feeding only."

Veterinary therapeutic/wellness foods are those products that are intended for use by, or under the supervision or direction of a veterinarian. These foods may contain the nutritional statement "Use only as directed by your veterinarian." In addition to this statement, the food must include a supplemental feeding statement or the appropriate AAFCO lifestage claim. The absence of a feeding test claim on the label does not necessarily mean a product has not passed feeding tests.

FEEDING GUIDELINES In the United States, dog and cat foods labeled as complete and balanced for any or all lifestages must list feeding directions on the product label. These directions must be expressed in common terms and must appear prominently on the label. Feeding directions

Table 5-2. How to interpret label claims of nutritional adequacy.*

Claim 1: "Good Things Dog Food is formulated to meet the AAFCO (Association of American Feed Control Officials) dog food nutrient profile for maintenance of adult dogs."

Interpretation: This food has been formulated to meet the nutrient levels in the AAFCO Dog Food Nutrient Profile for adult maintenance. This product does not meet the nutrient profile for growth/lactation and has probably not undergone AAFCO feeding tests.

Claim 2: "Good Things Cat Food meets the nutrient requirements established by the AAFCO Nutrient Profile for all stages of a cat's life."

Interpretation: This food has been formulated to meet the nutrient levels in the AAFCO Cat Food Nutrient Profile for growth/reproduction and adult maintenance. This product has probably not undergone AAFCO feeding tests.

Claim 3: "Animal feeding tests using the AAFCO procedures substantiate that Good Things Dog Food provides complete and balanced nutrition for the growth of puppies and maintenance of adult dogs."

Interpretation: This food has successfully completed an AAFCO minimum protocol feeding trial for growing puppies (10 weeks of feeding) or is a family member of a tested product.

Claim 4: "Good Things Cat Food provides complete and balanced nutrition for kittens and adult reproducing queens as substantiated by feeding tests performed in accordance with procedures established by the Association of American Feed Control Officials (AAFCO)."

Interpretation: This cat food (or a family member) has undergone AAFCO minimum protocol feeding studies for gestation/lactation and growth. This food would be nutritionally adequate for adult cats but has not undergone an adult maintenance feeding trial and is not recommended by this manufacturer for long-term maintenance of adult cats.

Claim 5: "Complete and balanced nutrition for adult maintenance based on AAFCO protocol feeding studies conducted at the Good Things Nutrition Center."

Interpretation: This food (or a family member) has undergone AAFCO minimum protocol feeding studies for adult maintenance only and has not been tested for gestation/lactation or growth.

Claim 6: "Complete and balanced nutrition for all lifestages of the dog, substantiated by testing performed in accordance with feeding protocols established by AAFCO."

Interpretation: This dog food (or a family member) has undergone AAFCO minimum protocol feeding trials for gestation/lactation and growth.

Claim 7: "Good Things Dog Food meets or exceeds the requirements of the National Academy of Sciences (USA) for the complete nutrition of your dog or puppy."

(Continued on next page.)

Table 5-2. How to interpret label claims of nutritional adequacy* (Continued.).

Interpretation: This food has been formulated to meet or exceed the nutrient levels established for growth and adult maintenance by the National Research Council (NRC) in the United States. This product has probably not undergone feeding tests. This nutrition statement would be illegal in the United States because the NRC nutrient profiles have been replaced by AAFCO Dog Food Nutrient Profiles. However, references to NRC are still made on pet foods sold in countries other than the United States.

Claim 8: "Meets or exceeds the nutritional levels established by the National Research Council specifications for all stages of a cat's life."

Interpretation: This cat food has been formulated to meet or exceed the nutrient levels established for growth, gestation/lactation and adult maintenance by the National Research Council (NRC) in the United States. This product has probably not undergone feeding tests. This nutrition statement would be illegal in the United States because the NRC nutrient profiles have been replaced by AAFCO Cat Food Nutrient Profiles. However, references to NRC are still made on pet foods sold in countries other than the United States.

Claim 9: "Good Things for Dogs: CVMA Certified; Certified by the Canadian Veterinary Medical Association to meet its nutritional standards on the basis of comprehensive feeding trials, chemical analysis and on going monitoring."

Interpretation: This dog food meets or exceeds the standards established by the CVMA Pet Food Certification Program for adult maintenance. The food meets or exceeds the CVMA standards for nutrient content, digestibility and labeling requirements. Nutrient digestibility is the only feeding test performed after the product is initially certified.

*NOTE: Claims 2, 4, 5 and 6 appear on pet food labels in the United States market, but Claim 3 is the preferred wording for products that have passed an AAFCO-protocol feeding trial and Claim 1 is the preferred wording for products that meet the profiles.

should, at a minimum, state "Feed (weight/unit of product) per (weight unit) of dog (or cat)" and frequency of feeding. These feeding statements are general guidelines at best. Because of individual variation, many animals will require more or less food than that recommended on the label to maintain optimal body condition and health.

PET FOOD LABELS IN CANADA (Pages 157-158)

Regulation in Canada (Page 157)
Canadian Government (Page 157)
The Canadian government has few pet food labeling regulations. Three basic mandatory statements must appear in English and French languages on a pet food label for food sold in Canada: 1) product identity, 2) product

net quantity (metric units first) and 3) the manufacturer's or dealer's name and address. A manufacturer or dealer may choose to include more information but the information must only conform to "truth in labeling."

Canadian Veterinary Medical Association Pet Food Certification
(Page 157)

The Canadian Veterinary Medical Association (CVMA) Pet Food Certification Program was established in 1976 as a voluntary, third-party, quality assurance program for pet foods sold in Canada. The CVMA Program establishes nutrient standards, lifestage feeding protocols and digestibility feeding protocols for dogs and cats. Involvement in the CVMA Pet Food Certification Program is not mandatory.

Principal Display Panel (Page 157)

Principal display panels on Canadian pet food containers may vary. The Canadian government requires that product identity and net quantity (net weight) be listed on all principal display panels of pet foods sold in Canada. Other elements of the principal display panel described under United States regulations may appear on the container depending on several factors.

The CVMA Pet Food Certification Program requires more extensive labeling requirements than does the Canadian government. The CVMA Program labeling requirements include product identity, designator and net quantity, which are usually found on the principal display panel. Nutritional claims can be stated but must be substantiated.

Information Panel (Pages 157-158)
Ingredient Statement (Page 157)

Ingredient statements on pet food containers in Canada also vary. Canadian government regulations do not require an ingredient statement. The CVMA Program states that ingredients should be listed on the label in decreasing order of concentration in the product.

Guaranteed Analysis *(Pages 157-158)*

The Canadian government does not require guarantees on pet food labels. Pet foods certified by the CVMA Program must include guarantees similar to those required for pet food labels in the United States. Ash maximums (not more than 6% dry matter) are required for cat foods certified by the CVMA Program, and magnesium maximums (not more than 0.1% dry matter) are required for cat foods that make a "low ash" claim.

Nutritional Adequacy Statement *(Page 158)*

The Canadian government does not have requirements for substantiation of nutritional claims on pet food labels. The CVMA Pet Food

Certification Program has published nutrient standards and protocols for digestibility feeding trials for dogs and cats. Nutrient digestibility, feeding protocols and feeding guideline standards have also been published for "special foods" including light (lite) foods, calorie-reduced foods, geriatric foods, growth foods, gestation/lactation foods and low-ash, low-magnesium cat foods. Feeding trials are incorporated into the standards for geriatric foods (three-month period) and growth foods (weaning to six months). Products that meet these standards can display the CVMA Seal of Certification and use the following words as a nutritional statement: "This product meets nutritional standards established by the Canadian Veterinary Medical Association (CVMA)." In addition to the CVMA certification logo, products certified as special foods may carry language to the effect that: "This product is formulated to provide (claim for level of nutrients)" or "This product meets the CVMA standard for a (type of special food)."

Other Items on Information Panel (Page 158)

In Canada, pet foods certified by the CVMA must provide feeding instructions on the label if they are sold as light, calorie-reduced or geriatric foods. Pet foods certified by the CVMA Program as light, calorie-reduced or geriatric foods have energy density (kcal per gram of dry matter gross energy) standards, but caloric density is not required on the label. No other feeding guideline requirements exist for pet foods sold in Canada.

PET FOOD LABELS IN EUROPE (Pages 158-161)

Regulation in Europe (Pages 158-159)

The regulations about pet food labeling for Europe, as discussed in this chapter, apply primarily to the European Union (EU). Legislation controlling pet food labels originates in EU institutions and is then implemented into national law. Outside the EU, individual countries have different structures and rules.

The European Commission is the main legislative body within the EU and is independent from the council of ministers and from the different member states.

Label Design (Pages 159-161)

Pet food labels in Europe are divided into a principal display panel and the statutory statement, although the distinction is less stringent than that for pet food labels in the United States.

Table 5-3. Information found in the statutory statements of European pet food labels.

Additives
Address of person (company) responsible for the accuracy of declarations
Complete/complementary food
Expiration date and reference to manufacturing date
Ingredient list
Instructions for use
Net weight and/or volume
Reference (batch) number
Species/category
Typical analysis

Principal Display Panel *(Page 159)*

As in the United States, this part of the label gives information about product identity, shows graphics and pictures, includes marketing claims to promote the product and contains descriptions of meat types and other information that companies may choose to convey outside of the statutory statements.

No specific rules apply to the principal display panel other than general legislation concerning misleading claims that applies to all advertising. Labels should not mislead the purchaser; the label must not suggest that the product possesses properties that it does not have, nor should the label imply that the product is special when similar properties are found in other products.

A pet food label must not claim that the product will prevent, treat or cure disease. The label should clearly differentiate between pet foods and foods for human consumption. In the case of moist food, the words "Animal Food" must be written on the can lid in the language(s) of the country.

Statutory Statement *(Pages 159-161)*

GENERAL The mandatory and optional declarations are encapsulated in what is called "The Statutory Statement" (United Kingdom) or "Cadre Réservé" (France). In addition to being visible, legible and indelible, the statutory statement must be separate from all other information on the label (**Table 5-3**). Some of this information may be outside the statutory statement, but the statutory statement must indicate where to find the information. Such information as the "best before" date, net weight and the name and address of the company responsible for the product are often found elsewhere on the label.

A pet food label must indicate whether the food is a complete or a complementary pet food (i.e., whether the food can satisfy all nutritional demands without an additional ration [complete] or whether it must be fed with another product [complementary]). For complementary foods, the other food or supplement should be stated. The description "complete" or "complementary" must be considered in relation to the intended purpose of the food or to the particular lifestage for which it is defined (i.e., adult, growth or all lifestages).

The species or category of animals must be stated with the indication complete or complementary (e.g., Brand X is a complete food for adult dogs). This statement of intent is often communicated on the principal display panel, but is repeated in the statutory statement.

INGREDIENT LIST In Europe, ingredients are declared by the individual name or grouped under various categories. These categories are designed to provide consumers with some indication of the source of raw materials used, while allowing the manufacturer some flexibility in the selection of the ingredients within a specific category. These categories are well defined and names and descriptions are officially published. Ingredients should be listed in descending order by weight of each individual ingredient or category.

Vitamins are considered additives and are not listed under ingredients. Water does not have to be declared as an ingredient even if added during processing.

TYPICAL ANALYSIS Contrary to pet food labels in the United States, where minimum and maximum guarantees are stated, the EU regulations dictate that the typical analysis must be declared for: 1) crude protein, 2) crude fat, 3) crude fiber and 4) ash. Moisture must be declared if it exceeds 14%. Typical analysis (percentage) is the average of the nutrient level calculated from several samples and should correspond with the target level of each nutrient for which precise limits of variation are defined. The typical analysis gives the percentages found in the actual food. Declaration of nutrients such as calcium, phosphorus, sodium, potassium and magnesium is optional. Energy declaration is forbidden in the EU except for some veterinary dietetic pet foods. Other nutrients must be declared if a manufacturer wants to draw attention to them by saying a food is "high in" or "low in" a particular nutrient.

ADDITIVES Five types of substances are commonly declared as additives: 1) vitamins, 2) copper, 3) preservatives, 4) antioxidants and 5) coloring agents. Vitamins A, D and E must be declared when added by the manufacturer. The added amount should be declared although some countries ask the manufacturer to declare the total amount of the vitamins found in the food. Vitamins are declared in IU or in mg/kg of food.

If a container has a net weight of up to 10 kg, the manufacturer can use the following statements: "Contains European Economic Community (EEC) permitted antioxidant(s)," "Contains EEC permitted preservative(s)" or "Contains EEC permitted colorant(s)." However, if a container has a net weight of more than 10 kg, the name of the additive must be stated in the following way: "with antioxidant X," "with preservative Y," or "preserved with Y" and "with colorant Z" or "colored with Z." Only those additives are declared that have been added during production of the food. Additives (e.g., preservatives or antioxidants) added during rendering in order to preserve raw materials (e.g., meat or fish meals) do not have to be declared on the label.

FEEDING INSTRUCTIONS Feeding instructions are compulsory on European pet food labels but are not as strictly regulated as in the United States. The manufacturer will usually list the weight of the food to feed per body weight of the animal.

LABELS

CHAPTER 6

Making Pet Foods at Home

For a review of the unabridged chapter, see Remillard RL, Paragon B-M, Crane SW, et al. Making Pet Foods at Home. In: Hand MS, Thatcher CD, Remillard RL, et al, eds. Small Animal Clinical Nutrition, 4th ed. Topeka, KS: Mark Morris Institute, 2000; 163-181.

INTRODUCTION (Page 163)*

Feeding commercially prepared pet foods offers several advantages over feeding homemade foods. Most commercial foods are easier to use, less expensive and provide better nutritional balance. Nevertheless, many owners prefer to prepare homemade foods. Therefore, veterinarians and their health care teams must also be able to provide good advice about home-cooked pet foods to clients who prefer this option.

It is important for veterinarians to understand why some clients want to prepare pet foods at home (**Table 6-1**). In many cases, it is possible to address their concerns and to recommend an appropriate commercial food. In one survey, at least 25% of pet owners said they would be influenced to purchase a specific brand of pet food based on a recommendation by their veterinarian. However, when owners strongly prefer to cook for their pet, it is better to provide them with a well-designed homemade recipe, rather than allow them to prepare food according to their own or a breeder's well-intentioned formulation that may have deficiencies and excesses.

COMMON PROBLEMS WITH HOMEMADE FOODS
(Pages 169-171)

It is possible to achieve the same nutrient balance with a homemade food as with a commercially prepared food. However, this largely depends on the accuracy and competence of the veterinarian or animal nutritionist formulating the food, and on the compliance and discipline of the owner. Unfortunately, some homemade recipes are flawed even when followed exactly and consistently. In one survey, 90% of the homemade elimination foods prescribed by 116 veterinarians in North America were not nutritionally adequate for adult canine or feline maintenance. Unlike most commercial foods, many printed homemade recipes are not com-

*Page numbers in headings refer to Small Animal Clinical Nutrition, 4th ed., where additional information may be found.

Table 6-1. Common reasons pet owners and veterinarians prefer homemade foods.

1. They wish to use ingredients that are fresh, wild grown, organic or natural.
2. They wish to avoid additives that are present in some commercial pet foods.
3. They wish to avoid contaminants thought to be present in prepared foods.
4. They are concerned that the ingredient list is an indecipherable list of chemicals.
5. They fear an ingredient in a commercial food, such as a "byproduct."
6. They wish to maintain adequate food intake in a finicky pet through exceptional palatability.
7. They desire to personally cook for the pet.
8. The pet is addicted to table foods or a single grocery item.
9. They wish to feed major quantities of an ingredient not found in commercial pet foods.
10. They hope to construct a nutritional profile for dietary management of a disease for which no commercial food is available.
11. They hope to restrict the allergens/causative substances during an elimination trial or for long-term feeding of animals with adverse reactions to food.
12. They wish to support a sick or terminally ill animal through home cooking and hand feeding.
13. They wish to provide food variety as a defense against malnutrition, or because of the popular idea that pets need variety.
14. They wish to lower feeding costs by using significant quantities of table food and leftovers.
15. They wish to feed a pet according to human nutritional guidelines (e.g., low fat, low cholesterol).

HOME FOODS

plete or balanced to fulfill animal requirements. Few of the numerous published homemade food recipes for dogs and cats have been tested to document performance over sustained periods. Additionally, making homemade foods requires knowledge, motivation, additional financial resources and careful, consistent attention to recipe detail to ensure a consistent, balanced intake of nutrients.

Formulations for homemade foods should not be assumed to be complete or balanced for any canine or feline lifestage until sufficiently tested (feeding tests, nutrient analysis, etc.). Most recipes have been crudely balanced using the average nutrient content of specific foods and computer assimilation. The palatability, digestibility and safety of these recipes have not been adequately or scientifically tested. Even formulations that are initially complete and balanced put pets at risk when pet owners make their own food substitutions, omit ingredients because of personal preferences or convenience or make preparation errors. Therefore, veterinarians and their health care teams should encourage regular dietary histories and patient monitoring for pets that belong to clients who feed home-made foods.

Common Nutrient Problems in Homemade Foods
(Pages 169-170)

Many formulations contain excessive protein, but are deficient in calories, calcium, vitamins and microminerals. Commonly used meat and carbohydrate sources contain more phosphorus than calcium; therefore, homemade foods may have inverse calcium-phosphorus ratios as high as 1:10. Most homemade foods for dogs contain excessive quantities of meat, often far exceeding the animal's protein and phosphorus requirements.

Feline foods designed by clients are commonly deficient in fat and energy density and/or contain an unpalatable fat source (vegetable oil). Homemade foods are rarely balanced for microminerals and vitamins because veterinary vitamin-mineral supplements are not complete nor are the nutrients well balanced within the product.

Vegetarian and Vegan Foods (Page 165)

Pet owners who want to feed a vegetarian food to their dog or cat may assume they must prepare the food at home. Commercially prepared vegetarian foods exist for dogs, and can be well balanced using egg and milk products. Vegan foods (no animal products) should be carefully checked because they may be deficient in arginine, lysine, methionine, tryptophan, taurine, iron, calcium, zinc, vitamin A and some B vitamins. Owners should be discouraged from preparing vegetarian or vegan foods at home for cats because cats are strict carnivores. Without adequate supplementation, cats fed vegetarian and vegan diets are at high risk for taurine, arginine, tryptophan, lysine and vitamin A deficiency.

Common Ingredient Problems in Homemade Foods
(Pages 170-171)

Many owners who make their pet's food according to published canine or feline recipes, over time, make their own ingredient substitutions that may or may not be correct. In addition, foods made at home are typically designed from a variety of table foods, and generally have no consistent ingredient composition. Inconsistency is the rule.

The second most common error made by pet owners who cook for their pets is to eliminate the vitamin-mineral supplement because of its inconvenience, expense or a failure to understand its importance. Foods made from recipes that were once crudely balanced become grossly unbalanced when owners eliminate supplements. Regular veterinary checkups are necessary to monitor the patient's progress and response to the food and to monitor the owner's level of compliance.

Some owners and breeders encourage the use of uncooked meat, liver and eggs in their homemade pet food recipes. This practice can be dangerous because uncooked animal ingredients can harbor pathogenic bacteria that normally would be killed during cooking. (See Chapter 7 and Small Animal Clinical Nutrition, 4th ed., pp 183-198.)

■ RECOMMENDING HOMEMADE FOODS (Pages 171-178)

Veterinarians should be willing to: 1) assess an existing recipe, 2) make appropriate formula substitutions for the client, if necessary or 3) offer nutritionally adequate recipes.

Assessing Recipes (Pages 171-173)
A method to correct an inadequate homemade formulation is to adjust the proportions or change the ingredients in the recipe. Homemade formulations can be checked for nutritional adequacy and adjusted using the "quick check" guidelines below.

1. Do Five Food Groups Appear in the Recipe? (Page 171)
- A carbohydrate/fiber source from a cooked cereal grain.
- A protein source, preferably of animal origin, or if more than one protein source is used, one source should be of animal origin.
- A fat source.
- A source of minerals, particularly calcium.
- A multivitamin and trace mineral source.

2. Is the Carbohydrate Source a Cooked Cereal and Present in a Higher or Equal Quantity than the Meat Source? (Page 171)
The carbohydrate to protein ratio should be at least 1:1 to 2:1 for cat foods and 2:1 to 3:1 for dog foods. Carbohydrate sources for dog and cat foods are used for energy and are usually a cereal such as cooked corn, rice, wheat, potato or barley. These carbohydrate sources have similar caloric contributions, but some carbohydrate sources also contribute a significant amount of protein, fiber and fat. A specific carbohydrate may be chosen based on specific changes in the patient's protein, fat and fiber requirements. For example, soybean may be substituted for corn if more protein is needed, or peas may be substituted if more fiber is needed.

3. What is the Type and Quantity of the Primary Protein Source? (Pages 171-172)
The overall protein quality of a homemade food can be improved by substituting an animal-source protein for a vegetable-source protein.

HOME FOODS

Skeletal muscle protein from different animal species has very similar amino acid profiles. The protein content of various mammalian and avian skeletal muscle tissues is generally equivalent on a water-free basis. Thus, there is no great advantage to feeding one meat source over another. Any cooked animal protein source should provide the majority of a dog's or cat's essential amino acids. The final food should contain 25 to 30% cooked meat for dogs and 35 to 50% cooked meat for cats.

Providing some liver in the meat portion is recommended once a week or no more than half of the meat portion on a regular basis. Liver corrects most potential amino acid deficiencies in homemade foods for dogs and cats. Liver not only improves the amino acid profile over that provided by vegetable and skeletal meat sources, but also contributes essential fatty acids, cholesterol, energy, vitamins and microminerals. If a pet owner requests an ovolacto vegetarian food, eggs are the best ingredient. If a vegan food is requested, soybeans provide the next best, but incomplete, amino acid profile.

4. Is the Primary Protein Source Lean or Fatty? *(Page 172)*

The fat content of different cuts of meat varies. When the specified protein source is "lean," an additional animal, vegetable or fish fat source should compose at least 2% of the formula weight for dogs, and 5% of the formula for cats to ensure adequate energy density and essential fatty acids. If a homemade food lacks sufficient caloric density (fat), the addition of cooked beef or chicken fat, poultry skins, vegetable or fish oils (tuna, mackerel, sardine) can markedly increase the caloric density without adding other nutrients. Changing the cut of meat can also markedly increase the fat content of a food.

5. Is a Source of Calcium and Other Minerals Provided? *(Page 172)*

A homemade food is almost never spontaneously balanced in minerals; an absolute calcium deficiency is common. Unfortunately, pet owners erroneously assume cottage cheese, cheese or milk added in small quantities to homemade pet foods provides adequate calcium. Most foods require a specific calcium supplement. When the protein fraction equals or is greater than the carbohydrate fraction, usually only calcium carbonate is added to the food (0.5 g/4.5-kg cat/day and at least 2.0 g/15-kg dog/day). Calcium carbonate, containing 40% calcium and <1% phosphorus, is available in various size tablets from most pharmacies, health food stores and grocery stores.

Calcium and phosphorus supplementation may be necessary when the protein fraction is less than the carbohydrate fraction. Steamed bone meal, dicalcium phosphate and certain proprietary mineral supplements contain approximately 27% calcium and 16% phosphorus (about 2:1) and microminerals. These supplements, fed at the same dose as calcium carbonate, usually correct the calcium and phosphorus content.

6. Is a Source of Vitamins and Other Nutrients Provided? (Pages 172-173)

Supplements providing vitamins, microminerals, fatty acids and specific nutrients of concern for cats and dogs can be obtained, but they may be cumbersome to feed and greatly increase the cost of the food. An adult over-the-counter vitamin-mineral tablet that contains no more than 200% of the recommended daily allowances for people works well for both dogs and cats at one-half to one tablet per day. One tablet per day of a human adult product will not oversupplement pets with calcium, phosphorus, magnesium, vitamins A, D and E, iron, copper, zinc, iodine and selenium, according to Association of American Feed Control Officials' (AAFCO) maximum allowances for canine and feline foods. In general, veterinary supplements contribute between 0 and 300% of the vitamin-mineral requirements of dogs and cats.

Cats fed a homemade formula exclusively should receive 200 to 500 mg taurine daily, depending on the calculated taurine content of the food. Iodized salt should be used whenever salt is added to the food. It is difficult to meet the iodine requirement without using the iodized form (400 μg of iodine/6 g [1 tsp] sodium chloride).

Making Ingredient Substitutions (Page 173)

When formulating a homemade recipe, proportions of carbohydrate, protein, fat and fiber must be maintained. **Table 6-2** suggests starch, meat, fat and fiber ingredient substitutions and their relative nutrient values. When substituting one ingredient for another, determine the relative nutrient value of the old ingredient and that of the replacement ingredient. If the old recipe recommended 75 g of rice, and the owner would like to use pumpkin instead of rice, 200 g of pumpkin will be needed to supply the same amount (15 g) of carbohydrate as 75 g of rice (**Table 6-2**). For more information about formulating homemade foods, see Small Animal Clinical Nutrition, 4th ed., pp 173-177.

Nutritionally Adequate Recipes (Pages 168-171)

Tables 6-3 and **6-4** are homemade food recipes for healthy adult dogs and cats, respectively. These recipes would be considered all-purpose foods. (See Small Animal Clinical Nutrition, 4th ed., p 117.)

There are also recipes for homemade dietary animal foods for clinical patients. **Table 6-5** contains recipes for reduced protein/low-phosphorus foods for dogs and cats with kidney disease. (See Chapter 19 and Small

HOME FOODS

Table 6-2. Ingredient substitution lists.

Ingredients	Major nutrient	18-kg dog	4.5-kg cat
Starch, cooked	Carbohydrate	60 g	12 g
Meat, cooked	Protein	28 g	9 g
Fat	Fat	10 g	10 g
Fiber	Dietary fiber	10 g (max)	5 g (max)

Starch: These foods in these amounts yield 15 g carbohydrate with 3 g protein, trace fat and 80 kcal

Bread	25 g	Breadsticks, raisin, rye, whole wheat, white
	30 g	Bagel, English muffins, buns, rolls, pita, tortilla
Cereal	20 g	Ready to eat cereals
	25 g	Bran cereals, shredded wheat
	30 g	Bran flakes, Chex
	100 g	Cooked cereals and grits
Grains	20 g	Cornmeal, flour, cornstarch, popcorn, tapioca
	75 g	Rice
	100 g	Barley, pasta
Vegetables	50 g	Baked beans, sweet potato
	75 g	Beans, peas, lentils, plantain
	80 g	Corn
	100 g	Corn on the cob, lima beans, green peas, potato, yam
	150 g	Squash, parsnips
	200 g	Pumpkin

Protein: Should be weighed after cooking and after bone, skin and excess fat have been removed
Low fat: These foods in these amounts yield 7 g protein with 3 g fat and 55 kcal

Beef	30 g	Baby beef, chipped beef, flank tenderloin, plate ribs, round (bottom, top), all rump cuts, lean spareribs, tripe, ground beef (>90% lean), USDA good and choice
Dairy	30 g	Cottage cheese
	45 g	Cheeses (low fat 3 g or less/oz.)
Fish	30 g	Fresh or frozen, tuna or mackerel canned in water
Mixed meats	30 g	Low-fat luncheon meats with 3 g fat or less/oz., >90% lean
Pork	30 g	Leg, tenderloin, ham, Canadian bacon
Poultry	30 g	Chicken or turkey meat without skin
	90 g	Egg whites
Seafood	30 g	Clams, crab, lobster
	50 g	Scallops
	60 g	Shrimp
	90 g	Oysters

(Continued on next page.)

Table 6-2. Ingredient substitution lists (Continued.).

Veal	30 g	Leg, loin, rib, shank, shoulder
Wild game	30 g	Venison, rabbit, squirrel, pheasant, goose (no skin)

Medium fat: These foods in these amounts yield 7 g protein with 5 g fat and 73 kcal

Beef	30 g	Ground beef (>80% lean), corned beef, rib eye
Dairy	30 g	Cheese: mozzarella, ricotta, farmer
Fish	30 g	Tuna, salmon canned in oil, drained
Lamb	30 g	Leg, rib, sirloin, loin, shank, shoulder
Mixed meats	30 g	Low-fat luncheon meats with 3-5 g fat/oz., 85-90% lean
Organ meats	30 g	Liver, kidney, heart, sweetbreads
Pork	30 g	Loin, shoulder arm and blade, butt
Poultry	30 g	Chicken or turkey meat with skin, duck and goose well drained of fat
	50 g	Egg whole
Veal	30 g	Cutlet
Vegetable	120 g	Tofu

High fat: These foods in these amounts yield 7 g protein with 8 g fat and 100 kcal

Beef	30 g	Ground beef (<80% lean), brisket, chuck, ribs, USDA prime
Dairy	30 g	Cheese spreads, all regular American, blue, cheddar, Monterey, Swiss
Lamb	30 g	Breast, ground
Mixed meats	30 g	Cold cuts, sausages
	45 g	Frankfurter
Pork	30 g	Spareribs, back ribs, ground, country style and deviled ham, sausage
Veal	30 g	Breast
Vegetable	30 g	Peanut butter

HOME FOODS

Fats: These foods in these amounts yield 5 g fat with 45 kcal

Monounsaturated	5 g	Margarine with soybean, cottonseed, partially hydrogenated oils, peanut oil, olive oil
Polyunsaturated	5 g	Soft tub margarine, oil (safflower, corn, sunflower, cottonseed, sesame)
	15 g	Diet margarine with safflower, corn, sunflower oil
Saturated	5 g	Chicken fat, beef fat, bacon fat, lard, butter, shortening
	15 g	Heavy cream, cream cheese
	30 g	Sour cream, nondairy substitutes, gravy
	45 g	Light cream, half & half

(Continued on next page.)

Table 6-2. Ingredient substitution lists (Continued.).

Fiber: Grams of dietary fiber per 100 g of these foods	
Low (0-2 g)	Asparagus, cucumber, lettuce, zucchini, alfalfa sprouts, eggplant, mushrooms, celery, green pepper, tomatoes
Medium (2-4 g)	Bamboo shoots, carrots, peas, string beans, bean sprouts, chickpeas, pinto beans, summer squash, broccoli, cauliflower, pumpkin, turnips, cabbage, kidney beans, spinach, watercress
High (5 g or more)	Beans (white, red, lima, black, broad, soy)

Animal Clinical Nutrition, 4th ed., pp 563-604.) **Table 6-6** contains recipes for low-fat/high-fiber foods for overweight dogs and cats. (See Chapter 13 and Small Animal Clinical Nutrition, 4th ed., pp 401-430.) **Table 6-7** contains recipes for low-sodium/low-mineral homemade foods for dogs and cats with heart disease. (See Chapter 18 and Small Animal Clinical Nutrition, 4th ed., pp 529-562.) **Table 6-8** contains a recipe for a low-protein/low-purine food for dogs with urinary calculi. (See Chapter 20 and Small Animal Clinical Nutrition, 4th ed., pp 605-688.) **Table 6-9** is a recipe for a low-residue food for dogs with gastrointestinal disease. (See Chapter 22 and Small Animal Clinical Nutrition, 4th ed., pp 725-810.)

Because these foods are for clinical patients, it is imperative that clients follow these recipes meticulously. Well-intentioned but ill-informed substitutions, or other modifications could result in ineffective, or even counterproductive, nutritional therapy. Furthermore, it is unlikely that these recipes will be as effective as commercial foods in managing clinical patients, even when followed closely.

Additional Instructions (Pages 173-178)

Explaining the importance of a balanced food and providing practical recommendations about how to mix and cook the food will increase compliance.

Some owners and breeders encourage the use of uncooked meat and eggs in their homemade pet food recipes. Pet owners should be informed that uncooked meat and eggs can harbor pathogenic bacteria that are normally killed during cooking. (See Chapter 7 and Small Animal Clinical Nutrition, 4th ed., pp 183-198.) Animal ingredients (meat and eggs) should be cooked for at least 10 minutes at 82°C (180°F). Vegetable ingredients should be washed or rinsed and cooked if increased digestibility is desired. Owners can make homemade foods that lack preservatives and antioxidants in three- to seven-day batches, but must refrigerate the food in airtight containers between meals (0 to 4°C [32 to 40°F]). Larger quantities of food can be frozen (-20°C [<0°F]). Because

Table 6-3. Balanced generic homemade formula for healthy adult dogs that meets AAFCO allowances.

Daily food formulation for an 18-kg dog

Ingredients	Grams	Percent	Nutrient content (% DMB)	
Carbohydrate, cooked*	240	58	Protein	21
Meat, cooked**	120	29	Fat	20
Fat***	10	2	Crude fiber	6.5
Fiber†	30	7	Calcium	0.66
Bone meal††	4.0		Phosphorus	0.59
Potassium chloride†††	1.0		Magnesium	0.1
			Sodium	0.2
			Potassium	0.6
			kcal (as fed)	820

*Examples include rice, cornmeal, oatmeal, potato, pasta and various infant cereals.
**Examples include all typical meats, poultry, fish and liver.
***Chicken fat, beef fat, vegetable oil or fish oil.
†Prepared high-fiber cereals (All Bran, Fiber One) or vegetables (raw or cooked).
††Dicalcium phosphate can replace bone meal.
†††Readily available as a salt substitute in grocery stores.
Human adult vitamin-mineral tablet (9 g/tablet, give 1 tablet/day).

Directions: Bake or microwave meat component and cook starch component separately. Grind or finely chop meat if necessary. Pulverize the bone meal or dicalcium phosphate. Mix with all other components except the vitamin-mineral supplement. Mix well and serve immediately or cover and refrigerate. Feed the vitamin-mineral supplement with the meal; give as a pill or pulverize and thoroughly mix with food before feeding.

HOME FOODS

homemade foods are relatively high in moisture and lack a preservative system, they are highly susceptible to bacterial and fungal growth when left at room temperatures for more than a few hours.

Pet owners should be encouraged to use a dietary gram scale to weigh ingredients until they become familiar with the approximate volumes of each food.

Cooking is necessary to improve the digestibility of starch in carbohydrate sources. Cooking also destroys anti-nutrient factors that may be present (e.g., antitrypsin in soybeans, thiaminase in some fish). However, carbohydrate and animal protein sources should be cooked separately. Carbohydrate sources need a longer cooking time to increase starch digestibility, due to swelling and gelatinizing of starch granules. Meat and liver, on the other hand, should not be overcooked to avoid protein denaturation.

After cooking, all ingredients should be thoroughly mixed (in a blender) to prevent the animal from picking out single food items.

Owners should be warned that although vitamins and minerals are present in only small quantities, they are very important and are not option-

Table 6-4. Balanced generic homemade formula for healthy adult cats that meets AAFCO allowances.

Daily food formulation for a 4.5-kg cat				
Ingredients	Grams	Percent	Nutrient content (% DMB)	
Carbohydrate, cooked*	60	50	Protein	31
Meat, cooked**	40	34	Fat	28
Fat***	10	8	Crude fiber	2.0
Bone meal†	1.2		Calcium	0.69
Salts (NaCl/KCl††)	1.0		Phosphorus	0.58
Taurine	0.5		Magnesium	0.1
			Sodium	0.4
			Potassium	0.75
			kcal (as fed)	250

*Examples include rice, cornmeal, oatmeal, potato, pasta and various infant cereals.
**Examples include all typical meats, poultry, fish and liver.
***Chicken fat, beef fat, vegetable oil or fish oil.
†Dicalcium phosphate can replace bone meal.
††Readily available as lite salt in grocery stores.
Human adult vitamin-mineral tablet (9 g/tablet, give 0.5 tablet/day).

Directions: Bake or microwave meat component and cook starch component. Grind or finely chop meat if necessary. Pulverize the bone meal and mix with other components except the vitamin-mineral supplement. Mix well and serve immediately or cover and refrigerate. Feed the vitamin-mineral supplement with the meal; give as a pill or pulverize and mix in food before feeding.

al. Vitamin-mineral supplements should not be cooked or heated or stored with the food. Vitamins may be destroyed by heat and oxidation. The vitamin-mineral supplement should be kept separate from the food, and administered just before, during or after a meal to ensure proper dosing.

The food should be warmed to just below body temperature before feeding. Clients should be advised to carefully check for "hot spots" that could burn a pet's mouth after food has been warmed in a microwave oven.

Wetting the food may improve palatability. The moisture content of homemade foods is approximately 70%, which is similar to that of moist food. Animals that favor dry forms may reject homemade foods.

When stored too long, the food mixture may separate and dry out, becoming less palatable. Therefore, it is best not to prepare large amounts of food that cannot be eaten in a few days. Mixing the food before warming improves palatability.

Vegetable and meat sources may be substituted for similar ingredients in a recipe (**Table 6-2**). Feeding a variety of foods decreases the risk that a particular nutrient might be below requirements long enough to cause clinical signs of deficiency. Clients should receive a list of possible substitutes, and be informed that inappropriate substitutions may jeopardize nutritional balance.

Table 6-5. Balanced reduced-protein/low-phosphorus homemade formulas for adult dogs and cats with kidney disease.*

Daily food formulation for an 18-kg dog (as fed)

Ingredients	Grams	Nutrient content (% DMB)†	
Rice, white, cooked**	237	Dry matter	41.0
Beef, regular, cooked***	78	Protein	21.1
Egg, large, boiled	20	Fat	13.7
Bread, white	50	Linoleic acid	1.8
Oil, vegetable	3	Crude fiber	1.4
Calcium carbonate	1.5	Calcium	0.43
Salt, iodized	0.5	Phosphorus	0.22
Total	390	Potassium	0.26
		Sodium	0.33
		Magnesium	0.09
		Energy (kcal/100 g)	445

Daily food formulation for a 4.5-kg cat (as fed)

Ingredients	Grams	Nutrient content (% DMB)†	
Liver, chicken, cooked	21	Dry matter	37.8
Rice, white, cooked**	98	Protein	24.4
Chicken, white, cooked	21	Fat	17.5
Oil, vegetable	7	Linoleic acid	7.9
Calcium carbonate	0.7	Crude fiber	0.85
Salt, iodized	0.5	Calcium	0.54
Salt, substitute (KCl)	0.5	Phosphorus	0.29
Total	149	Potassium	0.66
		Sodium	0.42
		Magnesium	0.09
		Energy (kcal/100 g)	458

*Also feed one human adult vitamin-mineral tablet daily to dogs and one-half tablet to cats to ensure all vitamins and trace minerals are included. Cats should be given one-half to one taurine tablet (500 mg/tablet) daily.
**May substitute rice baby cereal and flavor either selection with meat broth during cooking.
***Retain the fat.
†Nutrients of concern are italicized.

In practices where homemade foods are regularly recommended, the staff should have experience preparing homemade food recipes. Furthermore, it is worthwhile and may be cost effective to send the most commonly recommended formulas to a food analytical laboratory to confirm the calculated analysis. (See Appendix U, Small Animal Clinical Nutrition, 4th ed., pp 1134-1144.) In the United States, AAFCO provides valuable guidelines for minimum and maximum nutrient allowances within which a food for healthy dogs and cats should be formulated, if no feeding tests are done. These guidelines are also a useful target for formulating homemade foods. (See Appendix J, Small Animal Clinical Nutrition, 4th ed., pp 1048-1063.)

Table 6-6. Balanced low-fat/high-fiber homemade formulas for overweight adult dogs and cats.*

Daily food formulation for an 18-kg dog (as fed)

Ingredients	Grams	Nutrient content (% DMB)***	
Chicken, white meat	65	Dry matter	36.5
Egg, large, boiled	81	Protein	22.6
Rice, white, cooked**	325	*Fat*	8.0
Cereal, All Bran	26	Linoleic acid	1.1
Calcium carbonate	2	*Fiber*	5.8
Salt, iodized	1	Calcium	0.50
Salt substitute (KCl)	1	Phosphorus	0.37
Total	501	Potassium	0.63
		Sodium	0.45
		Magnesium	0.14
		Energy (kcal/100 g)	398

Daily food formulation for a 4.5-kg cat (as fed)

Ingredients	Grams	Nutrient content (% DMB)***	
Liver, chicken, cooked	125	Dry matter	33.8
Rice, white, cooked**	46	Protein	52.7
Cereal, All Bran	8	*Fat*	11.4
Calcium carbonate	1.2	Linoleic acid	1.2
Salt, iodized	0.3	*Fiber*	5.2
Salt, substitute (KCl)	0.3	Calcium	0.85
Total	180	Phosphorus	0.77
		Potassium	0.67
		Sodium	0.44
		Magnesium	0.11
		Energy (kcal/100 g)	420

*Also feed one human adult vitamin-mineral tablet daily to dogs and one-half tablet to cats to ensure all vitamins and trace minerals are included. Cats should be given one-half to one taurine tablet (500 mg/tablet) daily.
**May substitute rice baby cereal and flavor either selection with meat broth during cooking.
***Nutrients of concern are italicized.

▮ PATIENT ASSESSMENT AND MONITORING (Page 178)

Patients that eat homemade foods should be presented for veterinary examinations two to three times annually. Because the nutritional profile of homemade foods is quite variable, a nutritional review is recommended at least twice a year. If a dog or cat eats a homemade food exclusively for more than six months, the veterinarian should ask the client to record and submit a three- to five-day food history so that the nutrient profile and ingredient substitutions can be reevaluated.

The effectiveness of a food can be grossly evaluated by noting the

Table 6-7. Balanced low-sodium/low-mineral homemade formulas for adult dogs and cats with heart disease.*

Daily food formulation for an 18-kg dog (as fed)

Ingredients	Grams	Nutrient content (% DMB)†	
Beef, regular, cooked**	94	Dry matter	38.7
Rice, white, cooked***	330	Protein	20.8
Cereal, All Bran	9.0	Fat	12.4
Oil, vegetable	2.0	Linoleic acid	1.0
Calcium carbonate	2.0	Fiber	2.9
Salt, substitute (KCl)	1.0	*Calcium*	0.49
Total	438	*Phosphorus*	0.26
		Potassium	0.59
		Sodium	0.12
		Magnesium	0.11
		Energy (kcal/100 g)	431

Daily food formulation for a 4.5-kg cat (as fed)

Ingredients	Grams	Nutrient content (% DMB)†	
Beef, lean, cooked**	67	Dry matter	37.9
Rice, white, cooked***	67	Protein	36.4
Calcium carbonate	0.7	Fat	21.5
Salt, iodized	0.1	Linoleic acid	0.73
Salt, substitute (KCl)	0.1	Fiber	0.65
Total	135	*Calcium*	0.55
		Phosphorus	0.28
		Potassium	0.54
		Sodium	0.17
		Magnesium	0.07
		Energy (kcal/100 g)	500

*Also feed one human adult vitamin-mineral tablet daily to dogs and one-half tablet to cats to ensure all vitamins and trace minerals are included. Cats should be given one-half to one taurine tablet (500 mg/tablet) daily.
**Retain the fat.
***May substitute rice baby cereal and flavor either selection with meat broth during cooking.
†Nutrients of concern are italicized.

HOME FOODS

patient's body weight, body condition and activity level. Laboratory data such as albumin level, red blood cell number and size and hemoglobin concentration are gross estimations of the animal's nutritional status and can be used with other clinical observations to evaluate homemade foods. More specifically, the skin and hair should be examined closely and an ophthalmic examination, including evaluation of the lens and retina, should be performed. These tissues are more sensitive than others to nutritional status. Stool quality should also be assessed.

Veterinarians should always: 1) offer to have a homemade recipe evaluated by a nutritionist and 2) recommend the feeding of a complete and

Table 6-8. Balanced low-protein/low-purine homemade formula for adult dogs with urinary calculi.*

Daily food formulation for an 18-kg dog (as fed)

Ingredients	Grams	Nutrient content (% DMB)**	
Rice, white, cooked	431	Dry matter	29.5
Egg, large, boiled	49	*Protein*	9.8
Oil, vegetable	27	Fat	21.8
Calcium carbonate	1.2	Fiber	2.2
Salt, substitute (KCl)	1.2	Calcium	0.38
Total	509	*Phosphorus*	0.10
		Energy (kcal/100 g)	483

*Also feed one human adult vitamin-mineral tablet daily.
**Nutrients of concern are italicized.

Table 6-9. Balanced low-residue homemade formula for adult dogs with gastrointestinal disease.*

Food formulation for an 18-kg dog (as fed)

Ingredients	Grams	Nutrient content (% DMB)**	
Rice, white, cooked	232	Dry matter	27.7
Cottage cheese	232	Protein	30.4
Egg, large, boiled	116	Fat	15.6
Oil, vegetable	2.0	*Fiber*	0.71
Salt, substitute (KCl)	1.0	Calcium	0.42
Calcium carbonate	1.0	Phosphorus	0.39
Total	585	Energy (kcal/100 g)	450

*Also feed one human adult vitamin-mineral tablet daily.
**Nutrients of concern are italicized.

balanced commercial product as often as possible. This is especially true if the pet has a medical condition for which dietary management depends on the highest level of diet consistency and quality assurance.

Clinical cases that illustrate and reinforce the nutritional concepts presented in this chapter can be found in Small Animal Clinical Nutrition, 4th ed., pp 179-181.

CHAPTER 7

Food Safety

For a review of the unabridged chapter, see Miller EP, Cullor JS. Food Safety. In: Hand MS, Thatcher CD, Remillard RL, et al, eds. Small Animal Clinical Nutrition, 4th ed. Topeka, KS: Mark Morris Institute, 2000; 183-198.

CLINICAL IMPORTANCE (Pages 183-186)*

Foodborne diseases can be divided into: 1) food infections (usually bacterial) and 2) food intoxication (microbial toxicoses) (**Figure 7-1**). Food infections such as salmonellosis and salmon disease (*Neorickettsia helminthoeca*) result from the ingestion of infectious microbial cells that invade the host's tissues and produce the disease after an appropriate incubation period. Because it takes time for these cells to replicate to pathogenic numbers, clinical disease in food infections does not become evident until at least 12 to 24 hours after ingestion.

Food poisonings, or more specifically "food intoxications," do not depend on the ingestion of viable cells, but result from the ingestion of a food that already contains a microbial toxin. Because cell replication is not required, the signs of food poisoning appear rapidly, sometimes less than one hour after ingestion. The term "food poisoning" is often incorrectly used as a synonym for foodborne illness.

Today, foodborne disease in household pets is rare. The 1990 annual report of the American Association of Poison Control Centers (AAPCC) indicated that of the 41,854 dog and cat poisoning cases reported, foodborne illnesses accounted for only 1.7% of the total.

The low incidence of foodborne diseases can be attributed to two primary changes in the way domestic pets are fed. First, most pets in developed countries depend totally on commercial pet foods to meet their nutritional needs. Second, present-day commercial pet foods are much safer than in the past.

In addition to manufacturing quality control and storage improvements, pet foods and individual pet food ingredients are regulated by several governmental agencies to ensure safety. Most regulatory agencies, both domestic and international, use monitoring programs to maintain surveillance over pet food products.

The use of home-prepared pet foods also has clinical relevance to foodborne disease. Many breeders and individuals in the dog racing industry

*Page numbers in headings refer to Small Animal Clinical Nutrition, 4th ed., where additional information may be found.

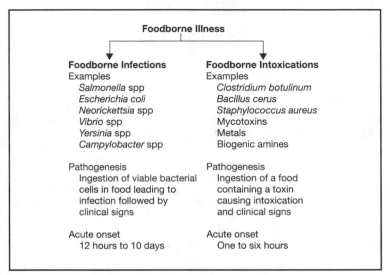

Figure 7-1. Classification of foodborne illnesses.

have their own special food formulas. Raw meat may make up 50 to 75% of the food consumed by racing greyhounds in the United States. These foods are a common source of foodborne illness because uncooked meat may contain large numbers of bacterial cells and the foods are mass produced without proper hygiene or adequate cooking.

ASSESSMENT (Pages 186-193)

Assess the Animal (Pages 186-190)

The most important goal in dealing with a case of suspected foodborne illness is to obtain an accurate diagnosis. One must adhere stubbornly to the principles of a proper toxicologic investigation, including the careful evaluation of information supplied by: 1) the history, 2) clinical signs, 3) postmortem findings, 4) chemical analyses and 5) laboratory animal tests.

History (Pages 186-187)

Although an adequate history is important in all clinical cases, it is especially important when foodborne illness is suspected because some of the critical facts in the case may be lacking.

First, ascertain when and what clinical signs first appeared. From here, veterinarians can annotate the sequence and relevant facts about the events that transpired before the patient's presentation. For example, it is important to know what day the neighborhood trash was left out for collection, especially if it was the day before the illness occurred. The recent application of a pesticide to the premises or yard coupled with signs typical of pesticide toxicity would be important, particularly if the pet owner applied the pesticide instead of a professional exterminator.

Exposure to other toxicants in the pet's environment, such as pesticides or cleaning chemicals, is a distinct possibility and should be thoroughly investigated. The AAPCC report noted that drugs and household products are responsible for 71% of canine and feline poisonings. This report also noted that 89% of pet poisonings are from sources within the home and its immediate surroundings.

Contamination of a commercial pet food will usually produce an epizootic of sick pets within a wide geographic area. The veterinarian should contact the pet food manufacturer to determine if similar cases have been reported. Company technical personnel can also help the differential diagnosis process by supplying key information about product testing, additional areas of investigation and beneficial laboratory tests. If other animals in the same household are eating food from the same bag or container and remain asymptomatic, then implication of the food is diminished and other possible causes should be investigated.

However, even if it appears that the commercial food has been exonerated, one should not end the investigation here because the commercial food could still be involved if the pet owner has compromised the product's integrity by improper storage or usage. Request that the pet owner bring the food in its container to the hospital for testing.

Physical Examination (Page 187)
Although signs of gastrointestinal (GI) disease may be obvious, one should also be alert for other clinical signs such as cutaneous lesions or signs that might signify central nervous system or hepatic disease (**Table 7-1**). Patients may have signs such as dehydration, seizures or high fever that require symptomatic or supportive treatment before a final diagnosis is made. If possible, samples of vomitus and feces should be obtained for laboratory testing.

Clinical Laboratory Testing (Page 187)
Clinical laboratory testing should be performed routinely in all suspected foodborne illness cases. Many such illnesses are short-lived so hematologic and serum biochemistry values may be within normal limits. However, clinical biochemistry values would be invaluable in establishing

Table 7-1. Clinical signs of various foodborne illnesses.

Clinical signs	Causative agents
Vomiting/diarrhea	S. aureus, Salmonella spp, Neorickettsia spp, E. coli, B. cereus, Yersinia spp, Campylobacter spp, biogenic amines, aflatoxins, vomitoxin, cyclopiazonic acid, lead, arsenic, zinc, cadmium
Liver disorders, jaundice	Aflatoxins, fumonisins, lead, arsenic, rubratoxin, Yersinia spp
Blood disorders, (anemia, hemorrhages)	Aflatoxins, Neorickettsia spp, lead, onions, garlic, rubratoxin, cyclopiazonic acid, mycotoxins
CNS/nervous disturbances	C. botulinum, fumonisins, penitrem A, lead, arsenic, mercury, chocolate
Kidney pathology	Ochratoxin, cyclopiazonic acid, E. coli, lead, arsenic, mercury, cadmium, chocolate
Skin lesions	E. coli, garlic, arsenic, cyclopiazonic acid

the diagnosis and prognosis in serious illnesses such as the mycotoxicoses.

Vomitus, feces and urine should be collected, labeled, frozen and tested for bacteria, viruses, biotoxins, metals, pesticides or chemicals as deemed appropriate by discussions between the veterinarian and testing laboratory personnel.

Risk Factors (Page 187)
Young, old, debilitated and immunocompromised animals are most susceptible to foodborne illnesses. Dogs are at greater risk than cats because they are more likely to forage spoiled foods (e.g., trash, garbage and carrion); dogs account for 75% of all animal poisonings. Historically, the risk of foodborne illnesses in pets is increased during warm weather, hunting seasons and two holidays: Thanksgiving and Christmas.

If the pet owner feeds a commercial pet food, follows label directions and follows proper storage recommendations, the likelihood of foodborne illness is low. However, if the same pet is allowed to roam freely outdoors, then the risk of exposure to foodborne disease agents is increased greatly. Home-prepared foods are riskier if the owners do not follow proper preparation and storage procedures discussed later in this chapter. Animals fed foods containing large amounts of uncooked meat and offal are at much greater risk. In general, the risk of contracting foodborne illness from various food sources increases as follows (from least to greatest risk): federally regulated canned pet foods <federally regulated dry pet foods <federally regulated semi-moist pet foods <individual homemade fresh foods <locally prepared commercial dry pet foods <mass-produced kennel foods <free access to garbage, trash and carrion.

Etiopathogenesis (Pages 187-190)
The following discussion involves the etiopathogenesis of bacteria and other agents that can cause foodborne disease in pets.

MYCOTOXINS Up to one-fourth of the world's annual food crop is affected by mycotoxins. These organisms can be highly toxic and are produced by a wide variety of saprophytic and pathologic fungi. Toxic syndromes range from mild GI discomfort and vomiting to an acute fulminating episode with death. Long-term, low-level exposure can produce vague signs such as chronic organ damage (e.g., hepatic cirrhosis), immunosuppression and decreased production and performance. Although cereal grains are most commonly associated with mycotoxins, a wide variety of foodstuffs including cheeses, nuts, forages, fruits and even beer can be contaminated.

AFLATOXINS Corn, peanuts, cottonseed and grains are potential sources of aflatoxins in pet foods. Dogs and cats are among the species most sensitive to the effects of aflatoxin, with LD_{50} values ranging from 0.5 to 1.0 mg/kg.
The principal target organ is the liver. Clinical signs include anorexia, severe GI disturbances, jaundice and hemorrhage, with a corresponding increase in hepatic enzyme activities and a decrease in serum protein values. The most frequently observed hepatic lesions are centrilobular necrosis, fibrosis and bile duct proliferation. Intravascular coagulation may also result from chronic aflatoxicosis.

VOMITOXIN Vomitoxin can be found in any grain but most commonly affects wheat and barley. Like most other mycotoxins, it is heat stable. Dogs and swine are most susceptible to the effects of vomitoxin; these species are affected at relatively low concentrations. Clinical signs include feed refusal, vomiting and diarrhea.

SALMONELLA SPECIES Healthy adult dogs and cats are resistant to the pathogenic effects of salmonellae but serve as important sources of infection for people and weak, debilitated animals. It has been estimated that 36% of healthy dogs and 17% of healthy cats harbor these organisms in their GI tracts.
The most common route of exposure is through the ingestion of fecal-contaminated food and water. Salmonellae in food or water indicate inadequate hygiene and improper cooking. Racing greyhounds are frequently infected when they consume foods largely composed of contaminated raw meat and offal from rendering plants. About 45% of the raw meat used in greyhound foods is contaminated with salmonellae.

SAFETY

Salmonellae elaborate a heat-labile endotoxin. Clinical syndromes can be divided into gastroenteritis, bacteremia/toxemia and organ localization. Infections can usually be treated successfully with antibacterial drugs and supportive treatment. Persistent carriers are common and can be a source of human exposure. Proper cooking of foods and boiling of water will kill the vegetative bacterial cells and inactivate the endotoxin.

CLOSTRIDIUM BOTULINUM These saprophytic bacilli are commonly found in soil and as contaminants in raw meat, carrion and vegetables. They are not considered dangerous to man or animals unless allowed to grow under anaerobic conditions in uncooked meats, improperly canned foods and the carcasses of dead animals. Under anaerobic conditions, *C. botulinum* produces the most potent biotoxin known. This powerful exotoxin blocks the release of acetylcholine. Dogs are less susceptible to the effects of the toxin than people, but naturally occurring botulism occurs in dogs. There have been no documented occurrences in cats.

Clinical signs occur as early as 12 hours or as late as five to six days after ingestion of the exotoxin. The primary clinical sign is generalized paralysis that starts in the posterior limbs and progresses to quadriplegia.

STAPHYLOCOCCUS AUREUS Staphylococci are ubiquitous and are common inhabitants of the skin and mucous membranes of people and other animals. *S. aureus* is the most common cause of foodborne illness in people. The typical GI signs result from a potent *S. aureus* enterotoxin. Dogs and cats are tolerant to staphylococcal enterotoxin and have remained asymptomatic after administration of oral doses as high as 100 μg/kg body weight.

NEORICKETTSIA SPECIES Two rickettsiae, *Neorickettsia helminthoeca* and *N. elokominica*, cause a serious systemic infection in dogs known as salmon poisoning. The disease is transmitted by the ingestion of raw salmon, which contain *Nanophyetus salmincola*. Clinical signs include vomiting, hemorrhagic diarrhea, high fever, dehydration and peripheral lymphadenopathy.

ESCHERICHIA COLI *Escherichia coli* strain O157:H7 has been incriminated in an unusual clinical syndrome in racing greyhounds termed "Alabama rot" or "Greenetrack disease." This disease is characterized by erythema, ulceration of the extremities and renal glomerular pathology.

BACILLUS CEREUS *Bacillus cereus* is a ubiquitous, spore-forming aerobic saprophyte found in soil, grains, cereal products and other foods. At room temperature, *B. cereus* flourishes, producing a potent endotoxin. As an example, it is commonly found in uncooked rice. *B. cereus* is a common isolate from samples of dry pet food. It has also been isolated from food packaging paper and materials.

The standard heat used to manufacture pet foods is not likely to destroy the spores of this organism. However, the number of organisms isolated from pet food samples (<10 cells/g of food) is unlikely to cause foodborne disease in pets unless the food is exposed to moisture and heat conditions conducive to bacterial proliferation. Therefore, pet owners should be warned not to add water to dry pet foods and leave them exposed to high ambient temperatures for prolonged periods.

BIOGENIC AMINES Excessive levels of histamine (around 500 ppm) in the flesh of spoiled fish in combination with the toxin called saurine are thought to be involved in the pathogenesis of a human foodborne illness called "scombroid fish poisoning." This common seafood-related illness is named for its association with the consumption of scombroid fishes, such as tuna, wahoo, mackerel and sardines, although other fishes and cheese have also been implicated. The disease produces clinical signs of an allergic nature, i.e., flushing, sweating, nausea, diarrhea, rash, dizziness, facial swelling, respiratory distress and occasionally vasodilatory shock. The disease is rarely fatal. The Food and Drug Administration has recognized histamine's role in scombroid poisoning by setting an action level maximum of 500-ppm histamine in canned fish.

Histamine and other biogenic amines such as putrescine and cadaverine have also been detected in pet foods. Their presence has been attributed to the use of poultry, fish and meat byproducts as raw ingredients.

METALS Pet foods can become contaminated by metals in several ways. First, metals tend to accumulate in plant and animal matter creating the possibility of toxic levels in food ingredients. Foods can also become contaminated during commercial manufacturing and home preparation by the inadvertent addition of metal shavings, grease, oils and other chemicals. Acidic foods can leach paint, soldered joints or plating agents from the food container. Young animals may ingest lead by chewing on painted wood, linoleum, metal toys, golf balls, roofing materials, drapery weights and ornaments.

Most foodborne metal toxicities in dogs and cats involve lead, zinc, cadmium and arsenic. These agents cause a variety of clinical syndromes depending on age, dose ingested and length of exposure. The specifics of metal toxicities are well described in several veterinary toxicology textbooks and are beyond the scope of this chapter.

SAFETY

OTHER SOURCES Many people supplement their own food with vitamins, herbal remedies and other items purchased at health food stores. Well-meaning pet owners likewise think that what is good for them is also good for their pets.

Owners may supplement a pet's food with onions or garlic. Onions may injure the lipid membranes of erythrocytes and irreversibly denature hemoglobin. These changes result in formation of Heinz bodies, hemolytic anemia and hemoglobinuria. Cattle are most susceptible but dogs are also susceptible. Heinz body anemia has occurred in dogs consuming relatively small amounts (5 to 10 g of onions/kg body weight) of raw, cooked or dehydrated onions. In one study, consumption of approximately 30 g of raw onions/kg body weight for three consecutive days produced severe anemia, erythrocyte Heinz bodies and hemoglobinuria in all dogs fed onions.

Cats are prone to developing erythrocyte Heinz bodies after exposure to many chemicals in food. Likewise, Heinz body anemia may occur in cats following the consumption of onions.

Garlic (*Allium sativum*) is also a member of the onion family. Chronic exposure to garlic and garlic extracts causes anemia, contact dermatitis and asthmatic attacks in dogs.

Chocolate products contain variable amounts of theobromine, a potent cardiovascular and central nervous system stimulant. Although pet owners might believe that chocolate is innocuous, one poison center documented six cases of chocolate poisoning in dogs during a single year. Signs such as vomiting, diarrhea, panting, nervousness, excitement, tremors, tachycardia, cardiac dysrhythmias, coma, convulsions and sudden death may appear four to 15 hours after ingestion. Renal damage may occur in severe cases.

The toxic dose of theobromine has been reported to be greater than 200 mg/kg body weight. However, a springer spaniel died after ingestion of 2 lb of milk chocolate, corresponding to a dose of only 92 mg of theobromine/kg body weight. Based on this case, consumption of one 1.55-oz. milk chocolate bar (93 mg theobromine)/kg body weight could produce clinical signs and possibly death. Unsweetened baking chocolate also contains high levels of theobromine (450 mg/oz.) and has been implicated in cases of toxicity. Dogs have also been poisoned by ingesting cocoa powder (1 to 3% theobromine).

Assess the Food and Feeding Method (Pages 190-193)

If a diagnosis of foodborne illness seems feasible, then the pet owner should be questioned extensively about the animal's food. First, the veterinarian should identify all possible food sources (including commercial foods, home-prepared foods and table scraps) and determine the feeding amounts and the availability of unintentional food sources. Common questions about commercial foods should include: 1) brand name, 2) manufacturer, 3) lot or date code, 4) form of food (i.e., dry, semi-moist, moist), 5) feeding method (i.e., meal fed, free choice), 6) the length of time the pet has been consuming the brand of food, 7) the length of time the pet has been fed from the present container of food (i.e., bag or can), 8) whether water is mixed with the food, 9) how long the food is left in the food bowl, 10) the ambient temperature at feeding and 11) the method of storing the food.

Questions about home-prepared foods should include: 1) the source of ingredients, 2) storage methods for the ingredients and the food, 3) method of preparation, 4) preparation temperatures, 5) method of measuring temperatures and 6) feeding method. Any recent change in either the food ingredients or preparation methods should be investigated further.

The amount of food consumed should be compared with the calculated amount typically consumed by an animal of similar size. If the amount consumed is markedly less than the calculated amount, it could mean that the animal doesn't like the food and may be foraging other food sources or garbage. Decreased intake may also indicate feed refusal typical of vomitoxin contamination.

Sampling Procedures *(Page 191)*

Most veterinary diagnostic laboratories can perform the tests necessary to facilitate a diagnosis of foodborne illness. See Appendix U, Small Animal Clinical Nutrition, 4th ed., pp 1134-1144 for a list of publicly funded and private laboratories. Many investigative tests and techniques are available to the diagnostic laboratory to help assess the case. In fact, the number is so overwhelming that only a few can be used on a particular sample. It is essential that the veterinarian discuss the likely diagnoses with laboratory personnel before test initiation to ensure the tests most critical to a correct diagnosis are performed. Veterinarians also need to determine the laboratory's preferred specimens and method of specimen preservation. (See Small Animal Clinical Nutrition, 4th ed., pp 191-192 for more information about pathogen levels in food and detection methods, including methods for collecting evidence [Table 7-3, p 191] and sampling procedures [Table 7-4, p 191].)

Control and Prevention *(Pages 192-193)*

Methods for control and prevention of foodborne illness in pets apply to commercial (after purchase) and home-prepared foods. Following the practices described in **Table 7-2** can help prevent foodborne illnesses in pets.

All commercial pet foods should be stored in the 4.4 to 15.6°C (40 to 60°F) temperature range. Fresh, home-prepared foods should be refrigerated at -1.6 to 15.5°C (29 to 60°F) before feeding. (Most household refrigerators hold foods at 4.4 to 7.2°C [40 to 45°F].) The length of time that a food

SAFETY

Table 7-2. Prevention of foodborne illness in animals.

1. Select only wholesome pet foods or pet food ingredients.
For home-prepared foods, use raw ingredients that meet human food grade
standards.
Discard all foods with an abnormal color, composition, odor or moldy
appearance.
Discard foods from bulging or leaking cans.
2. Control food contamination.
Use stainless steel utensils, feeding bowls, etc. whenever possible.
Keep food preparation areas, cooking utensils and food bowls spotlessly
clean. Wash and disinfect bowls and utensils daily.
Store dry commercial foods in a cool, dry environment, free from insects
and rodents.
Empty the feeding bowl of moist or moistened foods not consumed within
two to four hours if the ambient temperature is above 10°C (50°F).
Clean, wash and disinfect food utensils and food bowls after each feeding.
If feeding free choice, check food daily for mold and spoilage.
3. Control microorganisms in food using physical means.
Cook all home-prepared foods at 82°C (180°F) for at least 10 minutes.
Verify cooking temperatures with a cooking thermometer and internal meat
temperatures with a meat thermometer.
Validate thermometer accuracy periodically with boiling water.
Cover all perishable foods and opened cans of pet food and store in the
refrigerator at 4°C (40°F) when not being prepared, cooked or consumed.
4. Control the pet's access to unintentional foods.

can be kept refrigerated depends on its type and age. Fresh meats, fish and
poultry can be kept for two to 10 days whereas fruits and vegetables will
remain wholesome for weeks when refrigerated.

Moist products are sealed and therefore not affected by moisture or air;
control of these factors applies only to storage of bulk dry commercial
foods. Spoilage bacteria require at least 30% moisture for growth whereas
molds require 5 to 15%. Dry pet foods have a moisture content in the range
of 6 to 9%. Therefore, dry commercial pet foods will have a satisfactory shelf
life if stored in a cool dry place with the top of the bag or container closed.
These precautions limit the availability of moisture and air needed for
oxidative chemical degradation and microbial growth. Placing the food in a
canister or other closed container will extend the shelf life by further reducing the availability of moisture and oxygen. This method of storage has the
added advantage of preventing rodent and insect damage.

Clinical cases that illustrate and reinforce the nutritional concepts presented in this chapter can be found in Small Animal Clinical Nutrition,
4th ed., pp 195-198.

CHAPTER 8

Health Maintenance Programs for Dogs and Cats

For a review of the unabridged chapter, see Roudebush P, Goldston RT. Health Maintenance Programs for Dogs and Cats. In: Hand MS, Thatcher CD, Remillard RL, et al, eds. Small Animal Clinical Nutrition, 4th ed. Topeka, KS: Mark Morris Institute, 2000; 201-211.

CHAPTER 9

Normal Dogs

For a review of the unabridged chapter, see Debraekeleer J, Gross KL, Zicker SC. Normal Dogs. In Hand MS, Thatcher CD, Remillard RL, et al, eds. Small Animal Clinical Nutrition, 4th ed. Topeka, KS: Mark Morris Institute, 2000; 213-260.

DOGS AS OMNIVORES (Pages 216-219)*

The word carnivore can be used to indicate either a taxonomic classification or a type of feeding behavior. The order Carnivora is quite diverse and consists of 12 families containing more than 260 species. Omnivorous and carnivorous feeding behaviors are most common among members of the order Carnivora; however, the order also includes species that are herbivores (e.g., pandas). Dogs are opportunistic eaters and have developed anatomic and physiologic characteristics that permit digestion and usage of a varied diet.

LIFESTAGE NUTRITION (Page 219)

Lifestage nutrition is the practice of feeding animals foods designed to meet their optimal nutritional needs at a specific age or physiologic state (i.e., maintenance, reproduction, growth or senior) (**Table 9-1**). The concept of lifestage nutrition recognizes that feeding either below or above an optimal nutrient concentration can negatively affect biologic performance and/or health. This concept differs markedly from feeding a single product for all lifestages (all-purpose foods) whereby nutrients are added at levels to meet the highest potential need (usually growth and reproduction). Adult animals at maintenance are always provided nutrients well in excess of their biologic needs when fed all-purpose foods. Because the goals in nutrition are to feed for optimal health, performance and longevity, feeding foods designed to more closely meet individual needs is preferred. This philosophy is key to lifestage nutrition and preventive medicine. In addition to providing advice about animals' basic nutritional requirements, veterinarians should assess and minimize the nutrition-related health risks at each lifestage. For maximal benefit, risk assessment and prevention plans should begin well before the onset of disease.

---

*Page numbers in headings refer to Small Animal Clinical Nutrition, 4th ed., where additional information may be found.

Key Nutritional Factors–Normal Dogs (Page 215)

> **Table 9-3** lists key nutritional factors for adult dogs at maintenance (young to middle age, obese prone and older).
> **Table 9-6** lists key nutritional factors for reproduction (gestation and lactation).
> **Table 9-9** lists key nutritional factors for growing dogs.

Table 9-1. Feeding plan summary for normal dogs.

Adult Dogs

1. Body condition and other assessment criteria will determine the daily energy requirement (DER). Remember that DER calculations should be used as guidelines, starting points and estimates for individual animals and not as absolute requirements.
   Neutered adult = 1.6 x resting energy requirement (RER)
   Intact adult = 1.8 x RER
   Obese-prone adult = 1.2 to 1.4 x RER
   Working adult = 2.0 to 8.0 x RER
2. Select a food or foods with the following energy density.
   Ideal body condition = 3.5 to 4.5 kcal (14.6 to 18.8 kJ) metabolizable energy (ME)/g dry matter
   Obese prone = 3.0 to 3.5 kcal (12.5 to 14.6 kJ) ME/g dry matter
3. Select a food or foods with levels of key nutritional factors listed in **Table 9-3**.
4. Determine quantity of food based on DER calculations.
5. Determine preferred feeding method.
6. Monitor body weight, body condition and general health.

Reproducing Dogs

1. Body condition and other assessment criteria will determine the DER. Remember, DER calculations should be used as guidelines, starting points and estimates for individual animals and not as absolute requirements.
   Gestation = 1.8 to 2.0 x RER for the first four weeks, then 2.2 to 3.5 x RER for the last five weeks
   Lactation = 4.0 to 8.0 x RER (peak lactation = 1.9 x RER + 25% per puppy)
2. Select a food or foods with the following energy density.
   Gestation = 3.5 to 4.5 kcal (14.6 to 18.8 kJ) ME/g dry matter
   Lactation = 4.0 to 5.0 kcal (16.7 to 20.9 kJ) ME/g dry matter
   = 3.5 to 4.5 kcal (14.6 to 18.8 kJ) ME/g dry matter for bitches with fewer than four puppies
3. Select a food or foods with levels of key nutritional factors listed in **Table 9-6**.
4. Select a food or foods with above average digestibility.
5. In general, free-choice feeding is the preferred feeding method.
6. Monitor body weight, body condition and general health.

DOGS

(Continued on next page.)

Table 9-1. Feeding plan summary for normal dogs (Continued.).

Growing Dogs

1. Estimate adult body weight and size, which will determine key nutritional factors.
2. Determine the DER. Remember, DER calculations should be used as guidelines, starting points and estimates for individual animals and not as absolute requirements.
   Weaning to 50% of adult body weight (four to five months of age)
   = 3.0 x RER
   Four to five months of age to 80% of adult weight
   = 2.0 to 2.5 x RER
   When 80% of adult body weight is reached = 1.8 to 2.0 x RER
3. Select a food or foods with the following energy density.
   Adult body weight <25 kg = 3.5 to 4.5 kcal (14.6 to 18.8 kJ) ME/g dry matter
   Adult body weight >25 kg = 3.25 to 3.75 kcal (13.6 to 15.7 kJ) ME/g dry matter
4. Select a food or foods with levels of key nutritional factors listed in **Table 9-9**.
5. Select a food or foods with above average digestibility.
6. In general, free-choice feeding should be avoided; use a meal-restricted (preferred) or time-restricted feeding method.
7. Monitor body weight, body condition and general health.

The value of lifestage feeding is enhanced if risk factor management is incorporated into the feeding practice. In many instances, when the nutritional needs associated with a dog's age and physiologic state are combined with the nutritional goals of disease risk factor reduction, a more narrow, but optimal, range of nutrient recommendations results. Nearly all commercial dog foods meet or exceed the minimum nutrient requirements of dogs. However, certain nutrients may still be outside the desired range for optimal health or risk factor reduction. For dogs fed commercial foods, these nutrients require particular consideration and, thus, are referred to as nutrients of concern. Specific food factors such as digestibility and texture can also affect health and disease risk. Together, nutrients of concern and specific food factors are called key nutritional factors. In the following sections, the key nutritional factors for different lifestages of healthy dogs will be discussed, including those associated with reducing the risk of specific diseases and those involved with optimizing performance during different physiologic states. Dogs eating homemade foods are more at risk for developing nutrient deficiencies (e.g., too little calcium) and excesses (e.g., too much phosphorus); therefore, dogs fed homemade foods have a longer list of key nutritional factors than dogs fed commercial foods.

YOUNG TO MIDDLE-AGED ADULT DOGS
(Pages 219-229)

Depending on breed, young to middle-aged dogs are those that are full grown (about 12 months old) and not over five (large breeds) to seven years (small breeds) old. Dogs are often considered older when they reach half their life expectancy. The goals of nutritional management are to maximize longevity and quality of life (disease prevention). The most important health concerns that may benefit from proper nutritional management in this age group include dental disease, obesity and kidney disease. Also, many owners are concerned about outward appearances; thus, a "healthy coat" may be an additional goal.

Assessment (Pages 219-227)
Assess the Animal (Pages 219-225)
Animal assessment should be a structured process that includes: 1) obtaining an accurate and detailed history, 2) reviewing the medical record, 3) conducting a physical examination and 4) evaluating results of laboratory and other diagnostic tests. During assessment, the feeding goals should be established, risk factors for nutrition-related diseases considered and key nutritional factors identified.

HISTORY AND PHYSICAL EXAMINATION For normal, healthy adult dogs, there is usually little need to obtain detailed nutritional information and the time available to obtain a dietary history and conduct a physical examination is often limited. A minimum dietary database for all canine patients should include: 1) the type of food fed to the dog (homemade, commercial, dry, moist, semi-moist, etc.), 2) recipes if homemade food represents the majority of the diet, 3) brand names of commercial foods, if known, 4) names of supplements, treats and snacks and 5) method of feeding (free choice, meal feeding, etc.).

An extended dietary database includes: 6) quantities fed, 7) recent changes in food type, intake and preferences, 8) access to food for other pets or livestock, 9) who in the family buys food for the pet, 10) who in the family feeds the pet and 11) appetite changes with estimates of magnitude and duration. The general type and level of activity (e.g., house pet, confined to kennel, working dog, etc.) and neuter status should be noted because these factors are important determinants of energy requirements. The dietary history should be expanded if nutrition-related problems such as obesity are identified in the initial evaluation of the patient.

Body weight, body condition score (BCS), oral health and overall appearance of the skin and coat of all adult dogs should be assessed and

DOGS

recorded in the record. These parameters are general indicators of nutritional adequacy. An otherwise healthy young to middle-aged adult dog with normal body weight, skin and coat and BCS (3/5) and no evidence of significant dental disease is unlikely to need further assessment.

Gender and Neuter Status No controlled studies have been performed to delineate differences in nutritional requirements of intact male vs. intact female dogs. It may be presumed that, like other mammals, intact females require less caloric intake than males due to differences in lean body mass. Lean body mass accounts for nearly all of an animal's resting energy requirement (RER).

Obesity occurs twice as often in neutered dogs than in reproductively intact dogs. Neutering does not appear to have a marked impact on the resting energy expenditure of female dogs; however, it may significantly increase food intake. A decrease in physical activity is also assumed to occur in many dogs after neutering and may play a more important role in male dogs because of a decrease in roaming. The daily energy intake should be limited to prevent rapid weight gain in neutered dogs; 1.6 x RER is a good starting point. For some breeds and individual dogs, it may be necessary to reduce caloric intake of neutered dogs to 1.2 to 1.4 x RER.

Breed Newfoundland dogs have energy requirements about 15% less than average, whereas Great Danes and Dalmatians may have energy requirements up to 16% higher than average.

Activity Level Daily energy requirement (DER) represents the average energy expended by an animal each day; it is dependent on lifestage and activity. DER may range from RER for sedentary dogs to almost 15 x RER for endurance athletes under extreme conditions. A consistently higher level of physical activity would likely increase lean body mass, which would increase energy usage, even at rest.

Estimations for DER include enough energy to support spontaneous activity, such as eating, sleeping, going outside and up to three hours of play and exercise per day. However, many dogs are relatively inactive, particularly those in urban areas. Approximately 19% of owners never play with their dogs and 22% take their dogs out for exercise fewer than three hours per week. Chapter 10 covers the nutrient requirements of canine athletes.

Age Age-related changes occur between the onset of adulthood and five to seven years of age. The prevalence of dental disease, obesity and kidney disease will generally increase over this time span. (See Chapters 13, 16 and 19.) Apart from reproduction and imposed activity during work or

Table 9-2. Influence of age on daily energy requirements of dogs.*

Age (years)	kcal ME/BW$_{kg}^{0.75}$	Typical DER ranges** kJ ME/BW$_{kg}^{0.75}$	x RER
1-2	120-140	500-585	1.7-2.0
3-7	100-130	420-550	1.4-1.9
>7	80-120	335-500	1.1-1.7

Key: DER = daily energy requirement, ME = metabolizable energy, RER = resting energy requirement, kcal = kilocalories, kJ = kilojoules.
*See Appendix C, Small Animal Clinical Nutrition, 4th ed., pp 1006-1007.
**The energy requirements indicated in this table are only starting points and should be adapted for individual dogs.

sport, age may be the single most important factor influencing DER of most adult dogs (**Table 9-2**).

Environment Temperature, humidity, type of housing, stress level and the degree of acclimatization should be considered with respect to breed and lifestage nutrient requirements of animals. Animal factors including insulative characteristics of skin and coat (i.e., subcutaneous fat, hair length and coat density) and differences in stature, behavior and activity interact and affect DER. When kept outside in cold weather, dogs may need 10 to 90% more energy than during optimal weather conditions.

LABORATORY AND OTHER CLINICAL INFORMATION Healthy young to middle-aged adult dogs require few laboratory and other diagnostic tests as part of routine assessment.

KEY NUTRITIONAL FACTORS Table 9-3 summarizes the key nutritional factors for young to middle-aged adult dogs. The following section describes these key nutritional factors in more detail.

Water Total water intake (i.e., drinking and water from food) is influenced by several factors such as environment, physiologic state, activity, disease processes and food composition. Total water intake increases almost linearly with increasing salt levels in food. Switching from a moist to a dry food and vice versa markedly affects the amount of water taken with the food; however, dogs compensate well for this difference by changing the quantity of water they drink, thus keeping their total daily water intake constant. In general, dogs self-regulate water intake according to physiologic need. Requirements may be met by allowing free access to a source of potable water. Healthy adult dogs need roughly the equivalent of the energy requirement in kcal metabolizable energy (ME)/day, expressed in ml/day.

DOGS

Table 9-3. Key nutritional factors for adult dogs at maintenance.

Factors	Recommended levels in food (DM)		
	Young to middle age	Obese prone	Older
Energy density (kcal ME/g)*	3.5-4.5	3.0-3.5	3.0-4.0
Energy density (kJ ME/g)*	14.6-18.8	12.5-14.6	12.5-16.7
Crude protein (%)	15-30	15-30	15-23
Crude fat (%)	10-20	7-12	7-15
Crude fiber (%)**	≤5	≥5	≥2.0
Calcium (%)	0.5-1.0	0.5-1.0	0.5-1.0
Phosphorus (%)	0.4-0.9	0.4-0.9	0.25-0.75
Ca/P ratio	1:1-2:1	1:1-2:1	1:1-2:1
Sodium (%)	0.2-0.4	0.2-0.4	0.15-0.35
Chloride (%)	0.3-0.6	0.3-0.6	0.3-0.5

Key: DM = dry matter, kcal = kilocalories, kJ = kilojoules, ME = metabolizable energy.
*If the caloric density of the food is different, the nutrient content in the dry matter must be adapted accordingly.
**Crude fiber measurements underestimate total dietary fiber levels in food.

Energy Graphically, the DER for a population of dogs results in a bell-shaped curve; therefore, the energy intake of individual dogs may vary by about 50% above or below the average requirements, even within the same age group. The RER, however, is not markedly influenced by these factors, and is similar for all dogs, independent of breed or age. However, RER is not linearly related to body weight. RER is approximately 70 kcal (293 kJ)/$BW_{kg}^{0.75}$.

Because DER is the sum of RER plus all the above influences, it is better to use RER as the basis for calculating energy requirements of adult dogs and to assign different multipliers to account for differences in activity, age and environmental influences. When assigning multipliers to RER, it is important to account for neuter status because this variable can be an important factor in determining DER of household dogs. Neutered animals may have a lower DER than intact counterparts. Recent surveys have shown that the prevalence of obesity increases progressively and peaks in middle-aged dogs. Thus, prevention of obesity should be an important goal of feeding programs for young adult dogs. It is far more beneficial to an animal to set up an appropriate weight-maintenance program than to treat obesity.

Three groups of adult dogs can be distinguished based on DER: 1) young adults one to two years old, 2) middle-aged dogs (three to seven years old) and 3) older dogs (more than seven years old) (**Table 9-2**). The differences in DER probably reflect an age-related decrease in activity and lean body mass. A good starting point for estimating the DER of a neutered adult dog is 1.6 x RER (115 kcal [480 kJ] ME/$BW_{kg}^{0.75}$).

Alternatively, **Table 1-4** (See pages 14-15.) may be used to assign different multipliers of RER that may be used as estimates of DER variation in this age group. All initial estimates of energy needs must subsequently be evaluated by body condition assessment and adjusted as needed for individual dogs. Sedentary dogs may have a DER that approaches their RER. Sedentary dogs fed caloric intakes recommended for maintenance (1.6 x RER) will be overfed and are likely to become overweight.

Table 9-3 lists recommended energy density levels for foods intended for young to middle-aged adult dogs. The levels of dietary fat and fiber are important determinants of a food's energy density. Fat provides more than twice as much energy on a weight basis than carbohydrate or protein. High-fat foods have increased energy density; conversely, low-fat foods have decreased energy density. Fiber is a poor source of energy for dogs; thus, as the fiber content of foods increases, energy density decreases. Dietary fiber also helps promote satiety.

It is difficult to determine the optimal concentration of crude fiber in a complete food for dogs; however, up to 5% dry matter (DM) seems to be adequate. Obese-prone dogs may benefit from 10 to 15% crude fiber on a dry matter basis (DMB). Foods that are both low in fat and high in fiber tend to have the lowest energy density and are recommended for obese-prone dogs. **Table 9-3** lists recommended levels of dietary fat and fiber.

Phosphorus and Calcium Commercial foods contain adequate and sometimes excessive amounts of calcium and phosphorus and, therefore, should not be supplemented. However, calcium is often deficient and phosphorus may be excessive in homemade foods, especially when most of the diet comes from meat and leftovers from the table.

Based on endogenous losses, a daily intake of 100 mg calcium/kg body weight and 75 mg phosphorus/kg body weight is adequate. At an energy density of 3.5 kcal (14.6 kJ)/g DM this corresponds to an average dry matter content of about 0.5 to 0.8% calcium and 0.4 to 0.6% phosphorus. **Table 9-3** lists recommended levels of phosphorus and calcium for foods intended for young to middle-aged adult dogs and with energy densities varying from 3.5 to 4.5 kcal/g. The calcium-phosphorus ratio in dog foods should not be less than 1:1.

The above mentioned levels are adequate but not excessive; even a daily intake of 20 to 30% less is still sufficient. Therefore, it is unnecessary to feed foods with higher levels of phosphorus, or to supplement commercial foods with calcium-phosphorus supplements. Moreover, higher phosphorus levels are contraindicated for up to 25% of the young adult dog population that may already be affected by subclinical kidney disease.

DOGS

Protein The amount of protein in commercial foods for healthy dogs varies widely (15 to 60% DM). (See Appendix 2.) After the amino acid requirements are met for an individual animal, addition of more protein provides no known benefit. This fact often runs contrary to popular belief that more protein is better. Some pet food manufacturers advertise that extra protein in commercial dog foods is necessary for carnivores. This marketing ploy misrepresents the fact that dogs are omnivores. Excess dietary protein, above the amino acid requirement, is not stored as protein, but rather is deaminated by the liver. Subsequently, the kidneys excrete the byproducts of protein catabolism and the remaining keto acid analogues are used for energy or stored as fat.

Foods high in protein are often also high in phosphorus. Excess dietary phosphorus accelerates the progression of kidney disease in dogs.

The minimum crude protein content of food depends on digestibility and quality. For example, if the digestibility of an average quality protein is 75%, then about 12% DM crude protein is adequate. Foods containing less than 12% DM crude protein must be of higher biologic value. Biologic value becomes less important for healthy adult dogs if foods contain crude protein levels greater than 12%. **Table 9-3** lists recommended levels of crude protein for foods for young to middle-aged adult dogs.

Sodium and Chloride Essential hypertension is not a common problem in dogs; therefore, higher intakes of dietary sodium and chloride have not been considered harmful to young, healthy dogs. However, one study suggested that up to 10% of apparently healthy dogs may have high blood pressure.

High sodium and chloride intake is contraindicated in dogs with certain diseases that may have a hypertensive component such as obesity, renal disease and some endocrinopathies. Uncontrolled high blood pressure may lead to kidney, brain, eye and cardiovascular damage. Dietary sodium chloride restriction is the first step in and an important part of antihypertensive therapy.

It is prudent to meet but not greatly exceed sodium and chloride requirements when selecting foods for adult dogs. The best estimate for a minimum requirement of sodium is about 4 mg/kg body weight/day. In general, 25 to 50 mg/kg body weight/day is recommended for adult maintenance; these levels are six to 12 times more than the minimum. A content of 0.15 to 0.4% DM sodium will provide this recommended intake level. Sodium levels in commercial foods for adult dogs range from 0.11 to 2.2% DM and are higher in moist foods than in dry foods. (See Appendix 2.) In the absence of studies establishing chloride requirements in dogs, a value 1.5 times the sodium requirement is recommended. **Table 9-3** lists recommended levels of sodium and chloride in foods intended for young to middle-aged adult dogs.

Fat and Essential Fatty Acids Fats are an excellent source of energy, but the real requirement for fat is to supply essential fatty acids (EFA). In addition, fat serves as a carrier for the absorption of fat-soluble vitamins (A, D, E and K). Linoleic acid and α-linolenic acid are considered essential because dogs lack the enzymes to synthesize them. Ensuring an adequate intake of EFA is key to maintaining normal skin and coat.

Whether n-3 fatty acids are essential is less certain because of the inability of n-3 fatty acids to support all of the physiologic functions that are supported by n-6 fatty acids. Nevertheless, a source of dietary n-3 fatty acids is often recommended. N-3 fatty acids are abundant in membrane lipids, especially in retinal and neural tissues.

The minimum amount of fat in foods for normal, healthy adult dogs is at least 5% DMB, with at least 1% of the food as linoleic acid (DMB). A source of dietary n-3 fatty acids is also recommended. Increasing the amount of fat in foods increases palatability and EFA levels; however, energy content also increases. **Table 9-3** lists recommended levels of fat for foods intended for young to middle-aged adult dogs. Note that lower levels of dietary fat are recommended for obese-prone adult dogs.

OTHER NUTRITIONAL FACTORS/Food Texture Periodontal disease is the most common health problem of adult dogs and may predispose affected animals to systemic complications. Periodontal disease can be prevented in many animals with routine veterinary care and frequent plaque control at home. Feeding recommendations for oral health commonly include feeding a dry pet food. However, typical dry dog foods contribute little dental cleansing and the general statement that dry foods provide significant oral cleansing should be regarded with skepticism. Research has demonstrated that a maintenance dog food with specific textural properties and processing techniques can significantly decrease plaque and calculus accumulation and maintain gingival health. (See Chapter 16 and Small Animal Clinical Nutrition, 4th ed., pp 474-504.)

Assess the Food (Page 225)

After the nutritional status of the dog has been assessed and the key nutritional factors and their target levels determined, the adequacy of the food is assessed. The three most useful components when assessing foods for normal adult dogs are to: 1) ensure that the food has been tested or fed to dogs, 2) determine the food's DM nutrient content (especially for the key nutritional factors) and 3) compare the food's key nutritional factors with the recommended levels (**Table 9-3**).

Whether or not commercial foods for healthy pets have been animal tested can usually be determined from the nutritional adequacy statement on the product's label. (See Chapter 5 and Small Animal Clinical

DOGS

Nutrition, 4th ed., pp 147-161.) Few, if any, homemade recipes have been animal tested according to prescribed feeding protocols. Commercial pet foods that have undergone Association of American Feed Control Official's (AAFCO) prescribed or other feeding tests provide reasonable assurance of nutrient availability and sufficient palatability to ensure acceptability (i.e., food intake sufficient to meet nutrient needs). However, even controlled animal testing is not infallible. Passing such tests does not ensure the food will be effective in preventing long-term nutritional/health problems.

Thus, in addition to having passed feeding tests, the food should be evaluated to ensure that key nutritional factors are at levels appropriate for promoting long-term health and optimal performance for the intended lifestage.

Appendices 2 and 3 (Also see Small Animal Clinical Nutrition, 4th ed., pp 1084-1091.) provide partial nutrient profiles for selected commercial foods and treats sold in the United States, Canada and Europe. In most instances, these profiles provide the necessary information about a food's or treat's nutrient content. The manufacturer should be contacted if the food or treat in question cannot be found in the appendices. Manufacturers' addresses and toll-free phone numbers are listed on pet food labels.

Comparing a food's nutrient content with the animal's nutrient needs will help identify any significant nutritional imbalances in the food being fed. This comparison is fundamental to determining whether or not to feed a different food.

The energy density of a food is an important aspect to consider when assessing foods. For example, if each of two foods contains 20% protein (DMB) but one food has an energy density of 4.5 kcal ME/g DM and the other 3.5 kcal ME/g DM, the former will provide about 25% less protein on a daily basis than the latter. This comparison, however, is not valid for foods intended for obese or obese-prone dogs and foods for active, sporting dogs. Fat is the most important contributor to a food's energy density, conversely fiber decreases a food's energy density. Foods with a low fat content and a high fiber level tend to have the lowest energy density.

Commercial treats, snacks and table food should also be included in the food assessment step because they are part of the total food intake of an animal and, if misused, may create an imbalance in an otherwise balanced feeding plan.

Assess the Feeding Method (Pages 225-227)

It may not always be necessary to change the feeding method when managing healthy adult dogs in optimal body condition. However, a thorough evaluation includes verification that an appropriate feeding method is being used. In addition, future risk factors such as obesity should be

Table 9-4. Advantages and disadvantages of various feeding methods for dogs.

Feeding methods	Advantages	Disadvantages
Free-choice feeding	Less labor intensive Less knowledge required Quieting effect in a kennel Less dominant animals have a better chance to get their share	Less control over food intake Predisposes to obesity Less monitoring of individual changes in food intake
Meal-restricted feeding	Better control of food dose Early detection of altered appetite Better control of body weight	Intermediate labor intensive Most knowledge required for food dose calculation
Time-restricted feeding	Intermediate control of food dose Some monitoring of appetite possible	Inaccurate control of food intake Risk of obesity similar to free choice Most labor intensive

considered when evaluating the current method. Current feeding methods should have been obtained when the history was taken.

The objective of this part of the assessment process is to establish how much food has been fed and how it has been offered (i.e., when, where, by whom and how often). An important determinant of food intake in domestic dogs is the owner and family situation because these factors usually control the amounts and types of food fed. Dogs are usually fed either free choice or in a restricted (time restricted or food restricted) fashion. Each method has advantages and disadvantages (**Table 9-4**).

Nutrient requirements of animals are met by a combination of nutrient levels in food and amounts fed. Even if a food has an appropriate nutrient profile, significant malnutrition may result if excess or insufficient amounts are consumed. If the animal in question has a normal BCS (3/5), the amount being fed is probably appropriate. The amount fed can be estimated either by calculation (See Chapter 1 and Small Animal Clinical Nutrition, 4th ed., pp 1-19.) or by referring to feeding guides on product labels. Calculated estimates and feeding guides represent population averages and may need to be adjusted for individual dogs.

ALTERNATIVE EATING BEHAVIORS Owners are often concerned about alternative eating behaviors displayed by their dogs. In fact, these behaviors may be more offensive to the owner than detrimental to the dog. Alternative eating behaviors may be of nutritional or non-nutritional origin, and some may be indicators of underlying disease.

Response to Food Variety Dogs may display preferences for specific types of foods according to taste and texture. However, the notion that dogs need a variety of flavors or taste in their meals is incorrect and may be detrimental in some instances. Dogs prefer novel foods or flavors to familiar foods; therefore, feeding a variety of novel foods free choice may lead to overeating and obesity. Dogs may correct for excessive energy intake by decreasing or refusing food intake the next day(s). Reduction of food intake to maintain weight following engorgement may erroneously be interpreted as a dislike of the current food instead of an autoregulatory mechanism to achieve the previous set-point weight.

Garbage Eating Garbage eating is probably normal behavior. Many dogs prefer food in an advanced stage of decomposition. However, garbage eating is oftentimes unhealthy. Ingestion of garbage may cause brief, mild gastroenteritis or more serious intoxication. Because the etiology is complex and may involve bacterial toxins, mycotoxins and byproducts from putrefaction or decomposition, the clinical signs vary widely from vomiting, diarrhea, abdominal pain, weakness, incoordination and dyspnea, to shock, coma and death. Scavenging dogs may eat less of their regular meal; therefore, garbage eating may be mistaken for anorexia at home. Spraying garbage bags with a dog repellent usually will not stop the problem. Preventing access to garbage is the obvious best solution.

Grass Eating Plant and grass eating is normal behavior. Herbivores are the natural prey for wolves and most other canids. The viscera of prey is often eaten first and contains partially digested vegetable material. Because dogs' ancestors and close relatives in the wild regularly ingest plant material, some investigators have suggested that domestic dogs must also eat grass. Probably the better explanation is that, to date, no one knows for sure why dogs eat plants or grass, but they may simply like the way plants taste or prefer the texture.

Begging for Food Begging for food may be fun when dogs sit up or perform other tricks; however, the behavior can become annoying when whining, barking, persistent nudging and scratching take over. Begging for food was one of the most common complaints raised in a study involving more than 1,400 owners and was perceived as a problem in one-third of the dogs. In addition, begging may encourage owners to feed more of the dog's regular food. Begging tends to increase with age and may indicate that most owners don't realize that they reinforce begging by continuing to offer tidbits to their begging pet. All between-meal treats reinforce begging. Also, the fact that begging for food is directly proportional to the number of people in the family may be related to an increase in the number of tidbits fed. Treatment consists of ignoring behaviors such as begging, barking and whining. Owners should be prepared for a prolonged period of such behaviors before begging subsides completely. Intermittent reinforcement of begging when these behaviors become too much can be more powerful than continuous rewarding, even though the owner may have refused to provide snacks in the interim. It may also help to keep the dog out of the kitchen and dining areas when preparing and eating food and to feed the dog before or after the family has eaten.

Pica Pica is defined as perverted appetite with craving for and ingestion of non-food items. The etiology of true pica is unknown. Suggested causes include mineral deficiencies, permanent anxiety and psychological disturbances. A few cases of pica have been noted in relation to zinc intoxication and hepatic encephalopathy. Pica is common in dogs with exocrine pancreatic insufficiency, probably as a manifestation of polyphagia, and perhaps as a consequence of some specific nutritional deficiency. Sometimes, coprophagy and garbage eating are mistakenly considered forms of pica.

Pica can be treated with aversion therapy, by offering a counter attraction at the moment the dog begins to eat foreign material and by punishment if there is no response. Outdoors, the dog should be kept on a leash or even muzzled. Most treatments for pica are unrewarding. Physically preventing the animal from engaging in pica is sometimes the only solution.

Coprophagy Coprophagy is defined as eating feces and may involve consumption of the animal's own stools or the feces of other animals. Coprophagy is probably widespread among pet dogs and is probably more disturbing to owners than it is harmful to dogs. Bitches normally eat the feces of their puppies during the first three weeks of lactation. In rural areas and in the free-living state, the ingestion of large-animal feces by dogs is also considered normal behavior. In many cases, however, coprophagy is a behavioral problem and the etiology is unknown. Coprophagy can also be related to certain diseases.

The danger for transmission of parasitic diseases is probably the most important health reason for managing coprophagy; however, the associated halitosis is of primary concern to owners. The dog's motivation must be reduced to correct coprophagy. Several measures have been proposed. Punishment may deter the animal's behavior, but may violate the confidence between owner and pet. Punishment may also aggravate the coprophagic behavior. Thus, a good balance has to be found. Walking the dog on a leash and keeping it away from feces after the dog defecates is helpful.

Repulsive substances can be used to create aversion for feces. Many different products have been recommended including spices (e.g., pepper,

DOGS

sambal, hot pepper sauce), quinine, strong perfumes and specific products such as cythioate and meat tenderizers. Adding repulsive substances to feces can be time-consuming and has questionable efficacy.

Food changes to deter coprophagy have been recommended; however, most of these recommendations lack substantiation. Using foods with increased fiber levels has been reported to help. Free-choice feeding has also been recommended, whereas a strict schedule of two meals per day and avoiding all tidbits and table foods has worked for others.

Feeding Plan (Pages 227-229)

When done properly, assessment of the animal, the food and the feeding method should provide all the information necessary to develop a feeding plan. In some instances, the plan currently in effect may not need to be changed. In other cases, major changes may need to be made in foods and feeding methods. In either case it is necessary to reassess the animal at regular intervals as indicated in the specific lifestage sections below.

Select a Food (Page 227)

A new food should be selected if significant differences are seen between the recommended nutrient levels and those in the food currently fed. The new food should provide the key nutritional factors in amounts listed in **Table 9-3**. Appendices 2 and 3 list nutrient levels of a number of readily available commercial dog foods and treats (Also see Small Animal Clinical Nutrition, 4th ed., pp 1084-1091.) or manufacturers can be contacted directly for the same information. Foods selected should also have passed AAFCO or similar feeding trials for adult dogs. This information is usually on the pet food label.

Determine a Feeding Method (Pages 228-229)

The method of feeding may not need to be revised if it appears adequate. Free-choice feeding is popular and will suffice for healthy dogs unless obesity is an issue. If so, the amount of food offered should be limited. An understanding of the other pets in the home, which family member is responsible for selecting and purchasing the dog's food and who feeds the dog regularly are helpful to evaluate the feasibility of new dietary recommendations and will increase compliance, if a food or feeding method change is necessary. **Table 9-1** contains a feeding plan summary for healthy adult dogs.

Most healthy adult dogs adapt well to new foods. However, it is good practice to allow for a transition period to avoid digestive upsets. This is particularly true when switching from lower fat foods to higher fat foods or when changing forms of food (e.g., changing from dry to moist food). New food should be increased and old food decreased in progressive amounts over a three- to seven-day period until the changeover is completed. (See Chapter 1.)

Dogs may eat an insufficient amount or completely refuse new foods. Investigation of food refusal may reveal problems with owner compliance rather than a finicky appetite. The following guidelines may be useful when a food change must be made: 1) Explain clearly to the owners why a change in food is necessary or preferable. 2) Justify your recommendation to the owners (i.e., food profile vs. specific needs of the dog). 3) As a general rule, start with one or two meals per day, always presented at the same time. Uneaten food should be removed after 15 to 20 minutes. 4) Don't give treats or table foods between meals for the first few days. If a small snack is given, it should be given immediately (i.e., within seconds) after the new food is eaten. Most dogs will accept the new food within a few days.

Reassessment (Page 229)

Owners should be encouraged to weigh their dog every month or so, and should be trained to observe their dog and adapt the food intake according to its needs. Dogs whose nutrition is well managed are alert, have an ideal BCS (3/5) with a stable, normal body weight and a healthy coat. Stools should be firm, well formed and medium to dark brown.

Reassessment by a veterinarian should take place regularly. Healthy dogs should be reassessed every six to 12 months. Because few if any homemade recipes have been tested according to prescribed feeding protocols, dogs should be reassessed more frequently if homemade food is a significant part of their caloric intake. Reassessment should take place immediately if clinical signs arise indicating that the current feeding regimen is inappropriate, or if the needs of the dog change (e.g., reproduction or change in activity).

If expected results are not obtained, the owner should also be questioned in detail about compliance with the feeding regimen or the possibility that the dog has access to other food sources.

OLDER DOGS (Pages 229-232)

The overall goals of feeding older adult dogs are similar to those for feeding young and middle-aged adult dogs: optimize quality and longevity of life and minimize disease. Aging is characterized by progressive and irreversible change, and its rate and manifestations are determined by intrinsic and extrinsic factors, one of which is nutrition. Dogs often are considered older or likely to start having diseases associated with aging between seven and one-half and 13.5 years. Smaller dogs tend to live longer than large dogs; the life expectancy of smaller dogs may be more than 20 years. Because dogs are often considered older when they reach half of their life expectancy, a food change should be considered around the age of five years for large- and giant-breed dogs and around seven years for small dogs. At these ages, dogs may gradually start to gain weight and develop age-related physical and behavioral changes.

Assessment (Pages 230-232)
Assess the Animal (Pages 230-232)
HISTORY AND PHYSICAL EXAMINATION All of the considerations discussed above for young to middle-aged adult dogs (i.e., breed, gender and health status) should be considered when developing key nutritional factors for older dogs. Special attention should be directed to physiologic changes associated with aging and diseases that are more prevalent in older animals such as renal disease, cancer, degenerative joint disease, cardiac disease, endocrine disorders, periodontal disease and obesity.

LABORATORY AND OTHER CLINICAL INFORMATION All older dogs should be screened for renal disease and hypertension. Chronic renal disease is best diagnosed with a urinalysis and serum biochemistry profile. In general, indirect blood pressure measurements obtained routinely during hospital visits are reasonable estimates of a dog's true blood pressure. Fundic examination may detect changes associated with hypertension and other systemic diseases. Thoracic radiographs and echocardiography should be performed if a cardiac murmur is detected or if there is a history of coughing or an abnormal respiratory pattern.

KEY NUTRITIONAL FACTORS Table 9-3 summarizes key nutritional factors for older adult dogs. The following section describes these key nutritional factors in more detail.

Water Older dogs are more prone to dehydration due to possible osmoregulatory disturbances, medications (diuretics) and chronic renal disease, and may have compromised urine concentrating ability. Therefore, continuous access to a fresh, clean water supply is very important for older dogs. Owners of older dogs should monitor water intake closely.

Energy With increasing age, lean body mass decreases, subcutaneous fat increases, basal metabolic rate gradually declines and body temperature may decrease. Older dogs become slower and less active, and their thyroid function may be impaired. All these changes result in a 12 to 13% decrease in DER by around seven years of age (**Table 9-2**). For older dogs, a daily energy intake of 1.4 x RER (100 kcal [418 kJ] ME/BW$_{kg}^{0.75}$) is a good starting point. This amount should be modified if a dog tends to lose or gain weight when fed at the recommended level.

Very old dogs are often underweight and may have inadequate energy intake. Underweight, very elderly people show an increase in body weight when a food of higher caloric density is provided. Thus, it may be appropriate to feed a more energy-dense food to very old dogs. Because of the potential for older dogs to have different energy needs, energy densities in foods recommended for this age group may vary from 3.0 to 4.0 kcal (12.6 to 16.7 kJ)/g DM (**Table 9-3**).

Phosphorus and Calcium Some degree of clinical or subclinical renal disease is often present in older dogs; as many as 25% of all dogs may be affected. Excessive phosphorus intake should therefore be avoided. Researchers have observed that dogs with advanced renal disease had slowed progression and reduced severity of renal disease when phosphorus levels in foods were decreased, thereby improving survival time. Foods with 0.25 to 0.75% DM phosphorus are recommended for older dogs (**Table 9-3**).

Foods with 0.5 to 1.0% DM calcium (providing 75 to 150 mg calcium/BW_{kg}) are recommended for older dogs. The calcium-phosphorus ratio should not be less than 1:1 (**Table 9-3**).

Protein As with all lifestages, healthy older dogs should receive enough protein and energy to avoid protein-energy malnutrition. However, improving protein quality, rather than increasing its intake can provide sufficient protein. Commercial foods containing 15 to 23% DM protein will provide sufficient protein for healthy older dogs (**Table 9-3**).

Sodium and Chloride There is no nutritional need for higher levels of sodium and chloride found in some commercial dog foods, especially considering the increased prevalence of heart disease and renal disease in older dogs. High sodium chloride intake may be harmful in diseases that have a hypertensive component. Secondary hypertension is associated with obesity and chronic renal disease, which are frequently seen in older dogs.

Minimum sodium and chloride requirements in older dogs are probably not different from those of younger adults, which are estimated to be around 4 mg sodium/kg body weight. The recommended sodium levels for adult dogs (25 to 50 mg sodium/BW_{kg}) are still six to 12 times higher than the minimum requirement. Foods with 0.2 to 0.35% DM sodium provide recommended levels (**Table 9-3**). Although the chloride requirement of dogs has not been established, a chloride level 1.5 times the sodium requirement is a reasonable recommendation.

Fat A relatively low fat intake helps prevent obesity in healthy older dogs. However, some dogs may need different foods at seven years of age than

DOGS

they will at 13 years of age. Very old dogs may have a tendency to lose weight. For these dogs, increasing the fat content of the food increases energy intake, improves palatability and improves protein efficiency.

In general, fat levels between 7 and 15% DM are recommended for most older dogs (**Table 9-3**). The fat level should be selected as needed to meet the desired energy density to achieve ideal body weight and condition (BCS 3/5). EFA requirements should also be met as outlined for young to middle-aged adult dogs.

OTHER NUTRITIONAL FACTORS/Fiber Older dogs are prone to developing constipation, which may justify increased fiber intake. In addition, fiber added to foods for obese-prone older dogs dilutes calories. Fiber also decreases postprandial glycemic effects in older diabetic dogs. Very old dogs that tend to lose weight, however, should be offered a food with increased caloric density. **Table 9-3** lists recommended levels of crude fiber for foods intended for older dogs.

OTHER NUTRITIONAL FACTORS/Food Texture Oral disease is the most common health problem in older dogs. Both veterinary and home care are important in the treatment and prevention of periodontal disease. Foods designed to reduce the accumulation of dental substrates (i.e., plaque, calculus) and help control gingivitis and malodor are an important part of an oral home-care program for older dogs. (See Chapter 16 and Small Animal Clinical Nutrition, 4th ed., pp 475-504.)

Assess the Food and Feeding Method (Page 232)

Assessment of the food and feeding method for older dogs is similar to those procedures outlined for young and middle-aged dogs. Compare the current food's nutrient levels with key nutritional factors and nutrient requirements established during animal assessment, identify discrepancies between key nutritional factor levels and current intake and decide whether changes in the food are required.

It may not always be necessary to change the feeding method when managing healthy older dogs. However, a thorough evaluation includes verification that an appropriate feeding method is being used.

Feeding Plan (Page 232)

Older dogs are more prone to obesity, degenerative joint disease, cardiac disease, renal disease and metabolic aberrations than younger dogs. They are also usually less active than younger dogs. The feeding plan should be based on potential risk factors and information attained in the assessment. Nutritional surveillance is more important for older dogs than for young and middle-aged dogs; therefore, the number of veterinary assessments per year should be increased.

Select a Food (Page 232)

It may not be necessary to change the food that is currently fed. However, a new food should be selected if discrepancies are determined under the food assessment phase. This is accomplished by choosing a food that provides the appropriate key nutritional factors and amounts listed in **Table 9-3**. Appendices 2 and 3 list nutrient levels for a number of readily available commercial dog foods and treats. (Also see Small Animal Clinical Nutrition, 4th ed., pp 1084-1091.) Alternatively, pet food manufacturers can be contacted directly for information. Foods selected should also have passed AAFCO or similar feeding trials for adult dogs. This information may be found on the pet food label.

Determine a Feeding Method (Page 232)

The method of feeding should be monitored more closely in older than in younger dogs. Free-choice feeding should not be used for obese or overweight animals, but this method may be preferred for thinner, very old animals to allow increased food intake. It is very important to measure food intake of older dogs; this measurement may be more accurate when the dog is meal fed. Measures to stimulate food intake may be necessary for some very old dogs. Most older adult dogs adapt well to new foods, but some animals may have difficulty. It is always good practice to allow for a transition period to avoid digestive upsets. This is particularly true when switching from foods with a lower to a higher fat content. New foods should be increased and old foods decreased in progressive amounts over a three- to seven-day period until the changeover is completed. (See Chapter 1.) **Table 9-1** contains a feeding plan summary for healthy older dogs.

Reassessment (Page 232)

Nutritional status for healthy older dogs should be assessed at least every six to 12 months. Immediate reassessment should take place if clinical signs arise that indicate the current regimen is inappropriate or if the needs of the dog change due to a change in activity.

DOGS

REPRODUCING DOGS (Pages 232-241)

The objectives of a good reproductive feeding program are to optimize: 1) conception, 2) number of puppies per litter, 3) the ability of the bitch to deliver and 4) viability of prenatal and neonatal puppies. Appropriate feeding and management will increase the likelihood of successful reproductive performance, whereas improper nutrition can negatively affect reproductive performance in bitches.

Estrus and Mating (Pages 232-234)
Assessment (Pages 232-234)
ASSESS THE ANIMAL Optimal nutrition of reproducing animals should precede mating and conception. As a rule, only healthy dogs in a good nutritional state (BCS 3/5) should be used for breeding. A BCS of 2/5 is acceptable only for a bitch that is bred occasionally. Obese bitches may have a lower ovulation rate, smaller litter size and insufficient milk production. Obesity may also cause silent heat, prolonged interestrous intervals and anestrus. Therefore, to optimize fertility, overweight bitches should lose weight before breeding. A good history and general physical examination should precede breeding to document and correct problems that may interfere with successful breeding.

Key Nutritional Factors Compared with adult maintenance, there are no special nutritional requirements for bitches during estrus. Like breeding females, most sires do not have special nutritional needs beyond adult maintenance requirements and do well when fed foods for young to middle-aged adult dogs. However, intact males and females may require more energy intake than their neutered counterparts to maintain ideal body condition (BCS 3/5).

ASSESS THE FOOD AND FEEDING METHOD/Females If a breeding bitch is in poor body condition (BCS <2/5), it may be prudent to postpone mating and bring the bitch into good body condition for the next estrus. If breeding cannot be postponed, the bitch should be fed a well-balanced, energy-dense food (at least 4.0 kcal/g [16.7 kJ/g]), in sufficient quantities to improve body condition throughout gestation. The digestibility of the nutrients in the food should be above average to achieve this goal (**Table 9-5**), and the food should contain a minimum of 25% DM crude protein and 15% DM fat.

Bitches tend to have a depressed appetite during estrus; therefore, a decrease in food intake of as much as 17% can be expected during peak estrus. Occasional vomiting may occur in bitches due to hormonal changes, nervousness, travel and environmental changes associated with mating. To reduce these problems, it may be better to feed small meals or not to feed the bitch at all immediately before or after mating.

ASSESS THE FOOD AND FEEDING METHOD/Males Some males in heavy service may have decreased food consumption and lose weight. If weight loss is a problem in reproducing males, the amount of food provided should be increased or a more energy-dense food should be fed to help maintain condition. This assumes that other causes of weight loss have been ruled out.

Table 9-5. Apparent digestibility of nutrients in typical commercial pet foods.*

Nutrient	Average digestibility (%)	Above average digestibility (%)**
Protein	80	≥85
Fat	90	≥95
Carbohydrate (NFE)	85	≥90

Key: NFE = nitrogen-free extract.
*Apparent digestibility of a nutrient is the difference between the amount ingested with the food and the amount excreted in the feces, divided by the amount ingested and multiplied by 100.
**Above average digestibility is defined as 5% or more above the average nutrient digestibility in pet foods. During certain physiologic states (e.g., intense exercise, late gestation, lactation and early growth), the food digestibility should be above average.

Feeding Plan *(Page 234)*

Follow the guidelines discussed above for young to middle-aged adult dogs. **Table 9-1** contains a feeding guide summary for reproducing dogs.

Reassessment (Page 234)

Animals should be reassessed before every estrous cycle in which a pregnancy is planned. Breeders should be encouraged to present reproducing bitches for a checkup at least a month before the upcoming estrus. Problems detected by the assessment still may be corrected before breeding. See the young to middle-aged adult dog section above for methods of assessment.

Pregnancy (Pages 234-238)
Assessment (Pages 234-237)

ASSESS THE ANIMAL Bitches that are fed properly gain 15 to 25% more than their pre-breeding weight before whelping (**Figure 9-1**). After birth, bitches should weigh about 5 to 10% more than their pre-breeding weight. This weight gain corresponds with development of mammary tissue, extracellular water and some gain in extragenital tissue. Retention of more than 10% above pre-breeding weight may adversely affect whelping. Furthermore, dogs do not need to maintain a body fat reserve to provide energy for the subsequent lactation because they can increase their food intake during lactation.

Malnutrition, whether due to inadequate or excessive intake of nutrients, may affect pregnancy. Embryos may die at an early stage or, later, fetuses may develop incorrectly, die and be resorbed or expelled before term or carried to full term. Embryo loss and in utero resorption are manifested by smaller litter size. Malnutrition during pregnancy is also a cause of low birth weight puppies that are particularly prone to hypoglycemia, sepsis, pneumonia and hemorrhage.

DOGS

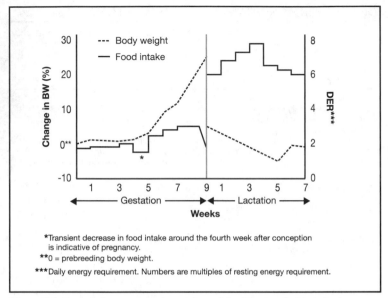

*Transient decrease in food intake around the fourth week after conception is indicative of pregnancy.
**0 = prebreeding body weight.
***Daily energy requirement. Numbers are multiples of resting energy requirement.

Figure 9-1. Typical changes in body weight and food intake of a bitch during gestation and lactation. A bitch only weighs 5 to 10% above pre-breeding weight after parturition, and should lose no more than 5% of its body weight during the first month of lactation. Food intake may drop precipitously during the last days of gestation.

Obesity at the end of pregnancy may increase dystocia, prolong labor and therefore predispose the puppies to hypoxia and hypoglycemia.

Key Nutritional Factors Table 9-6 summarizes key nutritional factors for pregnant dogs. The following section describes these key nutritional factors in more detail.

Energy Only 2% of total fetal mass is developed at 35 days of pregnancy and 5.5% at 40 days (**Figure 9-2**). Therefore, during the first two-thirds of gestation, energy requirements are not different from those of adult maintenance. After Day 40, fetal tissue grows exponentially; energy intake correspondingly increases during Week 5 and peaks between Weeks 6 and 8 of gestation. Energy requirements for gestation peak at about 30% above adult maintenance for bitches with smaller litters, whereas energy needs for bitches with larger litters can increase by 50 to 60% (**Table 9-7**).

Energy needs are highest during the last week of gestation; however, food intake is limited by the abdominal space occupied by the gravid uterus. Giant breeds may have difficulty ingesting enough food and maintaining body weight even before the last week of gestation. Food intake may decrease precipitously just before whelping with some bitches becoming completely anorectic. Enough energy should be provided to bitches during the earlier weeks of gestation, otherwise bitches might be under-

Table 9-6. Key nutritional factors for gestation and lactation.

Factors	Recommended levels in food (DM)	
	Gestation/lactation*	Gestation/lactation**
Energy density (kcal ME/g)***	3.5-4.5	4.0-5.0
Energy density (kJ ME/g)***	14.6-18.8	16.7-20.9
Crude protein (%)	22-32	25-35
Crude fat (%)	10-25	≥18
Soluble carbohydrate (%)	≥23	≥23
Crude fiber (%)	≤5	≤5
Calcium (%)	0.75-1.5	1.0-1.7
Phosphorus (%)	0.6-1.3	0.7-1.3
Ca/P ratio	1:1-1.5:1	1:1-2:1
Sodium (%)	0.35-0.60	0.35-0.6
Chloride (%)	0.50-0.90	0.50-0.9
Digestibility	Above average	Above average

Key: DM = dry matter, kcal = kilocalories, kJ = kilojoules.
*Gestation for all bitches and for lactation of bitches with four or fewer puppies.
**Lactation for bitches with litters of more than four puppies. Some giant-breed bitches may need this type of food during gestation in order to maintain body weight, particularly during late pregnancy.
***If the caloric density of the food is different, the nutrient content in the dry matter must be adjusted accordingly.

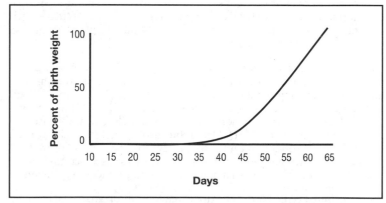

Figure 9-2. The development of fetal mass during pregnancy in beagle dogs.

Table 9-7. Practical recommendations for energy intake during gestation.

Week of gestation	Daily energy requirement	
	kcal ME/day*	kJ ME/day*
1-4	DER**	DER**
5	DER + 18 kcal ME/kg BW	DER + 75 kJ ME/kg BW
6-8	DER + 36 kcal ME/kg BW	DER + 150 kJ ME/kg BW
9	DER + 18 kcal ME/kg BW	DER + 75 kJ ME/kg BW

Key: DER = daily energy requirement, kcal = kilocalories, kJ = kilojoules, ME = metabolizable energy, BW = body weight.
*Energy requirements during gestation are the sum of the energy needed for normal adult maintenance of a non-pregnant dog plus what is needed for accretion of fetal and maternal tissue. Because accretion of fetal and maternal tissue is minimal during the first 35 days of gestation, the increase in energy requirement only becomes significant from Week 6 on. However, it is better to increase the food intake progressively during Week 5. This allows the bitch to build reserves for the last week of gestation, during which food intake is compromised by abdominal fill.
**During gestation DER is estimated at 1.9 x RER (DER = 132 kcal ME/BW$_{kg}^{0.75}$ or 550 kJ ME/BW$_{kg}^{0.75}$).

weight during mid and late gestation and have difficulty maintaining body condition and milk production after whelping. During the last few weeks of gestation, the food should be high in energy density (4.0 kcal/g [16.7 kJ/g]) to provide adequate energy, especially for large-breed bitches.

Protein Protein requirement increases 40 to 70% above maintenance during late gestation, and follows the increase in energy requirement. Thus, foods for dogs in late gestation should also have increased levels of protein. A food containing 20 to 25% DM crude protein and 4.0 kcal/g DM (16.7 kJ/g) is recommended (**Table 9-6**). The quality of the protein should also be higher to improve vigor of newborn puppies and minimize neonatal mortality. Protein deficiency during pregnancy may decrease birth weight, increase mortality during the first 48 hours of life and decrease immunocompetence of offspring.

Fat Increasing fat levels in foods improves digestibility and provides energy, which in turn improves nitrogen retention. Although adult maintenance foods are appropriate for the first two-thirds of pregnancy, a food with an energy density of approximately 4.0 kcal (16.7 kJ) ME/g is recommended for the last third of gestation. This is usually achieved by feeding a food containing 10 to 25% DM fat (**Table 9-6**). The target level may need to be altered depending on litter size, body condition of the bitch, food intake of the bitch and other extraneous factors as discussed previously. Giant-breed bitches may need a high-energy food throughout gestation.

Carbohydrate Feeding a carbohydrate-free food to pregnant bitches may result in weight loss, decreased food intake, reduced birth weight and decreased neonatal survival and may increase the risk of stillbirth. Because more than 50% of the energy for fetal development is supplied by glucose, bitches have a high metabolic requirement for glucose during the last weeks of gestation. Feeding a carbohydrate-free food to pregnant bitches increases the risk of hypoglycemia and ketosis during late pregnancy, and lactose concentration in the milk may decrease by 40% during peak lactation.

Providing approximately 20% of the energy from carbohydrate is sufficient to prevent the negative side effects of a carbohydrate-free diet. **Table 9-6** provides recommendations for minimum DM carbohydrate levels.

Clinical hypoglycemia (i.e., glucose values <45 mg/dl or 2.5 mmol/l) occurs infrequently in bitches, but when it develops it is usually observed during the last two to three weeks of gestation. Neurologic signs of hypoglycemia predominate and differentiation from eclampsia may be difficult. Elevated levels of serum ketones (mainly β-hydroxybutyrate) are characteristic in bitches with clinical disease. However, ketonemia may be missed when strips or tablets containing nitroprusside are used to detect ketones. Nitroprusside primarily detects acetone and acetoacetate. Risk factors that may predispose bitches to this syndrome include: 1) poor body condition during pregnancy, 2) malnutrition and 3) feeding a high-fat, carbohydrate-free food.

Treatment of clinical hypoglycemia during pregnancy should consist of intravenous administration of a bolus of 20 to 50% glucose solution, which can be followed by intravenous infusion of a 5% glucose solution at a rate of 2 ml/BW_{kg}/hour. During or soon after the infusion, a palatable food should be provided that has adequate soluble carbohydrate, protein and calories (**Table 9-6**) and has above average digestibility (**Table 9-5**).

Calcium and Phosphorus During the last 35 days of gestation, requirements for calcium and phosphorus increase by approximately 60% because of rapid skeletal growth of the fetuses. As occurs with some dairy cows, excessive intake of calcium during pregnancy may decrease activity of the parathyroid glands and predispose the bitch to eclampsia during lactation. Therefore, it has been recommended to feed a food during pregnancy that avoids large excesses of calcium (recommend 0.75 to 1.5% DM) and has a calcium-phosphorus ratio of 1.1 to 1.5:1 (**Table 9-6**).

Bitches are at the highest risk for developing eclampsia (puerperal tetany) during Weeks 3 and 4 of lactation (peak lactation), when calcium losses via secretion in milk are highest. The number of nursing puppies is the most important stimulus for milk production, so it is not surprising that eclampsia is commonly seen in bitches nursing large litters. Typically, affected bitches are primipara, are less than four years old, in Weeks 2 to 4 of lactation and are toy-breed dogs. Occasionally, bitches may be affect-

DOGS

ed at or just before whelping. Investigators have suggested that toy breeds may be more predisposed to developing eclampsia than large breeds because toy breeds tend to receive more meat-based homemade foods, which are low in calcium.

Although most bitches with eclampsia are hypocalcemic, some may be normocalcemic. Some bitches with hypocalcemia, on the other hand, may not exhibit clinical signs. This finding may indicate that factors other than calcemia determine whether tetany manifests clinically or not. Magnesium levels in bitches with eclampsia are normal.

Hypocalcemia leads to increased neuromuscular irritability resulting in restlessness and whining, stiffness of gait, ataxia and tonic-clonic convulsions.

Intravenous infusion of 10% calcium gluconate administered to effect (1 to 2 mg calcium/BW_{kg}) results in rapid clinical improvement. To lessen the risk of relapse, calcium may be injected subcutaneously or intramuscularly in addition to the immediate intravenous infusion. Subcutaneous injections, however, may cause skin necrosis and should be administered only when other routes are inaccessible. If possible, puppies should be separated from the bitch for the first 24 hours of treatment. If tetany recurs during the same lactation, the puppies should be weaned and fed a milk replacer. Administration of corticosteroids is contraindicated because they may further decrease plasma calcium levels.

Measures to prevent eclampsia should be initiated during gestation and include feeding a nutritionally balanced food (no excess calcium; **Table 9-6**). Vitamin D therapy (10,000 to 25,000 IU daily) has been proposed during the last week of gestation. This therapy is similar to recommendations for dairy cows to prevent postparturient paresis. However, this approach may be inappropriate for bitches because eclampsia and the highest calcium losses do not occur immediately after parturition.

Digestibility Although not a nutrient, this food factor is important in late gestation. Apparent digestibility is the difference between the amount of food ingested and that excreted in feces. During late gestation, the ability to ingest adequate amounts of food may be limited. This situation is exacerbated if the food is poorly digestible. Therefore, it is important to assess digestibility and recommend foods with above average digestibility for late gestation. Above average digestibility is more than 5% above average values (**Table 9-5**).

Other Nutritional Factors Hematocrit, hemoglobin and plasma iron values often decrease in bitches near the end of gestation. Iron requirements are particularly high during the last week of gestation because large quantities are stored in the liver of the fetuses and secreted in colostrum. Colostrum is rich in iron. However, iron concentrations are low in mature milk. Therefore, neonates must have an iron reserve to overcome the initial three-week nursing period, when milk is the only source of food. Latent iron deficiency may impair neutrophil function and cell-mediated immunity, increasing susceptibility to infections. During periods when requirements for tissue synthesis are greater than normal (e.g., pregnancy, lactation and growth), animals are particularly susceptible to zinc deprivation. Most commercial foods provide adequate zinc. However, if zinc deficiency does occur during pregnancy, it may lead to fetal resorption or less viable offspring.

ASSESS THE FOOD Food assessment includes a comparison of the current food's levels of key nutritional factors with those recommended in **Table 9-6**. Appendices 2 and 3 list levels of key nutritional factors for selected foods and treats. (Also see Small Animal Clinical Nutrition, 4th ed., pp 1084-1091.) In some cases, it may be necessary to contact the manufacturer for this information. The recommended food should also have passed a feeding test to prove it will support gestation (i.e., AAFCO or equivalent). This information should be available on the product label. The food assessment step determines the appropriateness of the current food. Gestation is a unique situation in which nutritional requirements increase markedly over a relatively short period of time. It is very important to provide the correct food.

ASSESS THE FEEDING METHOD The feeding method for pregnant dogs may need to be altered, especially in late gestation. Evaluation of current feeding methods with foreknowledge of the demands of gestation will allow for development of an appropriate feeding plan.

One or two meals per day will suffice for most bitches during the first half of pregnancy. At least two meals per day should be provided in the last half of pregnancy. Giant breeds may need to be fed free choice. Bitches pregnant with a large litter may also need to be fed free choice to help them compensate for the space occupied by the gravid uterus. Food restriction during gestation may lead to smaller litter size, lower birth weights and may compromise the subsequent lactation.

During the third or fourth week of gestation, bitches commonly experience a decrease in appetite that may result in as much as a 30% reduction in food intake. This decrease may be due to the effect of embryo implantation, which starts around 20 days of pregnancy (**Figure 9-1**). The bitch is usually fed the same amount of energy as an intact adult dog (approximately 1.8 x RER) during the first two-thirds of gestation. This amount is increased to approximately 3.0 x RER during the last three weeks of gestation. Energy intake may need to be increased further to maintain normal body condition in some dogs, especially larger breeds.

DOGS

Feeding Plan (Pages 237-238)

Information gleaned from the assessment step (i.e., animal, food and feeding method) sets the stage for developing the feeding plan; specifically which foods to feed and which feeding methods to use in providing the food. **Table 9-1** contains a feeding plan summary for pregnant dogs.

SELECT A FOOD In general, foods for non-reproducing adult dogs should suffice for the first four weeks of gestation, then the dog should receive a food for growth/reproduction. As mentioned above, giant-breed dogs may need a growth/reproduction-type food throughout gestation.

A different food should be selected if the current food does not provide the recommended levels of key nutritional factors as determined by the food assessment step (**Table 9-6**). Appendix 2 lists some appropriate foods. The manufacturer should be contacted for information if a food is considered that is not listed in Appendix 2. The selected food should have successfully passed AAFCO (or a comparable regulatory agency) feeding tests.

DETERMINE A FEEDING METHOD Because overfeeding during gestation may have similar side effects as underfeeding, small- and medium-sized bitches should be meal fed. Large- and giant-breed bitches may be fed free choice beginning with Week 5. Some giant-breed dogs may have difficulty maintaining weight and should be fed free choice throughout gestation. Follow the recommendations described in the feeding methods assessment step. Also, if changing foods, gradually transition the bitch to the new food over several days as described in Chapter 1.

Reassessment (Page 238)

There are two occasions during pregnancy when owners should present a bitch for assessment by a veterinarian. The first time is to confirm pregnancy with ultrasonography between 17 to 20 days after breeding, or by palpation between 25 to 36 days after breeding. A thorough physical examination should be conducted at the first visit. The owner should be encouraged to present the bitch again one week before parturition, or earlier if an abnormality is found during the first checkup. In addition to another physical examination, the following parameters should be assessed at the second checkup: a complete blood count and serum glucose, calcium and total protein concentrations.

Lactation (Pages 238-241)

Successful lactation depends on body condition before breeding, and adequate nutrition throughout gestation and lactation. During lactation, nutrient requirements are directly related to milk production, which in turn depend primarily on the number of nursing puppies. A bitch's nutrient requirement during lactation is greater than at any other adult lifestage and, in some cases, equal to or higher than for growth.

Assessment (*Pages 238-240*)

ASSESS THE ANIMAL A physical examination and anamnesis should be performed. After parturition, a bitch should weigh 5 to 10% above its pre-breeding body weight (**Figure 9-1**). Unlike cats, bitches do not need to maintain a body fat reserve to provide energy for lactation. During the first week of lactation, milk production is approximately 2.7% of body weight independent of litter size. Thereafter, milk production steadily increases and peaks during the third and fourth week of lactation.

After the first two to five days of lactation, the composition of the milk is stable and the bitch's nutrient requirements are primarily determined by the quantity of milk produced. During peak lactation, the quantity of milk produced depends primarily on the number of nursing puppies. The puppies' intake of solid food begins to increase around the fifth week, after which, milk production progressively declines. Therefore, the stage of lactation and the number of nursing puppies primarily determine the bitch's protein and energy requirements for lactation.

Key Nutritional Factors/*Water* Although often overlooked, water is an important nutrient for lactation. Water is needed in large quantities to produce milk and aid in thermoregulation. Water requirements in ml are roughly equal to energy requirements in kcal. A 35-kg bitch nursing a large litter may require 5 to 6 l/day at peak lactation. Therefore, it is critical that clean, fresh water be available at all times during lactation.

Key Nutritional Factors/*Digestibility* Nutrients in food should be highly available due to the considerable nutritional demands associated with lactation. Therefore, foods with above average digestibility are recommended for lactating bitches (**Table 9-5**).

Key Nutritional Factors/*Energy* After whelping, the bitch's energy requirement steadily increases and peaks between three and five weeks at a level two to four times higher than the DER for non-lactating, intact adults. The energy requirement returns to adult maintenance levels about eight weeks after whelping. Bitches are capable of increasing food intake during lactation; however, the energy density of the food is usually the limiting factor for meeting energy requirements of lactating dogs. If foods with low energy density are fed (<3.5 kcal [14.6 kJ]/g), the bitch may not be physically able to consume enough food and may lose weight, have decreased milk production and display signs of severe exhaustion. These signs are most pronounced in giant-breed dogs with large litters. Therefore, it is better to provide foods containing more than 4 kcal ME (16.6 kJ)/g DM (**Table 9-6**).

Energy requirements for lactating bitches can be subdivided into energy for maintenance and energy used for milk production. The DER, without allotment for milk production, may be slightly higher than that for average adults because of stress and increased activity associated with caring for puppies. The DER, without allotment for milk production, has been estimated to be 143 kcal (600 kJ) digestible energy/$BW_{kg}^{0.75}$. This is equivalent to 132 kcal (550 kJ) ME/$BW_{kg}^{0.75}$ or about 1.9 x RER.

Milk production is related to demand for milk, which is directly related to litter size and stage of lactation. Bitches with large litters will produce more milk than those with small litters and more milk is produced in mid-lactation than during late lactation. After five weeks of lactation, puppies begin eating more solid food and the number of puppies becomes less of a determinant. See Small Animal Clinical Nutrition, 4th ed., p 239 for a method to calculate milk production.

By combining the DER, without allotment for milk production, with the energy required for milk production, the total daily energy needed during lactation can be expressed as:

132 kcal ME per $BW_{kg}^{0.75}$ + (1.7 kcal x ml of milk/day).

In other words, during peak lactation, about 220 kcal (920 kJ) ME/kg litter weight are needed in addition to maintenance energy requirements. As mentioned above, lactating bitches need about 1.9 x RER for maintenance; this amount should be increased 25% for each nursing puppy. Because this is a rough estimate, body condition of bitches should be evaluated and adjustments made to feeding amounts as necessary to maintain an ideal body condition (BCS 3/5). Foods for lactation should provide between 4.0 to 5.0 kcal ME (16.7 to 20.9 kJ)/g DM (**Table 9-6**).

Key Nutritional Factors/*Protein* The requirement for protein increases more than the requirement for energy during lactation. Therefore, the protein-energy ratio must be higher in foods for lactation than in foods for adult maintenance. Ratios of 4.8 to 6.8 g digestible protein/100 kcal ME (10.5 to 15 g/MJ DE) have been recommended. This recommendation corresponds to about 19 to 27% DM digestible protein of an energy-dense food (4.0 kcal [16.7 kJ] ME/g). Generally, it is recommended to feed a food containing at least 25% or more crude protein (**Table 9-6**) from mixed sources with increased digestibility (**Table 9-5**) and sufficient energy density.

Key Nutritional Factors/*Fat* Fat provides EFA and energy and enhances fat-soluble vitamin absorption. The minimum level of fat in foods intended for lactating bitches (four or fewer puppies) is 10% DMB. However, an increase in fat intake results in better efficiency of food utilization during lactation. Increasing concentrations of fat will also increase the caloric density of foods and help meet the high energy requirements of bitches during lactation. An increase in fat should be balanced by increasing other nutrients proportionally to match energy density (**Table 9-6**).

Key Nutritional Factors/*Carbohydrate* Foods for lactation should provide at least 10 to 20% of the energy intake in the form of soluble carbohydrate to support normal lactose production. See **Table 9-6** for DM soluble carbohydrate recommendations.

Key Nutritional Factors/*Calcium and Phosphorus* Mineral requirements during lactation are determined by mineral secretion in milk and thus by the number of nursing puppies. A definite increase in calcium content is seen over the course of lactation; however, the calcium-phosphorus ratio is consistently maintained around 1.3:1. Bitches need two to five times more calcium during peak lactation than for adult maintenance. Depending on the number of puppies, bitches need 250 to 500 mg calcium and 175 to 335 mg of phosphorus/BW_{kg} per day. One investigator recommended that a food for lactation contain at least 0.8 to 1.1% calcium and 0.6 to 0.8% phosphorus; however, reducing these needs by 10 to 20% will not necessarily lead to disturbances in milk mineral content. **Table 9-6** lists practical recommendations for dietary calcium and phosphorus levels.

Other Nutritional Factors Requirements for most trace elements depend on litter size. Iron requirements increase only slightly during lactation when compared with adult maintenance requirements because mature milk is relatively low in iron. Colostrum is very rich in iron; however, levels decrease within 48 hours. Requirements for copper (17 to 20 mg/kg DM), on the other hand, increase more than for energy.

ASSESS THE FOOD Lactation represents an extreme test of a food's nutritional adequacy, because no other lifestage requires such a marked increase in energy density and nutrient content. The nutrient demands are directly related to the dam's ability to produce milk. Food assessment includes a comparison of the current food's levels of key nutritional factors with those recommended in **Table 9-6**. Appendices 2 and 3 list levels of key nutritional factors for selected foods and treats. (Also see Small Animal Clinical Nutrition, 4th ed., pp 1084-1091.) Alternatively, the manufacturer can be contacted for this information. Pet food labels usually lack information about soluble carbohydrate content, digestibility and energy density. The recommended food should also have passed an AAFCO or similar feeding test to prove it will support lactation. This information should be available on the product label. The food assessment step determines the appropriateness of the current food. Because nutritional requirements for lactation increase markedly over a relatively short period of time, it is very important to provide the correct food.

ASSESS THE FEEDING METHOD A lactating bitch's nutrient needs are met by a combination of the nutrient levels in the food and the amount fed. Even if the food has an appropriate nutrient profile, significant undernutrition may result if the bitch is fed an insufficient amount. If the bitch maintains normal body condition (BCS 3/5) and the puppies are growing at a normal rate, then the amount being fed is probably appropriate. The amount to feed can be estimated either by calculation (See Chapter 1.) or by referring to feeding guides on product labels. As a rough estimate, bitches should ingest their DER + 25% of DER for each nursing puppy. During peak lactation, a bitch's energy needs may be three to four times greater than its requirements for adult maintenance. The amount fed during lactation is usually offered either three times per day or free choice.

Feeding Plan (Pages 240-241)
    Table 9-1 contains a feeding plan summary for lactating bitches.

SELECT A FOOD A more appropriate food should be selected if food assessment indicates inadequacies or if lactation performance is suboptimal. Lactating bitches are best fed commercial foods. Dry foods are more nutrient dense, as fed, and have higher levels of carbohydrates than moist foods. These foods may benefit bitches experiencing weight loss and those spending little time eating. Conversely, moist foods are often higher in fat and provide additional water to support lactation. The added water also improves palatability; therefore, bitches may be more likely to eat. Because both food types have advantages, many breeders choose to feed both forms during reproduction. Appendix 2 lists several commercial foods appropriate for lactation. Select a food that has passed an AAFCO or a similar feeding trial for lactation and provides levels of key nutritional factors listed in **Table 9-6**.

DETERMINE A FEEDING METHOD In practice, it is best to feed bitches free choice during lactation, except when the bitch has only one puppy and may have a tendency to gain weight. Free-choice feeding is especially important for lactating bitches with more than four puppies. Some bitches are nervous throughout lactation and free-choice feeding will allow them to eat on their schedule. Meal-fed lactating bitches should receive at least three meals per day. Puppies may begin to eat the bitch's food at three weeks of age; therefore, it is important to allow them access to the food.

Reassessment (Page 241)

Body weight gain by puppies during early lactation provides an indication of milk production by the bitch (quantity and quality) and milk intake by puppies. Failure to gain weight for more than one day or continuous vocalization may indicate that the quantity or quality of milk production is insufficient due to mastitis, agalactia or inadequate nutrition.

Body weight and body condition scoring are important tools to assess nutritional adequacy. A bitch should not lose more than 5% of its body weight during the first month of lactation, and optimal body weight should again be reached within a month after lactation ceases. BCS should be maintained around 3/5 throughout lactation, otherwise adjustments should be made in the food or feeding method, assuming other potential causes of weight loss are ruled out.

WEANING (Page 241)

Weaning is a gradual process with two phases. The first phase begins when the puppies begin eating solid food between three and four weeks of age. This phase should be encouraged, especially if the bitch has a large litter. In addition, nursing is an important stimulus for milk production. Therefore, milk production will progressively decline as the puppies' intake of solid food increases, making complete weaning (second phase) less stressful. However, some bitches may continue to produce large quantities of milk and are at risk for development of mammary congestion at the time the puppies are completely separated from the bitch. The feeding schedule in **Table 9-8** may be helpful, particularly in cases of early weaning (around the fifth week of age).

Restricting food intake a day or two before weaning reduces nutrients available for milk production, thereby reducing mammary gland engorgement. Leaving one or two puppies to nurse will not alleviate mammary gland engorgement in bitches that are still producing a large amount of milk at weaning. This practice continues to stimulate milk production, and therefore prolongs the problem. When it is decided to completely separate the puppies from the mother, all puppies should be taken away at once.

GROWING DOGS (Pages 241-250)

Compared with the young of other species, newborn puppies are relatively immature at birth. For example, their skeletons have a low degree

Table 9-8. Recommended feeding schedule for reducing mammary congestion in bitches during weaning of puppies.

Day of weaning	No food
First day after weaning	One-fourth of DER for adult maintenance (i.e., 0.5 x RER)
Second day after weaning	One-half of DER for adult maintenance (i.e., RER)
Third day after weaning	Three-fourths of DER for adult maintenance (i.e., 1.4 x RER)

Key: DER = daily energy requirement, RER = resting energy requirement.

of mineralization. Large-breed puppies are more premature than small-breed puppies, which may be one of the reasons why they are more susceptible to malnutrition and developmental orthopedic diseases during the rapid growth phase.

Growing dogs progress through three critical phases in the first 12 months of life, during which nutrition is essential for survival and healthy development.

- A nursing period during which the transition is made from in utero nutrition to postpartum nutrition. This period is largely influenced by the nutrition of the bitch during gestation and early lactation.
- A weaning period, which is very stressful due to changes in food and environment. The transition from bitch's milk to solid food for further growth must therefore be handled properly.
- A postweaning period that occurs from two to 12 months of age and is a critical time for skeletal and other development. Proper feeding during this period is especially critical for large- and giant-breed puppies because nutrition has proved to be the most important non-genetic factor for healthy bone development.

Growing Dogs: Nursing Period (Pages 241-247)

Before weaning, mortality may be as high as 30%, with two-thirds of the deaths occurring during the first week of life. Nutrition during gestation and early lactation, behavior and health of the bitch and good neonatal care are factors critical to successful transition from fetal life to the nursing period. (See Appendix 1 and Small Animal Clinical Nutrition, 4th ed., pp 1012-1019.)

Assessment (Pages 241-247)

ASSESS THE ANIMAL/Physical Examination The three most important areas of evaluation of nursing puppies are assessment of body weight (especially with respect to temporal changes), body temperature and other physical parameters. Healthy puppies sleep and nurse; when a puppy continues to vocalize it is probably ill, malnourished, cold or dehydrated.

The syndrome of hypoglycemia, hypothermia and dehydration is by far the most common nutrition-related condition seen in neonates. Orphan puppies are at a much higher risk than nursing puppies, especially when deprived of colostrum. Low fat stores and the degree of poikilothermy make puppies very dependent on effective nursing and optimal environmental temperature during the first two weeks of life. The first three days of life, however, are the most critical. Rectal temperatures of newborn puppies may decrease up to 4 to 5°C (7 to 8°F) immediately after birth. Furthermore, healthy puppies may lose about 0.5 g of body weight every 30 minutes that they sleep without being fed.

When food intake is inadequate or when the environmental temperature is too low, newborn puppies rapidly deplete glycogen and fat stores and soon chill and become hypoglycemic, weak and dehydrated. Etiology includes inadequate milk production by the bitch (qualitative and/or quantitative), and all the causes of anorexia and reasons why a puppy refuses or is unable to nurse, including early maternal rejection, prematurity and low birth weight.

Infections, parasites and other illnesses lead to anorexia and may cause hypoglycemia, dehydration and hypothermia. Diarrhea rapidly causes dehydration in young puppies. Hypoxia is an important cause of anorexia and hypoglycemia. Hypoxia may result from dystocia, prolonged birth or trauma caused by the bitch. Neonates have significantly lower blood glucose levels during the first day of life when their dam did not eat during the last days of pregnancy.

Hypoglycemia, hypothermia and dehydration are interrelated; one can cause or worsen the others, starting a vicious cycle.

ASSESS THE ANIMAL/Physical Examination/*Body Weight* Birth weight of puppies is the single most important measure of their chances of survival, and reflects, among other factors, the adequacy of the bitch's nutrition during pregnancy. The evolution of a puppy's body weight gives useful information about food intake and general health. Body weight should be recorded within 24 hours after birth, and then daily or every other day for the first four weeks of life, using an accurate gram scale.

Due to variation in breed size, an exact optimal birth weight is difficult to estimate for individual puppies. Body weight at birth correlates primarily with the weight of the mother; birth weights range from 1% for some large and giant breeds to about 6.5% in Chihuahuas. Interestingly, investigators found a consistent ratio between the weight of the total litter and the body weight of the dam. Birth weight of the entire litter averages about 12 to 14% of adult body weight. The ratio can be slightly smaller in

DOGS

large breeds. Given the number of puppies and the ratio of litter to adult body weight, the birth weight of individual puppies can be evaluated in relation to the expected number of puppies per litter.

Daily weight gain averages about 5% of the puppy's current body weight during the first four weeks after birth. The absolute daily weight gain is lowest during the first week of life; however, the relative increase is largest (average 7.7% of body weight), and can reach 10% of body weight. In the first 48 hours, the increase in body weight is not related to the puppy's body weight, because healthy smaller puppies eat relatively more in an effort to replete body reserves.

The puppy's body weight often doubles by eight to 10 days after birth and it may triple by the third week. Although the relative weight gain gradually decreases, weight gain in g/day varies little from the second to the fourth week of life.

Daily gain can vary markedly. Although puppies should be weighed every day or every other day, a more precise evaluation should be based on the average weekly weight gain. Between one and two months of age, daily weight gain may average 3 g/BW$_{kg}$, and between 2 and 4 g/kg adult body weight until five months of age. These numbers may be used to help assess growth rates. However, dogs do not grow linearly; the growth curve has a sigmoid shape, with a fast exponential growth component first followed by slower growth (**Figure 9-3**). The exact timing of these phases differs from breed to breed. As a rule, small- and medium-sized dogs (up to 25 kg) reach about 50% of their adult weight around four months of age, whereas dogs with adult weights above 25 kg reach the 50% point at about five months of age. (See Appendix F, Small Animal Clinical Nutrition, 4th ed., pp 1020-1026.)

Low birth weight is highly correlated to neonatal mortality. Low birth weight puppies are susceptible to hypoglycemia and sepsis, and are less likely to survive without special care. Nursing puppies should be weighed daily or every other day on a gram scale. Monitoring the puppies' weight is a good way to evaluate the quality and quantity of milk the bitch is producing as well as the milk intake and health status of the puppies.

Puppies should neither lose weight nor fail to gain weight for more than one day. Weight loss or failure to gain weight in an individual puppy or the entire litter may indicate disease in the puppies or bitch, inadequate milk production or inability to suckle. It is essential to evaluate puppies' growth rate in relation to changes in behavior such as crying, restlessness and continuous vocalization.

ASSESS THE ANIMAL/Physical Examination/*Body Temperature* When examining a puppy, the clinician should determine whether the puppy is warm. Neonates show a certain degree of poikilothermy during the first

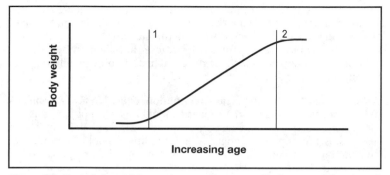

Figure 9-3. Typical sigmoidal growth curve of puppies. The beginning (line 1) and endpoint (line 2) of exponential growth can shift depending on the breed and individual variation.

two weeks of life, and have an extremely low level of body fat. Therefore, it is vital for newborn puppies to eat and be kept in a warm environment. During the first week, the immediate environment of the puppies should be kept between 29 and 32°C (84 to 90°F). This means that the temperature in the room with the bitch and its litter should be maintained between 24 and 27°C (75 to 81°F). (See Appendix 1 and Small Animal Clinical Nutrition, 4th ed., pp 1012-1019.) Marginal hypothermia can often be detected by palpation of the lower limbs. The behavior of the bitch may give some indication as to whether a puppy is hypothermic or ill. A bitch may start pushing a puppy away and neglect its cries when the puppy's skin temperature drops below a certain level.

Once a puppy's rectal temperature drops below 34.5°C (94°F) the puppy becomes less active and nurses ineffectually, digestion no longer occurs and bowel movements stop. When a puppy's skin feels cold, the dam will push the puppy away and ignore its cries. The puppy then becomes hypoglycemic and is too weak to nurse, initiating a vicious cycle from which the puppy will not survive without help. Tissue hypoxia and metabolic acidosis may reach profound proportions. Once the body temperature reaches the critical level of 32°C (90°F), hypothermia becomes severe and the puppy lies motionless, with a very slow respiratory rate and an occasional air hunger response. It has been reported that healthy newborn puppies can survive up to 12 hours of deep hypothermia and recover if warmed slowly. In practice, however, hypothermic puppies can be rescued only when the problem is detected early and treated correctly.

Hypothermia that develops in puppies kept at the correct environmental temperature may indicate insufficient milk intake. Insufficient milk intake can be caused by: disease or weakness, inability to reach the bitch's nipples, insufficient milk production and/or inadequate maternal behavior and poor milk quality or quantity due to insufficient nutrition of the dam, disease of the dam and/or inherited factors.

Orphan puppies are at greater risk because they are more sensitive to suboptimal temperatures without the dam. In addition, the milk replacer formula or feeding schedule may be inadequate.

ASSESS THE ANIMAL/Physical Examination/ *Other Physical Parameters*
When evaluating neonates, the clinician should hold each puppy to assess alertness, muscle tone and response to handling. The clinician should also assess the puppy's strength. Gastric fullness should be evaluated, and the owner asked if the puppies are nursing. Hungry puppies will start crying; however, healthy puppies will soon stop crying and sleep, even without nursing. Small and weak puppies may appear to nurse and develop abdominal fullness, yet fail to thrive. Weak puppies may also have an enlarged abdomen, but are often restless and continue vocalizing. This distention may result from aerophagia; however, more often it is caused by malnutrition or illness of the bitch or puppy.

ASSESS THE ANIMAL/Key Nutritional Factors/ *Colostrum and Milk*
The composition of the milk changes rapidly to become normal or "mature" milk between 24 hours after parturition and the end of the first week of lactation. Colostrum transfers immunoglobulins, provides a concentrated source of energy and selected nutrients and produces a laxative effect.

Colostrum has a very different composition than mature milk. Due to its high dry matter content, colostrum is somewhat sticky and viscous, which makes nursing more difficult, especially for weaker puppies. Dry matter content of colostrum decreases within 12 to 24 hours after whelping, primarily reflecting a decrease in protein.

The lactose concentration of colostrum is very low compared with that of mature milk (i.e., 1.0 vs. 3.4%). Levels of calcium, phosphorus and magnesium are very high in colostrum and decrease after two to three days to levels that are lower than in the subsequent mature milk.

Just after whelping, colostrum contains high levels of iron, copper and zinc, which decrease to the levels in mature milk within 48 hours postpartum. Colostrum is particularly high in vitamin A; colostrum levels increase the liver reserve of vitamin A in puppies by 25% within a week.

Milk is a complete food for neonates. The composition of milk (i.e., water, protein, fat, lactose, minerals and vitamins) is designed to support the normal growth rate of neonates.

Canine milk is higher in energy, protein and minerals than bovine milk. (See Appendix K, Small Animal Clinical Nutrition, 4th ed., pp 1064-

1072.) The composition of mature milk does not change significantly over the course of the lactation.

ASSESS THE ANIMAL/Key Nutritional Factors/*Energy* Bitch's milk is high in energy and provides about 146 kcal of gross energy (GE) (610 kJ)/100 g of milk. Total milk intake per puppy is lowest during the first week of life. However, expressed per kg body weight, puppies' milk intake is highest during the first week and decreases progressively. Daily intake averages about 239 kcal (1 MJ)/BW_{kg}/day during the first four weeks of life. Averages, however, may vary from as high as 287 kcal GE (1.2 MJ)/BW_{kg} during the first week of life to as low as 190 kcal GE (0.8 MJ)/BW_{kg} by Week 4.

Energy requirements of puppies are the sum of energy needed for maintenance and the requirements for growth. Because puppies sleep more than 80% of their time, and huddle together in a warm whelping box, they are able to decrease their energy requirements for maintenance to a level that approaches RER (70 kcal/$BW_{kg}^{0.75}$) during the first week of life. Therefore, all additional ingested energy can be used to grow and build body reserves. Puppies born with a lower body weight will ingest an amount of milk similar to that of their larger littermates during the first 48 hours of life to replete body reserves.

ASSESS THE ANIMAL/Key Nutritional Factors/*Water* Normal puppies need 130 to 220 ml of water/BW_{kg}/day.

ASSESS THE ANIMAL/Key Nutritional Factors/*Protein* Protein digestibility of bitch's milk is very high (up to 99%), and nitrogen retention is about 90% during the first week. Compared with cow's milk, bitch's milk contains about twice as much protein per 100 ml (7.5 vs. 3.3%). Bitch's milk also provides high levels of arginine, lysine and branched-chain amino acids. This nutrient profile is important when assessing and formulating milk replacers, and reflects the enormous anabolic activity of puppies at this young age. Protein requirements will be met if puppies ingest adequate amounts of energy as bitch's milk. (See Appendix K, Small Animal Clinical Nutrition, 4th ed., pp 1064-1072.)

ASSESS THE ANIMAL/Key Nutritional Factors/*Fat* Body fat is approximately 1.5% of total body mass at birth, which is very low when compared with the 22% body fat of non-obese adult dogs. Puppies increase body fat during the first month of life; accretion of body fat is about 50% of total weight gain. Body fat increases to about 10% of body weight by two weeks of age and to 17% after one month. The dam's milk, therefore, must contain enough energy (fat) to support development of these reserves.

DOGS

Milk fat and fatty acid composition are two of the most variable components of milk. The content and quality of the fat in milk depend on the food the bitch receives during lactation. Bitch's milk should contain 9 g or more fat/100 g of milk.

ASSESS THE ANIMAL/Key Nutritional Factors/*Carbohydrate* Lactose is the primary carbohydrate in milk. Lactose levels in bitch's milk vary between 3.0 and 3.5%, which are about 30% lower than those in cow's milk. (See Appendix K, Small Animal Clinical Nutrition, 4th ed., pp 1064-1072.) Although the lactose content of milk varies widely among animal species, it is very consistent and maintained within narrow limits within a species. Lactose and minerals in milk are the primary contributors to osmolarity. Any increase or decrease in lactose content is offset by changes in the content of the other soluble components.

To avoid diarrhea, lactose should be the main source of carbohydrate during the first weeks of life. The activity of pancreatic amylases is insignificant at four weeks of age and is still very low at eight weeks, whereas intestinal lactase activity is high until about four months of age.

ASSESS THE ANIMAL/Key Nutritional Factors/*Calcium and Phosphorus* Calcium levels are very high in colostrum; however, after two to three days, levels decrease to less than those found in subsequent mature milk. An increase in calcium content is seen over the course of lactation; however, the calcium-phosphorus ratio is consistently maintained around 1.3:1.

ASSESS THE ANIMAL/Key Nutritional Factors/*Iron* Iron deficiency may occur if insufficient stores are accumulated during the last week of pregnancy, or if excessive blood loss occurs due to severe hookworm infection or severe flea infestation. During the first three to four weeks of life, body iron stores and hematocrit and hemoglobin values decrease below levels at birth. The decrease is more pronounced in fast-growing, large-breed puppies.

Milk is a poor source of iron and puppy requirements are usually higher than intake. Iron reserves increase again when puppies receive additional food at weaning, and body iron stores normalize around four months of age. Therefore, puppies should receive additional food as soon as possible (around three weeks of age).

ASSESS THE ANIMAL/Key Nutritional Factors/*Digestibility* Data from two studies show that bitch's milk is extremely digestible. The high digestibility of milk maximizes its usage and helps puppies survive the critical first weeks.

ASSESS THE FOOD Because direct assessment of milk quality is difficult, indirect parameters should be evaluated. Such parameters in puppies include weakness, enlarged abdomens, abnormal behaviors such as restlessness and continuous vocalization and failure to grow. Presence of these signs after illness is ruled out may indicate insufficient milk production and/or deficient milk quality.

Milk intake can be roughly estimated by weighing puppies before and after they nurse. The ratio of weight gain to milk intake may provide some indication of milk quality. Weight gains reportedly range from about 1 g/2 g of milk intake to 1 g/almost 5 g of milk intake during the first weeks of life. This wide range results primarily from differences in ability to estimate milk intake. Also, an underweight bitch (BCS 1/5 or 2/5) may be at risk for producing inadequate amounts and/or poor quality milk. Therefore, the bitch's food should also be assessed.

ASSESS THE FEEDING METHOD Puppies should be encouraged to nurse as often as possible during the first week of life (eight to 12 times per day); after Week 1, they should be encouraged to nurse at least three to four times per day. Inexperienced bitches should be carefully observed to ensure that all puppies receive sufficient amounts of colostrum within 24 hours of birth, when they are still able to absorb intact proteins such as immunoglobulins. This involvement may include positioning the puppies on the bitch's nipples at feeding time or encouraging a nervous bitch to lie quietly as the puppies nurse. Handling the dam and puppies facilitates monitoring the progress of the litter.

Competition in large litters may prevent smaller, weaker puppies from nursing and predispose them to dehydration and hypoglycemia. Partial orphan rearing of the entire litter should be done in these cases. Partial orphan rearing allows the puppies to stay with the dam in their normal environment and permits proper socialization. Dividing the litter into two groups of equal number and size of puppies facilitates partial orphan rearing. One group is allowed to nurse the bitch, while the other group is given a milk replacer. The groups are exchanged three to four times per day. The group separated from the bitch should be fed before the puppies are returned to the bitch. Thus, these puppies will be full and less likely to nurse the bitch. Supplementing all puppies with milk replacer is better than complete orphan feeding a few puppies. (See Appendix 1 and Small Animal Clinical Nutrition, 4th ed., pp 1012-1019 for details about complete orphan feeding.)

Feeding Plan (Page 247)

Because nursing puppies depend completely on the bitch's milk, the feeding method used for the bitch should be evaluated and adapted if necessary. Most lactating bitches should be fed free choice. Puppies that fail to thrive when receiving bitch's milk should be fed immediately via

DOGS

partial or complete orphan feeding techniques to avoid the risk of hypoglycemia, hypothermia and dehydration. (See Appendix 1 and Small Animal Clinical Nutrition, 4th ed., pp 1012-1019.)

Reassessment (Page 247)

Nursing puppies should be reassessed daily. Puppy body weights should be obtained at birth, daily or every other day for the first four weeks and then weekly. Adequacy of the bitch's milk production can be assessed by growth rate of the puppies, puppy contentment and mammary gland distention. To determine whether an individual mammary gland is producing milk, gently express milk from the nipple while the bitch is relaxed.

Growing Dogs: Weaning Period (Page 247)

Puppies can be offered gruel to stimulate food intake at three weeks of age. Gruels can be made by blending a moist growth/reproduction-type food (**Tables 9-6** and **9-9**) with an equal volume of warm water. Alternatively, dry food can be ground and mixed one part with three parts of warm water (volume basis). Puppies should be encouraged to lap the gruel by touching their lips to the food or the owner can place a fingertip in the gruel and then into the puppies' mouth. The gruel should be very digestible, and contain at least 4.5 kcal (18.8 kJ) ME/g DM, and 25 to 30% highly digestible protein. Puppies are very prone to vomiting and diarrhea during this period. If gastrointestinal disturbances occur, gruel can be made from a highly digestible moist food intended for dietary management of diarrhea with a minimum of 25% DM protein. (See Chapter 22.)

As the puppies' interest in solid food increases, the water content of the gruel can be reduced progressively. Puppies should be eating sufficient quantities of solid food at five weeks of age because the bitch's milk production will probably start declining.

From three weeks of age on, puppies can be separated from their mother for short periods of time. The time away from the dam can be progressively increased to about four hours a day by around six weeks of age. Weaning should be effectively completed between six and seven weeks of age and puppies can be removed from the dam. After weaning, the puppies should be continued on the same food to minimize stress and the risk of diarrhea.

Growing Dogs: Postweaning to Adulthood (Pages 247-250)

The ultimate goal of a feeding plan for puppies is to create a healthy adult. The specific objectives of the puppy-feeding plan are to optimize growth and minimize obesity and developmental orthopedic disease.

Table 9-9. Key nutritional factors for growing dogs.

	Recommended levels in food (DM)	
Factors	Adult BW <25 kg*	Adult BW >25 kg**
Energy density (kcal ME/g)	3.5-4.5	3.2-3.8
Energy density (kJ ME/g)	14.6-18.8	13.6-15.7
Crude protein (%)	22-32	20-32
Crude fat (%)	10-25	8-12
Calcium (%)	0.7-1.7	0.7-1.2
Phosphorus (%)	0.6-1.3	0.6-1.1
Ca/P ratio	1:1-1.8:1	1:1-1.5:1
Digestibility	See **Table 9-5**	See **Table 9-5**

Key: DM = dry matter, BW = body weight, kcal = kilocalories, kJ = kilojoules, ME = metabolizable energy.
*These dogs can be fed the same food as recommended for gestation/lactation. Balanced commercial foods may match the profile for gestation/lactation of most bitches and growth of puppies whose adult weight does not exceed 25 kg.
**For large- and giant-breed dogs.

Assessment (Pages 247-250)

ASSESS THE ANIMAL Puppies should be assessed for risk factors before weaning so recommendations for appropriate nutrition can be implemented. As always, a thorough history and physical evaluation are required. Special attention should be paid to large- and giant-breed puppies (See Chapter 17.) and breeds and genders at risk for obesity. (See Chapter 13.) In addition to breed and gender risks, growth rates and body condition provide valuable information about nutritional risks.

Besides being breed dependent, growth rates of young dogs are dependent on the nutrient density of food and the amount of food fed. Puppies should be fed to grow at an average rather than at a maximum rate for their breed. Growing animals reach a similar adult weight whether growth rate is rapid or slow. Feeding for maximum growth increases the risk of skeletal deformities and decreases longevity in other species.

Appendix F (See Small Animal Clinical Nutrition, 4th ed., pp 1020-1026.) lists body weights of various breeds of puppies at different ages. Despite the variability between breeds and individual dogs, these values can be used as an approximate guide to evaluate growth rates. If the body weight of a dog is markedly less than the listed values, the animal may be undernourished. Puppies weighing markedly more than the values in Appendix F are likely to be overfed, growing too fast and at risk for obesity and/or skeletal disease. As a rule of thumb, small- and medium-sized dogs (adult weights up to 25 kg) reach about 50% of their adult weight around four months of age and dogs with adult weights greater than 25 kg at about five months of age. All puppies should have their body condition evaluated and reassessed every two weeks to allow for adjustments in amounts fed and growth rates. (See Chapter 1.)

Key Nutritional Factors The requirements for all nutrients increase during growth compared with requirements for adult dogs. However, energy and calcium levels are of special concern; energy for small- and medium-breed puppies (for obesity prevention) and energy and calcium for large- and giant-breed puppies (for skeletal health). In addition, the method of feeding is important especially in larger breeds.

Table 9-9 summarizes key nutritional factors for growing dogs. The following sections describe these key nutritional factors in more detail.

Energy Energy requirements of growing puppies consist of energy needed for maintenance and energy required for growth. During the first weeks after weaning, when body weight is relatively small and the growth rate is high, puppies use about 50% of their total energy intake for maintenance and 50% for growth. Gradually, the growth curves reach a plateau, as puppies become young adults (**Figure 9-3**). The proportion of energy needed for maintenance increases progressively, whereas the part for growth becomes less important. Energy needed for growth decreases to about 8 to 10% of the total energy requirement when puppies reach 80% or more of adult body weight.

Because of the shift in energy usage, total food intake of a typical German shepherd puppy (adult body weight 35 kg) may no longer increase after about four months of age. (See Appendix F, See Small Animal Clinical Nutrition, 4th ed., pp 1020-1026.) A puppy should receive 3 x RER until it reaches about 50% of its adult body weight (**Table 9-10**). Thereafter, energy intake should be about 2.5 x RER and can be reduced progressively to 2 x RER. When approximately 80% of adult size is reached, 1.8 to 2 x RER will suffice.

Great Dane puppies may have energy requirements 25% higher than those of other breeds. However, this finding should not be extrapolated to other giant-breed puppies. Appendix F (See Small Animal Clinical Nutrition, 4th ed., pp 1020-1026.) provides a complete guide for daily energy intake during growth; values in Appendix F can be used as an initial feeding guideline. Body condition scoring should be used to adapt this information to individual dogs.

Management of obesity should start at weaning. Excessive food intake during growth may contribute to overweight and obese body condition. Obesity increases load and stress on immature bones and may contribute to skeletal disorders.

Protein Protein requirements of growing dogs differ quantitatively and qualitatively from those of adults. An important difference is that arginine is an essential amino acid for puppies, whereas it is only conditionally

Table 9-10. Recommendations for energy intake of growing dogs.

Time frame	x RER	kcal/BW$_{kg}^{0.75}$	kJ/BW$_{kg}^{0.75}$
Weaning to 50% of adult BW*	3	210	880
50 to 80% of adult BW	2.5	175	735
>80% of adult BW	1.8-2.0	125-140	525-585

Key: RER = resting energy requirement, kcal = kilocalories, kJ = kilojoules, BW = body weight.
*Great Dane puppies may need 20% more energy during the first two months after weaning = 250 kcal or 1,050 kJ/BW$_{kg}^{0.75}$.

essential for adult dogs. Quantitatively, protein requirements are highest around weaning, and decrease progressively.

Protein requirements for growing dogs follow a pattern similar to energy requirements. Thus, the same nutrient profile is appropriate throughout this phase of growth (postweaning to adulthood). The relatively higher requirements for protein immediately after weaning are compensated for by the higher total food intake to meet energy needs.

Most foods for growing dogs contain more protein than is required just after weaning. See **Table 9-9** for protein recommendations for growing dogs. When average-quality protein sources are used, foods containing 18% DM digestible protein and 3.5 kcal/g DM, will support optimal growth from weaning to adulthood. This corresponds to 22% crude protein of high quality and 82% digestibility.

Fat Growing dogs have an estimated daily requirement for EFA (linoleic acid) of about 250 mg/BW$_{kg}$, which can be provided by a food containing between 5 to 10% fat (DMB). The fat source must be carefully chosen when low-fat foods (<10% DM fat) are fed so that sufficient amounts of linoleic acid are provided.

Fat contributes greatly to the energy density of a food and excessive energy intake can affect endochondral bone formation in growing large- and giant-breed dogs. The fat content of foods for large- and giant-breed puppies should be controlled to decrease the likelihood of excessive energy intake. This can be achieved by feeding foods with a target energy density of 3.5 kcal (14.6 kJ) ME/g (range 3.2 to 3.8 kcal ME/g [13.4 to 15.9 kJ]). **Table 9-9** lists optimal ranges of dietary fat for puppies of different breed sizes.

Calcium and Phosphorus Although growing dogs need more calcium and phosphorus than adult dogs, the minimum requirements are relatively low. Puppies have been successfully raised on foods containing 0.37 to 0.6% DM calcium and 0.33% DM phosphorus.

Foods for large- and giant-breed puppies should contain 0.7 to 1.2% DM calcium, and the food should provide about 3.5 kcal (14.6 kJ) ME/g

DOGS

DM. Foods with a calcium content of 1.1% DM provide more calcium to puppies just after weaning than if bitch's milk were fed exclusively. Because small dogs are less sensitive to slightly overfeeding or underfeeding with calcium, the level of calcium in foods for smaller breed puppies can range from 0.7 to 1.7% DM, without risk. The phosphorus intake is less critical than the calcium intake, provided the minimum requirements of 0.35% DM are met, and the calcium-phosphorus ratio is between 1:1 and 1.8:1. **Table 9-9** summarizes dietary calcium and phosphorus recommendations for growing dogs.

Digestibility Puppies fed foods low in energy density and digestibility need to eat large quantities of food, increasing the risk of flatulence, vomiting, diarrhea and the development of a "pot-bellied" appearance. Therefore, foods recommended for puppies should be more digestible than average (**Table 9-5**).

Other Nutritional Factors No specific recommendations for carbohydrate levels are available for growing dogs. It has been suggested that foods contain at least 20% carbohydrate until puppies are four months of age to ensure optimal health.

Zinc requirements on a body weight basis are highest during the first two months of life. However, zinc requirements on a per unit of energy basis are highest between four to six months of age. Puppies are most sensitive to deficiency during this latter period. Therefore, foods for growing dogs should contain zinc levels that meet the requirements between four to six months of age. Excessive calcium and phosphorus intake should be avoided to ensure zinc availability.

Most commercial pet foods should contain adequate levels of copper unless the availability is low (e.g., when sources such as copper oxide are used). Puppies with copper deficiency may have loss of hair pigmentation, with graying of black and dark brown hair. Hyperextension of the distal phalanges and splayed toes on the front feet and normochromic, normocytic anemia may develop in more extreme cases. The recommendation for copper intake in growing puppies is between 0.25 and 0.5 mg/BW_{kg}.

ASSESS THE FOOD The choice of an appropriate food can decrease the risks associated with rapid growth, especially for large- and giant-breed puppies. Therefore, a prerequisite for developing a feeding plan includes comparing the nutrient profile of the current food with the levels of key nutritional factors listed in **Table 9-9**. Puppies of small- to medium-sized breeds may continue to receive the same food as the bitch received during lactation. These puppies were probably transitioned to this food during weaning. Large- and giant-breed puppies should be fed a food that contains less energy and calcium to decrease the risk of developmental orthopedic disease. (See Chapter 17.) If possible, such foods should be fed during early weaning. The greatest influence of nutrition on the incidence of phenotypic hip dysplasia occurs when energy is restricted very early in life.

Appendices 2 and 3 list levels of key nutritional factors in selected commercial foods and treats. (Also see Small Animal Clinical Nutrition, 4th ed., pp 1084-1091.) Alternatively, pet food manufacturers can be contacted for this information. The guaranteed or typical analysis on pet food labels is of limited usefulness and will not contain information about digestibility. The product label indicates whether the food has successfully completed an AAFCO or similar feeding trial for growth. (See Chapter 5.) This information is important because foods with similar label declarations can have markedly different nutrient availabilities and growth performance. Information about digestibility and energy density should be obtained from the manufacturer; digestibility must be sufficiently high to avoid gastrointestinal problems.

Growing dogs should not receive vitamin-mineral supplements when fed complete, balanced commercial foods. Supplements may be justified to balance homemade foods.

ASSESS THE FEEDING METHOD Assessment of feeding methods is critical to successful management of growing puppies, especially those of large and giant breeds. Free-choice and time-restricted feeding (See Chapter 1.) should be avoided during periods of rapid growth.

Free-choice feeding may increase body fat, predispose to obesity and induce skeletal deformities at a young age. Breeders who want to maximize growth and skeletal development of large- and giant-breed puppies must be taught that overfeeding predisposes these animals to developmental orthopedic disease. Weight gain and body condition should be monitored closely if free-choice feeding is used.

It has been recommended to feed a puppy all it can eat in 20 minutes, twice daily. However, recent work has shown that puppies fed using this method had increased body weight, more body fat and increased bone mineral accretion than puppies receiving the same food free choice.

During the period of rapid growth, it is best for puppies to receive a specific amount of food. This amount can be fed in two to four meals per day. The energy recommendations in this section (**Table 9-9**) coupled with food dosage calculations (See Chapter 1 and Small Animal Clinical Nutrition, 4th ed., pp 1-19.) or the feeding guidelines listed in Appendix F (See Small Animal Clinical Nutrition, 4th ed., pp 1020-1026.) can be used as starting points.

Feeding Plan (Page 250)
SELECT A FOOD The food assessment phase will determine whether or not it is necessary to change foods. If a change is indicated, it is best to

select a food that has passed an AAFCO or similar feeding trial for growth. Even so, because AAFCO feeding trials only last 10 weeks, they probably will not expose problems related to excess calcium and energy consumption, especially in large- and giant-breed puppies. Therefore, foods selected for growth should have key nutrients in the ranges provided in **Table 9-9**. See Appendix 2 for levels of key nutritional factors in selected commercial foods, or contact the manufacturer.

DETERMINE A FEEDING METHOD Free-choice feeding is not generally recommended for puppies unless they are extremely thin (BCS 1/5) or have difficulty maintaining adequate body weight. Meal-restricted feeding is appropriate for most puppies to allow better control of body weight and growth rate. **Table 9-4** outlines advantages and disadvantages of these feeding methods. Chapter 17 outlines more specific techniques to minimize nutritional risk factors for developmental orthopedic disease in rapidly growing, large- and giant-breed puppies. **Table 9-1** contains a feeding plan summary for growing dogs.

Reassessment (Page 250)
Owners should weigh growing puppies every week and record body weights and food intake (including snacks and treats). Veterinarians can provide owners with expected body weights at different ages based on Appendix F to allow better assessment of body weight and food intake. (See Small Animal Clinical Nutrition, 4th ed., pp 1020-1026.) Owners should be taught body condition scoring techniques and be informed about early clinical signs of developmental orthopedic disease. Veterinarians should reassess puppies at the time of routine vaccinations and more frequently if any indication of under- or overnutrition is detected at that time. Reassessment should include body weight and body condition assessment, food assessment and determination of correct food dosage.

Clinical cases that illustrate and reinforce the nutritional concepts presented in this chapter can be found in Small Animal Clinical Nutrition, 4th ed., pp 255-260.

CHAPTER 10

The Canine Athlete

For a review of the unabridged chapter, see Toll PW, Reynolds AJ. The Canine Athlete. In: Hand MS, Thatcher CD, Remillard RL, et al, eds. Small Animal Clinical Nutrition, 4th ed. Topeka, KS: Mark Morris Institute, 2000; 261-289.

INTRODUCTION (Pages 261-263)*

Nutrition (Page 263)

Nutrition cannot overcome deficits in genetics and training. However, matching nutrition to exercise type (i.e., intensity, duration and frequency) allows a canine athlete to perform to its genetic potential and level of training. The goal of feeding canine athletes is to provide appropriate nutrition for health and optimal performance. As with feeding any animal, it is important that the food type, amount and feeding method be matched to the animal's needs.

EXERCISE PHYSIOLOGY (Pages 265-267)

Muscle Energetics (Pages 265-267)

Exercise requires the transfer of chemical energy into physical work. Chemical energy stored in high-energy phosphate bonds of adenosine triphosphate (ATP) is the sole source of energy for muscle contraction.

Although ATP is the high-energy compound that cells use as fuel to perform work, the energy required for exercise can ultimately come from a variety of sources. Because the concentration of ATP in muscle cells is relatively low in comparison to the cell's need during exercise, ATP must be replenished from other fuel sources. These metabolic fuels are stored in muscle (endogenous) and at other body sites (exogenous). The metabolism of these fuels occurs either with oxygen (aerobic) or without oxygen (anaerobic).

Resting muscle cells have only enough ATP to fuel muscle contraction for a few seconds. If work continues beyond this point, ATP must be regenerated from other metabolic fuels at a rate comparable to that at which it is being consumed. Glucose is a versatile metabolic fuel that is

*Page numbers in headings refer to Small Animal Clinical Nutrition, 4th ed., where additional information may be found.

Key Nutritional Factors–The Canine Athlete (Page 263)

Factors	Sprint athletes	Intermediate athletes (low/moderate duration and frequency)	Intermediate athletes (high duration and frequency)	Endurance athletes
Energy density	Use food with 3.5 to 4.0 kcal ME/g DM	Use food with 4.0 to 5.0 kcal ME/g DM	Use food with 4.5 to 5.5 kcal ME/g DM	Use food with >6.0 kcal ME/g DM
Fat	Use food with 8 to 10% fat DM or 20 to 24% of calories from fat	Use food with 15 to 30% fat DM or 30 to 55% of calories from fat	Use food with 25 to 40% fat DM or 45 to 65% of calories from fat	Use food with >50% fat DM or >75% of calories from fat
Soluble carbohydrate	Use food with 55 to 65% NFE DM or 50 to 60% of calories from NFE	Use food with 30 to 55% NFE DM or 20 to 50% of calories from NFE	Use food with 30 to 35% NFE DM or 15 to 30% of calories from NFE	Use food with <15% NFE DM or <10% of calories from NFE
Protein	Use food with 22 to 28% protein DM or 20 to 25% of calories from protein	Use food with 22 to 32% protein DM or 20 to 25% of calories from protein	Use food with 22 to 30% protein DM or 18 to 25% of calories from protein	Use food with 28 to 34% protein DM or 18 to 22% of calories from protein
Digestibility	DM digestibility >80%	DM digestibility >80%	DM digestibility >80%	DM digestibility >80%
Water	Unlimited access except just before a race	Unlimited access	Unlimited access	Unlimited access

Key: ME = metabolizable energy, DM = dry matter, NFE = nitrogen-free extract.

stored endogenously as muscle glycogen and exogenously as glycogen in the liver and to a much smaller extent as free glucose in the blood. Glucose can be metabolized to regenerate ATP by both anaerobic and aerobic pathways. Anaerobic metabolism of glucose results in very rapid ATP production, but only yields two ATP per molecule of glucose. Aerobic metabolism, the complete oxidation of glucose to CO_2 and water, regenerates ATP less rapidly, but results in a much greater yield (36 ATP per molecule of glucose). Because total body glucose stores (glycogen) are relatively small (1 to 2% of body weight), even aerobic metabolism of glucose cannot sustain exercise for extended periods of time. Fatty acids are stored in ample supply in adipose tissue and within muscle. They are the primary energy source for long-lasting exercise. Although small amounts of fatty acids are stored in muscle, this source may contribute up to 60% of the fatty acids oxidized during the first two to three hours of exercise.

Amino acids are usually not a primary energy source for exercise. Oxidation of amino acids may contribute up to 5 to 15% of the energy used during exercise, depending on the intensity and duration of the task.

The proportion of energy substrates and metabolic pathways used during exercise depends on the intensity and duration of the exercise. High-power activities (e.g., sprinting) rely heavily on anaerobic metabolism, whereas more prolonged activities require the higher energy yield provided by oxidation of glucose and fatty acids. As the duration of exercise increases, oxidation of fatty acids becomes more important.

Exercise Intensity and Duration (Page 268)

Energy and other nutrient requirements for canine athletes are determined by the intensity and duration of exercise. Exercise intensity can be described in a variety of ways depending on body weight and type of activity. Exercise intensity is a measure of work done per unit time. For dogs, the type of work done is usually running and the amount of work done depends on body weight, distance traveled and change in elevation. The amount of work done is directly proportional to the amount of energy used. Therefore, energy use describes work done.

For example, a 30-kg dog expends about 30 kcal to cover 1 km on a flat surface, regardless of how fast it walks or runs (minor differences may occur due to differences in efficiency of various gaits for running at a specific speed).

Exercise intensity dictates metabolic pathways and substrate use. High-intensity activities (sprinting) depend on anaerobic metabolism of carbohydrate (glucose and glycogen), which is supported by high-carbohydrate foods. The severe acidemia produced by high-intensity activities underscores the need for adequate electrolyte and water intake. Endurance events that take place at low to moderate intensity for long periods are

Table 10-1. Caloric cost of running 1 km for dogs of varying size.

Body weight (kg)	Cost of running (kcal) 1 km/kg body weight*	Cost of running 1 km (kcal)**
5	1.77	9
10	1.41	14
15	1.23	19
20	1.13	23
25	1.05	26
30	0.99	30
35	0.94	33
40	0.90	36
45	0.87	39
50	0.84	42
70	0.76	53

*Formula: $Energy_{(kcal)} /BW_{(kg)} = 1.77 \times distance_{(km)} \times BW_{(kg)}^{-0.4} + 1.25 \times BW_{(kg)}^{-0.25}$
**To convert to kJ, multiply kcal x 4.184.

completely aerobic and rely mostly on oxidation of fatty acids. Thus, as exercise duration increases, the fat fraction of the food becomes more important to supply energy needs. Intermediate exercise (as performed by most canine athletes) is usually of low to moderate intensity, but may include some short periods of high-intensity work. Both fats and carbohydrates are important fuel sources in intermediate exercise.

Energy Cost of Running (Page 269)

The energy cost of running depends on body size and distance traveled. **Table 10-1** shows the caloric cost of running 1 km for dogs of various sizes. This table also illustrates an important principle about the mass-specific caloric cost of running as size changes. As body size increases, the efficiency of running increases (i.e., larger animals use fewer kcal/kg to run the same distance). By using the data in **Table 10-1**, it is possible to estimate the energy requirement for a dog of a known size to run a given distance.

ASSESSMENT (Pages 269-282)

Assess the Animal (Pages 269-278)
History (Pages 269-271)

In addition to the normal historical information that is usually obtained about a patient, the following information should be gathered from owners of canine athletes: environmental/housing data, medications, dietary history and exercise type, amount, frequency and performance. Detailed information should be gathered about how the animal is housed, including: indoors or outdoors, size and type of housing, opportunity for spontaneous exercise, type of surface, number of animals housed together and access to food and water. All medications used should be recorded, including drugs used to suppress estrus and drugs used to enhance performance.

The dietary history should include all foods and supplements used. The amount fed, nutrient profile and timing of feeding in relation to exercise should be noted. The amount eaten should also be assessed (i.e., Does the dog have a normal appetite and is it actually consuming a reasonable amount?). In some cases, the composition of the overall diet (food plus supplements) may be complex and individual meals may vary in composition. It is also important to ascertain the duration of a particular feeding plan. Abrupt or frequent changes in food or feeding method may affect performance.

EXERCISE TYPE Functionally, exercise can be divided into three types based on intensity and duration: 1) sprint: high-intensity activities that can be sustained less than two minutes, 2) intermediate: activities lasting a few minutes to a few hours and 3) endurance: activities that last many hours. These definitions are arbitrary and vague, but are useful for assessing and developing feeding plans.

Most canine athletes participate in intermediate exercise activities. Most of these activities are of low to moderate intensity and last only a few hours. Intensity and duration of exercise vary widely within this category. For example, most guide dogs work at a low level of physical exertion for variable lengths of time throughout the day, whereas a search and rescue dog may work at a much higher level for many consecutive hours. Dogs that work at a relatively high intensity level for an extended time, such as sled dogs, have much greater metabolic requirements and are true endurance athletes.

Exercise amount can be quantified as hours per day or week. Frequency is how often the animal exercises: daily, weekly, weekends only or seasonally. Many hunting dogs only work hard on weekends during hunting season, whereas some livestock dogs may work several hours daily. Canine athletes should be categorized as either "full-time" or "part-time" athletes.

ENVIRONMENTAL INFLUENCES ON EXERCISE Ambient temperature, psychological stress and geography are environmental factors that may influence nutritional needs of canine athletes. Of these, ambient temperature exerts the greatest effect. Hot temperatures result in increased work

ATHLETES

and water loss (i.e., to excrete metabolic heat and maintain body temperature homeostasis). High humidity impairs evaporative cooling, thus adding to the work of heat excretion. Cold temperatures without exercise increase the energy requirement for thermogenesis. For working dogs, cold environmental temperatures aid in heat dissipation during exercise. Excitement or stress associated with some activities increases body temperature and respiration, leading to greater requirements for energy, water and perhaps electrolytes. Stress may also negatively affect food intake. Geographic factors such as high elevation, changing elevation (running up and down hills) and the presence of sand or tall grass underfoot can increase workload.

Physical Examination *(Page 271)*

During the physical examination, the veterinarian should evaluate the dog's general health, musculoskeletal soundness, hydration, cardiopulmonary function and body condition. A complete physical examination is crucial because disease affecting any body system can impair performance. For example, severe periodontal disease can cause sufficient pain to affect food intake, thus causing a retriever to retrieve poorly or a greyhound to run poorly. Likewise any injury or deterioration of the musculoskeletal system affects performance.

Energy balance can be evaluated by body condition scoring. The body condition score (BCS) is an indication of fat mass. If dietary energy intake is less than energy needs, fat mass declines and BCS decreases. A BCS of 3/5 is normal for most pets and for many canine athletes. However, a much leaner body composition is desirable for some canine athletes (e.g., racing greyhounds and sled dogs).

Key Nutritional Factors *(Pages 272-278)*

ENERGY Providing the right amount of energy from the right sources is central to feeding canine athletes. Providing the correct amount of energy is partly controlled by the amount fed and feeding method, whereas nutrient profile affects the maximum possible caloric consumption. Additionally, the preferred source of energy (fat vs. carbohydrate) depends on exercise type. Energy for exercise comes from three nutrients: fat, carbohydrate and protein. Fats and carbohydrates are the primary energy substrates for exercise. Fat is the preferred substrate for longer duration exercise, whereas sprinters depend more on carbohydrate. Under most conditions, the energy contribution of protein during exercise is small.

Energy required for exercise depends on the intensity, duration and frequency of exercise. The amount of energy required for exercise depends on total work done (intensity x duration x frequency). The preferred source of energy depends mostly on intensity. Sprinters, even though they work at a very high intensity, have relatively low energy requirements because the duration of their events is so short and frequency is usually only a few times each week. Generally, 1.6 to 2 x resting energy requirement (RER) is adequate for most sprint athletes (note the daily energy requirement [DER] for the average pet dog is 1.6 x RER).

For activities of very short duration and high intensity, the energy substrate source is the main determinant of the nutrient profile. Foods for sprint athletes should be high in carbohydrate and lower in fat, with a resulting energy density lower than that of many dog foods. Intensity, duration and frequency of exercise are variable for intermediate athletes; therefore, the energy requirement is highly variable. DER for these athletes ranges from 2 to 5 x RER. Foods with a higher fat content are typically fed to provide adequate dietary energy density. Endurance athletes require more than 5 x RER. For activities of long duration, providing adequate energy is a major determinant in the choice of a nutrient profile for exercising dogs. Foods that are very high in fat are required.

FAT Fat provides approximately 8.5 kcal (36 kJ) of metabolizable energy (ME) per g of dry matter (DM) or more than twice the amount provided by protein and carbohydrate. Because of these differences in caloric density, the only practical means of significantly increasing the energy density of a food is to increase its fat concentration. Reasonable increases in fat also increase palatability. Energy density and palatability make dietary fat levels an important consideration in the formulation of foods for canine athletes.

Ingesting adequate kcal to meet daily energy expenditure is often a serious problem for working dogs. In extreme cases, sled dogs in long-distance races expend from 6,000 to 10,000 kcal/day (25 to 42 MJ/day), in which case dry matter intake becomes a performance-limiting factor. Because the total daily dry matter intake is limited to about 3.5% of body weight, the energy density of a ration should be maximized. Under these circumstances, each nonessential gram of protein and carbohydrate ingested potentially robs the dog of 5 kcal (21 kJ). The calorie deficit is paid through mobilization of body fat stores. Over-reliance on these depots may lead to catabolism of more functionally crucial energy sources, such as muscle and plasma proteins. In addition to its role as an energy store, adipose tissue also functions as an insulator. Excessive adipose depletion may increase a dog's cost of maintaining its body temperature, especially at rest in cold environments.

Even under the less severe conditions of intermediate exercise, increased dietary fat levels provide needed kcal and other valuable benefits. Fatigue and dehydration may decrease appetite. Increasing dietary fat concentration increases energy intake and encourages stressed dogs to ingest more food because the higher fat content improves palatability.

ATHLETES

Feeding high levels of fat may also positively affect endurance. Training may elevate the carbohydrate threshold, thus increasing the proportion of energy supplied by free fatty acid (FFA) oxidation at all but the highest intensities of exercise. Increasing dietary fat concentration may augment this process by enhancing FFA availability. Working dogs fed high-fat foods have higher circulating levels of FFA at rest and respond to exercise stimuli by releasing more FFA than those fed isocaloric amounts of a high-carbohydrate food. This difference in FFA availability may be related to the decreased resting plasma concentration of insulin in animals fed high-fat foods, and the induction of key lipolytic enzymes.

The relationship between fat intake and canine endurance is well established. As the duration of the event performed by the athlete increases, so should the dietary fat intake.

Dogs can tolerate high levels of dietary fat if fat is gradually introduced and an adequate intake of non-fat nutrients is maintained. Steatorrhea and a decrease in food palatability are indicators that a ration has exceeded an animal's fat tolerance. Under conditions of extreme training, sled dogs may ingest up to 60% of their kcal as fat. During ultra-endurance events, such as the Iditarod or the Yukon Quest, fat intake may compose 80% of the kcal ingested. This "super fat loading" should be attempted only during the most strenuous periods of such events, when it is difficult or impossible for canine athletes to ingest as many kcal as they are expending.

The type of fat used must also be considered in the formulation of a performance ration. Essential fatty acids should make up at least 2% of the dry matter of a ration. The remainder of the fat may come from any of a number of plant or animal sources.

Sprint exercise is almost entirely dependent on carbohydrate; therefore, the fat requirement for sprinters is not different than that for other dogs. Total fat content should be 8 to 10% of dry matter or 20 to 24% of kcal. Dietary fat needs for intermediate athletes are directly proportional to the amount of work done. Part-time athletes during off season should be fed as other dogs. (See Chapter 9.) Dietary fat content should be increased as the amount of work increases: 15 to 30% dry matter (30 to 55% fat kcal) for moderate amounts of work and 25 to 40% dry matter (45 to 65% fat kcal) for large amounts of work. Endurance athletes require very high levels of dietary fat to meet their energy needs, in excess of 50% dry matter and 75% fat kcal. The Key Nutritional Factors table summarizes recommendations for fat and other nutrients by exercise type.

CARBOHYDRATE Canine athletes requiring less than twice maintenance levels of energy may derive a significant portion of their kcal from carbohydrate sources. This is an advantage for high-power athletes such as racing greyhounds that are highly dependent on anaerobic metabolism. Because carbohydrates contain only about 3.5 kcal (15 kJ) ME/g, they cannot be used to increase the energy density of a food. This limitation is an important consideration for endurance athletes that have difficulty ingesting a sufficient volume of food to meet caloric requirements.

Even endurance athletes may benefit from low levels of dietary carbohydrate. Studies involving sled dogs fed 0 or 17% of their kcal as carbohydrate showed that dogs were more susceptible to developing "stress" diarrhea when fed foods devoid of carbohydrate. There are other advantages associated with feeding carbohydrates to sprint athletes. Because these dogs derive more of their energy for exercise from glucose/glycogen, glycogen depletion may play a role in the onset of fatigue for athletes working at or above their anaerobic threshold.

Carbohydrate availability to working muscles is a limiting factor for prolonged exercise in people and other species. This finding has led to development of strategies for carbohydrate loading or glycogen super compensation. The classic (Åstrand) method uses a combination of exhaustive exercise and low-carbohydrate foods (≤10% kcal from carbohydrate) to deplete muscle glycogen. Glycogen depletion is followed by consumption of high-carbohydrate foods (80 to 90% kcal from carbohydrate) and little activity. This method dramatically increases muscle glycogen in people. An alternative carbohydrate-loading method (Sherman/Costill) simply requires consumption of 60 to 70% of kcal from carbohydrate consistently over time. In people, this method produces results similar to those achieved by the classic method.

The carbohydrates used in foods for canine athletes should be highly digestible to limit fecal bulk. Excessive amounts of undigested carbohydrates reaching the colon may increase water loss via the stool, increase colonic gas production and increase overall fecal bulk. These changes in fecal consistency have been proposed to increase an athlete's risk of developing "stress diarrhea," further increasing fecal water losses.

Metabolic power or a high rate of ATP generation is required for sprint performance. Consequently, anaerobic metabolism of glucose and glycogen is the dominant energy generation pathway. High-carbohydrate foods should be fed to maximize muscle glycogen. Dietary carbohydrates should compose 50 to 70% of total kcal to maximize muscle glycogen levels (based on research done with people).

The dietary carbohydrate recommendation for intermediate athletes is highly variable, depending on the intensity and duration of work. Dogs that perform relatively long bouts of low to moderate intensity work require more dietary energy (higher fat) and relatively low carbohydrate levels (as low as 15% of kcal). Dogs that perform short bursts of higher intensity work should be fed more carbohydrate, up to 50% of kcal.

Endurance athletes require very little carbohydrate. Endurance rations should contain less than 15% of kcal from carbohydrate to achieve the energy density required for the amount of work done by these dogs. Some carbohydrate and/or soluble fiber should be included in the ration to avoid loose stools. The Key Nutritional Factors table summarizes carbohydrate recommendations for canine athletes by exercise type.

PROTEIN Dietary protein is used to fulfill structural, biochemical and, to a lesser extent, energy requirements. Work increases the requirement for protein. The work-induced elevation in protein requirement is most pronounced when the intensity and/or duration of exercise performed is rapidly increased above an animal's present level of conditioning.

Several anabolic processes contribute to the exercise-induced increment in protein requirement. Protein demand is elevated due to an increase in the synthesis of structural and functional proteins. Training induces synthesis of many enzymes and transport proteins in each of the energy-generating pathways. Blood volume also expands during aerobic training. Such expansion necessitates an increase in plasma protein synthesis to maintain oncotic and osmotic balance between the plasma and the interstitial fluids. The increase in hematocrit sometimes observed during endurance conditioning programs may reflect an increase in tissue protein synthesis. Anaerobic training regimens may also induce muscle hypertrophy. Amino acids are used in the formation of new muscle tissue and in the repair of damage that may occur to muscle and connective tissue during intensive conditioning programs.

In addition to enhancing the rate of tissue protein synthesis, exercise increases the rate of amino acid catabolism. Amino acids may provide 5 to 15% of the energy used during exercise, depending on the intensity and duration of the task. Most of this energy comes from the oxidation of branched-chain amino acids.

The proportion of energy supplied by amino acids may be even greater in underfed athletes and those participating in ultra-endurance events in which there is a high risk for depletion of endogenous carbohydrate stores. In these instances, gluconeogenesis becomes the major pathway for maintaining blood glucose levels.

For endurance athletes, there may be some disadvantages inherent in exploiting dietary protein sources for energy. Protein has only about 3.5 kcal (15 kJ) ME/g dry matter. Thus, increasing the proportion of protein in the formulation cannot increase the energy density of a ration. The energy density of the food is one of the major determinants of endurance capacity when working dogs have difficulty ingesting as many kcal as they expend.

Excessive protein intake may predispose an animal to increased amino acid catabolism because dietary amino acids are not stored in labile protein depots, but are deaminated. The resulting ketoacids are either oxidized for energy directly or converted into fatty acids and/or glucose and then stored as adipose tissue or glycogen. The urea produced from amino acid breakdown is excreted from the body in urine. In healthy animals, the amount of water lost increases with increases in urea production.

An optimal food for canine athletes should contain enough high-quality protein to meet the dog's anabolic requirements and enough non-protein energy nutrients to meet its energy requirements. Such a food encourages the use of ingested protein in synthetic rather than energy-generating processes. As non-protein caloric intake increases, less dietary protein is used for energy and more is available for use in anabolic processes. Energy requirements should be met by fat and carbohydrate, leaving the majority of amino acids available for synthetic purposes. During long-duration exercise, DER may increase several-fold whereas protein requirement increases only a few percent. To meet the energy needs of hard-working dogs, either more food must be consumed (increasing both energy and protein intake equally) or a higher energy, lower protein food must be fed (increasing energy intake more than protein intake). Providing sufficient dietary kcal by increasing the fat content should limit the use of amino acids for energy production. Increased caloric requirements are better met from dietary fat, leaving amino acids available for synthetic purposes.

The protein requirement for exercise is only mildly increased (5 to 15%) regardless of exercise type. Protein is used for muscle hypertrophy and muscle maintenance/repair and the branched-chain amino acids can contribute to energy production. Dietary protein should be at least 24% of kcal. Because the energy requirement of some endurance athletes is so high (up to 10 or 11 x RER), it may not be feasible to feed even this level of protein and provide adequate kcal. Protein kcal at 16% should be viewed as an absolute minimum. Note that for endurance exercise, energy requirement increases many fold (up to 11-fold), whereas protein requirements increase much less (5 to 15%). For a given food, as intake increases to meet energy requirements, protein intake increases proportionally. Because of the disparity between the increase in need for energy and protein for exercise, as total dietary energy requirement increases, the percent protein kcal of the food may decrease. The Key Nutritional Factors table summarizes protein recommendations by exercise type.

DIGESTIBILITY Dry matter digestibility of food is important to canine athletes for two reasons. First, exercise may be limited by a dog's ability to obtain sufficient amounts of nutrients (usually energy). Enhanced digestibility increases the maximum possible delivery of nutrients to tissues. Second, lower digestibility means greater fecal bulk, and therefore a greater handicap. Although increased animal size results in greater running efficiency, increased fecal weight creates a greater energetic cost of running with no benefit. Total dry matter digestibility of any food for canine athletes should exceed 80%.

WATER Only about 20 to 30% of the energy consumed within muscle cells during exercise produces work; the remaining 70 to 80% is converted into heat. This heat must be dissipated to prevent performance impairments and perhaps life-threatening increases in body temperature.

During prolonged periods of exercise in warm and humid environments, heat dissipation leads to a decrease in total body water and plasma volume. Exercise in very cold, dry environments also increases evaporative fluid losses. Significant fluid loss during exercise may impair performance. Even mild dehydration can limit exercise performance. Several studies indicate that hydration status is the single most important determinant of endurance capacity.

Under most exercise situations, these athletes lose more water than electrolytes, causing a decrease in plasma volume and an increase in plasma osmolality. Efforts to return electrolyte values to normal should therefore concentrate on water replacement. Ideally, clean fresh water should be available at all times. There are occasions when such accommodations cannot be made due to the nature of the athletic event or the environmental conditions. Under these conditions, water should be offered at least three times a day and more often if possible. "Baiting" the water with a flavor enhancer such as meat juice can encourage water intake.

OTHER NUTRITIONAL FACTORS Vitamins, minerals and electrolytes play important roles in maintaining homeostasis and chemical reactions during exercise. However, they are of secondary concern when feeding canine athletes and are found in adequate amounts in most commercial foods. Likewise, the acid-base composition of the food and base loading may also affect performance; however, these effects are poorly understood in canine athletes. Because exercise increases metabolic rate and the opportunity for free radical formation, it is important to ensure adequate intake of antioxidant nutrients. Evidence that canine athletes have an increased need for these nutrients is inconclusive. For more information, see Small Animal Clinical Nutrition, 4th ed., pp 277-278.

Assess the Food (Pages 278-279)

All foods for canine athletes (performance foods) should share a few important characteristics. First, the food should be calorically dense so that canine athletes can consume enough food to meet their energy requirements. Second, the food must be acceptable and highly digestible. Dry matter digestibility should exceed 80%. High digestibility reduces fecal bulk and fecal water loss and may decrease the risk of stress diarrhea.

Finally, the food should be practical. Factors such as the cost of the food, the form of the food, the environment in which the food is stored and fed and the number of dogs being fed are all important considerations. What is practical for a single hunting dog at home may not be practical for a team of sled dogs hundreds of miles from civilization or racing greyhounds at a track.

Because the greatest nutritional demand of exercise is for energy, foods for canine athletes must provide sufficient kcal from the right sources. Increasing the fat content of the food usually enhances energy density. Energy need and exercise intensity dictate the appropriate fat content of the food. For example, dogs participating in short-duration, intense exercise may benefit from lower-fat, higher-carbohydrate foods.

Canine athletes are fed a wide variety of foods and supplements. When assessing the overall ration, it is important to assess all foods and supplements fed. The nutrient profile of the total daily ration should be evaluated for the key nutritional factors based on the type and amount of exercise performed by each dog. (See the Key Nutritional Factors table.)

Most intermediate athletes are fed commercial foods, whereas many sprint and endurance athletes are fed homemade foods or more commonly a mixture of commercial food and other ingredients. Supplements are commonly given to canine athletes of all types.

Commercial Foods (Page 279)

Appendix 2 lists the nutrients of concern for many commercial dog foods and **Table 10-2** lists selected commercial foods formulated for active dogs and canine athletes. For commercial foods not found in Appendix 2 or **Table 10-2**, minimum fat and protein levels are listed in the guaranteed or typical analysis on the pet food label. (See Chapter 1 for limitations of this information.)

Caloric density may also be listed on the label but usually must be obtained from the manufacturer or estimated from the guaranteed analysis. The carbohydrate (nitrogen-free extract [NFE]) content is estimated from the guaranteed analysis if a value is assumed for ash content (100% – % moisture – % fat – % protein – % fiber – 5% ash = NFE). Digestibility information is usually only available from the manufacturer.

Homemade Foods (Page 279)

Homemade foods can be very complicated mixtures of many ingredients. Chapter 6 discusses assessment of homemade foods in detail. Fortunately, most homemade rations for canine athletes use a commercial dry dog food as a base. Many racing greyhound rations contain dry dog food mixed with either raw or cooked meat, water, vitamin and mineral supplements and a variety of other ingredients. Likewise, many sled dog

ATHLETES

Table 10-2. Levels of key nutrients in selected commercial foods used for active and athletic dogs.*

	kcal (ME/g)	Protein (%)	Fat (%)	NFE (%)	Digestibility (%)
Diamond Professional Dog Food, dry	4.50	33.2	22.3	33.1	na
Eagle Pack Kennel Pack, dry	3.94	28.1	16.3	44.7	na
Eagle Pack Power Pack, dry	4.20	32.8	21.7	34.2	na
Hill's Science Diet Canine Active/Performance, moist	5.01	30.1	25.8	36.9	86
Hill's Science Diet Canine Active/Performance, dry	4.99	30.3	27.2	34.7	85
Hill's Science Diet Canine Maintenance, moist	4.33	25.4	15.7	53.0	na
Hill's Science Diet Canine Maintenance, dry	4.02	24.6	15.5	53.6	na
Iams Eukanuba Adult Maintenance Formula, dry	4.71	28.5	18.4	45.3	na
Iams Eukanuba Premium Performance, dry	4.83	34.0	23.0	35.1	na
Nutro's Natural Choice Plus, dry	4.17	32.2	20.0	38.4	na
Pedigree Chum Advance Formula Activity Plus, dry	4.28	34.8	21.7	32.1	na
Purina Hi Pro, dry	4.08	30.2	11.8	49.7	83
Purina ProPlan Performance, dry	4.71	34.1	21.6	36.9	84
Wafcol Energy Plus, dry	3.82	26.7	10.7	49.2	na

*This list represents products with the largest market share and for which published information is available. Manufacturers' published values or analyses performed in June 1996; expressed as % dry matter. See the Key Nutritional Factors table, p 176 for recommended nutrient levels according to exercise type.
To convert to kJ, multiply kcal x 4.184.
na = information not published by manufacturer.

mushers mix animal fat or both meat and fat with dry dog food and other ingredients. If the commercial dry food constitutes 50 to 75% of the total mixture on a weight basis and most of the added ingredients are wet ingredients or animal fat, it is unlikely that vitamin and mineral deficiencies will occur.

Because many elite canine athletes (racing greyhounds and sled dogs) are fed homemade foods containing meat and animal byproducts of variable quality, the safety of these foods should always be evaluated. Some raw meat sources contain abundant bacteria and bacterial toxins. These materials may pose a health hazard for the people that care for these animals and for the athletes themselves. Chapter 7 discusses food safety in detail.

Assess the Feeding Method (Pages 280-282)

Performance can be influenced by the composition of the food and how it is fed. It is possible to feed the right food in the wrong way and vice versa. Items to be assessed should include amount fed, frequency of feeding, timing of meals in relation to exercise, access to water and the use of supplements. If the animal is in appropriate body condition and hydration status, it is likely that the amount of food and water consumed is appropriate.

Amount to Feed (Page 280)

An increase in energy requirement is the hallmark of exercise. The wide variation in the intensity and duration of exercise and therefore energy requirements of different types of canine athletes emphasizes the need for food dose calculations. The basics of energy requirements and food dose calculation are covered in Chapter 1.

DER is the product of RER and a factor that accounts for normal activity. For the average neutered adult dog, DER is 1.6 x RER. DER for exercising dogs has a wide range of values from 1.8 x RER to 11 x RER, depending on the intensity and duration of exercise. As discussed earlier, the caloric cost of running is determined by the size of the animal (BW), weight carried or pulled and distance traveled.

Energy is also used to maintain body temperature. Extreme arctic and tropical temperatures will both increase a dog's RER. Dogs working in cold climates may require less energy than the sum of those determined for work and for thermoregulation because exercise generates significant quantities of heat. RER for non-working dogs in hot environments increases only marginally as a result of increased work of the respiratory muscles (panting).

Most working dogs expending fewer kcal than 3 x RER can adequately fulfill their energy needs by eating a commercial food formulated for performance (**Table 10-2**). These foods are acceptable, complete and convenient in most situations. Athletes working in extremely warm or extremely cold environments and those working for several hours a day for several days in a row may expend more calories than 3 x RER. Dry matter intake is limited to about 3.5% of body weight under most physiologic conditions. This quantity of a performance food (30% protein, 20% fat, 40% carbohydrate on a dry matter basis [DMB], >80% DM digestibility) provides a maximum of 5 x RER for a 25-kg dog.

Because true endurance athletes have a DER greater than 5 x RER, providing sufficient dietary energy becomes the focus of feeding these athletes. Long-distance sled dog drivers frequently encounter situations in which their 25- to 30-kg dogs require 6,000 to 10,000 kcal/day (25 to 42 MJ/day) (7 to 11 x RER). Under these extreme circumstances, dogs are fed 1,500 to 2,500 kcal (6 to 11 MJ) of a dry commercial food in an attempt to fulfill protein, carbohydrate, vitamin and mineral requirements. Fulfilling the rest of the dog's dry matter intake with fat or fatty meat then maximizes energy intake. Strategies that maximize fat intake have been successfully used in virtually all of the recent Iditarod, Quest and Alpirod victories and in sled dog expeditions to both poles. These extremely high-fat foods, which derive up to 80% of their kcal from lipid sources, should be fed only to dogs previously acclimated to high-fat intake (i.e., 30 to 60% fat kcal), through feeding and training. Also, there may be a limited amount of time that dogs can be maintained on such a food or at such a level of stress.

Feeding to Maintain Proper Body Condition (Pages 280-281)

Food-dose calculations are based on average energy needs for a population of animals and therefore will not be accurate for all animals in various circumstances. Variation in individual metabolic rate, environmental temperature and exercise affect energy requirement and food dose. Repeated or continual body condition assessment is clearly the best clinical measure of energy balance. Body condition scoring is primarily a measure of body fat. Increasing body fat indicates positive energy balance; therefore, food dosage should be decreased. If body fat falls below optimal, energy balance is negative and food dosage should be increased to ensure adequate energy for maximal performance. One method of body condition scoring is presented in Chapter 1. A BCS of 2/5 or 3/5 is normal for most canine athletes. Because fat in excess of what is needed for energy reserves during racing adds weight and may affect performance, many sight hounds and some sled dogs are kept very lean (BCS 1/5 to 2/5).

Most racing greyhounds normally have a BCS of 1/5. Being very lean may be an important physical characteristic for maximal sprint performance. Excess body fat adds weight, which is unnecessary, especially in light of the greyhound's very limited ability to use fat as an energy source for sprinting.

When to Feed (Pages 281-282)

It is not enough to feed the right amount of the right food. To gain maximum benefit from a specific food, meals must be fed at the right time in relation to exercise and ample time must be allowed for metabolism to adapt when changing foods.

After the amount to feed has been determined, an appropriate feeding schedule should be used. The temporal relationship between food intake and exercise greatly affects nutrient use. In one study, dogs fed within six hours of exercise developed a higher working body temperature than those fed 17 hours before exercise. The elevated body temperatures in dogs fed closest to the onset of exercise may have been caused by the heat released by the digestive process (specific dynamic action of food), and by vasodilatation of the splanchnic vessels. Such shunting may decrease cutaneous circulation and thus diminish heat dissipation. In performing the same task, dogs fed within six hours of exercise used more glucose and less fat than postabsorptive dogs. Higher circulating insulin levels in the more recently fed dogs may cause this alteration in substrate use. Because insulin tends to decrease free fatty acid mobilization from peripheral adipose depots, feeding too close to exercise may impair endurance by encouraging use and thus depletion of limited carbohydrate (glycogen) stores.

The importance of the temporal relationship between feeding and exercise is seen in the poorly documented syndrome known as hunting dog hypoglycemia. The exact etiology of this syndrome is unknown. It is often associated with hyperactive, under-conditioned hunting dogs. Elevated ambient temperature has also been implicated as a risk factor. Dogs experiencing this syndrome begin working normally and then develop signs of weakness and tremors that may progress to seizures and even death. Their purported inability to maintain normoglycemia has been attributed to inadequate glycogen mobilization (due to a lack of a glycogen debranching enzyme), excessive rates of glycogen mobilization or a combination of the two. Feeding these dogs several hours (four or more hours) before the onset of exercise may help decrease insulin levels at the onset of exercise. Exercise also dampens the insulin response to ingested carbohydrate. Providing exogenous carbohydrate via small amounts of food offered at the onset of and periodically during exercise may aid blood glucose homeostasis in these dogs.

Feeding long before exercise (more than four hours) may also aid endurance by allowing the dog to evacuate its bowels before it begins work. This decreases the weight carried by the dog and may decrease its risk of developing stress diarrhea. Although the cause of loose stools postexercise has not been determined, some researchers have attributed it to the presence of stool in the colon at the onset of exercise.

ATHLETES

As with pre-exercise feedings, the timing of postexercise meals also influences nutrient use. Glycogen synthesis postexercise occurs much more rapidly in human athletes given exogenous substrates within 30 minutes to two hours postexercise. Feeding within this time frame may aid repletion of glycogen stores in athletes who must perform strenuous exercise on several consecutive days.

The practical application of the above information is: 1) feed more than four hours before exercise, 2) feed within two hours after exercise and 3) small amounts of food may be fed during exercise. Feeding must be done during exercise or during short breaks. Feeding a hunting dog that has hypoglycemic tendencies at the beginning of a 45-minute lunch break may contribute to exercise-induced hypoglycemia.

Canine athletes may become significantly dehydrated after prolonged exercise and under relatively warm or humid conditions. Attempts should not be made to replace the entire fluid deficit orally or at once. Gradual oral replacement can be supplemented with subcutaneous (or in severe cases intravenous) isotonic solutions. Body temperature should be monitored because dehydrated animals are less capable of regulating this parameter.

Food Adaptation (Page 282)

Animals require some time to adapt to a new food whenever a dietary change is made. When dramatic changes in proportion of fat and carbohydrate are made, gastrointestinal (GI) and metabolic adaptations occur. The GI adjustments usually happen in a few days provided the transition to the new food is gradual. The metabolic changes generally take more time. Muscle glycogen responds to feeding a high-carbohydrate food in a few days to a few weeks. Changes in muscle enzymes and oxidative capacity occur in response to high-fat rations in six to eight weeks. Allowing appropriate time for these adaptations to occur is especially important for seasonal athletes that may be fed a high-fat performance food only part of the year and a maintenance food the remainder of the year. Both training and dietary change should occur six weeks before exercise season.

DETERMINE A FEEDING PLAN (Pages 282-283)

The feeding plan (**Table 10-3**) should be formulated based on realistic and quantifiable nutritional objectives after the animal, food and feeding method have been assessed.

Select a Food (Page 282)

Comparing the nutritional content of the current food to the key nutritional factors for the canine athlete allows decisions to be made about the

Table 10-3. Feeding plan summary for canine athletes.

Sprint athletes
1. Feed a highly digestible, high-carbohydrate, low-fat, moderate-protein food. (See Key Nutritional Factors table.)
2. Feed the right amount of food (DER = 1.6 to 2 x RER).
3. Check body condition frequently to assess energy balance and food dose.
4. Time meals and snacks correctly. Provide food or snack more than four hours before racing; offer high-carbohydrate snack within 30 minutes after racing to enhance glycogen repletion.
5. Allow free access to water except just before racing.
6. Monitor hydration status frequently.

Intermediate athletes (working)
1. Feed a highly digestible, moderate-carbohydrate, moderate- to high-fat, moderate-protein food. The energy density of the food should be high enough to allow consumption of DER. (See Key Nutritional Factors table.)
2. Feed the right amount of food. The food dose will be highly variable depending on duration and frequency of exercise (DER = 2 to 5 x RER) and should be calculated after assessing the amount of exercise performed.
3. Check body condition frequently to assess energy balance and food dose.
4. Time meals and snacks correctly. Feed after exercise or more than four hours before exercise. Snacks should be given during exercise or at the end of breaks (within 15 minutes of resuming exercise).
5. Allow free access to water.
6. Monitor hydration status frequently.

Intermediate athletes (training)
1. Feed the same as for work (see above).
2. Allow adequate time to adapt to new food (more than six weeks) before seasonal work.
3. Begin training and new food at least six weeks before seasonal work begins.
4. Allow free access to water.

Intermediate athletes (idle)
1. Feed as typical adult dog. (See Chapter 9.)
2. Feed a performance food (smaller amount) or typical adult maintenance food as needed to maintain optimal body condition.
3. Allow free access to water.

(Continued on next page.)

adequacy of the food for that individual. If the current food is appropriate (key nutritional factors in balance with the athlete's needs) then that food can continue to be fed. If discrepancies exist between the key nutritional factors for the animal and the content of the food, the food should be changed or "balanced" to meet the athlete's needs (**Table 10-2**).

Determine a Feeding Method (Pages 282-283)

As discussed above (See Assess the Feeding Method.), several aspects of the feeding method are important to canine athletic performance. These

ATHLETES

Table 10-3. Feeding plan summary for canine athletes (Continued.).

Endurance athletes

1. Feed a highly digestible, high-fat, low-carbohydrate, moderate-protein food. The energy density of the food should be high enough to allow consumption of DER. (See Key Nutritional Factors table.)
2. Feed the right amount of food. The food dose will be highly variable depending on duration and frequency of exercise (DER = 5 to 11 x RER) and should be calculated after assessing amount of exercise performed.
3. Check body condition frequently to assess energy balance and food dose.
4. Time meals and snacks correctly. Feed after exercise or more than four hours before exercise. If snacks are used they should be given during or after exercise.
5. Allow free access to water.
6. Monitor hydration status frequently.

include the amount fed, timing of feeding in relation to exercise and food adaptation. All of these factors should be matched to the individual athlete and the type of exercise performed (intensity, duration, frequency, season). If the current feeding method matches the individual's needs based on the assessment, no changes are necessary. Changes should be made if the assessment reveals discrepancies in the feeding method.

The amount of a new food to feed a canine athlete can be estimated several ways. Feeding guidelines from the manufacturer and those on pet food labels are seldom correct for canine athletes. Energy needs and food doses usually must be calculated. If the amount of the previous food was correct (i.e., appropriate body condition was maintained) and the ME of the food is known, simply feed the same amount of the new food to supply the same ME. If this method isn't feasible, the food dose should be calculated based on the animal's needs as shown above. (See Assess the Feeding Method.) In all cases, the animal should be assessed frequently and adjustments should be made to maintain correct body condition.

Timing of feeding and timing of food changes are important for canine athletes. (See Assess the Feeding Method.) Timing of feeding in relation to exercise influences hormonal status, substrate availability and performance. When changing foods, adequate time must be allowed for the animal to adapt to the new food type to take full advantage of its nutrient profile.

REASSESSMENT (Page 283)

After the feeding plan has been implemented, the animal should be monitored to evaluate its appropriateness. This process is identical to the original assessment of the animal. Frequent physical examinations are important for early detection of injuries or illnesses. Daily monitoring of food consumption can detect problems before they become serious. Frequent evaluation of stool quality can help assess how well the animal is tolerating the food. Weekly measurements of body condition and body weight allow assessment of energy balance (i.e., whether food intake matches energy expenditure). Appropriate body condition is also important for optimal performance. Excess fat represents an unneeded handicap, whereas excessively lean animals may not have sufficient energy stores.

Hydration status should be monitored frequently. Water plays a vital role in supporting cardiovascular function, transport of metabolic substrates and wastes and thermoregulation. Respiratory water losses can be large, particularly during lengthy exercise or under hot or cold environmental conditions.

Ultimately, athletic performance is the best means of monitoring the feeding plan for canine athletes.

Clinical cases that illustrate and reinforce the nutritional concepts presented in this chapter can be found in Small Animal Clinical Nutrition, 4th ed., pp 285-289.

ATHLETES

CHAPTER 11

Normal Cats

For a review of the unabridged chapter, see Kirk CA, Debraekeleer J, Armstrong PJ. Normal Cats. In: Hand MS, Thatcher CD, Remillard RL, et al, eds. Small Animal Clinical Nutrition, 4th ed. Topeka, KS: Mark Morris Institute, 2000; 291-347.

CATS AS CARNIVORES (Pages 294-303)*

Taxonomically, cats and dogs are members of the order Carnivora and are therefore classified as carnivores. However, from a dietary perspective, dogs are omnivores (See Chapter 9.) and cats and other members of the superfamily Feloidea are strict or true carnivores. The difference in food usage is evident from the unique anatomic, physiologic, metabolic and behavioral adaptations of cats to a strictly carnivorous diet. The anatomic, physiologic, metabolic and behavioral adaptations are discussed in detail in Small Animal Clinical Nutrition, 4th ed., pp 294-303.

Nutrient Requirements and Metabolism (Pages 299-303)

The evolution of cats as strict carnivores has resulted in notable nutritional and metabolic adaptations. Although these adaptations have resulted in nutritional requirements peculiar to cats, the basic physiologic and metabolic systems are relatively similar among all mammals. Nevertheless, adaptations to foods composed strictly of animal tissues obligate cats to be "meat eaters."

Energy Metabolism (Page 299)

Adult cats have very low liver glucokinase activity and a limited ability to metabolize large amounts of simple carbohydrates. Kittens naturally ingest soluble carbohydrates (i.e., lactose or milk sugar) before weaning; however, as adults they must rely primarily on gluconeogenesis from glucogenic amino acids (ketoacids), lactic acid and glycerol for maintenance of blood glucose concentration.

Protein Metabolism (Pages 300-303)

Protein metabolism is unique in cats and is manifested by an unusually high maintenance requirement for protein as compared with canine requirements. The protein requirement for growth in kittens is only 50%

*Page numbers in headings refer to Small Animal Clinical Nutrition, 4th ed., where additional information may be found.

Key Nutritional Factors–Normal Cats (Page 293)

Table 11-1 summarizes feeding plans for normal cats.
Table 11-3 lists key nutritional factors for adult cats at maintenance.
Table 11-5 lists key nutritional factors for reproducing cats (mating, gestation, lactation).
Table 11-6 lists energy requirements of pregnant queens.
Table 11-7 lists daily energy requirements of lactating queens over the lactation period.
Table 11-8 lists daily energy requirements of growing kittens.
Table 11-9 lists key nutritional factors for growing kittens.

Table 11-1. Feeding plan summary for normal cats.

Adult Cats

1. Body condition and other assessment criteria will determine the daily energy requirement (DER). DER is calculated by multiplying resting energy requirement (RER) by an appropriate factor. Remember, DER calculations should be used as guidelines, starting points and estimates for individual animals and not as absolute requirements.

    Neutered adult = 1.2 to 1.4 x RER
    Intact adult = 1.4 to 1.6 x RER
    Obese-prone adult = 1.0 x RER
    Senior adult (seven to 11 years) = 1.1 to 1.4 x RER
    Very old adult (≥12 years) = 1.1 to 1.6 x RER

2. Select a food with an appropriate energy density.

    Young to middle-aged adult = 4.0 to 5.0 kcal (17 to 21 kJ)
      metabolizable energy (ME)/g dry matter
    Obese-prone adult = 3.3 to 3.8 kcal (13.8 to 15.9 kJ) ME/g dry matter
    Senior adult = 3.5 to 4.5 kcal (14.6 to 18.8 kJ) ME/g dry matter
    Very old adult = 4.0 to 4.5 kcal (17 to 18.8 kJ) ME/g dry matter

3. Select a food with levels of key nutritional factors and desired urinary pH values listed in **Table 11-3**.

4. Determine quantity of food based on DER calculation (DER ÷ food energy density).

5. Determine preferred feeding method (**Table 11-4**).

6. Monitor body condition, body weight and general health.

Reproducing Cats

1. Body condition and other assessment criteria will determine the DER. Remember, DER calculations should be used as guidelines, starting points and estimates for individual animals and not as absolute requirements. See **Tables 11-6** and **11-7** for detailed DER information.

    Breeding male = 1.4 to 1.6 x RER
    Breeding female = 1.6 x RER
    Gestation = 1.6 to 2.0 x RER
    Lactation = 2.0 to 6.0 x RER

(Continued on next page.)

Table 11-1. Feeding plan summary for normal cats (Continued.).

Reproducing Cats (Continued.)
2. Select a food with an appropriate energy density.
   Breeding male and female = 4.5 to 5.0 kcal (18.8 to 21 kJ) ME/g
      dry matter
   Gestation = 4.0 to 5.0 kcal (17 to 21 kJ) ME/g dry matter
   Lactation = 4.0 to 5.0 kcal (17 to 21 kJ) ME/g dry matter
3. Select a food with levels of key nutritional factors, high digestibility and
   desired urinary pH values listed in **Table 11-5** (gestation and lactation) or
   Table 11-3 (breeding males).
4. Determine quantity of food based on DER calculation (DER ÷ food energy
   density) or provide food free choice.
5. Determine an appropriate feeding method (**Table 11-**4). Free choice is the
   preferred method of feeding gestating/lactating queens.
6. Monitor body condition, body weight, general health, reproductive performance and kitten growth rates.

Growing Cats
1. Age, body condition and other assessment criteria will determine the DER.
   Remember, DER calculations are only guidelines, starting points and estimates for individual animals and do not represent absolute requirements.
      Neonatal and postweaning kittens: See **Table 11-8** for details about calcu-
      lating DER for growing kittens
2. Select a food with an appropriate energy density.
   Postweaning to adult = 4.0 to 5.0 kcal (17 to 21 kJ) ME/g dry matter
3. Select a food with above average digestibility (>85%).
4. Select a food with levels of key nutritional factors and desired urinary pH
   values listed in **Table 11-9**. Remember urinary pH values are lower in kittens compared to those of adult cats and highly acidified foods should be
   avoided.
5. Determine quantity of food based on DER calculation (DER ÷ food energy
   density) or provide food free choice.
6. Determine preferred feeding method. In general, free-choice feeding is
   preferred for kittens less than five months of age (**Table 11-4**).
7. Monitor body condition, weight gain and general health.

higher than that of puppies, whereas the protein requirement for feline
maintenance is twice that of adult dogs. The higher protein requirement
of cats is not due to an exceptionally high requirement for any specific
amino acid; instead, it is caused by a high activity of hepatic enzymes (i.e.,
transaminases and deaminases) that remove amino groups from amino
acids so the resulting ketoacids can be used for energy or glucose production. Unlike omnivores (e.g., dogs) and herbivores, cats cannot decrease
the activity of these enzymes when fed low-protein foods.

Hepatic enzyme systems are constantly active; therefore, a fixed amount
of dietary protein is always catabolized for energy. However, cats do have

a special need for four amino acids: arginine, taurine, methionine and cystine.

ARGININE Cats cannot synthesize sufficient ornithine or citrulline for conversion to arginine, which is needed for the urea cycle. Without arginine, the urea cycle cannot convert ammonia to urea; therefore, feeding foods devoid of arginine may result in hyperammonemia in less than one hour. Affected cats exhibit severe signs of ammonia toxicity and may die within two to five hours.

TAURINE Taurine is β-amino sulfonic acid, abundant in the natural food of cats. In cats, dietary taurine is essential and clinical disease results if insufficient amounts are present. Many species can use glycine or taurine to conjugate bile acids into bile salts before they are secreted into bile. Cats can only conjugate bile acids with taurine. The loss of taurine in bile coupled with a low rate of taurine synthesis contributes to the obligatory taurine requirement of cats.

In addition to its importance in normal bile salt function, taurine is essential for normal retinal, cardiac, neurologic, reproductive, immune and platelet function. Taurine is needed for normal fetal development and may function as an antioxidant, osmolyte and neuromodulator. Most animal tissues, particularly muscle, viscera and brain, contain high levels of taurine; plants contain none.

Thus, homemade vegetarian diets and cereal-based dog foods have long been known to cause taurine deficiency when fed to cats. However, in 1987, taurine deficiency was reported to occur in cats fed commercial foods containing the National Research Council (NRC) recommended levels of taurine. This finding underscored the food-dependent nature of the taurine requirement and prompted the Association of American Feed Control Officials (AAFCO) to recommend an increase in taurine recommendations to 1,000 mg/kg (ppm) and 2,000 mg/kg food in commercial dry and moist foods, respectively. However, taurine levels of 2,500 ppm are often recommended for moist products. Taurine adequacy is best established through feeding trials.

Because taurine functions throughout the body, signs of deficiency have been demonstrated in virtually all body systems. Three syndromes of taurine deficiency in cats have been well established: 1) central retinal degeneration, 2) reproductive failure and impaired fetal development and 3) dilated cardiomyopathy. Hearing loss, platelet hyperaggregation and impaired immune function have also been demonstrated although specific clinical disorders have not been recognized.

METHIONINE AND CYSTINE The sulfur-containing amino acids methionine and cystine are required in higher amounts by cats than most other species especially during growth. Methionine and cystine are considered together because cystine can replace up to half of the methionine requirement of cats. Although these amino acids are present in high amounts in animal flesh, methionine tends to be the first limiting amino acid in many food ingredients. Nutritional deficiencies are possible, especially in cats fed home-prepared or vegetable-based foods. Clinical signs of methionine deficiency include poor growth and a crusting dermatitis at the mucocutaneous junctions of the mouth and nose. Approximately 19% of a food must be composed of animal protein to meet the methionine requirement of kittens. Foods high in plant proteins require additional methionine, which can be supplied as DL-methionine, a crystalline form of the amino acid.

Fat Metabolism (Page 303)

Cats have the ability to digest and use high levels of dietary fat (as is present in animal tissue). Cats have a special need for arachidonic acid (20:4n6) because they cannot synthesize it from linoleic acid (18:2n6), as can dogs. Arachidonic acid is abundant in animal tissues, particularly in organ meats and neural tissues. Thus, the dietary requirement for arachidonic acid has little consequence if cats consume animal tissues. Plants, on the other hand, do not contain arachidonic acid. Therefore, foods composed predominantly of plant-based ingredients may not meet the arachidonic acid requirement of cats.

Vitamin Metabolism (Page 303)

The vitamin needs of cats differ from those of dogs in several ways. Cats do not convert sufficient amounts of tryptophan to niacin. Although cats possess all the enzymes needed for niacin synthesis, the high activity of enzymes in the catabolic pathway (picolinic carboxylase) prevents niacin synthesis. As a result, the niacin requirement of cats is four times higher than that of dogs. Animal tissue is high in niacin.

The prosthetic group of all transaminases is pyridoxine (vitamin B_6). As discussed, flesh-eaters (e.g., cats) derive considerable energy from dietary protein. Cats have high transaminase activity. Therefore, it is logical to expect that their pyridoxine turnover and requirement would be higher than that of omnivores. The pyridoxine requirement of cats is about four times higher than that of dogs.

Vitamin A occurs naturally only in animal tissue. Plants synthesize vitamin A precursors (e.g., β-carotene). Omnivorous and herbivorous animals can convert β-carotene to vitamin A; cats cannot because they lack intestinal dioxygenase that cleaves β-carotene to retinol.

In addition, cats have insufficient 7-dehydrocholesterol in the skin to meet the metabolic need for vitamin D photosynthesis; therefore, they require a dietary source of vitamin D. Vitamin D is fairly ubiquitous in animal fats; therefore, primary vitamin D deficiency has been identified only in cats fed experimental diets.

Water *(Page 303)*

Water needs of cats also differ from those of dogs, not because of feline feeding behaviors (i.e., carnivorous vs. omnivorous) but because of their ancestors' adaptation to environmental extremes. Domestic cats are thought to have descended from the African wildcat (*Felis silvestris libyca*), a desert dweller. Several unique features of water balance in cats may be explained by adaptation to a dry environment. Cats appear to have a less sensitive stimulus for thirst than dogs. Cats compensate for reduced water intake, in part, by forming highly concentrated urine. Unfortunately, this strong concentrating ability coupled with a weak thirst drive may result in highly saturated urine, increasing the risk of crystalluria and/or urolithiasis, both components of the feline lower urinary tract disease (FLUTD) complex.

Eating Behaviors (Pages 294-296)

Domestic cats share several feeding behaviors with their wild counterparts. Unlike most mammals, cats do not display a regular daily rhythmicity in feeding and drinking. Cats typically eat 10 to 20 small meals throughout the day and night.

The predatory drive is so strong in cats that they will stop eating to make a kill. This strategy allows for multiple kills, which optimizes food availability. Unfortunately, this behavior and others may frustrate owners who confuse predatory behavior with hunger. Many owners reason that a fed cat will not hunt and are disappointed when their housecat kills songbirds. Supplemental feeding may reduce hunting time, but otherwise does not alter hunting behavior.

Cats are very sensitive to the physical form, odor and taste of foods. Cats prefer solid, moist foods typical of flesh. Cats reluctantly accept food with powdery, sticky and very greasy textures.

Cats accustomed to a specific texture or type of food (i.e., moist, dry, semi-moist) may refuse foods with different texture. This becomes an important consideration when feeding cats novel diets.

Despite the cat's reputation as a finicky eater, many cats will choose a new food over a currently fed food. Thus, cats are considered to be neophilic. The reverse is true in new or stressful situations in which cats tend to refuse novel foods (i.e., neophobic). These features of cat feeding behavior should be considered when feeding hospitalized cats, but are often overlooked as a cause of food refusal.

Food temperature also influences food acceptance by cats. Cats do not readily accept food served at temperature extremes, whereas foods offered near body temperature (38.5°C [101.5°F]) are most preferred.

Alternative Eating Behaviors (*Pages 295-296*)

Although there are such things as aberrant eating behaviors, many of the behaviors observed are actually normal behaviors that owners happen to find objectionable.

COPROPHAGIA Coprophagia, or consumption of excreta, is normal behavior in queens with kittens less than 30 days of age. The queen stimulates the kittens' urogenital reflexes and elimination by grooming the kittens' perineum. Then, the queen consumes the products of elimination. This process is important as an aid to elimination in young kittens. In addition, coprophagia maintains sanitation and reduces odors in the nest box. Thus, coprophagia has important survival value in wild and feral cats by reducing factors that might attract predators to the nest site. It is very uncommon for cats to continue coprophagia after the kittens are weaned.

CANNIBALISM/INFANTICIDE Cannibalism or infanticide is often normal behavior in male and female cats. Queens typically cannibalize aborted, dead and weak kittens. This behavior may reduce the spread of disease to healthy kittens, conserve maternal resources, optimize survival of the fittest kittens and help keep the nest box clean. In addition, the queen derives nutritional benefits from consuming dead kittens. Occasionally, queens will kill an apparently healthy litter. Environmental factors that cause kittens to mimic early signs of illness (e.g., inactivity, hyperthermia or hypothermia) may trigger infanticide and cannibalism. Maternal stress, malnourishment and hormonal insufficiency may contribute to unexplained cannibalism as well. Maternal experience or parity does not appear to be a contributing factor.

Tomcats may indiscriminately kill unrelated kittens. This behavior usually occurs when a strange male enters a new territory and encounters a lactating queen and kittens. A queen rapidly returns to estrus after the loss of its kittens. Thus, infanticide optimizes a male's genetic potential, in that it now has an opportunity to sire subsequent litters. Infanticide is an uncommon behavior by resident male cats.

The health status, dietary management and husbandry practices should be reviewed in queens or catteries experiencing persistent problems with cannibalism. Males should not have access to young kittens to reduce the chance of infanticide. Although resident male cats rarely pose a problem, it is prudent to err on the side of safety.

PLANT AND GRASS EATING Plant and grass eating is a natural behavior of cats. A variety of explanations have been advanced for grass eating. Grass is undigested within the cat's gastrointestinal (GI) tract; it acts as a local irritant and sometimes stimulates vomiting. Thus, ingested grass may serve as a purgative to eliminate hair and other indigestible material. However, many cats readily eat grass but do not vomit. Other explanations for the behavior include a response to nutritional deficiencies, boredom or a taste preference. Despite a wealth of theories, scientific support is lacking and the cause remains unknown.

RESPONSE TO CATNIP The smell or ingestion of catnip (*Nepeta cataria*) can invoke wild behavior for five to 15 minutes after exposure. Cats may become refractory for an hour or more after cessation of the initial response. The active ingredient, cis-trans-nepetalactone, is thought to act as a hallucinogen although stimulation of neurologic centers associated with estrous behaviors has also been suggested. Cats may respond to catnip by head rubbing and shaking, salivating, gazing, skin twitches, rolling and animated leaping. Only 50 to 70% of cats exhibit a behavioral response, which may have a genetic basis. Prolonged exposure may lead to a chronic state of partial unawareness.

WOOL CHEWING A commonly reported behavioral abnormality in cats is wool chewing. The behavior first appears near puberty when cats begin to lick, suck, chew or eat wool or other clothing articles. Although the cause is poorly understood, nutritional deficiencies are unlikely. Affected cats may be seeking the odor of lanolin or human sweat or the behavior may be a manifestation of prolonged nursing. Siamese, Siamese-cross and Burmese cats are primarily affected. Therefore, a strong genetic link is probable. Wool chewing is managed by limiting access to attractive items and through behavior modification. Feeding a high-fiber food or providing a continuous supply of dry food reduces fabric eating in some cats.

PROLONGED NURSING Prolonged nursing may occur in kittens that strive to satisfy a desire for non-nutritional sucking. Non-nutritional sucking normally subsides near weaning. Kittens may develop nursing vices when they are deprived of normal nursing behavior because they were orphaned, prematurely weaned or required bottle feeding. Within the litter, kittens will often nurse tails, ears, skin folds and/or the genitalia of their littermates. After a kitten is separated from its litter, it may transfer sucking vices to people, stuffed toys, clothing or other pets.

ANOREXIA Although a few days of inappetence is not particularly detrimental to an otherwise healthy cat, prolonged inanition results in malnutrition, reduced immune function and increased risk for hepatic lipidosis. Anorexia may be caused by stress, unacceptable foods or concurrent disease. Most commonly, cats presented to veterinarians for anorexia have a concurrent disease. Cats may endure prolonged starvation rather than eat an unpalatable food. Therefore, advising owners that a cat will "eat when it gets hungry enough" can have deadly results.

A thorough history is useful in differentiating potential causes of anorexia. To determine if inadequate food acceptance is the cause, offer a small selection of highly palatable foods along with the typical food. Because improperly stored foods may develop off flavors, bacterial contamination or fungal growth, confirm that the product is fresh and wholesome. Environmental or emotional factors reported to result in stress-mediated anorexia include hospitalization, boarding, travel, introduction of new people or pets to the household, loss of a companion, overcrowding, high temperatures and excessive handling. Stress-mediated anorexia is usually diagnosed from the history and by ruling out other diseases. Providing a quiet secluded area will often allow a cat to relax sufficiently enough to begin eating. Often, increasing the food's palatability will improve food intake. Warming, adding water or choosing foods high in animal protein and fat can enhance food palatability. If cats are highly stressed or appropriate feeding sites are unavailable, mild tranquilizers or appetite stimulants (e.g., diazepam, oxazepam or cyproheptadine) may be beneficial. (See Chapter 12.) Force feeding may stimulate taste receptors and appetite in some cats. Other cats find the process so stressful that any benefit is far outweighed by the additional stress.

FIXED-FOOD PREFERENCES The food type fed by the owner during a kitten's first six months influences the pattern of food preferences throughout life. Although uncommon, kittens exposed to a very limited number of foods may develop a food fixation, refusing to eat anything but a single food. Adult cats fed highly palatable, single-item foods have been reported to develop fixed-food preferences as well.

Cats with food fixations can be particularly troublesome if dietary modifications are necessary. Cats with strong food preferences should be transitioned to the new food over a prolonged period. Convert to the desired food by replacing 10 to 20% of the old food with an equal amount of the new food on Day 1, then gradually increase the ratio of the new food to the old over the next 14 days. A more gradual transition may be required if food intake drops below 70% of maintenance levels. Cats should be monitored to ensure they are not selecting the preferred food from the food dish and that food intake remains adequate. Feeding cats a complex ration-type commercial food and not feeding single-item foods can avoid food fixations.

LEARNED TASTE AVERSIONS Cats may develop learned aversions to certain foods when feeding is paired with a negative GI experience. The negative experience can be physical, emotional or physiologic. Typically, aversions occur when cats are fed before an episode of nausea or vomiting. Foods that were readily consumed before a negative incident will be avoided subsequently. Clinically, aversions may develop when GI upset is induced by various diseases, drugs or treatments protocols. Foods with high salience (i.e., strong odors or high protein levels) are more likely to become aversive and should not be fed within 24 hours of anticipated GI upsets. Aversions have been documented to last up to 40 days in cats. Learned aversions are considered an adaptive response. By avoiding foods that previously caused gastric distress, cats will avoid eating foods likely to be spoiled or tainted. Learned aversions have also been suggested to reduce repeated bouts of food allergy. Although this response may occur when feeding single-food items, it does not appear to be of particular benefit to cats fed commercial foods.

POLYPHAGIA Various diseases, drugs and psychological stresses can mediate excessive food consumption. Rarely, polyphagia (hyperphagia) may occur with CNS disease, particularly with lesions of the ventromedial hypothalamus. Presence of weight loss or gain is of key diagnostic importance. Polyphagia with weight loss is almost always associated with an underlying disease process or simple underfeeding. Caloric intake should always be calculated because underfeeding can result in a ravenous appetite that may be misinterpreted as abnormal. Nutritional management of polyphagia requires an accurate diagnosis because treatment is aimed at the primary disease.

LIFESTAGE NUTRITION (Pages 303-306)

Lifestage nutrition is the practice of feeding foods designed to meet an animal's optimal nutritional needs at a specific age or physiologic state (e.g., maintenance, reproduction or growth) (**Table 11-1**). The concept of lifestage nutrition recognizes that feeding either below or above an optimal nutrient concentration can negatively affect biologic performance and/or health. This concept differs markedly from feeding a single product for all lifestages (i.e., all-purpose foods), whereby nutrients are added at levels to meet the highest potential need (i.e., usually growth and reproduction). The goal in nutrition is to feed for optimal health, performance and longevity, thus feeding foods optimized toward individual needs is preferred. This philosophy is key to lifestage nutrition and preventive medicine. In addition to meeting the basic nutritional requirements of an animal, the nutrition-related health risks at each lifestage should be assessed and minimized. For maximum benefit, risk assessment and prevention plans should begin well before the onset of disease.

The value of lifestage feeding is greater if risk factor management is also incorporated into the feeding practice. A more narrow but optimal range of nutrient recommendations often emerges when age and physiologic needs are reviewed in conjunction with reducing disease risk factors. Nearly all commercial cat foods meet or exceed the minimum nutrient requirements of cats. However, certain nutrients may still be outside of the desired nutrient range for optimal health. For cats fed commercial foods, these nutrients require particular consideration and, thus, are referred to as nutrients of concern. Specific food factors (e.g., digestibility, texture and effect on urinary pH) can also affect health and risk of disease. (See Chapter 1.) Together, nutrients of concern and specific food factors are called key nutritional factors. The key nutritional factors for different lifestages of healthy cats will be discussed in the following sections, including those associated with reducing the risk of specific diseases and those involved with optimizing performance during different physiologic states. Cats eating homemade foods are at greater risk for nutrient deficiencies (e.g., calcium) and excesses (e.g., phosphorus) than those eating commercial foods. Therefore, these cats have a broader list of key nutritional factors, which are discussed in Chapter 6.

In sequence, this chapter covers feeding adult cats, reproducing cats and growing kittens. The adult cat section is divided into young to middle-aged cats and older (senior) cats. The chapter begins with feeding adult cats because most pet cats are adults and the nutrient needs of adults serve as a good basis for comparing the nutrient needs for reproduction, lactation and growth. Also, note that most of the general information about feeding cats is included in the adult cat section and is not repeated in the other sections of this chapter.

YOUNG TO MIDDLE-AGED ADULT CATS (Pages 306-314)

Adult maintenance refers to the needs of non-reproducing cats one to seven years of age. Optimal nutritional requirements for adult cats tend to be the most broadly defined of any lifestage. This is due, in part, because young to middle-aged adult cats have the greatest ability to tolerate or compensate for metabolic and physiologic perturbations. The goals of feeding adult cats are to maximize health, longevity and quality of life.

Assessment (Pages 306-314)
Assess the Animal (Pages 306-312)
Animal assessment encompasses the complete evaluation of the animal, its environment and risk factors for disease (**Table 11-2**). The purpose of

Table 11-2. Factors to consider during nutritional assessment.

Signalment		
Activity level	Disease status	Reproductive status
Age	Environment	Use
Breed	Gender	
Dietary history		
Adverse food reactions	Feeding method	Previous foods
Amount eaten	Feeding schedule	Supplements
Amount offered	Food aversions	Treats
Appetite (interest)	Food storage	Type
Brand fed	Nutritional losses	Water availability
Weight history		
Current weight	Percent weight change	Usual weight
Ideal weight	Rate of change	
Physical examination		
Body condition	Eyes	Oral health
Bone structure	Hydration	Skin condition
Coat condition	Muscle mass	Strength/activity
Diagnostic studies		
Albumin	Hemoglobin	Prothrombin time
Creatine kinase	Lymphocyte count	Serum urea nitrogen
Hematocrit	Potassium	Sodium

first assessing the animal is to establish feeding goals, recognize risk factors for diseases and identify key nutritional factors for individual cats.

HISTORY AND PHYSICAL EXAMINATION A detailed dietary history should evaluate the factors listed in **Table 11-2**. If the dietary history is perceived to be incomplete, it may prove useful to have owners continue to feed and medicate their pet as usual and record amounts, types and brands of all foods and supplements given for one to two weeks. Such dietary records help define nutrient intake, nutritional problems and errors in feeding management.

A thorough physical examination should include a systematic evaluation of each body system by observation and palpation and auscultation where possible. Special attention should be given to the oral cavity, hydration status, skin and coat condition, body weight and body condition score (BCS). (See Chapter 1.)

Gender Neutering reduces the daily energy requirement (DER) of adult cats by 24 to 33%, compared with the DER of intact cats. The decrease in DER does not appear to be influenced by age at neutering. The reduction in energy requirement is most likely attributable to a reduction in basal metabolic rate because obvious changes in behavior and activity are not observed after neutering.

Because a reduction in DER without a reduction in food intake leads to obesity, nutritional counseling should be provided to owners who present cats for neutering. Kittens neutered at less than six months of age should be fed foods designed for growth until they reach skeletal maturity between eight and 10 months of age. Many foods designed for growing kittens are energy dense; therefore, portion control and regular monitoring of body condition is advised.

Activity Level Activity level is one of the key determinants of DER. Two-fold differences in energy requirement have been observed between active and sedentary cats. Therefore, food intake should be adjusted according to activity level to maintain optimal body condition (BCS 3/5).

Sedentary, inactive, caged and older cats often have energy requirements very near or even below the average resting energy requirement (RER) (0.8 to 1.2 x RER). A reduction in food intake is expected in normally active cats that are subsequently confined.

Cats with unlimited activity may have energy needs 10 to 15% above normal. Very active or "high strung" cats may expend markedly more energy than other cats. For example, the energy requirement of Abyssinian cats has been reported as 79 kcal/BW_{kg}/day (330 kJ/BW_{kg}/day), or 1.6 x RER, which is 30% greater than that required for average adult housecats.

Age Young to middle-aged cats may have specific nutrient concerns especially with respect to weight control, lower urinary tract health, dental health and subclinical kidney disease.

Environment DER for cats may be markedly altered when ambient temperatures deviate significantly from a thermoneutral environment. Behavioral responses usually compensate for minor deviations in temperature with little effect on a cat's water or energy needs. However, temperatures low enough to cause shivering (5 to 8°C [41 to 46.4°F]) can increase the resting caloric requirement from two to five times normal. Cats kept in very hot environments (>38°C [>100.4°F]) may initially reduce food intake by 15 to 40%; however, as respiratory rate and grooming behavior increase and panting begins, the requirement for calories and water increases. Water is critically important to prevent heat stress in hot environments.

Multi-cat environments refer to individual households with two or more cats. Challenges associated with feeding cats in a multi-cat environment include difficulty in monitoring food and water intake, ensuring all cats have unfettered access to food and providing specialized foods to individual cats. Multiple feeding stations and individual feed pans, particularly if placed at different levels, allow timid and low-status cats to eat alone or away from dominant cats.

CATS

Cats that are under stress may develop diminished appetites or complete anorexia. A prolonged reduction in food intake in healthy cats or short-term food deprivation in sick cats can lead to malnourishment and increased risk of hepatic lipidosis.

Breed Certain breeds of cats (e.g., Abyssinians) are noted for their lively and rambunctious disposition, whereas others (e.g., Persians or ragdolls) tend to be quiet and tranquil. Thus, disposition affects energy requirements among breeds. Other nutritional variances may be elucidated with continued research into specific requirements of different cat breeds.

LABORATORY AND OTHER CLINICAL INFORMATION Laboratory analyses may help develop a complete picture of nutritional status. Results should always be interpreted in relation to the physical examination and historical findings.

KEY NUTRITIONAL FACTORS Table 11-3 summarizes key nutritional factors for young to middle-aged adult cats. The following section describes these key nutritional factors in more detail.

Water Cats should generally drink 1 ml water/kcal metabolizable energy (ME) requirement. In practice, adult cats should have unlimited access to potable water. Increased water intake is thought to be useful for managing urolithiasis by reducing the urinary concentration of urolith-forming minerals. Feeding moist foods (vs. dry foods) increases water intake and urine volume in most cats. When allowed free access to water, the total water intake of cats eating dry food is only half that of cats eating moist food.

Energy The DER of average young to middle-aged adult cats is generally between 60 to 80 kcal/BW_{kg}/day (251 to 335 kJ/BW_{kg}/day) or approximately 1.2 to 1.6 x RER, where RER in kcal = $70(BW_{kg})^{0.75}$ or RER in kJ = $293(BW_{kg})^{0.75}$. Caloric requirements for neutered or inactive cats are calculated using the lower end of the range (1.2 x RER), whereas the upper end of the range (1.4 to 1.6 x RER) is used for active sexually intact cats. Most neutered housecats require between 1.2 to 1.4 x RER.

Individual cats may have energy requirements that vary by up to 50% or more above or below the average requirement. This range is not surprising considering that the DER of a particular cat is influenced by differences in lean body mass, gender, activity, environmental temperature and genetic traits. Thus, it is important to remember that calculated energy requirements are only estimates for individual cats. The true caloric requirement for an individual cat is the amount of food that will maintain an ideal body condition (BCS 3/5) and stable weight.

Table 11-3. Key nutritional factors for adult cats at maintenance.

Factors	Recommended food levels*		
	Young to middle aged	Obese prone	Older
Energy density (kcal ME/g)	4.0-5.0	3.3-3.8	3.5-4.5
Energy density (kJ ME/g)	16.7-20.9	13.8-15.9	14.6-18.8
Protein (%)	30-45	30-45	30-45
Fat (%)	10-30	8-17	10-25
Crude fiber (%)	<5	5-15	<10
Calcium (%)	0.5-1.0	0.5-1.0	0.6-1.0
Phosphorus (%)	0.5-0.8	0.5-0.9	0.5-0.7
Ca/P ratio	0.9:1-1.5:1	0.9:1-1.5:1	0.9:1-1.5:1
Sodium (%)	0.2-0.6	0.2-0.6	0.2-0.6
Potassium (%)	0.6-1.0	0.6-1.0	0.6-1.0
Magnesium (%)	0.04-0.1	0.04-0.1	0.05-0.1
Chloride (%)	>0.3	>0.3	>0.3
Average urinary pH	6.2-6.5	6.2-6.5	6.2-6.6

*Dry matter basis. Concentrations presume an energy density of 4.0 kcal/g. Levels should be corrected for foods with higher energy densities. Adjustment is unnecessary for foods with lower energy densities.

Controlling energy intake is important in managing and preventing obesity. Approximately 25% of pet cats in the United States are overweight. The prevalence is highest in middle-aged cats (seven to eight years); nearly 50% of this age group are overweight or obese (BCS 4/5 or 5/5). Obesity increases the risk of death in middle-aged cats 2.7 times above that of lean cats, thus preventing obesity has important consequences for long-term health. (See Chapter 13.) Risk factors associated with obesity include: 1) middle age, 2) neuter status, 3) low activity and 4) high-fat, high-calorie foods. Some cats that are inactive, confined or obese prone may require markedly fewer calories than predicted by equations to determine DER. Obese cats may require as few as 0.8 x RER to achieve an average weight loss of 1% of body weight per week.

Food digestibility and energy density may influence the risk for FLUTD. Energy-dense foods reduce overall dry matter intake. Lower dry matter intake decreases stool volume, which subsequently reduces fecal water loss. Both features are beneficial in reducing total magnesium intake and increasing urine volume.

See **Table 11-3** for recommended levels of fat, fiber and energy density in foods for obese-prone cats.

Protein AAFCO has suggested a minimum dietary protein level of 26% dry matter (DM) for adult maintenance. Protein and amino acid requirements vary with the energy content of a food. The minimum protein allowance suggested by AAFCO is based on foods containing 4.0 kcal/g

(16.7 kJ/g) DM and should be corrected in foods with energy densities greater than 4.5 kcal/g (18.8 kJ/g). (See Chapter 1 for the correction method.)

Meeting the minimum protein needs of cats is critical because they have minimal capacity to adapt to low levels of dietary protein. However, protein in excess of the requirement is rapidly catabolized and used to provide energy and maintain blood glucose levels. Any excess energy will be stored as fat. Therefore, there appears to be little benefit to feeding large excesses of protein to cats. Conversely, dietary protein excess may increase proteinuria and the progression of subclinical renal disease.

Although cats can be fed vegetable-based foods, most protein in the food should be derived from animal tissues. The amino acid profile of most animal tissues better reflects the nutritional requirements of cats. The recommended protein allowance for normal adult cats is 30 to 45% of the dry matter (**Table 11-3**).

Taurine Taurine is a key nutrient for all feline lifestages. Sporadic cases of taurine depletion continue to be diagnosed. Therefore, dietary taurine concentrations should be evaluated in cats with signs of deficiency or disease.

Fats and Essential Fatty Acids Cats use dietary fat for energy, as a source of essential fatty acids and to facilitate absorption of fat-soluble vitamins. Linoleic acid and α-linolenic acid are essential for normal membrane structure and function, including growth, lipid transport, maintenance of the epidermal permeability barrier and normal skin and coat. Arachidonic acid, on the other hand, is important for functions that rely on eicosanoid synthesis.

Fatty acids of the n-3 series (linolenic acid, 18:3n-3) are probably required in the diet of all animals. Studies indicate n-3 fatty acids are essential for normal neural development in neonates. Cats appear particularly susceptible to the deleterious effects of lipid oxidation. Foods high in polyunsaturated fatty acids have been associated with the development of feline pansteatitis when not adequately supplemented with vitamin E.

Fat levels above 9.0% DM are recommended for most cats. When both fatty acids are present, linoleic acid is required at 0.5% of the food and arachidonic acid at 0.02% of the food, or 5% and 0.04% of the dietary energy as linoleic acid and arachidonic acid, respectively. Fat enhances the palatability of food; cats prefer foods with levels near 25% DM vs. foods at 10 or 50% DM.

High-fat foods have been associated with an increased incidence of obesity in cats. Most cats do well on foods with 10 to 30% fat. However, cats prone to obesity should be fed foods with lower levels of dietary fat (8 to 17% DM) (**Table 11-3**).

Fiber Although cats do not require dietary fiber, small amounts in commercial foods enhance stool quality and promote normal GI function. The natural foods of cats typically contain less than 1% dietary fiber although cats tolerate much higher levels. Fiber concentrations less than 5% DM are recommended for normal adult cats.

Fiber supplementation may benefit cats with frequent hairballs. Clinical evidence and field trials have demonstrated a reduction in the frequency of hairball vomiting with fiber supplementation.

Calcium and Phosphorus Deficiencies of calcium and phosphorus are uncommon in cats fed commercial foods. Most cases of calcium deficiency have occurred in cats eating only unsupplemented meats, in which the calcium concentration is excessively low. Phosphorus excess appears to be of greater concern for adult cats fed commercial foods, especially as related to lower urinary tract and renal disease. (See Chapters 19 and 21.)

The calcium requirement for growing kittens is 0.5% (DM) of the diet. Adult needs are typically less than those for growth. Calcium and phosphorus requirements of adult cats have recently been determined. The minimum calcium requirement is 132 mg/MJ (60 mg/100 kcal or 0.25% DM at an energy density of 4.5 kcal/g). The minimum phosphorus requirement is 143 mg/MJ (55 mg/100 kcal or 0.27% DM at an energy density of 4.5 kcal/g). Both values are nearly half the AAFCO minimum allowance (i.e., 0.6% DM for calcium and 0.5% DM for phosphorus). Typical commercial foods contain calcium and phosphorus levels well in excess of these guidelines. (See Appendix 2.) Calcium-phosphorus ratios between 0.9:1 to 1.5:1 appear optimal for most feline foods.

Dietary phosphorus is a key nutrient in the management of two common feline diseases: struvite-mediated FLUTD and renal disease. The mineral constituents of struvite are magnesium, ammonium and phosphate. Although the primary objectives for preventing FLUTD due to struvite precipitates are to reduce urinary pH and, to a lesser extent, restrict dietary magnesium, limiting dietary phosphorus may be beneficial. (See Chapter 21.) The risk of clinically apparent struvite crystalluria and urolithiasis appears highest in adult cats from two to five years of age. Controlling phosphorus intake in combination with appropriate reductions in dietary magnesium concentrations and urinary pH reduces the risk of struvite-associated FLUTD in cats of this age group.

Excess dietary phosphorus is not considered a cause of renal damage but accelerates the progression of renal disease toward failure and death. High levels of dietary phosphorus (1.2 to 1.8% DM) lower creatinine clearance values and possibly reduce renal function in young, healthy cats. Phosphorus reduction is advised in the early nutritional management of renal disease in cats to decrease the renal excretory workload and avoid phosphorus retention. Cats with renal insufficiency are often not diagnosed until three-fourths or more of kidney function has been lost. Furthermore, older cats have an increased prevalence of kidney disease. Generalized phosphorus reduction may slow progression of renal disease in cats with subclinical or undiagnosed disease. Dietary phosphorus may be reduced as low as 0.3% DM in cats with overt renal disease, otherwise levels of 0.5 to 0.8% are recommended (**Table 11-3**).

Sodium and Chloride In people, limiting sodium intake to levels that meet the requirement without significant excess reduces the risk of hypertension and is considered important to long-term health. This same nutritional practice has been advocated for cats. In a study involving feline hypertension, nearly 50% of hypertensive cats fed a low-sodium food had a significant reduction in blood pressure. This response is similar to that seen in people in that not all people are "salt-sensitive." In cats, hypertension is commonly associated with renal failure, hyperthyroidism, cardiac disease and possibly obesity. High blood pressure has been associated with significant end-organ damage in hypertensive cats. Blindness, retinal hemorrhage, stroke, cardiac dilatation and murmurs and renal damage are common findings. Avoiding excess sodium chloride seems prudent because: 1) hypertension has significant deleterious health effects, 2) diagnostics to detect hypertension are not commonly performed and 3) the medical conditions associated with hypertension are common in cats. Chloride has been implicated more recently as a major determinant in the development of hypertension in salt-sensitive people. The interaction of sodium with chloride appears to cause the greatest increase in blood pressure compared with sodium combined with other anions. The minimum chloride requirement is not known for cats, but the NRC recommended concentration (i.e., 0.19% DM) appears sufficient.

The minimum sodium requirement for adult cats is estimated at 9.2 mg/BW_{kg}/day or approximately 0.08% DM. Sodium concentrations from 0.2 to 0.6% DM will satisfy the needs of healthy adult cats without providing excessive levels (**Table 11-3**).

Potassium The potassium requirement of cats varies with the dietary protein concentration and the effect of the food on urinary pH. High-protein foods and foods that result in an acidic urinary pH increase the potassium requirement of cats. Dietary potassium levels in foods for normal adult cats should be greater than 0.5% DM and ideally between 0.6 to 1.0% DM to prevent hypokalemia (**Table 11-3**). Increased potassium losses occur in cats with certain metabolic abnormalities (e.g., renal insufficiency, renal tubular acidosis, diabetes mellitus and enteritis). Potassium supplementation may be necessary to maintain normal potassium balance in cats with these conditions.

Magnesium Magnesium is an essential nutrient, but is also a major constituent of struvite crystals (magnesium ammonium phosphate). Struvite precipitation in the urinary tract contributes significantly to the development of FLUTD. To reduce the risk of FLUTD due to struvite, dietary magnesium concentrations should be less than 20 mg/100 kcal of food (or <0.10% DM) in conjunction with a reduced urinary pH (i.e., average pH values <6.5 pH units). These levels are similar to those found in the natural food of cats. Excessive magnesium restriction may be associated with the increasing prevalence of calcium oxalate uroliths in cats. Therefore, excessive restriction of magnesium (i.e., <0.04% DM) is not recommended (**Table 11-3**).

Urinary pH Food ingredients and feeding methods contribute to the urinary pH produced by cats. The risk of struvite precipitation and FLUTD is greatly reduced at urinary pH values less than 6.5. Many cats develop metabolic acidosis when the urinary pH is less than 6.0. Metabolic acidosis may promote bone demineralization, urinary calcium and potassium loss and increase the risk of calcium oxalate urolithiasis. Free-choice food intake modulates urinary pH by dampening the postprandial alkaline tide that occurs three to six hours following larger meals. Meal feeding promotes a much greater alkaline tide and higher average urinary pH.

Foods that produce average urinary pH values of 6.2 to 6.5 when fed free choice reduce the risk of struvite-mediated FLUTD and avoid metabolic acidosis in most adult cats. The normal urinary pH of cats eating mice and rats is 6.2 to 6.4. Thus, 6.2 to 6.4 is the "normal acidic urinary pH" of cats fed a wild-type food and the recommended range for healthy adult cats (**Table 11-3**).

Texture Food texture influences oral health. (See Chapter 16.) Dental calculus and periodontal disease are the most prevalent diseases in cats one year old and older. Dry foods specifically designed to promote oral health are beneficial in reducing accumulation of dental plaque and calculus and reducing the severity of gingivitis. Generally, dry foods result in less plaque and calculus accumulation in cats than do moist and semi-moist foods.

Recently, very hard foods were implicated as a potential factor in the etiology of feline odontoclastic resorptive lesions (i.e., neck lesions). Chronic tooth trauma may result in abfraction (fractures) of the enamel at the gum line. This trauma alone or combined with poor oral hygiene may initiate odontoclastic resorptive lesions in cats. Although a direct association has yet to be demonstrated, it seems prudent to avoid feeding very hard foods (e.g., bones) whenever possible, especially to cats with dental disease.

Food texture also influences the palatability and acceptability of foods for cats. A sudden change in texture may result in reduced food intake or food refusal. Cats accustomed to eating only dry foods may refuse moist foods and vice versa. Recognizing the preferred food texture from the dietary history will help identify textural change as a cause of inappetence.

Assess the Food (Pages 312-313)

After the nutritional status of the cat has been assessed and the key nutritional factors and their target levels have been determined, the adequacy of the food being fed can be assessed. The three steps to food assessment include: 1) ensuring the food has undergone feeding trials with cats, 2) determining the food's dry matter nutrient content (especially for the key nutritional factors) and 3) comparing the food's key nutritional factors with the recommended levels (**Table 11-3**).

In the United States, commercial foods that have been animal tested will usually have a nutritional adequacy statement on the label. (See Chapter 5.) Few if any homemade recipes have been animal tested according to prescribed feeding protocols. Commercial cat foods that have passed AAFCO-prescribed or other feeding tests provide reasonable assurance of nutrient availability and sufficient palatability to ensure that food intake will be sufficient to meet nutrient needs. However, even controlled animal testing is not infallible. Passing such tests does not ensure the food will be effective in preventing long-term nutrition and health problems. Therefore, in addition to having passed feeding tests, the food should be evaluated to ensure that the key nutritional factors are at levels appropriate for promoting long-term health.

Appendices 2 and 3 provide partial nutrient profiles for selected commercial foods and treats sold in the United States, Canada and Europe. (Also see Appendix M, Small Animal Clinical Nutrition, 4th ed., pp 1084-1091.) The manufacturer should be contacted if the food or treat in question cannot be found in these appendices. Manufacturers' addresses and toll-free customer service numbers are listed on pet food labels. If the manufacturer cannot provide the necessary information, consider switching to a food for which this information is available.

Comparing a food's nutrient content with the cat's nutrient needs (**Table 11-3**) will help identify any significant nutritional imbalances in the food being fed. This comparison is fundamental to deciding whether or not to feed a different food.

The impact of treats on daily nutrient intake depends on three factors: 1) the nutrient profile of the treat, 2) the number of treats provided daily and 3) the nutrient composition of the cat's regular food. Meeting nutrient requirements is not the primary goal of feeding treats; therefore, many commercial treats are not complete and balanced. Appendix 3 lists common commercially available treats and their nutrient content. (Also see Appendix M, Small Animal Clinical Nutrition, 4th ed., pp 1084-1091.) Similarly, most table foods are nutritionally incomplete and unbalanced and may contain high levels of fat, minerals and/or protein. If snacks are fed, it is simplest to recommend those that best match the nutritional profile recommended for a particular lifestage. Otherwise, the nutritional composition of the treat and food must be combined and assessed as the entire diet.

Assess the Feeding Method *(Pages 313-314)*

The veterinarian should evaluate the feeding method, feeding frequency and the amount of food offered. It is also useful to know how the food is prepared (e.g., heated, water added, etc.) and by whom and where the cat is fed. It may not always be necessary to change the feeding method when managing healthy cats. However, a thorough evaluation includes verification that an appropriate feeding method is being used. (See Chapter 1.)

No single feeding method is optimal for all cats. The preferred method of feeding an individual cat is often determined by non-nutritional factors (i.e., food type, owner preference, owner schedule and feeding environment). Nutritional considerations for selecting an appropriate feeding regimen include the cat's body condition, health status/disease risk factors and the food's energy density and palatability.

Two methods are typically used to feed cats: 1) free choice in which the food is continuously available and the cat eats as much as it wants whenever it wants and 2) meal feeding in which a discrete amount of food is offered one or more times per day. Many owners use a combination of free-choice and meal-feeding methods. Typically, dry food is available throughout the day and supplemented with one or more meals of moist food. Free-choice or combination feeding accommodates the normal feeding behavior of cats by allowing the animal to eat several small meals spaced irregularly throughout the day and night. Each feeding method has advantages and disadvantages that should be considered when making feeding recommendations (**Table 11-4**). The major disadvantages to combination feeding are the inability to accurately monitor and control food intake.

The amount fed is important because nutrient requirements are met by a combination of nutrient levels in the food and the amount of food eaten. Even if a food has an appropriate profile of key nutritional factors, significant malnutrition could result from feeding excessive or insufficient amounts. The amount fed is appropriate if the cat has an optimal BCS (3/5) (See Chapter 1.) and its body weight is stable. The amount fed can be estimated by calculation (See Chapter 1.) or by referring to feeding guides on product labels or product information. These guides, however, usually represent population averages and thus may not be optimal for individual cats.

CATS

Table 11-4. Advantages and disadvantages of various feeding methods for cats.

Methods	Advantages	Disadvantages	Food types
Free choice	Convenient Ensures adequate food availability Mimics natural feeding behavior Dampens postprandial alkaline tide (lower mean urinary pH)	Overconsumption leads to weight gain and/or obesity Difficult to monitor appetite/food intake Moist food may spoil Less owner contact	Dry Semi moist
Meal fed*	Enhances human-animal bond Facilitates monitoring of appetite/food intake Enhances control of food intake	Enhances postprandial alkaline tide (higher mean urinary pH) Large meals may result in vomiting Less convenient Three or more meals are needed for pregnant/nursing queens, kittens/debilitated cats	Dry Semi moist Moist
Combination**	Enhances human-animal bond (vs. free choice) Variable effect on urinary pH	Poor monitoring of appetite/food intake Poor control of food intake Less convenient than free choice Variable effect on urinary pH	Dry Semi moist Moist

*One or more individual feedings per day, one to two hour availability per feeding.
**Dry foods available free choice, moist foods meal fed one or more times daily.

Determine a Feeding Plan (Page 314)

The aforementioned assessment of the cat, current food and feeding method should provide the necessary information to develop a feeding plan. Sometimes, the assessment steps determine that the current plan is optimal and no changes are necessary. However, changes in food, feeding method or both will need to be made if results of the assessment indicate that the food or feeding method is inappropriate. In either case, the cat should be reassessed regularly.

Select a Food (Page 314)

It is not necessary to change foods if the food currently fed supplies the correct amounts of the key nutritional factors and the food has been test fed to cats at similar lifestages. However, a new food should be selected if discrepancies were determined during the food assessment phase. The new food should provide levels of the key nutritional factors listed in **Table 11-3**. Appendices 2 and 3 list nutrient levels for selected, readily available commercial cat foods and treats. (Also see Appendix M, Small Animal Clinical Nutrition, 4th ed., pp 1084-1091.) The same information can be obtained directly from the manufacturer. Be sure to confirm that the new food is complete and balanced by verifying it has passed animal feeding tests.

Determine a Feeding Method (Page 314)

The method of feeding may not need to be revised if it appears adequate. Adult cats may be meal fed, free choice fed or fed using a combination of methods. The method of choice largely depends on the food form, the ability of the cat to regulate food intake and owner preference. Most obese-prone cats should be fed a measured quantity of food; however, some cats can be fed low-calorie foods free choice. Most cats tolerate once daily feeding with no problems; however, meal feeding at least twice daily is preferred. Cats should be allowed one to two hours to complete a measured meal; many cats will return for several small feedings before finishing the entire offering. Food should be available at all times for underweight cats to encourage sufficient food intake. In general, clean drinking water should be continuously available.

Cats are not typically bothered by food changes and food variety stimulates increased food intake. Unfortunately, in a few cats, rapid changes in the food or feeding method can cause GI upsets or food refusal. Transitioning to a new food over four to seven days may be required to avoid food intolerances. To change to a new food, replace 25% of the old food with the new food on Day 1 and continue this incremental change daily until the change is complete on Day 4. A slower transition may be required for cats that have been historically sensitive to dietary changes, those with GI diseases and when the new food differs markedly from the old (e.g., low fat vs. high fat or raw meat vs. dry food).

Food and water bowls should be cleaned regularly with warm soapy water and rinsed well. Pans used for moist foods need daily cleaning, whereas dry food feeders should be cleaned at least once weekly. Many cats prefer shallow dishes, especially "flat-faced" breeds such as Persians. Contamination of the chin and face with food particles can exacerbate feline acne. Using shallow pans and regular pan cleaning can help reduce the severity of the problem.

Reassessment (Page 314)

Cats provided proper nutritional management are healthy and alert, have ideal body condition and stable weight and have a clean, glossy coat. The owner should evaluate body condition every two to four weeks. In addition, owners should monitor daily food and water intake and observe the cat's interest in its food and its appetite. Stools should be evaluated regularly because changes in frequency or character may signify nutritional problems or disease. Normal stools should be firm, well-formed and medium to dark brown. Any abnormalities should be investigated. The veterinarian should also conduct a nutritional assessment as part of the annual physical examination and vaccination visit.

OLDER CATS (Pages 314-320)

For nutritional purposes, cats are often considered senior at seven to eight years of age and geriatric or very old at 10 to 12 years of age. There is an increasing prevalence of age-related diseases around seven years of age coupled with the gradual onset of behavioral, physical and metabolic changes related to aging.

Outwardly, a seven-year-old cat may not appear old, but changes in nutritional management and preventive care are important to reduce risk factors for common age-associated diseases, maintain good health and maximize longevity.

Older cats become less active and have reduced lean body mass. Together, these changes reduce the basal metabolic rate.

Aging cats become less adaptable and have reduced physiologic reserve to withstand perturbations in their health and environment, including changes in their food. Different rates of aging occur among older cats, thus there is greater diversity in individual animal needs than at any other lifestage. Individualization of nutritional management becomes even more important because of the poor adaptability of older cats. The goals for the nutritional management of older cats are:

- Maintenance of optimal nutrition (i.e., maintenance of ideal body condition and weight, adequate intake of a nutritious food and good hydration)
- Risk factor management (i.e., minimization of associated disease risks)
- Disease management (i.e., amelioration of clinical signs of common diseases, slowing progression of certain chronic diseases)
- Improvement in the quality and longevity of life.

This section describes how to assess older cats and meet their nutritional needs.

Assessment (Pages 316-320)
Assess the Animal (Pages 316-319)
HISTORY AND PHYSICAL EXAMINATION A complete history should be taken and physical examination performed, as described for young to mid-

dle-aged adult cats. Of particular interest are physiologic changes associated with aging and age-related diseases. Note any changes in appetite, food or water intake, activity, oral health and body condition. Abnormalities in these parameters are often early indicators of underlying disease.

Oral disease is the most prevalent disease in older cats; however, weight loss, renal disease, cardiac disorders, diabetes mellitus and hyperthyroidism are frequently diagnosed in this age category. Kidney disease may affect nearly 30% of older cats and is a major cause of feline death. Physical evaluation of renal size, shape and firmness may uncover kidney abnormalities, whereas thoracic auscultation may expose cardiac disease. Hyperthyroidism may be suspected based on the history and other physical findings and may be detected by palpating enlarged thyroid glands. A fundic examination may help detect hypertension, which is often secondary to renal, cardiac or thyroid disease in older cats. Retinal hemorrhage is a common finding in older hypertensive cats.

LABORATORY AND OTHER CLINICAL INFORMATION Specific abnormalities in the physical and historical examination should be pursued further using appropriate diagnostic procedures. A geriatric-type blood panel to screen for common age-associated diseases should be performed annually. Feline leukemia virus and feline immunodeficiency virus testing should be current and repeated if potential exposure has occurred or suspicious clinical signs are present. Specialized diagnostics may be indicated by physical or biochemical findings (e.g., electrocardiography, ultrasonography, radiography, blood pressure monitoring).

KEY NUTRITIONAL FACTORS The recommended range of nutrient allowances can be optimized to support changes in physiologic function and reduce risk factors for common age-related diseases. **Table 11-3** summarizes key nutritional factors for older adult cats.

Water Water is an often overlooked but critical nutrient in the health of older cats. Aging impairs thirst sensitivity, which is already low in cats compared with that in other species. In addition, the decline in renal function observed in many older cats may increase water losses due to impaired urine concentrating ability. Together, these characteristics predispose older cats to dehydration. Chronic dehydration can impair normal metabolic processes and exacerbate subclinical disease. In addition, dehydration reduces a cat's ability to thermoregulate. Water intake in healthy cats without increased losses is 200 to 250 ml per day. This intake comes from a combination of free water, metabolic water and water contained in food.

Changing to a moist food or adding water to the food (moist or dry) increases the water content of food. Offering low-salt broth, meat juices or "pet drinks" has been advocated as a means to enhance water consumption; however, the long-term effectiveness of this strategy is unknown. Clean fresh water should be continuously available and readily accessible to further encourage increased water intake.

Energy Reductions in lean body mass, basal metabolic rate and physical activity are factors that decrease energy requirements as animals age. In many species, the decline in lean body mass is counterbalanced by an increase in total body fat such that obesity becomes more prevalent with age. However, studies reveal the prevalence of obesity plateaus and then declines in cats after seven years of age, whereas the prevalence of underweight conditions increases dramatically after 11 years of age. Although the prevalence of obesity declines after seven years, a significant proportion of older cats remain overweight. Both obesity and cachexia significantly increase mortality in cats over eight years of age; obese cats are nearly three times as likely to die as cats of optimal weight. Therefore, it is critical to recommend foods and feeding methods that will achieve optimal weight and body condition in individual older cats.

The energy density of foods formulated for senior cats should be between 3.5 to 4.5 kcal/g (14.6 to 18.8 kJ/g) on a dry matter basis (DMB). Very old cats should be fed energy-dense foods (4.0 to 4.5 kcal/g [16.7 to 18.8 kJ/g] DMB) and caloric intake should not be restricted, except to prevent or treat obesity.

Reasonable estimates of caloric needs in senior cats are 1.1 to 1.4 x RER (55 to 70 kcal/BW_{kg} [230 to 293 kJ/BW_{kg}]), with very old cats needing up to 1.6 x RER (80 kcal/BW_{kg} [344 kJ/BW_{kg}]). Obese cats can be managed with standard weight-control programs appropriate for adult maintenance. (See Chapter 13.)

Protein The long-term effects of feeding foods with high dietary protein levels to healthy cats are still largely unknown. High-protein foods have been implicated in the progression of renal failure. Protein restriction in foods for older cats has been advocated because of the high prevalence of renal disease in this age group and the knowledge that renal failure is rarely diagnosed until at least three-fourths of renal function is lost. The potential benefits of protein restriction include a delay in age-related renal impairment and slowed progression of subclinical renal disease.

Healthy older cats should receive sufficient protein to adequately meet protein needs and avoid protein-calorie malnutrition. Improving protein quality can fulfill any additional protein needs of older cats without increasing protein intake. Until further research defines an optimal range of dietary protein for older cats, moderate levels of dietary protein (30 to 45%) are recommended (**Table 11-3**).

Fat Although weight loss is prevalent in very old cats, obesity still affects a large proportion of the older cat population. Certain diseases associated with obesity are also common in older cats (e.g., diabetes mellitus, hypertension and heart disease). In addition, the risk of death increases nearly threefold in older obese cats (i.e., eight to 12 years). Moderate to low levels of fat are indicated to reduce the risk of obesity. However, very old cats need energy-dense foods and ample levels of essential fatty acids. Essential fatty acids (i.e., linoleic, arachidonic and possibly linolenic acid) help maintain normal skin and coat condition. As animals age, they tend to lose skin elasticity, develop epidermal and follicular atrophy and have reduced sebum secretion. Marked reduction in dietary fat (i.e., calorie-restricted or "light" foods) is not ideal for older cats unless they are obese prone. Fat should be highly digestible in foods intended for older cats. As discussed above, fat digestion declines as cats age, which may account for the weight decline noted in very old cats. Dietary fat improves the palatability of food and contributes significantly to the energy density.

Table 11-3 lists recommendations for the fat content in foods for older cats. Foods with lower fat levels are recommended for obese-prone cats, and foods with higher fat levels should be fed to thin cats (BCS <3/5) and cats with poor appetites. Essential fatty acids should be provided at or above levels recommended for young to middle-aged adults.

Fiber Dietary fiber promotes normal intestinal motility and provides fuel for colonocytes via volatile fatty acid production as a result of fiber fermentation by colonic microbes. Feeding small amounts (i.e., <5%) of soluble and insoluble fiber results in these effects. Promoting intestinal motility may benefit older cats with constipation.

High levels of dietary fiber (>10%) reduce food dry matter digestibility and dilute caloric density. Very old cats appear to need energy-dense foods; therefore, high levels of dietary fiber are not recommended except to manage obesity and fiber-responsive diseases (i.e., diabetes mellitus, colitis and constipation).

Calcium and Phosphorus After skeletal growth is complete, the nutritional requirement for calcium and phosphorus declines to levels needed by adult cats and is thought to remain relatively constant for life. Unlike the situation in people, osteoporosis is not commonly diagnosed in very old cats.

Older cats should receive foods with moderate levels of available dietary calcium (**Table 11-3**) to help maintain bone mass and possibly reduce the risk of calcium oxalate urolithiasis.

In contrast to the moderate calcium needs during aging, reduction of dietary phosphorus is commonly recommended in foods designed for older cats. The recommendation is predicated on the fact that nearly 30% of older cats may have kidney disease.

Renal insufficiency is rarely diagnosed until significant loss of renal function has occurred. Thus, a large proportion of older cats has subclinical renal damage and may benefit from reduced dietary phosphorus. It is commonly accepted that phosphorus restriction slows the progression of renal disease in cats. (See Chapter 19.) Phosphorus reduction helps decrease: 1) the renal excretory workload, 2) phosphorus retention, 3) renal secondary hyperparathyroidism and 4) the subsequent renal mineralization in cats with chronic renal insufficiency. Therefore, phosphorus levels should be reduced from levels typically found in commercial foods (See Appendix 2.) in the early nutritional management of renal disease in dogs and cats. Slowing progression of early renal disease in affected older cats will likely extend longevity. Phosphorus may be reduced to as low as 0.3% of the food (DM) for cats with overt renal disease, otherwise the general population of older cats should be fed foods containing 0.5 to 0.7% DM phosphorus. **Table 11-3** lists recommended levels of calcium and phosphorus for foods intended for older cats. Although adult cats appear to be remarkably tolerant to perturbations in dietary calcium-phosphorus ratios, a ratio between 0.9:1 to 1.1:1 maximizes availability and ratios between 0.9:1 to 1.5:1 are recommended.

Potassium The potassium requirement for older cats is thought to be greater than that for young to middle-aged cats. Factors common in older cats that support the need for increased dietary potassium include: 1) kaliuresis as a result of kidney disease, high dietary protein or high metabolic and/or dietary acid load, 2) reduced food intake and 3) increased intestinal loss. Older cats with normal appetite and renal function probably do not benefit significantly from increased dietary potassium levels. However, hypokalemia can cause signs ranging from mild lethargy to marked polymyopathy or nephropathy. Thus, increasing dietary potassium to support moderate losses may benefit some older cats. Levels as low as 0.3% result in hypokalemia when provided in high-protein or acidified foods. Dietary potassium levels for older cats should be at least 0.6% of the diet dry matter (**Table 11-3**).

Magnesium Increased losses of magnesium, similar to those seen with potassium, may affect magnesium balance in older cats. Hypomagnesemia has also been associated with refractory hypokalemia, particularly in cats with diabetes mellitus. The benefit of limiting dietary magnesium in cats is a reduced risk of struvite-mediated lower urinary tract disease. However, the risk of struvite-mediated disease is low in older cats. Furthermore, foods containing very low levels of magnesium have been associated with the development of calcium oxalate uroliths in an epidemiologic survey of cats and deficiency is known to increase urolith formation in rats. Therefore, magnesium should be provided at moderate levels (**Table 11-3**) and severe magnesium restriction should be avoided (<0.04% DM).

Sodium and Chloride Avoiding excessive sodium intake to reduce risk factors appears even more important in older cats than in young to middle-aged cats. Although the sodium and chloride requirements of older cats are not likely to be different from those of young to middle-aged adults, the prevalence of chronic diseases associated with hypertension (e.g., renal disease, hyperthyroidism, cardiac disease) increases with age. The exact prevalence of secondary hypertension in the feline population is unknown, but it appears highest in older cats. In one study, systolic arterial pressures were significantly higher in older cats than in middle-aged or younger cats. Furthermore, hypertension affects 60 to 65% of cats with renal disease and 23% of cats with hyperthyroidism. Chronic hypertension results in end-organ damage and progression of renal and cardiac disease; therefore, control of risk factors for salt-sensitive individuals is desirable. Unfortunately, accurate monitoring of blood pressure in all feline patients is uncommon and hypertension is rarely diagnosed until clinical signs are evident. Therefore, nutritional needs for sodium and chloride should be met, but excesses should be avoided.

The minimum dietary requirement of sodium for adult cats is 0.08% DM. AAFCO recommends an intake of 0.2% of the diet dry matter, or 2.5 times the minimum requirement. Some commercial moist foods exceed 1.0% dietary sodium (DM) or 12.5 times the requirement. (See Appendix 2.) Sodium intake at this level is markedly above that needed for optimal health. Chloride is now recognized as a co-determinant in salt-sensitive hypertension, thus control of dietary excess is equally important. Unfortunately, little information is available about the chloride requirement of cats. An intake of 0.2 to 0.6% DM sodium is recommended to ensure sodium adequacy and simultaneously avoid excess in older cats (**Table 11-3**). Minimum chloride levels of 0.19% are suggested; however, more typically, chloride values are approximately 1.5 times the concentration of sodium.

Urinary pH Older cats frequently have clinical or subclinical renal disease that can impair their ability to compensate for acid-base alterations resulting from metabolic and/or dietary influences. In a study in which cats were fed a food with higher urinary acidifying potential (pH 6.39 vs. pH 6.6 in the control food), older cats lost more weight, had lower red cell counts and had greater systemic acid loads than younger cats. This observation, combined with the reduced risk of struvite urolithiasis, increased risk of calcium oxalate urolithiasis and high frequency of kidney disease in older cats, supports the idea that foods fed to older cats should have a lower urine acidifying potential (i.e., higher published urinary pH averages) than foods for young and middle-aged adults. A safe range of measured urinary pH values in older cats is still between 6.2 to 6.5.

The acidifying potential of commercial foods is not typically tested for in older cats, despite the fact that older cats generate a significantly lower urinary pH than younger cats when fed the same foods. To achieve a normal urinary pH, the acidifying potential of foods for older cats should be lower than that of foods for young to middle-aged cats. Published urinary pH averages should be greater for foods for older cats than for foods for young to middle-aged adults, unless the foods have been specifically tested in old (senior) or very old cats and found to be safe. Providing food with less acidifying potential helps avoid metabolic acidosis and its complications in older cats.

Palatability and Digestibility Reduced smell or taste, the presence of oral disease or metabolic disturbances, the use of medications or a combination of factors can impair appetite and food intake in older cats. Foods for very old cats should be highly palatable and highly digestible to lessen concerns about weight loss and inadequate food intake.

Texture Oral disease is the most common disease of older cats. Age-related changes include an increased prevalence of dental calculus, periodontal disease, loss of teeth and oral neoplasia. Cats with poor oral health have more difficulty eating, and pathologic lesions may act as a portal for bacteria into the body.

Dry foods designed with dental cleansing benefits improve oral health by reducing accumulation of dental substrates (i.e., plaque and calculus) and reducing the severity of gingivitis. (See Chapter 16.) Conversely, hard dry foods may cause oral pain if fed to cats with gingivitis or periodontitis. Dry foods with softer texture, semi-moist foods or moist foods may be easier to chew. The optimal texture depends on the oral health and food texture preference of individual older cats.

Assess the Food and Feeding Method (Pages 319-320)

After assessing the cat and identifying key nutritional factors, the food and feeding method should be assessed as described for young and middle-aged cats. Foods currently being fed should be evaluated as discussed previously. (See Chapter 1.)

- Ensure feeding tests have been conducted (i.e., review package label for statement).
- Compare the nutrient content of the current food with the cat's nutrient needs and the key nutritional factors. (See Table 11-3 and Appendix 2.)

- Identify discrepancies between the key nutritional factors and the food currently fed.

It may not always be necessary to change the food and feeding method when managing healthy geriatric cats. However, a thorough evaluation includes verification that an appropriate food and feeding method are being used. Older cats should be reevaluated at each examination because nutrition and health needs change with disease status, risk factors and overall health.

Determine a Feeding Plan (Page 320)

Older cats are more prone to weight loss, cardiac disease, renal disease and metabolic aberrations and usually have a decreased activity level than younger cats. The feeding plan should be based on the information obtained in the assessment as well as detected risk factors. Nutritional surveillance and therefore the number of contacts per year should be increased for older cats. Although goals remain the same as those listed at the beginning of this section, each animal should be evaluated individually.

Select a Food (Page 320)

Several nutrients are of particular interest because of their role in the management of health risks or age-related diseases or because older cats poorly tolerate and adapt to wider variations in nutrient concentrations. The nutrient profile of the current food should be compared with the appropriate key nutritional factors to determine if the food is satisfactory.

A different food should be selected if discrepancies are found between the recommended levels of key nutritional factors and those in the current food (**Table 11-3**). Appendix 2 lists nutrient levels for several readily available commercial cat foods, or the product manufacturer can be contacted for the same information. Foods that have passed AAFCO or similar feeding trials for adult cats are recommended.

An important goal when managing the nutrition of older cats is to ensure proper food intake. There is little need to change the form of food a cat eats well simply because of age. In fact, some cats will refuse to eat a new form or texture of food. However, cats with poor intake may benefit from changing food forms if the new food is more palatable and easier to chew.

Determine a Feeding Method (Page 320)

As mentioned above, healthy senior cats may be fed free choice, meal fed or fed by a combination of methods. Obese cats should be offered measured amounts of food. The measured quantity may be fed in meals or dispensed at one time to allow continuous access throughout the day. Underweight cats should be fed free choice. Only dry and semi-moist foods should be fed free choice and these foods are typically less palatable than moist foods. Older cats may have reduced olfaction and taste perception; therefore, it may be preferable to feed moist and warm foods to encourage food intake. Providing dry foods for free-choice consumption and moist foods in several meals throughout the day may optimize food intake. Adding broth or canned meat juices to dry foods may enhance food and water intake in geriatric cats.

Although most cats do not experience digestive upsets with typical food changes, a gradual transition to a new food may benefit older cats. Progressively exchanging the new food for the usual food over four to seven days will minimize untoward effects and food refusal.

Reassessment (Page 320)

Veterinarians should examine and conduct a nutritional assessment of geriatric cats regularly. The frequency of monitoring depends on the overall health of the cat and the presence or absence of chronic diseases. Annual veterinary examinations are usually recommended for older cats, whereas biannual check-ups are recommended for very old cats.

The owner should evaluate body condition every two to four weeks. Although lean body mass tends to decline as cats reach extreme geriatric age (older than 16 years), significant loss of muscle mass or body weight warrants immediate evaluation by a veterinarian. Owners should also monitor daily food and water intake and stools and urination. Any persistent change, whether increased or decreased, should prompt the veterinarian to assess the cat and perform diagnostics as indicated.

Dental disease is the most frequent diagnosis made in geriatric cats. Therefore, a dental health program should be part of every older cat's preventive health care plan. (See Chapter 16.)

REPRODUCING CATS (Pages 320-328)

Domestic cats generally reach puberty by six to nine months of age. However, the best age for breeding is between one and one-half to seven years of age. Before 10 to 12 months of age, queens are still growing and must meet nutritional demands for their own growth as well as for their kittens, if pregnant. Queens older than seven years should not be bred due to reproductive complications, irregular estrous cycles and reduced litter size. The reproductive stage of the queen can be divided into four periods: 1) estrus and mating, 2) gestation, 3) lactation and 4) weaning. Reproducing cats have significantly altered nutritional needs compared with maintenance requirements, especially during late pregnancy and lactation. During reproduction, energy requirements increase and the minimum requirements for certain nutrients exceed even those required for growth.

The objectives of a good feeding program for feline reproduction are to optimize: 1) the health and body condition of the queen throughout the various reproductive periods, 2) reproductive performance and 3) kitten health and development through the weaning period. Key indicators of optimal reproduction are ease of conception, a low rate of fetal and neonatal death, normal parturition, maximum litter size, adequate lactation and an optimal rate of growth of healthy kittens. Providing adequate nutrition throughout reproduction has long-range health implications for the offspring. Immune function is impaired for life in animals born to nutritionally deficient dams. Meeting the nutritional needs of reproducing queens is critical to successful conception, delivery and weaning of healthy kittens.

Lactation begins at parturition and lasts six to 12 weeks depending on breed, kitten growth rates and management practices. Most kittens are sufficiently mature at eight weeks of age to maintain adequate food intake for optimal development. Purebred kittens are typically weaned later than domestic shorthair kittens. Lactation is the most demanding stage of reproduction. The queen must maintain its own nutritional needs and provide nutritionally complete, energy-dense milk to support the needs of growing kittens. Consequently, queens should enter lactation with sufficient energy stores to support needs above those supplied by daily food intake. Poor lactation performance is common without these reserves. Thus, successful lactation depends on appropriate nutritional management during the pre-breeding period, gestation and lactation.

Assessment (Pages 321-327)
Assess the Animal (Pages 321-326)
ESTRUS AND MATING A history and physical examination should precede breeding to assess problems that may interfere with conception, parturition and lactation. Queens should be at ideal body weight at mating (BCS 3/5). Small variations in body condition can be corrected during pregnancy; however, breeding should be delayed in cats that are significantly under- or overweight (BCS <2/5 or >4/5). Both obesity and undernourishment can be detrimental to reproduction.

Tomcats should also be healthy and in optimal body condition (BCS 3/5); however, abnormalities associated with moderate deviations from ideal (BCS 2/5 to 4/5) have not been reported. The level of activity required during the breeding period should be ascertained. Single matings result in minimal changes in energy needs, whereas multiple matings may require an increase in the amount of food fed or an adjustment in feeding method, based on body condition.

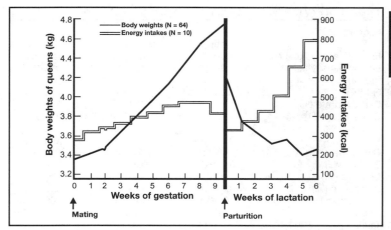

Figure 11-1. Body weight and energy intake during gestation and lactation in queens. Unlike bitches, which have a dramatic increase in energy intake and body weight during the last trimester, queens have a regular linear increase in both body weight and energy intake throughout gestation. Mobilized stores of body fat provide needed energy during lactation, which accounts for weight loss during this period. Food intake parallels lactation and peaks during the sixth to seventh week.

PREGNANCY One of the early indicators of successful breeding and conception is a steady gain in body weight. Weight gain increases linearly from conception to parturition in queens (**Figure 11-1**). This pattern is different from that of most other species, which experience small increases in body weight until the last third of gestation when weight gain and energy intake greatly increase. Weight gain in early pregnancy is not associated with significant growth of reproductive tissues or conceptuses but appears to be stored in energy depots (presumably as fat) to support lactation. Mean weight gain during gestation is approximately 40% of the pre-mating weight (900 to 1,200 g for a litter of average size).

At parturition, only 40% of the weight gained by queens during gestation will be lost, whereas bitches should return to pre-breeding weight. The remaining 60% of prepartum weight gain will be used during lactation to sustain milk production. Poor nutrition may lead to failure to conceive, fetal death, fetal malformations and underweight kittens. Queens underweight at parturition may subsequently experience poor lactation performance and inability to maintain body condition.

Overnutrition or obesity (BCS 5/5) has an equally negative effect on pregnancy outcome. Stillbirths, dystocia and cesarean sections occur more frequently in obese queens than in cats at ideal body condition.

Table 11-5. Key nutritional factors for reproducing cats.*

Factors	Recommended food levels		
	Mating	**Gestation**	**Lactation**
Energy density (kcal ME/g)	4.5-5.0	4.0-5.0	4.0-5.0
Energy density (kJ ME/g)	19-21	17-21	17-21
Protein (%)	30-45	35-50	35-50
Fat (%)	10-30	18-35	18-35
Carbohydrate (%)	–	≥10	≥10
Crude fiber (%)	<5	<5	<5
Calcium (%)	0.6-1.0	1.0-1.6	1.0-1.6
Phosphorus (%)	0.5-1.0	0.8-1.4	0.8-1.4
Ca/P ratio	1:1-1.5:1	1:1-1.5:1	1:1-1.5:1
Sodium (%)	0.2-0.6	0.3-0.6	0.3-0.6
Chloride (%)	≥0.3	≥0.45	≥0.45
Potassium (%)	0.6-1.0	0.6-1.2	0.6-1.2
Magnesium (%)	0.04-0.1	0.08-0.15	0.08-0.15
Copper (ppm)	≥5	≥15	≥15
Taurine (ppm) (dry kibble)	1,000	1,000	1,000
Taurine (ppm) (moist)	2,500	2,500	2,500
Average urinary pH	6.2-6.5	6.2-6.5	6.2-6.5

*Dry matter basis. Concentrations presume an energy density of 4.0 kcal/g. Levels should be corrected for foods with higher energy densities. Adjustment is unnecessary for foods with lower energy densities.

LACTATION The queen and kittens should be weighed within 24 hours after parturition. The queen should weigh 700 to 900 g above the prebreeding weight and each kitten should weigh approximately 100 g.

The queen's appetite, which is reduced 24 to 48 hours before parturition, should return to normal or to an increased level within 24 hours of parturition. All kittens should nurse soon after parturition and within the first six to eight hours to ensure transfer of colostral antibodies. Neonatal kittens may not absorb immunoglobulins after 12 hours postpartum.

Milk yield depends on litter size and stage of lactation, with peak lactation occurring at three to four weeks.

Continuous weight gain by the kittens is the best indicator of the queen's lactation performance. Neonatal kittens should gain between 10 to 15 g daily. Gains less than 7 g/day are inadequate.

KEY NUTRITIONAL FACTORS Although most foods appropriate for growing kittens are deemed adequate for female reproduction, complete and balanced foods specifically designed to support gestation/lactation should be fed. **Table 11-5** summarizes key nutritional factors for reproducing cats. The following section describes these key nutritional factors in more detail.

CATS

Water Water is important for normal reproduction. Expansion of extra-cellular fluid compartments and maternal and fetal tissues during pregnancy increases the need for water. Water is particularly important for milk production during lactation. Water needs for lactating queens vary according to maintenance needs, type of food (moist vs. dry) and the rate of milk production. Although specific levels of water intake have not been established, reproducing queens should be provided with ample potable water at all times. Some queens are reluctant to leave the nest box during the first few days after parturition. Water intake should be encouraged by placing water very near the enclosure to allow easy access. Feeding moist foods or adding water to food can improve water intake.

Energy/*Energy Requirements During Estrus and Mating* The energy requirements of queens during mating do not appear to be significantly different from those of young to middle-aged adults. However, during behavioral estrus, queens typically reduce food intake and body weight may decline. Food intake and body weight rebound upon cessation of estrus. A nutrient-dense (4.5 to 5.0 kcal/g food [18.8 to 20.9 kJ/g food]), palatable food is appropriate to ensure optimal body condition at conception. Intact female cats typically require more calories than neutered housecats. The DER for sexually intact cats is 1.4 to 1.6 x RER.

Breeding male cats that are used infrequently or in small catteries have energy requirements similar to those of intact young to middle-aged cats (1.4 to 1.6 x RER). Tomcats that are used extensively for breeding may have difficulty maintaining proper body condition due to increased energy expenditure or, more often, reduced food intake. The stress of travel, new environments, social interactions and preoccupation with breeding may contribute to inappetence. These tomcats should be managed similarly to cats that are very active or under stress. Energy-dense (4.5 to 5.0 kcal/g food [18.8 to 20.9 kJ/g food]), highly digestible foods (>85%) with above average palatability should help these cats maintain ideal body condition (BCS 3/5) and activity.

Energy/*Energy Requirements During Pregnancy* One of the most important changes in nutrient requirements of gestating cats is an increase in energy requirement. Although many essential nutrients are required at increased levels during gestation, dietary energy is often the most limiting "nutrient."

As mentioned previously, energy intake and weight gain increase linearly from conception to parturition in queens (**Figure 11-1**). However, food intake normally fluctuates slightly throughout gestation. Reduced food intake occurs approximately two weeks after mating and is thought to occur in association with fetal implantation at about Day 15 postconception. Energy intake increases then peaks between six to seven weeks of gestation. A second decline in food intake occurs during the last week

Table 11-6. Energy requirements of pregnant queens.

Body weights		kcal ME per day		kJ ME per day	
kg	lb	At 90 kcal/ BW_{kg}	At 100 kcal/ BW_{kg}	At 375 kJ/ BW_{kg}	At 420 kJ/ BW_{kg}
2	4.4	180	200	750	840
3	6.6	270	300	1,125	1,260
4	8.8	360	400	1,500	1,680
5	11.0	450	500	1,875	2,100
6	13.2	540	600	2,250	2,520
7	15.4	630	700	2,625	2,940
8	17.6	720	800	3,000	3,360

Key: ME = metabolizable energy, BW_{kg} = body weight in kilograms.

of gestation. These transient declines in food intake do not appear harmful. However, inadequate food intake over the course of gestation may impair weight gain, the subsequent lactation and kitten health. The recommended energy allowance for gestation is 25 to 50% above maintenance levels or approximately 90 to 110 kcal/BW_{kg}/day (376 to 460 kJ/BW_{kg}/day), although total caloric intake may increase as much as 70% above maintenance. The increased need for energy can be met by providing 1.6 x RER at breeding with a gradual increase to 2 x RER at parturition (**Table 11-6**). Energy requirements sometimes exceed the recommended energy allowance due to individual cat variation and increased energy needs of queens with large litters. Therefore, free-choice feeding allows queens to adjust food intake as needed to meet the energy requirement for gestation.

Feeding energy-dense foods (ME = 4.0 to 5.0 kcal/g DM [16.74 to 20.9 kJ/g DM]) helps meet the energy needs of pregnant queens, especially during late gestation when the gravid uterus reduces stomach capacity.

Energy/*Energy Requirements During Lactation* Lactation is the most energy-demanding stage of a cat's life. Peak milk production typically occurs at three to four weeks of lactation and, theoretically, peak energy demand should occur concurrently. However, actual peak energy demand occurs at six to seven weeks postpartum when energy requirements may exceed 250 kcal/BW_{kg}/day (1.05 MJ/BW_{kg}/day) or 2 to 6 x RER (**Table 11-7**). Observed energy intakes of queens and their litters during lactation increase from 90 kcal/BW_{kg}/day (376 kJ/BW_{kg}/day) at parturition to 270 kcal/BW_{kg}/day (1.13 MJ/BW_{kg}/day) at Week 7. The discrepancy in the timing of peak lactation and peak energy demand is due to combined food consumption by kittens and the queen. Kittens begin eating the queen's food in increasing amounts from three weeks of age until weaning. Therefore, the above estimates of energy requirement for the lactating

Table 11-7. Daily energy requirements of lactating queens over the lactation period.*

Weeks of lactation	Daily energy requirements		
	Factor x RER	kcal/BW$_{kg}$	kJ/BW$_{kg}$
1	2.3	115	481
2	2.5	125	523
3	3.0	150	628
4	3.5	175	732
5	4.0	200	837
6	5.0	250	1,046

Key: RER = resting energy requirement, $70(BW_{kg})^{0.75}$ or $30(BW_{kg}) + 70$.
*Based on average queen at parturition (3.8 kg) nursing four to five kittens. These values represent average energy requirements for lactating queens. Individual animal variation and litter size may alter total daily energy needs. (See Appendix G, Small Animal Clinical Nutrition, 4th ed., pp 1027-1036.)

queen include energy consumed by the queen and its kittens. When energy intake was measured for the queen alone, the energy requirement at Week 6 of lactation was 229 kcal/BW$_{kg}$/day (962 kJ/BW$_{kg}$/day). Within large litters, up to 50% of the total energy was consumed by kittens, increasing the total energy consumption (i.e., kittens and queens) to as high as 306 kcal/BW$_{kg}$/day (1.28 MJ/BW$_{kg}$/day). Even with these large increases in energy intake, queens will continue to lose weight during lactation and return to pre-mating weight by weaning. Queens that lose excessive weight are prone to lactation failure. **Table 11-7** is a guide to estimate the energy requirements of lactating queens. However, it is preferable to feed lactating queens free choice because the wide variation in energy needs makes accurate prediction difficult.

The high energy demands during lactation require a marked increase in total food intake. Feeding an energy-dense food (4.0 to 5.0 kcal ME/g DM, [16.74 to 20.9 kJ ME/g]) helps meet these demands without overwhelming gastric capacity.

If kittens are encouraged to eat a solid food beginning at three weeks of age, the energy demands placed on the lactating queen will decline as kittens increasingly obtain nutrition from solid food. Maintenance energy levels are sufficient for queens at ideal body condition after the kittens are weaned. Queens that have lost excess body weight during lactation should be provided additional food to restore ideal body condition (BCS 3/5).

Protein Protein synthesis in the queen is greatly increased during gestation. In addition, protein quality and quantity are important to provide essential amino acids for growth and development of the fetuses.

Considering the varying nutrient availability in typical pet food ingredients compared with purified foods, protein levels at or above 35% DM are recommended for gestating queens (**Table 11-5**). Animal-based proteins are preferred as the major contributor to dietary protein because they generally have greater digestibility and more desirable amino-acid profiles for cats. Deficiency of protein during pregnancy may result in lower birth weights, higher neonatal mortality and impaired immunocompetency in the kittens. In addition, feeding queens protein-restricted foods during late gestation and lactation results in delayed home orientation (i.e., ability of kittens to orient to and return to the nest), aberrant locomotor development and decreased emotional responsiveness in the kittens.

During lactation, queens increase protein synthesis to supply milk with protein concentrations suitable for growth. Near optimal performance is achieved with foods containing 30% DM crude protein. Additionally, food intake and kitten growth rates are higher at dietary protein levels of 30% DM. Because of variations in food digestibility and ingredient quality and the goal to promote optimal reproductive performance, the recommended crude protein allowance for lactation is at least 35% DM (**Table 11-5**). The protein sources in commercial foods should be highly digestible and have high biologic value. Animal-based proteins should provide the major source of amino acids and protein for lactating queens.

Inadequate protein concentrations result in poor lactation and kitten growth. Queens fed foods containing 20% DM protein had lower hematocrit values at Week 6 of lactation compared with queens fed foods with higher protein levels.

Taurine Taurine is required for normal reproduction and fetal development. Taurine deficiency in gestating queens may result in fetal death near the 25th day of gestation, abortions throughout gestation, fetal deformities and delayed growth and development. However, the taurine requirement for gestation is similar to that for other lifestages (i.e., a minimum of 0.1% DM taurine in dry foods and 0.2% DM in moist foods).

Fats and Essential Fatty Acids High-energy foods are beneficial because of the increased energy demand during gestation. Fat delivers 2.25 times the number of calories as the same amount of protein or carbohydrate; therefore, fat represents an important source of calories.

Increasing dietary fat from 15 to 27% of the food dry matter will: 1) increase the number of kittens per litter, 2) decrease kitten mortality from more than 20 to 9% and 3) improve reproductive efficiency in queens (more litters/year). For optimal reproductive performance, foods for gestating queens should contain at least 18% DM fat, although foods with lower levels of fat have been successfully fed during gestation.

Arachidonic acid at 0.04% of the dietary energy supports normal reproduction in queens. However, lower levels have been used when interference from n-3 fatty acids is avoided. Current AAFCO allowances for linoleic and arachidonic acid are appropriate for gestating cats. (See Appendix J, Small Animal Clinical Nutrition, 4th ed., pp 1048-1063.)

As in gestation, the high energy demands of lactation are best met by feeding energy-dense foods. Moderate- to high-fat foods enhance lactation performance in queens. Queens have improved reproductive performance when switched from lower fat foods (12 to 15% DM) to moderately high-fat foods (21 to 27% DM). Kitten survival and growth rate and reproductive efficiency of queens are improved if queens are fed higher fat foods throughout lactation. For optimal reproductive performance, foods for lactating queens should contain between 18 and 35% fat (**Table 11-5**).

In addition to meeting the essential fatty acid needs of lactation and aiding in absorption of fat-soluble vitamins, higher levels of fat in foods for lactation increase the food's energy density. Thus, smaller amounts of food can be consumed to meet the queen's energy demands. Other nutrients in the food should be balanced to the higher energy content of energy-dense foods (>4.5 kcal/g DM [18.8 kJ/g DM]).

Minimum essential fatty acid requirements for lactation do not appear to differ significantly from those of gestation. However, a dietary source of docosahexaenoic acid (DHA, 22:6n-3) is required for normal development of retinal function in nursing kittens. Milk concentrations of DHA parallel dietary intake. Therefore, DHA should be included in foods fed to lactating queens. Common ingredients such as fish and poultry meal represent a source of DHA in the diet of queens.

Calcium and Phosphorus Calcium and phosphorus are required at levels greater than maintenance to support fetal skeletal development and lactation. Recommended dietary calcium levels are 1.0 to 1.6% DM while maintaining a normal calcium-phosphorus ratio (1:1 to 1.5:1). These levels are typically found in commercial cat foods; therefore, supplementation is rarely indicated except for homemade foods.

Queens with a history of eclampsia may benefit from calcium supplementation during subsequent lactations; however, the efficacy of this treatment in preventing recurrence has not been reported. Occasionally, pregnant queens may develop hypocalcemia, hypoglycemia or both one to two weeks before parturition. Initially, affected queens should be treated with intravenous calcium and glucose solutions, followed by oral calcium supplements throughout the remainder of gestation. After treatment, gestation proceeds normally and calcium supplementation is not typically required during lactation.

Carbohydrate Although a true carbohydrate requirement for cats has not been demonstrated, carbohydrates apparently protect against weight loss in queens during lactation. Carbohydrates spare protein necessary to sustain blood glucose concentrations in queens and provide a substrate for lactose during milk production. Even with an abundant supply of dietary protein, providing some dietary carbohydrate improves lactation performance. Until further studies define optimal levels of carbohydrates for lactation, at least 10% DM carbohydrate should be included in foods for lactating queens.

OTHER NUTRITIONAL FACTORS/Magnesium Dietary magnesium levels of 0.08 to 0.15% DM are recommended in foods for reproducing female cats.

OTHER NUTRITIONAL FACTORS/Copper Copper requirements for growth and reproduction are thought to be approximately 5 mg/kg food. However, copper deficiency has been reported to occur in queens fed a food containing 15 mg copper/kg food, supplied, in part, by copper oxide. Copper from copper oxide is poorly available. The combination of poorly available copper from the diet and competition from high levels of dietary zinc, iron, calcium and phytate significantly impair copper availability. Clinical signs of copper deficiency include fetal death and abortions, achromotrichia, arthrogryposis, fusion of digits, craniofacial deformities and cerebral dysgenesis. For this reason, copper levels of 15 mg/kg food (DM) from an available source have been recommended for queens eating dry foods.

OTHER NUTRITIONAL FACTORS/Digestibility Foods with dry matter digestibility greater than 85% are better suited than less digestible foods for pregnant cats because: 1) nutrient needs increase as pregnancy progresses and 2) increased abdominal fullness as the pregnancy progresses may impair the queen's ability to ingest adequate amounts of nutrients, especially if the food is poorly digestible.

Dry matter digestibility and availability of foods intended for lactation should also be above average because of the high nutritional demands of lactation. Apparent dry matter digestibility of 85% or greater is desirable. By comparison, the digestibility of fresh meat (uncooked) is 96%.

OTHER NUTRITIONAL FACTORS/Urinary pH Highly acidified foods should be avoided during gestation because metabolic acidosis may impair bone mineralization in adult cats and kittens, which can be especially detrimental to developing fetuses. Foods designed to produce average urinary pH values between 6.2 to 6.5 appear to be safe.

Assess the Food (Page 326)

Assessment of the food and the individual animal allows for development of an appropriate feeding plan. The following general steps should be followed: 1) Compare the current food's nutrient profile with the nutrient recommendations for reproduction (**Table 11-5**). Appendices 2 and 3 list the key nutritional factors in selected commercial foods and treats. (Also see Appendix M, Small Animal Clinical Nutrition, 4th ed., pp 1084-1091.) The same information can be obtained from the pet food manufacturer. 2) Identify discrepancies between the key nutritional factors and the cat's current intake. 3) Ensure the food is appropriate for reproduction based on AAFCO-type feeding trials.

The nutritional adequacy statement on the pet food label should indicate if the food is complete and balanced for reproduction, gestation and lactation, or all lifestages. The statement of nutritional adequacy should also be based on animal feeding trials. Lactation is the most nutritionally demanding lifestage for cats; therefore, the manufacturer must prove the food's nutrients are available through successful animal feeding. After it has been established that the minimum standards have been met, the food's nutrient levels should be compared with the key nutritional factors for lactation (**Table 11-5**). Any discrepancies between target levels and the levels in the food are then noted. Particular attention should be devoted to the protein source, food digestibility and energy density. Foods with an energy density in the upper range should be chosen (4.7 to 5.0 kcal ME/g DM [19.7 to 20.9 kJ ME/g DM]) for queens with large litters and those with marginal weight gain during pregnancy. Queens with small litters and those prone to obesity may benefit from foods with a lower caloric density (4.0 to 4.5 kcal ME/g DM [16.74 to 18.8 kJ ME/g DM]) to avoid excessive weight gain and obesity.

Food assessment during lactation also includes assessment of lactation performance. Evaluation of kitten growth rate and queen weight loss can point to nutritional inadequacies. Nursing kittens should gain approximately 100 g/week or 10 to 15 g/day. Weight gains less than 7 g/day require immediate evaluation of the food, the queen and the kittens. Queens normally lose weight during lactation and return to within 2% of their pre-breeding body weight by weaning (**Figure 11-1**).

If either the queen's rate of loss or the kittens' growth rate is inadequate, the food and feeding method should be carefully reviewed. If inadequacies exist, a more appropriate food should be selected. Supplements should not be given to improve lactation performance. Supplements, unless carefully balanced to the nutrients in the food, can unbalance a food or impair availability of other nutrients.

Assess the Feeding Method (Pages 326-327)

It may be necessary to alter the feeding method when managing reproducing cats, especially in late-term pregnancy, when a queen is carrying a very large litter and during lactation. Evaluation of current feeding meth ods with foreknowledge of reproductive demands will allow for development of a rational feeding plan.

Free-choice feeding is the preferred method for reproducing female cats. Meal size and therefore caloric intake may be limited as the uterus and fetal mass occupy much of the abdominal cavity and limit gastric capacity. The queen's energy needs may increase fourfold over maintenance requirements during peak lactation. Providing food free choice allows reproducing female cats to consume sufficient calories in multiple small feedings. Cats may also be fed multiple meals (three to four/day) using the recommended energy allowances in **Table 11-6**. However, food intake should not be limited unless obesity becomes a problem. **Table 11-7** lists estimates of average energy intake during lactation.

Obese queens (i.e., those with heavy fat accumulations over the ribs and bony prominences [BCS 4/5]) should be fed controlled amounts of food during gestation; however, they should not be fed to reduce weight. Obesity increases the risk of dystocia and kitten mortality; therefore, careful weight management before breeding and monitoring during gestation is important.

Clean water should be available at all times. Food and water should be placed within easy reach for the queen. Food should be placed directly in or very near the nest box during the first few days after parturition, when many queens refuse to leave the nest box. Methods to improve food intake include adding water or moist food to dry food to enhance palatability and increase water intake.

Determine a Feeding Plan (Page 327)

The feeding plan should be tailored to meet the needs of the individual cat based on unique variations in genetics, environment, litter size and health status.

Select a Food (Page 327)

Queens should be fed a food appropriate for gestation and lactation at or before mating. This recommendation is generally accomplished by: 1) selecting a food that has been proved suitable for reproduction by animal feeding trials (i.e., see the product's label) and 2) choosing a food that provides recommended levels of key nutritional factors (**Table 11-5**). Although, nutritional demands are greatest in the last one-half to one-third of pregnancy, conception rate and in utero fetal viability are markedly impaired in queens fed foods with marginal nutrient content and availability at breeding and early gestation. Changing to a new food more suitable for gestation and lactation before conception avoids any reduction in food intake or GI upsets during the critical time of conception and implantation, improves any marginal nutrient stores and typically increases energy intake.

The food form selected for reproducing female cats bears consideration. Semi-moist products should be selected cautiously because many
produce urinary pH values below desired levels. Dry foods are more
nutrient dense as fed and have higher carbohydrate levels than moist
foods. Dry foods may benefit queens undergoing rapid weight loss and
those spending little time eating. Conversely, moist foods often have higher fat levels and provide additional water to support lactation. The added
water also improves palatability; therefore, queens may spend more time
eating. Dry and moist food types each have advantages; therefore, many
breeders choose to feed both forms during reproduction.

Intact male cats in heavy service and those stressed during breeding
should be fed foods with high energy density (4.5 to 5.0 kcal/g DM [18.8
to 20.9 kJ/g DM]). Otherwise, foods appropriate for young to middle-aged
cats are adequate (**Table 11-3**). Male cats used in harem-breeding programs are typically fed the same foods as the queens. Although the vitamin and mineral levels of these foods are typically well in excess of the
male cat's needs, the high energy density may be beneficial.

Determine a Feeding Method (Page 327)

Free-choice feeding is preferred for most queens during reproduction.
Overweight or obese-prone queens should be fed measured portions
adequate for weight maintenance. If both dry and moist foods are fed, it
may be desirable to feed dry foods free choice and provide multiple meals
of moist foods. Obese-prone cats should be fed three to four meals per
day in controlled portions. Kittens should be allowed access to the
queen's food, which they typically begin eating at three weeks of age.
Kittens may need to be fed away from the queen if the queen is fed portion-controlled amounts of food.

Some queens with strong maternal instincts are reluctant to leave the nest
box. When this occurs, food and water should be placed in the immediate
vicinity of the nest box. Use care when placing water bowls near neonates to
avoid accidental drowning. If food intake does not improve, the kittens may
be removed from the queen for short periods three to four times a day.

Reassessment (Pages 327-328)

Male and female cats should be reassessed before every reproductive
cycle. Females should have returned to optimal body weight and condition
(BCS 3/5) before the next breeding. Oral health should be optimal and vaccinations and parasite control should be completed before the next reproductive cycle. The last reproductive performance should be evaluated and
compared with previous performances and the cattery average. If performance was suboptimal, a detailed review of genetic selection, husbandry and
nutritional management should be completed to identify deficiencies.

Modifications can be then be incorporated to improve subsequent reproductive outcomes.

Monitoring the queen during gestation should include weekly assessment of food intake and body weight. Body condition scoring is particularly important in assessing weight gain during gestation. Inadequate nutrition and poor weight gain may be overlooked if total body weight and the queen's expanding abdomen are the only criteria used to monitor weight gain. If underfed, the queen may continue to gain weight as the kittens grow, but fail to develop the energy reserves needed for lactation. Body condition scoring during gestation should ignore the abdominal component of the scoring process and allow for slight increases in body fat. When assigning body condition scores to pregnant queens, the areas of focus include muscle mass and fat covering the ribs and bony prominences. The queen and each kitten should be thoroughly evaluated at parturition. Average weight loss at parturition is 6 to 14% (254 to 638 g) of the prepartum weight, depending on litter size. The remaining 700 to 850 g of gain will be used to sustain normal lactation. Evaluation of gestational performance should include: 1) the queen's weight record, 2) litter size, 3) kitten birth weights, 4) kitten growth rates, 5) kitten vigor, 6) mortality rates and 7) congenital defects. Although stools may normally vary from soft to firm during reproduction, stool quality should be monitored. Constipation and diarrhea are always considered abnormal and should be evaluated and treated as needed.

Reassessment of lactating queens is similar to that of pregnant queens. Body weight and condition should be evaluated after parturition and weekly thereafter. Kittens should exhibit steady weight gain, have good muscle tone and suckle vigorously. Young kittens are quiet between feedings. Kittens are often restless and cry excessively if milk production is inadequate. Gastric distention is not a good indicator of adequate nursing. Aerophagia can give the appearance of gastric fullness in kittens, despite inadequate milk intake.

Kitten mortality reportedly varies from 9 to 63% depending on the source of cats and the cattery. Breeders should compare reproductive performance of each queen to the cattery standard. Several genetic, husbandry and nutritional factors may cause high kitten mortality. If kitten death or cannibalism rates are high, all three areas should be investigated thoroughly.

WEANING (Pages 328-329)

Weaning is usually a gradual process that begins with the queen avoiding the kittens and kittens eating increasing amounts of solid food. This phase begins when kittens are three to four weeks old and is complete at six to 10 weeks of age. At three to four weeks of age, kittens begin to eat solid foods although approximately 95% of their caloric intake is still provided by the queen's milk. By five to six weeks of age, kittens eat nearly 30% of their caloric requirement as solid food and the remainder as milk. A progressive intake of solid food continues until the kittens are completely independent of the queen. Most domestic shorthair kittens are weaned by six weeks of age, whereas purebred kittens are usually weaned around eight to nine weeks of age. Later weaning allows more time for kitten growth and immune system maturation, which may help reduce kitten mortality in the postweaning period.

Restricting the food of queens that are abruptly removed from their kittens and those that are heavy milk producers a day or two before weaning reduces energy available for milk production, thereby minimizing mammary gland engorgement. A commonly used weaning schedule follows: kittens and food are withheld from the queen the day before weaning. The kittens are returned at the end of the day and allowed to nurse. The following day, the kittens are removed and the queen is given one-fourth of its ration. Food amounts are then gradually increased over the next three days to pre-breeding levels.

Weaning can be a stressful event in a kitten's life. Transition to independent feeding, greater environmental exposure and waning maternal antibodies result in reduced immune defense. These factors contribute to increased morbidity and mortality in the postweaning period. Proper nutrition and careful husbandry can reduce these rates markedly.

Recommended nutrient allowances for weanling kittens are similar to those for older growing kittens. Energy requirements for weanling kittens are between 200 to 250 kcal/BW$_{kg}$ (837 to 1,046 kJ/BW$_{kg}$). The stomach volume of kittens is small; therefore, feeding energy-dense foods helps meet the higher energy needs of weanling kittens without exceeding gastric capacity.

At the onset of weaning, kittens should be offered moist foods or dry foods moistened with water or milk replacer. The food should be moistened until it forms a soft but not liquid gruel. Kittens at this stage lap at food but do not prehend food. By six to eight weeks of age, most kittens have learned to eat solid, unmoistened foods; therefore, a gruel is no longer necessary. The food should be highly digestible (i.e., apparent dry matter >80% and protein digestibility >85%), and complete and balanced for growth and reproduction. Semi-moist foods that promote a highly acidic urinary pH should not be fed as the sole food source for growing kittens. High levels of dietary acid may lead to metabolic acidosis and impaired bone mineralization. Limited amounts of semi-moist treats are acceptable.

The weaning process will be less stressful if kittens are initially offered the same food that will be fed after weaning. Not only will kittens readily recognize this diet as food, but GI upsets associated with food transitions will be avoided. Water should be accessible to kittens when they are about three weeks old.

Kittens should have water and food available at all times in addition to free access to the queen. Food and water should be easily accessible and offered in broad shallow pans. Food should be replenished three to four times daily. High-moisture foods begin to spoil and harbor high levels of bacteria when left at room temperature for prolonged periods. Ideally, food should be warmed to about 38°C (100°F) or at least brought to room temperature. Kittens first eat by accident, as they step into food and then ingest it during grooming. This process can be hastened by smearing small quantities of food around a kitten's mouth.

Monitoring consists of daily evaluation of physical appearance, activity, stool quality and food intake. Normal kittens should be weighed and their body condition assessed weekly; they should continue to grow at approximately 100 g/week. Gender differences in growth rate are now evident; female kittens are normally smaller than males. Kittens should demonstrate increasing activity and social and exploratory behavior. After a meal, the kittens' abdomen should be well rounded but not overly distended. Crying in neonates and older kittens usually indicates discomfort (e.g., cold, hunger, pain, disease or isolation).

The queen still consumes the kittens' feces to keep the nest box clean early during this phase. At about four weeks, the kittens begin to defecate outside the nest box and stools can be readily monitored. Kittens eating solid foods should have soft-formed stools, whereas those eating predominantly milk will have pasty yellow to light brown stools. It is vital during this phase to practice good cattery husbandry and monitor kittens closely for disease. Weaning is a stressful event and outbreaks of diarrhea and disease are very common. Growth rate is universally impaired in sick and malnourished kittens.

GROWING KITTENS (Pages 329-337)

Kittens usually depend on the queen to provide food during the neonatal or nursing period (discussed first below). Specific information about raising orphan kittens and using milk replacers is found in Appendix 1 and Small Animal Clinical Nutrition, 4th ed. (See Appendices E and K, pp 1012-1019 and 1064-1072.) The transition from queen's milk to solid food (weaning period) was discussed above. Feeding kittens from two to 12 months of age is discussed below, after the section on kittens in the neonatal period.

Growing Kittens: Neonatal Period (Pages 329-333)

Proper nutrition of the queen during gestation and lactation, the behavior and health of the queen and good neonatal care are important to achieving a successful transition from fetal life to the nursing period.

Assessment (Pages 329-333)

ASSESS THE ANIMAL/History Persons who raise kittens (i.e., orphaned, fostered and normal) should be encouraged to keep logbooks of all data that may provide information about the health and nutritional status of kittens and the reproductive performance of the queen. Records should include food intake, body weight, body temperature and stool characteristics, especially during the first two weeks postpartum. Changes in kitten behavior, activity and other indicators of normal development (e.g., opening of eyes, eruption of teeth and coat quality) may prove useful as well. In some instances, it may be helpful to mark kittens (e.g., with nail polish or nontoxic dyes) to differentiate individual kittens.

It is particularly important that good records be maintained for orphaned and foster kittens. Orphaned kittens are hand-raised kittens, whereas foster kittens are those raised by a queen other than their mother. Successful management of these kittens depends on the quick recognition and correction of health and management problems. Parameters such as weight gain, daily food intake, stool characteristics and kitten vigor (i.e., muscle tone, activity and alertness) should be recorded. Kittens should be observed for suckling activity in addition to the above parameters. Orphaned kittens should have consistent weight gains similar to those of suckling kittens. Queens should also be monitored for signs of impending cannibalism (e.g., extreme nervousness, aggressiveness toward the kitten and kitten rejection). Unfortunately, cannibalism often occurs without warning. In addition, housing and environmental hygiene should also be evaluated. Improper housing and hygiene are important risk factors for poor kitten development and impaired health.

ASSESS THE ANIMAL/Physical Examination The goals of the neonatal physical examination are to: 1) establish baseline data for future reference, 2) assess overall health and development of the kittens and 3) detect abnormalities that may impair normal development and health. During the physical examination, particular attention should be given to kitten behavior, body weight, body temperature and oral cavity health.

ASSESS THE ANIMAL/Physical Examination/*Kitten Behavior* Normal kittens are vigorous and have good muscle tone. They should nurse immediately or soon after parturition and have a strong sucking reflex. Well-fed kittens should have a distended abdomen and are quiet after feeding. Kittens that are hungry, cold, hot or in discomfort will cry continuously and should be closely monitored. Nursing behavior and milk intake should be carefully observed because some kittens develop rounded abdomens as a result of aerophagia. Kittens may have difficulty nursing queens of longhaired breeds due to hair accumulation or matting around the nipples. In these cases, abdominal hair can be clipped to allow kittens easy access to the queen's nipples. Care should be taken not to damage nipples during this process.

The behavioral response of kittens to the litter and queen is also important. Poor maternal-kitten interaction may result in cannibalism or neglect. Kittens depend on the queen for food, antibodies, warmth and hygiene; therefore, serious metabolic alterations (e.g., hypoglycemia, hypothermia, dehydration and malnutrition), infectious disease and death are common sequelae to abnormal behavior and maternal neglect.

ASSESS THE ANIMAL/Physical Examination/*Body Weight* Monitoring initial and subsequent body weight is a good way to evaluate milk intake and health status of nursing and orphaned kittens. Daily weighing is important to evaluate the queen's milk production and to help assess sick, weak and underweight kittens. Weight loss or slow weight gain in individuals or entire litters may indicate one or more of the following: 1) disease in kittens or the queen, 2) inability of kittens to suckle or 3) inadequate milk production. Healthy nursing kittens should be weighed at birth and weekly thereafter using a gram scale.

Birth weights are normally between 85 to 120 g with mean weights of approximately 100 g. Kittens weighing less than 75 g have very high mortality rates and require extra care and monitoring if they are to survive. Low birth weight kittens should be weighed every 24 to 48 hours for the first one to three weeks of life to ensure proper weight gain. Kittens gain an average of 100 g/week for the first six months of life. Minimally, they should gain 7 g/day.

ASSESS THE ANIMAL/Physical Examination/*Body Temperature* Kittens regulate body temperature poorly during the first four weeks of life. Normal body temperature is approximately 36.0°C (96.8°F) at birth and increases to 37.5°C (100.0°F) by one week of age. Extreme environmental conditions or abandonment by the queen may lead to hypothermia, which may quickly result in circulatory failure and death. See Small Animal Clinical Nutrition, 4th ed., p 330 for information about managing environmental temperature and humidity.

ASSESS THE ANIMAL/Physical Examination/*Oral Cavity* Examination of the oral cavity should include careful evaluation of the mucous membranes and hard palate. The mucous membranes should be light pink and moist. Cleft palates are relatively common defects in kittens. Vitamin A toxicity and trace mineral deficiencies (e.g., copper and zinc) during gestation have been associated with the development of cleft palates in kittens. However, in most cases, a cause is not identified. Most kittens with a cleft palate are unable to nurse effectively. Affected kittens must either be tube fed until the time of surgical correction or spontaneous closure, or they should be humanely euthanatized.

ASSESS THE ANIMAL/Laboratory Evaluation Laboratory tests should be performed as needed to assess any abnormalities noted during the physical examination. Particular attention should be given to hydration status and serum glucose and electrolyte concentrations. When evaluating laboratory data in kittens, age-appropriate reference values should be used because concentrations of certain analytes (e.g., phosphorus, hematocrit, serum proteins) vary markedly from adult values.

ASSESS THE ANIMAL/Key Nutritional Factors/*Colostrum and Milk* Colostrum is milk provided by the queen during the first 24 to 72 hours after parturition. Colostrum provides nutrients, water, growth factors, digestive enzymes and maternal immunoglobulins, all of which are critical to survival of neonatal kittens. Colostrum differs from mature milk in water and nutrient composition. The dry matter content of colostrum is high, which accounts for its sticky concentrated appearance compared with mature milk. The dry matter concentration declines as water content increases from Day 1 to 3 of lactation.

In addition to providing complete nutrition for nursing kittens, queen's milk also supplies non-nutritive factors that enhance food digestion, neonatal development and immune protection.

Kittens acquire passive systemic and local immunity from consuming either colostrum or mature milk. Kittens should receive colostrum within the first 12 hours of life to obtain adequate systemic immunity; after 16 hours, passive immunoglobulin transfer does not occur in kittens. During this time, kittens absorb intact immunoglobulins across the intestine. Failure to ingest colostrum or queen's milk during this absorptive window leaves kittens immunologically compromised and susceptible to infections and sepsis. Passive transfer of systemic immunity is particularly important to orphaned and hand-raised kittens that are fed only milk replacers. Consumption of queen's milk provides local concentrations of immunoglobulins within the GI tract and helps prevent invasion of microorganisms into the bloodstream (passive local immunity). Local immunity persists as long as kittens receive queen's milk. Both systemic and local immunity are important in maintaining kitten health until immune system maturation.

Mature milk is a complete food for nursing kittens. Water, protein, fat, lactose, minerals and vitamins are provided in amounts sufficient for normal growth and development.

Nutrient recommendations for neonates are based on the composition of queen's milk and growth studies in weaned kittens. Despite discrepancies in published nutrient values, queen's milk varies markedly from milk of other species. (See Appendix K, Small Animal Clinical Nutrition, 4th ed., pp 1064-1072.) Consequently, milk from other species is not suitable for nursing kittens. Replacement formulas with a nutrient profile similar to that of mature milk from queens should be used for orphans and supplemental feedings. For nutrients in which the concentration in mature milk is unknown, values recommended by AAFCO for growth should suffice.

ASSESS THE ANIMAL/Key Nutritional Factors/*Energy* Queen's milk typically meets the energy requirements of nursing kittens. Estimated caloric intake is approximately 200 kcal/BW_{kg} (837 kJ/BW_{kg}) for kittens up to four weeks of age, and 250 kcal/BW_{kg} (1,046 kJ/BW_{kg}) for kittens over four weeks of age (**Table 11-8**). As a rule, milk contains from 0.85 to 1.6 kcal/ml (3.6 to 6.7 kJ/ml) and milk replacers contain approximately 1 kcal/ml (4.2 kJ/ml) as fed. (See Appendix K, Small Animal Clinical Nutrition, 4th ed., pp 1064-1072.) By six weeks of age, male kittens are significantly heavier than female kittens and consume a proportionately larger quantity of food.

ASSESS THE ANIMAL/Key Nutritional Factors/*Water* Kittens contain 78.8% body water at one week of age. Total body water decreases to 70.1% at weaning. By comparison, adult cats are composed of only 61.7% water. The water intake of kittens is relatively high. A normal kitten needs about 155 to 230 ml water/BW_{kg}/day (i.e., 4.4 to 6.5 ml water/oz.).

ASSESS THE ANIMAL/Key Nutritional Factors/*Protein* The minimum protein requirement of nursing kittens has not been established. However, it is assumed to be comparable to that of weanling kittens, which is approximately 18 to 20% DM. These requirements were established using purified diets and may not accurately reflect the needs of kittens fed commercial foods made from typical ingredients. The AAFCO recommendation of 30% DM appears adequate; however, the protein content of queen's milk ranges from 33 to 44% DM.

ASSESS THE ANIMAL/Key Nutritional Factors/*Taurine* Taurine is important for normal growth and development of kittens. Fortunately, dietary taurine is more available to kittens than adult cats, presumably because of reduced bacterial destruction in the GI tract. Normal plasma taurine concentrations are maintained in 12- and 18-week-old kittens fed taurine at 150 to 197 mg/BW_{kg}/day. Queen's milk supplies about 300 mg taurine/liter. Queens fed low-taurine foods have significantly lower milk taurine levels, which may impair normal growth and development. Milk taurine concentrations are influenced by dietary taurine intake, thus it is not surprising that cow's milk is a poor source of taurine (i.e., only 1.3 mg/l). Therefore, homemade milk replacers based on cow's milk should be supplemented with taurine.

Table 11-8. Daily energy requirements of growing kittens.

Age (months)	kcal/BW$_{kg}$/day	kJ/BW$_{kg}$/day
Birth	250	1,045
1	240	1,005
2	210	880
3	200	840
4*	175	730
5	145	610
6**	135	565
7	120	500
8	110	460
9***	100	420
10	95	400
11	90	375
12	85	355

Key: BW = body weight, RER = resting energy requirement = 70 x (BW$_{kg}$)$^{0.75}$.
*Up to 50% of adult BW (at about four months of age) or 3.0 x RER.
**Between 50 and 70% of adult BW (around six months of age) or 2.5 x RER.
***Between 70 and 100% of adult BW (around nine to 12 months of age) or 2 x RER.
See Appendices C and F, Small Animal Clinical Nutrition, 4th ed., pp 1006-1007 and 1020-1026.

ASSESS THE ANIMAL/Key Nutritional Factors/*Fat* Milk fat is an important source of energy and essential fatty acids for nursing kittens. The composition of the queen's diet can significantly influence milk fat quantity and quality, which translates into fat composition of the offspring. The fat content of queen's milk increases throughout lactation. Average fat concentrations of 28% DM or 86 g/l are typical. Queen's milk provides the essential fatty acids linoleic and arachidonic acid at 5.8 and 0.5% DM, respectively. DHA is also essential for normal retinal development and function in kittens. Milk concentrations of DHA reflect the dietary intake of the queen.

ASSESS THE ANIMAL/Key Nutritional Factors/*Carbohydrate* No carbohydrate requirements have been established for nursing and growing kittens. However, the lactose concentration of queen's milk ranges from 14 to 26% DM. Intestinal lactase activity declines to adult levels very soon after weaning. Overfeeding cow's milk causes diarrhea, bloating and abdominal discomfort in kittens due to bacterial metabolism of undigested lactose in the large intestine. Owners who wish to offer cow's milk should be advised to limit the quantities given and to discontinue feeding cow's milk if intolerance occurs.

ASSESS THE ANIMAL/Key Nutritional Factors/*Calcium and Phosphorus* Calcium concentrations are low in colostrum (0.22% DM) and increase significantly to approximately 1% DM by mid to late lactation. Thus, requirements appear limited early on and increase with bone mineralization and growth. Milk phosphorus concentrations do not vary to the same extent. Thus, calcium-phosphorus ratios change from 0.4:1 on Day 1 of lactation to approximately 1.2:1 at one week and remain as such throughout lactation.

ASSESS THE ANIMAL/Key Nutritional Factors/*Trace Minerals* Queen's milk contains iron, copper and zinc concentrations markedly higher than those in human and bovine milk but similar to those in canine milk. Copper and iron levels tend to gradually decline throughout lactation, whereas zinc concentrations remain constant. Thus, mineral deficiencies are rarely reported to occur in nursing kittens fed queen's milk. However, milk replacers made from cow's milk should be supplemented to levels typically found in queen's milk to avoid deficiency.

ASSESS THE FOOD Foods should be liquid until kittens are three to four weeks old, then semi-solid to solid foods may be introduced. Foods may consist of queen's milk, commercial milk replacers and homemade milk replacers (including supplemented human enteral formulas). See Appendix K (Small Animal Clinical Nutrition, 4th ed., pp 1064-1072.) for more information about milk and milk replacers.

Queen's milk is considered ideal, providing all essential nutrients, antibodies, enzymes and hormones.

The quality of queen's milk and milk replacers is difficult to assess without analysis. Indirect measurement of kitten growth is probably the most practical method of assessment. In addition, the queen's food should be assessed if the queen is losing excessive weight. A thin queen (BCS 1/5 to 2/5) may not produce enough milk or may produce poor-quality milk. If milk analysis is required, a sample can be collected by manually expressing milk from the queen after preventing the kittens from nursing for a short time. Parenteral oxytocin (5 IU/queen) facilitates milk collection. Small samples (1 to 3 ml) are easily collected during normal lactation and should be frozen until analysis. Commercial laboratories do not routinely analyze such small milk samples; therefore, an appropriate research facility should be contacted for specific information about sample size, preservation and shipping instructions.

ASSESS THE FEEDING METHOD It may be necessary to alter the feeding method when managing neonatal kittens, especially if they are hand reared. Evaluation of current feeding methods with foreknowledge of growth demands will allow for development of a rational feeding plan.

Nursing kittens should be allowed free access to the queen. Kittens should be observed to ensure they have received colostrum by 12 hours after birth. Most neonatal kittens require feeding every two to four hours during the first week of life then every four to six hours until weaning. Weak kittens may need to be placed on the queen and held to facilitate nursing. Chilled kittens will not suckle and have reduced GI function. Thus, it is imperative to adequately warm weak kittens before they are fed. Hypoglycemia and hypothermia may occur simultaneously in neonates and have similar clinical signs. If kittens fail to respond to warming, a dilute glucose solution (2.5% glucose) may be given orally. This should be repeated until kittens are able to initiate a strong sucking reflex. Orphaned kittens and those too weak to nurse may be fed with a stomach tube, pet nurser or small syringe. (See Appendix 1.) At three to five weeks of age, kittens begin to eat semi-solid foods and enter the transition period from nursing to weaning.

Determine a Feeding Plan (Page 333)

Nursing kittens should be allowed to suckle the queen as the preferred feeding method. Nursing kittens depend completely on queen's milk; therefore, the feeding plan for the queen should be evaluated and modified if necessary. (See Reproducing Cats.)

Kittens should receive colostrum within the first 12 hours after birth. After this time, immunoglobulins are no longer absorbed from the GI tract and passive transfer will not occur. If colostrum is unavailable, milk collected from queens at any stage of lactation may be substituted. Antibody levels in non-colostral milk appear to adequately transfer passive immunity to kittens. Alternatively, sterile serum may be given to kittens subcutaneously if milk is unavailable. After the first 24 hours, kittens should be fed queen's milk or a complete and balanced milk replacer.

If circumstances require alternate feeding methods, the next best option is to attempt to foster kittens to a surrogate queen. Finally, feeding by bottle, syringe or stomach tube may be required. The method of choice largely depends on the age, vitality and adequacy of the suckling reflex of the kitten and the handler's expertise. (See Appendix 1.)

Reassessment (Page 333)

Kittens should be reassessed daily. Body weights should be obtained at birth then once weekly, if no complications are present. Poor weight gain or failure to thrive should prompt the breeder/owner to seek an immediate evaluation by a veterinarian.

Adequacy of the queen's milk production can be assessed by the growth rate of the kittens, kitten contentment and, to some extent, the degree of mammary gland distention. Expressing milk from a queen's nipples demonstrates the functionality of individual mammary glands, but does not indicate adequate milk production.

Growing Kittens: Postweaning to Adulthood (Pages 333-337)

The postweaning growth period includes kittens from about eight weeks of age until adulthood (i.e., 10 to 12 months). The nutritional needs of growing kittens include maintenance needs similar to those of adult cats and energy and substrates necessary for rapid tissue accretion. Growth rate slows if nutritional deficiencies exist. Thus, nutritional requirements are easiest to determine in growing animals using growth rates as a nutritional marker. The ultimate goal of feeding kittens is to ensure a healthy adult. The specific objectives, however, are to optimize growth, minimize risk factors for disease and achieve optimal health.

Assessment (Pages 334-336)

ASSESS THE ANIMAL/History and Physical Examination The general health and risk factors should be determined for every kitten early in the growth phase. A thorough history and physical examination, including determination of body weight and body condition, are generally sufficient. Ideally, a veterinarian should assess the kitten at weaning and monthly thereafter until the kitten is four months old. This schedule coincides with typical vaccination protocols for young kittens. The veterinary health care team should educate the owner about nutrition, weight management, neutering and dental care during these examinations. The owner can then evaluate stool quality and appetite daily and body condition weekly or biweekly.

Kittens should continue to grow at approximately 100 g/week until about 20 weeks of age. At 20 weeks, males typically gain 20 g/day whereas females gain 11 g/day. Growth rate slows as kittens approach 80% of adult size at 30 weeks and reach adult body weight at Week 40 (10 months). Most cats will achieve skeletal maturity at 10 months of age although some growth plates have yet to close. Additional weight gain may occur after 12 months of age and represents a phase of maturation and muscle development.

There is no evidence that the age at neutering alters the rate of growth. Investigators found kittens neutered at 12 weeks of age reached similar size as adults neutered at the more typical ages of six to nine months. Unfortunately, energy requirements decline with neutering and the risk for obesity increases. Neutering, however, is not the only risk factor for obesity. Practitioners have noted an alarming increase in the number of young cats with marked abdominal fat accumulation before neutering. An indoor lifestyle, feeding high-fat foods, overfeeding and certain feeding practices (e.g., free-choice feeding) are additional risk factors for obesity. Obesity should be prevented at an early age because it significantly affects the health and longevity of cats. Therefore, the risks for obesity should be determined as part of each cat's health evaluation at each veterinary visit.

ASSESS THE ANIMAL/Key Nutritional Factors Many of the key nutritional factors were discussed in the Neonatal Kittens section above and are outlined in **Tables 11-8** and **11-9**. Key nutritional factors as they relate to postweaning, growing kittens are reviewed here.

ASSESS THE ANIMAL/Key Nutritional Factors/*Energy* Growing kittens have high energy requirements to meet the needs of a rapid growth rate, thermoregulation and maintenance. Kittens may grow at rates from 14 to 30 g/day during the rapid growth phase. Although ensuring optimal growth is desired, excessive energy intake may lead to obesity. Ten-week-old kittens have a DER of approximately 200 kcal/BW_{kg} (837 kJ/BW_{kg}), which declines to adult levels (80 kcal/BW_{kg} [335 kJ/BW_{kg}]) at 10 months of age. After 10 months, age-related changes in energy requirement have not been observed.

Neutering reduces energy requirements by 24 to 33% regardless of the age at neutering. After neutering, limiting food intake or decreasing dietary energy may be required to prevent excessive weight gain. The energy density of the food fed to rapidly growing kittens should be between 4.0 to 5.0 kcal ME/g (16.7 to 20.9 kJ ME/g) (**Table 11-9**). A higher energy density allows smaller volumes of food intake to satisfy caloric needs. However, foods with energy densities at the lower end of this range should be fed to neutered kittens and those with a BCS of 4/5 or greater. The prevalence of obesity increases dramatically after one year of age.

ASSESS THE ANIMAL/Key Nutritional Factors/*Protein* Protein requirements are high at weaning then decrease gradually to adult levels. Kittens fed purified foods meeting all essential amino acids at or above the requirement have minimum protein needs of 20% DM in food containing 4.7 kcal ME/g (19.7 kJ ME/g). Protein biologic value and amino acid digestibility in practical cat foods are typically lower than in purified foods. AAFCO recommends a minimum crude protein level of 30% DM. To provide sufficient sulfur-containing amino acids without additional supplementation, at least 19% of the food must come from animal protein. **Table 11-9** lists crude protein recommendations in practical foods for growing kittens.

ASSESS THE ANIMAL/Key Nutritional Factors/*Fat* As kittens grow, body composition changes dramatically. Body fat composes only 5.5% of body weight in eight-week-old kittens and increases to 14.6% of body weight by 18 weeks. Dietary fat serves three primary functions in growing kittens: 1) it supplies essential fatty acids, 2) it acts as a carrier for fat-soluble vitamins and 3) it provides a concentrated source of energy in the food.

Kittens, like adults, require linoleic and arachidonic acid. Kittens also require fatty acids of the n-3 series (DHA, 20:6n-3) for normal neural

Table 11-9. Key nutritional factors for growing kittens.*

Factors	Recommended food levels**
Energy density (kcal ME/g)	4.0-5.0
Energy density (kJ ME/g)	17-21
Protein (%)	35-50
Fat (%)	18-35
Carbohydrate (%)	10
Crude fiber (%)	<5
Calcium (%)	0.8-1.6
Phosphorus (%)	0.6-1.4
Ca/P ratio	1:1-1.5:1
Sodium (%)	0.3-0.6
Chloride (%)	≥0.45
Potassium (%)	0.6-1.2
Magnesium (%)	0.08-0.15
Taurine (ppm) (extruded)	1,000
Taurine (ppm) (moist)	2,500
Average urinary pH***	6.2-6.5

*Concentrations presume an energy density of 4.0 kcal/g. Levels should be corrected for foods with higher energy densities. Adjustment is unnecessary for foods with lower energy densities.
**Dry matter basis.
***As determined in growing kittens.

development. Although true fat requirements are much lower, AAFCO recommendations for growth are 9.0, 0.5 and 0.02% for total fat, linoleic acid and arachidonic acid, respectively. These levels will sustain adequate growth; however, optimal growth rates are achieved with higher fat intake. Unless excessive growth or weight gain is evident, feeding foods with 18 to 35% fat is preferred to enhance palatability, meet essential fatty acid needs and maintain the energy density of the food above 4.5 kcal ME/g (18.8 kJ ME/g). Overweight and neutered kittens may require foods with dietary fat levels well below these to achieve ideal body condition (BCS 3/5).

ASSESS THE ANIMAL/Key Nutritional Factors/*Calcium and Phosphorus*
Weaned kittens appear to be fairly insensitive to inverse calcium-phosphorus ratios (e.g., kittens have been fed foods with ratios as low as 0.38:1 with no deleterious effects). The minimum requirement for dietary calcium in growing kittens is approximately 5 g/kg food (0.5% DM). Thus, AAFCO minimum allowances for calcium (0.8% DM) and phosphorus (0.6% DM) are appropriate for nursing, weanling and postweaning kittens. The AAFCO calcium allowance is higher than the established minimum requirement to compensate for foods with poor calcium availability. Unlike the situation with puppies, calcium excess in kittens is not

associated with developmental orthopedic disease. However, very high concentrations of calcium significantly reduce magnesium availability. Dietary calcium concentrations of 2% result in a nearly twofold increase in the magnesium requirement of growing kittens. Calcium levels in **Table 11-9** are sufficient to meet the needs of growing kittens and avoid impairing the availability of other nutrients.

Calcium deficiency coupled with phosphorus excess is most commonly observed in kittens fed unsupplemented all-meat diets. Nutritional secondary hyperparathyroidism results in osteitis fibrosa and is manifested by limping, pain and reluctance to move. Kittens fed such foods should immediately be fed a commercial food that meets the recommended minimum requirements with a calcium-phosphorus ratio of 1.2:1 to 2:1. Additional supplementation of calcium is not recommended and may lead to hypercalcemia as a result of serum parathyroid hormone excess.

ASSESS THE ANIMAL/Key Nutritional Factors/*Potassium* The potassium requirement of kittens is highly dependent on the protein content of the food and the effect of the food on acid-base balance. Urinary potassium loss is markedly increased when kittens are fed high-protein, acidified foods. To avoid syndromes associated with hypokalemia, postweaning kittens should not be fed highly acidifying foods and potassium allowances should be at least 0.6% of the dry matter intake. Chloride levels of 0.1% DM also cause hypokalemia despite adequate potassium levels. Some foods intended for lifestage feeding target urinary pH levels more appropriate for adult cats. These foods should be carefully assessed to ensure potassium is provided at appropriate levels (**Table 11-9**).

ASSESS THE ANIMAL/Other Nutritional Factors/*Urinary pH* The urinary pH of growing kittens is lower than that of adult cats fed similar foods. Presumably, the lower pH is caused by hydrogen ions released during bone formation, which are excreted into the urine. This increased response to dietary acidification continues until kittens are about 12 months old. Kittens fed highly acidifying foods (e.g., free-choice fed, urinary pH at or below 6.0) grow more slowly and plateau at lower body weights than kittens fed more basic foods. In addition to contributing to slow growth rates, feeding highly acidified foods results in poor bone mineralization in growing kittens. To reduce the risk of acidification on bone mineralization and growth, kittens should not be fed foods that produce urinary pH values less than 6.2 when fed free choice. Because growing kittens have a reduced risk for developing struvite-mediated lower urinary tract disease, an upper maximum for urinary pH is poorly defined. A maximum urinary pH of 6.5 will reduce the risk of struvite precipitates in cats at risk for lower urinary tract disease and avoid overacidification.

ASSESS THE ANIMAL/Other Nutritional Factors/*Digestibility* The food should be palatable with apparent dry matter digestibility greater than 80% and protein digestibility greater than 85%. The small stomach capacity of young kittens limits food intake in the face of relatively high energy demands. Providing highly digestible foods maximizes usage of the nutrients consumed.

ASSESS THE ANIMAL/Other Nutritional Factors/*Carbohydrates* Carbohydrates are not required in the food of growing kittens as long as an adequate supply of glucogenic amino acids is available. Nevertheless, cats can readily digest starch in cereal grains (i.e., >95% digestible). Excessive feeding of poorly digestible carbohydrates may result in bloating, gas and diarrhea. **Table 11-9** lists carbohydrate recommendations in foods for growing kittens.

ASSESS THE FOOD The profile of key nutritional factors for the food being fed should be compared to the levels recommended in **Table 11-9**. Appendix 2 lists nutrient information for selected foods. The manufacturer should be contacted for additional information. Alternatively, laboratories can analyze food samples. (See Appendix U, Small Animal Clinical Nutrition, 4th ed., pp 1134-1144.) This approach is usually necessary to obtain information about homemade foods. A food that is complete and balanced as demonstrated by AAFCO or similar animal feeding trials should be fed to kittens until they reach adulthood (10 to 12 months).

Unmoistened dry foods and moist foods are appropriate for weaned kittens. Dry foods are more energy dense per volume of food, which benefits small kittens needing increased caloric intake. Moist foods are often more palatable and thus encourage food intake. Semi-moist foods that excessively acidify the urine (i.e., <6.0 pH units) should be avoided until skeletal growth is completed. Identification of health risks such as obesity or overacidification necessitates a scrupulous review of foods provided for growth.

Treats are unnecessary but may be fed in small quantities (i.e., <10% of the daily intake). Milk is commonly offered to kittens as a treat. Amounts offered should be limited because intestinal lactase levels decline shortly after weaning. Appendix 3 lists nutritional profiles of common treats. (Also see Appendix M, Small Animal Clinical Nutrition, 4th ed., pp 1084-1091.)

ASSESS THE FEEDING METHOD Several feeding methods are appropriate for growing kittens. However, the method should be tailored to the individual animal's needs, the type of food being offered and the owner's preference (**Table 11-4**). Free-choice feeding is often preferred because it

reduces the risk of underfeeding and reduces the marked gastric distention that sometimes accompanies rapid meal feeding in young kittens.
The feeding frequency should be three to four times daily for meal-fed kittens less than six months old. This frequency ensures sufficient food
intake to meet the high nutritional demands of kittens without encouraging engorgement. By six months of age, most kittens will tolerate twice
daily feeding.

If kittens are thriving on their current regimen, alterations in the feeding method are unnecessary. A more appropriate feeding method should
be considered for kittens with poor growth rates and those with excessive
weight gain and obesity. Free-choice feeding methods should be used for
underweight and slow-growing kittens. Providing unlimited food for free-
choice intake is inappropriate for overweight and obese kittens. A defined
food quantity should be measured then offered as meals or fed free choice
until gone. Neutering increases the risk for obesity; therefore, free-choice
feeding of high-fat foods to neutered kittens should be done very cautiously. **Table 11-8** outlines the recommended daily ME intake for growing kittens.

To determine the amount to feed, the DER may be calculated based on
the age-appropriate energy requirements listed in **Table 11-8**, divided by
the caloric content of the food. The caloric content of many foods is not
readily available; therefore, feeding guides on food labels and the manufacturers' literature are useful starting places. After an initial food amount
is chosen, weight gain and body condition can be evaluated to tailor the
feeding amounts to individual cats. Young postweaning kittens should be
evaluated weekly. Biweekly evaluations are appropriate after kittens are
about four months old. Owners can easily evaluate body weight and condition; however, a veterinarian should confirm their findings during vaccination and wellness visits.

Determine a Feeding Plan (Page 336)

Assessment of the kitten, the current food and the feeding method
should provide the necessary information to develop a feeding plan. If it
is concluded that the food or feeding method is inappropriate, then
changes in the food, feeding method or both will need to be made. Often,
the assessment steps will determine that the current plan is optimal and
no changes are necessary.

SELECT A FOOD A more appropriate food should be selected if the current food does not adequately address the key nutritional factors listed in
Table 11-9. It is better to change to a food formulated specifically for kittens than to try balancing an inappropriate food. Appendix 2 lists the key
nutritional factors in selected commercial foods. The manufacturer can

also be contacted for additional information and for foods not listed in Appendix 2. Of key importance, foods for kittens should have undergone feeding trials appropriate for growth. Foods labeled for all lifestages, for growth and reproduction or for growth are appropriate for kittens, if animal feeding tests support the label claim.

SELECT A FEEDING METHOD As described previously, all types of feeding methods are appropriate for kittens. The method should be tailored to the individual animal's needs, the food type and the owner's preference. Free-choice feeding is preferred for kittens younger than five months. Fresh water should be provided daily and be available at all times. Young kittens should be fed in shallow pans to facilitate access to food. Food should be offered at room temperature; however, moist foods should not be left out for prolonged periods to reduce spoilage.

Reassessment (Pages 336-337)

After weaning, kittens should be weighed once a month until they are four to five months old. Weighing is usually performed at the time of vaccinations or veterinary examinations. Owners should continue to monitor daily food and water consumption to ensure normal appetite. Determination of total food intake is necessary only if inappetence, illness or poor growth rate is evident. Body condition scoring every one to two weeks is a better means to assess growth and adequacy of food intake. Results of body condition assessment allow owners to monitor kitten growth and adjust food offerings as needed to maintain ideal body condition (BCS 3/5). Kittens provided proper nutrition are healthy and alert and have ideal body condition, steady weight gain and a clean, glossy coat. Normal stools are firm, well-formed and medium to dark brown. The veterinarian should conduct a nutritional assessment at each visit, or approximately monthly for kittens between six to 16 weeks of age, and then annually. Instructions for nutritional modifications and dental care can be given at that time.

SUMMARY (Page 337)

The nutritional peculiarities of cats are easily understood if one considers the consistency of the natural diet to which cats evolved. Failure to feed nutritionally complete and balance foods appropriate for the age, activity and use of the cat can impair health and vitality. To ensure the best possible nutrition for cats and to alleviate problems caused by feeding errors follow these guidelines:

CATS

- Feed a good-quality food proven nutritionally adequate for cats.
- Choose a food for which nutritional adequacy was established through animal feeding trials.
- Select a food formulated for the purpose for which it is being fed.
- Provide continuous access to fresh clean water.
- Limit treats to 10% or less of the daily intake.
- Provide supplements only when medically indicated.
- Regularly monitor the cat's body condition, appetite and food intake.
- Adjust food offerings to maintain ideal body condition (BCS 3/5).

Clinical cases that illustrate and reinforce the nutritional concepts presented in this chapter can be found in Small Animal Clinical Nutrition, 4th ed., pp 340-347.

CHAPTER 12

Enteral and Parenteral Nutrition

For a review of the unabridged chapter, see Remillard RL, Armstrong PJ, Davenport DJ. Assisted Feeding in Hospitalized Patients: Enteral and Parenteral Nutrition. In: Hand MS, Thatcher CD, Remillard RL, et al, eds. Small Animal Clinical Nutrition, 4th ed. Topeka, KS: Mark Morris Institute, 2000; 351-399.

CLINICAL IMPORTANCE (Pages 351-353)*

The major consequences of malnutrition in all patients, but more prominently in sick or injured patients, are decreased immunocompetence, decreased tissue synthesis and repair and altered intermediary drug metabolism.

A malnourished animal is more susceptible to infections and a septic patient is more likely to be anorectic, which results in malnutrition. Nutrient imbalances suppress immune function, which increases the risk of disease; conversely, certain diseases alter some nutrient requirements.

Immunoglobulins and circulating antibodies are maintained at relatively low levels during malnutrition, but are highly responsive to appropriate refeeding stimuli.

Tissue synthesis and wound healing are a function of local and whole body nutritional status. Locally, amino acids and carbohydrates are needed for collagen and ground substance synthesis. Fibroblasts require energy to synthesize the RNA, DNA and ATP necessary for protein anabolism. Migration of fibroblasts and epithelial and endothelial cells also requires energy.

At distant sites, the liver has energy and protein needs specifically for synthesis of fibronectin, complement and glucose. The bone marrow requires nutrients for production of platelets, leukocytes and monocytes. Transportation of these necessary components and oxygen to wound sites requires the muscular activities of respiration and cardiac work. Tissue trauma and healing alter the normal continuous cycle of protein turnover (synthesis and degradation) in the body.

Proper nutrition for local tissue synthesis and wound healing depends on adequate whole body nutrition.

Cellular activities are dependent upon and regulated by the coordinated actions of peptides, lipids, vitamins and minerals as substrates, enzymes, coenzymes and cofactors of intermediary metabolism. All nutrients are

*Page numbers in headings refer to Small Animal Clinical Nutrition, 4th ed., where additional information may be found.

Key Nutritional Factors–Enteral and Parenteral Nutrition (Page 353)

(Continued on next page.)

EN/PN

Factors	Associated conditions	Enteral recommendations	Parenteral recommendations
Water	Dehydration	Correct dehydration with parenteral fluid therapy before starting assisted feeding.	Recalculate or increase intravenous fluid rate if dehydration persists or occurs.
	Edema		Recalculate or decrease intravenous fluid rate if edema occurs.
Electrolytes	Cardiac dysrhythmias Muscle weakness Hypo- or hyperkalemia Hypo- or hyper-phosphatemia	Correct electrolyte abnormalities with fluid therapy before starting enteral feeding. Use food that contains macrominerals.	Correct electrolyte abnormalities with fluid therapy before starting parenteral feeding.
Energy	Weight loss Depressed attitude and lethargy	If the patient is not eating at least resting energy requirement (RER) per os, provide nutritional support by assisted-feeding techniques to meet this requirement. By the fifth day of food deprivation or longer, patients should receive the majority (60 to 90%) of their calculated RER as lipid. If using a liquid or blended food, select a product that provides 1.0 to 2.0 kcal/ml (1.0 to 2.0 kcal/g) as fed.	Provide at least RER each day with parenteral feeding. By the fifth day of food deprivation or longer, patients should receive the majority (60 to 90%) of their calculated RER as lipid.

KNFs–EN/PN (Page 353)
(Continued.)

Factors	Associated conditions	Enteral recommendations	Parenteral recommendations
Protein	Weight loss Muscle wasting Decreased serum protein levels Depressed immune response Poor wound healing	Provide essential and conditionally essential (arginine, glutamine) amino acids. Dogs: Use a food that provides at least 4 to 6 g protein/100 kcal. Cats: Use a food that provides at least 6 to 8 g protein/100 kcal.	Dogs: Use a solution that provides at least 2 to 3 g protein/100 kcal. Cats: Use a solution that provides at least 3 to 4 g protein/100 kcal.
B-complex vitamins	Poor energy metabolism Anorexia	Use a food that is fortified with B-complex vitamins.	Add B-complex vitamins to parenteral feeding solutions.

essential for the maintenance of normal cellular structure and function. Nutrient deprivation alters the normal metabolic synergy responsible for ion gradients, membrane potentials, production of high-energy phosphate compounds and antioxidant defenses.

Protein-calorie deficiencies may result in decreased: 1) hepatic biotransformation of certain antibiotics, 2) concentrations of serum proteins that bind and transport drugs throughout the body and 3) renal blood flow, which decreases the rate of drug elimination and increases the possibility of drug overdose. Therefore, protein-calorie malnutrition may alter the normal or expected metabolism of certain drugs, which may increase or decrease their therapeutic effect even when given at recommended dosages. (See Chapter 27.) Animals receiving sufficient calories and protein are expected to have better or near expected drug distribution, metabolism and elimination than animals with protein-calorie malnutrition.

ASSESSMENT
(Pages 354-367)

Assess the Animal (Pages 354-367)

Nutritional assessment helps identify those patients that require assisted feeding to avoid or reduce nutrient deficiencies and the associated complications. Although inadequate nutrient intake may complicate many disorders, anorexia has been traditionally viewed as a secondary problem that will improve when the primary disease problem has resolved (i.e., "They'll eat when they feel better."). Conversely, it is better to be proactive and recognize the value of administering nutrients to veterinary patients and realize that "they'll feel better when they eat."

Diseased and debilitated patients (hospitalized or not) need to be assessed frequently, regardless of their age or lifestage. A veterinary nutritional assessment protocol should include history, physical examination with special attention given to certain risk factors, body condition assessment (body condition score [BCS]) and laboratory tests.

History and Physical Examination (Pages 354-356)

All patients should receive a physical examination including an accurate determination of body weight and an estimate of body condition. Weight changes must be viewed as a proportion or percentage of "normal, usual or optimal" weight within a certain time period as opposed to absolute changes in units (e.g., g or kg lost). Weight loss of more than 10% within a week is clinically significant and warrants further assessment. A 10% (5 kg) weight loss within seven days for a 50-kg dog is easily recognized as significant, but a similar weight loss within seven days for a 5-kg cat (i.e., 0.5 kg) is not easily recognized. This weight loss should be considered as serious as the same percentage weight loss in the dog example.

Body condition scoring adds valuable information to the breed and body weight data. Fat stores indicate previous energy intake (i.e., decreasing fat stores indicate low energy intake and vice versa). Muscle wasting implies protein intake has been insufficient because skeletal muscle mass supports hepatic protein synthesis when dietary intake is inadequate.

Decreased muscle mass may occur before serum protein levels drop below normal in chronic states because overall muscle wasting is less life threatening than decreased serum protein concentrations. Muscle atrophy due to protein malnutrition occurs bilaterally and should involve several muscle groups.

Laboratory Data and Other Clinical Information (Pages 356-357)

The changes in most laboratory data due to malnutrition are indistinguishable from those occurring in some disease processes; however, malnutrition should be considered when examining the animal and reviewing the data. Red blood cells, hemoglobin, albumin and total protein have moderately long half-lives of one to eight weeks and are an indication of the energy and protein status of the animal over the preceding weeks to months.

Decreased serum protein levels may occur in more acute states of inadequate protein intake relative to a large protein loss (e.g., protein-losing enteropathies, open abdomen). In starving animals, the loss of muscle mass decreases the body's protein reserves and, together with a slower rate of protein turnover in the remaining muscle, decreases the body's ability to synthesize proteins in response to metabolic needs. Such patients are poor surgical candidates because the body's protein reserves (muscle mass) have been catabolized to maintain the higher priority protein pools. If surgery can be safely postponed, several days of preoperative nutritional support in such patients is advisable. Evidence suggests that only one to three days of adequate energy and protein intake are required to up-regulate hepatic and muscle anabolic enzymes.

Serum potassium and urea nitrogen concentrations may also be decreased in anorectic animals because these variables are largely affected by daily food intake. Urea nitrogen, however, tends to increase in endstage starvation because muscle is catabolized for energy when fat stores are depleted.

Risk Factors (Page 357)

The physiologic status of the patient should be noted. Knowing the gender, reproductive status, age and activity level of a patient aids in the nutritional evaluation. Reproductive status (intact vs. neutered) alters metabolic rate and energy needs in animals. The metabolic processes of growth, gestation and lactation do not necessarily cease when an animal becomes acutely ill. Several days of inadequate energy intake may be necessary before the hormonal milieu for growth, gestation or lactation is down regulated.

Animals fed homemade food, table food, vegetarian or single-item foods are at greater risk for developing subclinical nutritional imbalances and warrant further nutritional assessment. Foods designed, formulated or prepared by owners are rarely nutritionally complete, balanced or consistent. (See Chapter 6.) These patients may not only have protein-calorie malnutrition, but are more likely to have several vitamin and mineral imbalances concurrently (e.g., calcium and micromineral deficiencies and/or subclinical vitamin A and D toxicoses).

Patients with a history of nausea, vomiting and diarrhea are at increased risk of malnutrition because nutritional intake probably has been less than optimal before admission. Nutrient intake may be voluntarily decreased with nausea, whereas vomiting and diarrhea can compromise nutrient digestion and absorption. Such clinical signs are also associated with additional losses of body protein.

Anorexia, Cachexia and Accommodation (Pages 357-361)

Anorexia is the loss of desire for food before caloric needs have been satisfied.

Disorders of taste or smell can impair appetite and occur because of:
- Old age. The number of taste buds declines with age.
- Damage to neural connections due to surgery or traumatic head injury.

- Impaired renewal of taste buds and olfactory epithelium. Decreased chemosensory cell turnover is consistent with the decreased cell renewal that has been reported to occur in the small intestinal epithelium as a result of food deprivation, radiation therapy, uremia, vitamin B_{12} deficiency and therapy with methotrexate. Many endocrine factors also depress cell proliferation. These factors and many conditions and drugs (**Table 12-1**) probably impair regeneration and function of taste buds and olfactory cells in the same manner that they impair regeneration of intestinal epithelium. The turnover time of taste bud and olfactory cells is about 10 days. Therefore, a return to normal taste function after mitosis is interrupted requires at least 10 days and usually longer.
- Modification of receptor cells as a result of a chronic change in local environment due to drugs or metabolic agents such as urea.

Numerous medical problems including organic disease, inflammation, trauma and neoplasia can cause anorexia. In addition, pain, fear and other components of emotional stress inhibit the desire for food. If anorexia persists, nutritional depletion occurs. Nutritional depletion may also result from facial or oral injuries, or obstruction or dysfunction of the gastrointestinal (GI) tract, liver or pancreas so that the animal is incapable of ingesting, chewing, swallowing, digesting or absorbing food. In general, animals not eating for more than 48 hours or those consuming less than 50% of normal intake for more than three days should be of concern and noted as having a form of anorexia. Cats and dogs with a history of complete anorexia for three or more days or animals with a history of partial anorexia for several weeks warrant further nutritional assessment.

Cachexia is a state of general illness, malnutrition and profound disability. Loss of peripheral (skeletal) and central (visceral) proteins can have adverse anatomic and functional consequences in food-deprived animals. These adverse effects include anemia, reduced heart muscle mass and function, decreased pulmonary mechanical function and diminished respiratory drive, altered intestinal morphology and mildly impaired absorptive abilities. Cachexia may affect dogs and cats with longstanding cardiac disease, renal disease or cancer. A state of metabolic "accommodation" that prolongs survival has been recognized in people with chronic diseases. A similar state of metabolic accommodation probably occurs in dogs and cats.

Accommodation occurs when the energy equilibrium is re-established at a constant but lower food intake and lean-tissue wasting is arrested before protein deficiency becomes fatal. Accommodation with the exception of an intact delayed cutaneous hypersensitivity response, accurately describes the condition of some chronically ill animals (i.e., those with chronic renal, hepatic or cardiac insufficiency). Some chronically ill, cachectic cats and dogs may be maintained at a less than optimal body

Table 12-1. Disorders and drugs that affect taste and smell in people.

Disorders

Adrenocortical insufficiency	Diabetes mellitus	Nasal polyposis
Allergic rhinitis	Head trauma	Niacin deficiency
Bronchial asthma	Hepatic cirrhosis	Radiation therapy
Burns	Hypertension	Sinusitis
Cancer	Hypothyroidism	Viral hepatitis
Chronic renal failure	Influenza-like	(acute)
Cobalam's syndrome	infections	Zinc deficiency
Cushing's syndrome		

Drugs

Drug classification	Examples
Amebicides	Metronidazole
Antiepileptic drugs	Phenytoin
Anesthetics (local)	Benzocaine, procaine hydrochloride, tetracaine hydrochloride
Antihistamines	Chlorpheniramine maleate
Antimicrobial agents	Amphotericin B, ampicillin, cephalosporins, chloramphenicol, gentamicin, griseofulvin, kanamycin, lincomycin, neomycin, nitrofurantoin, sulfonamides, streptomycin, tetracyclines
Antineoplastic agents	Doxorubicin, methotrexate, vincristine sulfate
Antirheumatic, analgesic, antipyretic, antiinflammatory, immunosuppressive agents	Allopurinol, azathioprine, colchicine, levamisole, D-penicillamine, phenylbutazone
Antithyroid agents	Propylthiouracil, thiouracil
Diuretics and antihypertensive agents	Captopril, furosemide, thiazides
Opiates	Codeine, morphine
Sympathomimetic drugs	Amphetamines, ephedrine
Others	Digitalis glycosides, estrogens, iron sorbitex, oral antidiabetic agents, vitamin D

weight and condition for some time, even though important organ function deficits are apparent. In these cases, metabolic rate has been down regulated and protein turnover has been altered to establish a fragile homeostasis. This homeostasis can be maintained until a new stress supervenes. Affected animals very often do not survive additional stresses such as trauma, surgery, infection or tumors, as might a previously healthy animal.

METABOLIC CHANGES/Simple Starvation The metabolic changes that occur with food deprivation in the absence of disease are often referred to as simple starvation. The time course of these metabolic changes should be the basis by which hospitalized patients are fed. In the immediate postprandial period, exogenous dietary nutrients are first used to meet immediate metabolic needs, sparing endogenous fuels. The second priority is to replenish glycogen reserves in the liver, fat and muscle, and to replace proteins catabolized since the last meal. The third priority is to convert any excess energy consumed as carbohydrate, fat or protein into triacylglycerides and to store that energy as fat in adipose tissue, muscle and liver. In the fed state when blood glucose concentrations are high, the liver is a net importer of glucose.

Priorities are reversed in food deprivation and, under the influence of endocrine changes, energy is drawn from endogenous stores. Animals use different proportions of stored body carbohydrate, fat and protein to maintain blood glucose concentrations throughout the course of starvation to maintain vital functions as long as possible (**Figure 12-1**). The adaptation from the fed to starved state is one in which fuel usage by the animal shifts from primarily a mixture of fuels, to one in which the primary fuel is fatty acids. Carbohydrate metabolism is profoundly altered during the first week of starvation. An understanding of these metabolic changes, which occur primarily in the liver, during simple starvation is essential to understanding the underlying metabolic alterations present during anorexia, illness and cachexia.

As blood glucose levels fall below 120 mg/dl in simple, uncomplicated starvation, the liver becomes a net exporter of glucose to maintain blood glucose concentrations. Omnivores (e.g., dogs) maintain blood glucose levels during the first two days of food deprivation through glycogenolysis and gluconeogenesis. The glycogenolysis will begin intraprandially but cease at the next meal. After four to five hours of fasting, however, the liver will begin to export glucose from the breakdown of glycogen stores to maintain blood glucose concentrations. Glycogenolysis maintains blood glucose levels for only another 12 to 28 hours. Thereafter, gluconeogenesis must maintain blood glucose concentrations because hepatic glycogen stores will have been depleted. Carnivores (e.g., cats) maintain blood glucose levels via gluconeogenesis beginning intraprandially due to minimal hepatic glycogen stores. Carnivores have relatively small hepatic glycogen stores compared with those in omnivores.

Gluconeogenesis by the liver and kidneys using glycerol, lactate and glucogenic amino acids is initiated by glucagon but later maintained by glucocorticoids as blood glucose levels decrease. Adipose tissue supplies glycerol for glucose production and fatty acids for oxidation to supply energy, whereas muscle catabolism releases glucogenic amino acids, lac-

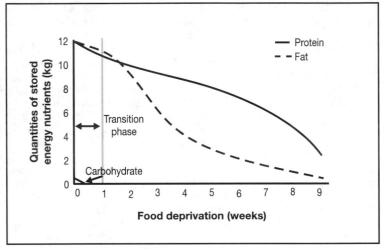

Figure 12-1. Disappearance of nutrient stores during starvation.

tic acid and pyruvate for glucose production by the liver. Extrahepatic tissues can trap glucose intracellularly even at very low blood glucose levels. By the third day of food deprivation in all mammals, there is a reduction in metabolic rate that continues for several weeks to slow fat and muscle catabolism in an effort to survive long-term starvation. Food deprivation in normal animals causes a decrease in blood glucose levels, which results in lower serum levels of insulin. The conversion of thyroxine (T_4) to triiodothyronine (T_3) is responsive to insulin; therefore, as insulin levels decline, conversion to T_3 declines. T_3 levels decrease within 24 hours of fasting and may be 40 to 50% less than levels in fed animals within three days. Thus, animals receiving less than their daily energy requirement (DER), regardless of the circumstances (e.g., food deprivation or anorexia), have a decreased metabolic rate to conserve body functions until appropriate caloric intake resumes.

The liver releases ketone bodies from fatty acids as an alternate fuel for non-glucose-dependent tissues within the first few days of food deprivation. Ketone bodies are essential to maintaining an energy supply to all tissues because the distribution of fatty acids (ketone precursors) is limited. Fatty acids have a limited distribution because they are water insoluble and must be carried in blood bound to albumin. This insolubility severely limits the serum concentration of fatty acids and their rate of diffusion into cells. Thus, fatty acids are restricted from serving directly as an alternate energy source for vital organs such as the brain.

The advantages of converting fatty acids to ketone bodies are threefold: 1) ketones are water soluble, 2) they do not require albumin for transport and 3) they provide a lipid fuel source to cells at much higher blood and interstitial fluid concentrations. The increase in levels of blood ketones causes enzymatic changes in peripheral tissues and in the brain to promote ketone use and decrease glucose demand, which then conserves body protein. These changes effectively spare glucose for the non-adaptive, glucose-dependent tissues (i.e., erythrocytes and renal medullary cells).

Ketosis in food deprivation is an appropriate physiologic response and may not lead to severe ketoacidosis except in diabetic dogs and cats. Thus, ketone bodies serve as a readily diffusible lipid fuel for muscles, kidney cortex, peripheral nerves and the brain during periods of starvation. Ketone body production is usually maintained until adipose tissue is depleted.

By Day 5 of food deprivation in all mammals, endocrine changes have mandated a metabolic shift from exogenous fuel usage in the fed state to endogenous fatty acid and ketone body usage. Blood glucose concentrations decrease, thus insulin levels decrease, which allows lipolysis. Fatty acids and glycerol from adipose tissue become important fuels for the body in starvation after three to five days.

Muscles also begin using ketone bodies for energy, sparing glucose. Muscles catabolize branched-chain amino acids and export the nitrogen in combination with pyruvate as alanine, and to a lesser extent as glutamine. These amino acids are deaminated and transaminated for hepatic protein synthesis and their ketoacid analogs are used for hepatic and renal glucose synthesis. Fatty acid oxidation partially spares oxidation of amino acids for energy, which helps maintain muscle protein stores throughout starvation until the endstages. Overall, blood glucose levels are maintained within the normal range throughout starvation for non-adaptive, glucose-dependent tissues.

As described, animals use different proportions of stored body carbohydrate, fat and protein to maintain blood glucose levels and for other vital functions at different times throughout the first week of food deprivation. A patient's dietary history will indicate the number of "no food" days. Matching the number of no food days with the pattern of fuel use (i.e., adaptation from using a mixture of fuels to primarily using fat) throughout the course of food deprivation is important in selecting the first refeeding formula. Animals deprived of food are in a state of catabolism, which can often be reversed by refeeding, if body protein losses have not exceeded 25 to 30%. To minimize the metabolic complications of refeeding, the refeeding formula should contain a complete balance of nutrients, and should have a carbohydrate, fat and protein profile similar to that which the liver has become adapted, or is estimated to be using from body stores.

METABOLIC CHANGES/Disease State Diseased patients have similar metabolic changes as patients in simple starvation (i.e., insulin resistance with decreased glucose use and increased proteolysis and lipolysis). General neuroendocrine responses and local mediators common to stressful stimuli mediate these metabolic and physiologic alterations. The specific hormonal and subsequent physiologic changes that occur in a stressed patient are unique to that disease condition, the time course of the disease and other complicating diseases or conditions. Hormonal and tissue substrate changes have best been characterized in disease conditions with a known acute onset such as trauma or infection. The response has been described as having an acute (ebb) phase and then an adaptive (flow) phase. The acute and adaptive phases can vary in duration and intensity depending on the specifics of the condition. The response to infection is qualitatively similar to that occurring with trauma except that the response to trauma is usually limited in time, whereas the response to infection will continue until the infection has been eliminated.

The acute phase generally occurs within the first 24 to 72 hours and is associated with catabolism. The response is similar to simple starvation except that the response is driven by a milieu of neuroendocrine and locally activated mediators rather than decreased T_3 concentrations. Simultaneous sympathetic nervous system stimulation and release of catecholamines, cortisol, adrenocorticoids, glucagon, growth hormones and antidiuretic hormones induce metabolic and physiologic responses:

- Suppression of insulin release.
- Hyperglycemia from glycogenolysis and gluconeogenesis, which provides an energy source for the "flight or fight" response.
- Increased proteolysis and net protein catabolism from albumin and skeletal muscle sources to supply glucogenic amino acids for glucose production.
- Increased lipolysis to provide fatty acids and glycerol for energy and glucose production.
- Increased rate and depth of respiration, and increased cardiac work to deliver oxygen, remove carbon dioxide and supply blood components to muscles and wound sites.

During an infection, a toxic response results from the invasive organisms and resorption of necrotic tissue. Lysosomal enzymes and specific chemical mediators are released including histamine, kinins, prostaglandins and serotonin. These substances amplify the previously mentioned responses to trauma. Fever increases energy expenditure 10 to 13% per degree Celsius increase in body temperature. Cytokine release also increases degradation of muscle protein and stimulates production of hepatic acute-phase proteins. Endotoxins released from dead gram-negative bacteria trigger coagulation cascades and profoundly affect carbohydrate metabolism.

In the acute phase, glucose is the preferred fuel; however, muscles are insulin resistant and hyperglycemia is maintained for net hepatic and immune tissue anabolism. As in starvation, these metabolic changes benefit the animal. Hyperglycemia ensures an energy source in the face of hypotension or poor organ perfusion. Tissue sequestration of minerals and trace elements (e.g., iron) slows bacterial growth. General feelings of malaise and anorexia signal the animal to reduce activity levels of skeletal muscle and the GI tract to conserve energy for essential tissue maintenance and repair.

The catabolic phase will continue until the neuroendocrine stimuli and cytokine mediators are removed. During severe head trauma, multiple trauma, burns and sepsis, weight and lean-tissue loss are rapid and unremitting in the absence of feeding. Providing maintenance or even supraphysiologic quantities of nutrients to patients with these injuries and conditions will not reverse the ongoing catabolism, and may not achieve nitrogen balance. The goal in providing nutrition to these catabolic patients is to feed the catabolism with exogenous sources of protein and fat. Feeding spares endogenous protein, which is critical because loss of 25 to 30% body protein stores has been associated with reduced heart muscle mass and function, decreased pulmonary function and diminished respiratory drive, compromised immune function and, therefore, increased mortality.

There is a definite turning point in which clinicians note a subjective improvement in their patients. This noted improvement is associated with the adaptive phase in which net anabolism occurs. The adaptive phase is characterized by increased metabolic rate, nitrogen gain and normal body temperature and may last for several days, weeks or years. The convalescent anabolic phase rebuilds damaged and catabolized lean tissue, and therefore, requires exogenous energy and protein sources. Investigators have measured the respiratory quotient (RQ) in resting postoperative and severely traumatized dogs at 0.76, indicating that fat was the preferred energy fuel.

Cats have very low glucokinase activity and cannot effectively transport glucose given in high intravenous concentrations into hepatic cells. This phenomenon is probably one of many metabolic reasons why hyperglycemia is often seen in cats as a refeeding complication, but rarely in dogs. In the adaptive or recovery phase of disease, however, fat is well used by dogs and cats as the primary fuel. The recovery phase is different for each patient, which underscores the need for continuous monitoring.

Metabolic changes occur in animals with cancer. Neoplastic tissues can be very aggressive and compete with the host for energy and nitrogen-supplying nutrients. Some tumors use glucose anaerobically and therefore are metabolically inefficient, placing a disproportionate energy burden on the host and in the process generating increased amounts of lactic acid. (See Chapter 25.)

Key Nutritional Factors (*Pages 361-367*)

Nutrient requirements for critically ill patients are more often expressed on an energy basis rather than on a dry matter basis (DMB). (See Chapter 1.) This designation is primarily an extension of the units used in actual clinical metabolic trials. In addition, nutrient profiles of oral liquid products and parenteral solutions used in nutritional support/recovery are more commonly expressed on an energy basis rather than a DMB.

FLUID AND ELECTROLYTE THERAPY Initial support often involves management of fluid, electrolyte and acid-base disorders. The water requirements of normal healthy animals approximate their DER (i.e., 1 kcal [4.2 kJ] = 1 ml of water). Fresh, clean water should be available to patients at all times, unless the patient requires a period of nothing per os. Most patients in an intensive care unit have venous catheters in place and are receiving crystalloid fluid therapy. These patients may have fluid restrictions or conversely may require diuresis. In these cases, the water or fluid administered will not be equal to the patient's DER. Daily maintenance fluid requirements are approximately 60 ml/BW_{kg}/day.

Nutritional support should not be initiated until the patient is hemodynamically stable because administering enteral or parenteral nutrition may further compromise the patient. Nutritional support should not be initiated as a "last ditch" effort in unstable patients. Major electrolyte disorders, acid-base abnormalities and blood glucose levels should be corrected before instituting enteral or parenteral nutritional support. It is also desirable to correct severe tachycardia, hypotension and colloid and volume deficits before starting assisted feeding. A practical goal is to begin nutritional assessment and support within 24 hours of the injury, illness or presentation.

ENERGY Knowing the patient's approximate caloric requirement is important because feeding more of any food than is necessary may cause metabolic complications. Overfeeding patients is possible through a feeding tube or with parenteral nutritional support.

The proportion of fat, carbohydrate and protein in foods fed to hospitalized patients should be similar to that which the liver is estimated to be using from body stores. Caloric density is important in both enteral and parenteral feedings when volume is limited. Enterally fed patients may be volume restricted due to gastric or intestinal sensitivities. Parenterally fed patients may be fluid restricted due to cardiorespiratory diseases and functional disabilities. In general, most dogs and cats tolerate the volume of food or solution that meets resting energy requirement (RER).

Patients with disease have metabolic rates and energy requirements that are less than those of comparable normal healthy individuals. In malnutrition, without disease or injury, decreased T_3 concentrations decrease the metabolic rate in an effort to conserve body functions. However, with an ongoing disease process or traumatic injury, the neuroendocrine responses to stress increase the metabolic rate above that found in simple starvation. Hospitalized veterinary patients have metabolic rates very near their RER.

Estimating the RER of hospitalized patients can be relatively simple using the equation RER = $70(BW_{kg})^{0.75}$. Most hospitalized veterinary patients should be fed at their calculated RER, realizing their actual energy requirement is likely to change over the course of the disease process and recovery. Initially feeding patients at RER, or 50% of RER if 100% is not possible, is a rational and safe recommendation that decreases the probability of metabolic complications. Regular nutritional assessment of the patient is strongly recommended to guide adjustments to initial feeding rates.

There are exceptions when the caloric requirement will be greater than RER. Particular cases have energy requirements 1.3 to 2.1 x RER. For example, people with severe closed-head and brain injury have energy requirements 40 to 60% above their calculated RER. Brain injury apparently increases oxygen consumption and acute-phase protein synthesis, which increases caloric and protein requirements significantly above RER. Energy requirements of twice RER appear to be the upper limit in the most severe head injuries. Energy expenditure may be 30 to 50% above RER in patients with multisystem trauma. Severely burned patients also have energy and protein requirements 80 to 100% above RER, relative to the extent of skin damage and surface area exposed. The body loses heat, moisture and protein through wounds with little or no epithelial covering. The patient's actual metabolic rate and resultant energy requirement are related to the degree of trauma, disease and/or complications and can only be approximated in a clinical setting.

PROTEIN Protein in the body is always in a flux between synthesis and catabolism. Protein synthesis requires that amino acids be present within cells at the correct time and ratio so that a protein may be constructed successfully. Protein degradation involves the release of amino acids, and if the amino acid is deaminated, the ketoacid analog is converted to glucose or fat and the amino group enters the hepatic urea cycle and is ultimately excreted in the urine. Under most circumstances, about 15% of the RER comes from the oxidation of amino acids. Providing a protein source to animals in catabolic states spares endogenous skeletal muscle protein and supplies essential amino acids and amino groups for acute-phase proteins and the immune response. Excessive dietary protein should be avoided in patients with liver or kidney disease. (See Chapters 19 and 23.)

Excessive protein feeding requires energy expenditure to rid the body of excess nitrogen, which may or may not be handled efficiently by the liver (urea cycle) and kidneys and may result in hyperammonemia with accompanying clinical signs of encephalopathy. Conversely, insufficient protein intake has been linked to low albumin concentrations, poor immune response, impaired healing and increased risk of wound dehiscence and muscle wasting.

In formulating nutritional support, it is prudent to first provide for total caloric needs with fat and carbohydrate, and then meet the protein requirement with protein. It is important to remember that if sufficient calories are supplied to patients as either fat or carbohydrate, then most of the protein will be used for protein synthesis and not burned for energy. A starting point of 4 to 6 g protein/100 kcal enterally and 2 to 3 g protein/100 kcal parenterally can be used for most dogs that do not have an extraordinary protein loss and that can excrete protein waste products. Higher ranges (6 to 8 g/100 kcal enterally and 3 to 4 g/100 kcal parenterally) are a more reasonable estimate for cats because of their constant state of gluconeogenesis and higher protein requirement. Protein intake can then be adjusted based on the patient's needs and ability to metabolize the amount of protein provided initially (e.g., decreasing serum albumin concentration or encephalopathic signs).

B VITAMINS Folic acid, thiamin, riboflavin, niacin, pantothenic acid, pyridoxine and B_{12} are essential for hepatic metabolism of glucose, fat and protein. These vitamins function as coenzymes for the tricarboxylic acid (TCA) cycle, ATP production and red blood cell metabolism. B vitamins are required in small amounts relative to other nutrients, but they are required daily and are necessary for efficient energy metabolism. B vitamins should be added to the fluids of all animals that are not eating but receiving fluid therapy. B vitamins are easily and inexpensively replaced and should be included in all forms of assisted feeding. Most pet foods contain adequate amounts of these nutrients; therefore, deficiency should not be of concern if the patient is eating enough food to meet its resting energy requirement.

MICROMINERALS Zinc, copper, manganese, chromium and selenium are vital cofactors for optimal hepatic and peripheral metabolism of energy substrates. Microminerals are important cofactors (metalloenzymes) and participate in tissue repair and albumin synthesis; therefore, they should be included in all food forms used for assisted feeding. Most pet foods contain adequate amounts of these nutrients, thus deficiency should not be of concern if the patient is eating enough food to meet its resting energy requirement.

OTHER FACTORS/Fat-Soluble Vitamins and Macrominerals Hospitalized patients rarely need fat-soluble vitamins and minerals. Most patients have fat and hepatic stores of fat-soluble vitamins sufficient to meet metabolic needs for months to years. However, administering fat-soluble vitamins should be considered in cases of prolonged malnutrition in which the patient is severely underweight with little to no fat stores (i.e., BCS 1/5). Patients rarely require macrominerals (i.e., calcium, phosphorus, magnesium, sodium and potassium) in excess of quantities needed to maintain serum electrolyte levels. Whole body stores of these minerals are usually not the problem. However, correct distribution between the intracellular and extracellular fluid space can be a problem and imbalances should be corrected before assisted feeding is begun. Sodium, potassium and magnesium levels may become a concern in patients experiencing excessive urinary loss of those minerals due to intensive diuretic therapy.

OTHER FACTORS/Palatability Many patients have poor appetites in the hospital for numerous reasons (**Table 12-1**). Therefore, highly palatable foods should be offered whenever possible. In general, foods that are moist (i.e., dry matter [DM] <40%), warmed to room temperature with higher percentages of fat and protein are more palatable than dry, cold foods with very low levels of fat and protein. A dietary history should also clarify the type of foods (i.e., moist, dry, homemade) the patient normally consumes. Many cats and small dogs prefer one form of food (e.g., moist vs. dry); therefore, food acceptance in the hospital can be improved by offering the same form fed at home. Owners are usually willing to bring in the pet's own food in order to improve food intake during hospitalization.

OTHER FACTORS/N-3 Fatty Acids Dietary fats are thought to affect the immune system by three mechanisms: 1) altered eicosanoid synthesis, 2) changes in cell membranes that affect membrane-associated protein and receptor function and 3) changes in intracellular nonesterified fatty acid pools (NEFA) that affect cytokine production. Generally, n-3 (omega-3) fatty acids produce less inflammatory cytokines, whereas n-6 fatty acids produce more pro-inflammatory cytokines.

Eicosanoids and lipoxygenase products, when produced in excess (e.g., trauma, injury, postoperative states), are generally immunosuppressive. The capacity of tissues and white blood cells to produce prostaglandins and lipoxygenase products is largely determined by the amount of arachidonic acid present, which is mostly determined by dietary linoleic acid. A significant reduction in n-6 polyunsaturated fatty acids (PUFA) appears prudent in the diet of immunocompromised, traumatized, postoperative or infected patients. The inclusion of n-3 PUFA in such diets would seem to be beneficial in decreasing eicosanoid production. Recent findings suggest that marked improvement can be made in foods by adjusting the n-6 and n-3 components to ensure optimal immune function.

OTHER FACTORS/Nucleotides Nucleotides are precursors of DNA and RNA, but they also participate in a number of metabolic reactions fundamental to cellular activity. Dietary nucleotides appear to be important for normal cellular immunity and are vital for maintenance of host defenses against bacterial and fungal pathogens. Dietary nucleotides are essential to the normal maturation of lymphocytes.

All commercially available parenteral and nearly all human enteral products are devoid of nucleotides. Pet foods that use meats and cereal grains as ingredients should provide adequate levels of dietary nucleotides. Dietary nucleotides are a vital component of regimens to maintain or restore immune function and host defense.

OTHER FACTORS/Arginine Arginine is essential to traumatized patients. Arginine enrichment stimulates the immune system, improves wound healing and decreases morbidity and mortality in burn patients. As a nutrient substrate, arginine may benefit surgical patients at increased risk for infection. Numerous studies demonstrate the efficacy of arginine-supplemented foods in reducing the catabolic response to major trauma, sepsis and injury and in improving the immune response after a variety of adverse stimuli. Further, exogenous arginine supplementation consistently improves nitrogen retention, protein turnover and wound healing.

Arginine has been recognized as an essential amino acid in dogs and cats for more than a decade. Therefore, most pet foods meeting Association of American Feed Control Officials (AAFCO) nutrient profiles should contain at least 146 mg arginine/100 kcal for adult dogs and 250 mg arginine/100 kcal for adult cats (approximately 80 to 200 mg/BW_{kg}). The arginine content of human enteral products varies, but is usually listed on the label. Human enteral products must contain at least adequate amounts of arginine if they are administered to dogs and cats for more than a few days.

OTHER FACTORS/Glutamine Glutamine concentrations in whole blood and skeletal muscle decrease markedly following injury and other catabolic states, thus making it "conditionally" essential. Replicating cells such as fibroblasts, lymphocytes and intestinal epithelial cells consume glutamine. These cells have high glutaminase activity and low levels of intracellular glutamine. Glutamine is important in stimulating immune function, possibly through an effect on gut-associated lymphoid tissue or through stimulation of macrophage function. Glutamine is the preferred fuel for rapidly dividing tissues such as white blood cells and intestinal mucosa.

Historically glutamine has been considered a nonessential amino acid in dogs and cats. Intestinal uptake of glutamine increases with surgery or trauma, most likely because glutamine is the preferred fuel for enterocytes. At least 80% of the published data in animals demonstrate a positive effect with glutamine-enriched feedings. Feeding a glutamine-enriched enteral diet has positive effects on: 1) protein metabolism, 2) intestinal and pancreatic repair and regeneration, 3) nutrient absorption, 4) gut-barrier function, 5) systemic and intestinal immune function and 6) animal survival. Glutamine is now thought to be a conditionally essential amino acid and probably need only be administered during early periods of physiologic stress to stimulate DNA synthesis and increase mucosal mass.

Assess the Food and Feeding Method (Page 367)

Ideally, all hospitalized patients should consume at least enough balanced food to meet their daily RER and sufficient fluid to meet their water requirement. Requirements for all other nutrients need not be calculated when the food contains non-energy nutrients properly balanced to the caloric density of the product. When the patient consumes the proper amount of a balanced food, all other nutrient needs have been met, unless known losses of particular nutrients occur (e.g., protein and electrolytes). When it becomes certain the patient is not eating enough food to meet at least RER, then assisted feeding should be instituted and the feeding plan revised.

Evaluating the diet of a pet with medical problems is more involved and requires more frequent and detailed attention than when dealing with a healthy pet. Case reviews may occur several times daily and should include a food-intake assessment. Feeding orders of hospitalized patients should be clear and complete. Properly written hospital feeding orders identify a specific food product with the amount, frequency and the route of intake specified, if not per os (**Table 12-2**).

Recording the food intake of hospitalized patients is essential to determining whether or not assisted feeding is necessary. In addition to having complete feeding orders, the medical record should also contain the time of day and amount of food actually consumed by the patient. Consumption can be simply recorded as some percentage of the food offered (e.g., 0%, 50%, 100%). If feeding orders are properly written and food consumption is recorded, it will be apparent after 24 hours of hospitalization whether or not the patient is consuming sufficient food to meet its RER, and whether assisted feeding is necessary.

Table 12-2. Examples of hospital feeding orders.

1. Offer 2 cans of product XX every 6 hr PO.
2. Give 100 ml of product YY gruel every 6 hr via PEG tube.
3. Administer 300 ml of parenteral solution IV every 8 hr.

Sometimes the feeding orders should contain special conditions, such as:

4. Begin feeding liquid product ZZ at 10 ml/hr via NG tube. D/C (discontinue) all feeding if vomiting begins.
5. Administer 300 ml of parenteral solution IV every 8 hr. Check urine glucose and decrease rate to 150 ml every 8 hr if urine is positive. Recheck serum K daily and increase to 40 mEq/l if below normal.
6. Give 30 ml of product YY gruel every 6 hr via PEG tube. Increase meal volume fed by 10 ml every 24 hr. Decrease volume by 50% if vomiting begins.

◼ FEEDING PLAN (Pages 367-385)

Devising a feeding plan for hospitalized patients requires complete knowledge of the case and the plan often needs to be individually tailored because of each case's unique circumstances. Nutritional plans require an understanding of the patient's metabolic state relative to changes in metabolism resulting from ongoing food deprivation. Refeeding patients in the early phase vs. refeeding in the later phases of food deprivation dictates the proportion of fat and carbohydrate in the refeeding formula (**Figure 12-1**). For example, the refeeding formula for a patient that has not eaten in seven days or more should contain predominantly fat as the energy fuel, as opposed to higher levels of carbohydrate (e.g., glucose).

Pre-existing conditions requiring specific nutritional modifications (e.g., renal insufficiency) or dietary modifications (e.g., adverse reactions to foods) must be understood and incorporated into the new feeding plan for the patient. Prior knowledge that a patient requires other medical and surgical procedures should also be taken into account when formulating an assisted-feeding plan. For example, feeding tubes can easily be placed at the end of a procedure requiring anesthesia or tranquilization. Nutritional plans should also consider the treatment plan and owners' expectations. Some nutritional plans can only be implemented while the patient is in the hospital, whereas the owner can implement others at home. A feeding tube may be placed differently depending on whether it will be used by trained personnel in the hospital or by owners at home.

Patient and Feeding Method Selection (Pages 369-371)

Any patient with a suspected or documented food intake below the calculated daily RER for more than three days is a good candidate for assisted feeding. Nutritional support should initially deliver sufficient amounts of a nutritionally balanced food to provide enough calories and protein to meet the RER of the patient at its current weight when the BCS is 3/5 or less. RER is primarily determined by total weight of metabolically active tissues such as skeletal and smooth muscle and visceral organs. BCS is primarily a measure of body fat stores. RER and BCS taken together are used to initially estimate the patient's daily caloric requirement. Animals with a BCS of 4/5 or 5/5 generally have the same muscle and organ mass as those with a BCS of 3/5; however, these animals have increased fat stores, which does not increase RER. It may be prudent, therefore, to calculate RER on an estimate of optimal weight in overweight patients to prevent overfeeding. (See Chapter 13.) After several days, the food intake may be increased as warranted on a case-by-case basis.

There are only two methods by which nutrients can be supplied to the body: enterally and parenterally (**Figure 12-2**). Enteral feeding provides adequate nutrition simply and cost effectively whether done orally or by feeding tube. Enteral feeding is preferred to parenteral feeding in most clinical cases because using the GI tract is less expensive, stimulates the immune system and avoids most metabolic complications. However, nutrients must be administered parenterally when the small intestine is not accessible or functioning adequately to meet at least the patient's nutrient requirements enterally. The two methods are not mutually exclusive; supplementing what the patient consumes voluntarily with a parenteral caloric and protein infusion is possible in most veterinary practices. Therefore, overall patient assessment, paying particular attention to GI function, is essential when deciding how assisted feeding should be provided.

Parenteral nutrition (PN) is valuable in meeting a patient's daily resting energy and amino acid requirements. However, patient selection is very important to the successful use of PN support. Patients with small intestinal impairment that is unlikely to be resolved within the next three days are candidates for PN support. PN can be used initially to meet energy and amino acid requirements for cases in which enteral access cannot be safely acquired for several days. Depending on patient size, it may not be cost effective to use PN as the only method of assisted feeding for less than a three-day course. There is a substantial startup cost to preparing parenteral solutions. The procedure becomes cost-effective when the cost is spread over several days or more than one animal.

There is evidence that between 48 to 72 hours are required to reverse a catabolic state and begin anabolism. Thus, proper patient selection mandates that the patient be hospitalized for at least three days because instituting PN for only one or two days is of questionable cost benefit. However, when PN is done in conjunction with some enteral intake, a course shorter than three days may be cost-effective and of great nutri-

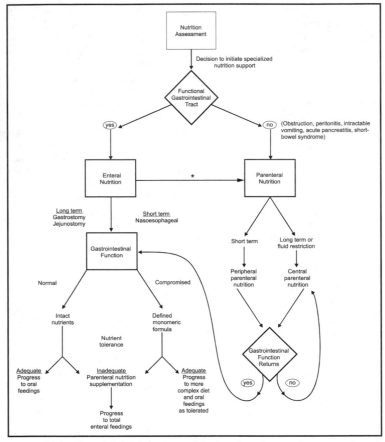

Figure 12-2. Clinical decision-making algorithm for selecting the route of nutritional support.
*Nasoesophageal tube not tolerated or anesthesia not possible.

tional benefit. PN support should not begin until the patient is hemodynamically stable and electrolyte and acid-base abnormalities, severe tachycardia, hypotension and volume deficits have been corrected (**Table 12-3**).

Parenteral administration of nutrients has value in patients with inflammatory (small and large) bowel disease, parvoviral enteritis and other causes of impaired GI motility, peritonitis, pancreatitis, intestinal lymphosarcoma and short-bowel syndrome (extensive small bowel resec-

Table 12-3. Patient criteria for administration of parenteral nutrition.

1. The patient is hemodynamically stable and major electrolyte and acid-base abnormalities, severe tachycardia, hypotension and volume deficits have been corrected.
2. Actual or anticipated food intake is less than calculated resting energy requirement for more than three days.
3. Concurrent small intestinal disorder is known or suspected to be present, or a safe enteral nutritional access cannot be established.
4. The patient is expected to be hospitalized for at least the next three days.

EN/PN

tion). Neurologic patients and those that are comatose or receiving large doses of pain-control medications that cause lateral recumbency with questionable swallowing reflexes and/or risk of aspiration when oral or tube feedings are attempted also benefit from PN. PN has been used successfully in complicated cases of feline hepatic lipidosis, in animals with facial fractures, pneumonia, lung lobe torsion or contusion and diaphragmatic hernias, and in other patients that are high anesthetic risks.

PN can be administered until patients are more stable and can tolerate placement of a feeding tube. Septic and anemic patients and those with severe upper respiratory infections that cause persistently poor appetites may benefit from PN. Sometimes PN support may be as simple as augmenting oral intake with intravenous lipids to meet RER. Any patient with a poor appetite that also has large heat and/or protein losses (e.g., continuous-suction chest or abdominal drains, large areas of skin loss due to burns, degloving injuries) benefit from PN in addition to any voluntary oral food intake.

The goal of assisted feeding is to provide adequate nutrition to meet the patient's RER. The logistics of that support should be determined on a case-by-case basis.

Enteral-Assisted Feeding (Pages 371-375)

Enteral-assisted feeding is providing nutrients to the patient using some portion of the GI tract. Patients that can digest and absorb nutrients from the small intestine but cannot or will not eat should receive enteral-assisted feeding. Feeding via the GI tract can be the simplest, fastest, easiest, safest, least expensive and most physiologic method of feeding patients.

Methods of Enteral Delivery (Pages 371-373)

ORAL FEEDING There are several methods of enteral feeding, but the first attempt usually should be oral feeding. Placing a bolus of food in the proximal portion of the mouth may stimulate the swallowing reflex and, if the patient offers no resistance, is a plausible method.

Some patients refuse to swallow boluses of food; therefore, forced feeding may increase the risk of food aspiration. Oral feeding should be discontinued if the patient does not swallow food voluntarily. Appetite stimulants may be used to induce food consumption in some patients; however, voluntary food intake rarely continues and the animal's RER is often not met (**Table 12-4**).

Orogastric tubes require placement at each feeding but may provide a useful option for one or two days of feeding. (See Appendix 4.) Neonates appear to tolerate multiple daily oral tube feedings better than adults. A red rubber or polyvinyl chloride tube (8 to 24 Fr.) may be used with the tip placed in the caudal esophagus or stomach. An indwelling feeding tube is the method of choice if enteral-assisted feeding is necessary for more than two days.

After an indwelling feeding tube has been placed, feeding is easier and less stressful on the patient than forced feeding or placing an orogastric tube. Nasoesophageal, pharyngostomy, esophagostomy, gastrostomy and enterostomy are potential placement sites. (See Appendix 4.) Tubes should be placed in the most proximal functioning portion of the GI tract possible via the least invasive method, and whenever possible, the stomach should be used.

NASOESOPHAGEAL TUBES Nasoesophageal tubes are generally used for three to seven days, but are occasionally used longer (weeks). Polyurethane tubes (6 to 8 Fr., 90 to 100 cm) with or without a tungsten-weighted tip and silicone feeding tubes (3.5 to 10 Fr., 20 to 105 cm) may be placed in the caudal esophagus or stomach. The preferred placement of all tubes originating cranial to the stomach is in the caudal esophagus to minimize gastric reflux and subsequent esophagitis. An 8-Fr. tube will pass through the nasal cavity of most dogs. A 5-Fr. tube is more comfortable in cats. Either orogastric or nasoesophageal feedings may be used in anorectic patients that do not have nasal, oral or orpharyngeal disease or trauma. Anesthesia or tranquilization is not necessary to place an orogastric or nasoesophageal tube. Therefore, these tubes provide enteral access to patients considered anesthetic risks. These tubes are most often used in the hospital, although conscientious owners can use nasoesophageal tubes at home. See Appendix 4 for information about tube placement.

PHARYNGOSTOMY/ESOPHAGOSTOMY/GASTROSTOMY TUBES Pharyngostomy or esophagostomy tubes (8 to 16 Fr.) may be placed in patients with disease or trauma to the nasal or oral cavity. The tip of the tube is placed in the caudal esophagus and the tube can be used for long-term (weeks to months) in-hospital or home feedings.

For patients in which the pharynx and esophagus must be bypassed, gastrostomy tubes (mushroom-tipped, 16 to 22 Fr.) can be placed either

Table 12-4. Pharmacologic appetite stimulants.

Pharmaceutical agents	Effects
Cyproheptadine	Antihistamine and antiserotonin effects. Dose cats at: a) 2-4 mg/cat per os once or twice daily, b) 2 mg/cat every 12 hr. May take up to 24 hr for a response. Give at this dosage for one week and then taper.
Diazepam	Short-lived appetite stimulant with sedative properties. Dose in cats varies: a) 0.05-0.15 mg/kg IV once daily to every other day or 1 mg per os once daily, b) 0.05-0.4 mg/kg IV, IM or per os. Eating may begin within a few seconds after IV administration. Have food readily available.
Oxazepam	Short-lived appetite stimulant with sedative properties. Dose cats at 2 mg/cat (total dose) every 12 hr.
Prednisolone	Dose glucocorticoids at 0.25-0.5 mg/kg per os every day, every other day or intermittently as needed in dogs.

intraoperatively or percutaneously using an endoscope or a gastrostomy tube introduction device. Gastrostomy tubes are also recommended for long-term feeding (weeks to months) if needed, and have generally replaced pharyngostomy tubes, even when the esophagus is normal. Gastrostomy tubes are convenient and safe for in-hospital and at-home feedings. See Appendix 4 for descriptions of tube placement.

Any tube that has been placed into the esophagus or stomach allows bolus or meal-type feeding schedules because the stomach acts as a food reservoir. The exception to this rule is the patient that cannot tolerate bolus feeding to the stomach without vomiting. Such patients benefit from a slow, continuous drip administration (by pump or gravity flow) of food to the stomach. Most veterinary patients receive bolus feedings of enteral nutritional support via nasoesophageal or gastrostomy feeding tubes.

JEJUNOSTOMY TUBES Jejunostomy tubes (J-tubes, 5 to 8 Fr.) are placed within the small intestine, ideally at the time of exploratory celiotomy, to bypass the proximal GI tract. Ideally, food should be administered through jejunostomy tubes at a slow, continuous drip delivered by a pump. Some patients, however, will tolerate frequent small-bolus feedings. Consult surgical texts for descriptions of tube placement.

Selecting an Enteral Food (Pages 373-375)

Food selection depends on tube size and location within the GI tract, the availability and cost of products and the experience of the clinician.

Table 12-5. Selected commercial products used for enteral feeding of veterinary patients (Continued on next page.).*

	Form	Energy (kcal/ ml/g)	Protein (g/100 kcal)	Fat (kcal %)	CHO (kcal %)
Abbott CliniCare Canine^Ü	L	1	5.5	55	25
Abbott CliniCare Feline^{ÜÜ}	L	1	8.6	45	25
Abbott CliniCare RF Feline^{ÜÜÜ}	L	1	5.6	57	21
Hill's Prescription Diet Canine/Feline a/d^á	MH	1.2	8.9	53	12
Iams Eukanuba Maximum-Calorie/Canine & Feline	MH	2.1	7.8	54	15
Purina CNM CV-Formula Feline	M	1.4	8.7	50	18
Select Care Canine Development Formula	M	0.9	8	30	42
Select Care Feline Development Formula	M	1	10.4	54	10
Waltham/Pedigree Concentration Diet/Canine	M	1.4	11.4	48	23
Waltham/Whiskas Concentration Diet/Feline	M	1.2	11	64	3
Waltham/Pedigree Concentration Instant Diet/Canine	L	1.5	14.3	37	25
Waltham/Whiskas Concentration Instant Diet/Feline	L	1.3	11	48	15

Key: L = liquid, M = moist, MH = moist homogenized, FA = fatty acid, CHO = soluble carbohydrate, na = information not available or not applicable.
*These products meet or exceed AAFCO nutrient profiles for intended species.
**5/8 French is the smallest tube size using a catheter-tip syringe and comfortable pressure.
***18 French is the smallest tube size using a catheter-tip syringe and comfortable pressure.

Commercial foods available for enteral use in veterinary patients can be divided into two major types: 1) liquid or modular products and 2) blended pet foods. Nasal and jejunostomy tubes usually have a small diameter (<8 Fr.), which necessitates use of liquid foods. Orogastric, pharyngostomy, esophagostomy and gastrostomy tubes have large diameters (>8 Fr.) and are suitable for blended pet foods. **Tables 12-5** and **12-6** list commercial foods available for enteral use. **Table 12-7** lists selected dry commercial foods marketed for recovery of critically ill patients. These foods may be used when patients are able to eat sufficiently on their own; alternatively, the patient may be fed the food it was accustomed to eating before the injury or illness. The latter approach reduces the number of food changes that ultimately will need to be made.

LIQUID FOODS AND MODULES In general, human liquid foods cost more than veterinary liquid products. Most human liquid foods are ade-

Table 12-5. Selected commercial products used for enteral feeding of veterinary patients (Continued.).*

	Fatty acid ratio (n-6:n-3)	Arginine (mg/100 kcal)	Glutamine (mg/100 kcal)	Tube feeding (Fr.)
Abbott CliniCare Canine†	6.4:1	249	550	5/8**
Abbott CliniCare Feline††	6.4:1	350	720	5/8**
Abbott CliniCare RF Feline†††	6.4:1	350	600	5/8***
Hill's Prescription Diet Canine/Feline a/d‡	2.2:1	398	1,180	18***
Iams Eukanuba Maximum-Calorie/Canine & Feline	5.1:1	460	na	18***
Purina CNM CV-Formula Feline	na	na	na	na
Select Care Canine Development Formula	10.3:1	na	na	na
Select Care Feline Development Formula	5.3:1	na	na	na
Waltham/Pedigree Concentration Diet/Canine	na	na	na	na
Waltham/Whiskas Concentration Diet/Feline	na	na	na	na
Waltham/Pedigree Concentration Instant Diet/Canine	na	na	na	5/8**
Waltham/Whiskas Concentration Instant Diet/Feline	na	na	na	5/8**

†Osmolality = 230 mOsm/kg.
††Osmolality = 235 mOsm/kg.
†††Osmolality = 165 mOsm/kg.
‡Carnitine = 1.10 mg/100 kcal.

EN/PN

quate for adult dogs but are too low in protein for cats, puppies and adult dogs with increased protein losses (e.g., protein-losing enteropathies, drains, etc.). Human liquid enteral products may not contain adequate concentrations of protein, taurine, arginine and arachidonic acid for long-term feeding of cats, but are satisfactory for fewer than seven days.

Liquid foods are of two basic types: 1) elemental or monomeric and 2) polymeric. Foods said to be "elemental" are not truly elemental, but contain nutrients in small hydrolyzed absorbable forms and are best described as monomeric. The proteins are usually present as free amino acids, small dipeptides or tripeptides or larger hydrolyzed protein fractions. The fat source is often an oil of mixed (medium- and long-chain) fatty acids and the carbohydrate sources are mono-, di- and trisaccharides. There are several liquid foods on the human medical market that are positioned as monomeric or hydrolyzed diets and are suitable when initially refeeding dogs and cats. These monomeric products are homogenized liquids that can be fed through any feeding tube including J-tubes. Monomeric foods

Table 12-6. Selected commercial human products used for enteral feeding of veterinary patients.*

	Energy (kcal/ml)	Protein (g/100 kcal)	Fat (kcal %)	CHO (kcal %)	Arginine (mg/100 kcal)	Glutamine (mg/100 kcal)	Comments	Tube feeding (Fr.)
Polymeric liquid products								
Ensure HN (Ross Laboratories)†	1.1	4.2	30	53	163	452	No MCTs	5/8**
Impact (Sandoz Nutrition)	1	5.6	25	53	na	350	27% MCTs; menhaden fish oil	8***
Jevity (Ross Laboratories)††	1.1	4.2	29	54	146	484	20% MCTs; soluble (soy) fiber	8***
Osmolite HN (Ross Laboratories)††	1.1	4.2	29	54	155	436	19% MCTs	5/8**
Pulmocare (Ross Laboratories)†††	1.5	4.2	55	28	220	333	20% MCTs	8**
Sustacal (Mead Johnson)	1	6.1	21	55	220	1,280	No MCTs	8***
Monomeric liquid products								
Peptamen (CliniTec Nutrition)	1	4.4	33	51	118	730	70% MCTs	5/8**
Perative (Ross Laboratories)†††	1.3	5.2	25	55	1,130	415	40% MCTs	5/8**
Vital HN (Ross Laboratories)†	1	4.1	9	74	200	210	32% MCTs	5/8**
Vivonex (Sandoz Nutrition)	1	2.1	2	83	292	491	No MCTs	5/8**

Key: MCTs = medium-chain triglycerides, CHO = soluble carbohydrate, na = information not available.
*Note: Veterinary enteral products should be routinely used rather than the commercial human enteral products listed in this table. Veterinary products are complete and balanced for the intended species, which might not be the case for human products. Polymeric and monomeric products listed in this table are appropriate for short term or emergency use in dogs and cats. See Small Animal Clinical Nutrition, 4th ed., p 372 for osmolality information.
**5/8 French is the smallest sideport red rubber tube size using a pump; 8 French is the smallest sideport red rubber tube using a 35-ml Luer-tip syringe and comfortable pressure.
***8 French is the smallest sideport red rubber tube size using a pump or a 35-ml Luer-tip syringe and comfortable pressure.
† Contains no carnitine.
††Contains 11 mg carnitine/100 kcal.
†††Contains 10 mg carnitine/100 kcal.

Table 12-7. Selected dry commercial veterinary products used for feeding hospitalized veterinary patients.

	Energy (kcal/g)	Protein (g/100 kcal)	Protein (kcal %)	Fat (kcal %)	CHO (kcal %)	FA ratio (n-6:n-3)	Arginine (mg/100 kcal)
Canine products							
Hill's Prescription Diet Canine p/d	4.2	7.0	26	46	23	13.2:1	458
Iams Eukanuba Maximum-Calorie/Canine	4.7	7.8	31	54	15	5.1:1	460
Select Care Canine Development Formula	3.6	6.6	23	34	43	7.0:1	na
Feline products							
Hill's Prescription Diet Feline p/d	4.4	8.1	28	46	26	16.2:1	453
Iams Eukanuba Maximum-Calorie/Feline	4.9	8.2	31	54	15	4.9:1	na
Select Care Feline Development Formula	4	8.1	28	47	25	5.2:1	na

Key: CHO = soluble carbohydrate, FA = fatty acid, na = information not available.

are indicated in disease conditions such as inflammatory bowel disease, lymphangiectasia, refeeding parvoviral enteritis and pancreatitis cases and any other condition in which a patient's digestive capabilities are questionable (**Table 12-6**).

Polymeric products contain mixtures of more complex nutrients. Protein is supplied in the form of large peptides (e.g., casein or whey). Carbohydrates are usually supplied as cornstarch or syrup, and fats are provided by medium-chain triglycerides (MCT) or vegetable oil. These foods require normal digestive processes and are appropriate for most veterinary clinical situations, especially when a small tube (<8 Fr.) has been placed and particular nutrient profiles are needed (e.g., low sodium, high protein, soluble fiber) (**Table 12-6**).

One of the leading liquid veterinary foods is a polymeric form that meets the current AAFCO nutrient allowances for adult dogs and cats. This product is a homogenized liquid containing 1 kcal/ml (4.2 kJ/ml) and is usually accepted better than human liquid products containing MCT oil. This liquid food is the best option currently available in North America when small-diameter nasogastric and jejunostomy feeding tubes have been placed, or when continuous drip feedings are necessary.

Several liquid milk replacer products are available; however, these products are not appropriate to feed to adult dogs and cats. They typically contain lactose, have high osmolarity, are lower in caloric density and do not meet AAFCO nutrient allowances for adult animals.

EN/PN

Module products are concentrated powdered or liquid forms of nutrients and are primarily supplemental. These products may be added to a liquid product to increase the concentration of a specific nutrient. There are protein, fat and carbohydrate modules (e.g., casein powder, vegetable oil or corn syrup). For example, a protein modular product may be added to a human liquid product for an animal with high protein requirements. Soluble fiber can be added to these foods using psyllium husk fiber or pectin; however, these fibers may block the small side ports in 8-Fr. and smaller tubes.

BLENDED PET FOODS Blended pet foods refer to commercial products nutritionally complete and balanced according to AAFCO allowances for dogs and cats. These products can easily be blended with a liquid to make a consistency that flows through a feeding tube. Some products have a blended texture, a high water content and very small particle size, whereas others must be blenderized with water and may need to be strained to remove particulate matter.

The best recommendation when using the blended pet food method is to use a product that has been tested in feeding trials and is proven to be balanced and complete for dogs or cats. These products are more readily available, better tolerated and less expensive than the human liquid foods. These pet food products contain essential amino acids and essential micronutrients properly balanced to the caloric density of the food. Fewer medical complications (e.g., diarrhea) are likely to result. However, blended products are more likely to plug the feeding tube if the tube is not properly flushed after feeding. Patients may later consume the pet food orally, eliminating a diet change when the patient's appetite returns and the tube has been removed. These products are appropriate for patients in catabolic states that are using fat and protein substrates from body stores. When using small-diameter (<8 Fr.) feeding tubes, it will be necessary to dilute the pet food with water, which dilutes the caloric density.

Blenderized moist veterinary therapeutic foods may have a place in assisted feeding of patients with specific disease conditions. Moist veterinary therapeutic foods are available with nutrient profiles that assist in the management of different disease conditions in dogs and cats. Examples of patients that may benefit from these blended therapeutic formulas include those with renal failure, hepatic failure, diabetes mellitus, hyperlipidemia and heart failure. These moist foods can be blended with water and strained to produce slurries or gruels that are administered through medium- and large-bore feeding tubes (i.e., 14 Fr. or larger). Blended pet food is prepared by blending 400 g of moist food with 1.5 to 2.5 cups of water at high speed for one minute and then straining the mixture once through a kitchen strainer (approximately 1-mm mesh).

Appropriate moist foods are listed in the respective disease chapters of this pocket companion.

HUMAN BABY FOODS Veterinarians have fed human baby foods packed in jars because some canine and feline patients voluntarily eat these products. In general, the meat and/or egg baby foods are high in protein (30 to 70% DM) and fat (20 to 60% DM), which compares favorably with blended pet food products. However, baby foods are more costly, contain only one or two food types (protein, protein/grain) and do not contain a balanced mixture of other essential nutrients (amino acids, vitamins and minerals). For example, these products contain only 10% of the calcium required by dogs and cats, and therefore have a large inverse calcium-phosphorus ratio. Some products contain onion powder, which has resulted in Heinz body formation in cats. The human and veterinary liquid products have a better nutritional profile than do the human baby food products.

Enteral Feeding Schedule (Pages 374-375)

The feeding schedule is often determined by the patient's ability to tolerate food and the logistics of feeding. Feeding an amount equal to the patient's RER during the first 24 hours of food reintroduction, if physically tolerated, is recommended. Initially feeding one-third of RER and then increasing the amount by one-third every 24 hours is a more cautious approach to initial feeding, but isn't always necessary. Foods should be warmed to room temperature, but not higher than body temperature before feeding.

Food boluses must be infused slowly (over approximately one minute) to allow gastric expansion. Daily food dosage should be divided into several meals according to the expected stomach capacity. Capacities for cats and dogs are 5 to 10 ml/BW_{kg} during initial food reintroduction. Maximum capacities as high as 45 to 90 ml/BW_{kg} have been measured in cats and dogs when fully re-alimented. Most often, meeting the patient's RER can be done in volumes far less than these maximum stomach capacities. Salivating, gulping, retching and even vomiting may occur when too much food has been infused or when the infusion rate is too fast.

Feeding should be stopped at the first sign of retching or salivating, the meal size reduced by 50% for 24 hours and then increased by 25% gradually. Foods provided via J-tubes must be infused slowly and often in either very small quantities or by a slow gravity drip or enteral pump with an hourly rate equal to RER/24 hours because the jejunum is volume sensitive.

Each meal must be followed by a water flush to clear the feeding tube of food residue. When the patient is volume sensitive, it is important to know the minimum volume required to flush the tube. The patient's daily fluid requirement must also be met and additional tap water may be administered through the feeding tube to meet that requirement. Liquid oral medications may also be administered easily through feeding tubes. Plugged feeding tubes can be cleared by filling the tube with water or a nonalcoholic carbonated beverage and allowing time for the food plug to dissolve. In general, endport tubes are easier to maintain than sideport tubes because food tends to become trapped in the blind end of sideport tubes. All tubes except orogastric and nasoesophageal tubes require standard every-other-day bandage care.

Parenteral-Assisted Feeding (Pages 375-385)

The term "parenteral" indicates administration of nutrients in a manner other than through the GI tract. Parenteral could therefore be administration by intravenous, intramuscular, subcutaneous, intraosseous or intraperitoneal routes, but not via the alimentary canal.

In veterinary medicine, clinicians attempt to meet the patient's estimated RER and most (but not all) of the patient's requirements for essential amino and fatty acids, and supply some water-soluble vitamins and selected macro and trace elements. The shorter time frame of assisted feeding of pets allows omission of less immediately essential nutrients (e.g., fat-soluble vitamins).

Parenteral Products (Pages 375-382)

Most veterinary practices can administer PN to patients. Individual PN solutions of dextrose, lipid and amino acids can be combined as a "three-in-one" solution, also called a total nutrient admixture (TNA). TNA in veterinary medicine refers to a one-fluid bag containing a sufficient mixture of parenteral solutions to meet a particular patient's fluid, energy, amino acid, electrolyte and B-vitamin needs for a 24-hour period. This is a very convenient method requiring only one bag, one infusion pump and one administration set. Almost any infusion pump can be used. The formulation is designed specifically for the patient based on the current RER, daily fluid and electrolyte requirements, approximate protein need and ability to handle dextrose vs. lipid.

The TNA solution should be calculated to first meet the patient's RER and protein needs with water-soluble vitamins and trace minerals (if available) added. Second, the total fluid volume is adjusted with standard crystalloid solution (e.g., lactated Ringer's solution, Plasmalyte A) to meet the patient's daily fluid requirement. Then electrolytes are adjusted, if necessary (**Table 12-8**). Alternatively, the additional crystalloid fluids with added potassium may be administered via a separate intravenous line piggybacked into the same catheter.

Table 12-8. Standard total nutrient admixture (TNA) formulations.

PART A. CALCULATION WORKSHEET

Patient data needed	Feline example
1. Current body weight in kg	4.1 kg
2. Calculate resting energy requirement (RER) as kcal/day	200 kcal/day
3. Expected fluid volume in ml/kg/day	70 ml/kg
4. Calories from fat as a percent	80%
5. Protein-calorie ratio as g/100 kcal RER	4 g/100 kcal RER
6. Potassium concentration as mEq/l	30 mEq/l

Parenteral solution formula

1. Determine volume of fat and dextrose needed daily

Calculate RER calories from fat	200 x 0.80 = 160 kcal
Calculate volume of 20% lipid needed	160 kcal ÷ 2 kcal/ml = **80 ml/day**
Calculate RER calories from dextrose	200 − 160 = 40 kcal
Calculate volume of 50% dextrose needed	40 kcal ÷ 1.7 kcal/ml = **24 ml/day**

2. Determine volume of amino acid solution needed daily

Calculate g of protein needed	RER x 4 g/100 kcal = 8 g protein/day
Calculate volume of 8.5% amino acid needed	8 g ÷ 0.085 g/ml = **95 ml/day**

3. Determine volume of B vitamins and trace minerals needed daily

Calculate B vitamins needed	RER x 1 ml/100 kcal = **2 ml/day**
Calculate trace minerals needed	RER x 1 ml/100 kcal = **2 ml/day**
Daily parenteral nutrition formula	80 ml of 20% lipid emulsion
	24 ml of 50% dextrose
	95 ml of 8.5% amino acid with electrolytes
	2 ml of vitamin-B complex
	2 ml of trace elements
	Total = 203 ml

4. Determine volume of crystalloid solution needed to meet daily fluid requirement

Daily fluid volume requested	4.1 kg x 70 ml/kg = 287 ml/day
Volume required is daily total − PN total	**287 − 203 = 84 ml**

5. Determine phosphorus supplementation

Phosphorus from amino acids	95 x 30 mM/l = 2.9 Mm
Desired final phosphorus concentration in the TNA	= 10 mM/l x 287 = 2.9 mM
	(no phosphorus is needed)

6. Determine potassium supplementation

K$^+$ from lactated Ringer's solution	84 ml x 4 mEq/l = 0.3 mEq
K$^+$ from amino acid solution	95 ml x 60 mEq/l = 5.7 mEq
Total K$^+$ in TNA solution	0.3 mEq + 5.7 mEq = 6.0 mEq
Desired final K$^+$ concentration in TNA	30 mEq/l x 287 ml = 8.6 mEq
KCl (2.0 mEq/ml) required	8.6 mEq − 6.0 mEq = 2.6 mEq ÷ 2.0 = 1.3 ml

(Continued on next page.)

Table 12-8. Standard total nutrient admixture (TNA) formulations (Continued.).

PART B. FELINE FORMULA EXAMPLE
Animal data

Body weight	4.1 kg
RER	200 kcal/day
Calories from fat	80%
Calories from glucose	20%
Protein-calorie ratio	4 g/100 kcal (adequate for most cats)
Fluid volume	70 ml/kg (maintenance fluid volume)
Potassium concentration	30 mEq/l

Parenteral solution

50% dextrose	24 ml providing 40 kcal
20% lipid emulsion	80 ml providing 160 kcal
8.5% amino acids with electrolytes	95 ml providing 8 g of amino acids
Potassium chloride	1.3 ml
Vitamin-B complex	2 ml
Trace elements	2 ml
Lactated Ringer's solution	84 ml
Total fluid volume	288 ml

This final solution is a 500-ml bag containing 200 kcal (80% from fat), adequate nitrogen, major B vitamins with the following electrolyte profile

Sodium	61.6 mEq/l
Potassium	29.5 mEq/l
Magnesium	3.3 mEq/l
Phosphorus	9.8 mM/l
Chloride	55.4 mEq/l
Calcium	0.8 mEq/l
Zinc	2 mg
Copper	1 mg
Manganese	0.2 mg
Chromium	8 mg
Final osmolarity	755 mOsm/l
Approximate cost = $100 for a three-day supply	

(Continued on next page.)

ENERGY SOLUTIONS A TNA solution should supply sufficient energy to meet, but not exceed, the patient's daily RER. It is becoming increasingly evident that the negative consequences of PN administration (i.e., metabolic complications) are due to administering energy in excess of patient expenditure. Energy is routinely provided to veterinary patients receiving PN as a combination of dextrose and lipid. Several companies manufacture dextrose and lipid products of various strengths and attributes (**Table 12-9**). The vast majority of TNA solutions formulated for veterinary patients use 50% dextrose and 20% lipid. Dextrose solutions

Table 12-8. Standard total nutrient admixture (TNA) formulations (Continued.).

PART C. CANINE FORMULA EXAMPLE
Animal data

Body weight	14 kg
RER	507 kcal/day
Calories from fat	90%
Calories from glucose	10%
Protein-calorie ratio	3 g/100 kcal (adequate for most dogs)
Fluid volume	70 ml/kg (maintenance fluid volume requested)
Potassium concentration	20 mEq/l

Parenteral solution

50% dextrose	30 ml providing 51 kcal
20% lipid emulsion	227 ml providing 454 kcal
8.5% amino acids/electrolytes	178 ml providing 15 g of amino acids
Potassium phosphate	1.4 ml
Vitamin-B complex	5 ml
Trace elements	5 ml
Plasmalyte A	516 ml
Total fluid volume	1,014 ml

This final solution is a 1-liter bag containing 507 kcal (90% from fat), adequate nitrogen, major B vitamins with the following electrolyte profile

Sodium	88.0 mEq/l
Potassium	20.3 mEq/l
Magnesium	3.5 mEq/l
Phosphorus	9.9 mM/l
Chloride	65.5 mEq/l
Calcium	0 mEq/l
Zinc	5 mg
Copper	2 mg
Chromium	20 mg
Manganese	0.5 mg
Final osmolarity	528 mOsm/l
Approximate cost = $130 for a three-day supply	

range from 2.5 to 70% glucose, which is usually derived from hydrolyzed cornstarch. Osmolarity ranges from 126 to 3,530 mOsm/l and is directly proportional to the glucose concentration. Dextrose solutions are maintained in the pH range of 3.5 to 5.5 and are sterilized by autoclave to prolong shelf life at room temperature.

Lipid (10, 20 or 30%) products (**Table 12-9**) contain emulsified fat particles (0.5 μm) of soybean oil and/or safflower oil, glycerin and linoleic and linolenic acids. Lipid emulsions are maintained in a pH range of 6.0 to 8.9 and have an osmolarity range of 260 to 310 mOsm/l, which effectively decreases the final osmolarity of the TNA. Dextrose and lipids are

Table 12-9. Nutritional comparison of parenteral products.

Products	Caloric density (kcal/ml)	Osmolarity (mOsm/l)	Amino acids (g/100 ml)	Fat (g/100 ml)	Carbohydrate (g/100 ml)	Electrolytes
Amino acids 8.5% without electrolytes (Abbott, Baxter)*	na	785-860	8.5	0	0	Few
Amino acids 8.5% with electrolytes (Abbott, Baxter)**	na	1,160	8.5	0	0	Yes
ProcalAmine (McGaw)***	0.25	735	3	0	3	Yes
Lipid 10% (Abbott, Baxter)†	1.1	268	0	10	0	No
Lipid 20% (Abbott, Baxter)†	2	268	0	20	0	No
Dextrose 2.5-70% (various)††	0.09-2.4	126-3,535	0	0	2.5-70	No

Key: na = not available.
*Contains all essential amino acids except taurine, nitrogen 1.3 g/100 ml, pH 5.3-6.5, available in 500- and 1,000-ml sizes.
**Contains all essential amino acids except taurine, nitrogen 1.3 g/100 ml, pH 5.3-6.5, electrolytes Na, K, Mg, Cl, PO₄, available in 500- and 1,000-ml sizes.
***Contains 3% glycerol and 3% amino acids, nitrogen 4.6 g/1,000 ml, pH 6.5-7.0, electrolytes Na, K, Mg, Cl, PO₄, available in 500- and 1,000-ml sizes. Contains 13 nonprotein kcal/ml and 22.5 g amino acids/100 nonprotein kcal as is. Mix 775 ml ProcalAmine with 300 ml 20% lipid to prepare 1,075 ml of a 3.2 g protein/100 nonprotein kcal solution.
†Contains soybean and/or safflower oil, glycerin, linoleic and linolenic acids, egg yolk as phospholipid emulsifier, pH 6.0-8.9, available in 50-, 100-, 250- and 500-ml sizes.
††Contains hydrolyzed cornstarch, pH 3.5-5.5, available in 50- and 500-ml sizes.

readily available and both are strongly recommended as sources of energy in a TNA solution.

Dextrose-Fat Ratio When PN is begun, most patients have not consumed their daily RER for at least three days, and are more likely even further along in the course of food deprivation. The proportion of glucose to lipid in the PN solution should mirror the current metabolic condition of the liver. Fewer metabolic complications will arise if the glucose-lipid ratio in the PN solution is well tolerated by the liver.

PN is rarely instituted during the early phases of food deprivation (less than three days of anorexia); however, if PN is indicated, canine patients should tolerate a moderate percentage of their calculated RER as dextrose. For example, dogs in this early phase of food deprivation maintain blood glucose levels by glycogenolysis and therefore should receive 60 to 90% of the RER as dextrose. However, feline patients in the early phases of food deprivation maintain blood glucose levels by lipolysis and gluconeogenesis, and should receive 60 to 90% of their RER from lipid.

By Day 5 of food deprivation or longer, patients should receive the majority (60 to 90%) of their calculated RER as lipid because the liver is using glycerol from endogenous fat for gluconeogenesis. Giving high doses of glucose at a time when the animal's natural response is to minimize glucose usage is unlikely to result in optimal glucose use. This is the most likely cause of the hyperglycemia often seen in such patients. There is evidence to suggest the proportion of calories needed from fat increases markedly (>60%) in starving and diseased states.

In people, fat is well oxidized in the septic state, and as the sepsis worsens the amount of fat oxidized increases and the glucose oxidative capacity decreases. Dogs with a septic abdomen receiving PN with both glucose and lipid maintain nitrogen balance better than dogs receiving glucose-only PN solutions.

The optimal caloric source is a mixture of glucose and fat; however, the precise ratio is unknown. A mixed fuel source should decrease the possibility of fat deposition in the liver when any metabolic pathway that handles either fat or glucose becomes overloaded.

To date, there appears to be little in the veterinary literature documenting why lipids could not or should not provide more than 60% of a dog's or cat's RER. In fact, when central venous access is limited and the patient requires fluid therapy at or below maintenance rates, administering a lipid emulsion via peripheral access (providing 100% of the caloric intake as fat) is not only possible but well tolerated.

Glycemia is better controlled in cases of diabetes mellitus, pancreatitis and septicemia with a TNA solution that provides most of the calories as fat. Intravenous infusion of lipid emulsion routinely causes a transient

EN/PN

increase in plasma triglyceride levels. However, this should not be considered a true hyperlipidemia because most patients can clear these chylomicron-size lipid particles within 30 minutes. The half-life of chylomicrons from the diet or intravenous infusion of soybean oil and safflower oil emulsions in the plasma of dogs is from seven to 16 minutes. Therefore, it is sometimes necessary to shut off the TNA infusion pump 20 to 30 minutes before blood is drawn if hyperlipidemia is a problem.

Dogs and cats at risk for developing hypertriglyceridemia (e.g., those with pancreatitis) have received a high-fat TNA at their RER with no additional problems.

PROTEIN SOLUTIONS Patients must receive a source of essential and nonessential amino acids. Solutions are available containing 3.5 to 15% amino acids. These solutions are maintained in the pH range of 5.3 to 6.5, have an osmolarity between 300 and 1,400 mOsm/l and may also contain various combinations of electrolytes and/or dextrose.

The most commonly used product in veterinary medicine is the conventional 8.5% amino acid solution either with or without electrolytes (**Table 12-9**). Most amino acid solutions contain all the essential amino acids for dogs and cats, except taurine. However, some specialized pediatric amino acid products contain taurine.

Protein should be provided to the patient within a ratio of 1 to 6 g/100 kcal of nonprotein energy provided. Adult dogs do well on 2 to 3 g/100 kcal, whereas, adult cats do well on 3 to 4 g/100 kcal. The lower protein-calorie ratios are recommended for patients with renal or hepatic insufficiency. The higher protein intakes are recommended for patients with increased protein needs (e.g., albumin losses, chest-tube drains). It is important to use the levels recommended here as guidelines only. A reasonable estimate of a patient's protein need should be made, the patient's response to that particular protein intake should be monitored and the intake should be adjusted accordingly.

ELECTROLYTE SOLUTIONS The more common electrolyte abnormalities associated with PN occur with the major intracellular cation potassium and the anion phosphorus. Potassium and phosphorus rapidly move intracellularly with refeeding by either enteral or parenteral methods or by the administration of glucose or insulin. Potassium moves intracellularly with the correction of acidosis and in response to insulin release. A TNA solution composed of 8.5% amino acids with electrolytes and lactated Ringer's solution contains approximately 12 mEq potassium/l, which is not adequate to maintain normal serum potassium levels. Potassium can be added to the PN solution using either a 2 mEq/ml potassium chloride solution or a 4.4 mEq/ml potassium phosphate solution.

If the patient is normokalemic when PN is initiated, 20 to 30 mEq potassium/l will usually maintain normokalemia. However, if the patient is hypokalemic when PN is started, 40 or more mEq potassium/l will be required. If the patient is hyperkalemic when PN is initiated, no additional potassium is recommended, but serum potassium concentrations should be monitored daily. Crystalloid solutions containing potassium and administered via a second intravenous line are a convenient method of regulating serum potassium levels in difficult cases.

Phosphorus moves intracellularly with refeeding because of increased production of high-energy phosphate compounds. Animals receiving PN rarely become hypophosphatemic. Sufficient quantities of phosphorus (5 to 10 mM/l) appear to be available in the TNA from lipids (15 mM/l) and amino acid/electrolyte (30 mM/l) solutions. However, adding a potassium phosphate solution containing 4.4 mEq potassium and 3 mM phosphorus/ml will increase the potassium and phosphorus content of the TNA. In cases of hyperphosphatemia, the quantity of amino acids, electrolytes and fat must be reduced to lower TNA phosphate concentrations in the TNA. Alternatively, an amino acid solution without electrolytes and potassium chloride can be used.

VITAMIN SOLUTIONS Very few veterinary patients receiving PN require fat-soluble vitamins unless there is a history of prolonged weight loss, inappetence and decreased fat absorption (diarrhea/steatorrhea). Dogs and cats usually have sufficient body stores of vitamins A, D, E and K to last several months. Fat-soluble vitamin supplementation is warranted in cases of long-term fat malabsorption (months). One-time administration of 1 ml of a vitamin A, D and E product, divided into two intramuscular sites, is simple, cost-effective and supplies fat-soluble vitamins for about three months. Vitamin K_1 injections (3 to 5 mg/cat, b.i.d., subcutaneously) improve abnormal coagulation times in cases of severe idiopathic hepatic lipidosis.

Water-soluble vitamins, however, must be supplied daily by either the enteral or parenteral route. Most veterinary vitamin B-complex products do not contain all 11 B-complex vitamins, because some B vitamins are not compatible (e.g., folic acid is incompatible with riboflavin in the same solution). Folic acid, therefore, must be administered separately if needed. Based on National Research Council daily vitamin recommendations for healthy dogs and cats and given the vitamin concentrations available in most solutions, the recommended dose of 1 ml of B vitamins per 100 kcal exceeds daily B-vitamin requirements by several-fold, except for B_{12}. Most previously healthy pets and people, however, have sufficient hepatic stores of B_{12}.

Some B vitamins are light labile; therefore, most B-vitamin preparations should be kept in a light-resistant bottle and stored between 15 to 30°C

(59 to 86°F). B vitamins are incompatible with some drugs commonly administered to veterinary patients by continuous intravenous infusion (**Table 12-10**).

TRACE-ELEMENT SOLUTIONS PN solutions may contain 1 mg zinc and 0.5 mg copper/100 kcal. These elements can be added to the PN solution most economically (pennies per day) using a multiple trace-element combination available in multiple-dose vials. Approximately 1 ml of the trace-element solution should be added per 100 kcal daily.

Drug Additions (Page 382)

Although it is very convenient to administer drugs intravenously with the PN solution, *extreme caution must be taken before any medications are added to the TNA.* Drug and TNA solution compatibility studies are ongoing, and there are published lists of drugs known to be compatible and safe. **Table 12-11** lists the drugs of most interest to veterinarians that can be incorporated into a three-in-one mixture. The *Handbook on Injectable Drugs* is updated and published every two years and is a good source for current information about drug compatibility with PN solutions. After a medication has been added to the day's PN solution, a decision to discontinue that medication can be costly because a new bag of PN solution must be compounded. Therefore, use of a second peripheral catheter or a double-lumen central catheter may be preferable to adding drugs to PN solutions.

Compounding (Page 382)

PN solutions can be obtained from several sources. Some human hospitals and independent pharmaceutical companies will compound TNA solutions for veterinarians. A prescription must be written indicating the volume or final concentration of each nutrient (fat, dextrose, amino acids and each electrolyte), and the person preparing the TNA is likely to refer to the solution as "TPN." Some veterinary schools, large referral practices and private veterinary hospitals maintain parenteral solution compounders and supplies for their own use and will compound and sell TNA bags directly to veterinary practitioners. Several bags of PN solution (up to 10 days' worth) can be sent by overnight mail services directly to the practice. This is often the safest, most convenient and economical method of obtaining an all-in-one PN solution for the occasional patient in most practices.

TNA solutions can be compounded by one of three basic methods: 1) syringe, 2) gravity flow or 3) computerized flow. Several variants of each method exist but will not be discussed here. All-in-one PN or TNA supplies can be purchased from the same sources that provide the PN solutions. The least desirable method uses a 35- or 60-ml syringe to transfer

Table 12-10. Drug incompatibility with B-complex vitamins.

Known incompatible	Suspected incompatible
2-PAM (pralidoxime chloride)	4-methylpyrazole
Aminophylline	Adriamycin
Asparaginase	Carboplatin
Bicarbonate	Cisplatin
Calcium versenate	Dobutamine
Cefazolin	Dopamine
Diazepam	Fentanyl
Digoxin (injectable)	Propranolol
Mannitol	
Nitroprusside	
Penicillin G	
Quinidine	

Table 12-11. Drugs compatible with total nutrient admixtures (TNA).

Aminophylline	Diphenhydramine	Lidocaine
Ampicillin	Dopamine	Metoclopramide
Cefazolin	Erythromycin	Penicillin G
Chloramphenicol	Furosemide	Phytonadione
Cimetidine	Gentamicin	Ranitidine
Clindamycin	Heparin	Ticarcillin
Digoxin	Insulin (regular)	

each nutrient solution (dextrose, amino acid and lipid) into a sterile, empty fluid bag. This method is the most time-consuming and carries the greatest risk of contamination because of the multiple transfers required. Transfers are ideally done under a laminar flow hood.

The second method uses a closed-circuit fluid system in which the all-in-one bag comes with a pre-attached three-lead transfer set. Each lead, with a vented filter spike, is inserted directly into the individual nutrient solutions (dextrose, amino acid and lipid), and the nutrients are transferred directly into the all-in-one bag by gravity flow. This method is faster and safer than the syringe method, but transfer of exact quantities is not possible. This method may be most economical when few patients require PN. Both syringe and gravity feed methods usually leave partially unused bottles of dextrose, fat and amino acids.

The third and best method, used by most human hospitals and some large referral veterinary hospitals, employs a high-speed, closed-circuit fluid compounder that pumps three or four solutions (dextrose, amino acid, lipid and fluid) directly into one TNA bag within 60 seconds. Each solution is accurately transferred to within 1 ml. The method has a mean error of less than 2%. Multiple bags of TNA for several patients can be efficiently made at one time using partial bottles of dextrose, fat and amino acids. Making TNA bags with a compounder is safe, fast, accurate and efficient. Veterinary technicians can routinely accomplish this task.

Administration (Page 382)

PN solutions can be delivered to patients by a central, peripheral, intraosseous or intraperitoneal catheter. PN solutions with an osmolarity greater than 600 mOsm/l should be administered into a central vein to avoid thrombophlebitis. Solutions less than 600 mOsm/l may be administered via a peripheral vein.

PERIPHERAL VEIN INFUSION Calories can easily be administered peripherally to dogs and cats using a TNA of 400 to 600 mOsm/l or an isomolar 20% lipid solution piggybacked with standard fluid therapy at volumes sufficient to meet RER.

TNA solutions can be administered to dogs and cats through peripherally inserted central lines. Cats, compared with most dogs, are smaller, have higher protein requirements and sometimes have restrictive fluid allowances. Therefore, for cats (more than for dogs), the final osmolarity of the PN solution will be between 600 and 800 mOsm/l and the solution should be administered into a large vein. Placing a 10- to 20-cm polyurethane or silicone catheter (with a break-away needle) into the medial saphenous vein at the level of the tarsus and advancing the catheter up the vein places the tip of the catheter into the caudal vena cava of the cat. A similar, but longer (20- to 30-cm) polyurethane or silicone catheter, placed in the lateral saphenous vein, is more useful in dogs weighing less than 20 kg. For long-term applications (more than three days), silicone and polyurethane remain the only acceptable catheter materials, and they need not be removed at a predetermined time.

CENTRAL VEIN INFUSION A single- or multiple-lumen polyurethane or silicone elastomer catheter can be placed by a percutaneous procedure (or rarely a cut-down procedure) in the external jugular vein of most dogs and cats. The tip of the catheter should be located in the cranial vena cava or right atrium. Catheters made of silicone elastomer and polyurethane are softer and less irritating, though more expensive, than the polytetrafluoroethylene (Teflon) catheters often used for fluid administration in veterinary patients. Silicone and polyurethane are likely to result in fewer mechanical and septic complications and are less thrombogenic. Again, for long-term applications (more than three days), silicone and polyurethane are the only acceptable catheter materials.

Central catheters are changed as indicated by particular problems (e.g., infection, subcutaneous migration, thrombosis) and not at predetermined intervals. Multiple-lumen catheters allow multipurpose venous access for administering incompatible fluid/drug therapies or different fluids at different rates. Although use of central catheters has not been adequately evaluated in veterinary patients, their use for PN administration in people is associated with an increase in septic complications.

CATHETER COMPLICATIONS TNA solutions should be administered at room temperature. Also, it is prudent not to extend the delivery of any one bag more than 24 hours. The most clinically significant problem in administering TNA solutions involves the catheter, including loss of access, thrombophlebitis and infection. Catheter kinking, catheter tip migration or catheter blockage may cause loss of venous access. In these instances, the catheter should be removed and a second catheter placed in another vein.

Thrombophlebitis Thrombophlebitis is a response of the vein intima to the unique combination of the infusate, the catheter material and placement and the ratio of catheter to vessel size. Intravenous catheters reduce blood flow and frequently induce thrombophlebitis within 72 hours; therefore, they are routinely changed every three days. The catheter material is thought to be the single most important factor in the severity of infusion thrombophlebitis.

Teflon intravenous catheters are commonly used in veterinary medicine, and can be used routinely as short-term (two to three days) peripheral catheters. Vialon catheters are easier to insert, have lower rates of phlebitis and are less thrombogenic than Teflon catheters.

The list of catheter materials in order of increasing thrombogenicity is: Vialon (lowest), polyurethane, silicone and Teflon (highest).

Infection Infectious complications with intravenous infusions are primarily associated with substandard catheter care. Most catheter-related septicemias are due to microbial invasion at the catheter wound, either during or after insertion. Hospitals that use iodine solutions to disinfect the skin and emphasize catheter site asepsis have significantly lower rates of positive catheter cultures and septicemias. Catheters for PN administration must be placed using meticulous aseptic technique. The catheter bandage and administration sets should be changed at least every other day, if not daily. When the bandage is changed, the venipuncture site should be cleaned with an iodine solution and examined for redness, edema or swelling. A topical antibiotic ointment (e.g., povidone iodine) that contains antifungal properties should be applied at the catheter-skin junction. When redness, edema or swelling is noticed, the catheter should be removed and cultured and the site should be kept clean and hot packed, if necessary, to reduce swelling. Appropriate antibiotics should be given if the catheter or TNA solution is shown to be contaminated by culture and antimicrobial sensitivity testing.

Second, thrombi at the catheter tip may be seeded hematogenously with organisms from urinary tract infections, abscesses, pneumonia or other infected sites.

Infusion of contaminated fluid is a third source of infection.

Fungi can still proliferate in admixtures, though refrigeration at 4°C suppresses all microbial growth. Lipid emulsions alone, on the other hand, support gram-positive and gram-negative bacterial growth or fungal growth if contaminated. The Centers for Disease Control has recommended that lipid emulsions be administered for no longer than 12 hours, except in TNA systems, which can be administered over a 24-hour period.

Combined Enteral and Parenteral Feeding (Page 385)

Prolonged fasting (more than three days) results in enterocyte deterioration and decreased GI immunity. A possible source of infection with parenteral administration is translocation of enteric bacteria due to a compromised intestinal mucosal barrier. A combination of enteral and parenteral administration has been suggested because enteral infusion of small quantities of a liquid diet has helped prevent intestinal mucosal deterioration during parenteral nutrition in piglets, human infants and adults. Intestinal adaptations after disease and intestinal hypertrophy after surgery require the presence of intraluminal nutrients. Food intake promotes intestinal hyperplasia and brush border enzyme activity. Therefore, most recent recommendations encourage some enteral feeding to patients receiving parenteral nutritional support, if possible. Feeding both the small bowel and the patient is important.

REASSESSMENT (Pages 385-386)

Regular reassessment is a critical step in successful nutritional management of hospitalized patients, regardless of whether the enteral route, the parenteral route or both routes are used. Malnutrition in the form of insufficient nutrient intake to support tissue metabolism undermines appropriate medical and/or surgical therapeutic management of a case. Malnutrition is far more common in veterinary patients than is currently recognized. Patients resting in a cage have been mistakenly assumed to require little or no nutrition when, in fact, the nutrient costs of tissue repair, immunocompetence and drug metabolism are significant. Therefore, reassessment of nutritional status is important whether the animal remains in the hospital or recovers at home.

Monitoring Parameters (Page 385)

Food intake or administration of nutritional support for hospitalized patients should be reviewed at least daily. Body weight should be record-

Table 12-12. Metabolic complications of parenteral-nutrition (PN) administration, treatment and potential patient considerations.

Complications are listed in descending order of likely occurrence and treatments are listed from immediate to longer-term solutions. To minimize complications, patients should be hemodynamically stable and any electrolyte and acid-base abnormalities, severe tachycardia, hypotension and volume deficits should be corrected before starting PN.

EN/PN

Complications	Treatments	Patient considerations

Hyperglycemia	Stop infusion, recheck in two to to four hours, decrease PN infusion by 50% until normal, then increase infusion rate slowly
	Subcutaneous insulin therapy
	Change caloric sources:
	  Increase lipid fraction of calories
	  Decrease glucose fraction of calories
		Glucose intolerance

Hypokalemia	Add KCl or KPO_4 to PN bag
	Correct serum Mg as needed
	Change caloric sources:
	  Increase lipid fraction of calories
	  Decrease glucose fraction of calories
		GI or renal losses
		Drug therapies that increase urinary excretion
		Insulin therapy

Hypophosphatemia	Add $NaPO_4$ or KPO_4 to PN bag	Diabetic ketoacidosis

Hyperlipidemia	Stop infusion, recheck in two to four hours, decrease PN infusion by 50% until normal, then increase infusion rate slowly
	Change caloric sources:
	  Decrease lipid fraction of calories
	  Increase glucose fraction of calories
		Decreased lipid clearance

(Continued on next page.)

ed daily. Body condition should be noted; however, an animal's BCS is unlikely to change during the course of a hospital stay. Laboratory assessments specifically for patients receiving nutritional support are generally not necessary beyond those tests already routinely performed for critically ill patients. The most common alterations that occur in laboratory parameters associated with nutrient administration are decreases in serum potassium and phosphate levels, increases in serum glucose concentrations and triglyceridemia (**Table 12-12**).

Table 12-12. Metabolic complications of parenteral-nutrition (PN) administration, treatment and potential patient considerations (Continued.).

Complications	Treatments	Patient considerations

Phlebitis	Change catheter and infusion site
	Lower PN osmolality:
	  Increase lipid fraction
	    of calories
	  Decrease glucose fraction
	    of calories
	Add heparin to PN bag
		Properly hydrated
		Endogenous site
		  of infection

Hyperkalemia	Change PN bag and
	  decrease K^+
		Acidosis, renal failure,
		  sepsis
		Drug therapies that
		  decrease urinary
		  excretion

Hyperammonemia	Decrease PN infusion by
	  50% until normal
	Change PN bag, decrease
	  amino acid concentration
	Use branched-chain amino
	  acid sources
		Liver dysfunction
		GI bleeding

Hypomagnesemia	Add $MgSO_4$ to PN bag	GI or renal losses
		Drug therapies that
		  increase urinary
		  excretion

Hypoglycemia	Piggyback 50% dextrose drip
	  until normal
	Change caloric sources:
	  Decrease lipid fraction
	    of calories
	  Increase glucose fraction
	    of calories
		Sepsis
		Insulin therapy
		Insulinoma

Infected catheter site	Change catheter and infusion
	  site
	Culture catheter and PN
	  solution
	Give antibiotics based on
	  culture and antimicrobial
	  sensitivity tests
	Hot pack site
		Substandard catheter
		  care
		Endogenous site
		  of infection
		Properly hydrated

Changing Foods (Pages 385-386)

Parenterally fed patients should be fed enterally as soon as possible, but may continue to receive PN as enteral intake increases to meet RER. The food offered enterally may be a fixed-formula therapeutic food intended as the food to be fed to the patient at home because of an ongoing disease condition. When the patient has a decreased appetite, a highly palatable, fixed-formula food may be offered initially to stimulate oral consumption. This food may then be mixed in gradually decreasing proportions with the food to be fed on a long-term basis. (See Chapter 1.) Vomiting and diarrhea are the most common problems seen when refeeding patients orally. Foods should be introduced in amounts equal to RER in small frequent meals, and the amounts increased if well tolerated over the course of several days.

Clinical cases that illustrate and reinforce the nutritional concepts presented in this chapter can be found in Small Animal Clinical Nutrition, 4th ed., pp 391-399.

EN/PN

CHAPTER 13

Obesity

For a review of the unabridged chapter, see Burkholder WJ, Toll PW. Obesity. In: Hand MS, Thatcher CD, Remillard RL, et al, eds. Small Animal Clinical Nutrition, 4th ed. Topeka, KS: Mark Morris Institute, 2000; 401-430.

CLINICAL IMPORTANCE (Pages 401-403)*

Overconsumption of calories resulting in excess body fat is believed to be the most prevalent form of malnutrition in pets of westernized societies. The most extensive surveys estimate that approximately one-fourth of the dogs and cats presented to small animal practitioners in westernized societies are overweight to grossly obese.

Overweight dogs and cats are divided into three categories: 1) those animals 1 to 9% above optimal weight are simply above optimal, 2) those 10 to 19% above optimal are considered overweight and 3) those 20% above optimal are considered obese.

Health Risks of Obesity (Pages 402-403)

Obesity has detrimental effects on the health of dogs and cats. **Table 13-1** lists abnormalities associated with or exacerbated by excess body weight.

ASSESSMENT (Pages 403-414)

Assess the Animal (Pages 403-409)

Clinically, it is useful to assess body condition of cats and dogs as objectively as possible. The ability to assess body condition is necessary to determine when a dog or cat is likely to benefit from weight loss, and to substantiate a diagnosis of obesity to pet owners and convince them that their pet needs to lose weight. Additionally, quantifying excess body weight and determining ideal body weight are essential to a weight-loss program. Several clinical methods can be used to differentiate between optimal, overweight and obese body conditions. Relative body weight (RBW), body condition score (BCS) and morphometric analysis are tools that can be used to substantiate a diagnosis of obesity. Morphometric analysis is discussed in Small Animal Clinical Nutrition, 4th ed., pp 406-407.

*Page numbers in headings refer to Small Animal Clinical Nutrition, 4th ed., where additional information may be found.

Key Nutritional Factors–Obesity (Page 403)

Factors	Dietary recommendations
Energy	Calorie-restricted dog foods should contain <3.4 kcal (14.23 kJ) ME/g DM for weight loss Calorie-restricted cat foods should contain <3.6 kcal (15.06 kJ) ME/g DM for weight loss Weight loss in dogs: 1.0 to 1.2 x RER for optimal weight Weight loss in cats: 0.8 to 1.0 x RER for optimal weight
Fat	Dogs: calorie-restricted foods should contain 5 to 12% fat (DMB) for weight loss and prevention of weight gain in obese-prone dogs Cats: calorie-restricted foods should contain 7 to 14% fat (DMB) for weight loss and prevention of weight gain in obese-prone cats
Fiber	Weight loss: 12 to 30% crude fiber (DMB) may be helpful, indicated and required Prevention of weight gain: 6 to 30% crude fiber (DMB) may be helpful, indicated and required
Protein	Dogs: foods for weight loss should contain >25% crude protein (DMB) to help prevent loss of lean body tissues Cats: foods for weight loss should contain >35% crude protein (DMB) to help prevent loss of lean body tissues

Table 13-1. Diseases associated with or exacerbated by obesity.

Endocrinopathies
Diabetes mellitus
Hyperadrenocorticism
Hypopituitarism
Hypothalamic lesions
Hypothyroidism
Insulinoma
Pituitary chromophobe adenoma

Functional alterations
Decreased immune function
Dyspnea
Dystocia
Exercise intolerance
Heat intolerance
Hypertension
Joint stress/musculoskeletal pain

Metabolic alterations
Anesthetic complications
Glucose intolerance
Hepatic lipidosis (cats)
Hyperlipidemia
Insulin resistance

Other diseases
Cardiovascular disease
Degenerative joint and orthopedic disease
Transitional cell carcinoma (bladder)

Body condition scoring is probably the single most useful of the three methods. Body condition scoring can be used to demonstrate to the owner what contours are absent that otherwise should be present and what bony prominences should be easily felt but are not readily palpable.

Relative Body Weight (Pages 404-405)

RBW is simply an animal's current weight divided by its estimated optimal weight. Animals that are at their optimal weight have an RBW of 1.00 or 100%. Animals weighing less than optimal have an RBW less than 1 and animals weighing more than optimal have an RBW greater than 1. The values of 1.10 and 1.20 have been suggested as division points for placing people, and by extrapolation dogs and cats, into overweight and obese categories, respectively (**Table 13-2**).

Deciding on an optimal weight can be problematic for the veterinarian and the pet owner, especially if the two disagree. An individual dog's or cat's optimal weight can be estimated from several sources. The best estimate is a recorded mature adult weight at a time when the pet's body condition was simultaneously assessed as optimal. Weight at the time the dog or cat reached adult age is often a good indicator of the optimal weight if body condition assessments are unavailable.

Average weights determined by the American Kennel Club for individual breeds and both genders may be used if there is no record of an animal's past weight(s). (See Appendix H, Small Animal Clinical Nutrition, 4th ed., pp 1037-1046.) However, optimal weights among individual dogs of a given breed may vary 25% or more, as indicated by the range of optimal weights listed for different breeds in Appendix H. Thus, frame size of the individual becomes important for selecting or determining optimal weight from published averages.

Establishing absolute weights to indicate obesity for cats is somewhat easier than for dogs. Most domestic cats have an optimal body weight in the range of 3.2 to 4.5 kg. Some cats may weigh up to 5.5 kg without being overweight. However, any domestic cat weighing more than 6.4 kg is likely to be obese unless it has an exceptionally large skeletal frame. These ranges are similar to average body weights of 4.8 and 6.5 kg for optimal and overweight, respectively, as determined in a recent assessment of prevalence and risk factors for obesity in cats.

Body Condition Scoring (Pages 405-406)

The BCS is a subjective assessment of an animal's body fat, and to a lesser extent protein stores, that takes into account the animal's frame size independent of its weight. Chapter 1 presents a 5-point body condition scoring system in detail.

Table 13-2. Relationships between actual, optimal and relative weight, anticipated body condition score and percent body fat.

| Ideal weight (kg) | Relative weight | | | | |
| | 0.8 or 80% | 0.9 or 90% | 1.0 or 100% | 1.1 or 110% | 1.2 or 120% |
	Actual body weight (kg)				
1	0.8	0.9	1.0	1.1	1.2
2	1.6	1.8	2.0	2.2	2.4
3	2.4	2.7	3.0	3.3	3.6
4	3.2	3.6	4.0	4.4	4.8
5	4.0	4.5	5.0	5.5	6.0
10	8.0	9.0	10.0	11.0	12.0
15	12.0	13.5	15.0	16.5	18.0
20	16.0	18.0	20.0	22.0	24.0
25	20.0	22.5	25.0	27.5	30.0
30	24.0	27.0	30.0	33.0	36.0
35	28.0	31.5	35.0	38.5	42.0
40	32.0	36.0	40.0	44.0	48.0
45	36.0	40.5	45.0	49.5	54.0
50	40.0	45.0	50.0	55.0	60.0
55	44.0	49.5	55.0	60.5	66.0
60	48.0	54.0	60.0	66.0	72.0
65	52.0	58.5	65.0	71.5	78.0
70	56.0	63.0	70.0	77.0	84.0
75	60.0	67.5	75.0	82.5	90.0
80	64.0	72.0	80.0	88.0	96.0
Anticipated body condition score					
	1	2	3	4	5
Anticipated % body fat					
	<5	5 to 15	16 to 25	26 to 35	>35

A 5-point system scored to the nearest half score subdivides into three categories each for insufficient, optimal and excess body conditions, with a score of 3.0 falling in the middle of the optimal range.

In general, dogs and cats in optimal body condition have: 1) normal body contours and silhouettes, 2) bony prominences that can be readily palpated but not seen or felt above skin surfaces and 3) intra-abdominal fat insufficient to obscure or interfere with abdominal palpation. The most critical division points in a 5-point system are between the scores of 2.0 vs. 2.5 and 3.5 vs. 4.0 because assignment of a BCS <2.5 or >3.5 suggests action should be taken to return the animal's BCS to the optimal range.

In addition to body weight, the BCS should always be recorded in the hospital record whenever a veterinarian examines an animal. Body weight alone does not indicate how appropriate the weight is for an individual animal. A Labrador retriever weighing 30 kg or a domestic shorthair cat weighing 4 kg may be underweight, at optimal weight or overweight. The BCS puts body weight in perspective for what individual dogs and cats ought to weigh.

Body condition can be formally defined as the ratio of fat to nonfat tissues in the body and thus can be used to estimate percent body fat (%BF). If 15 to 25% body fat is accepted as optimal for dogs and cats, then an animal with a BCS of 3.0 out of 5.0 (3/5) should have about 20% body fat. Research to critically assess the capability of BCS to predict body composition suggests that %BF changes by 10% for each change in BCS on a 5-point scale. For example, as BCS increases from 3 to 5, the corresponding body fat increases from 20 to 40%. Therefore, a BCS of 4.0 correlates with an average of 30% body fat, which is similar to the critical %BF for assessing when people are at risk for ill effects from being overweight (**Table 13-2**).

Risk Factors for Obesity (Pages 407-409)

Obesity develops when animals are in positive energy balance for an extended period of time. This occurs when energy intake increases, energy expenditure decreases or both happen. Under most circumstances, homeostatic mechanisms control energy intake and maintain body composition at or near some "set point." Several risk factors contribute to positive energy balance and/or affect the body's compositional set point. Genetics, gender, age, physical activity and caloric composition of foods are risk factors for positive energy balance, weight gain and obesity. These risk factors must be understood if obesity is to be prevented or treated effectively.

GENETICS Labrador retrievers, Cairn terriers, cocker spaniels, longhaired dachshunds, Shetland sheepdogs, basset hounds, cavalier King Charles spaniels and beagles have a greater prevalence of obesity than other breeds. In contrast to dogs, cats of mixed breeding are more likely to be obese than are purebred cats.

GENDER AND GONADECTOMY Gonadectomy increases the risk of obesity in dogs and cats. Neutered cats are more likely to be overweight than intact cats of either gender. Neutered female dogs are about twice as likely to be overweight than intact female dogs. A similar trend occurs in castrated male dogs. Gonadectomy predisposes dogs and cats to weight gain and eventual obesity for several reasons. Gonadectomized cats have resting metabolic rates 20 to 25% below those of intact cats of similar age, as measured by indirect calorimetry. In practical terms, this finding indicates that neutered cats require only 75 to 80% of the food required by intact animals to maintain optimal body weight. These measurements confirm the previously suspected decrease in metabolic rate caused by the loss of estrogens and androgens from gonadectomy. This reduction in resting metabolic rate appears to be in addition to any decrease in physical activity that might occur from decreased roaming and sexual behavior.

Furthermore, estrogens suppress appetite in several species of animals. Removal of the metabolic effects of estrogens and androgens by gonadectomy may lead to increased food consumption, when at the same time, the animal's energy requirement is lower because of its decreased metabolic rate and physical activity.

AGE Age has been correlated with prevalence of excess body weight in dogs and cats. Very few animals younger than two years of age are classified as overweight. After two years of age the prevalence of overweight dogs and cats increases and reaches a maximum at around six to eight years of age. Studies show a plateau or slight decrease in prevalence of overweight dogs and cats until about 12 years of age when the prevalence tends to decrease markedly in most cross-sectional studies.

ACTIVITY Activity or exercise can contribute markedly to daily energy expenditure. Thus, it is not surprising that animals with decreased activity or restricted opportunities for exercise are at greater risk for becoming overweight.

FOOD AND FEEDING Feeding very palatable foods free choice to dogs and cats may encourage consumption that exceeds requirements. Likewise, excessive use of treats or substitution of food for other types of interaction between the owner and pet may also encourage excessive consumption of food.

Energy Requirements (Page 409)

Daily consumption of calories must exceed daily energy expenditure for a sustained period in order for overweight or obese body conditions to develop. An understanding of the components that contribute to the daily energy requirement (DER) is useful to appreciate why animals of similar body weight and frame size can have different caloric requirements independent of genetics or neuter status. An understanding of the components that contribute to DER is also important to understanding the rationale behind recommendations and alterations made to correct obesity.

The DER to maintain body weight of an animal can be subdivided into: 1) resting energy requirement (RER), 2) exercise energy requirement (EER), 3) thermic effect of food (TEF) and 4) adaptive thermogenesis (AT). RER correlates closely with lean body mass in people, and accounts for 60 to 80% of the total DER for adult maintenance. The RER represents energy used to maintain normal physiologic functions at rest in a thermoneutral environment several hours after eating. Lean tissues perform these physiologic functions with very little energy required to maintain adipose tissue.

EER is the energy expended for muscular activity. The contribution of EER to DER is determined by the animal's body weight plus the time and intensity of muscular activity. Certainly, animals that are less active or have little opportunity to exercise expend less energy compared with active animals of similar size. The EER can account for 10 to 20% of total daily energy expended by nonathletic people.

TEF is the obligatory cost of digesting and absorbing food. TEF constitutes approximately 10% of total expenditure and is affected by food composition and the number of meals eaten per day. The obligatory cost associated with digesting and absorbing each meal is the reason weight-reduction programs recommend multiple small meals per day rather than one or two large meals. RER, EER and TEF make up the majority of DER; thus, these are the components that can be manipulated to affect the amount and rate of weight loss.

AT makes up the smallest proportion of the DER for most pets. AT is the energy expended to regulate body temperature during exposure to ambient temperatures below or above the thermoneutral zone or during transient periods of excess caloric consumption.

Imbalances favoring caloric intake relative to caloric expenditure (positive energy balance) can occur several different ways. Gonadectomy may suddenly make the number of calories fed before surgery excessive for the animal's subsequent rate of metabolism. Determining amounts to feed based on manufacturer recommendations may also lead to excessive caloric intake. This results not because manufacturers make inappropriate or self-serving recommendations, but rather because manufacturers base recommendations on ranges and average caloric requirements for a given body weight. Recommendations often list a minimum and maximum amount of food to feed within a given range of body weights (e.g., two to four cups for a 5.9- to 11.4-kg dog). The maximum amount can be one and one-half to four times the minimum amount listed for a given range of body weights. Excess caloric intake can occur if the pet owner interprets that a smaller dog should be fed the larger amount.

When manufacturers recommend only one amount of food for a specific body weight, the amount is generally based on the average caloric requirement for that body weight. However, averages are population means and energy requirements vary widely around the population mean for any specified body weight. A recommended amount of food to supply the average caloric requirement for a 10-kg dog will be inappropriate for a 10-kg dog that requires only 60 to 80% of the average because of a lower metabolic rate (decreased RER) or less activity (lower EER).

Intake of calories can become excessive if changes occur in a pet's lifestyle or daily routine that markedly reduce activity without reducing calories. Such changes include moving to smaller dwellings, musculoskeletal injuries and diseases that require persistent long-term use of central nervous system depressants or corticosteroids. Finally, caloric requirements will decrease as some animals age. Certainly, requirements for a given weight are less for maintenance of adults than for growing individuals of similar weight. Some older animals may also require slightly less energy than younger adults as a result of reduced activity and possibly a reduction in lean body mass as a normal consequence of aging.

Assess the Food (Pages 409-414)

A pet's body weight and condition are determined by the nutrient composition of the food and the amount of food eaten daily. An accurate, complete history of the types and amounts of all foods eaten by the pet is necessary for making feeding recommendations in general, but especially for weight loss. Besides total quantity of food and calories, some consideration should be given to which nutrients (protein, fat and soluble carbohydrate) supply what proportion of the calories. The proportion of these nutrients in a food determines the food's caloric density and to some extent the acceptability of the food for reducing or maintaining body weight and condition of the pet. The Key Nutritional Factors table lists guidelines and ranges for energy, fat, fiber and protein content in foods suitable for producing weight loss or maintaining weight in dogs and cats that have lost weight or are predisposed to being overweight.

Food Amount (Pages 409-410)

A quantitative food record is an essential diagnostic step in developing a weight-loss program. To be quantitative, the food record must include amounts of all foods and account for all calories the patient consumes. Caloric content of commercial pet foods and treats can either be obtained from manufacturers or calculated. (See Chapter 1.) Appendices 2 and 3 list caloric content for many other foods and treats. (Also see Appendix M, Small Animal Clinical Nutrition, 4th ed., pp 1084-1091.) Most packaged human foods include caloric content on the label.

A quantitative food record helps determine how efficiently an animal uses calories and how easy it will be to return the animal to optimal weight. The food record can be indispensable for determining how severe caloric restriction will need to be to produce weight loss, both in total calories and in product selection. Equations to estimate energy requirements for cats or dogs often produce erroneous results because of wide variation between individual animals. In practice, individual animals are encountered that need precisely the same, markedly fewer and, occasionally, markedly more calories than the calculations suggest. Caloric restriction may be insufficient to produce weight loss or may even produce weight gain in some animals if calculations for caloric restriction are applied without taking into account the calories being eaten to maintain the animal's current weight.

Food Type (Pages 410-413)

Decreasing the caloric density of the food fed to overweight pets is the primary strategy for producing weight loss. Pet foods marketed as restricted in calories can vary widely in caloric content, proportion of nutrients contributing calories, fiber and digestibility. **Tables 13-3** and **13-4** list selected veterinary therapeutic and commercial pet foods marketed specifically for loss or control of body weight in dogs and cats, respectively. Regulatory definitions for the terms light, lean, reduced calorie and reduced fat have been implemented in the United States (**Table 13-5**).

Pet food manufacturers decrease the caloric density of foods by reducing fat and simultaneously increasing the fiber, air or moisture content of the food. Pet owners facing the challenge of getting their pet to lose weight often ask whether feeding less of the pet's current food would produce weight loss rather than having to switch to a calorie-restricted food. This approach is usually unsuccessful for three reasons.

First, most calorically dense pet foods contain more fat than do calorie-restricted foods. Fat has about 2.25 times the calories of an equivalent weight of carbohydrate or protein. The TEF from absorption of fat has been found to be less than the TEF from absorption of carbohydrate or protein in obese people. Studies in people have also determined that fat stored in the body comes primarily from dietary fat, whereas TEF is more closely correlated with carbohydrate intake. Thus, a food with more calories supplied from fat will tend to support retention of body weight and body fat even when total calories consumed are reduced.

Second, digestibility of a food (i.e., the amount eaten that is digested and absorbed) is inversely proportional to the total amount of food eaten. When less of a calorically dense food is fed, the proportion digested and thus the proportion of total energy extracted from the food will increase slightly. This is a secondary contributor to the caloric efficiency that occurs when less of a calorically dense food is fed. However, the amount of energy derived from the food is slightly greater than expected based on calculations using the reported energy density of the food and the amount eaten.

Third, all nutrients are reduced when amounts of a calorically dense food are decreased to restrict calories. A food for weight loss ideally should

Table 13-3. Levels of key nutrients in selected commercial foods marketed for caloric restriction and weight loss in dogs.

Recommended levels (wt. loss)***	kcal/kg* <3,400	MJ/kg <14.23	Fat** 5-12	Fiber 12-30	Protein >25
Dry canine products					
Hill's Prescription Diet Canine r/d	2,966	12.4	8.5	23.4	24.7
Hill's Prescription Diet Canine w/d	3,281	13.7	8.7	16.9	18.9
Hill's Science Diet Canine Maintenance Light	3,293	13.8	8.9	14.3	24.4
Iams Eukanuba Reduced Fat Adult Formula	4,189	17.5	10.5	1.9	21.3
Iams Eukanuba Restricted-Calorie	4,002	16.7	7.7	2.0	25.0
Iams Less Active	4,170	17.4	12.5	2.8	22.2
Leo Specific Fitness CRD	3,633	15.2	5.6	8.9	24.4
Medi-Cal Canine Fibre Formula	3,078	12.9	10.1	15.8	24.3
Medi-Cal Canine Weight Control/ Geriatric	3,434	14.4	8.3	5.9	19.6
Purina CNM OM-Formula	3,088	12.9	6.7	10.7	32.0
Purina Fit & Trim	3,104	13.0	6.5	8.0	28.4
Purina O.N.E. Reduced Calorie Formula	3,863	16.1	9.6	3.0	28.6
Purina ProPlan Reduced Calorie Formula	3,976	16.6	9.4	2.8	28.3
Skippy Cycle Custom Fitness Lite	3,217	13.5	10.1	4.8	18.7
Waltham/Pedigree Calorie Control/ Low Calorie	3,682	15.4	11.4	1.9	29.5
Moist canine products					
Hill's Prescription Diet Canine r/d	2,995	12.5	8.4	21.8	25.5
Hill's Prescription Diet Canine w/d	3,423	14.3	13.0	12.2	18.2
Hill's Science Diet Canine Maintenance Light	3,470	14.5	8.8	10.0	18.7
Iams Less Active (Beef, Liver & Rice Formula)	4,295	18.0	17.2	1.5	34.8
Leo Specific Fitness CRW (foil pack)	3,511	14.7	8.9	17.7	31.6
Medi-Cal Canine Fibre Formula	3,309	13.8	7.9	15.9	24.7
Medi-Cal Canine Weight Control	na	na	12.3	6.1	21.8
Purina CNM OM-Formula	2,468	10.3	8.4	19.2	44.1
Skippy Cycle Custom Fitness Lite	4,160	17.4	19.0	1.8	27.2
Waltham/Pedigree Calorie Control/ Low Calorie	3,662	15.4	16.9	2.1	52.8

Key: na = information not published by manufacturer.
*Energy expressed on dry matter basis. From manufacturers' published information or calculated from manufacturers' published as fed values.
**Nutrients expressed as % dry matter.
***Recommended levels for prevention of weight gain are the same except 6 to 30% crude fiber is indicated.

Table 13-4. Levels of key nutrients in selected commercial foods marketed for caloric restriction and weight loss in cats.

	kcal/kg*	MJ/kg	Fat**	Fiber	Protein
Recommended levels (wt. loss)*	<3,600	<15.06	7-14	12-30	>35
Dry feline products					
Hill's Prescription Diet Feline r/d	3,316	13.9	9.5	14.6	38.1
Hill's Prescription Diet Feline w/d	3,546	14.8	9.3	7.9	38.8
Hill's Science Diet Feline Maintenance Light	3,523	14.7	9.3	6.4	36.8
Iams Eukanuba Restricted-Calorie	4,246	17.8	10.1	2.1	35.4
Iams Less Active Formula	4,298	18.0	13.1	2.4	30.9
Leo Specific Fitness FRD	3,290	13.8	7.2	18.7	36.4
Medi-Cal Feline Fibre Control	2,813	11.8	13.8	7.3	34.1
Medi-Cal Feline Weight Control	3,831	16.0	12.3	3.8	34.4
Purina CNM OM-Formula	3,369	14.1	8.5	8.0	38.0
Purina ProPlan Reduced Calorie Formula	4,129	17.3	10.1	2.5	34.9
Select Care Feline Hifactor Formula	3,778	15.8	12.9	5.0	38.2
Select Care Feline Weight Formula	3,800	15.9	12.2	3.6	36.2
Waltham Calorie Control for Cats	3,375	14.1	7.7	4.8	44.0
Moist feline products					
Hill's Prescription Diet Feline r/d	3,153	13.2	8.9	17.4	36.0
Hill's Prescription Diet Feline w/d	3,758	15.7	16.7	10.7	41.3
Hill's Science Diet Feline Maintenance Light	3,550	14.9	12.5	10.0	47.5
Leo Specific Fitness FRW (foil pack)	3,458	14.5	9.3	18.7	42.1
Medi-Cal Feline Fibre Formula	3,873	16.2	17.7	5.5	40.2
Medi-Cal Feline Weight Control	3,391	14.2	14.4	8.6	40.2
Select Care Feline Hifactor Formula	3,857	16.1	16.1	7.2	34.1
Waltham/Whiskas Feline Calorie Control/Low Calorie	3,797	15.9	18.4	1.9	50.0

Key: na = information not published by manufacturer.
*Energy expressed on dry matter basis. From manufacturers' published information or calculated from manufacturers' published as fed values.
**Nutrients expressed as % dry matter.
***Recommended levels for prevention of weight gain are the same except 6 to 30% crude fiber is indicated.

be replete in all nutrients except energy so that protein, essential fatty acids, vitamins and minerals are present in amounts sufficient to support normal physiologic processes and retention of lean body tissue. Most pet foods are balanced for all other nutrients based on the energy content of the food and the expected intake required to support a given body weight. A deficiency in energy and other nutrients will occur if the amount of a maintenance food is markedly decreased to produce weight loss. Calorie-restricted foods generally have proportional increases in protein, vitamins and minerals to avert or minimize the creation of other nutrient deficiencies when fewer total calories are fed. The goal of a reducing food should be to restrict only energy, not other nutrients.

In addition to decreasing fat content to decrease caloric density, increasing the air in dry extruded foods, water in moist formulas and indigestible fiber in either moist or dry products can further dilute calories per volume of food on an as fed basis. The primary reason for diluting calories by one of these methods is to allow the animal to eat a larger volume of food without consuming additional calories. This is an attempt to accomplish two things. First, feeding larger volumes of food attempts to avert perceptions by some owners that the pet is being fed too little. Second, volume is increased in an effort to maximize the effect of bulk fill in the gastrointestinal (GI) tract for satiety (i.e., the transient postprandial interruption in hunger sensations and food-seeking activity of animals).

Use of Fiber in Foods for Weight Reduction (Pages 413-414)

Potential calorie-diluting agents are dietary fiber, dietary water and air added to dry foods. Water and air are quickly removed from the GI tract and contribute only transiently to GI fill.

Dietary fiber helps produce weight loss by diluting calories, increasing satiety and limiting food consumption as a result of more bulk being present in the GI tract. Fiber may also help produce weight loss by decreasing the availability of calories by interfering with the digestion and absorption of fat, protein and soluble carbohydrate. Many of the effects of dietary fiber depend on the specific type, form and amount of fiber used.

Studies suggest that increased levels of dietary fiber contribute to satiety via prolonged distention of the GI tract. Fiber types affect duration of gastric and intestinal distention differently. Insoluble fibers have little effect on gastric emptying, whereas soluble fibers slow gastric emptying. Although both soluble and insoluble fibers slow intestinal transit, insoluble fiber (purified cellulose) produces the greater effect. Thus, even though the type of fiber affects the two segments of the GI tract differently, total transit time through the entire GI tract is increased and is approximately the same for soluble and insoluble fibers.

Actual documentation of increased satiety from dietary fiber is difficult to prove in people and more so in other animals, because satiety is a subjective feeling of fullness and a lack of desire to eat. Indirect evidence for satiety can be obtained from animals by measuring decreases in food consumption and food-seeking activities. In one study, dogs offered maintenance calories from food containing 21% insoluble fiber consumed significantly less food and calories than when offered equivalent calories from foods containing less fiber. These same dogs also ate less food when subsequent meals were offered 30 to 45 minutes after consuming the high-fiber food, indicating a satiety effect.

Table 13-5. "Light," "lite" and "lean."

	Dry foods (<20% moisture)	Semi-moist foods (20 to 65% moisture)	Moist foods (>65% moisture)
		Dogs	
Light	3,100 kcal ME/kg food	2,500 kcal ME/kg food	900 kcal ME/kg food
Lean	9% fat as fed	7% fat as fed	4% fat as fed
		Cats	
Light	3,250 kcal ME/kg food	2,650 kcal ME/kg food	950 kcal ME/kg food
Lean	10% fat as fed	8% fat as fed	5% fat as fed

Key: ME = metabolizable energy.

Another factor regarding the value of dietary fiber in weight-management foods is fiber decreases the apparent digestibility of energy-providing nutrients in the food by 2 to 8%.

The effect of dietary fiber on mineral availability depends on the specific fiber(s) and mineral(s). In general, insoluble fibers such as cellulose are less likely to reduce mineral availability than are soluble fibers. (See Chapter 2, Small Animal Clinical Nutrition, 4th ed., pp 42-48.)

Dietary fiber increases the amount of fecal material and frequency of defecation. Dogs fed soluble fiber produced more feces than dogs fed similar amounts of predominantly insoluble fiber.

Pet owners should be informed that the quantity of feces their cat or dog produces will probably increase when fed foods containing more than 10% dry matter (DM) from fiber. Excessive flatus can also be an unwelcome side effect of feeding high-fiber foods. Fiber solubility roughly equates with fiber fermentability. Increased amounts of highly fermentable fiber in a food are more likely to result in flatulence.

Tables 13-3 and **13-4** show that the crude fiber content of selected commercial products marketed as reduced-calorie or weight-loss products can be similar to the crude fiber content of non-calorie-restricted pet foods (<5% of DM), mildly increased (5 to 10% of DM), moderately increased (10 to 15% of DM) or greatly increased (15 to 30% of DM). For the reasons noted above, most commercial calorie-restricted foods with increased fiber contain primarily insoluble fiber.

Assess the Feeding Method (Page 414)

Free-choice feeding rarely works for weight loss or for maintenance of reduced body weight even with the most calorie-restricted foods. The owner's quantitative description of how much pet food, treats, table food, and consumable chew toys are provided for the dog or cat must be assessed. "Bowls," "cups" and "handfuls" reported by owners come in all sizes; thus, the amount of food and calories these objects contain varies as well. The veterinary nutritionist's "cup" is a standard 8-oz. volume measure. The amount of dry dog or cat food reportedly fed by owners needs to be converted to this standard or some other usable quantity (i.e., weight) for determining calories, if the food record is to be quantitative. Treats, consumable chew toys and table food can supply significant calories, especially if the owner is unaware of their caloric content or how many the animal eats daily.

Whether the pet has access to any other sources of food also needs to be determined. Other sources include other pets' food in multi-pet households. Having multiple people feed the pet can result in multiple sources of food, particularly if different people have different opinions about the body condition of the pet. The previous two situations can condemn a weight-reduction program to failure before it ever begins if the owner cannot, or will not, feed the overweight pet separately and prevent the overweight pet from eating other pets' food. Dogs and cats that roam unsupervised also have the opportunity to eat at other locations.

FEEDING PLAN (Pages 414-419)

A successful weight-reduction program is a multi-step process that requires pet owner commitment, a feeding plan, an exercise plan, pet owner communication and patient monitoring.

Pet Owner Commitment (Pages 414-415)

The first step, and the foundation for weight loss, is for everyone involved in feeding the pet to recognize, accept and understand the reason why the pet should lose weight, and to make a commitment to accomplish that goal. Weight loss will not occur unless the pet owners recognize the problem and are willing to take corrective steps.

Several methods and techniques can be used to help owners recognize and accept that their pet is overweight and not just "stocky." Some of these techniques have already been discussed. Past body weights and BCS in the patient's medical record can be used to show an owner how excessive present body weight relates to the animal's frame size and optimal body weight. References to breed standards for adults in optimal weight can be used if no records of the patient's body weight are available. The owner can be shown and made to feel where bony structures on the patient should be readily palpable but are not, and where body contours of the patient differ from optimal. The BCS can be used to estimate %BF for additional body composition information. Estimates of %BF should be interpreted for the owner in relation to ranges of %BF found in animals of optimal body condition (15 to 25%), with obesity being defined as greater than 30% body fat.

If thoracic or abdominal radiographs have been taken, a side-by-side comparison with similar radiographic views from an animal of similar size at optimal body weight can effectively demonstrate to the owner the excess subcutaneous or intra-abdominal fat on the pet. Practitioners should consider keeping a reference set of radiographs for cats and dogs in optimal body condition for this specific purpose. Also, a side-by-side comparison of the overweight animal and one of the same breed and frame size in optimal body condition can serve the same purpose if an animal of optimal weight is available.

After the owners recognize and accept that the pet is overweight, the next step is for them to commit to a weight-loss program. There are several strategies to help owners make this commitment. Owners can be informed about documented problems associated with obesity and how returning the animal to optimal weight will reduce the risk of one or more of these problems. The risks can be quantified economically for animals likely to suffer orthopedic or metabolic problems because of their degree of obesity.

Often a strong motivating factor for commitment is to improve or palliate a problem caused or exacerbated by obesity. Weight loss in these cases becomes part of the overall therapeutic plan and can be crucial for realizing clinical improvement and benefit from other treatments.

If the pet owner commits to weight loss for the pet, the veterinarian's responsibility is to use knowledge of the physiology and nutrition involved with weight loss to prescribe a program with the best chance of succeeding. The veterinarian also has the responsibility to adjust the initial plan based on monitoring of the patient's progress and to reinforce the owner's commitment during the difficult periods that almost always occur when less weight is lost than was expected. Ideally, pet owners are responsible for the daily tasks that will produce weight loss because then they will modify the way they feed and exercise the pet.

Reports and position statements addressing obesity in people indicate the combination of reduced-calorie foods, regular exercise and behavior modification has the greatest chance of achieving and maintaining weight loss. Formulation of a program for achieving weight reduction consists of: 1) setting a goal for the amount of weight to lose, 2) setting an amount for daily caloric intake, 3) selecting a specific food and feeding method, 4) selecting a specific amount of exercise, 5) monitoring the progress of weight loss, 6) adjusting calories, food and exercise as necessary and 7) stabilizing caloric intake of the animal at its reduced weight to ensure that weight is not regained.

Select a Food (Pages 415-416)

Formulating a dietary plan involves: 1) setting the amount of calories the animal is allowed daily, 2) selecting and specifying the food and amount to supply daily calories and 3) selecting the way the food is to be fed. Development of an appropriate dietary plan must include calculation of the current caloric intake (DER for obese weight), determination of the animal's ideal body weight, calculation of a food dose (DER for weight loss) and adjustments to the plan based on monitoring.

Calculation of Caloric Intake (Page 415)

A detailed quantitative food record or feeding history completed by the owner is the only way of estimating the DER for maintaining a given body weight of an individual animal. It is crucial to know the number of calories required to maintain a pet's obese weight in order to specify the amount of food that will produce calorie-restricted weight loss. Although some obese patients may actually have "average" caloric requirements and simply overeat, many have caloric requirements that are below average. The only way to differentiate between the two is to compare the individual's current intake with the expected average DER for the present body weight. For the food record to be useful, all calories consumed must be accounted for as discussed in the food assessment section.

OBESITY

Determination of Ideal Body Weight (Page 415)

Determination of ideal body weight is important to set a goal for weight loss and provide one reference point for calculating amounts of food for caloric restriction. Because adipose tissue requires very little energy for maintenance, lean body mass primarily determines DER. Therefore, a given animal will have similar DER in ideal and obese body conditions (assuming similar lean body mass). Using ideal weight for energy requirement calculations will usually avoid overestimation of DER. Ideal weight is determined using the methods described above. **Table 13-2** lists ideal weight in comparison to current weight and BCS.

Food Dose Calculation (Pages 415-416)

If triglycerides were the only tissue component lost during weight reduction, then simply starving dogs, but not cats, would be an acceptable option for weight loss from a physiologic perspective. There are several disadvantages, however, to using starvation for weight reduction in dogs. Unfortunately, when body weight is lost under the best of circumstances, 10 to 25% of the loss comes from lean tissues. Loss of lean body mass ultimately decreases an animal's RER and the number of calories required for DER, unless the level of activity is increased to that associated with athletic training. Therefore, the underlying objective in setting the number of daily calories for weight loss is to restrict calories enough to produce weight loss, but still provide enough calories, protein, vitamins and minerals to prevent or minimize nutrient deficiencies and loss of lean body tissue.

Most calories consumed are used to maintain lean body tissues and physiologic functions. Therefore, if lean body mass is similar at ideal and obese body conditions, the energy required to maintain each body condition should also be similar. This is one reason why standard energy calculations may overestimate energy requirements for most obese patients. Daily caloric intake must be less than calories needed to maintain lean body tissues in order to use adipose tissue to supply energy to meet DER. The recommendation for dogs is to use 1.0 to 1.2 x RER (50 to 60% of DER) for optimal weight as an initial calculated estimate of calories required to produce appropriate weight loss. Theoretically this level of restriction will make caloric intake nearly equal to the calories required to support lean body mass at optimal weight. Energy for exercise and digestion of food must subsequently be supplied by catabolizing fat stores.

Recommendations for caloric restriction to produce weight loss in cats range from 50 to 80% of maintenance calories for optimal weight. Restricting calories for maintenance of optimal weight of a cat by more than 70% effectively makes caloric intake less than RER because DER for cats is only 1.2 to 1.4 x RER. RER represents a theoretical minimum for daily energy consumption. Restrictions to less than RER might seem problematic especially in overweight cats in which severe caloric restriction has been demonstrated to produce hepatic lipidosis. However, experimental and clinical trials using caloric restrictions between 59 and 80% of RER produced acceptable rates of weight loss in overweight cats with no biochemical evidence of hepatic lipidosis.

The number of calories estimated for weight loss needs to be compared with the number of calories currently being consumed as indicated in the feeding record. This comparison is necessary to ensure the calculated caloric restriction is less than what the animal is currently eating. More severe caloric restriction will be required than calculations suggest if calories calculated to produce weight loss are greater than or equal to calories being consumed to maintain the animal's excess body weight. In such cases, caloric restriction markedly less than the calculated RER may be required to achieve weight loss. Initially, 80% of present consumption is a reasonable starting point if an animal is maintaining excess weight on fewer calories than 1.2 x RER for optimal weight.

Ideally treats, snacks and human foods should be eliminated from reducing diets for dogs and cats to maximize the chances for successful weight loss. A portion of the total daily calories for weight loss can be reserved for treats if the owner must feed treats or snacks. Treats should be low-calorie foods such as the dry form of the reducing food, popcorn popped without butter, low-fat, low-starch vegetables or low-fat commercial treats. The calories supplied by the treats must be accounted for within the total calories allowed in the feeding plan.

Determine a Feeding Method (Pages 416-417)

Pet owners must quantitatively account for the foods they feed their pets. As little as one-fourth cup of a calorie-restricted food per day can be the difference between achieving or not achieving weight loss. Similarly, treats can be the difference between achieving or not achieving weight loss, if they are not accounted for within the total allowed calories or their specified number is exceeded. If the owner must feed treats, the number can be controlled by placing a specific quantity of treats containing the number of calories reserved for treats in a "treat container" each day. No additional treats are allowed for that day after the treat container is empty. In multi-pet households, the obese pet must be fed separately to prevent access to other pets' food.

Dogs and cats on weight-reduction programs should be fed multiple small meals during the day rather than a single large meal in order to take advantage of the obligatory energy cost for digesting and absorbing food. The optimal number of meals for maximizing caloric expenditure from thermic effect of food has not been determined. However, the total daily food should be divided into at least two portions fed eight to 12 hours apart. Most pet owners can feed two meals per day without disrupting their schedules. Clients who can conveniently feed three or more meals per day should do so.

Meal sizes should be in portions that are practical to measure (i.e., to the nearest one-fourth cup or can). If the daily amount of food does not divide evenly into portions that are readily measurable, some meals will contain less and others more food. The meals containing more food should be fed when the owner will be with the pet for the longest time between meals. The pet should be kept out of the kitchen and dining areas during preparation and consumption of family meals. These practices can help reduce the pet's begging and the owner's urge to give the pet additional food or treats.

Exercise (Pages 417-419)

Exercise is the only practical means of increasing energy expenditure to create or widen a deficit between energy consumed and energy expended for patients fed calorie-restricted foods. Exercise may also benefit obese patients by reducing the loss of lean body mass and maintaining or improving RER. In some cases, pets fail to lose weight unless exercise is part of the weight-reduction plan, regardless of the severity of caloric restriction.

Exercise should be implemented gradually, starting with amounts the patient can comfortably tolerate, especially if orthopedic, cardiovascular or pulmonary disease is also present. It is more important that the animal increase its activity by some amount each day even if it initially is able to walk only out the door to the sidewalk and back inside. The goal should be to work up to 20 minutes of uninterrupted walking if the animal cannot do this initially. Exercise may need to be omitted initially for patients recovering from orthopedic surgery because walking may exacerbate joint pain. Swimming is an alternative to walking that sometimes works for orthopedic patients if facilities are available to the owner. Because swimming requires more calories per minute than walking, the same number of calories can be expended in less time.

Some creativity is often required to increase the activity of an overweight cat. Although cats are not typically trained to walk on a leash, they can be if the owner is patient and persistent. Sometimes a cat will walk back home on a leash if an exceptionally dedicated owner is willing to carry the cat on the out-bound half of the walk. Less extraordinary ways of increasing a cat's daily activity are to engage the cat in supervised play with string, balls, laser "mice," other toys or other pets.

Pet Owner Communication (Page 419)

The specific recommendations for feeding, exercising and rechecking the animal need to be provided in a clear, concise, written format. Several pet food companies provide brochures that briefly explain obesity, its consequences and provide space to write individual instructions for feeding, exercise and recheck appointments. These forms usually have a space to document progress on a graph or to record weights on specific dates. Several computer software programs also generate forms incorporating recommendations and progress updates for a particular pet and owner. A practice may also elect to design and distribute its own printed or computer-generated material for this purpose. Information for a weight-reduction program should be clear, concise and fit on both sides of an 8.5- x 11-in. sheet of paper. See Small Animal Clinical Nutrition, 4th ed., p 418 for an example of such a form.

REASSESSMENT (Pages 419-423)

Regular monitoring of patient weight loss is important to ensure the prescribed program is effective and to motivate the owner. Simply telling a pet's owner to feed a certain quantity of a calorie-restricted food and increase the pet's activity is unlikely to produce weight loss for several reasons. Office rechecks or weigh-ins to monitor patient progress throughout weight loss are an integral component of a weight-reduction program, equal in importance to diet and exercise. There are three critical times during a weight-reduction program when rechecks can prevent the program from failing. These are at the very beginning, the very end and anytime in between when weight loss slows or stops.

Rechecks are used to accomplish several things necessary to ensure success of the weight-loss program. First, rechecks reinforce the importance of weight loss for the pet and the veterinary practice's commitment to helping the pet owner get the animal to lose weight. Rechecks also reinforce the owner's commitment to get the animal to lose weight. Rechecks become a moment of truth for both pet owner and veterinarian. Rechecks give pet owners an opportunity to see the results of their efforts or to see the impact of inadvertently or purposefully feeding extra calories or not ensuring that the pet performed the specified amount of exercise since the last recheck. Rechecks allow the veterinary health care team to adjust the caloric intake, feeding plan and exercise recommendations to get or keep weight loss proceeding at a reasonable rate. The opportunity to make these adjustments is an extremely important iterative step in a weight-loss program. The initial considerations and calculations for caloric restriction and the feeding plan, no matter how carefully or scientifically made, are only an educated guess at what the caloric restriction and food should be for a safe and reasonable rate of weight loss for an individual animal. The success and appropriateness of this educated guess are ultimately determined by changes in the body weight measured on scales, the BCS and morphometric analyses. For more information about using morphometric measurements, see Small Animal Clinical Nutrition, 4th ed., pp 406-407.

The amount of caloric restriction may be insufficient to accomplish weight loss for an individual animal despite what any calculation would suggest and despite 100% compliance by the owner. Problems with caloric restriction can occur initially or after some period of weight loss, perhaps because of a decreasing metabolic rate from the weight loss. If monitoring and counseling in the form of rechecks are not being done, these problems will not be detected until the animal is seen some time in the future weighing the same or more than when the weight-loss program was started. The opportunity to promote weight loss in such animals will probably be lost because the pet owner will conclude that switching the food and tolerating undesirable behaviors did not produce results and was not worth the trouble or expense involved.

The best reinforcement and encouragement come initially from seeing the pet's body weight decrease, and later from seeing the return of normal body contours and resolution of clinical signs (e.g., better exercise tolerance, reduced lameness or decreased insulin doses). Often, weight loss is imperceptible to the owner and requires objective documentation with scales or tapes. Change in body weight is the ultimate criterion for judging success or failure of the weight-loss program. However, if the period of

OBESITY

time between rechecks is short, or the rate of weight loss is particularly slow, progress based on body weight alone may not be readily apparent.

Rechecks should be continued after the pet attains its optimal or target weight. Simply feeding the animal its previous food, even at reduced amounts, may lead to weight gain, negating the effort required to produce weight loss and the resulting benefits. Recommendations for the specific food and calories required to maintain reduced body weight need to be individualized for each patient. The number of calories needed to maintain the reduced weight can be gauged from how many calories the pet ate to maintain its obese weight compared with standard estimates for that weight. The required calories can range from somewhat less than the estimated RER to the largest estimates for the DER.

Rechecks should be scheduled to allow enough time for detectable progress, but not so much time that the pet owner becomes dismayed at the lack of progress when problems are finally detected. Shorter intervals between rechecks are needed at the beginning and end of a weight-reduction program when the caloric content and amounts of food are changed. At the minimum, a week will be required before any progress can be detected. Generally two weeks is a safe and reasonable interval for most animals. Cats and some small dogs may take three weeks to lose enough weight for scales to measure the loss.

Three body weights are usually required to establish a trend and rate for changes in weight. Thus, a determination that initial caloric restriction is insufficient to produce weight loss can be made sooner with a two-week recheck interval than with a four-week interval, saving at least two and perhaps six weeks, during which the animal is not losing weight. Intervals between rechecks can be increased to every four to six weeks after weight loss is documented to occur at a steady rate acceptable to the pet owner and veterinary health care team. If the animal fails to lose weight during a four- to six-week interval with no apparent explanation (i.e., more calories or less exercise) then the rechecks need to be more frequent to determine if weight loss has stopped and to assess the degree of caloric restriction needed for weight loss to recur.

More frequent rechecks are needed when the animal reaches its target weight and calories are increased to maintain that weight. Rechecks should occur every one to two weeks to assess the appropriateness of caloric intake in conjunction with continued exercise. During this stage of the weight-loss program, no more than two weeks should elapse between weighings because consumption of too little or too much food has undesirable consequences.

It is better to focus on acceptable rates of weight loss instead of calculating a specific number of days that pet owners view as the time it will take to complete the weight-loss program. Minimum and maximum acceptable times for a cat or dog to complete a weight-loss program can be calculated. The loss of 2% of initial body weight per week can be used as the maximum desired rate of weight loss in typical obese patients and a loss of 0.5% of initial body weight per week can be used as the minimum desired rate of weight loss. These two weight-loss rates can be used to calculate the minimum and maximum time expected for a dog or cat to reach its ideal or target body weight.

Obese weight – desired weight = A (kg)
Obese weight x 2% = B (kg/week)
A ÷ B = C (number of weeks necessary for weight loss at 2% rate)
C x 4 = D (number of weeks necessary for weight loss at 0.5% rate)
Desired weight loss should occur within these two time frames.

See Small Animal Clinical Nutrition, 4th ed., pp 422 and 426-430 for examples.

A greater proportion of lean body tissue is lost when more than 2% of body weight is lost per week. This ultimately reduces the RER and works against the goal of maintaining the greatest metabolic rate possible in a weight-reduced animal. A 2% loss of initial body weight per week is a reasonable estimate of the maximum acceptable rate of weight loss.

At the other extreme, a rate of at least 0.5% of the initial body weight per week is needed to maintain owner interest and complete the weight-reduction program in a reasonable period. Realistically, eight to 12 months will be required to complete weight reduction of most dogs and cats that are truly obese and that have metabolic rates slower than predicted by standard equations.

OBESITY PREVENTION (Pages 423-424)

Successful prevention of obesity requires risk factor assessment, body composition evaluation and appropriate feeding recommendations. Pets that have increased risk or are beginning to increase body fat above optimal amounts should be fed fewer calories. Calculated DER is about 1.6 x RER for most dogs and 1.2 x RER for most cats. Obese-prone pets should be fed about 15% fewer calories (or 1.4 x RER for dogs and 1.0 x RER for cats). These are guidelines or starting points. Individual energy requirements can vary markedly from average estimates. Food amounts and feeding methods need to be evaluated and adjusted on an individual basis to maintain optimal body condition.

Growing animals that are overweight or obese have a greater risk of becoming overweight or obese adults. Pet owners who present new puppies and kittens for vaccinations should be counseled on how to feed their pets to prevent excessive rates of growth and weight gain. No single feeding method and type of food will work for all pets and exceptions can be found to refute any standard recommendation. Body condition and rate of growth are the ultimate criteria for determining whether the type of food, amount of food and feeding method are appropriate for a given individual. Pet owners should be told the characteristics of optimal body condition and shown how to assess the body condition of their pets. Owners can also be given target weights for age based on reasonable guidelines for appropriate rates of growth for the breed of dog. (See Appendix F, Small Animal Clinical Nutrition, 4th ed., pp 1020-1026.)

When pets are presented for gonadectomy, their owners should also be counseled that the amount of food given to cats and dogs that have been neutered will probably need to be reduced to maintain optimal body condition. Body condition should also be closely monitored in animals with orthopedic or physiologic conditions that decrease their physical activity or metabolic rate.

Clinical cases that illustrate and reinforce the nutritional concepts presented in this chapter can be found in Small Animal Clinical Nutrition, 4th ed., pp 426-430.

CHAPTER 14

Adverse Reactions to Food

For a review of the unabridged chapter, see Roudebush P, Guilford WG, Shanley KJ. Adverse Reactions to Food. In: Hand MS, Thatcher CD, Remillard RL, et al, eds. Small Animal Clinical Nutrition, 4th ed. Topeka, KS: Mark Morris Institute, 2000; 431-453.

CLINICAL IMPORTANCE (Pages 431-433)*

An adverse reaction to food is an abnormal response to an ingested food or food additive. Adverse reactions to food are composed of a variety of sub-classifications based on pathomechanisms (**Figure 14-1**). The terms food allergy and food hypersensitivity should be reserved for those adverse reactions to food that have an immunologic basis. Food intolerance refers to a large category of adverse food reactions due to nonimmunologic mechanisms.

Adverse food reactions mimic other diseases, especially pruritic dermatoses, and they often coexist with other allergic conditions. Veterinary dermatologists suggest that adverse food reactions account for 1 to 6% of all dermatoses in general practice and that food allergy constitutes 10 to 20% of allergic responses in dogs and cats. Food allergy is probably the third most common hypersensitivity skin disease in dogs and cats after arthropod (flea) hypersensitivity and atopy. Adverse food reactions can cause a wide variety of cutaneous lesions and should be considered in any pruritic dog or cat.

Plasmacytic-lymphocytic enteritis and eosinophilic enteritis are the most common forms of inflammatory bowel disease (IBD) identified in dogs and cats, and are the most common cause of chronic vomiting and diarrhea in these two species. Plasmacytic-lymphocytic and eosinophilic intestinal infiltrates may be associated with many conditions, but no obvious cause can be identified in most cases. Food sensitivity seems to be involved in some cases. Clinical response to a modification in the feeding plan suggests that hypersensitivity to food antigens plays a role in dogs with chronic idiopathic or plasmacytic-lymphocytic colitis. It is not known if chronic colitis or other forms of inflammatory disease of the small bowel are a direct manifestation of an adverse food reaction or if modifying the feeding plan is merely palliative in some animals.

Food additives, preservatives and dyes are frequently mentioned by veterinarians as ingredients that may be associated with adverse food reactions,

*Page numbers in headings refer to Small Animal Clinical Nutrition, 4th ed., where additional information may be found.

Key Nutritional Factors–Adverse Reactions to Food (Page 433)

Factors (Dogs)	Associated conditions	Dietary recommendations
Protein	Nonseasonal pruritic dermatitis Pruritic bilateral otitis externa Recurrent bacterial pyoderma Angioedema Atopy Flea-allergy dermatitis Vomiting Small bowel diarrhea Colitis	Limit dietary protein to one or two sources Use novel protein sources Avoid excess levels of dietary protein (dermatologic cases only) Protein 16 to 20% (DMB) Use a food with protein digestibility >87% or one containing a protein hydrolysate Use a food that is nutritionally balanced for dogs Avoid foods that contain wheat, barley or rye (dogs with diarrhea)
Vasoactive amines	Nonseasonal pruritic dermatitis Pruritic bilateral otitis externa Recurrent bacterial pyoderma Angiodema Atopy Flea-allergy dermatitis Vomiting Small bowel diarrhea Colitis	Avoid foods that contain certain fish ingredients (tuna, mackerel)
Food additives	Nonseasonal pruritic dermatitis Pruritic bilateral otitis externa Recurrent bacterial pyoderma Angiodema Atopy Flea-allergy dermatitis Vomiting Small bowel diarrhea Colitis Erythema multiforme	Use a food that is free of or has reduced numbers of food additives

(Continued on next page.)

Key Nutritional Factors–Adverse Reactions to Food (Page 433)
(Continued.)

Factors (Cats)	Associated conditions	Dietary recommendations
Protein	Severe, generalized pruritus	Limit dietary protein to one or two sources
	Miliary dermatitis	Use protein sources to which the cat has not been
	Pruritus with self trauma	exposed previously
	Self-inflicted alopecia	Avoid excess levels of dietary protein
	Eosinophilic plaque	(dermatologic cases only)
	Angioedema, urticaria or	Protein 30 to 45% (DMB)
	conjunctivitis	Use a food with protein digestibility >87% or one
	Indolent ulcer of lip	containing a protein hydrolysate
	Vomiting	Use a food that is nutritionally balanced for cats
	Small bowel diarrhea	
	Colitis	
Vasoactive amines	Same as above	Avoid foods that contain certain fish ingredients
		(tuna, mackerel)
Food additives	Same as above	Use a food that is free of or has reduced numbers
	Erythema multiforme	of food additives

FOOD RXNS

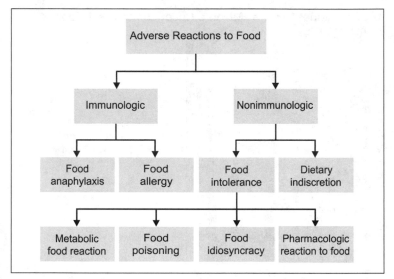

Figure 14-1. Classification of adverse reactions to food.

specifically dermatologic and gastrointestinal (GI) signs due to food allergy. Although additives are frequently incriminated as causing problems in dogs and cats, few data confirm this perception.

ASSESSMENT (Pages 433-440)

Assess the Animal (Pages 433-439)
Nutritional History (Page 433)

Dogs and cats may develop food allergy after prolonged exposure to one brand, type or form of food. In contrast, adverse reactions due to food intolerance may occur after a single exposure to a food ingredient because immune amplification is not necessary.

The nutritional history of the patient should be reviewed carefully for ingredients thought to be commonly associated with adverse food reactions. The nutritional history should include a complete list of the foods used in the pet's regular feeding plan or as treats including: 1) specific commercial foods, 2) commercial snacks and treats, 3) supplements, 4) chewable medications, 5) chew toys, 6) human foods and 7) access to other sources of food. It is often helpful to have the pet owner keep a diary for several weeks documenting the types of food and other items the pet ingests daily.

History and Physical Examination (Pages 433-435)

DERMATOLOGIC RESPONSES TO ADVERSE FOOD REACTIONS IN DOGS Reports of adverse food reactions in dogs with cutaneous disease do not document a gender predisposition and ages range from four months to 14 years. Up to one-third of canine cases, however, may occur in dogs less than one year of age. Because many adverse food reactions occur in young dogs, the index of suspicion for food allergy may rise above that for atopic disease when pruritic dermatoses occur in dogs less than six months old. Most investigators have not found a breed predilection, whereas others have found that cocker spaniels, springer spaniels, Labrador retrievers, collies, miniature schnauzers, Chinese Shar-Pei, West Highland white terriers, wheaten terriers, boxers, dachshunds, Dalmatians, Lhasa apsos, German shepherd dogs and golden retrievers are at increased risk.

Adverse food reactions in dogs typically occur as nonseasonal pruritic dermatitis, occasionally accompanied by GI signs. The pruritus varies in severity. Lesion distribution is often indistinguishable from that seen with atopy; feet, face, axillae, perineal region, inguinal region, rump and ears are often affected. One-fourth of dogs with adverse food reactions have lesions only in the region of the ears. This finding suggests that adverse food reactions should always be suspected in dogs with pruritic, bilateral otitis externa, even if accompanied by secondary bacterial or *Malassezia* infections.

Adverse food reactions in dogs produce no set of pathognomonic cutaneous signs. A variety of primary and secondary skin lesions occur and include: 1) papules, 2) erythroderma, 3) excoriations, 4) hyperpigmentation, 5) epidermal collarettes, 6) pododermatitis, 7) seborrhea sicca and 8) otitis externa. Adverse food reactions often mimic other common canine skin disorders including pyoderma, pruritic seborrheic dermatoses, folliculitis and ectoparasitism. Twenty to 30% or more of dogs with suspected adverse food reactions may have concurrent allergic disease, such as flea-allergic dermatitis or atopy. Some dogs present with only recurrent bacterial pyoderma, with or without pruritus, wherein all clinical signs resolve temporarily with antibiotic therapy.

Food anaphylaxis is an acute reaction to food or food additives with systemic consequences. The most common clinical manifestation in dogs occurs in localized form referred to as angioedema or facioconjunctival edema. Angioedema is typically manifested by large edematous swellings of the lips, face, eyelids, ears, conjunctiva and/or tongue, with or without pruritus.

FOOD RXNS

DERMATOLOGIC RESPONSES TO ADVERSE FOOD REACTIONS IN CATS Gender predisposition has not been documented in adverse food reactions in cats and ages have ranged from six months to 12 years. In one study, almost half of the cats developed the disease by two years of age. Siamese and Siamese cross cats may be at increased risk because they accounted for nearly one-third of cases in two studies.

Dermatologic signs include several different clinical reaction patterns such as: 1) severe, generalized pruritus without lesions, 2) miliary dermatitis, 3) pruritus with self trauma centered around the head, neck and ears, 4) traumatic alopecia, 5) moist dermatitis and 6) scaling dermatoses. In one study, angioedema, urticaria or conjunctivitis occurred in one-third of cats with adverse food reactions. Adverse reactions to food may also cause self-inflicted alopecia (i.e., psychogenic alopecia, neurodermatitis), eosinophilic plaques and indolent ulcers of the lips in some cats. Concurrent flea-allergy dermatitis or atopy may occur in up to 30% of cats with suspected adverse food reactions.

Moderate to marked peripheral lymphadomegaly is found in up to one-third of cats with dermatologic manifestations of food allergy. Absolute peripheral eosinophilia occurs in 20 to 50% of feline cases.

GI RESPONSES TO ADVERSE FOOD REACTIONS IN DOGS AND CATS Gender predilections have not been established for GI disease resulting from adverse reactions to foods. Similarly, there are no well-documented breed predispositions to GI food allergy, but Chinese Shar-Pei and German shepherd dogs are commonly affected. Furthermore, gluten-sensitive enteropathy has been well documented in Irish setter dogs. A wide age range of patients can be affected, including dogs and cats as young as weaning age.

Every level of the GI tract can be damaged by food allergies. Clinical signs usually relate to gastric and small bowel dysfunction, but colitis can also occur. Vomiting and diarrhea are prominent features. The diarrhea can be profuse and watery, mucoid or hemorrhagic. Intermittent abdominal pain is occasionally seen. Concurrent cutaneous signs may be seen. GI disturbances occur in 10 to 15% of dogs and cats with cutaneous manifestations of food sensitivity. In experimentally induced food hypersensitivity, the most common clinical signs are diarrhea, an increase in the number of bowel movements and occasional vomiting.

The role of food allergy in canine and feline IBD is unknown. Hypersensitivity to food is probably involved in the pathogenesis of this syndrome; at least some affected animals could be more appropriately diagnosed as suffering from food protein-induced enterocolitis. The subclassification of IBD in which hypersensitivity to food antigens is most likely to play a causal role is eosinophilic enterocolitis.

Risk Factors *(Page 435)*

Risk factors for adverse food reactions in animals are poorly documented but may include: 1) certain foods or food ingredients, 2) poorly digestible proteins, 3) any disease that increases intestinal mucosal permeability (e.g., viral enteritis), 4) selective IgA deficiency, 5) certain breeds, 6) age (less than one year old) and 7) concurrent allergic disease.

Etiopathogenesis *(Pages 435-439)*

IMMUNOLOGIC REACTIONS TO FOOD/Food Allergens The specific food allergens or ingredients that cause problems in animals have been poorly documented (**Table 14-1**). In general, the major food allergens that have been identified in people are water-soluble glycoproteins that have molecular weights ranging from 10,000 to 70,000 daltons and are stable to treatment with heat, acid and proteases. Other physiochemical properties that account for their unique allergenicity are poorly understood.

Abnormalities in the GI defense mechanisms may predispose patients to food allergies. Predisposing factors for food allergy include: 1) mucosal barrier failure (poorly digestible proteins, incomplete protein digestion, increased intestinal mucosal permeability, age-related changes in microvillous cell membrane composition, inflammatory-induced changes in mucus composition) and 2) defective immunoregulation (decreased IgA secretion, deranged cell-mediated responses of gut-associated lymphoid tissue (GALT), monocyte-macrophage system dysfunction). Which of these pathomechanisms are important predisposing factors in dogs and cats awaits further investigation. Selected IgA deficiency may be common in dogs with chronic dermatopathies and enteropathies, and may occur in some allergic cats.

IMMUNOLOGIC REACTIONS TO FOOD/Gluten (Gliadin) Enteropathy Gluten-induced enteropathy (celiac disease) is an important chronic inflammatory disease of the small intestine of people. The prevalence of gluten intolerance in dogs and cats is unknown. Recent work has conclusively demonstrated that an analogous disorder affects the Irish setter dog, and clinical experience suggests that other breeds may also be affected.

NONIMMUNOLOGIC REACTIONS TO FOOD Nonimmunologic, abnormal physiologic reactions to food include food intolerance and dietary indiscretion (**Figure 14-1**). Food intolerance mimics food allergy except that it can occur on the first exposure to a food or food additive, because nonimmunologic mechanisms are involved. The incidence of food intolerance vs. food hypersensitivity or food allergy is unknown.

NONIMMUNOLOGIC REACTIONS TO FOOD/Food Poisoning Food poisoning or food toxicosis is an adverse effect caused by the direct action of

Table 14-1. Ingredients commonly associated with adverse food reactions.

Dogs		Cats	
Beef	68% of	Beef	89% of
Dairy products	reported	Dairy products	reported
Wheat	cases	Fish	cases
Lamb	25% of	Chicken/poultry*	
Chicken egg	reported	Food additives*	
Chicken	cases		
Soy			
Corn*			
Food additives*			

*These pet food ingredients are commonly incriminated as causing adverse food reactions but are rarely reported in the literature.

a food or food additive on the host. Examples of food poisoning include ingestion of: 1) nutrient excesses (vitamin A or vitamin D toxicosis), 2) food contaminated with microorganisms or their toxic metabolites (e.g., scavenging putrefied material, vomitoxin), 3) specific foods (onions, chocolate) or 4) toxic food preservatives (benzoic acid or propylene glycol in cats). (See Chapter 7.) Food poisoning is a frequent cause of GI disease in dogs and cats. In addition to ingestion of pathogenic microorganisms and/or their toxins, food poisoning can result from the ingestion of plant-derived toxins or irritants.

NONIMMUNOLOGIC REACTIONS TO FOOD/Reactions to Food Additives
Although food additives are frequently incriminated as causing problems in dogs and cats, few data confirm this perception. Propylene glycol has been documented to cause hematologic abnormalities in cats and subsequently has been eliminated from cat foods sold in the United States and some other countries. Disulfides found in onions (e.g., onion powder, onion-based broth and baby foods containing onion) promote oxidative damage to hemoglobin in canine and feline red blood cells. The result is Heinz body production and red cell destruction.

NONIMMUNOLOGIC REACTIONS TO FOOD/Reactions to Vasoactive Amines in Food Another cause of food intolerance is pharmacologic reactions to substances found in food. Vasoactive or biogenic amines such as histamine cause clinical signs in people when present in excessive levels in food. Scombroid fish such as tuna, mackerel, skipjack and bonito that spoil before consumption are a frequent cause of histamine toxicosis in people. Clinical signs usually include diarrhea, flushing, sweating, nausea, vomiting, urticaria, facial swelling and erythroderma.

The role of histamine and other vasoactive amines in food intolerance in animals is unknown. Adverse reactions to ingested scombroid fish have been observed in cats and dogs. Recent surveys to detect histamine in pet foods found the highest levels of histamine in moist fish-based cat foods and those cat foods containing fish solubles. Vasoactive amines such as cadaverine may also exacerbate adverse reactions to spoiled fish by inhibiting histamine metabolism. Tyramine, spermine, spermidine, phenethylamine, putrescine and cadaverine are other vasoactive amines found in pet foods. Vasoactive or biogenic amines may not be present in levels high enough to cause clinical signs, but could lower the threshold levels for allergens in individual dogs and cats. Idiosyncratic intolerances to small quantities of histamine have been reported to occur in people and animals.

NONIMMUNOLOGIC REACTIONS TO FOOD/Carbohydrate Intolerance
The diarrhea, bloating and abdominal discomfort that occur when animals with lactose intolerance ingest milk are relatively common metabolic adverse reactions in dogs and cats. Puppies and kittens normally have adequate levels of intestinal lactase to permit digestion of lactose in the dam's milk. In many subjects, brush border disaccharidase activity decreases after weaning to a fraction of the activity found in young animals. Osmotic diarrhea will often occur when excessive levels of lactose are consumed. Puppies, kittens and adult animals may develop diarrhea when given cow's or goat's milk because these milk sources contain more lactose than either bitch's or queen's milk. (See Appendix K, Small Animal Clinical Nutrition, 4th ed., pp 1064-1072.)

Intolerance to disaccharides commonly occurs secondary to enteritis or rapid food changes. Loss of intestinal brush border disaccharidase activity contributes to the diarrhea associated with enteritis. Inadequate intestinal disaccharidase activity is also one of the factors responsible for diarrhea subsequent to rapid food changes. Several days are required for intestinal disaccharidase activity to adapt to changes in food carbohydrate sources.

NONIMMUNOLOGIC REACTIONS TO FOOD/Dietary Indiscretion Dietary indiscretions such as gluttony, pica and garbage ingestion usually cause GI signs and can be suspected based on the environmental and nutritional history. The clinical signs may be caused by ingestion of excessive fat, bacterial or fungal toxins, vasoactive amines or indigestible materials such as bone, plastic, wood and aluminum foil.

Key Nutritional Factors *(Page 439)*

Because most food allergens are thought to be glycoproteins, protein in food is the nutrient of most concern in patients with suspected adverse food reactions. The number of different proteins in the food, amount of protein, digestibility of the protein and whether the patient has been exposed previously to the protein are all important factors. Pet food additives

FOOD RXNS

such as antimicrobial preservatives, colorants, antioxidant preservatives and emulsifying agents may cause either food intolerance or food allergy.

Assess the Food and Feeding Method (Pages 439-440)

Ingredient statements on commercial pet food labels in the United States are sources of information for identifying all the food ingredients that might cause adverse reactions. An individual animal may develop an adverse reaction to virtually any pet food ingredient. However, particular attention should be directed at those ingredients that contain protein and food additives. When the ingredient statement is incomplete, the manufacturer or distributor should be contacted for detailed ingredient information.

In addition to questioning the pet owner about the regular food eaten by the animal, the nutritional assessment should include information about other items the pet may have ingested that contain protein or food additives. Examples include commercial snacks and treats, chew toys, human foods routinely fed to the pet and access to other sources of food (e.g., a dog with access to cat food in the same household). Supplements such as chewable vitamin tablets, chewable medications and fatty acid capsules, liquids and powders may also contain allergenic proteins or additives. It is often helpful to have the pet owner keep a diary for several weeks to document what types of food, supplements and other items are ingested daily by the animal.

Although pet food additives are frequently incriminated as causing food allergy or food intolerance, there have been no published case reports to date of an adverse food reaction in dogs or cats specifically caused by a pet food additive. Additives are found least often in moist pet foods and most commonly in semi-moist foods, treats, snacks and dry foods. Many moist commercial pet foods are free of additives.

Nutritional assessment should also evaluate foods or pet food ingredients for excessive levels of vasoactive or biogenic amines such as histamine. As mentioned above, the highest levels of histamine occur in moist fish-based cat foods and cat foods containing fish solubles. Human foods that may contain excessive levels of vasoactive or biogenic amines include tomato, avocado, cheese, liver, processed meats such as sausage and certain fish.

It may not always be necessary to change the feeding method when managing a patient with an adverse food reaction, but a thorough assessment includes verification that an appropriate feeding method is being used. Items to consider include feeding route, amount fed, how the food is offered, access to other food and who feeds the animal. All of this information should have been gathered when the history of the animal was obtained. If the animal has a normal body condition score (BCS 3/5), the amount of food it was fed previously (energy basis) was probably appropriate.

Table 14-2. Characteristics of an ideal elimination food.

Avoids excess levels of dietary protein
Free of excessive levels of vasoactive amines
Free of food additives
High protein digestibility (>87%) or contains a protein hydrolysate
Limited number of protein sources
Novel protein sources
Nutritionally adequate for the intended species, age and lifestyle

FEEDING PLAN (Pages 440-445)

The Ideal Elimination Food (Page 440)

Dietary elimination trials are the main diagnostic method used in dogs and cats with suspected adverse food reactions. At the present time, intradermal skin testing, radioallergosorbent tests (RAST) and enzyme-linked immunosorbent assays (ELISA) for food hypersensitivity are considered unreliable in animals with dermatologic disease.

The ideal elimination food is described in **Table 14-2**. Ingredients in an ideal elimination food should provide a limited number of novel protein sources; preferably one to two different types of protein to which the animal has not been previously exposed. This recommendation often includes a commercial or homemade food with one animal protein source and one vegetable protein source. Excess protein levels should be avoided to reduce the amount of potential allergens to which the dermatologic patient is exposed. A higher protein level may be necessary to counteract protein losses from the GI tract or impaired absorption in patients with hypoproteinemia and weight loss associated with severe GI disease.

Protein digestibility is also an important factor when assessing an elimination food. Complete digestion of food protein results in free amino acids and small peptides that are poor antigens. Thus, an incompletely digested food protein has the potential to incite an allergic response because of residual antigenic proteins and large polypeptides. Protein digestibility has been documented for some commercial pet foods marketed as hypoallergenic or elimination foods. A protein digestibility exceeding 87% is recommended for such foods. An alternate strategy is to use a food containing protein hydrolysates. Protein hydrolysates have molecular weights below levels that commonly elicit an allergic response.

Although specific pet food additives have not been documented to cause adverse food reactions, food additives generally should be avoided in elimination foods. The ideal elimination food should avoid ingredients such as certain kinds of fish that are known to contain higher levels of vasoactive amines than do other pet food ingredients.

FOOD RXNS

Finally, although elimination trials are only performed for several weeks to months, the food used in the trial should be nutritionally complete and balanced for the intended species, age and lifestyle of the animal. Elimination trials are often performed with young animals in which nutritionally inadequate foods are more likely to result in clinical disease. Homemade foods typically contain excess protein, but are deficient in calories, calcium, vitamins and microminerals. (See Chapter 6.)

Commercial Elimination Foods (Pages 442-443)

Several companies manufacture foods with limited and different protein sources (**Tables 14-3** through **14-5**). These commercial products are attractive because they are convenient, often contain novel protein sources and are nutritionally complete and balanced for either dogs or cats. Protein digestibility among these products varies markedly. Furthermore, few of these commercial foods have been adequately tested in dogs and cats with known adverse food reactions; only a few commercial foods have undergone the scrutiny of clinical trials using patients with dermatologic or GI disease (**Tables 14-3** and **14-4**). In published clinical trials, two-thirds to three-fourths of patients with suspected adverse food reactions showed significant improvement in clinical signs when fed commercial elimination-type foods. When selecting a commercial elimination food, use the criteria outlined in **Table 14-2**. Also, foods selected should have passed an Association of American Feed Control Officials (AAFCO) or similar feeding trial.

Commercial Foods Containing Protein Hydrolysates

The newest concept for managing veterinary patients with suspected adverse food reactions is use of commercial foods containing hydrolyzed protein ingredients. Veterinary therapeutic foods containing protein hydrolysates offer several hypothetical advantages to traditional commercial or homemade elimination foods. Protein hydrolysates of appropriate molecular weight (<10,000 daltons) will not elicit an immune-mediated response and may be regarded as truly "hypoallergenic" ingredients. Foods containing protein hydrolysates may also benefit patients that have increased GI permeability, in which enhanced protein absorption contributes to the pathogenesis of the disease. As an example, patients with IBD may benefit from use of protein hydrolysate-based foods. Protein hydrolysates have been used for many years in human infant formulas and for human patients with various GI diseases.

Current veterinary products containing protein hydrolysates as major ingredients are listed in **Table 14-5**. Some of these products contain only hydrolyzed protein sources, whereas others contain hydrolyzed and intact protein sources. Novel or unique protein sources are not important with protein hydrolysates. Total protein content, average molecular weight of the hydrolyzed protein and digestibility of nutrients vary among these products. Protein hydrolysates in some commercial veterinary therapeutic foods have markedly lower immunogenicity than the parent proteins. These protein hydrolysates are excellent ingredients for elimination foods. Palatability of veterinary therapeutic foods containing protein hydrolysates is often better than or at parity with traditional products containing novel protein sources such as duck, venison, fish and potato.

To date, few published clinical trials document use of foods containing protein hydrolysates in veterinary patients. One trial found positive responses in a small number of dogs (n = 6) with IBD.

Homemade Elimination Foods (Pages 440-442)

Results of a survey of veterinarians in the American Academy of Veterinary Dermatology (AAVD) showed homemade foods were recommended most often as the initial test food for dogs and cats with suspected food allergy. Homemade test foods usually include a single protein source or a combination of a single protein source and a single carbohydrate source. Ingredients recommended most often for homemade feline foods include lamb baby food, lamb, rice and rabbit. Ingredients recommended most often for homemade canine foods include lamb, rice, potato, fish, rabbit, venison and tofu.

Most of the homemade foods recommended in the AAVD survey for initial management of dogs and cats with suspected food allergy were nutritionally inadequate for growth or adult maintenance. Most homemade foods fail to meet nutritional requirements because they are made from a minimum of ingredients. In general, homemade foods lack a source of calcium, essential fatty acids, certain vitamins and other micronutrients and contain excessive levels of protein, which are contraindicated in food allergy cases.

Feeding nutritionally inadequate homemade foods to young dogs and cats for more than three weeks may result in clinical disease. Clinical signs of anorexia and poor growth occur in puppies within 10 to 20 days of feeding a thiamin-deficient food. Anorexia and emesis also appear within one to two weeks of feeding a thiamin-deficient food to cats. Many previously recommended homemade elimination foods have a severe inverse calcium-phosphorus ratio of 1:10. Foods with severe mineral imbalances can cause skeletal disease in young dogs within four weeks and should not be fed for longer than three weeks.

Complete and balanced homemade food recipes are available in this book (See Chapter 6.) and in other references. Nonflavored vitamin and mineral supplements are not perceived as causes of adverse food reactions. Additive-free supplements that do not contain animal or vegetable proteins are unlikely to be sources of ingested allergens. Intolerance to calcium supplements in

Table 14-3. Selected commercial products marketed or recommended as elimination foods for dogs.

Products	Protein sources*	Digestibility >87%	AAFCO approved	Clinical trials
Moist canine products				
Hill's Prescription Diet Canine d/d Lamb & Rice	rice, lamb, lamb liver	Y	Y	Y
Hill's Prescription Diet Canine d/d Whitefish & Rice	whitefish, rice	na	Y	N
Iams Eukanuba Veterinary Diets Response FP/Canine	catfish, herring meal, potato starch, beet pulp	na	N	N
IVD Limited Ingredient Diets Duck Formula	potato, duck, duck stock, duck byproducts	na	na	N
IVD Limited Ingredient Diets Lamb Formula	potato, lamb, lamb stock, lamb byproducts	na	na	N
IVD Limited Ingredient Diets Rabbit Formula	potato, rabbit, rabbit stock, rabbit byproducts	na	na	N
IVD Limited Ingredient Diets Venison Formula	potato, venison stock, venison, venison byproducts	na	na	N
IVD Limited Ingredient Diets Whitefish Formula	potato, whitefish, fish stock	na	Y	N
Leo Specific Dermil CDW	mutton, rice	na	na	N
Waltham/Pedigree Selected Protein Diet-Chicken & Rice	chicken byproducts, chicken, rice, natural flavors	na	na	Y
Waltham/Pedigree Selected Protein Diet-Lamb & Rice	lamb byproducts, lamb, rice, natural flavors	na	na	N
Waltham/Pedigree Selected Protein Diet-Venison	venison, venison byproducts, rice, natural flavors	na	na	N
Wysong Canine Anergen	lamb, lamb liver, brown rice, flax, yeast	na	na	N

(Continued on next page.)

Table 14-3. Selected commercial products marketed or recommended as elimination foods for dogs (Continued.).

Products	Protein sources*	Digestibility >87%	AAFCO approved	Clinical trials
Dry canine products				
Hill's Prescription Diet Canine d/d Rice & Duck	rice, duck byproducts, rice protein concentrate	Y	Y	N
Hill's Prescription Diet Canine d/d Rice & Egg	rice, egg	Y	Y	Y
Hill's Prescription Diet Canine d/d Rice & Salmon	rice, salmon, rice protein concentrate	Y	Y	N
Iams Eukanuba Veterinary Diets Response FP/Canine	potato, herring meal, catfish, beet pulp	na	N	Y
IVD Limited Ingredient Duck & Potato	potato, duck, duck meal, potato fiber, natural flavor	na	Y	N
IVD Limited Ingredient Lamb & Potato	potato, lamb, lamb meal, lamb digest	na	na	N
IVD Limited Ingredient Venison & Potato	potato, venison, venison meal, venison liver	na	na	N
IVD Select Care Vegetarian Formula	oat flour, rice, potato, flaxseed, beet pulp, tomato, yeast, carrot	na	N	N
Leo Specific Dermil CDD	egg, rice	na	na	N
Medi-Cal Canine Hypoallergenic Formula	oat flour, duck meal, oat bran, yeast, potato protein, duck digest	na	na	N

(Continued on next page.)

FOOD RXNS

Table 14-3. Selected commercial products marketed or recommended as elimination foods for dogs (Continued.).

Products	Protein sources*	Digestibility >87%	AAFCO approved	Clinical trials
Purina Veterinary Diets CNM LA-Formula	rice, salmon meal, trout, canola meal, yeast	N	Y	N
Waltham/Pedigree Selected Protein Diet-Capelin & Tapioca	tapioca, capelin	na	na	Y
Waltham/Pedigree Selected Protein Diet-Chicken & Rice	chicken byproducts, chicken, rice, natural flavors	na	na	N
Waltham/Pedigree Selected Protein Diet-Rice & Catfish	rice, catfish meal, rice gluten, catfish, natural flavor	na	na	N
Wysong Canine Anergen	lamb meal, chicken, brown rice, flax, quinoa, yeast, kelp	na	na	N

Key: na = information not available from manufacturer, Y = yes, N = no.
*Sources obtained from ingredient list on information panel of package or manufacturer's technical information.

Table 14-4. Selected commercial products marketed or recommended as elimination foods for cats.

Products	Protein sources*	Digestibility >87%	AAFCO approved	Clinical trials
Moist feline products				
Hill's Prescription Diet Feline d/d	lamb lungs, lamb liver, rice	Y	Y	Y
Iams Eukanuba Veterinary Diets Response LB/Feline	lamb liver, lamb tripe, barley, lamb meal, beet pulp	na	N	N
IVD Limited Ingredient Diets Lamb & Green Peas Formula	lamb byproducts, lamb stock, lamb, peas	na	na	N
IVD Limited Ingredient Diets Rabbit Formula	rabbit byproducts, rabbit stock, rabbit, potato	na	na	N
IVD Limited Ingredient Diets Venison Formula	venison byproducts, venison stock, venison, potato	na	na	N
Waltham/Whiskas Selected Protein Diet-Chicken & Rice	chicken, rice	na	na	Y
Waltham/Whiskas Selected Protein Diet-with Venison & Rice	venison, venison byproducts, rice, natural flavors	na	na	Y
Wysong Feline Anergen	lamb, lamb liver, brown rice, kelp, quinoa	na	na	N
Dry feline products				
IVD Limited Ingredient Diets Green Peas & Duck	peas, duck meal, duck, natural flavors	na	na	N
IVD Limited Ingredient Diets Lamb & Potato	potato, lamb, lamb meal, lamb digest	na	na	N
IVD Limited Ingredient Diets Venison & Potato	potato, venison, venison meal, venison liver, venison digest	na	na	N

(Continued on next page.)

FOOD RXNS

Table 14-4. Selected commercial products marketed or recommended as elimination foods for cats (Continued.).

Products	Protein sources*	Digestibility >87%	AAFCO approved	Clinical trials
Dry feline products				
Medi-Cal Feline Hypoallergenic/ Gastro Formula	oat flour, duck meal, potato protein, duck digest, yeast, oat bran	na	na	N
Waltham Selected Protein Diet-Rice & Duck	rice, duck byproduct meal, rice gluten, duck digest, duck	na	na	N
Waltham/Whiskas Selected Protein Diet-Capelin & Tapioca	tapicoa, capelin	na	na	Y
Wysong Feline Anergen	poultry, poultry meal, rice, oats, lamb meal, liver digest, flax, kelp, yeast	na	na	N

Key: na = information not available from manufacturer, Y = yes, N = no.
*Sources obtained from ingredient list on information panel of package or manufacturer's technical information.

Table 14-5. Selected characteristics of veterinary therapeutic foods containing protein hydrolysates.

Product names	Intact animal protein sources (Y/N) (ingredients)	Intact plant protein sources (Y/N) (ingredients)	Hydrolysate (av. mol. wt)	Protein (% DMB)
Hill's Prescription Diet Canine z/d ULTRA Allergen Free	N (hydrolyzed chicken liver, hydrolyzed chicken)	N (starch)	6,000	18.1
Hill's Prescription Diet Canine z/d Low Allergen	N (hydrolyzed chicken liver, hydrolyzed chicken)	Y (potato)	6,000	19.5
Hill's Prescription Diet Feline z/d Low Allergen	N (hydrolyzed chicken liver, hydrolyzed chicken)	Y (rice)	6,000	38.2
Purina CNM HA-Formula	N (modified isolated soy protein)	N (cornstarch)	12,200	21.3
DVM EXclude Veterinary Exclusion Diet	N (hydrolyzed casein and liver)	Y (oats, pinto beans)	5,000	18.3

Key: DMB = dry matter basis, av. mol. wt. = average molecular weight (daltons), Y = yes, N = no.

atopic children has been reported but is rare. Homemade rations should also contain a source of essential fatty acids, such as vegetable oil. Vegetable oils are not a routine source of ingested allergens; studies show that people allergic to peanuts and soybeans can safely ingest peanut oil or soybean oil. Homemade food recipes should provide an optimal amount of protein and foods for cats should be supplemented with taurine.

Performing an Elimination Trial in Patients with Dermatologic Disease (Page 443)

Before an elimination trial is initiated, the client should feed the dog or cat its usual food for seven to 14 days. During this time the client should record the type and amount of food ingested, any other ingested food items, such as table foods, treats and snacks, and the occurrence and character of adverse reactions (**Figure 14-2**). The patient is then fed a controlled elimination food for four to 12 weeks. In addition to the feeding change, no other substances should be ingested including treats, flavored vitamin supplements, chewable medications, fatty acid supplements and chew toys. During the elimination trial, the client should document daily the type and amount of food ingested and the occurrence and character of adverse reactions. A daily food diary helps document progression of clinical signs during the elimination trial and whether a strict elimination trial was performed in the home environment. See Small Animal Clinical Nutrition, 4th ed., pp 444-445 for a daily food diary that can be used in dietary elimination trials.

A tentative diagnosis of an adverse food reaction in dermatologic patients is made if the level of pruritus markedly decreases. This improvement may be gradual and may take four to 12 weeks to become evident. A diagnosis of an adverse food reaction is confirmed if clinical signs reappear within 10 to 14 days after the animal's former food and other ingested substances are offered as a challenge. Reinstituting the elimination food should resolve the clinical signs induced by the food challenge.

Provocation involves introducing single ingredients until as many positive reactions as possible can be documented. Clients and veterinarians are often reluctant to pursue challenge and provocation after clinical signs have improved or been eliminated. Provocation may also be difficult to perform in many dogs and cats because commercial pet foods contain large numbers of ingredients and feeding the same ingredients often cannot be duplicated in challenge studies.

Elimination trials are often difficult to interpret because of concurrent allergic skin disease. In several studies, at least 20 to 30% of dogs and cats with adverse food reactions had concurrent hypersensitivities. These patients may only partially respond to an elimination trial. Flea-allergy dermatitis and atopy are the most common canine and feline allergies and should be eliminated through other diagnostic testing.

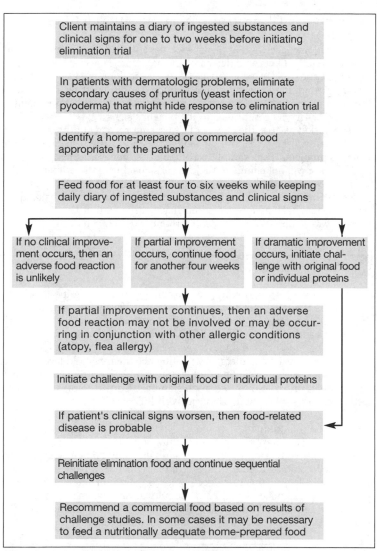

FOOD RXNS

Figure 14-2. Protocol for elimination-challenge trials for the diagnosis of adverse reactions to food.

Performing an Elimination Trial in Patients with GI Disease
(Page 443)

Elimination-challenge trial designs for patients with GI disease are similar to those for patients with dermatologic problems. However, shorter elimination periods are usually satisfactory (two to four weeks). In chronic relapsing conditions, the elimination period chosen must be greater than the usual symptom-free period of the patient to allow reliable assessment of how food sensitivity contributes to the patient's signs.

As with skin disease, the degree of clinical improvement during the elimination trial will be 100% only if food sensitivity is the sole cause of the patient's problems. For example, resolution of allergies acquired as a result of GI disease will not eliminate the clinical signs due to the primary GI disease process. Recrudescence of GI signs after challenge of a food-sensitive patient with the responsible allergen will usually occur within the first three days, but may take as long as seven days, particularly if the responsible allergen was removed from the food for longer than one month.

REASSESSMENT (Page 446)

For most adverse food reactions, avoiding the offending foods or food additives is the most effective treatment. How selective or meticulous an avoidance diet must be depends on the individual animal's sensitivity. Some dogs and cats may suffer adverse reactions to even trace quantities of an offending food or food additive, whereas others may have a higher tolerance level. Concurrent allergies will influence the threshold level of clinical signs in some animals. Symptomatic therapy for pruritic animals may also include corticosteroids and antihistamines. Corticosteroids along with feeding changes are often used in cats with IBD.

One-third of people fed a strict avoidance food for one to two years have tolerated the reintroduction of food allergens. This finding suggests that strict avoidance of food allergens may allow some dogs and cats to tolerate exposure to the same food allergens later in life. Both homemade and commercial foods can be used for long-term maintenance of patients with suspected food allergy. Homemade recipes for long-term maintenance must be nutritionally adequate. An attempt should always be made to find an acceptable commercial food that will increase owner compliance with the feeding change and ensure a nutritionally adequate ration.

Clinical cases that illustrate and reinforce the nutritional concepts presented in this chapter can be found in Small Animal Clinical Nutrition, 4th ed., pp 448-453.

CHAPTER 15

Skin and Hair Disorders

For a review of the unabridged chapter, see Roudebush P, Sousa CA, Logas DE. Skin and Hair Disorders. In: Hand MS, Thatcher CD, Remillard RL, et al, eds. Small Animal Clinical Nutrition, 4th ed. Topeka, KS: Mark Morris Institute, 2000; 455-474.

CLINICAL IMPORTANCE (Pages 455-456)*

Surveys indicate that 15 to 25% of all small animal practice activity is involved with the diagnosis and treatment of problems with the skin and coat.

The most commonly diagnosed canine skin disorders are: 1) allergy (flea-bite hypersensitivity, atopy), 2) cutaneous neoplasms, 3) bacterial pyoderma, 4) seborrhea, 5) parasitic dermatoses, 6) adverse reactions to food (food hypersensitivity or food intolerance), 7) immune-mediated dermatoses and 8) endocrine dermatoses. The most common feline skin disorders are: 1) abscesses, 2) parasitic dermatoses, 3) allergy (flea-bite hypersensitivity, atopy), 4) miliary dermatitis, 5) eosinophilic granuloma complex, 6) fungal infections, 7) adverse reactions to food, 8) psychogenic dermatoses, 9) seborrheic conditions, 10) neoplastic tumors and 11) immune-mediated dermatoses.

The skin and coat can be affected by many nutritional factors (See the Key Nutritional Factors–Skin and Hair Disorders table.). In addition, many pet owners are anxious to improve the quality and appearance of their animal's coat. This emphasizes the importance of understanding the nutritional factors that affect normal skin and hair and the nutritional factors that should be investigated in animals with skin disorders.

ASSESSMENT (Pages 456-467)

Assess the Animal (Pages 456-466)
History (Pages 456-458)
The signalment (species, breed, age, gender, reproductive status, hair color) is an important part of the historical information that should be obtained for patients with dermatologic problems, especially those with

---

*Page numbers in headings refer to Small Animal Clinical Nutrition, 4th ed., where additional information may be found.

Key Nutritional Factors–Skin and Hair Disorders (Page 457)

Factors	Associated conditions	Nutritional recommendations
Protein and energy	Keratinization abnormalities Loss of normal hair color Secondary bacterial or yeast infection Impaired wound healing Decubital ulcers Telogen defluxion Anagen defluxion	Avoid protein and energy deficiency Adult maintenance Dog: Protein = 25 to 30% (DMB*) 　Fat = 10 to 15% (DMB) Cat: Protein = 30 to 45% (DMB) 　Fat = 15 to 25% (DMB) Growth/lactation Dog: Protein = 30 to 35% (DMB) 　Fat = 15 to 30% (DMB) Cat: Protein = 35 to 50% (DMB) 　Fat = 20 to 35% (DMB) Use food with dry matter digestibility >80%
Essential fatty acids	Excessive scales (seborrhea sicca) Alopecia Dry, dull coat Lack of normal hair growth Erythroderma Interdigital exudation	Avoid fatty acid deficiency Dog: Linoleic acid >1.0% (DMB) Cat: Linoleic acid >0.5% (DMB) Some dogs and cats respond to levels in excess of those listed above Provide adequate levels and availability of zinc, B-complex vitamins and vitamin E
Zinc	Alopecia Skin ulceration Dermatitis Paronychia Foot pad disease Slow hair growth Buccal margin ulceration Hyperkeratotic plaques	Avoid zinc deficiency Dog: 100 to 200 mg/kg food (DMB) Cat: 50 to 150 mg/kg food (DMB) Avoid excess calcium Higher levels of zinc should be present in foods with calcium >1.5% (DMB) Avoid excess copper (copper <200 mg/kg food, DMB)

(Continued on next page.)

Key Nutritional Factors–Skin and Hair Disorders (Page 457)
(Continued.)

Factors	Associated conditions	Nutritional recommendations
Zinc	Secondary bacterial or yeast infection	Avoid essential fatty acid deficiency (See previous page.) Zinc supplementation (Do not give with food.) Zinc sulfate 10 mg/BW_{kg}/day per os 10 to 15 mg/BW_{kg}/week IV Zinc methionine 2 mg/BW_{kg}/day
Copper	Loss of normal color Dull or rough coat Reduced density of hair Alopecia	Avoid copper deficiency Dog: >5 to 10 mg/kg food (DMB) Cat: >15 mg/kg food (DMB) Avoid excess zinc (zinc <1,000 mg/kg food, DMB) Avoid ingredients that have low copper availability Copper oxide Liver from simple-stomached mammals Avoid excess calcium Higher levels of copper should be present in foods with calcium >1.5% (DMB)
Vitamin A	Seborrheic skin disease (mainly cocker spaniels) Keratinization disorders Chin acne Nasodigital hyperkeratosis Ear margin seborrhea/dermatosis Callus	Treatment with retinoids: See **Table 15-4** for details Vitamin A alcohol 625 to 1,000 U/BW_{kg}, q24h, per os 10,000 U q24h, per os cocker spaniel, miniature schnauzer 50,000 U q24h, per os Labrador retriever Tretinoin Apply topically q12 to 24h

SKIN/HAIR

(Continued on next page.)

Key Nutritional Factors–Skin and Hair Disorders (Page 457)
(Continued.)

Factors	Associated conditions	Nutritional recommendations
Vitamin A	Actinic keratosis Cutaneous neoplasms Schnauzer comedo syndrome Sebaceous adenitis Lamellar ichthyosis	Isotretinoin 1 to 3 mg/BW$_{kg}$, q24h, per os Etretinate 0.75 to 1.0 mg/BW$_{kg}$, q24h, per os
Vitamin E	Discoid lupus erythematosus Systemic lupus erythematosus Pemphigus erythematosus Sterile panniculitis Acanthosis nigricans Dermatomyositis Ear margin vasculitis	Treatment with vitamin E Dog: 200 to 800 IU twice daily, per os

*Dry matter basis.

possible nutritional disorders (**Table 15-1**). Both dogs and cats develop nutritionally related skin and hair disorders, although certain conditions such as zinc-responsive dermatoses are best characterized in dogs.

The age of the animal is important; most skin and coat changes due to nutritional deficiencies occur in young growing animals or adult females during gestation and lactation. The requirement for most nutrients is highest during growth and lactation, which accounts for the increase in nutritionally related skin and hair problems seen during these lifestages. As an example, a biotin-deficient food will cause dermatitis, alopecia, dull fur and achromotrichia when fed to young growing kittens but will not cause similar clinical signs when fed to nonlactating adult cats. There are many other examples of nutritional skin diseases that occur during periods of increased nutritional demand but do not occur during the normal adult lifestage. This age-related phenomenon is complicated by the fact that congenital defects of the integument and certain parasitic, fungal and bacterial infections of the skin are also more common in dogs and cats younger than six months. Gender and reproductive status affect the prevalence of certain skin problems, but they are not usual risk factors in nutritional skin disorders, unless the increased nutritional demands of pregnancy or lactation are present.

Table 15-1. Breed predilection for non-neoplastic skin diseases often managed by food changes or supplementation.*

Breed	Disease
Airedale terrier	Atopy
Akita	Sebaceous adenitis
Basenji	Atopy
Basset hound	Atopy
Beagle	Atopy
Boston terrier	Atopy
Boxer	Atopy, adverse reactions to food
Bull terrier	Atopy, acrodermatitis, zinc-responsive dermatosis
Chesapeake Bay retriever	Atopy
Dalmatian	Atopy
English bulldog	Atopy
German shepherd dog	Atopy, adverse reactions to food, seborrhea (primary)
Golden retriever	Atopy
Gordon setter	Atopy
Irish setter	Atopy, seborrhea (primary)
Labrador retriever	Atopy, adverse reactions to food, seborrhea (primary)
Lhasa apso	Atopy
Malamute	Zinc-responsive dermatosis
Old English sheepdog	Atopy
Poodle, standard	Sebaceous adenitis
Pug	Atopy
Schnauzer, miniature	Atopy
Shar Pei	Atopy, adverse reactions to food
Shih Tzu	Atopy
Siberian husky	Zinc-responsive dermatosis
Spaniels	Atopy (American cocker), adverse reactions to food, seborrhea (primary)
Terriers	Atopy
Vizsla	Sebaceous adenitis

*Atopy is often managed with fatty acid supplementation, sebaceous adenitis and primary seborrhea with retinoid supplementation, zinc-responsive dermatosis with zinc supplementation and adverse reactions to food with dietary changes. Specific nutrient deficiencies are usually not breed-specific.

The clinician should obtain a complete medical history in all cases. The nutritional history should focus on the adequacy of the specific food for the animal's lifestage, and types and dosages of nutritional supplements. The veterinarian or a veterinary nutritionist should evaluate home-prepared foods for nutritional adequacy (See Chapter 6. For a list of laboratories that can analyze homemade foods, see Appendix U, Small Animal Clinical Nutrition, 4th ed., pp 1134-1144.) because nutrient defi-

SKIN/HAIR

ciencies or imbalances are more likely to occur in animals eating homemade vs. commercial foods. Excessive nutrient levels in food can cause skin disease due to direct toxicosis or interaction with other nutrients in the food.

Physical Examination (Page 458)

A comprehensive physical examination that evaluates all body systems should be performed on patients with skin or hair disease. Internal disease is often manifested as skin and coat disease; this diagnostic possibility should not be overlooked by concentrating on the integumentary changes alone.

The skin can be affected by many nutritional factors, but usually responds in a limited number of ways. The cutaneous changes associated with nutritional abnormalities are often indistinguishable from those caused by other more common skin diseases. Changes that raise the suspicion for nutritional abnormalities include: 1) a sparse, dry, dull and brittle coat with hairs that epilate easily, 2) slow hair growth or slow regrowth from areas that have been clipped, 3) abnormal scale accumulation (seborrhea sicca), 4) loss of hair, erythema or crusting in areas of friction or stretch such as the distal extremities, 5) decubital ulcers and poor wound healing and 6) loss of normal hair color. Primary lesions such as papules and pustules rarely occur with nutritional abnormalities, but can occur with bacterial pyoderma secondary to nutritional, allergic or other underlying problems.

Laboratory/Other Clinical Information (Pages 458-459)

Routine laboratory evaluations including a complete blood count, serum biochemistry profile, urinalysis and thyroid panel are rarely helpful in evaluating nutritional skin disease. However, these tests can be used to rule out internal or metabolic diseases as causes of cutaneous problems.

Routine laboratory procedures for patients with dermatologic problems include skin scrapings for parasites, hair examination, cytologic examination of tissue or fluids, fungal culture, bacterial culture and biopsy for dermatohistopathologic examination. Of these procedures, hair examination and dermatohistopathology are most helpful for evaluation of potential nutritional problems.

HAIR EXAMINATION Normal adult animals have a mixture of anagen and telogen hairs, the ratio of which varies with the season and other factors. Estimation of the ratio of anagen to telogen hair bulbs can be useful. All the hair of normal animals should not be in telogen; this finding suggests a diagnosis of telogen defluxion or follicular arrest. Inappropriate numbers of telogen hairs (e.g., mostly telogen hairs during the summer when the ratio should be about 50:50) suggest a diagnosis of nutritional, endocrine or metabolic disease.

Hairs that are inappropriately curled, misshapen and malformed suggest an underlying nutritional or metabolic disease. When unusual pigmentation is observed, external sources (salivary staining, chemicals, topical medications), nutritional disorders, color dilution/color mutant disorders and endocrine disorders should be considered.

Breakage of hairs with abnormal shafts suggests nutritional disorders, dermatophytosis or congenitohereditary disorders such as color dilution alopecia. Morphologic changes in the hair bulb and hair diameter are sensitive indicators of overall protein status. Hair bulb atrophy, constriction and hair depigmentation may be seen in people after as little as two weeks of protein deprivation. Protein deprivation may not produce changes as rapidly in dog and cat hair because the hair in these species spends more time in telogen and less time in anagen. For a more complete discussion of hair examination, see Small Animal Clinical Nutrition, 4th ed., pp 458-459.

BIOPSY AND DERMATOHISTOPATHOLOGY The following are general guidelines for when a skin biopsy should be performed: 1) all obviously neoplastic or suspected neoplastic lesions, 2) all persistent ulcerations, 3) any case involving a major disease that is most readily diagnosed by biopsy (e.g., immune-mediated skin disease), 4) a dermatosis that is unresponsive to conventional therapy, 5) any unusual or serious dermatosis and 6) vesicular dermatitis. Some nutritional skin diseases, such as zinc-responsive dermatosis, have clearly delineated histopathologic lesions that are easily recognized during microscopic examination of a skin biopsy specimen. In general, skin biopsy should be performed within three weeks for any dermatosis that does not respond to appropriate therapy. This includes those dermatoses that do not respond to initial management with a food change or supplementation.

Risk Factors for Nutritionally Related Skin Disease *(Pages 459-460)*

Genotype, lifestage, food type and food supplementation are risk factors for nutritionally related skin disease. Breed predilection determines the prevalence of some skin disorders. Tables of common skin diseases categorized by breed are readily available (**Table 15-1**). In general, more than 30 dog breeds are at increased risk for skin diseases. The nutrient-sensitive skin diseases such as zinc-responsive dermatoses and retinoid-responsive dermatoses often occur in specific breeds. As an example, one form of zinc-responsive dermatosis is frequently seen in arctic-type dog breeds such as malamutes and Siberian husky dogs.

As mentioned before, nutrient deficiencies that cause skin disease are more likely to occur during growth, gestation, lactation and some illnesses when nutritional requirements are highest.

SKIN/HAIR

Some dry, commercial, generic, private label brand and grocery pet foods have lower fat content, lower nutrient digestibility and higher mineral content than other grocery and specialty brands. Low amounts of fat and poor-quality fat are risk factors for essential fatty acid deficiency; poor nutrient digestibility contributes to protein-energy malnutrition, especially during growth and lactation; and high levels of minerals such as calcium inhibit the absorption of nutrients such as zinc, which are essential for normal, healthy skin.

An animal that obtains most of its nutrients from home-prepared foods is at risk for several nutritional problems. (See Chapter 6.) In general, homemade foods lack adequate calcium, essential fatty acids, certain vitamins and other micronutrients. Homemade foods should include: 1) a calcium source such as bone meal, oyster shell or dicalcium phosphate, 2) a source of essential fatty acids such as corn oil, safflower oil or some other vegetable oil and 3) a multivitamin-trace mineral supplement. Also, homemade cat foods should be supplemented with taurine.

The final risk factor for nutritionally related skin disease is oversupplementation with naturally occurring foods or commercial supplements. Vitamin A toxicosis is associated with excessive use of liver as a supplement. High levels of minerals such as calcium in commercial supplements can interfere with absorption of essential trace elements such as zinc.

The Skin as a Metabolic Organ (Page 460)

The skin and coat significantly influence nutrient requirements. The ability of an animal's coat to regulate body temperature and energy requirements in cold environments correlates closely with hair length, thickness, density and with the medullation of individual hair fibers. The skin also influences water requirements by minimizing transepidermal moisture loss. Loss of this normal barrier function as a result of fatty acid deficiency can increase an animal's water requirement, which is clinically manifested as polydipsia.

The hair cycle, and thus the coat, is influenced by the general state of health, genetics, photoperiod, ambient temperature, hormones, nutrition and poorly understood intrinsic factors.

Dog breeds can be classified as having high, moderate and low weights of hair. Longhaired breeds with relatively large body surface areas per body weight, such as Pomeranians, have the largest relative amount of hair. Estimates indicate that as much as 30% of protein in food is needed to maintain daily hair growth in small breeds with long coats. On the other hand, larger dogs with short coats may use less than 10% of food protein to maintain daily hair growth.

The normal canine epidermis has a very slowly renewing cell population. Only 1.5% of epidermal basal cells undergo DNA replication at any point in time. In dogs, it takes approximately 22 days for cells to migrate from the basal layer to, but not through, the stratum corneum.

The upper external root sheath of the hair follicle and sebaceous gland has essentially the same cell kinetic growth characteristics as the surface epidermis. Conversely, the root matrix of anagen hairs is one of the most rapidly renewing cell populations of the body. In actively growing hair, up to 24% of cells are undergoing DNA replication.

Key Nutritional Factors (Pages 460-466)

PROTEIN AND ENERGY As mentioned previously, the integument is a metabolically active organ that is affected by the nutritional status of the animal. Protein and energy are required for the development of new hair and skin. Developing hair requires sulfur containing and other amino acids. Therefore, for normal skin and hair, it is important for the animal's food to provide optimal protein quantity, quality (appropriate levels of essential amino acids) and digestibility. Animals have increased protein and energy requirements during growth, gestation, lactation and some illnesses. Abnormal skin and hair will often be noted if nutritionally inadequate foods are fed during these stages. Optimal nutrient profiles for various lifestages of dogs and cats are listed in Chapters 9 and 11.

Foods inadequate in protein and energy can cause keratinization abnormalities, depigmentation of hair and changes in epidermal and sebaceous lipids. The skin loses its protective barrier function in animals with protein-energy malnutrition and becomes more susceptible to secondary bacterial or yeast infection. Impaired wound healing and decubital ulcers are also sequelae to protein-energy malnutrition. Protein-deficient animals have patchy alopecia and coats that are dry, dull and brittle.

Telogen defluxion is usually recognized as hair loss associated with a stressful event (e.g., pregnancy, severe illness, surgery) that causes the abrupt, premature cessation of growth of many anagen hair follicles and the synchronization of these hair follicles in catagen, then in telogen. Short-term increased requirements of energy, protein and other nutrients during growth, gestation, lactation and illness may cause telogen defluxion if appropriate nutritional changes are not instituted. Bitches and queens in late gestation and lactation, and growing puppies and kittens are at risk unless they are fed nutritionally balanced, highly digestible foods that meet their increased nutritional requirements.

Anagen defluxion is a sudden loss of hair due to an unusual event (e.g., antimitotic drugs, infectious disease, metabolic disease) that interferes with anagen, resulting in abnormalities of hair follicles and shafts. Animals suffering from the stress of illness, injury and surgery often require increased amounts of energy, protein, specific amino acids and other nutrients. Animals with severe illness that do not receive adequate nutritional sup-

SKIN/HAIR

port are at risk for telogen defluxion, anagen defluxion or other coat abnormalities.

Dogs with severe primary seborrhea may have increased protein and other nutrient requirements. The calculated epidermal cell renewal time is approximately seven to eight days for dogs with primary seborrhea. The hyperproliferative nature of the skin of dogs with primary seborrhea, with at least a threefold increase in epidermal cell renewal, may change the nutrient requirements of these dogs. However, no studies to date have evaluated the specific nutrient requirements of dogs with severe primary seborrhea vs. age- and breed-matched controls. Some authors suggest that primary seborrhea worsens markedly in dogs with nutritional inadequacies. Dogs with severe deep pyoderma secondary to generalized demodicosis or other underlying diseases may have increased nutrient requirements above those found in normal adult animals.

ESSENTIAL FATTY ACIDS The essential fatty acids (EFA) are polyunsaturated fatty acids derived from and including the parent EFA, cis-linoleic acid and α-linolenic acid. The epidermis depends on food to supply EFA or the continuous formation of γ-linolenic acid, arachidonic acid and eicosapentaenoic acid by the liver, with subsequent transportation to the skin by the blood.

One of the most important skin-related functions of EFA is the incorporation of linoleic acid into the ceramides of the lipid portion of the epidermal cornified envelope. This envelope serves an essential barrier function to prevent loss of water and other nutrients. EFA are a source of energy for the skin and serve as precursors to a variety of potent, short-lived molecules including prostaglandins, leukotrienes and their metabolites.

When mammals are deprived of fats, they develop characteristic signs of deficiency.

Cutaneous changes have been described in fatty acid deficiency in dogs and cats. These cutaneous abnormalities include scaliness (seborrhea sicca), matting of hair, loss of skin elasticity, alopecia, a dry and dull coat, erythroderma, hyperkeratosis, epidermal peeling, interdigital exudation, otitis externa and lack of hair regrowth following plucking. These changes are associated with epidermal and dermal metabolic effects leading to: 1) increased transepidermal water loss, 2) increased epidermal cell turnover, 3) sebaceous gland hypertrophy, 4) increased sebum viscosity, 5) poor wound healing, 6) increased susceptibility to infection and 7) weakening of cutaneous capillaries.

Dogs with cutaneous abnormalities due to low-fat foods have lower levels of fatty acids in serum, skin, liver, kidneys and heart muscle than do animals with healthy skin. Cats fed an EFA-deficient food develop moderate seborrhea sicca and mild hair loss after six months. Severe seborrhea sicca with large scales develops in EFA-deficient cats when the environmental relative humidity decreases from approximately 75 to 55%. Hair loss is extensive and stroking causes clumps of hair to fall out.

Clinical signs resembling those caused by experimental EFA deficiency can also be induced by deficiency of other nutrients, particularly zinc, vitamin E and pyridoxine. EFA intake influences the requirement of these nutrients. In rodents, clinical signs of zinc deficiency can be largely prevented by EFA supplementation.

Fatty acid deficiency is rapidly reversible if EFA are introduced orally, parenterally or topically. Various abnormal cutaneous parameters of dogs and cats are restored within a few days by linoleic acid supplementation. Supplementation with EFA will increase fatty acid levels in serum of dogs and cats, and in the skin of both normal and seborrheic dogs.

Use of Fatty Acids for Seborrhea Dogs with seborrhea have abnormally low cutaneous levels of linoleic acid and increased cutaneous levels of oleic acid. These low cutaneous levels are found despite normal food and serum fatty acid concentrations. Following supplementation for 30 days with a vegetable oil high in linoleic acid (sunflower oil), the cutaneous fatty acid concentrations return to near normal and clinical signs of seborrhea improve. This suggests that clinical signs of seborrhea in dogs may be partly attributable to a localized deficiency of linoleic acid, elevated levels of arachidonic acid in the skin or both.

Seborrhea sicca is also associated with increased transepidermal water loss, which can be reversed with cutaneous administration of vegetable oils rich in linoleic acid. Supplementing the food with α-linolenic acid can also decrease transepidermal water loss. Further studies are needed to determine the effects of supplementation of food with other fatty acids and to determine the optimal dose of fatty acid supplements for patients with seborrhea.

MINERALS Minerals in the food interact with one another and this interaction must be kept in mind when assessing integumentary problems that might be associated with certain homemade foods, commercial foods or nutritional supplements. Skin manifestations of mineral imbalances are seen most commonly with primary (nutritional inadequacy) or secondary (nutrient interaction) deficiencies of copper and zinc.

Copper Cutaneous manifestations of copper deficiency include achromotrichia or loss of normal hair coloration, reduced density or lack of hair and a dull or rough coat. Pigmented hair on the head and face loses its normal color, develops a "washed out" appearance and becomes gray. This change may extend to the entire body. In dogs and cats with cuta-

SKIN/HAIR

neous manifestations of copper deficiency, copper concentrations are significantly reduced in plasma, hair, liver, kidney and heart muscle. Copper deficiency is seen most commonly in young puppies and kittens.

Dogs and cats can develop copper deficiency due to: 1) a lack of copper in food, 2) poor availability of copper in food or 3) an excess of competing minerals. Zinc, in particular, can adversely affect copper homeostasis. Zinc is thought to inhibit copper absorption by its action on intestinal metallothioneins, which sequester copper in the intestinal epithelial cells making copper unavailable for use elsewhere in the body. The greater the intake of zinc and the lower the intake of copper (absolute or relative), the greater the potential for copper sequestration and ultimately, copper deficiency.

Copper availability varies widely among feed ingredients. Copper availability is relatively high in poultry byproduct meal, avian liver (chicken and turkey) and ruminant liver (beef and sheep); copper from soybean meal and corn gluten meal is moderately available; copper from monogastric mammalian liver (pork) and copper oxide is poorly available. Risk factors for copper deficiency in dogs and cats include: 1) rapid growth, 2) unsupplemented homemade foods, 3) commercial or homemade foods supplemented with copper oxide and 4) homemade or commercial foods supplemented with excessive levels of zinc, calcium or iron.

Zinc/*Zinc-Related Dermatoses* Over the past 20 years, a variety of cutaneous diseases in dogs have been described that are thought to be primary or secondary zinc deficiency, or that respond to zinc supplementation. The classification of these skin diseases is confusing and often overlaps (**Table 15-2**). A crusted dermatosis, termed dry juvenile pyoderma or juvenile hyperkeratosis, has been reported to occur in young shorthaired dog breeds. Many cases were not caused by primary bacterial infection and often resolved spontaneously at sexual maturity. In retrospect, these case reports most likely represent the first clinical descriptions of cutaneous disease caused by zinc deficiency in young dogs. A classification scheme was proposed in 1980 for zinc-responsive dermatoses that included two syndromes. Syndrome I included Siberian husky and malamute dogs, which usually developed lesions in early adulthood and responded to zinc supplementation. Syndrome II included rapidly growing puppies that developed lesions due to zinc deficiency and responded to food change, zinc supplementation or both. Several years later, a generic dog food syndrome was described in adult dogs and rapidly growing puppies consuming a poor-quality food. These animals had lesions consistent with zinc deficiency and responded to a food change. At the same time, acrodermatitis was described in bull terriers and linked to abnormal zinc absorption and metabolism.

Table 15-2. Classification scheme for zinc-related cutaneous disorders in dogs.

Previous classification schemes
  Acrodermatitis of bull terriers
  Dry juvenile pyoderma
  Generic dog food syndrome
  Syndrome I (Siberian husky, malamute, other breeds)
  Syndrome II (growing dogs)
Proposed classification
  Animal abnormalities
    Acrodermatitis of bull terriers
    Zinc malabsorption (Siberian husky, malamute)
  Nutritional abnormalities
    Essential fatty acid deficiency
    Primary zinc deficiency
    Secondary zinc deficiency

In all these syndromes, the dermatoses are clearly associated with zinc deficiency and possibly a deficiency or abnormal metabolism of other nutrients. A more practical classification scheme for zinc-related cutaneous changes in dogs includes clinical syndromes due to nutritional abnormalities (primary or secondary zinc deficiency) or animal abnormalities of zinc metabolism (**Table 15-2**).

Zinc/Zinc-Related Dermatoses Associated with Nutritional Abnormalities

Zinc-responsive cutaneous lesions have been frequently described in rapidly growing puppies and less frequently in adult dogs. Many breeds may be affected, but Great Danes, Doberman pinschers, German shepherd dogs, German shorthaired pointers, beagles, standard poodles, Rhodesian ridgebacks and Labrador retrievers are reportedly affected more often. Lesions somewhat resemble those of experimental zinc deficiency in puppies and include erythroderma, alopecia and hyperkeratotic plaques (exudative crusts) on the face, head, distal extremities and mucocutaneous junctions. Thickened, fissured footpads are also frequently seen. Severely affected animals have systemic signs of lymphadomegaly, poor growth, fever, depression and anorexia. Microscopic examination of skin biopsy specimens shows hyperplastic superficial perivascular dermatitis with diffuse parakeratotic hyperkeratosis.

Risk factors for development of zinc-responsive cutaneous disease are listed in **Table 15-3**. Foods with high mineral levels (calcium, phosphorus, magnesium), poor digestibility, high levels of phytate and/or low levels of total fat and EFA are significant risk factors, especially when fed to puppies during the rapid growth phase. Foods high in calcium, phosphorus and magnesium adversely affect absorption of zinc. Excessive use

SKIN/HAIR

Table 15-3. Risk factors for zinc-related skin disease in dogs.

Certain breeds
Bull terrier
Great Dane
Labrador retriever
Malamute
Siberian husky
Other rapidly growing large and giant breeds
Food
High mineral levels (calcium, phosphorus, magnesium)
High phytate levels (high levels of cereal ingredients)
Low essential fatty acid levels
Dietary supplements
Calcium and/or other mineral supplements
Cottage cheese or other dairy products
Small intestinal disease
Malassimilation (malabsorption, maldigestion)
Viral enteritis

of mineral supplements containing calcium in large- and giant-breed puppies is common and can inhibit zinc absorption.

Foods high in cereal ingredients often have excessive levels of phytate that complex with and prevent normal absorption of zinc. Phytate and calcium also interact to affect zinc absorption. The relative effect of phytate on zinc absorption increases with the calcium level in the food. Thus, foods high in phytate and calcium have an even greater negative impact on zinc absorption.

Low-cost, commercial generic or private label brand dry pet foods are often low in total fat and EFA because fat is an expensive ingredient. Zinc and EFA metabolism interact and foods with marginal concentrations of zinc and EFA may be more likely to cause clinical disease.

Viral enteritis and prolonged diarrhea adversely affect zinc absorption in swine and similar changes may occur in other animals.

Zinc/Zinc Deficiency Associated with Metabolic Abnormalities Lesions attributed to zinc deficiency develop early in adulthood in Siberian huskies, malamutes and bull terriers and progress at a variable rate. Skin lesions develop in these breeds despite consumption of well-balanced foods containing adequate levels of zinc. Lesions include erythroderma, alopecia, crusting, scaling and suppuration involving the head, extremities and mucocutaneous junctions. The footpads may become hyperkeratotic. Secondary *Malassezia* and bacterial infections are common. Some malamute and Siberian husky dogs have a decreased capability for zinc absorption. Bull terriers that are siblings of dogs with acrodermatitis may also be affected and probably have abnormal zinc absorption and metabolism. These animals probably require zinc supplementation for life to maintain normal tissue zinc concentrations and avert clinical disease.

Zinc/Diagnosis and Management of Zinc-Responsive Skin Disease Diagnosis of zinc-responsive cutaneous disease is based on the history, physical examination and results of skin biopsy evaluation. Hyperplastic superficial perivascular dermatitis with marked diffuse and follicular parakeratotic hyperkeratosis is suggestive of zinc deficiency. In general, zinc concentrations in serum, leukocytes and hair are not a good indicator of zinc status in dogs. Age, seasonal variation and many diseases affect serum zinc concentrations.

Treatment generally includes changing to a food that avoids excess minerals and contains adequate amounts of zinc and EFA. This type of change will usually result in rapid improvement in puppies and some adult dogs. Zinc supplementation will be necessary in those breeds in which decreased ability to absorb zinc is suspected. Oral supplementation with zinc sulfate (10 mg/BW_{kg}/day) or zinc methionine (2 mg/BW_{kg}/day) is adequate in most cases. Maximal zinc absorption occurs if supplements are given between rather than with meals. Supplemental zinc from zinc amino acid chelates may be more available to dogs than are inorganic zinc sources. Some dogs, especially Siberian huskies, do not respond to oral zinc supplementation. Intravenous injection of sterile zinc sulfate solutions at dosages of 10 to 15 mg/BW_{kg} is effective in these dogs. Weekly injections for at least four weeks are necessary to resolve the lesions, and maintenance injections every one to six months may be necessary to prevent relapses.

Existing skin lesions can be improved by hydrating the crusts with wet dressings, applying petrolatum or petrolatum-based topical agents or whole-body warm water soakings. Dogs with evidence of superficial pyoderma or *Malassezia* infections should be treated with appropriate antimicrobials. Some authors also recommend low doses of oral, short-acting glucocorticoids.

VITAMINS/Vitamin A Skin lesions and focal atrophy of the skin have been reported with experimental vitamin A deficiency in dogs and cats, although it is seldom encountered clinically. It is unlikely that vitamin A deficiency would occur in dogs and cats eating typical commercial pet foods because these foods contain several times the minimum daily requirement of vitamin A.

Vitamin A/Retinoid-Responsive Dermatoses The term "retinoids" refers to the entire group of naturally occurring and synthetic vitamin A derivatives. These therapeutic agents should be reserved for cases in which there are clinical and histopathologic abnormalities most consistent with primary keratinization disorders of the surface and/or follicular epithelium or abnormalities of the sebaceous glands. Other causes of clinical scaling (ectoparasitism, allergies, infections, endocrinopathies) should first be eliminated through other diagnostic testing.

A vitamin A-responsive dermatosis has been described primarily in cocker spaniels but also recognized in a Labrador retriever and a miniature schnauzer. The condition is characterized by adult-onset, medically refractory seborrheic skin disease with marked follicular plugging and hyperkeratotic plaques, primarily on the ventral and lateral thorax and abdomen. A ceruminous otitis externa and unthrifty appearing coat are often present. The clinical lesions are characterized histologically by marked follicular orthokeratotic hyperkeratosis. Improvement is noted within three to four weeks of starting oral vitamin A alcohol (retinol) with complete remission by eight to 10 weeks. It is important to remember that this syndrome represents only a small subset of seborrheic disease in cocker spaniels. However, it is logical to try a four- to eight-week course of vitamin A in dogs with ventral hyperkeratotic plaques that do not respond well to other therapy.

The most widely available and commonly used synthetic retinoids include tretinoin, isotretinoin and etretinate. Tretinoin is effective topically as therapy for localized follicular and epidermal keratinization disorders such as chin acne, nasodigital hyperkeratosis, calluses and ear margin seborrhea/dermatosis. Isotretinoin and etretinate are given orally and may be useful in managing primary idiopathic seborrhea in cocker spaniels, keratinization disorders in other breeds, schnauzer comedo syndrome, sebaceous adenitis, lamellar ichthyosis, actinic keratosis (solar-induced precancerous lesions) and various cutaneous neoplastic disorders (squamous cell carcinoma, cutaneous T-cell lymphoma, multiple keratoacanthomas).

Dosages commonly recommended by veterinary dermatologists are outlined in **Table 15-4**. Side effects that occur commonly with retinoids include conjunctivitis, decreased tear production, arthralgia/myalgia, moderate to marked elevations in serum triglyceride levels, elevations in liver enzyme activity and teratogenic effects.

VITAMINS/Vitamin E Naturally occurring vitamin E deficiency has only been reported to occur in cats. Steatitis occurs when sources of highly unsaturated fatty acids, such as red meat tuna, are fed to cats without adequate vitamin E. Clinical signs and laboratory findings include anorexia, fever, hyperesthesia, hemolytic anemia, leukocytosis and firm subcutaneous nodules. Diagnosis is confirmed by microscopic examination of biopsy specimens from adipose tissue. Typical lesions are firm, yellow to orange-brown fat with lobular panniculitis and ceroid within lipocytes, macrophages and giant cells. Treatment includes a change of food to a complete and balanced ration, supplemental vitamin E (25 to 75 mg/BW$_{kg}$/day), corticosteroids

Table 15-4. Indications and dosages for retinoids in primary keratinization disorders.

Vitamin A alcohol (retinol)
Subset of seborrheic skin disease, primarily in cocker spaniels
Dosage: 625 to 1,000 IU/kg q24h per os
10,000 IU q24h per os in cocker spaniels and miniature schnauzers
50,000 IU q24h per os in Labrador retrievers
Tretinoin (all-trans retinoic acid)
Chin acne of dogs and cats
Ear margin seborrhea/dermatoses
Nasodigital hyperkeratosis
Dosage: Apply topically q12h to q24h to control; then decrease frequency
for maintenance
Isotretinoin (13-cis retinoic acid)
Lamellar ichthyosis
Schnauzer comedo syndrome
Sebaceous adenitis
Dosage: 1 to 3 mg/BW$_{kg}$ q24h per os for control; then attempt to decrease
to alternate-day therapy
Etretinate (analogue of retinoic acid ethyl ester)
Actinic keratosis
Idiopathic seborrhea, especially of cocker spaniels
Lamellar ichthyosis
Sebaceous adenitis
Dosage: 0.75 to 1.0 mg/BW$_{kg}$ q24h per os for control; then attempt to
decrease to alternate-day therapy

SKIN/HAIR

and supportive care. Appendix A in Small Animal Clinical Nutrition, 4th ed., pp 997-1001 lists factors for converting mg/kg to IU.

Naturally occurring vitamin E deficiency has not been reported to occur in dogs, but experimentally induced vitamin E deficiency does produce skin lesions. Initial lesions consist of a keratinization defect (seborrhea sicca), followed by a greasy and inflammatory stage (erythroderma and seborrhea oleosa) and secondary bacterial pyoderma. The dermatosis rapidly responds to vitamin E supplementation. All lesions respond within eight to 10 weeks. It is unlikely that vitamin E deficiency would occur in dogs and cats that eat typical commercial pet foods because such foods contain three to five times the minimum daily requirement of vitamin E.

Vitamin E/*Vitamin E-Responsive Dermatoses* A number of inflammatory dermatoses in animals have been treated with oral vitamin E, including discoid lupus erythematosus, systemic lupus erythematosus, pemphigus erythematosus, sterile panniculitis, acanthosis nigricans, dermatomyositis and ear margin vasculitis. Vitamin E is often used in conjunction with systemic glucocorticoids, topical steroids and other immunosuppressive agents.

Large doses of vitamin E may stabilize cell and lysomal membranes against damage induced by free radicals and peroxides, modulate arachidonic acid and prostaglandin metabolism, inhibit proteolytic enzymes, enhance phagocytic activity and enhance humoral and cellular immunity.

The oral dosage of vitamin E for inflammatory dermatoses is 200 to 800 IU twice daily. This dose is 20 to 100 times the daily requirement for a 10-kg dog. Anecdotal reports suggest that topical vitamin E may help resolve discoid lupus erythematosus lesions.

VITAMINS/B-Complex Vitamins Experimental deficiencies of biotin, riboflavin, niacin and pyridoxine cause cutaneous lesions in dogs and cats. The most common clinical signs include anorexia, weight loss, diarrhea, alopecia and dry, flaky seborrhea. Clinical lesions are more likely to occur in young, growing animals than in adults.

It is unlikely that B-complex vitamin deficiency would occur in dogs and cats that eat typical commercial pet foods, because most foods contain several times the minimum daily requirement of these vitamins. B-complex deficiency could occur in animals eating unusual homemade foods that are not supplemented with vitamins.

Assess the Food (Pages 466-467)

Nutrients of concern for skin and hair include protein, EFA, zinc, copper and various fat-soluble and water-soluble vitamins. (See the Key Nutritional Factors–Skin and Hair Disorders table.) The food should include optimal levels of these nutrients, and the nutrients should be available to the animal. Digestibility and assimilation of nutrients are especially important during periods of increased nutrient demand such as growth, gestation and lactation. Use of maintenance-type foods (which are lower in protein, fat, minerals, vitamins and digestibility than growth/lactation foods or foods for repletion/recovery) may be a risk factor for nutritional skin disease during these lifestages.

Levels of these nutrients in specific foods can be obtained by contacting the manufacturer or distributor of the food. (Selected foods are listed in Appendix 2.) A detailed assessment of nutritional supplements is also important. Vitamin supplements are rarely indicated except in those nutrient-sensitive disorders that respond to high levels of vitamin A or vitamin E. Excessive use of mineral supplements can interfere with assimilation of zinc and copper.

Assess the Feeding Method (Page 467)

It may not always be necessary to change the feeding method when managing a patient with skin or hair disease, but a thorough assessment includes verification that an appropriate feeding method is being used.

Items to consider include feeding route, amount fed, how the food is offered, access to other food and who feeds the animal. All of this information should have been gathered when the history of the animal was obtained. If the animal has a normal body condition score (BCS 3/5), the amount of food it was fed previously (energy basis) was probably appropriate.

The food dosage and feeding method should be altered if the animal's body weight and condition are not optimal. For clinical nutrition to be effective, there needs to be good compliance. Enabling compliance includes limiting access to other foods and knowing who feeds the animal. If the animal comes from a household with multiple pets, it should be determined whether the pet with skin disease has access to the other pets' food.

REASSESSMENT (Page 467)

Cutaneous disease due to a nutrient deficiency will usually respond rapidly and dramatically to appropriate nutritional change or supplementation. Patients will usually improve within a few days to weeks. Nutrient-sensitive disorders usually respond to supplements more slowly, over several weeks to several months. Once a nutritional change or supplementation has been started, the patient should be examined monthly for significant changes in skin lesions and hair quality.

SKIN/HAIR

Clinical cases that illustrate and reinforce the nutritional concepts presented in this chapter can be found in Small Animal Clinical Nutrition, 4th ed., pp 469-474.

CHAPTER 16

Dental Disease

For a review of the unabridged chapter, see Logan EI, Wiggs RB, Zetner K, et al. Dental Disease. In: Hand MS, Thatcher CD, Remillard RL, et al, eds. Small Animal Clinical Nutrition, 4th ed. Topeka, KS: Mark Morris Institute, 2000; 475-504.

INTRODUCTION (Pages 475-478)*

The goals of promoting oral health and nutritional management of oral disease in dogs and cats are to: 1) control plaque, the cause of periodontal disease, 2) assess the level of plaque control necessary to prevent gingivitis in each patient, 3) determine each pet owner's ability to control substrate accumulation and select methods most likely to ensure compliance, 4) feed a food with an appropriate texture and nutritional profile and 5) recognize that oral health may affect systemic health; therefore, a healthy oral cavity may affect longevity and quality of life.

CLINICAL IMPORTANCE (Pages 478-480)

Prevalence of Periodontal Disease (Page 478)

Periodontal disease is the most common disease of adult dogs and cats. Periodontal disease has been observed in dogs and cats of varying breed, gender and age. Surveys from several countries report prevalence rates of periodontal disease that range from 60 to more than 80% of dogs and cats examined.

According to the National Companion Animal Study (representing 54 veterinary practices across the United States), oral disease was the most frequent diagnosis in all age categories of 39,556 dogs and 13,924 cats.

Odontoclastic Resorptive Lesions (Page 478)

Resorptive lesions are defined as noncarious defects of enamel, cementum and dentin, and have been commonly referred to as neck or cervical line lesions because they occur most often in the cervical or neck region of the tooth at the cementoenamel junction (CEJ). Resorptive lesions reportedly affect more than 50% of domestic cats examined. Odontoclastic resorptive lesions occur infrequently in dogs.

*Page numbers in headings refer to Small Animal Clinical Nutrition, 4th ed., where additional information may be found.

Key Nutritional Factors–Dental Disease (Page 477)

Factors	Associated conditions	Dietary recommendations
Protein	Plaque accumulation Periodontitis Dystrophic periodontium changes (protein deficiency)	Avoid protein deficiency or excess Adult maintenance Dogs: 16 to 25% protein (dry matter basis [DMB]) Cats: 30 to 40% protein (DMB) Growth/lactation Dogs: 30 to 35% protein (DMB) Cats: 35 to 50% protein (DMB) Use food with dry matter digestibility >80%
Calcium/ phosphorus	Calcium deficiency Nutritional secondary hyper- parathyroidism Significant loss of alveolar bone Calcium and/or phosphorus excess promotes calculus formation	Avoid calcium deficiency Ensure homemade food recipes contain a source of calcium Ensure food has proper calcium levels Adult maintenance (dog/cats) 0.5 to 1.0% calcium (DMB) 0.4 to 0.9% phosphorus (DMB) Growth/lactation (dogs/cats) 1.0 to 1.5% calcium (DMB) 0.8 to 1.3% phosphorus (DMB)
Food texture	Plaque and calculus accumulation	Provide food that promotes chewing and mechanical cleansing of teeth Dogs: Prescription Diet Canine t/d Cats: Prescription Diet Feline t/d Add dental treat (with proven efficacy) without changing the food Add new toy (appropriate for individual) without changing the food Add toothbrushing or other type of oral hygiene

TAL

Dental Caries (Page 478)

Dental caries is decay of the tooth structure. Dental caries occurs infrequently in dogs and cats.

Inflammatory and Immune-Mediated Diseases (Pages 478-479)

Several oral conditions have an inflammatory or immune-mediated component different from the typical inflammatory-mediated response seen in periodontal disease. Although the specific etiology of many of these conditions remains unknown, many are associated with oral bacteria and concurrent periodontal disease.

Ulcerative stomatitis occurs in dogs in which the mucosal surfaces contact plaque-covered teeth. Immune-mediated ulcerative gingivostomatitis has been reported to occur in Maltese dogs with minimal periodontal disease and has been reported to occur in other dogs with varying degrees of periodontal disease. A condition reported as lymphocytic-plasmacytic stomatitis occurs frequently in cats and is often refractory to treatment. Cats with this condition often exhibit severe oral pain and anorexia.

Malocclusion (Page 479)

Malocclusion is any occlusal abnormality and may affect individual teeth, groups or quadrants of teeth or the entire dental arch. Domesticated dogs and cats have been selectively bred for specific head types that result in a variety of malocclusions. Occlusal abnormalities affect the relationship of the teeth to one other, and to other oral structures including the periodontium, palates, tongue, oral mucosa and lips. Potential ramifications include: 1) a compromise in oral function, 2) self-induced oral trauma and 3) an increased risk of incidence and severity of plaque-associated dental diseases.

Attrition and Abrasion (Pages 479-480)

Dental attrition is abnormal wear of tooth surfaces due to contact with occluding teeth during mastication. Attrition may result from excessive chewing on inappropriate materials such as rocks or other hard objects. The incisor teeth, particularly in dogs, may show marked wear due to excessive chewing behaviors.

Abrasion is abnormal wear of tooth surfaces due to application of an external force, such as excessive toothbrushing or inappropriate use of power instruments. Excessive wear is not immediately pathologic as long as the wear rate does not exceed the rate of reparative dentin formation. However, rapid wear can lead to pulpal exposure and infection as well as compromised tooth strength, predisposing the tooth to fracture.

Research has indicated that occlusal trauma in the absence of periodontal disease can damage the periodontal ligament and affect periodontal health. Occlusal trauma can exacerbate periodontal disease.

ASSESSMENT (Pages 480-493)

Assess the Animal (Pages 480-489)
History (Page 480)

Inquiries specific to nutrition and oral care should include past and present information about: 1) oral hygiene, 2) chewing behavior, 3) access to dental treats and toys, 4) access to rocks and other materials that may cause occlusal trauma, 5) presence of any signs that may be related to oral dysfunction, 6) eating behavior and 7) foods eaten, with special attention given to texture and other factors. (See the Key Nutritional Factors–Dental Disease table.)

Physical Examination (Pages 480-481)

INITIAL ORAL EXAMINATION Examination of the skull and the oral cavity should be a regular part of every pet's physical examination. An extraoral examination should be done before opening the mouth to inspect the skull and facial areas for any abnormalities, such as muscle atrophy, swelling, draining tracts and ocular or nasal discharge. Extraoral examination should also include inspection for facial symmetry, palpation of the temporomandibular joints, regional lymph nodes and salivary glands and thorough inspection of the skin and lips. Extraoral abnormalities related to oral dysfunction may include mucopurulent discharge from the eyes or nostrils, soft or hard swellings, crepitus, salivation and an inability to open or close the mouth.

After the extraoral examination, the lips should be gently parted or retracted to allow inspection of the oral mucosa. However, animals experiencing severe oral pain may not tolerate even a cursory oral examination without sedation. The facial surfaces of the teeth and gingivae should be examined for substrate accumulation (i.e., plaque, calculus and stain [See Etiopathogenesis.]), inflammation, trauma and capillary refill time. Tooth position and occlusion should be evaluated. The lingual surfaces of the teeth and gingivae should be inspected, as well as the palates, tongue (ventral and dorsal), frenulum, oropharyngeal area and tonsils.

COMPREHENSIVE ORAL EXAMINATION A definitive oral examination must be done with the patient heavily sedated or anesthetized, and is often done immediately before periodontal therapy. The general examination should be used as a starting point in client communication with the understanding that the definitive oral examination may uncover other lesions that require treatment.

The examination should begin with a thorough inspection of all oral tissues. An overall assessment of oral health should consider the amount and location of substrate accumulation, which provides valuable information about the frequency and effectiveness of oral hygiene.

The other periodontal indices (e.g., probe depth, attachment loss, furcation exposure and tooth mobility) are usually charted after prophylaxis or periodontal therapy to ensure accurate assessment after subgingival debris is removed. Each tooth and its associated periodontium should be evaluated using a dental explorer-probe to examine the tooth for defects, lesions or both. The same instrument should be used to evaluate periodontal health by measuring the extent of gingival inflammation, attachment loss and alveolar bone loss. Any abnormalities in tooth or periodontal structures should be noted on the dental chart. Detailed dental charting allows for disease assessment and provides a record for future reference. The results should become part of the patient's permanent health record.

Radiographic Examination (Page 481)

Oral radiography may be indicated to identify lesions that cannot be detected visually or manually, and to determine the extent of pathology. Root fractures, periapical abscesses, alveolar bone loss, acute resorptive lesions and anatomic anomalies are difficult to assess without radiography. Additionally, oral radiographs are useful in selecting a definitive treatment plan and assessing the outcome of a dental procedure.

Laboratory Studies (Page 481)

A complete blood count, serum biochemistry profile, bacterial culture, virus isolation and examination of cytologic and biopsy specimens may add useful information. Other diagnostic tests such as urinalyses and cardiac examinations may be indicated, particularly in animals with suspected renal or cardiac disease. Such animals may be compromised by bacteremia associated with dental manipulations.

Risk Factors (Pages 481-482)

The primary etiologic agent associated with periodontal disease is bacterial plaque and bacterial byproducts. Bacterial plaque is also directly involved in the pathogenesis of enamel caries, and may be a contributing factor in the development and progression of odontoclastic resorptive and other oral inflammatory lesions. Any factor that enhances bacterial accumulation or affects the resistance of the periodontium may influence the disease process. Specific risk factors that contribute to the severity and progression of dental diseases include: 1) species, 2) breed, 3) age, 4) immunocompetence, 5) nutrition and food characteristics, 6) chewing behavior and 7) systemic health.

SPECIES Periodontal disease occurs frequently in dogs and cats. Other dental diseases appear to be more species dependent. Inflammation of the gin-

gival tissues and oral mucosa (ulcerative stomatitis, pharyngitis-gingivostom-
atitis complex) occurs as a separate entity from periodontal disease in dogs
and cats. Odontoclastic resorptive lesions occur more frequently in cats but
have been reported to occur in dogs. Attrition is observed more commonly
in dogs. Enamel caries occurs infrequently in dogs and cats.

BREED Breed plays a major role in the development of dental disease.
Small, toy and brachycephalic breeds are prone to malocclusive disorders
including overcrowding and rotation of teeth, retained deciduous teeth
and supernumerary teeth. Occlusal abnormalities provide plaque reten-
tive areas and increase the difficulty of oral hygiene procedures. Brachycephalic breeds are also predisposed to mouth breathing, which tends to
dry and irritate oral tissues. Periodontal disease, resorptive lesions and
gingivostomatitis have been reported to occur with relatively greater frequency in purebred cats, particularly Asian breeds such as Siamese and
Abyssinians.

AGE Several surveys have reported that older animals have a greater frequency and an increased severity of dental disease, including calculus
deposition, gingival inflammation, tooth mobility, furcation exposure,
attachment loss and missing teeth. In an evaluation of 4,776 cats aged
seven to 25 years and 8,692 dogs aged 10 to 25 years, oral disease was the
most frequent diagnosis reported.

IMMUNOCOMPETENCE The host immune response protects against
systemic infection from periodontal pathogens. An exaggerated immune
response can cause severe local periodontal destruction. An inadequate
immune response may predispose the animal to opportunistic or overwhelming systemic infections.

NUTRITION AND FOOD CHARACTERISTICS The dramatic difference
in food form represented by commercial dog and cat foods as compared
with the natural prey of wild canids and felids is often implicated as a significant cause of the degree of periodontal disease diagnosed in domestic
dogs and cats.

Commercial foods have been suggested as a causative factor in the
prevalence of resorptive lesions in cats. These foods were suspected
because of the common practice of applying an acid coating to dry cat
foods (e.g., feline digest) to enhance palatability, and the possibility that
hard dry cat foods cause microfractures that predispose teeth to infection
and initiate the inflammatory cascade leading to odontoclastic activation.

Research findings indicate that consumption of commercial dry food
does not contribute to the pathogenesis of resorptive lesions. In other

long-term studies to determine the effects of dry cat food on tooth and periodontal health, researchers found no correlation between the consumption of acid-coated food and increased prevalence of periodontal disease and resorptive lesions.

CHEWING BEHAVIOR Chewing behavior can adversely affect dental and oral health. Chewing on hard materials such as rocks, hard bones, fences and inappropriate chew toys may affect dental health by causing attrition, fractures, avulsions and gingival lacerations, with the potential to infect exposed pulpal tissue and exacerbate periodontal disease.

Etiopathogenesis (Pages 482-485)
TOOTH-ACCUMULATED MATERIALS Several materials accumulate on tooth surfaces and participate in the pathophysiology of dental and periodontal disease. These substances are commonly referred to as tooth-accumulated materials or dental substrates and are categorized as: 1) acquired enamel pellicle, 2) microbial plaque, 3) materia alba/debris, 4) calculus and 5) stain. These substrates accumulate in a dynamic continuum, initiated by the adsorption of salivary constituents onto tooth surfaces.

Enamel Pellicle Enamel pellicle is a thin film or cuticle. Early enamel pellicle is composed of proteins and glycoproteins deposited from saliva and gingival crevicular fluid. Early enamel pellicle protects and lubricates.

As pellicle ages, existing constituents are modified and additional salivary, crevicular and bacterial components are incorporated. Enamel pellicle and its components provide a framework for initial bacterial colonization and also function in the maturation of dental plaque.

Dental Plaque Pellicle deposition and subsequent bacterial colonization occur almost immediately after a dental prophylaxis. Studies have demonstrated that within minutes after polishing, approximately one million organisms are deposited per mm^2 of enamel surface. Aggregates of bacteria combine with salivary glycoproteins, extracellular polysaccharides and occasionally epithelial and inflammatory cells to form a soft adherent plaque that covers tooth surfaces. Dental plaque is not easily removed by normal tongue actions, water drinking or forced water spray, but can be affected by mechanical and chemical means.

Dental plaque has a specific composition and structure that changes with time. Supragingival dental plaque forms above and along the free gingival margin; subgingival dental plaque forms entirely within the gingival sulcus. Growth and maturation of supragingival plaque are necessary for subsequent colonization of subgingival surfaces by dental plaque. Supragingival and subgingival plaque are distinct compositional masses that influence the inflammatory reaction of gingival tissues.

Supragingival plaque in dogs with clinically healthy gingivae is primarily composed of gram-positive aerobic organisms. As plaque matures, the bacterial composition shifts to a predominately gram-negative anaerobic flora. The inflammation and destruction that accompanies periodontal disease results from the direct action of bacteria and their byproducts on periodontal tissues as well as the indirect activation of the host immune response. Thus, bacterial plaque is the most important substrate in the development of periodontal disease.

Materia Alba/Oral Debris Materia alba is a soft mixture of salivary proteins, bacteria, desquamated epithelial cells and leukocyte fragments. Materia alba and dental plaque are two distinct materials. Materia alba does not have the organized bacterial structure or the adherence properties of dental plaque; it can generally be removed by forced water spray. The role of materia alba in the etiopathogenesis of plaque accumulation and periodontal disease remains unclear.

Other debris commonly observed in the oral cavity of dogs and to some extent in cats includes food, impacted hair and miscellaneous foreign materials acquired through chewing behaviors. Food debris retained in the mouth after eating can usually be removed by the action of the tongue and saliva.

Dogs and cats fed soft, sticky foods, particularly those breeds compromised by occlusal abnormalities, may retain more food debris. Retained or impacted debris may act as a nidus for plaque accumulation and exacerbate gingival inflammation. The role of food type and texture in oral health and disease is discussed below.

Dental Calculus Dental calculus is mineralized plaque. Calculus is a hard substrate formed by the interactions of salivary and crevicular calcium and phosphate salts with existing plaque. Dental calculus is observed frequently in dogs and cats. Calculus accumulates supragingivally and subgingivally; calculus deposits thicken with time. Undisturbed calculus is always covered by vital dental plaque. Aged calculus may chip or break off with mastication; however, a film of plaque remains that is rapidly mineralized. Calculus provides a roughened surface to enhance plaque attachment and accumulation and chronically irritates gingival tissues.

Dental Stain Acquired dental stain (extrinsic stain) is initially stained pellicle that becomes part of the mineralized, layered laminate of pellicle, plaque and calculus. Dental stain occurs frequently in dogs. Various nutritional, chemical and bacterial factors affect the presence and intensity of stain.

Although nonpathogenic, dental stain is of aesthetic concern to pet owners and may signal teeth abnormalities.

PATHOPHYSIOLOGIC BASIS OF CLINICAL SIGNS/Periodontal Disease
Plaque accumulation along the gingival margin induces inflammation in adjacent gingival tissues. Without plaque removal or control, gingivitis progresses in severity, and local changes occur allowing subsequent bacterial colonization of subgingival sites. Inflammatory mediators damage the integrity of the gingival margin and sulcular epithelium, allowing infiltration of bacteria that release factors that interfere with normal host cell function or contribute to destruction or lysis of cells or cellular components. The immune response of the host attempts to localize the invasion to the periodontal tissues; the result may be further destruction of local tissues due to cytokines released from inflammatory cells.

Periodontal disease is cyclic with bursts of tissue destruction followed by periods of healing and relative quiescence. Four stages in the pathogenesis of periodontal disease have been proposed: 1) Microbial colonization. Salivary pellicle is deposited on the enamel surface and is soon colonized by oral bacteria that multiply forming plaque. 2) Microbial invasion. Plaque bacteria and their byproducts invade the gingival tissues and initiate a host inflammatory response. 3) Tissue destruction. Direct toxic effects of bacteria and their byproducts and indirect host-mediated toxic responses lead to destruction of periodontal tissue. 4) Healing. Periods of disease remission are characterized by a reduction in the inflammatory response and gingival healing. Additionally, not all teeth are affected at the same rate nor to the same degree. Periodontal disease begins with gingivitis and progresses through increased destruction of the periodontal apparatus, resulting in tooth mobility and eventual tooth loss. Generally, a stage classification system is used, beginning with a healthy periodontium and ending with tooth exfoliation (**Table 16-1**).

Periodontal disease is often a silent process that progresses without detection. Even in cases of severe disease, dogs and cats may not demonstrate obvious discomfort. One signal often noticed by pet owners is oral malodor, but even then pet owners may not link bad breath to periodontal disease. Oral disease is a primary cause of offensive breath odor, but other metabolic processes may be involved. A positive correlation between periodontal disease and malodor has been reported to occur in beagles.

Other signs of periodontal disease are listed in **Table 16-2**.

PATHOPHYSIOLOGIC BASIS OF CLINICAL SIGNS/Systemic Complications of Periodontal Disease Periodontal disease may predispose affected animals to systemic complications. A positive correlation has been found between the severity of periodontal disease and histopatho-

Table 16-1. Stages of periodontal disease.

Stage 1	**Gingivitis**
	Inflammatory changes affect the gingivae only
Stage 2	**Mild periodontal disease**
	Gingival inflammation with early destructive changes affect the periodontium (<25% attachment loss)
Stage 3	**Moderate periodontal disease**
	Gingival inflammation with progressive destruction of the periodontium (25 to 50% attachment loss)
Stage 4	**Severe periodontal disease**
	Gingival inflammation with severe destruction of the periodontium (>50% attachment loss)

Table 16-2. Clinical signs associated with periodontal disease.

Anorexia	Ptyalism
Behavioral changes	Red, swollen or bleeding gingivae
Difficulty eating	Substrate accumulation (plaque, calculus, stain)
Halitosis	Tooth mobility
Head shaking	Ulcerations on gingivae or oral mucosa

logic changes in the kidneys, myocardium and liver. Numerous reports speculate on the association between chronic periodontal disease and conditions affecting the heart valves and pulmonary airways of dogs. Periodontal infections allow bacterial migration into lymphatic and blood vessels, resulting in bacteremia. The host defenses of normal healthy animals can effectively clear transient bacteremia; however, blood-borne bacteria may colonize distant sites in animals with impaired immune function or organ compromise.

PATHOPHYSIOLOGIC BASIS OF CLINICAL SIGNS/Odontoclastic Resorptive Lesions Odontoclastic resorptive lesions present initially as shallow defects typically located at the CEJ. Early lesions are concavities lined by odontoclasts that do not involve the pulpal tissue. Moderate and advanced lesions are characterized by progressive destruction of crown and root with eventual involvement of the pulp cavity. These lesions are typically observed as external lesions, but may occur internally or apically. Odontoclastic resorptive lesions are often detected in cats with concurrent periodontal disease, stomatitis or both.

PATHOPHYSIOLOGIC BASIS OF CLINICAL SIGNS/Stomatitis This condition in cats is often referred to as lymphocytic-plasmacytic stomatitis because of the histologic appearance of the lesions. Inflammation may be limited to specific areas, such as the gingival margin, pharyngeal area or the glossopalatine arches, or may be more generalized, resulting in extensive stomatitis. The specific etiology is unknown, but several bacterial and viral factors may be involved. An excessive host immune response may also be involved. Elevated IgE levels against various ingredients of food may be found in the sera of cats with chronic gingivitis/periodontitis. These findings support clinical evidence that adverse reactions to food ingredients (food hypersensitivity) may occur in cats with chronic inflammatory oral disease. Changing the food (e.g., to a hypoallergenic type) may lessen the painful gingivitis/periodontitis response. Cats often have severe oral pain, ptyalism, dysphagia, anorexia and depression. Affected tissues appear bright red, swollen or proliferative, ulcerative and may be very friable.

Stomatitis in dogs usually results from direct contact of oral mucosa with plaque-laden teeth. These lesions are characteristically round, well-demarcated ulcerations that may coalesce in severe cases. Affected dogs generally have periodontal disease, halitosis, ptyalism, anorexia and behavioral changes associated with oral pain and discomfort such as lethargy and withdrawal.

Key Nutritional Factors *(Pages 485-489)*

Food texture and composition can directly affect the oral environment through: 1) maintenance of tissue integrity, 2) metabolism of plaque bacteria, 3) stimulation of salivary flow and 4) contact with tooth and oral surfaces. Nutritional factors have the potential to affect all of the various oral tissues during development, maturation and maintenance. (See the Key Nutritional Factors–Dental Disease table.) Mature enamel is a static tissue, but nutrition may affect its growth and maturation. The periodontal apparatus surrounds, protects and supports the teeth. Any negative influences affecting these structures may progress to tooth mobility and exfoliation. Oral mucosa has a high turnover rate; adequate nutrition is necessary to maintain tissue integrity.

Most commercial foods provide adequate levels of nutrients to prevent deficiency diseases provided the food meets levels recommended by the Association of American Feed Control Officials (AAFCO) and adequate amounts are fed to meet daily energy requirements.

WATER Water is a critical nutrient. Fresh, palatable water should be available at all times. Food is a source of water for dogs and cats; the moisture content of a pet food dictates its form. Food form can directly influence plaque and calculus accumulation as will be discussed later.

PROTEIN Protein deficiencies cause degenerative changes in the periodontium, specifically the gingivae, periodontal ligament and alveolar bone in laboratory animals. Protein deficiency, however, rarely occurs in dogs and cats and is not a practical consideration as a typical cause of periodontal disease in these species. The effects of protein excess are unknown; however, protein and its components provide nutrients for oral bacteria, which may contribute to increased substrate accumulation.

CARBOHYDRATES Dental caries occurs infrequently in dogs and cats. Fiber-containing foods have long been viewed as "nature's toothbrush." Investigators have theorized that fibrous foods: 1) exercise the gums, 2) promote gingival keratinization and 3) clean the teeth. Fiber in foods, especially as it relates to texture, affects plaque and calculus accumulation and gingival health in dogs and cats and will be discussed below.

CALCIUM AND PHOSPHORUS Foods deficient in calcium and excessive in phosphorus may lead to secondary nutritional hyperparathyroidism and significant loss of alveolar bone. Experiments in dogs have demonstrated resorption of alveolar bone following consumption of a low-calcium, high-phosphorus food.

Calcium deficiency rarely occurs in dogs and cats that consume commercial pet foods containing calcium levels that meet AAFCO allowances. A more realistic concern is the excessive levels of calcium and phosphorus present in many commercial pet foods. High levels of phosphate and calcium are calculogenic in rats.

Many homemade formulations, however, contain excessive protein, but are deficient in calcium. Commonly used meat and carbohydrate sources contain more phosphorus than calcium; therefore, homemade foods may have inverse calcium-phosphorus ratios as high as 1:10. Most homemade foods for dogs contain excessive quantities of meat, often far exceeding the animal's protein and phosphorus requirements. Pets fed homemade foods may be at increased risk for problems arising from calcium-phosphorus imbalances, unless the owner provides a daily source of calcium. (See Chapter 6.)

Further research clarifying the role of dietary calcium and phosphorus is warranted; however, the role these minerals have in calculus formation should be kept in mind when recommending a food as part of an oral care regimen.

VITAMINS Vitamins A, B, C and D have been studied in relation to periodontal disease. Deficiencies in vitamin A reportedly cause marginal gingivitis, gingival hypoplasia and resorption of alveolar bone. Vitamin B-complex (including folic acid, niacin, pantothenic acid and riboflavin) deficiencies have been associated with gingival inflammation, epithelial necrosis and resorption of alveolar bone. Vitamin C plays a key role in collagen synthesis. Deficiencies in ascorbic acid in people adversely affect periodontal tissues. Vitamin D helps regulate serum calcium concentrations. Deficiencies of vitamin D affect calcium homeostasis and reportedly affect the gingivae, periodontal ligament and alveolar bone. Almost all commercial pet foods contain adequate levels of these vitamins. Many homemade formulations, however, are deficient in vitamins and microminerals. Pets fed such formulations are at increased risk for vitamin deficiencies, and potentially periodontal disease, unless the owner provides daily vitamin-mineral supplementation. (See Chapter 6.)

Assess the Food (Pages 489-493)

Foods influence the growth and maturation of the oral microflora. Thus, they affect substrate accumulation and periodontal health, by direct nutrient availability from food/particle retention, as well as by systemic absorption and distribution of nutrients through serum, and salivary and crevicular fluids. Therefore, it is prudent to assess nutrient composition, particularly protein, carbohydrates, calcium and phosphorus, when evaluating a food as a component of an oral care regimen. (See the Key Nutritional Factors–Dental Disease table.)

A dental food should provide optimal nutritional balance for dogs and cats and be orally and systemically safe. The specific key nutritional factors mentioned above should be evaluated. Optimal nutrient balance is critical to tissue integrity and should not be overlooked when assessing dental foods.

Assessment of the food to preserve oral health or to help manage periodontal disease in dogs and cats includes two components: 1) ensuring the food meets AAFCO recommendations for the age/use of the pet evaluated, without providing excess nutrients that might cause imbalances or might contribute to substrate formation (See the Key Nutritional Factors–Dental Disease table.) and 2) determining whether the client's plaque control protocol effectively prevents gingivitis.

Most dogs and cats enter adulthood with healthy mouths. In many cases, periodontal disease can be prevented with appropriate plaque control. The level of plaque control necessary to maintain oral health must be assessed for each animal. Frequent plaque removal (daily, if possible) is widely recommended in the veterinary dental literature. Brushing, when done correctly and conscientiously, is a very effective method for achieving the level of plaque control necessary to control gingivitis. For more information, see Small Animal Clinical Nutrition, 4th ed., p 486.

If the pet owner is able to provide effective plaque control through toothbrushing, then the form of the food and other methods of plaque control may be of less concern. Realistically, however, compliance with toothbrushing may be a problem for many pet owners. Some animals may require aggressive plaque control combined with frequent professional care to maintain optimal oral health.

Plaque is the key substrate that must be controlled in order to control periodontal disease. This section reviews: 1) various methods of plaque control available to veterinarians and pet owners, 2) research that supports or dispels the dental benefit claims made by manufacturers of treats, foods and other products and 3) research that proves the effectiveness of textural fiber as a means of controlling accumulation of plaque and improving gingivitis when used in maintenance foods for adult cats and dogs (**Tables 16-3** and **16-4**).

Food Forms: Facts and Fallacies *(Pages 490-491)*

The physical consistency of foods has been reported to affect oral health of dogs and cats; feeding recommendations for oral health commonly include feeding a dry pet food. Hard food purportedly increases mastication, which aids oral health by exercising the gums, increasing keratinization of the gingivae and reducing accumulation of plaque and calculus.

NATURAL DIETS Early literature reported that the natural diets of wild canids and felids had a plaque-retardant effect and that wild canids and felids were not afflicted with the generalized form of periodontal disease seen in domesticated pets. Pet food commercialization is often implicated as a contributing factor to the increased prevalence and severity of periodontal disease in domestic dogs and cats.

There are emerging reports on the oral condition of small populations of dogs and cats consuming a natural diet. One study involved 67 English foxhounds, one to nine years of age, that were routinely fed raw carcasses consisting of the bony skeleton, muscle and associated tissues. Oral examinations revealed that all dogs had varying signs of periodontal disease as well as a high prevalence of tooth fractures. Another study examined 45 small feral cats from an Australian national park and reported conditions including calculus deposits, periodontal disease, fractured teeth, attrition and odontoclastic resorptive lesions. Examination of gastrointestinal (GI) contents of these cats revealed the presence of a natural diet including small mammals, birds, lizards and insects. These findings cast skepticism on the long-held view that a natural diet prevents development of oral disease, particularly periodontal disease, in dogs and cats.

DENTAL BENEFITS OF SOFT VS. HARD FOODS Consumption of soft foods may promote plaque accumulation. However, the general belief that dry foods provide significant oral cleansing should be regarded with skepticism. A moist food may perform similarly to a typical dry food in affecting plaque, stain and calculus accumulation. In a large epidemiologic survey, dogs consuming dry food alone did not consistently demonstrate improved periodontal health when compared with dogs eating moist foods.

DENTAL

Table 16-3. Dental benefits of selected commercial foods (fed as adult maintenance food).

Products*	Prot.	Fat	Fiber	Cal.	Phos.	Manufacturer's claims**
		(% dry matter)				
Dog food						
Hill's Prescription Diet Canine t/d*** (Original Bites and Small Bites)	16.5	16.1	10.5	0.56	0.40	Reduces the accumulation of plaque, stain and calculus on the teeth of adult dogs. Reduces oral malodor. Maintains gingival health.
Cat foods						
Friskies Dental Diet†	32.0	11.0	5.3	1.79	1.53	Helps maintain dental health. Reduces plaque and tartar buildup by 25% vs. leading brands.
Hill's Prescription Diet Feline t/d††	34.8	16.5	9.2	1.01	0.76	Reduces the accumulation of plaque, stain and calculus on the teeth of adult cats. Maintains gingival health.
Hill's Science Diet Plan Feline Oral Care Formula†††	34.8	16.5	8.0	1.01	0.76	Reduces the accumulation of plaque, stain and calculus on the teeth of adult cats.

Key: Prot = protein, Cal = calcium, Phos = phosphorus, VOHC = Veterinary Oral Health Council.
*This list represents products with the largest market share and for which published information is available.
**Obtained from product label or packaging.
***Published research supports efficacy claims for control of plaque, stain, calculus, malodor and gingivitis. VOHC Seal of Acceptance.
†VOHC Seal of Acceptance.
††Published research supports efficacy claims for control of plaque, stain, calculus and gingivitis. VOHC Seal of Acceptance.
†††Manufacturer's data support efficacy claims for control of plaque, stain and calculus.

Table 16-4. Dental benefits of selected commercial treats.

Products*	Recommended feeding method**	Prot.	Fat	Fiber	Cal.	Phos.	Manufacturer's claims**
			(% dry matter)				
Dog treats or snacks							
Friskies Cheew-eez Treats (Rawhide chew)***	Give at least two treats daily	89.0	2.4	0.6	na	na	Helps fight tartar and gives cleaner, healthier teeth. 50% less tartar than dogs fed dry food alone.
Heinz Tartar Check (Hexametaphosphate-coated biscuit)†	Give two biscuits each day	14.0	8.8	2.7	0.84	0.36	Reduces tartar buildup by 45%. Cleans teeth and freshens breath.
Iams Original Formula (Small biscuit)††	Feed as a treat; not more than two biscuits per cup of dry food	28.3	7.1	3.3	1.30	0.98	Promotes clean teeth, fresh breath and healthy gums.
Nabisco Milk-Bone Biscuit††	Feed five biscuits (small) twice a day	23.6	6.8	3.0	1.51	1.14	Fights tartar buildup above the gum line. Helps remove plaque.
Nutro Tartar Control Dog Biscuit††	Feed two to three biscuits throughout the day	29.2	9.0	3.4	1.12	1.12	They scrub away tartar buildup to help keep teeth and gums healthy.
Pedigree Dentabone/ Rask†††	Offer as a chewing snack	26.0	5.5	0.1	1.71	0.76	Reduces tartar and plaque buildup.
VRx C.E.T. Chews (Enzymatic beefhide chew)††	Feed at least one chew on those days when toothbrushing does not occur	na	na	na	na	na	When combined with brushing they can be an effective method of removing plaque and food debris on a daily basis.

(Continued on next page.)

DENTAL

Table 16-4. Dental benefits of selected commercial treats (Continued).

Products*	Recommended feeding method**	Prot.	Fat	Fiber	Cal.	Phos.	Manufacturer's claims**
		(% dry matter)					
Dog treats or snacks							
Waltham Formula Tartar Chew[ÜÜ]	Feed daily as a treat; not suitable for dogs under 5 kg	14.3	2.6	0.1	na	na	Clinically proven to reduce tartar and plaque buildup.
Cat treats or snacks							
Pounce Tartar Control (Dry cat treat)[ÜÜ]	Feed 10 to 26 pieces per day (approximately 12.5% of daily food intake)	33.1	12.3	2.0	1.33	1.20	Reduces tartar buildup and helps maintain the overall health of the teeth.
Whisker Lickin's Tartar Control (Dry cat treat)[ÜÜ]	Feed as a daily snack to your cat	35.0	11.2	4.5	1.24	1.16	Helps reduce tartar buildup and maintain overall health of the teeth. It's our crunchy texture that scrapes away the plaque and tartar on your cat's teeth.
VRx C.E.T. Forte Chews[â]	Feed one chew on days toothbrushing does not occur	na	na	na	na	na	Uniquely formulated to provide abrasive cleansing action. Specially treated with an antibacterial enzyme system to help maintain oral health.

Key: Prot = protein, Cal = calcium, Phos = phosphorus, na = information not available.
*This list represents products with the largest market share and for which published information is available.
**Obtained from product label or packaging.
***Published research supports claims for supragingival calculus control compared to feeding dry food or plain biscuits when fed two to three times/day.
[Ü]Published data support hexametaphosphate as a calculus-control agent.
[ÜÜ]No published data.
[ÜÜÜ]Published research supports efficacy claims when fed daily.
[â]One clinical trial indicated a reduction in plaque and calculus accumulation when cats ate one treat/day.

Foods with Textural Characteristics (Page 491)

Typical dry dog and cat foods contribute little dental cleansing. As a tooth penetrates a kibble or treat the initial contact causes the food to shatter and crumble with contact only at the coronal tip of the tooth surface (**Figure 16-1**). To provide effective mechanical cleansing, a food should promote chewing and maintain contact with the tooth surface (**Figure 16-1**).

Foods with enhanced textural characteristics promote oral health. Several maintenance pet foods are available (**Table 16-3**) that provide significant oral cleansing compared with commercial treats and typical dry and moist foods. Numerous studies have demonstrated that some of these foods provide significant plaque, calculus and stain control in dogs and cats. A six-month study investigating the effects of food on plaque accumulation and gingival inflammation in 40 adult mongrel dogs reported that dogs fed the test food (Prescription Diet Canine t/d, Hill's Pet Nutrition, Inc., Topeka, KS, USA) had 39% less plaque accumulation and 36% less gingival inflammation than dogs fed the control food (Purina Dog Chow, Ralston Purina Co., St. Louis, MO, USA). These studies used a clean-tooth model in which plaque, calculus and stain were evaluated at a specified time following a dental prophylaxis. (See Product Claims below.)

One study reported that feeding Prescription Diet Canine t/d to beagles with existing plaque, calculus and gingivitis resulted in a significant decrease in mean plaque and calculus indices after two weeks and in the gingival index after six weeks. Beagles eating the control food (Purina Dog Chow) had a significant increase in plaque and calculus accumulation and no change in gingival inflammation over the 16-week test period.

Combining an increased fiber content with a size and shape pattern (texture) that promotes chewing and maximizes contact with teeth is critical to obtaining dental benefits. A typical dry food does not possess the textural characteristics for adequate dental cleansing. Simply enlarging the kibble or varying the shape of the product is likewise inadequate. In the absence of effective plaque control through other measures or in cases demanding adjunctive plaque control, mechanical control of plaque and calculus accumulation daily with a maintenance dental food is reasonable. Given the prevalence of periodontal disease in dogs and cats, and the need for effective home care products that improve owner compliance, it is likely dental benefit technology will be added to more foods in the future.

Other Methods of Substrate Control: Facts and Fallacies (Pages 491-493)

TREATS Many treats that claim a wide variety of dental benefits are available to pet owners. Rawhide strips have been reported to control calculus accumulation, provided the dog actively chews the strips daily. Two rawhide

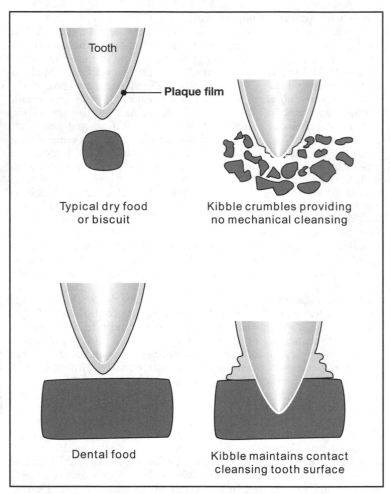

Tooth

Plaque film

Typical dry food
or biscuit

Kibble crumbles providing
no mechanical cleansing

Dental food

Kibble maintains contact
cleansing tooth surface

Figure 16-1. This illustration depicts the mechanical cleansing properties of commercial dog and cat foods. The top illustration demonstrates what happens when a dog or cat chews a typical dry food. The kibble crumbles providing little to no mechanical cleansing. The bottom illustration demonstrates what happens when a dog or cat chews a dental food. The kibble stays together, maintaining contact with the tooth surface and providing mechanical cleansing.

chews each day are typically recommended. Compacted rawhide treats in the shape of balls and bones can cause tooth fractures if chewed aggressively or if used as "catch" toys. Flat rawhide chews coated with an enzymatic system (C.E.T. Chews, VRx Products, Harbor City, CA, USA) are also available, but there are no published data demonstrating that these are any more effective than plain rawhide strips (**Table 16-4**).

A treat made of rice and whey and formed into a bone shape (Pedigree Chum Rask or Dentabone, Kal Kan Foods, Vernon, CA, USA) to promote chewing activity reportedly reduced accumulation of plaque and calculus in small dogs over a four-week period. The disadvantages of these products may include pet acceptance, potential for GI side effects, cost of the recommended feeding dosage and nutritional influences such as caloric excess and nutrient imbalances.

BISCUITS Plain baked biscuits, although long thought of as "dental" treats, provide little additional plaque and calculus reduction when compared with the feeding of dry dog food alone (**Table 16-4**). Additionally, manufacturers of some feline treats (Pounce Tartar Control Treats, Quaker Oats, Chicago, IL, USA; Whisker Lickin's Tartar Control Treats, Ralston Purina Co., St. Louis, MO, USA) make a tartar control claim; however, two studies have failed to demonstrate an effect on plaque and calculus accumulation compared with feeding dry or moist foods alone.

The addition of hexametaphosphate (HMP) to the surface of baked biscuits (Tartar Check, Heinz Pet Products, Newport, KY, USA) significantly reduced calculus accumulation in beagles over a four-week period compared with a regimen of plain baked biscuits and dry food alone. One three-week study, however, demonstrated no significant differences in plaque and calculus accumulation in dogs fed dry food, dry food plus baked biscuits or dry food plus HMP-coated biscuits.

CHEW TOYS Many varieties of chew toys are available with claims ranging from "flosses teeth" to "reduces harmful plaque;" however, few data in the literature substantiate these claims. One report claimed less calculus accumulation in 14 of 20 client-owned dogs when dogs were allowed access to a urethane chewing device (Nylabone, Nylabone Products, Neptune, NJ, USA) for one month. Anecdotal reports of oral trauma (e.g., gingival lacerations and tooth fractures) resulting from aggressive chewing of some dental toys can also be found in the veterinary dental literature.

PRODUCT CLAIMS It can be very confusing for veterinarians, and particularly for pet owners, to discern which products provide a significant dental benefit and thus warrant use as oral hygiene agents. The Center for Veterinary Medicine of the Food and Drug Administration (CVM-FDA) monitors and regulates dental health claims in the United States. Cosmetic claims are not objectionable and structure-function claims are not stringently regulated; thus, the wide availability of products that make some type of plaque or calculus claim with little or no research to document effectiveness. Phrases such as "cleans teeth, freshens breath" are commonplace on commercial treats packages. Because "crunchy" foods provide little dental benefit, the purported ability of these types of products to provide any significant level of oral hygiene is a misrepresentation to pet owners (**Table 16-4**).

The veterinary dental community has recognized the need for standardized and scientifically sound methods by which to measure substrate accumulation in dogs and cats and evaluate product efficacy.

The Veterinary Oral Health Council (VOHC) was established in 1997. The purpose of the VOHC is to provide an independent, objective and credible means of recognizing veterinary dental products that effectively control accumulation of plaque and/or calculus (tartar). The VOHC system is similar to the American Dental Association (ADA) Seal of Acceptance system. The VOHC does not conduct efficacy testing; the council reviews results of tests performed in accordance with approved protocols set by the VOHC. The first canine and feline dental products to receive the VOHC Seal of Acceptance were Prescription Diet Canine t/d and Feline t/d, respectively (**Figure 16-2**).

Products that increase chewing time and salivation and that maintain contact with the tooth surface have the potential to provide an oral health benefit through reduced accumulation of plaque and calculus. It is the veterinarian's responsibility to be aware of dental claims made by manufacturers and recommend oral hygiene aids that are safe and effective for clients' pets.

Assess the Feeding Method (Page 493)
Changing the feeding method in the management of dental disease may not always be necessary, but verifying that an appropriate feeding method is being used is part of a thorough assessment.

FEEDING PLAN (Page 493)

Select a Food (Page 493)
Oral disease should be treated with appropriate professional therapy. However, aftercare, or ongoing dental hygiene provided by the pet owner, will determine the overall success of professional therapy. Many pets can resume their normal nutritional regimen immediately after receiving professional care, provided the client has been instructed in appropriate

Figure 16-2. The Veterinary Oral Health Council Seal of Acceptance recognizes products that effectively control accumulation of plaque and/or calculus.

plaque control procedures. (See Small Animal Clinical Nutrition, 4th ed., p 486.) If the client isn't physically able to brush the pet's teeth often enough to control gingivitis or if the pet's temperament precludes such procedures, then nutritional recommendations should include feeding dental foods with textural characteristics sufficient to control plaque, calculus and gingivitis daily (**Table 16-3**).

A regimen of soft food may be recommended after invasive or advanced procedures during the initial healing phase. Chemical plaque control should be provided in these instances until mechanical plaque control can be resumed.

Determine a Feeding Method (Page 493)

The method of feeding is often not altered in the nutritional management of oral disease. If a new food is fed, the amount to feed can be determined from the product label or other supporting materials.

REASSESSMENT (Page 494)

The degree of monitoring required will depend on the: 1) degree of oral pathology, 2) level of periodontal therapy provided and 3) ability of the owner to provide routine oral hygiene. An annual oral examination and professional prophylaxis should be adequate for adult dogs and cats with good oral health and normal occlusion. As the severity of oral disease increases, the degree of periodontal therapy required to treat the condition will also increase. An increased level of oral hygiene will be necessary to prevent disease progression for advanced stages of periodontal disease (i.e., worse than gingivitis).

Initially, patients should be rechecked weekly to monitor healing and oral hygiene. If both are satisfactory, the time between recalls can increase to three-month intervals. If the animal has severe pathology that promotes plaque retention or if the owner is unable to provide effective plaque control, the time between periodontal therapy will need to be adjusted to maintain oral health. These recommendations are initial guidelines. Veterinarians must decide appropriate recall for each case, depending on the degree of oral pathology, periodontal therapy and owner compliance.

Clinical cases that illustrate and reinforce the nutritional concepts presented in this chapter can be found in Small Animal Clinical Nutrition, 4th ed., pp 498-504.

CHAPTER 17

Developmental Orthopedic Disease of Dogs

For a review of the unabridged chapter, see Richardson DC, Zentek J, Hazewinkel HA, et al. Developmental Orthopedic Disease of Dogs. In: Hand MS, Thatcher CD, Remillard RL, et al, eds. Small Animal Clinical Nutrition, 4th ed. Topeka, KS: Mark Morris Institute, 2000; 505-528.

CLINICAL IMPORTANCE (Pages 505-506)*

The prevalence of musculoskeletal problems in dogs less than one year old in all breeds is about 22%, with 20% possibly having a nutrition-related etiology. Developmental orthopedic disease (DOD) includes a diverse group of musculoskeletal disorders that occur in growing animals (most commonly fast-growing, large- and giant-breed dogs) and that are sometimes related to nutrition. Canine hip dysplasia (CHD) and osteochondrosis make up the overwhelming majority of the musculoskeletal problems with a possible nutrition-related etiology.

Canine Hip Dysplasia (Page 506)

CHD is the most frequently encountered orthopedic disease in veterinary medicine with heritability and a potential nutrition-related etiology. CHD is an abnormal development, or growth, of the hip joint manifested by varying degrees of laxity of surrounding soft tissues, instability of the joint and malformation of the femoral head and acetabulum with osteoarthrosis. The actual number of cases of CHD is estimated to be in the millions worldwide.

Osteochondrosis (Page 506)

Generally, osteochondrosis is a disruption in endochondral ossification that results in a focal lesion. Osteochondrosis occurs in the physis and/or epiphysis of growth cartilage, and may be considered a generalized or systemic disease. Clinical signs of osteochondrosis are related to the severity and location of disease.

When osteochondrosis affects physeal cartilage, it may cause growth abnormalities such as angular limb deformities in long bones. Osteochondrosis of articular epiphyseal cartilage commonly occurs in the shoulder (proximal humerus), stifle (distal femur), hock (talus) and elbow (distal humerus). Acute inflammatory joint disease (or degenerative joint disease)

ORTHOPEDIC

*Page numbers in headings refer to Small Animal Clinical Nutrition, 4th ed., where additional information may be found.

Key Nutritional Factors–Developmental Orthopedic Disease
(Page 507)

Factors	Associated conditions	Dietary recommendations
Energy	Canine hip dysplasia Osteochondrosis Obesity	Initially, calculate energy intake as follows: 0 to 4 months of age = 3 x resting energy requirement 4 to 12 months of age = 2 x resting energy requirement Feed a food with metabolizable energy of 3.2 to 3.8 kcal/g dry matter (DM) (13.4 to 15.9 kJ/g DM)
Fat	See above	Feed a food that does not exceed 12% fat (dry matter basis [DMB]), especially if feeding free choice
Calcium	Canine hip dysplasia Osteochondrosis	Feed a food with 0.7 to 1.2% calcium (DMB)
Supplements	See above	None recommended if a commercial food is fed

may ensue subsequent to development of osteochondrosis when the cartilage surface is disrupted and subchondral bone is exposed to synovial fluid. Inflammatory mediators and cartilage fragments are released into the joint (osteochondritis dissecans), which perpetuates the cycle of degenerative joint disease.

ASSESSMENT (Pages 507-520)

Assess the Animal (Pages 507-518)
History and Physical Examination (Page 507)

Breed and familial history are important predisposing factors for DOD. For mixed-breed dogs, it is useful to know the breed of the stud and bitch as well as historical wellness of offspring. If possible, it may be helpful to gather information pertaining to skeletal abnormalities of previous litters from the same bitch and stud.

Food intake and history should be evaluated as described in Chapter 1. Any treats and supplements fed to the patient should be scrutinized closely, paying special attention to the calcium and energy intake. It is critical to calculate, or closely estimate, metabolizable energy (ME) and calcium intake in order to provide rational advice for feeding growing large- and giant-breed dogs.

Dogs should be weighed during the initial visit and all subsequent visits to help monitor growth rate. A body condition score (BCS) should be determined and recorded at each visit. (See Chapter 1.) Attention to abnormal changes in weight or BCS will help in assessing and managing growing dogs. In some cases, graphs of body weight and BCS may prove useful in recognizing variances from desired goals.

Before a physical examination is conducted, historical information should be gathered about the degree, if any, of perceived lameness, the affected limb(s), duration of lameness and any peculiarities regarding the lameness. Following historical evaluation, the animal should be observed at rest for any gross conformational abnormalities. Next, the animal should be observed in motion to ascertain the degree of lameness and location of involvement.

If a locomotor defect is confirmed, the etiology should be determined. To determine the cause, the examination should include: 1) palpation of limbs for asymmetry, swelling, heat and sensitivity, 2) deep palpation of long bones, 3) flexion/extension of joints to determine range of motion, crepitation, instability and sensitivity and 4) neurologic evaluation. Even after a thorough physical examination, the exact cause of the lameness may remain undetermined.

Radiography *(Page 507)*

Radiographs should be taken to further define the clinical diagnosis. However, inability to identify lesions by survey radiography does not always negate the presence of disease.

Laboratory Information *(Pages 507-510)*

Uncomplicated cases of DOD rarely have altered complete blood counts. Severe elevations or decreases in white blood cell counts usually indicate other disease processes.

On occasion, serum concentrations of calcium and phosphorus may be elevated or decreased during the genesis of DOD. However, absence of calcium or phosphorus perturbations does not rule out a diagnosis of DOD. Conversely, many other disease processes may result in altered calcium or phosphorus homeostasis, which indicates abnormal values are not pathognomonic for a diagnosis of DOD.

Increased bone remodeling may result in increased serum alkaline phosphatase activity. This parameter is already high in young, growing animals and may not be a very sensitive indicator of ongoing metabolic bone disease.

ORTHOPEDIC

Table 17-1. Parathyroid hormone (PTH), ionized calcium and 1,25-dihydroxyvitamin D_3 concentrations in different physiologic/disease states.

States	PTH	Ionized calcium	1,25-dihydroxy-vitamin D_3
Apocrine gland tumors of the anal sacs	Low	High	Low
Chronic renal failure	High	Low/normal	Normal/low
High calcium intake	Low	High	Normal/low
Hypervitaminosis D	Low	High	Normal/high
Hypoparathyroidism	Low	Low	Low
Lymphosarcoma	Low	High	Low
Primary hyperparathyroidism	High	High	Normal/high

MEASURING AND INTERPRETING SPECIFIC LABORATORY TESTS/ Parathyroid Hormone Interpretation of serum parathyroid hormone (PTH) concentrations from other species has proven that evaluation must be made in conjunction with presenting signs and other biochemical tests such as concentrations of ionized calcium and 1,25-dihydroxyvitamin D_3 (**Table 17-1**). PTH values may be increased, decreased or normal in DOD depending on the etiology. Increased PTH concentrations may be observed in association with renal disease, vitamin D deficiency and states in which insufficient calcium is present in foods. Decreased PTH concentrations may be observed when excess calcium or vitamin D is present in foods, and in other metabolic diseases.

MEASURING AND INTERPRETING SPECIFIC LABORATORY TESTS/ Vitamin D Vitamin D_3 may be required in foods for dogs because endogenous synthesis may be limited. Because commercial foods contain added vitamin D_3 and in light of potentially limited endogenous synthesis, measurement of vitamin D_3 in serum may reflect dietary changes rather than specific disease states. 25-hydroxyvitamin D_3 is produced in the liver from vitamin D_3 and is a good indicator of general vitamin D_3 deficiency or excess. Another useful indicator of vitamin D_3 status is measurement of the most biologically active metabolite of vitamin D_3, 1,25-dihydroxyvitamin D_3, which is produced in the kidneys via the 1-α-hydroxylase enzyme. The concentration of 1,25-dihydroxyvitamin D_3 in serum is not a good indicator of vitamin D_3 toxicity; however, it is a more sensitive indicator of deficiency than serum concentrations of 25-hydroxyvitamin D_3.

MEASURING AND INTERPRETING SPECIFIC LABORATORY TESTS/ Calcium Calcium homeostasis is maintained by the sum of physiochemical and calciotropic hormonal processes. Calcium in blood is in equilibrium between the ionized state (45 to 50%), a protein-bound state (40 to 45%) and a complexed or chelated state (5 to 10%). The concentration of ionized calcium is the most important determinant of calciotropic homeostatic regulation initiated by the parathyroid gland and the C-cells of the thyroid gland. Sudden increases in ionized calcium concentrations stimulate release of calcitonin from the thyroid gland, whereas decreases in concentrations of ionized calcium stimulate release of PTH from the parathyroid gland. The total concentration of calcium in serum is affected by the interplay of the homeostatic mechanisms involving influx (gastrointestinal [GI] absorption and bone resorption), efflux (GI and renal losses) and skeletal mineralization of the less labile bone pool as outlined below.

When concentrations of ionized calcium are below the normal range:

1. PTH secretion is stimulated, which in turn stimulates conversion of 25-hydroxyvitamin D_3 to the biologically more potent 1,25-dihydroxyvitamin D_3 in the kidneys.
2. 1,25-dihydroxyvitamin D_3 stimulates calcium uptake in the gut via receptor-mediated mechanisms.
3. 1,25-dihydroxyvitamin D_3, in conjunction with PTH, stimulates bone resorption and calcium reabsorption in the kidneys.
4. PTH induces phosphaturia.

When concentrations of ionized calcium are above the normal range:

1. Calcitonin secretion is stimulated, PTH secretion is suppressed and 1,25-dihydroxyvitamin D_3 production is not stimulated. Instead, the kidneys produce 24,25-dihydroxyvitamin D_3, which is generally considered biologically inactive.
2. Gut absorption and bone resorption of calcium are not stimulated.
3. Calcitonin decreases osteoclastic activity.
4. Renal calcium excretion is increased.

Risk Factors *(Page 510)*

The genesis of DOD is presumed to be a multifactorial process. It appears that management (environmental), genetic and nutritional interactions have significant influence in young, growing animals. The most critical period for development of DOD is during the growth phase, before epiphyseal closure. Specific factors that are currently thought to increase the risk of DOD in young dogs include: 1) belonging to a large or giant breed (genetics), 2) free-choice feeding (management/genetics), 3) feeding of high-energy foods (nutrition) and 4) excessive intake of calcium from food, treats and supplements (nutrition).

Etiopathogenesis *(Pages 510-513)*

A variety of mechanisms are plausible in considering the pathogenesis of DOD. No one specific etiology is considered ultimately responsible for all observed clinical manifestations of DOD. Historically, feeding dogs imbalanced foods, especially those deficient in calcium, phosphorus or vitamin

D_3, was the main risk factor predisposing to skeletal diseases such as secondary hyperparathyroidism with subsequent development of osteodystrophia fibrosa. Dietary deficiencies are rare in young, growing dogs fed commercial growth foods because most foods are formulated to meet or exceed allowances for specific nutrients. Two popular, current theories for the pathogenesis of some types of DOD are discussed in the following sections. Specific nutrients are addressed in the Key Nutritional Factors section.

THEORY 1: ENERGY/GROWTH/BIOMECHANICAL STRESS The musculoskeletal system changes constantly throughout life with the most rapid changes occurring during the first few months. The skeletal system apparently is most susceptible to physical, nutritional and metabolic insults during the first 12 months of life because of heightened metabolic activity. Large- and giant-breed dogs are most susceptible to DOD, presumably because of their genetic propensity for rapid growth.

High energy intake directly affects growth velocity via nutrient supply and indirectly through changes in concentrations of growth hormone, insulin-like growth factor 1 (IGF-1), triiodothyronine (T_3), thyroxine (T_4) and insulin. Dysregulation of these endocrine factors, whether attributable to nutrition, feeding management or genetics, during this critical period of skeletal growth may be responsible for producing an environment in which DOD develops.

Growth hormone and IGF-1 stimulate chondrocyte proliferation and differentiation.

IGF-1 was found in significantly higher concentrations in growing dogs fed free choice compared with animals on restricted feed allowance.

Free-choice feeding of dogs is also accompanied by higher circulating concentrations of T_3 and T_4 compared with levels in food-restricted controls, reflecting a general stimulation of metabolic processes. Thyroid hormones are not only general stimuli for metabolic processes, including increasing the rate of bone formation and resorption, but are also important for capillary penetration of degenerating cartilage cells and the final stage of endochondral bone formation.

The result of these hormonal influences is enhanced mitotic activity of proliferative cartilage cells, which may enlarge the width of the inherently mechanically unstable zone of chondrocyte growth.

Histologic examinations have revealed articular cartilage is less well supported by solid bone plates in rapidly growing dogs, compared with smaller breeds or to littermates fed restricted amounts after weaning. The epiphyseal spongiosa of giant-breed dogs is inherently less dense and therefore assumed to be weaker than the spongiosa in small breeds, a tendency that may be exaggerated by overnutrition. Free-choice feeding may lead to a mismatch between bone growth and body growth, resulting in a lower ratio of long bone diaphyseal shaft cross-sectional area to body weight and also a less dense epiphyseal spongiosa.

The biomechanical stress induced by rapid weight gain during growth as discussed above has been cited as an etiology for DOD. It is unknown whether small focal cartilaginous lesions occur first and are then exacerbated by biomechanical stress, or if biomechanical stress first induces cartilaginous lesions. In either case, increased static forces (weight load) and dynamic forces (muscle pull) may damage immature skeletons, especially in large- and giant-breed dogs. These dysregulations of nutrient supply, bone formation and endocrine regulation may interfere with skeletal maturation, thus increasing the risk for DOD in young animals.

THEORY 2: EXCESS CALCIUM AND HYPERCALCITONINISM A contrasting theory to the preceding theory about high energy intake and rapid growth rate stems from the observation that the rate of DOD is increased in dogs with high calcium intakes. Young Great Danes fed a food high in energy and minerals free choice, or high in calcium alone, developed osteochondrosis lesions with overt clinical signs of disease. These lesions appeared at both weight-bearing sites and sites where weight bearing was of no influence, such as the growth plates of ribs.

Feeding high-calcium foods to growing small-breed dogs results in histologic lesions but no clinical manifestations of DOD. Large-breed dogs raised on food with a high calcium content or high calcium and phosphorus content had disturbed endochondral ossification, retained cartilaginous cores in the distal radius and ulna and delayed skeletal maturation and growth of bone length. Calcium intake, therefore, seems to be a significant determining factor in DOD.

Key Nutritional Factors (Pages 513-516)

Nutrients must be provided in appropriate amounts and balances for optimal bone development. Excesses of calcium and energy, together with rapid growth, appear to predispose dogs to certain musculoskeletal disorders such as osteochondrosis and CHD. However, severe excesses, deficits and imbalances of any nutrient may affect bone development.

ENERGY/FAT Energy intake, which depends on a variety of physiologic factors, is the main nutritional factor that determines growth intensity, as long as other nutrients are supplied in adequate and balanced amounts. The risk of DOD appears to be increased in large- and giant-breed dogs fed well-balanced, highly palatable, energy-dense foods, free choice as discussed previously. The detrimental influence of high energy intake on skeletal development during growth has been demonstrated in dogs and other animal species (e.g., turkeys, pigs). Lesions appear in physeal or

ORTHOPEDIC

articular epiphyseal cartilages as disturbances of endochondral ossification.

Fat must be considered when assessing the energy density of foods. As the fat content of food increases, the energy density also increases unless fiber is substituted for other metabolizable nutrients. Dogs grow slower and have less fat deposition when they are fed a food with low energy density free choice (3.16 kcal [13.22 kJ]/g ME, 8.0% fat dry matter basis [DMB]) compared with a high energy density food (3.98 kcal [16.65 kJ]/g ME, 23.9% fat DMB).

Because an increased growth rate is a risk factor for DOD in fast-growing, large- and giant-breed dogs, it follows that increased fat content (>12% DMB), or increased ME/g of dry matter (DM) of foods, must be considered a risk factor. However, some fat in the food is important for absorption of fat-soluble vitamins and for palatability. Concentrations of essential nutrients may need to be increased to meet nutritional requirements when the energy density of foods is altered.

It is difficult to determine the appropriate daily energy requirement (DER) for growing dogs because few well-controlled studies have been conducted. Energy intake reaches a maximum, as related to body weight, in the second to fourth month of life (<45% adult body weight). In general, a good starting point is to feed dogs three times the resting energy requirement (RER) from weaning to four months of age, followed by two times the RER until about one year of age.

The body condition of dogs should be evaluated every two weeks and the food dose adjusted as necessary (**Table 17-2**). It is usually best to keep the energy content of a growth food for larger breed puppies below 3.8 kcal [15.9 kJ]/g, which approximates 12% fat on a DMB, especially if fed free choice.

CALCIUM The amount of true calcium absorption in dogs ranges from 25 to 90%, depending on the amount of intake and the age of the animal. Calcium is absorbed via three mechanisms: 1) active absorption, 2) facilitated absorption and 3) passive diffusion. Passive diffusion is especially important in young dogs. Active absorption is most important in the proximal GI tract. Passive diffusion and facilitated absorption, however, are important in the distal GI tract, primarily because of prolonged transit time and increased calcium concentration through that section. Vitamin D_3 metabolites, especially 1,25-dihydroxyvitamin D_3, are the most important hormonal regulators of GI calcium absorption.

In the face of adequate levels of calcium in the food, the absolute level of calcium, rather than an imbalance in the calcium-phosphorus ratio, influences skeletal development. In one study, the prevalence of DOD was significantly increased in young, giant-breed dogs fed a food contain-

Table 17-2. Worksheet for calculating the initial feeding plan for large- and giant-breed dogs.

I. **Weigh (determine weight in kg)**

$\boxed{\text{WEIGH}}$

II. **Feed (determine daily energy requirement [DER])**
   A. Caloric requirement formulas
      Linear method
      RER $= 30(BW_{kg}) + 70 =$ initial estimate in kcal

$\boxed{\text{FEED}}$

      DER $=$ RER x 3.0 (two to four months of age)
            $=$ RER x 2.0 (four to 12 months of age)
            $=$ RER x 1.8 (intact adult)
            $=$ RER x 1.6 (neutered adult)
      Exponential method
      RER $= 70(BW_{kg})^{0.75} =$ initial estimate in kcal
      DER $=$ RER x 3.0 (2 to 4 months of age)
            $=$ RER x 2.0 (4 to 12 months of age)
            $=$ RER x 1.8 (intact adult)
            $=$ RER x 1.6 (neutered adult)
   B. Convert kcal to cups or grams
      Contact manufacturer to obtain energy density

III. **Evaluate (reassess every two weeks)**
   A. Weigh
   B. Body condition score (BCS)
   C. Clinical judgment

$\boxed{\text{EVALUATE}}$

IV. **Adjust (as needed)**
   A. If BCS >3/5, decrease intake by 10%
   B. If BCS <2/5, increase intake by 10%

$\boxed{\text{ADJUST}}$

Key: RER = resting energy requirement. To determine kJ, multiply kcal by 4.184.

ing excess calcium (3.3% DMB) with either normal phosphorus (0.9% DMB) or high phosphorus (3% DMB, to maintain a normal calcium-phosphorus ratio). These puppies apparently were unable to protect themselves against the negative effects of long-term calcium excess. Furthermore, long-term calcium intake increases the frequency and severity of osteochondrosis. Because the previously discussed studies have demonstrated the safety and adequacy of 1.1% calcium (DMB), it is recommended that growth foods for at-risk puppies contain calcium levels between 0.7 and 1.2% with no supplementation.

Other Nutritional Factors (*Pages 516-518*)

DIGESTIBILITY Digestibility is a nutritional factor that becomes important in certain physiologic states such as growth. During the growth period, the ability to ingest and absorb adequate amounts of various nutrients

ORTHOPEDIC

depends on food intake capacity and the quality of ingredients. It is especially important to consider quality of ingredients when trying to limit energy intake for at-risk dogs. The goal of energy restriction is not to provide low-quality foods that are poorly digestible, but to provide high-quality foods in a low energy density package that will promote appropriate growth. It is important to assess digestibility and recommend foods with above average digestibility for growth. Above average digestibility values for protein, fat and carbohydrate are listed in **Table 9-5**.

OTHER MINERALS/Phosphorus Excessive as well as inadequate phosphorus intake may affect calcium homeostasis and thus bone development. Chronic, inadequate phosphorus intake, to a lesser degree than calcium depletion, may stimulate 1,25-dihydroxyvitamin D_3 synthesis, which stimulates calcium and phosphorus resorption from bone and absorption in the gut. Mobilization of calcium and phosphorus decreases PTH secretion, increases the renal threshold for phosphorus and eliminates excess calcium in the urine. The result is an increase in serum phosphorus concentration while maintaining serum calcium levels.

Conversely, excessive phosphorus intake with inadequate calcium intake may result in nutritional secondary hyperparathyroidism. The excess phosphorus in food reduces the ionized calcium concentration in serum via mass action equilibrium, thus resulting in hypersecretion of PTH. The end result is a decreased renal threshold for phosphorus and excessive osteoclasia and pathologic fractures of growing bone.

The level of phosphorus recommended must be considered in conjunction with calcium recommendations. The calcium-phosphorus ratio should be maintained at 1.1:1 to 2:1; however, 1.1:1 is preferred. The absolute amount of calcium in the food is more important than the calcium-phosphorus ratio in young growing dogs. (See the calcium discussion above.) When calcium intake is set at 0.7 to 1.2% of the food, as recommended previously for large breeds at risk for DOD, the phosphorus content should be 0.6 to 1.1% of the food DM.

OTHER MINERALS/Copper In several animal species and in people, copper deficiency induces severe skeletal disease. Dietary copper levels less than 1 mg/kg DM are associated with severe growth deformities, fractures, wide "knotty" epiphyses and especially severe hyperextension of the limb axis in growing dogs. In young beagles, clinical signs of copper deficiency are less severe than those previously listed, but hyperextension of the forelegs is a characteristic feature. Feeding a low-copper food (1.2 mg/kg DM) vs. a normal copper food (14.1 mg/kg DM) depletes plasma (1.4 vs. 9.7 µmol/l) and liver copper stores (19 vs. 246 mg copper/kg DM). Secondary copper deficiency results in osteoporotic lesions in growing Great Dane puppies, which may be attributed to impaired osteoblastic function. These dogs were fed an experimental food containing high concentrations of molybdate, which strongly impairs copper absorption and induces secondary copper deficiency.

Some unsupplemented homemade foods (made of rice, dairy products, fat, starch) may contain low or suboptimal copper concentrations. Under certain circumstances, these foods may contribute to the development of skeletal disease, even if copper levels are higher than in deficient experimental foods. A suboptimal copper supply could evoke negative effects especially if combined with high growth intensity or other dietary imbalances (e.g., calcium, zinc or carbohydrates).

Impaired copper absorption may also occur with high dietary calcium or zinc levels; the latter induces copper binding metallothionein in the gut mucosa. High amounts of poorly digestible carbohydrates or foods that are rich in certain types of dietary fiber may also reduce copper absorption.

The recommendation for copper in canine growth foods has been 7.3 ppm; however, to achieve a safety margin for at-risk dogs, a minimum level of 10 ppm is encouraged. Most commercial canine growth foods deliver copper in a range from 10 to 20 mg/kg DM to meet this minimum recommendation.

OTHER MINERALS/Zinc Inadequate zinc supply, especially in growing animals, will lead to severe clinical signs within days, resulting mainly in growth depression, skin defects, impaired immune function and growth disorders of the skeleton. These disorders may be linked to the role of zinc as a cofactor in enzymes that are important for connective tissue metabolism. A low activity of alkaline phosphatase (<300 IU/l) is a good indicator of low zinc status (e.g., deficient zinc intake) in growing and young dogs. There are no reports that suggest excessive zinc intake is detrimental to skeletal development in dogs; however, excess zinc may be presumed to be toxic at higher levels, as observed in other species.

Skeletal abnormalities have been described in Alaskan malamutes with an inborn error in zinc metabolism and skeletal malformation in bull terriers with lethal acrodermatitis enteropathica, a genetically determined defect of zinc metabolism.

It is not known to what extent marginal zinc intake, due to either subnormal dietary zinc concentrations or high concentrations of interacting substances (e.g., phytic acid, calcium, copper, low digestible carbohydrates), contributes to DOD. Foods for growing dogs should contain enough zinc to compensate for negative interactions with other dietary ingredients, especially if the originally balanced food is "improved" by dog owners who add large amounts of calcium carbonate or other calci-

ORTHOPEDIC

um salts. Canine growth foods should contain 120 to 130 mg zinc/kg DM. Most commercial canine growth foods contain 200 to 300 mg/kg dry matter zinc to ensure this minimum recommendation is met.

OTHER MINERALS/Iodine T_3 and T_4 influence normal degeneration of growing cartilage, penetration of capillaries and mineralization of newly formed bone. Thyroid hormones stimulate formation and resorption of bone, which results in remodeling of the skeleton.

Low dietary iodine induces dysfunction of the thyroid glands. Goiter (enlarged thyroid glands) develops with extreme deficiency. In some regions of the world, goiter still occurs in dogs fed unbalanced, homemade rations. Stunted limb development, hyperplasia of the thyroid glands and myxedema with no loss of hair typically occur in young pups born to bitches that are iodine deficient during pregnancy. Most commercial foods meet the AAFCO-recommended iodine level of 1.5 mg/kg DM.

PROTEIN Protein deficiency may affect the general health of developing puppies, decrease plasma growth hormone levels and reduce skeletal growth. In Great Dane puppies, a protein level of 14.6% (DMB) with 13% of the dietary energy derived from protein results in significant decreases in body weight and plasma albumin and urea concentrations with no increased frequency of osteochondrosis. A growth food with average energy density should contain 22 to 32% protein (DMB) of high biologic value.

VITAMINS/Vitamin D Metabolites of vitamin D_3 act in concert with other hormones to regulate calcium metabolism and therefore skeletal development in dogs. Vitamin D_3 metabolites aid in calcium and phosphorus absorption from the gut and influence bone cell activity. Dogs may require vitamin D in food sources from plants (vitamin D_2) or animals (vitamin D_3).

Clinical cases of vitamin D_3 deficiency (rickets) are extremely rare in animals fed commercial foods. Measuring circulating levels of vitamin D_3 metabolites aids in diagnosing vitamin D_3 deficiency. Increased growth plate width and thin bone cortices are not associated with low-calcium, high-phosphorus foods, but are strong indicators of rickets.

Excess vitamin D can cause hypercalcemia, hyperphosphatemia, anorexia, polydipsia, polyuria, vomiting, muscle weakness, generalized soft tissue mineralization and lameness. In growing dogs, supplementation with excess vitamin D can markedly disturb normal skeletal development because of increased calcium and phosphorus absorption. Minimum recommendations are 500 IU vitamin D/kg food DM (143 IU/1,000 kcal [34.2 IU/MJ] ME). Commercial pet foods contain from two to 10 times the minimum amount recommended by AAFCO.

VITAMINS/Vitamin A Vitamin A is an essential factor in bone metabolism, especially osteoclastic activity. Deficiency or excess may lead to severe metabolic bone disease in growing dogs. Concentrations of vitamin A in canine serum range from 1,800 to 18,000 IU/l.

Hypervitaminosis A may result in anorexia, decreased weight gain, hyperesthesia, narrowing of long bone epiphyseal cartilage, ankylosis, new bone formation without osteolysis and thin bone cortices.

Hypovitaminosis A results in a variety of clinical signs including anorexia, weight loss, ataxia, xerophthalmia, metaplasia of bronchiolar epithelium, conjunctivitis and increased susceptibility to infection. In addition, faulty bone remodeling may constrict nerves passing through bone foramina resulting in neural degeneration.

The recommended concentration of vitamin A in dog foods is 5,000 IU/kg DM (1,429 IU/1,000 kcal [342 IU/MJ] ME). Most commercial dog foods are supplemented well above the minimum requirement for vitamin A.

VITAMINS/Vitamin C The relationship between vitamin C and DOD in dogs is unproved; therefore, no supplementation is recommended. There are no known dietary requirements for vitamin C in dogs.

Assess the Food (Pages 518-519)

The food should consist of a commercial food, or well-balanced homemade food, specific for the unique nutrient requirements of fast-growing, large- and giant-breed dogs. Recommended nutrient intake for such dogs is similar to that of other breeds (See Chapter 9.), except fat, energy and calcium intake should be more stringently restricted. (See the Key Nutritional Factors–Developmental Orthopedic Disease table.) Several commercial foods are available that have been formulated for fast-growing, large- and giant-breed dogs (**Table 17-3**). Large-breed growth formulations have marked differences in key nutrients (i.e., calcium and energy density) considered risk factors for skeletal disease in large- and giant-breed puppies. The energy density of the food depends on the components that make up the food. In general, as the fat content of the food increases, so does the energy density. The higher the energy content, the more likely dogs are to consume excess energy unless they are fed in a food-restricted manner. In addition, other nutrients may be under-consumed, if the formulation is not adjusted to compensate for the increased energy content.

Often puppies are switched from growth to adult maintenance-type foods under the pretense it will help avoid calcium excess and skeletal disease. However, because some maintenance foods have much lower energy density than most growth foods, the puppy must consume more dry matter volume to meet its energy requirement. If the calcium levels

ORTHOPEDIC

are similar (DMB) between the two foods, the puppy may actually consume more calcium when fed the maintenance food.

This point is exemplified in the case of switching a 15-week-old, 15-kg, male rottweiler puppy from a growth food containing, on an as fed basis, 4.0 kcal (16.74 kJ)/g ME and 1.35% calcium (1.5% on a DMB) to a maintenance food containing the same amount of calcium but at a lower energy density (3.2 kcal [13.4 kJ]/g). The puppy would require approximately 1,600 kcal/day (6.69 MJ/day). To meet this energy need, the puppy would consume approximately 400 g of the growth food (containing 5.4 g of calcium) vs. 500 g of the maintenance food (containing approximately 6.7 g of calcium).

Feeding dogs treats that contain calcium or providing calcium supplements further increases daily calcium intake. Two level teaspoons of a typical calcium supplement (calcium carbonate) added to the growth food of a 15-week-old, 15-kg, rottweiler puppy would more than double its daily calcium intake. This calcium intake is well beyond levels shown to increase the risk for DOD.

Assess the Feeding Method (Pages 519-520)

Assessment of the feeding methods requires owner knowledge of current feeding practices, which includes the amount being fed. If owners do not know how much food their puppy is consuming, they should measure the amount ingested under the current feeding regimen for several days. This information will help when making recommendations for future feeding plans. Both the nutrient profile of a food and how it is fed are risk factors for DOD.

The aim of feeding programs for large- and giant-breed puppies is to achieve moderate energy restriction. This energy restriction may be as high as 10 to 25% of free-choice intake of foods with higher energy density. This recommendation does not mean starving a dog or feeding it a weight-control food formulated for obese adult dogs. If healthy growing dogs eat to satiety, foods formulated for weight control may result in insufficient mineral, protein and/or vitamin intake. Rather, these puppies should be fed correct dietary allowances to satisfy physiologic needs for optimal skeletal growth in conjunction with moderate energy restriction. Slow growth during the first year does not deleteriously affect final adult body size.

However, the average ME intake for Great Dane puppies ranges from 311 kcal (1,300 kJ)/$(BW_{kg})^{0.75}$ at weaning to 263 kcal (1,100 kJ)/$(BW_{kg})^{0.75}$ at six months of age. These values are approximately 20% higher than those for other large-breed puppies and are consistent with reports of higher requirements for Great Dane puppies. Marked restriction of ME intake (191 kcal [800 kJ]/$(BW_{kg})^{0.75}$) for Great Dane puppies may lead to unacceptable body composition.

Table 17-3. Recommended levels of key nutrients for dogs at risk for developmental orthopedic disease vs. levels in selected commercial dog foods.*

	Calcium (%) 0.7-1.2	Phosphorus (%) 0.6-1.1	Energy (kcal ME/g) 3.2-3.8	Fat (%) 8-12	Protein (%) 22-32
Recommended levels					
Specific-purpose growth foods					
Bench & Field Puppy Plus, dry	3.1	1.8	3.7	13.5	31.5
Hill's Prescription Diet Canine p/d, dry	1.8	1.3	4.2	23.1	31.8
Hill's Prescription Diet Canine p/d Large Breed, dry	1.0	0.8	3.2	9.0	29.5
Hill's Science Diet Canine Growth, dry	1.5	1.2	3.8	19.4	29.4
Hill's Science Diet Lamb Meal & Rice Canine Growth, dry	1.7	1.2	4.0	20.0	28.9
Hill's Science Diet Large Breed Canine Growth, dry	1.2	0.8	3.4	10.7	30.0
Iams Eukanuba Lamb & Rice for Puppies, dry	1.3	1.0	4.1	17.3	29.9
Iams Eukanuba Large Breed Puppy, dry	0.9	0.7	4.1	17.2	29.5
Iams Eukanuba Medium Breed Puppy, dry	1.3	1.0	4.3	20.5	32.8
Iams Puppy Lamb Meal & Rice Formula, dry	1.3	1.0	4.1	19.0	30.6
Iams Puppy Original Formula, dry	1.4	1.0	4.3	19.9	32.1
NutroMax Puppy, dry	1.6	1.3	3.8	19.2	31.4
Nutro Natural Choice Puppy, dry	2.4	1.5	3.5	14.3	29.5
Purina ProPlan Beef & Rice Formula for Puppies, dry	1.3	1.1	4.1	20.6	31.2
Purina Puppy Chow, dry	1.3	1.1	3.9	15.6	29.8
Select Balance Puppy, dry	1.3	1.1	3.8	19.0	30.4
Solid Gold, dry	3.4	1.9	3.7	10.1	31.1
All-purpose foods					
Alpo Chunky with Liver, moist	2.2	1.5	4.4	25.7	50.4
Mighty Dog Beef, moist	4.5	2.5	5.2	40.7	41.5
Pedigree Chopped Combo, moist	4.1	2.4	5.1	38.0	40.3
Pedigree with Chunky Chicken, moist	2.8	1.7	5.3	43.3	38.1
Purina Dog Chow, dry	1.3	1.0	3.7	13.0	23.4

*Nutrients expressed on dry matter basis except energy, which is expressed on an as fed basis. To convert to kJ, multiply kcal x 4.184.

ORTHOPEDIC

There are three basic methods of feeding growing dogs: 1) free choice (ad libitum), 2) time limited or 3) food limited. Any feeding regimen requires an initial estimate of the amount to be fed. **Table 17-2** provides formulas for estimating initial food intake for at-risk breeds. It may be necessary to contact the food manufacturer to obtain the energy density of a particular food.

Free-Choice Feeding *(Pages 519-520)*

This feeding method increases the risk of overconsumption of food by large- and giant-breed puppies, and therefore the risk of DOD. It is especially important to recommend a food with an energy density less than 3.8 kcal/g (15.9 kJ/g) (<12% fat DMB) to decrease the risk of excess energy intake when dog owners use this feeding method.

Time-Restricted Meal Feeding *(Page 520)*

Time-limited feeding is a method in which dogs are allowed free access to food for a defined period, usually 10 to 15 minutes, once or twice daily. However, feeding 15 minutes twice a day does not reduce food intake between dogs fed free choice and those fed by time-restricted methods. It is important in this type of feeding program to recommend foods with a lower energy density (<12% fat DMB) to decrease the risk of overconsumption.

Food-Restricted Meal Feeding *(Page 520)*

The method of choice for feeding puppies at risk for DOD is limiting food intake to maintain optimal growth rate and body condition. Food-limited feeding requires feeding a measured amount of food based on the dog's calculated DER (**Table 17-2**), or as recommended by the manufacturer, divided into two or three meals per day. Energy requirement is most easily calculated by using RER as a base to build on. RER can be calculated using either of the following equations:

$$RER\ (kcal/day) = 70(BW_{kg})^{0.75}\ or$$
$$RER\ (kcal/day) = 30(BW_{kg}) + 70$$

To convert to kJ, multiplying kcal by 4.184. As a starting point, recommend 3 x RER for the first four months of life and 2 x RER from four months of age to 80% of expected mature weight (about 10 to 12 months for most breeds). Most large- and giant-breed dogs will increase body weight and muscle mass after 12 months; however, the growth rate will be reduced and most if not all growth plates will be closed. At 12 months, these puppies can be fed as adults (1.6 x RER for neutered dogs and 1.8 x RER for intact dogs).

FEEDING PLAN (Pages 520-521)

Prevention of DOD (Page 520)

1. Determine if the dog is at risk for DOD. (As are all large- and giant-breed dogs.)
2. If the dog is at risk, control the nutrients of concern through food composition (**Table 17-3**) and feeding method. (See above.)
3. Counsel owners not to add vitamin or mineral supplements to balanced foods, particularly those containing calcium, phosphorus, vitamin D and vitamin A. If a nutritionally adequate growth food is being fed, supplementation is contraindicated.
4. Determine the dog's BCS every two weeks. (See Chapter 1.) Dogs should have a BCS of 2/5 to 3/5.

Treatment of Affected Dogs (Pages 520-521)

1. If possible, determine if a nutritional imbalance is causing the skeletal disease observed. The feeding history, clinical signs, radiographic changes and laboratory values may be helpful.
2. To correct either deficiencies or excesses, recommend the pet owner feed a nutritionally adequate growth food designed for large-breed puppies. (See the Key Nutritional Factors–Developmental Orthopedic Disease table.)
3. If a well-balanced growth food is being fed and skeletal diseases occur, reduce food intake up to 25%.
4. Counsel owners not to give vitamin or mineral supplements to dogs eating commercial foods, particularly those containing calcium, phosphorus, vitamin D and vitamin A. If a nutritionally adequate commercial growth food is being fed, supplementation is contraindicated.
5. Provide appropriate treatment for specific problems, such as pathologic fractures.

Remember, dietary recommendations are inferred from limited group/ breed observations and applied to individual animals. All feeding programs need to be tailored to individual animal and client situations. Initial dietary recommendations are a generalized starting point for veterinary/client interactions. Monitoring BCS is necessary for assessing dietary adequacy, which necessitates veterinarian-client interaction at regular intervals.

Select a Food (Page 521)

In general, dry foods are more economical and less energy dense than moist foods. Considering most DOD occurs in large- and giant-breed dogs, the usual type of food selected is a dry formulation. However, moist foods may be fed as long as special attention is paid to key nutritional factors.

ORTHOPEDIC

Foods for growth should at the very least have passed an AAFCO or similar feeding trial specific for that lifestage. However, feeding trials do not ensure adequacy or safety for every breed. When dealing with dogs at risk for DOD, veterinarians should focus special attention on the energy density and calcium content of the food and the feeding method. (See the Key Nutritional Factors–Developmental Orthopedic Disease table.)

Determine a Feeding Method (Page 521)

In general, free-choice feeding is contraindicated in at-risk dogs until they have reached skeletal maturity (about 12 months of age or at least 80 to 90% of adult weight). If time-limited feeding is used, five- to 10-minute feeding periods may be required to decrease food intake in some puppies (three times per day for the first month after weaning, then twice per day). In some cases, the feeding periods may need to be even shorter.

Food-limited feeding is the recommended feeding method for rapidly growing puppies. Energy intake should be restricted up to 25% less than free-choice intake, if the puppy is fed a food with high energy density (>3.8 kcal/g [15.9 kJ/g]). A plausible starting point for energy intake is 3 x RER from weaning to four months of age for fast-growing, large- and giant-breed dogs. Great Dane puppies are the exception to this recommendation. They may require 20% more energy per metabolic body weight than other dogs. After the daily caloric requirement has been calculated, divide this number by the energy density of the food (kcal or kJ) to determine the number of cups or cans to feed per day. Remember, these calculations and the manufacturer recommendations are only starting points.

▮ REASSESSMENT (Page 521)

Regular clinical evaluation of growing puppies and adjustments in the food offered are crucial. Rapidly growing, large- and giant-breed dogs have a very steep growth curve and their intake requirements can change dramatically over short periods. These puppies should be weighed, their body condition evaluated and their daily feeding amount adjusted at least once every two weeks (**Table 17-2**). The veterinarian or another member of the veterinary health care team can perform this evaluation in the hospital, or owners can be taught to perform this evaluation at home.

Clinical cases that illustrate and reinforce the nutritional concepts presented in this chapter can be found in Small Animal Clinical Nutrition, 4th ed., pp 524-528.

Atlas Contents

(Continued on next page.)

Atlas Contents

Urogenital System

Special Senses

Parasite Life Cycles

Normal Feline Heart

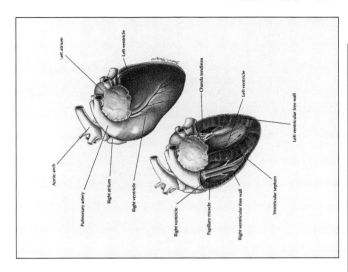

Normal Canine Heart

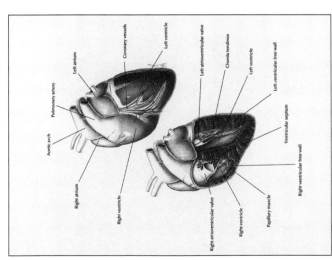

Cardiovascular System

Canine Dilated Cardiomyopathy

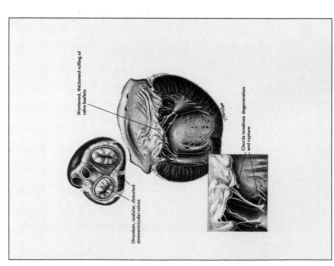

A globular-shaped heart with severe dilation of both atria and ventricles

Abnormally thin ventricular walls

Atrophied papillary muscle

Chronic Valvular Disease

Shortened, thickened rolling of valve leaflets

Shrunken, nodular, distorted atrioventricular valves

Chorda tendinea degeneration and rupture

Cardiovascular System

Feline Dilated Cardiomyopathy

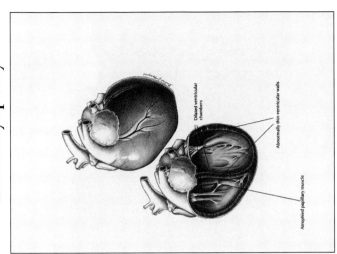

Feline Hypertrophic Cardiomyopathy

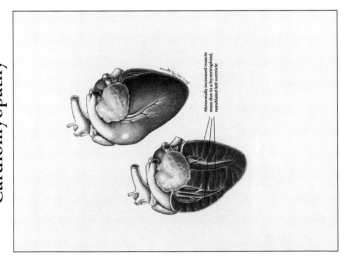

Periodontal Disease

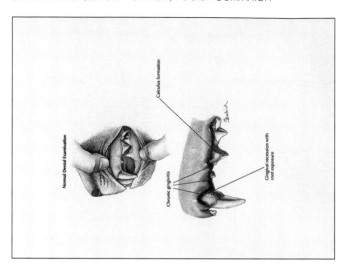

Normal Feline Dentition

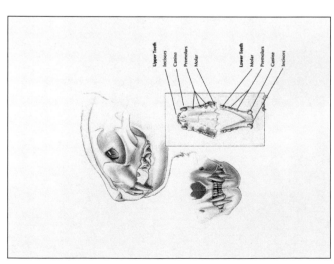

Digestive System

Carnassial Tooth Abscess

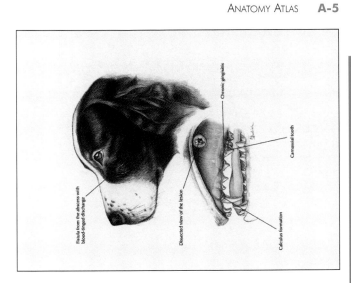

- Chronic gingivitis
- Carnassial tooth
- Fistula from the abscess with blood-tinged discharge
- Dissected view of the lesion
- Calculus formation

Normal Canine Dentition

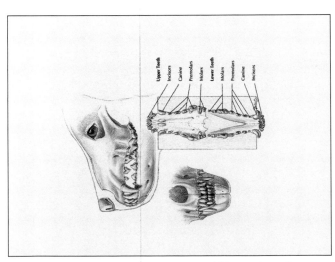

- Upper Teeth
  - Incisors
  - Canine
  - Premolars
  - Molars
- Lower Teeth
  - Molars
  - Premolars
  - Canine
  - Incisors

Digestive System

Hemorrhagic Gastritis with Ulcers

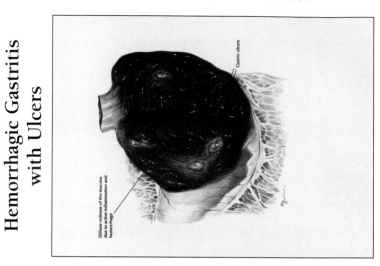

Gastric ulcers

Diffuse redness of the mucosa due to active inflammation and hemorrhage

Normal Stomach

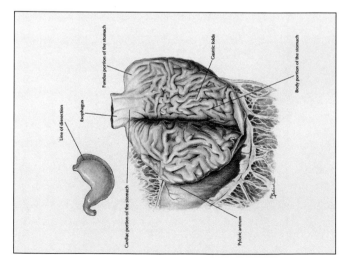

Fundus portion of the stomach

Gastric folds

Body portion of the stomach

Line of dissection

Esophagus

Cardiac portion of the stomach

Pyloric antrum

Digestive System

Normal Stomach

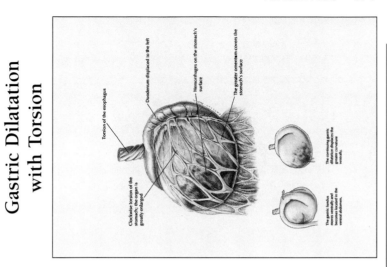

Esophagus

Fundus

Body

Omentum

Pyloric antrum

Pylorus

Sequence of Gastric Dilatation with Torsion

Clockwise rotation as viewed from a ventral position

The pyloric antrum is displaced downward.

The pylorus crosses the midline, passes underneath the distended/proximal part of stomach, and moves upward along the left abdominal wall.

Gastric Dilatation with Torsion

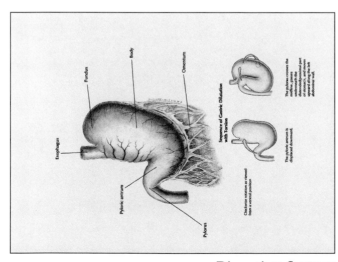

Torsion of the esophagus

Duodenum displaced to the left

Hemorrhages on the stomach's surface

The greater omentum covers the stomach's surface

Clockwise torsion of the stomach; the organ is greatly enlarged

The gastric fundus moves ventrally and becomes located in the ventral abdomen.

The continuing gastric dilatation displaces the greater curvature ventrally.

Digestive System

Foreign Bodies

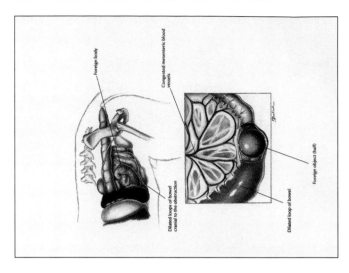

Normal Small Intestine

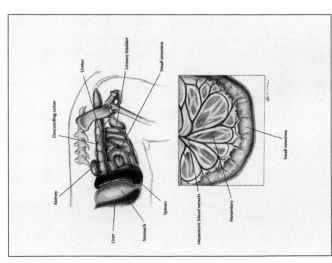

Digestive System

Intussusception

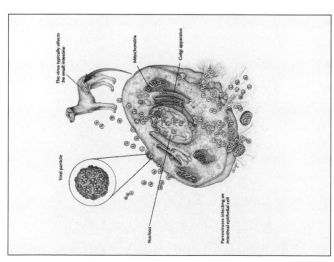

Congested mesenteric blood vessels

The mesentery and blood vessels supporting the invaginating segment of bowel are included in the intussusception

Obstruction of the small intestine caused by the telescoping of a segment of intestine into an adjacent segment

A loop of intestine within an adjacent segment of intestine

Parvoviral Enteritis

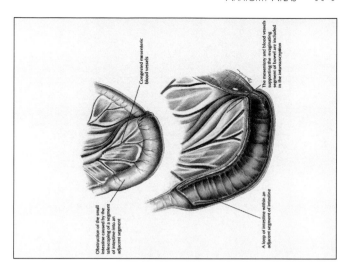

The virus typically affects the small intestine

Mitochondria

Golgi apparatus

Viral particle

Nucleus

Parvovirus infecting an intestinal epithelial cell

Digestive System

Chronic Colitis

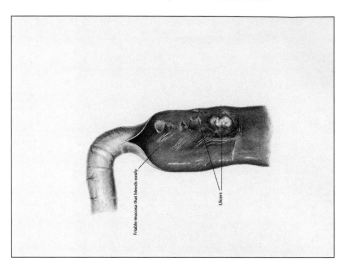

Normal Canine Colon

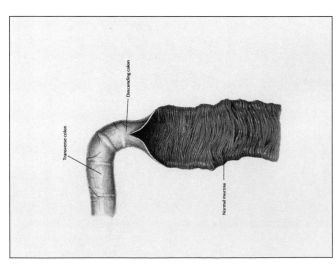

Digestive System

Constipation/
Colonic Impaction

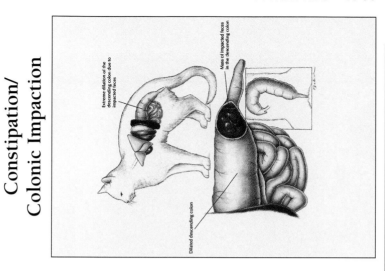

Extreme dilation of the descending colon due to impacted feces

Mass of impacted feces in the descending colon

Dilated descending colon

Normal Feline Colon

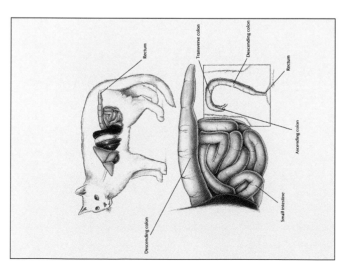

Rectum

Transverse colon

Descending colon

Rectum

Ascending colon

Small intestine

Descending colon

Digestive System

Acute Pancreatitis

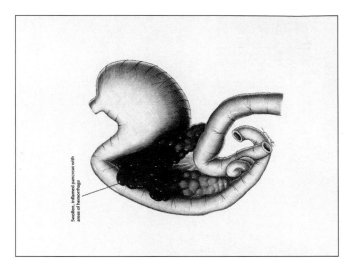

Swollen, inflamed pancreas with areas of hemorrhage

Normal Pancreas

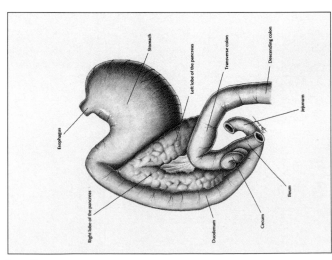

Esophagus

Stomach

Left lobe of the pancreas

Transverse colon

Descending colon

Jejunum

Ileum

Cecum

Duodenum

Right lobe of the pancreas

Digestive System

Exocrine Pancreatic Insufficiency

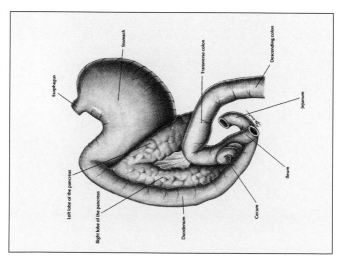

Shrunken pancreatic lobes with reduced production of digestive enzymes

Normal Pancreas

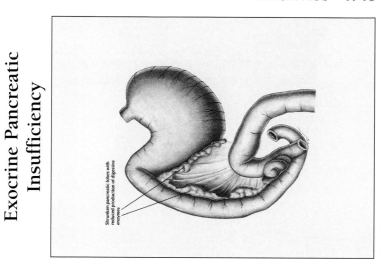

Esophagus

Stomach

Transverse colon

Descending colon

Jejunum

Ileum

Cecum

Duodenum

Right lobe of the pancreas

Left lobe of the pancreas

Digestive System

End-Stage Liver Disease

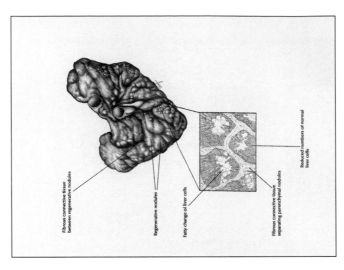

Fibrous connective tissue between regenerative nodules

Regenerative nodules

Fatty change of liver cells

Fibrous connective tissue separating parenchymal nodules

Reduced numbers of normal liver cells

Normal Liver

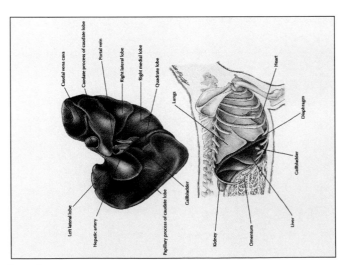

Caudal vena cava

Caudate process of caudate lobe

Portal vein

Right lateral lobe

Right medial lobe

Quadrate lobe

Left lateral lobe

Hepatic artery

Papillary process of caudate lobe

Gallbladder

Lungs

Heart

Diaphragm

Gallbladder

Liver

Omentum

Kidney

Digestive System

Hepatic Neoplasia

Interlobular connective tissue

Central vein

Tumors

Disruption of normal liver tissue by sheets of neoplastic cells

Normal Liver

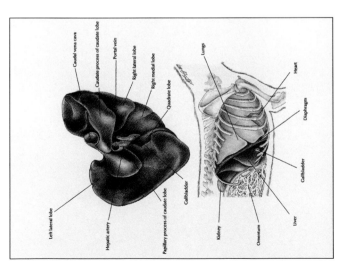

Caudal vena cava

Caudate process of caudate lobe

Portal vein

Right lateral lobe

Right medial lobe

Quadrate lobe

Left lateral lobe

Hepatic artery

Papillary process of caudate lobe

Gallbladder

Lungs

Heart

Diaphragm

Gallbladder

Liver

Omentum

Kidney

Digestive System

Anal Sac Abscess

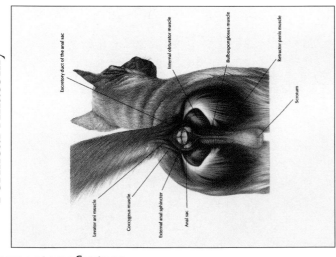

Ruptured anal sac abscess

Enlarged, inflamed anal sac

Normal Skin/ Perineal Anatomy

Excretory duct of the anal sac

Internal obturator muscle

Bulbospongiosus muscle

Retractor penis muscle

Scrotum

Levator ani muscle

Coccygeus muscle

External anal sphincter

Anal sac

Integumentary System

Flea-Allergy Dermatitis

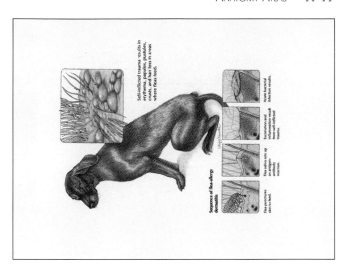

Skin Abscess

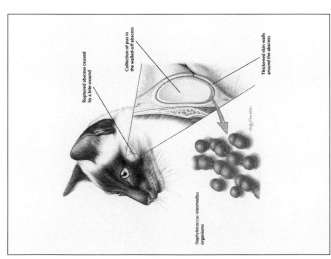

Integumentary System

Intervertebral Disk Disease

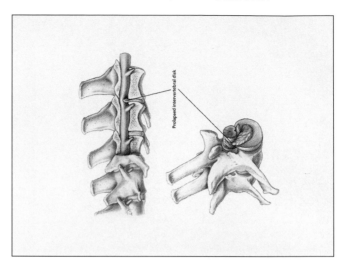

Normal Vertebrae/ Spinal Cord

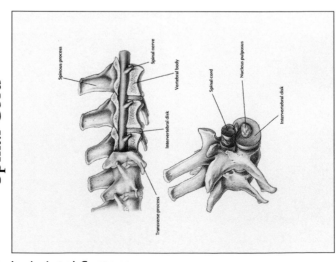

Musculoskeletal System

Osteochondritis Dissecans

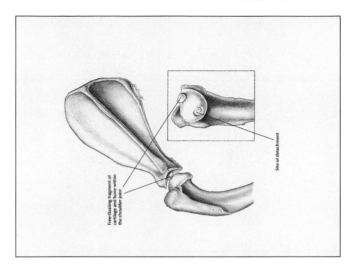

Free-floating fragment of cartilage and bone within the shoulder joint

Site of detachment

Normal Shoulder

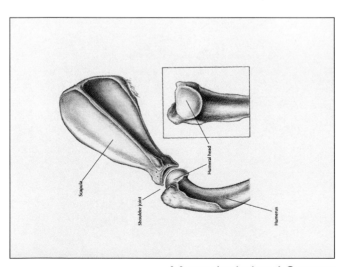

Scapula

Shoulder joint

Humeral head

Humerus

Musculoskeletal System

Ununited Anconeal Process/Panosteitis

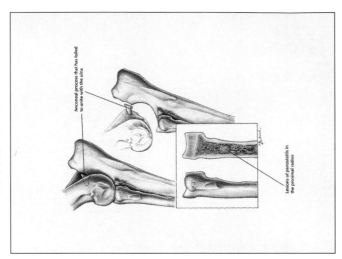

Normal Elbow

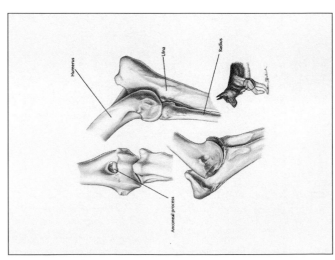

Musculoskeletal System

Hip Dysplasia

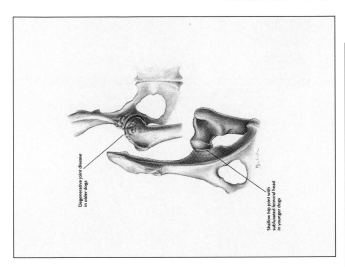

Degenerative joint disease in older dogs

Shallow hip joint with subluxated femoral head in younger dogs

Normal Hip Joint

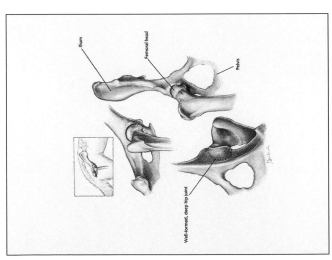

Ilium

Femoral head

Pelvis

Well-formed, deep hip joint

Musculoskeletal System

Femoral Fracture

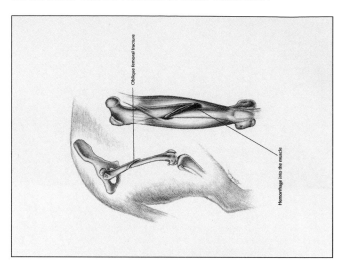

Normal Rear Leg

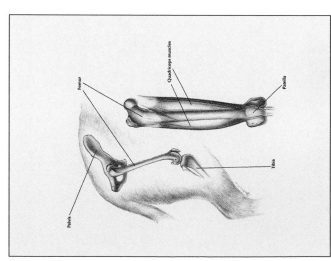

Musculoskeletal System

Ruptured Cranial Cruciate Ligament

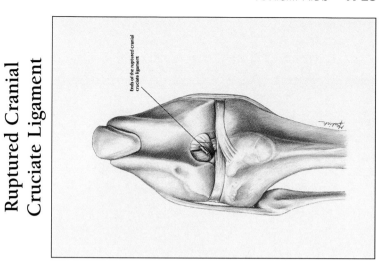

Ends of the ruptured cranial cruciate ligament

Normal Stifle

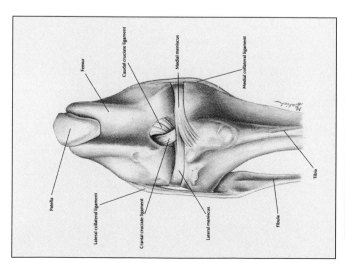

Femur

Caudal cruciate ligament

Medial meniscus

Medial collateral ligament

Patella

Tibia

Lateral collateral ligament

Cranial cruciate ligament

Lateral meniscus

Fibula

Musculoskeletal System

Patellar Luxation

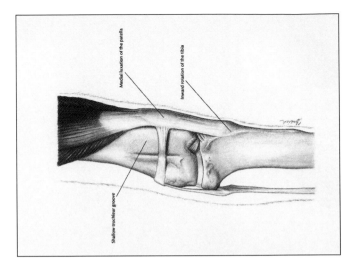

Medial luxation of the patella

Inward rotation of the tibia

Shallow trochlear groove

Normal Stifle

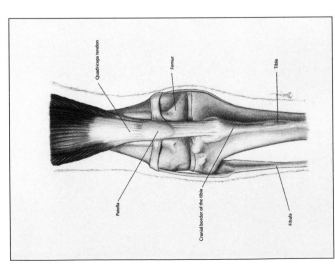

Quadriceps tendon

Femur

Tibia

Patella

Cranial border of the tibia

Fibula

Musculoskeletal System

Tonsillitis

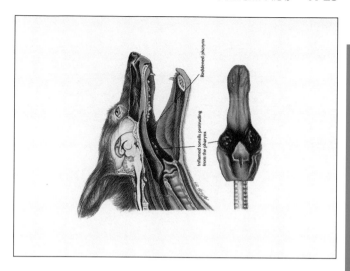

Reddened pharynx

Inflamed tonsils protruding from the pharynx

Normal Mouth/ Upper Airway

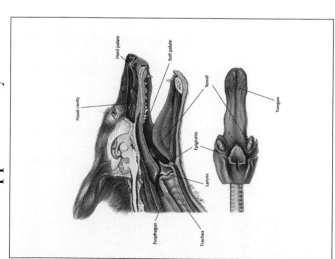

Hard palate

Soft palate

Nasal cavity

Tonsil

Tongue

Epiglottis

Larynx

Esophagus

Trachea

Respiratory System

Collapsing Trachea

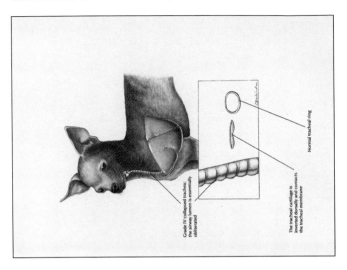

Grade IV collapsed trachea: the airway lumen is essentially obliterated

The tracheal cartilage is inverted dorsally and contacts the tracheal membrane

Normal tracheal ring

Normal Canine Thorax

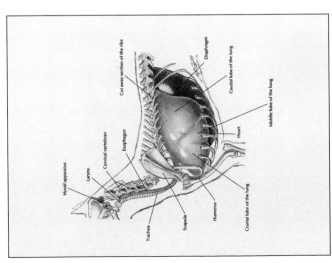

Hyoid apparatus
Larynx
Cervical vertebrae
Esophagus
Cut away section of the ribs
Diaphragm
Caudal lobe of the lung
Heart
Middle lobe of the lung
Cranial lobe of the lung
Humerus
Scapula
Trachea

Respiratory System

Pulmonary Edema

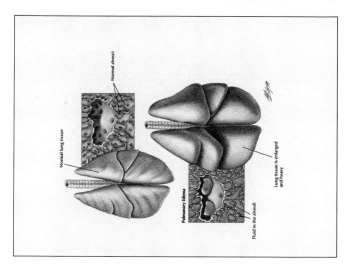

Normal lung tissue

Normal alveoli

Pulmonary Edema

Lung tissue is enlarged and heavy

Fluid in the alveoli

Normal Feline Thorax

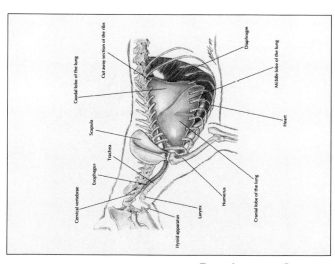

Cut away section of the ribs

Caudal lobe of the lung

Diaphragm

Middle lobe of the lung

Scapula

Heart

Trachea

Esophagus

Cervical vertebrae

Humerus

Larynx

Cranial lobe of the lung

Hyoid apparatus

Respiratory System

Chronic Renal Disease

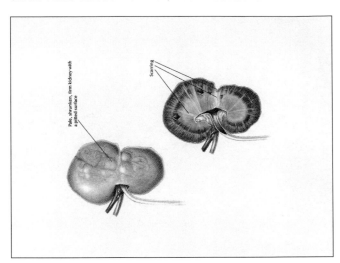

Pale, shrunken, firm kidney with a pitted surface

Scarring

Normal Canine Kidney

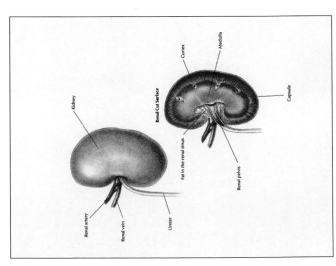

Kidney

Renal Cut Surface

Cortex

Medulla

Capsule

Fat in the renal sinus

Renal pelvis

Renal artery

Renal vein

Ureter

Urogenital System

Acute Renal Failure

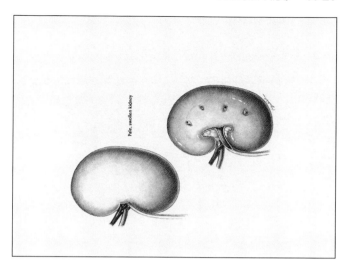

Normal Canine Kidney

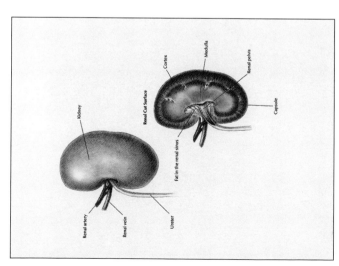

Urogenital System

Bladder Stones

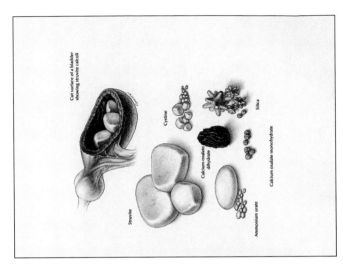

Normal Urinary Bladder

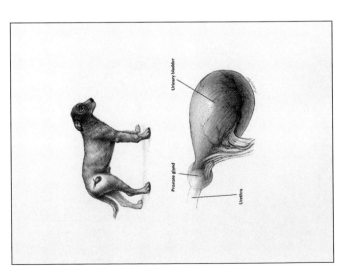

Urogenital System

Canine Urethral Obstruction

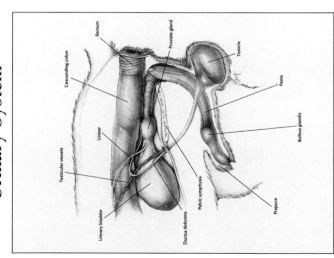

Distended urinary bladder caused by an obstructing urethral calculus

Urethral calculus immediately behind the os penis; the calculus is obstructing the outflow of urine from the bladder

Hemorrhages on the surface of the bladder

Normal Canine Lower Urinary System

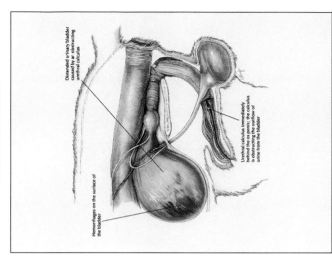

Rectum

Descending colon

Prostate gland

Testicle

Penis

Testicular vessels

Ureter

Bulbus glandis

Urinary bladder

Pelvic symphysis

Prepuce

Ductus deferens

Urogenital System

Feline Urologic Syndrome

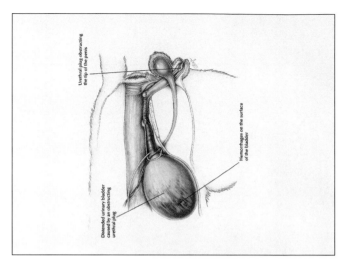

Urethral plug obstructing the tip of the penis

Hemorrhages on the surface of the bladder

Distended urinary bladder caused by an obstructing urethral plug

Normal Feline Lower Urinary System

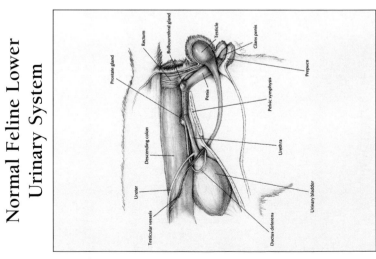

Rectum

Bulbourethral gland

Testicle

Glans penis

Prostate gland

Prepuce

Penis

Pelvic symphysis

Descending colon

Urethra

Ureter

Testicular vessels

Ductus deferens

Urinary bladder

Urogenital System

Benign Prostatic Hyperplasia

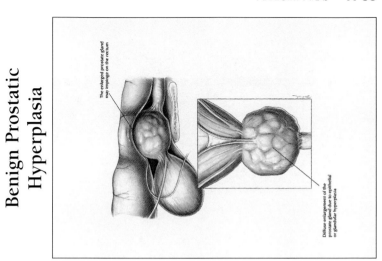

The enlarged prostate gland may impinge on the rectum

Diffuse enlargement of the prostate gland due to epithelial or glandular hyperplasia

Normal Prostate Gland

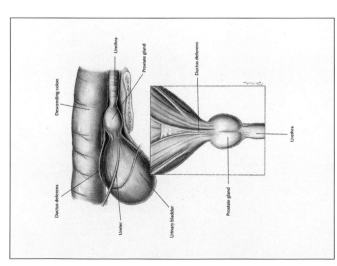

Descending colon

Urethra

Prostate gland

Ductus deferens

Ductus deferens

Ureter

Urinary bladder

Prostate gland

Urethra

Urogenital System

Pyometra

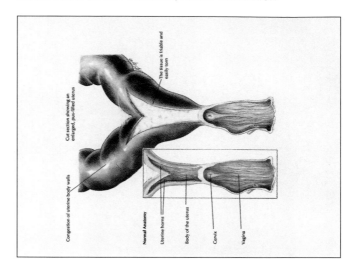

Ovariohysterectomy

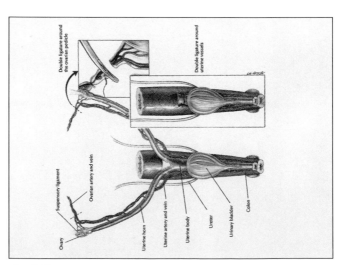

Urogenital System

Testicular Tumors

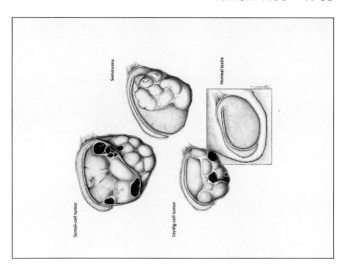

Seminoma

Normal testis

Sertoli-cell tumor

Leydig-cell tumor

Canine Castration

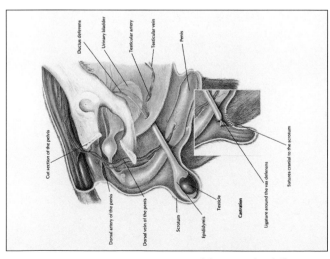

Ductus deferens

Urinary bladder

Testicular artery

Testicular vein

Penis

Cut section of the pelvis

Dorsal artery of the penis

Dorsal vein of the penis

Scrotum

Epididymis

Testicle

Castration

Ligature around the vas deferens

Sutures cranial to the scrotum

Urogenital System

Nuclear Sclerosis/Cataracts

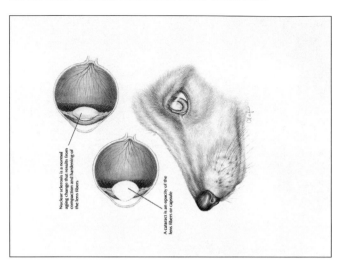

Nuclear sclerosis is a normal aging change that results from compaction and hardening of the lens fibers

A cataract is an opacity of the lens fibers or capsule

Normal Canine Eye

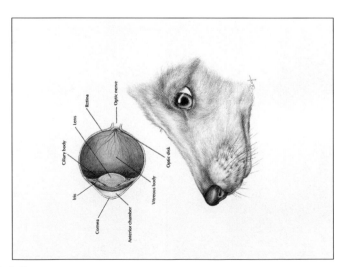

Retina
Optic nerve
Lens
Ciliary body
Optic disk
Iris
Vitreous body
Cornea
Anterior chamber

Special Senses

Glaucoma

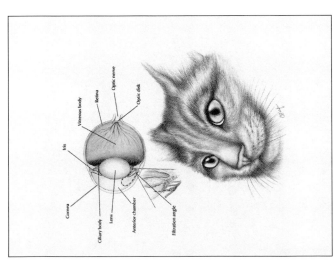

Increase in intraocular pressure

The globe is enlarged, pain may be present, the episcleral vessels are congested, and vision loss occurs.

Cloudy, edematous, insensitive cornea

Intraocular pressure is increased due to a disorder of the drainage angle

Normal Feline Eye

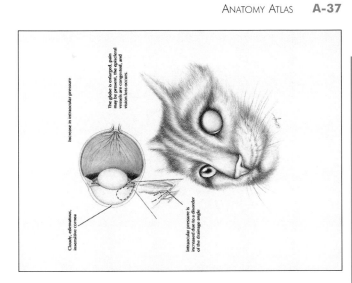

Vitreous body

Retina

Optic nerve

Optic disk

Iris

Cornea

Ciliary body

Lens

Anterior chamber

Filtration angle

Special Senses

Corneal Ulceration

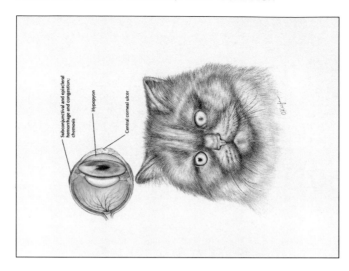

Subconjunctival and episcleral hemorrhage and congestion; chemosis

Hypopyon

Central corneal ulcer

Normal Feline Eye

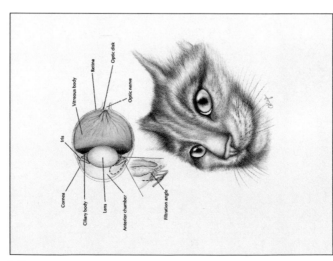

Vitreous body

Retina

Optic disk

Optic nerve

Iris

Cornea

Ciliary body

Lens

Anterior chamber

Filtration angle

Special Senses

Otitis Externa/Media/Interna

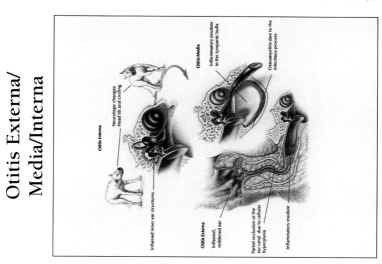

Otitis Interna

Neurologic changes
Head tilt and circling

Inflamed inner ear structures

Otitis Media

Inflammatory exudate
in the tympanic bulla

Osteomyelitis due to the
infectious process

Otitis Externa

Inflamed,
reddened ear

Partial occlusion of the
ear canal due to cellular
hyperplasia

Inflammatory exudate

Normal Hearing Apparatus

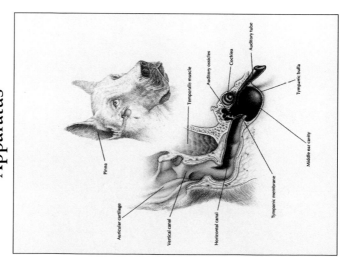

Pinna

Temporalis muscle

Auditory ossicles

Cochlea

Auditory tube

Tympanic bulla

Auricular cartilage

Vertical canal

Horizontal canal

Tympanic membrane

Middle ear cavity

Giardia

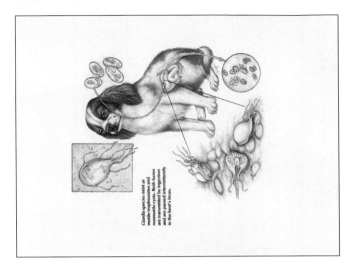

Giardia species exist as motile trophozoites and nonmotile cysts. Both forms are transmitted by ingestion and are passed intermittently in the host's feces.

Heartworms

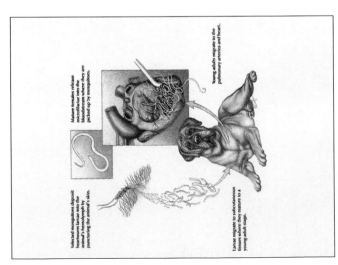

Mature females release microfilariae into the bloodstream where they are picked up by mosquitoes.

Young adults migrate to the pulmonary arteries and heart.

Infected mosquitoes deposit heartworm larvae into the animal's hemolymph by puncturing the animal's skin.

Larvae migrate to subcutaneous tissues where they mature to a young adult stage.

Parasite Life Cycles

Whipworms

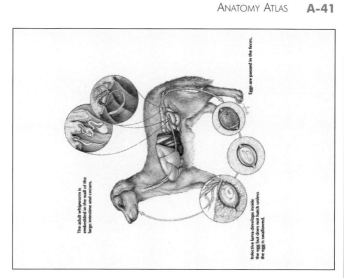

Hookworms

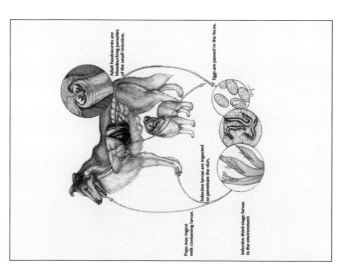

Parasite Life Cycles

Tapeworms (*Taenia*)

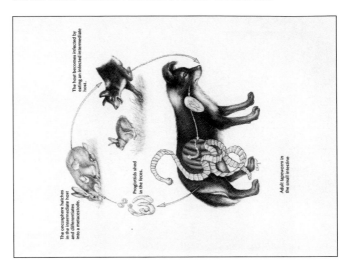

Roundworms

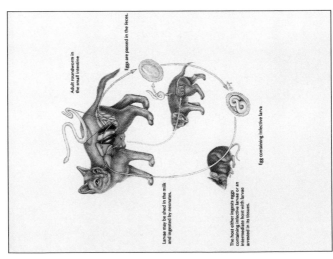

Parasite Life Cycles

Fleas

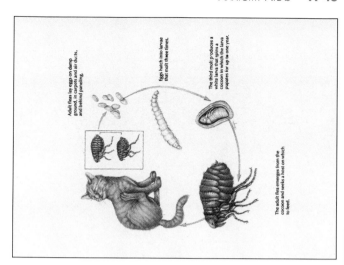

Adult fleas lay eggs on damp ground, in carpets and air ducts, and behind paneling.

Eggs hatch into larvae that molt three times.

The third molt produces a white larva that spins a cocoon in which the larva pupates for up to one year.

The adult flea emerges from the cocoon and seeks a host on which to feed.

Tapeworms
(*Dipylidium caninum*)

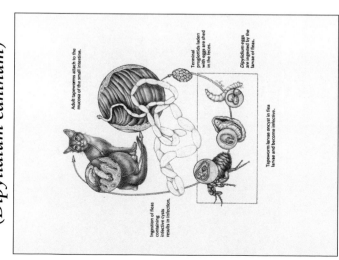

Adult tapeworms attach to the mucosa of the small intestine.

Terminal proglottids laden with eggs are shed in the feces.

Dipylidium eggs are ingested by the larvae of fleas.

Tapeworm larvae encyst in flea larvae and become infective.

Ingestion of fleas containing infective cysts results in infection.

Parasite Life Cycles

Sarcoptes

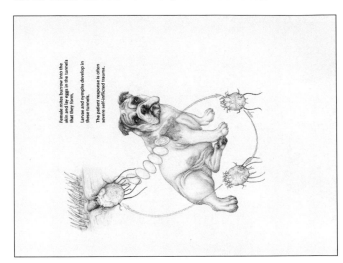

Female mites burrow into the skin and lay eggs in the tunnels that they form.

Larvae and nymphs develop in these tunnels.

The patient response is often severe self-inflicted trauma.

Ticks

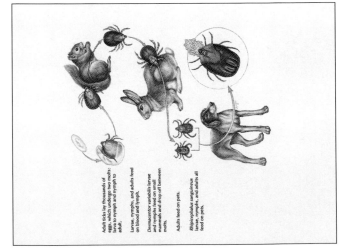

Adult ticks lay thousands of eggs, which undergo two molts: larva to nymph and nymph to adult.

Larvae, nymphs, and adults feed on blood and lymph.

Dermacentor variabilis larvae and nymphs feed on small mammals and drop off between molts.

Adults feed on pets.

Rhipicephalus sanguineus larvae, nymphs, and adults all feed on pets.

Parasite Life Cycles

Cheyletiella

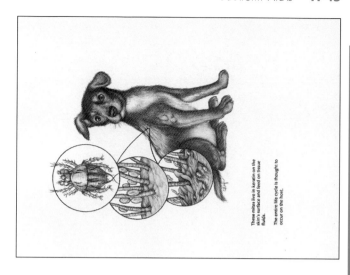

These mites live in keratin on the skin's surface and feed on tissue fluids.

The entire life cycle is thought to occur on the host.

Demodex

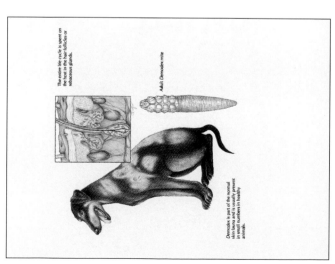

The entire life cycle is spent on the host in the hair follicles or sebaceous glands.

Adult *Demodex* mite

Demodex is part of the normal skin fauna and is usually present in small numbers in healthy animals.

Parasite Life Cycles

Ear Mites

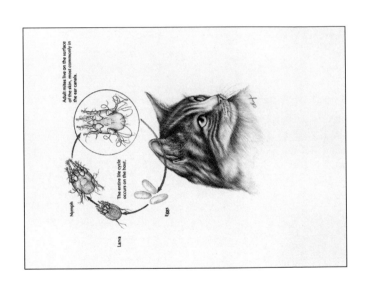

Parasite Life Cycles

Cardiovascular Disease

For a review of the unabridged chapter, see Roudebush P, Keene BW, Mizelle HL. Cardiovascular Disease. In: Hand MS, Thatcher CD, Remillard RL, et al, eds. Small Animal Clinical Nutrition, 4th ed. Topeka, KS: Mark Morris Institute, 2000; 529-562.

CLINICAL IMPORTANCE (Pages 529-531)*

Chronic mitral valvular disease (endocardiosis) is the most common acquired cardiac abnormality in dogs, affecting more than one-third of those over 10 years of age. The tricuspid valve is also frequently involved (approximately 30% of cases) but disease is less severe. Chronic valvular disease occurs with relatively greater frequency in small dogs. Acquired valvular disease in cats is rare.

Since 1987, dilated (congestive) cardiomyopathy in cats has decreased markedly following the discovery that taurine deficiency was the principal cause, and the subsequent supplementation of most commercial feline foods with taurine. Hypertrophic and restrictive cardiomyopathies are now the most common causes of myocardial failure in cats.

Various types of myocardial disease that were not recognized 30 years ago are now seen commonly in dogs. Large-breed dogs, especially males, are predisposed to dilated cardiomyopathy. Doberman pinscher dogs are particularly susceptible. Hypertrophic cardiomyopathy occurs rarely in dogs.

Pulmonary vascular disease with secondary cor pulmonale is most commonly associated with *Dirofilaria immitis* infection (heartworm disease). The prevalence of this disease is high in endemic areas and in those dogs that do not routinely receive preventive medication. Pulmonary hypertension unrelated to heartworm disease appears to be uncommon. Pulmonary thromboembolism is most commonly associated with renal disease, hyperadrenocorticism, corticosteroid therapy, neoplasia, nephrotic syndrome, pancreatitis and immune-mediated hemolytic anemia. Primary systemic vascular disease is uncommon; however, atherosclerosis and aortic or coronary thrombosis are occasionally recognized, particularly in dogs with hypothyroidism and elevated serum cholesterol concentrations. Secondary aortic thromboembolism in cats may occur with any of the forms of cardiomyopathy and is the most frequently acquired feline vascular abnormality.

*Page numbers in headings refer to Small Animal Clinical Nutrition, 4th ed., where additional information may be found.

Key Nutritional Factors–Cardiovascular Disease (Page 531)

Factors	Associated conditions	Nutritional recommendations
Water	Sodium, chloride and fluid retention Systemic hypertension Congestive heart failure	Offer water free choice at all times Recommend distilled water or water with <150 ppm sodium for patients with advanced heart disease
Sodium, chloride	Sodium, chloride and fluid retention Systemic hypertension Congestive heart failure	Avoid excess sodium chloride Restrict sodium to 0.07 to 0.25% dry matter (DM) in dogs and <0.35% DM in cats Chloride levels are typically 1.5 times sodium levels
Potassium	Hypokalemia associated with use of loop or thiazide diuretics, especially in patients receiving low levels of dietary potassium; associated with chronic renal failure in cats; may predispose to cardiac dysrhythmias and taurine deficiency	Oral potassium supplementation (3-5 mEq or mmol/BW_{kg}/day) Change to a food with a higher potassium concentration
	Hyperkalemia associated with use of angiotensin-converting enzyme inhibitors and/or potassium-sparing diuretics, especially in patients fed high levels of dietary potassium or those with an acute uremic crisis	Withdraw potassium supplements Switch from potassium-sparing diuretic to a loop or thiazide diuretic Change to a food with a lower potassium concentration

(Continued on next page.)

Key Nutritional Factors–Cardiovascular Disease (Page 531) (Continued.)

Factors	Associated conditions	Nutritional recommendations
Magnesium	Hypomagnesemia associated with use of diuretics, especially in patients receiving low levels of dietary magnesium; may predispose to cardiac dysrhythmias	Oral magnesium supplementation 20 to 40 mg/BW_{kg}/day per os (magnesium oxide) Change to a food with a higher magnesium concentration
Energy	Cachexia associated with protein-energy malnutrition and altered metabolism	Ensure adequate energy intake
	Obesity associated with neurohumoral activation, hypertension and/or blood volume expansion	Treatment: reduce body fat while maintaining lean body mass Prevention: caloric restriction if animal is at risk
Protein	Cachexia associated with protein-energy malnutrition and altered metabolism	Ensure adequate protein intake
	Uremia, metabolic acidosis and/or proteinuria associated with chronic renal failure	Avoid excess dietary protein See Chapter 19
Phosphorus	Hyperphosphatemia associated with chronic renal failure	Avoid excess phosphorus Restrict phosphorus to 0.2 to 0.3% DM in dogs and 0.5% in cats Consider use of phosphate binders See Chapter 19

(Continued on next page.)

HEART

Key Nutritional Factors–Cardiovascular Disease (Page 531) (Continued.)

Factors	Associated conditions	Nutritional recommendations
Taurine	Cat: dilated cardiomyopathy associated with taurine deficiency	Prevention: provide adequate taurine 1,000 to 2,000 mg/kg dry weight (dry extruded food) 2,000 to 3,000 mg/kg dry weight (moist food) Treatment: 250- to 500-mg taurine daily per os Use veterinary therapeutic food (cardiac food) containing increased levels of taurine
	Dog: dilated cardiomyopathy in American cocker spaniels	Treatment: 500-mg taurine twice daily per os Use veterinary therapeutic food (cardiac food) containing increased levels of taurine
Carnitine	Dog: dilated cardiomyopathy	Treatment: 50- to 100-mg L-carnitine/BW_{kg} three times daily

Hypertension in dogs and cats appears to be more common than studies indicated 30 years ago. Because blood pressure is not yet routinely measured, the prevalence of systemic hypertension in dogs and cats is still unknown. Spontaneous essential hypertension occurs in dogs; however, hypertension most commonly develops secondary to chronic renal disease in both dogs and cats, hyperadrenocorticism in dogs and hyperthyroidism in cats.

The most frequently encountered problems associated with cardiovascular disease that require nutritional modification are fluid retention states associated with chronic congestive heart failure (CHF), primary or secondary hypertension, obesity, cachexia and myocardial diseases related to a specific nutrient deficiency (taurine- and carnitine-associated cardiomyopathy and electrolyte disorders that potentiate cardiac dysrhythmias).

ASSESSMENT (Pages 531-543)

Assess the Animal (Pages 531-543)
History and Physical Examination (Pages 531-532)
Heart failure is a condition characterized by inadequate cardiac output and insufficient delivery of nutrients relative to tissue needs. Heart failure is not a specific disease, but a clinical syndrome caused by a variety of structural and functional disorders of the heart or great vessels. Clinical manifestations of heart failure are due to reduced cardiac output (weakness, exercise intolerance, syncope), pulmonary congestion (dyspnea, orthopnea, cough, abnormal breath sounds with crackles and wheezes), systemic fluid retention (jugular venous distention, hepatomegaly, ascites, pleural effusion) or a combination of these conditions.

In general, the clinical manifestations of heart failure are similar irrespective of the underlying cause, although the onset may vary. In many instances, heart failure becomes evident gradually; a long period of mild clinical signs that worsen precedes its diagnosis. Most of the clinical signs used as the basis for diagnosing chronic heart failure occur in other conditions.

Obesity and chronic bronchitis often occur in dogs and cats with heart disease and cause clinical manifestations similar to those of heart failure, thereby complicating the diagnosis.

Members of the International Small Animal Cardiac Health Council have developed a classification scheme that is applicable to veterinary patients (**Table 18-1**). This classification system is based on anatomic diagnosis and the severity of clinical signs at rest.

Although hypertensive dogs and cats may present at any age, most often they are middle aged to geriatric (mean age: dogs nine years; cats 15

HEART

Table 18-1. Functional classes of heart failure.

Class I. The asymptomatic patient
Heart disease is detectable (cardiac murmur, dysrhythmia), but the patient is not overtly affected and does not demonstrate clinical signs of heart failure.
a. Heart disease is detectable but no signs of compensation are evident, such as volume or pressure overload ventricular hypertrophy.
b. Heart disease is detectable in conjunction with radiographic or echocardiographic evidence of compensation, such as volume or pressure overload ventricular hypertrophy.

Class II. Mild to moderate heart failure
Clinical signs of heart failure are evident at rest or with mild exercise and adversely affect the quality of life. Typical clinical signs include exercise intolerance, cough, tachypnea, mild respiratory distress and mild to moderate ascites. Hypoperfusion at rest is generally not present.

Class III. Advanced heart failure
Clinical signs of CHF are immediately evident. These clinical signs include respiratory distress (dyspnea), marked ascites, profound exercise intolerance and hypoperfusion at rest. In the most severe cases, the patient is moribund and suffers from cardiogenic shock.

years). The strength of the peripheral pulses does not help detect systemic hypertension; absolute blood pressure numbers need to be determined. Retinal hemorrhages and detachments are common end-organ changes in patients with moderate to severe hypertension. These ocular signs are often the first evidence of cardiovascular disease, which suggests that a fundic examination should be included in the routine evaluation of all dogs and cats. Other clinical signs are related to the underlying disease that causes systemic hypertension.

Body condition is the most important assessment of the nutritional status of animals with cardiovascular disease. As will be discussed later, obesity causes cardiovascular changes that can complicate cardiovascular diseases. Treatment of the underlying cardiovascular disease may not be successful without first managing obesity.

Cachexia is a syndrome of severe wasting seen clinically in a variety of diseases, especially chronic heart failure, cancer and acquired immunodeficiency syndrome. Cachexia is an additional risk factor in people with heart failure; loss of lean body mass is a negative predictor of survival. Systems for accurately assessing and scoring body condition are available for dogs and cats. (See Chapter 1.) The body condition of dogs and cats with cardiovascular disease should be followed closely as part of reassessment.

Laboratory and Other Clinical Information (Pages 532-533)
MEASUREMENT OF SYSTEMIC BLOOD PRESSURE The systemic systolic/diastolic blood pressure in awake, untrained dogs and cats normal-

ly should not exceed 180/100 millimeters of mercury (mm Hg), with values up to 200/110 mm Hg considered borderline or mild hypertension.

SCREENING FOR CONCOMITANT DISEASE Cardiovascular disease is frequently associated with or exacerbates underlying chronic renal disease in dogs and cats. All patients with cardiovascular disease should be screened for concomitant renal disease. This is best accomplished with a urinalysis and a serum biochemistry profile, which includes urea nitrogen, creatinine, electrolyte, calcium and phosphorus concentrations.

Hyperthyroidism in cats is a risk factor for secondary hypertrophic cardiomyopathy and systemic hypertension. Older cats with evidence of cardiovascular disease should be screened for hyperthyroidism. (See Chapter 24.)

MEASURING AND INTERPRETING TISSUE NUTRIENTS AND HORMONES/Electrolytes and Magnesium Serum electrolyte and magnesium concentrations are important factors to assess in patients with cardiovascular disease. Abnormalities in electrolyte and magnesium homeostasis can cause cardiac dysrhythmias, decreased myocardial contractility and profound muscle weakness. Electrolyte and magnesium abnormalities can also potentiate adverse effects from cardiac glycosides and other cardiac drugs. Unfortunately, the precise diagnosis of potassium and magnesium depletion can be difficult to make because these are primarily intracellular constituents. Normal serum potassium and magnesium concentrations can occur in the presence of total body depletion of these elements; therefore, serum potassium and magnesium concentrations do not always reflect total body stores.

MEASURING AND INTERPRETING TISSUE NUTRIENTS AND HORMONES/Taurine Plasma and whole blood taurine concentrations are routinely measured to evaluate the taurine status of cats and dogs. Values for plasma taurine of less than 20 to 30 nmol/ml (μmol/l) have been associated with deficiency in clinical studies involving client-owned cats and dogs.

Studies involving laboratory cats have shown that plasma, but not whole blood taurine concentrations, are affected by meals and food deprivation. Therefore, whole blood taurine concentration is a more reliable index of taurine status in cats. In general, the whole blood taurine pool is remarkably stable and declines only during prolonged depletion, whereas plasma taurine concentrations fluctuate acutely depending on availability in food. Cats with whole blood taurine concentrations consistently less than 150 nmol/ml should be considered taurine deficient.

HEART

Table 18-2. Normal plasma carnitine concentrations of healthy dogs and cats eating commercial dry pet foods.

Plasma carnitine (μmol/l)	Dog	Cat
Total	12-40	9-35
Free	9-36	7-30
Esters	<7	<8

MEASURING AND INTERPRETING TISSUE NUTRIENTS AND HOR-MONES/Carnitine The lowest levels of carnitine are usually found in serum; in contrast, heart and skeletal muscle contain very high levels of carnitine, which underscores its importance in these tissues. Normal canine and feline values are similar for total, free and esterified carnitine concentrations in plasma based on measurements from a small number of healthy animals fed a standard dry commercial food (**Table 18-2**). Total carnitine concentrations in plasma are influenced by intake of carnitine in the food. Plasma concentrations will be elevated in animals that eat foods high in carnitine (e.g., raw meat or moist foods high in skeletal muscle content).

Risk Factors (Pages 533-535)
Risk factors for causing or complicating cardiovascular disease include breed, obesity, renal disease, drug therapy, endocrinopathies and heartworm infection. Breed is the most important risk factor for cardiovascular disease in dogs. A number of breeds are at increased risk for several different congenital cardiovascular malformations, including patent ductus arteriosus, portacaval shunts, aortic stenosis, pulmonic stenosis, ventricular septal defects, tricuspid dysplasia and persistent aortic arch and related vascular abnormalities.

Chronic valvular heart disease occurs with relatively greater frequency in small dogs, whereas large dogs, especially males, are predisposed to dilated cardiomyopathy. Certain canine breeds also have characteristic cardiac dysrhythmias that may occur with or without significant cardiomegaly or signs of CHF. Finally, an increased risk of pericardial effusion is noted in golden retrievers, Labrador retrievers, German shepherd dogs, German shorthaired pointers and Akitas.

Obesity occurs frequently in dogs and cats with cardiovascular disease. Obesity not only produces clinical signs that mimic those of early heart failure (i.e., exercise intolerance, tachypnea, weakness), but also causes cardiovascular changes that can exacerbate underlying cardiovascular disease.

Chronic progressive renal disease and failure often occur in dogs and cats with cardiovascular disease, especially in older patients. Cardiac disease often exacerbates underlying renal disease because a large proportion of the cardiac output is normally destined for the kidneys. Renal disease influences the types and dosages of medications that are used to treat patients with cardiovascular disease. Chronic renal disease is also a risk factor for secondary hypertension in dogs and cats.

Therapy for CHF often includes: 1) diuretics and salt restriction to reduce preload and venous congestion, 2) cardiac glycosides to increase contractility and control supraventricular dysrhythmias and 3) vasodilators to reduce venous congestion, preload and afterload. Dehydration, systemic hypotension, renal insufficiency, electrolyte abnormalities and acid-base disturbances are all potential complications of combined pharmacologic and nutritional therapy for patients with CHF.

Hyperthyroidism is a risk factor for both hypertrophic cardiomyopathy and secondary hypertension in older cats. Hyperadrenocorticism is a risk factor for pulmonary thromboembolism. Heartworm infection is a risk factor for pulmonary vascular disease, cor pulmonale, right-sided CHF and pulmonary thromboembolism.

Etiopathogenesis *(Pages 535-542)*
COMPENSATORY MECHANISMS IN HEART FAILURE The first priority of the cardiovascular system is to provide oxygen and nutrients to critical organs such as the brain, kidneys and heart. The next priority is to supply nutrients to all other tissues; a final priority is to maintain normal venous pressure. In heart failure, these cardiovascular priorities are lost in reverse order. The body will sacrifice normal venous pressure to provide nutrients to tissues. Increased venous pressure values above normal often result in clinical signs of CHF. The first and second cardiovascular priorities are maintained through compensatory responses from several neurohumoral mechanisms. In some animals, these compensatory changes ultimately result in: 1) sodium and water retention, 2) expanded extracellular fluid volume, 3) increased venous filling pressure and 4) clinical signs of cough, dyspnea, orthopnea, tachypnea, hepatomegaly and ascites. See Small Animal Clinical Nutrition, 4th ed., pp 535-538 for a detailed discussion of compensatory mechanisms in heart failure.

OBESITY Obesity has profound cardiovascular consequences. From a cardiovascular perspective, obesity is a disease of blood volume expansion. Increases in blood pressure, heart rate, cardiac output, left atrial pressure and extracellular fluid volume occur in dogs with experimentally induced obesity. Blood pressure always increases with increasing weight in dogs, regardless of the initial blood pressure.

The tendency toward blood volume expansion and neurohumoral activation in obese animals parallels the compensatory changes that often occur in animals with cardiac disease. Obesity, therefore, may have profound adverse effects in animals with concomitant cardiovascular disease.

HEART

CACHEXIA Cachexia is a syndrome of weight loss (unintentional loss of more than 10% body weight), lean tissue wasting and anorexia seen clinically in a variety of diseases, including chronic heart failure. The loss of lean body mass seen in cachexia is caused by a mismatch between food intake and nutritional requirements, resulting in negative nitrogen and energy balances. These imbalances may be due to inadequate intake, excessive losses or altered metabolism.

In patients with heart failure, anorexia may be due to the heart failure itself (dyspnea, fatigue), concomitant disease (nausea associated with renal failure), use of drugs that cause nausea (cardiac glycosides, angiotensin-converting enzyme [ACE] inhibitors) or sudden nutritional changes. However, the rate of loss of lean body mass with cardiac cachexia exceeds that attributable to anorexia alone. With simple starvation, most weight loss is loss of fat mass, whereas lean tissue is relatively spared, at least early on. Cachexia involves depletion of lean body mass. Physical inactivity may also contribute to loss of lean body mass because exercise is routinely restricted in patients with moderate to severe heart failure.

Altered metabolism appears to play a role in the pathogenesis of cachexia. Elevated serum tumor necrosis factor (TNF, cachectin) concentrations occur in people, dogs and cats with CHF. Both TNF and interleukin-1 (IL-1) cause cachexia by suppressing food intake and altering metabolism. TNF suppresses the expression of several genes that encode for essential lipogenic enzymes, including lipoprotein lipase, and promotes the breakdown of adipose tissue and skeletal muscle. TNF may also change the normal metabolic adaptation that accompanies caloric restriction and thus contributes to the nutritional imbalances observed in cachectic patients.

As heart failure worsens, tissue perfusion and renal blood flow decline progressively. The kidneys release renin and prostaglandins, particularly prostaglandin E_2, into the circulation in response to decreased renal blood flow. Prostaglandin E_2 stimulates production of TNF from monocytes in vitro.

RELATIONSHIP OF TAURINE DEFICIENCY TO MYOCARDIAL DISEASE

Taurine is an essential amino acid in cats. Cats have a limited ability to synthesize taurine from cysteine and methionine because their tissues contain low concentrations of cysteine dioxygenase and cysteine sulfinate acid decarboxylase, key enzymes in the synthesis of taurine. Cats must also use taurine exclusively for conjugation of bile acids, which contributes to an obligatory loss of taurine. The decreased ability to synthesize taurine and the continuous obligatory losses predispose cats to taurine deficiency when they eat foods with low taurine concentrations.

Treatment with oral taurine supplements significantly improves clinical signs, restores myocardial function and improves survival of cats with dilated cardiomyopathy. Supplementation of most commercial cat foods with taurine has markedly reduced the number of feline dilated cardiomyopathy cases. Idiopathic dilated cardiomyopathy is occasionally diagnosed in cats with no evidence of taurine deficiency; this condition does not improve with taurine supplementation. Dilated cardiomyopathy has also been associated with plasma taurine deficiency and low myocardial taurine concentrations in a small number of dogs.

The mechanism of heart failure in taurine-deficient cats and dogs is poorly understood. Taurine has been suggested to function in osmoregulation, calcium modulation and inactivation of free radicals. Other unidentified factors may be involved in the development of myocardial failure in animals with taurine deficiency.

Several studies have demonstrated an association between taurine and potassium balance in cats. Inadequate potassium intake may be sufficient to induce significant taurine depletion and cardiovascular disease in healthy cats. Female cats with dilated cardiomyopathy have significantly lower plasma taurine concentrations than do similarly affected male cats. This finding suggests that male cats are more prone to developing taurine-dependent dilated cardiomyopathy than are female cats, or they are more prone to developing clinical signs associated with cardiac decompensation at higher plasma taurine concentrations.

RELATIONSHIP OF CARNITINE DEFICIENCY TO MYOCARDIAL DISEASE L-carnitine is a small, water-soluble, vitamin-like quaternary amine found in high concentrations in mammalian heart and skeletal muscle. In dogs, L-carnitine is synthesized from the amino acids lysine and methionine, primarily in the liver. A poorly understood transport mechanism concentrates L-carnitine in cardiac and skeletal myocytes.

Although the heart uses various metabolic substrates to maintain the constant energy supply needed to sustain effective contraction and relaxation, it is well established that long-chain fatty acids are quantitatively the most important. Carnitine is a critical component of the mitochondrial membrane enzymes that transport activated fatty acids in the form of acylcarnitine esters across the mitochondrial membranes to the matrix, where β-oxidation and subsequent high-energy phosphate generation occur. In addition to its role in fatty acid transport, free carnitine serves as a mitochondrial detoxifying agent by accepting (or "scavenging") acyl groups and other potentially toxic metabolites and transporting them out of the mitochondria as carnitine esters.

A subset of dogs with dilated cardiomyopathy apparently suffers from myopathic carnitine deficiency and may respond to carnitine supplemen-

HEART

tation. Plasma carnitine deficiency appears to be a specific but insensitive marker for myocardial carnitine deficiency in dogs with dilated cardiomyopathy; unfortunately, dogs with myocardial carnitine deficiency do not always have low plasma carnitine concentrations. Most dogs in which myocardial carnitine deficiency has been associated with dilated cardiomyopathy fall into the classification of myopathic carnitine deficiency (i.e., decreased myocardial carnitine concentrations in the presence of normal or elevated plasma carnitine concentrations). Many of these dogs may suffer from a membrane transport defect that prevents adequate amounts of carnitine from moving into the myocardium from the plasma at plasma carnitine concentrations found in dogs fed most commercial foods. Systemic carnitine deficiency (decreased plasma and myocardial carnitine concentrations) accounts for approximately 20% of the cases.

HYPERTENSION Regulation of systemic blood pressure involves complex relationships between central and peripheral nervous, renal, endocrine and vascular systems. Most people with hypertension have essential hypertension (i.e., hypertension without a discernible organic cause [primary or idiopathic hypertension]). Hypertension secondary to an obvious underlying cause is more common in dogs and cats.

The kidneys ultimately provide long-term control of blood pressure because they are able to excrete sodium and water. This control is accomplished by manipulating the determinants of systemic blood pressure: cardiac output and total peripheral resistance (BP = CO x TPR). Cardiac output is related to heart rate and stroke volume (CO = HR x SV). Diseases associated with hypertension that increase heart rate include: 1) hyperthyroidism, 2) anemia, 3) hyperviscosity, 4) polycythemia and 5) pheochromocytoma. Increased stroke volume may occur during hypervolemic states, but is usually due to increased retention of sodium, chloride and water. Renal failure, hyperadrenocorticism and hyperaldosteronism may cause increased total body sodium, chloride and water.

Activation of the renin-angiotensin-aldosterone (RAA) pathway may elevate blood pressure by increasing stroke volume and total peripheral resistance. Angiotensin II is a potent vasoconstrictor, and angiotensin II and aldosterone stimulate renal sodium and chloride retention. Increased arteriolar tone, sensitivity to circulating vasopressors and levels of circulating catecholamines and decreased arteriolar elasticity may also increase total peripheral resistance.

Common causes of secondary hypertension include: 1) chronic progressive renal disease in dogs and cats (glomerulonephritis, amyloidosis, chronic interstitial nephritis, pyelonephritis, polycystic renal disease), 2) hyperadrenocorticism in dogs and 3) hyperthyroidism in cats. The "target organs" or end-organs or systems that appear most sensitive to increased blood pressure include the eyes, kidneys, cardiovascular system and cerebrovascular system. Clinical signs related to end-organ damage are usually the reason an animal with hypertension is brought to a veterinarian for examination.

PLEURAL EFFUSION Hydrostatic and oncotic forces (Starling's forces) are balanced within the pleurae and pleural space. Hydrostatic and oncotic pressures within the systemic circulation, pulmonary circulation and intrapleural space favor transudation of pleural fluid from the parietal pleura (pleura covering the inner chest wall) into the pleural space with subsequent absorption of the fluid into the visceral pleura's vasculature. The result is a continuous flow of fluid through the pleural space. This delicate balance can be disrupted by any disorder that alters oncotic pressure, systemic or pulmonary capillary pressure, lymphatic compliance, capillary permeability or effective pleural surface area.

Biventricular CHF with systemic and pulmonary venous hypertension is a primary cause of pleural effusion. However, other causes of pleural effusion can masquerade as heart failure and may occur in patients with known heart disease, especially older dogs and cats. Other common causes of pleural effusion include diseases that increase capillary permeability and alter the normal flow and absorption of pleural fluid (e.g., primary intrathoracic or metastatic malignancy, pleural space infection, traumatic diaphragmatic hernia with incarceration of abdominal viscera).

Key Nutritional Factors (Pages 542-543)

Because CHF is associated with retention of sodium, chloride and water, these nutrients are of primary importance in patients with cardiovascular disease. Within a few hours of ingesting high levels of sodium, normal dogs and cats easily excrete the excess in their urine. Early in the course of cardiac disease, animals may lose this ability to excrete excess sodium due to the neurohumoral compensatory mechanisms mentioned earlier.

In the past, retention of sodium was primarily implicated in the pathogenesis of CHF and some forms of hypertension. The full expression of sodium chloride-sensitive hypertension in people depends on the concomitant administration of both sodium and chloride. In experimental models of sodium chloride-sensitive hypertension in rodents and in clinical studies with small numbers of hypertensive people, blood pressure or volume was not increased by a high sodium intake provided with anions other than chloride, and high chloride intake without sodium affected blood pressure less than the intake of sodium chloride. The failure of nonchloride sodium salts to produce hypertension or hypervolemia may be related to their failure to expand plasma volume; renin release occurs in response to renal tubular chloride concentration. Chloride may also act

HEART

as a direct renal vasoconstrictor. These findings suggest that both sodium and chloride are nutrients of concern in patients with hypertension and heart disease.

Potassium and magnesium are nutrients of concern in patients with cardiovascular disease. Abnormalities in their homeostasis can: 1) cause cardiac dysrhythmias, 2) decrease myocardial contractility, 3) produce profound muscle weakness and 4) potentiate adverse effects from cardiac glycosides and other cardiac drugs. Nutrients that contribute to energy intake are important to consider in patients with either obesity or cachexia. Phosphorus and protein are nutrients of concern in patients with concurrent chronic renal disease. (See Chapter 19.) Taurine and carnitine are of importance in dogs and cats with myocardial failure. The Key Nutritional Factors–Cardiovascular Disease table summarizes nutrients of concern in cardiovascular disease.

Assess the Food (Page 543)

Foods for animals with cardiovascular disease should be evaluated for all the key nutritional factors mentioned above. The protein and fat content of the food can be estimated from the information provided by the guaranteed or typical analysis on the commercial pet food label. (See Chapter 5.) Some commercial pet food labels may also include information about the energy density of the food, although this information is not required and is prohibited in some countries. Levels of sodium, chloride, potassium, magnesium and phosphorus are not required on the guaranteed or typical analysis of pet food labels and are usually not listed. Levels of these nutrients must be obtained by contacting the manufacturer or consulting published product information. (See Appendix 2.) Levels of protein, sodium, chloride, potassium, magnesium and phosphorus in commercial pet foods intended for normal animals usually greatly exceed the minimum requirements for these nutrients.

In general, most commercial cat foods should have taurine added as an ingredient. Taurine is naturally high in many species of fish; therefore, fish-based moist cat foods often provide adequate levels of taurine without supplementation. For pet foods sold in the United States, taurine must appear in the ingredient list of foods if a pet food manufacturer adds it. Taurine may appear on labels of foods sold in other countries. Cat foods that have passed an Association of American Feed Control Officials (AAFCO) protocol feeding trial (See Chapter 5.) can be assumed to maintain adequate tissue taurine concentrations. Information about carnitine levels in pet foods is not widely available.

Nutrient sources other than commercial pet foods should be investigated. Water quality varies considerably, even within the same community. Water can be a significant source of sodium, chloride and other minerals. Veterinarians should be familiar with the mineral levels in their local water supply. Water samples can be submitted to laboratories for analysis, municipal water companies can be contacted or private companies that market water conditioning systems can be asked about mineral levels in local water supplies.

Other sources of nutrients include commercial treats and snacks for pets, and human foods offered as snacks or part of the pet's food. Commercial pet treats (See Appendix 3 and Small Animal Clinical Nutrition, 4th ed., pp 1084-1091.) and processed human foods are often high in sodium (**Table 18-3**), phosphorus and other minerals.

The caloric intake should be determined for patients with obesity and cachexia. A diary maintained by the client is helpful in documenting what types and quantities of foods and supplements are being offered and eaten by the animal. This caloric intake can be compared with the number of calories that are usually needed to maintain ideal body weight and condition in that animal.

Assess the Feeding Plan (Page 543)

It may not always be necessary to change the feeding method when managing a patient with cardiovascular disease, but a thorough assessment includes verification that an appropriate feeding method is being used. Items to consider include amount fed, how the food is offered, access to other food and who feeds the animal. All of this information should have been gathered when the history of the animal was obtained. If the animal has a normal body condition score (BCS 3/5), the amount of food it was fed previously (energy basis) was probably appropriate.

FEEDING PLAN (Pages 543-547)

Select a Food (Pages 543-547)
Avoiding Excess Sodium Chloride (Pages 543-545)

Table 18-4 classifies commercial dog foods by sodium content. This classification assumes that most of the foods in these categories have an energy density of 3.5 to 4.5 kcal (14.7 to 18.9 kJ) metabolizable energy per gram dry matter.

Low sodium intake can be achieved with most foods specifically formulated for geriatric dogs and veterinary therapeutic foods formulated for dogs and cats with renal disease. Veterinary therapeutic foods formulated for dogs with cardiovascular disease provide the lowest sodium levels. The levels of sodium in foods for dogs with cardiovascular disease still exceed the AAFCO minimum allowance.

HEART

Table 18-3. Sodium content of selected foods.

Foods	Amounts	Sodium (mg)
Bread, cereals and potatoes		
Recommended		
Macaroni	1 cup	1-10
Potato	1 (medium)	<5
Puffed wheat	1 oz	1-10
Rice (polished)	½ cup	1-10
Spaghetti	1 cup	1-10
Not recommended		
Bread	1 slice	200
Corn chips	1 oz	230
Potato chips	1 oz	300
Pretzel	1	275
Margarine and oil		
Recommended		
Unsalted margarine	1 tsp	0-1
Vegetable shortening	1 tbs	0-1
Not recommended		
Mayonnaise	1 tbs	60-90
Dairy products		
Not recommended		
American cheese	1 oz	200-300
Butter	1 tsp	50
Cottage cheese	3 oz	200-300
Cream cheese	1½ oz	100-120
Milk (regular and skim)	1 cup	122
Meats, poultry, fish		
Recommended		
Beef (fresh)	3½ oz	50
Chicken (no skin)		
Light meat	3½ oz	64
Dark meat	3½ oz	86
Lamb (fresh)	3½ oz	84
Pork (fresh)	3½ oz	62
Turkey (no skin)		
Light meat	3½ oz	82
Dark meat	3½ oz	98
Not recommended		
Bacon	2 slices	385
Egg	1	70
Frankfurter	1	560
Ham (processed)	3 oz	940
Tuna (canned)	1 can	320

(Continued on next page.)

Table 18-3. Sodium content of selected foods (Continued.).

Foods	Amounts	Sodium (mg)
Vegetables (fresh or dietetic canned)		
Recommended		
Corn	½ cup	<5
Cucumber	½ cup	<5
Green beans	½ cup	<5
Green pepper	¼ cup	<5
Lettuce	¼ cup	<5
Peas	½ cup	<5
Tomato	1	<5
Not recommended		
Most canned vegetables	½ cup	190-450
Fruits		
Most fresh and canned fruits are low in sodium and are permitted		
Other food items		
Not recommended		
Macaroni with cheese	1 cup	1,000
Peanut butter	1 tbs	81
Pizza (cheese)	1 slice	650
Dessert		
Recommended		
Sherbet	½ cup	15-25
Not recommended		
Cookies	1	35-100
Gelatins	½ cup	60-85
Ice cream	½ cup	60-85
Puddings	½ cup	100-200

Although there is currently little or no evidence that proves foods low in sodium chloride delay disease progression in the initial stages of heart disease in dogs, a prudent recommendation is to begin avoiding excess sodium chloride early in the disease process. At the first sign of heart disease without cardiac dilatation (Class Ia), foods in the moderate- to low-sodium category should be introduced. Early intervention may help the patient accept foods when more restricted sodium chloride levels are necessary later, and reminds the owners to remain vigilant for signs of disease progression. Furthermore, avoiding excess sodium chloride early in heart disease has not been shown to be harmful.

When cardiac dilatation becomes evident on radiographs or echocardiograms (Class Ib), then foods in the low sodium chloride category are appropriate. Cardiac dilatation implies abnormal sodium chloride handling and intravascular volume expansion. Cardiac dilatation is a prelude to congestion. The presence of moderate to severe cardiac dilatation, con-

Table 18-4. Classification of commercial pet foods by sodium content.[*]

Category	Dog food Average sodium content (% dry matter basis)
High-sodium foods	
Grocery moist foods (14)	1.13
Moderate-sodium foods	
Grocery dry foods (17)	0.41
Specialty dry foods (19)	0.41
Specialty moist foods (9)	0.40
Low-sodium foods	
Geriatric/senior foods (8)	0.20
Renal foods (8)	0.22
Very low-sodium foods	
Cardiac foods (5)	0.10
AAFCO minimum allowance	**0.06**

Category	Cat food Average sodium content (% dry matter basis)
High-sodium foods	
Grocery moist foods (17)	1.09
Moderate-sodium foods	
Grocery dry foods (12)	0.43
Specialty dry foods (9)	0.38
Specialty moist foods (11)	0.34
Low-sodium foods	
Renal foods (5)	0.25
Cardiac foods (3)	0.28
AAFCO minimum allowance	**0.20**

*Sodium contents are based on published or analytical values obtained for the leading products in the United States based on market share. The number in parentheses is the number of products used to determine the average value for each product category. Grocery foods are those pet foods usually purchased in grocery or mass merchandising outlets. Specialty and geriatric/senior foods are usually purchased in pet stores, veterinary hospitals or mass merchandising outlets. Renal and cardiac foods are veterinary therapeutic foods purchased in veterinary hospitals.

gestion or both conditions (Class II or III) indicates that foods low to very low in sodium chloride are appropriate. Many veterinary cardiologists prescribe foods in the low sodium chloride category when ACE inhibitors are used, especially when used in combination with diuretics. Nutrient profiles of selected veterinary therapeutic foods commonly used for patients with cardiovascular disease are listed in **Tables 18-5** and **18-6**.

Avoiding excess sodium chloride may also be important in patients with some forms of chronic respiratory disease such as chronic bronchitis or asthma. Epidemiologic and experimental evidence in people suggests that high sodium chloride intake may increase airway responsiveness and exacerbate clinical signs associated with asthma. When people with asthma are subjected to salt loading, clinical signs worsen, lung function deteriorates and the need for antiasthmatic drugs increases. Although similar canine and feline studies have not been completed, low-salt foods may be helpful in conjunction with other forms of therapy in animals with chronic bronchitis or asthma-like clinical signs.

Table 18-4 also classifies commercial cat foods by sodium content. Avoiding excess sodium chloride in cats is more difficult than in dogs because cat foods contain more sodium chloride than do dog foods. Ingredients used to meet the higher protein requirement of cats also contain moderate to high levels of sodium chloride and thus increase the sodium chloride content of cat food.

Normal animals are able to eliminate the excess levels of sodium found in many commercial foods, but patients with heart disease and failure have an impaired ability to handle these sodium levels.

Water, commercial treats and snacks and human foods offered as snacks are also sources of sodium, chloride and other minerals. Distilled water or water with less than 150 ppm sodium is recommended for patients with advanced heart disease and failure. Some commercial snacks or treats and processed human foods are often high in sodium chloride and should also be avoided (**Table 18-3**).

Palatability of Low-Salt Foods *(Page 545)*

Many foods with reduced sodium chloride levels for use in patients with cardiovascular disease have comparable or better palatability than grocery or specialty brand pet foods.

Occasionally, it is difficult to get a patient to accept a change to a lower salt commercial food. This can occur because of: 1) advanced illness associated with heart failure, 2) established feeding habits of older patients and their owners, 3) anorexia associated with concurrent renal failure and some cardiac drugs and 4) an "all or nothing" approach to feeding, rather than slowly changing to the new food. Changing the eating habits of most dogs is relatively easy, but changing the feeding habits and preconceptions (e.g., low-salt food is always unpalatable) of some pet owners and veterinarians is often much more difficult. Results of feeding studies using hospitalized dogs have shown that most dogs will readily accept a food that is very low in sodium chloride by the third day. For individual dogs, these foods can be made more palatable by warming the food or adding flavor enhancers (low-sodium soup or tomato sauce; sweeteners such as honey or syrup). Use of foods that are very low in sodium chloride in advanced heart disease and failure will be much easier if the dog

Table 18-5. Levels of key nutrients in selected commercial foods for dogs with cardiovascular disease.*

	Protein	Fat	Na	Cl	K	Mg	P	Taurine	Carnitine
Moist canine products									
Recommended levels	**	***	0.07-0.25	0.11-0.38	†	††	**	††	††
Hill's Prescription Diet Canine g/d	18.9	10.8	0.22	0.85	0.78	0.09	0.41	0.11	na
Hill's Prescription Diet Canine h/d	17.3	28.8	0.07	0.32	0.83	0.13	0.54	0.22	339
Hill's Prescription Diet Canine k/d	14.9	26.1	0.18	0.47	0.40	0.13	0.22	0.11	na
Hill's Prescription Diet Canine w/d	18.2	13.0	0.27	0.73	0.62	0.08	0.50	0.12	351
Leo Specific Cardil CHW	17.0	27.2	0.10	na	1.05	0.07	0.61	na	na
Purina CNM Canine CV-Formula	17.8	31.9	0.12	1.27	1.21	0.06	0.40	0.24	150
Purina CNM Canine NF-Formula	16.5	27.4	0.24	0.43	0.72	0.08	0.30	na	na
Select Care Canine Modified Formula	16.8	21.8	0.24	na	0.96	0.07	0.35	na	na
Waltham/Pedigree Low Sodium Diet	26.3	38.0	0.10	na	0.83	0.10	0.67	na	na
Dry canine products									
Hill's Prescription Diet Canine g/d	18.7	11.0	0.18	0.65	0.63	0.06	0.41	0.10	na
Hill's Prescription Diet Canine h/d	17.1	19.7	0.06	0.34	0.77	0.14	0.61	0.14	299
Hill's Prescription Diet Canine k/d	14.5	19.0	0.23	0.69	0.66	0.09	0.24	0.10	na

(Continued on next page.)

Table 18-5. Levels of key nutrients in selected commercial foods for dogs with cardiovascular disease* (Continued).

Dry canine products	Protein	Fat	Na	Cl	K	Mg	P	Taurine	Carnitine
Recommended levels	**	***	0.07-0.25	0.11-0.38	†	††	**	††	††
Hill's Prescription Diet Canine w/d	18.9	8.7	0.22	0.54	0.68	0.10	0.58	0.13	319
Purina CNM Canine NF-Formula	15.9	15.7	0.22	0.57	0.86	0.07	0.29	na	na
Select Care Canine Modified Formula	14.4	19.7	0.28	na	0.88	0.09	0.34	na	na

Key: na = information not available, Na = sodium, Cl = chloride, K = potassium, Mg = magnesium, P = phosphorus.
*This list represents products with the largest market share and for which published information is available. Manufacturers' published values; carnitine expressed as ppm.
**Protein levels must be adequate if cardiac cachexia is present. Protein and phosphorus levels should be restricted if chronic renal failure is present. (See Chapter 19.)
***Providing the right amount of energy is vital for managing patients with cardiovascular diseases. Overweight patients should be fed foods with reduced energy as part of a weight-reduction program. (See Chapter 13.) Patients suffering from cardiac cachexia may need more energy than otherwise normal pets. Because most pet food labels don't list energy, dietary fat may be used as a guide to a product's energy content. Body condition scoring should be used frequently to determine the patient's response to a food's energy/fat content.
†Hypokalemia, normokalemia or hyperkalemia may be present. See the Key Nutritional Factors–Cardiovascular Disease table for recommendations.
††See the Key Nutritional Factors–Cardiovascular Disease table if additional supplementation is required beyond that present in cardiac foods.

HEART

Table 18-6. Levels of key nutrients in selected commercial foods for cats with cardiovascular disease.*

	Protein	Fat	Na	Cl	K	Mg	P	Taurine
Moist feline products								
Recommended levels	**	***	<0.35	<0.75	†	††	**	††
Hill's Prescription Diet Feline g/d	35.1	19.4	0.29	0.58	0.74	0.08	0.54	0.45
Hill's Prescription Diet Feline h/d	43.6	27.0	0.31	0.72	0.91	0.07	0.78	0.41
Hill's Prescription Diet Feline k/d	29.5	26.7	0.32	0.56	1.05	0.04	0.39	0.39
Purina CNM Feline CV-Formula	42.5	26.8	0.20	1.09	1.33	0.07	0.92	0.31
Select Care Feline Modified Formula	35.0	53.0	0.23	na	1.07	0.06	0.49	na
Dry feline products								
Hill's Prescription Diet Feline g/d	33.4	18.9	0.34	0.68	0.75	0.05	0.55	0.12
Hill's Prescription Diet Feline k/d	28.2	22.3	0.25	0.50	0.76	0.04	0.46	0.14
Purina CNM Feline NF-Formula	30.8	12.9	0.20	0.64	0.88	0.10	0.41	0.18
Select Care Feline Modified Formula	28.3	22.1	0.27	na	0.92	0.07	0.92	na

Key: na = information not available, Na = sodium, Cl = chloride, K = potassium, Mg = magnesium, P = phosphorus.
*This list represents products with the largest market share and for which published information is available. Manufacturers' published values; expressed as % dry matter.
**Protein levels must be adequate if cardiac cachexia is present. Protein and phosphorus levels should be restricted if chronic renal failure is present. (See Chapter 19.)
***Providing the right amount of energy is vital for managing patients with cardiovascular disease. Overweight patients should be fed foods with reduced energy as part of a weight-reduction program. (See Chapter 13.) Patients suffering from cardiac cachexia may need more energy than otherwise normal pets. Because most pet food labels don't list energy, dietary fat may be used as a guide to a product's energy content. Body condition scoring should be used frequently to determine the patient's response to a food's energy/fat content.
†Hypokalemia, normokalemia or hyperkalemia may be present. See the Key Nutritional Factors–Cardiovascular Disease table for recommendations.
††See the Key Nutritional Factors–Cardiovascular Disease table if additional supplementation is required beyond that present in cardiac foods.

has already been fed a food in the low-sodium category. Cats may respond to warming moist food to body temperature and a more gradual transition (weeks instead of days) to the new food.

A common mistake is to insist that an owner feed only a salt-restricted food even if caloric intake is inadequate. Although avoiding excess sodium chloride is important in CHF patients, offering only a salt-restricted food should not be imposed to the detriment of overall nutrient intake. Changing to a different commercial food or homemade food may be a more beneficial solution for some patients. Appetite may be cyclical in patients with advanced heart failure, both in respect to overall appetite and food preferences. A dedicated owner is often required and a trial and error approach must be used with different foods and feeding methods.

Ensuring Optimal Intake of Potassium, Magnesium and Phosphorus (Pages 545-546)

Electrolyte abnormalities, including hypokalemia, hyperkalemia and hypomagnesemia, are potential complications of drug therapy in patients with cardiovascular disease. Patients receiving diuretic therapy should receive adequate amounts of potassium and magnesium. Patients treated with ACE inhibitors may be predisposed to mild hyperkalemia; therefore, their food should not contain excessive levels of potassium. Chronic renal disease often occurs concomitantly in patients with cardiovascular disorders. Patients with chronic renal disease should be fed foods without excess levels of phosphorus. (See Chapter 19.)

Protein and Energy Intake (Page 546)

Obesity causes profound changes that can complicate cardiovascular disorders. Obese patients should undergo management with a calorie-restricted food and client education should focus on the importance of the pet achieving an ideal body weight and condition. (See Chapter 13.) The veterinary health care team should emphasize the potentially damaging effects of obesity in patients with heart disease to clients to enlist their active participation in a successful weight-management program.

The protein requirements of patients with cardiac cachexia have not been investigated and it is unknown how the metabolic changes associated with cachexia affect overall nutrient requirements. Many patients with cachexia have concomitant disease, such as chronic renal failure, which affects nutrient requirements. Profound anorexia enhances protein-energy malnutrition in animals with cachexia. Animals with cachexia should be encouraged to eat a complete and balanced food that contains adequate calories and high-quality protein.

The cytokines TNF and IL-1 have been implicated as pathogenic mediators in cardiac cachexia. Fish oil, which is high in omega-3 (n-3) fatty acids, alters cytokine production. Preliminary results suggest that fish-oil-mediated alterations in cytokine production may benefit dogs with CHF. Circulating TNF and IL-1 concentrations decreased significantly in a group of dogs with CHF secondary to idiopathic dilated cardiomyopathy when they were treated with fish oil supplements (65 mg/kg). Dogs receiving fish oil tended to be judged as less cachectic when compared with those in the placebo group. Ventricular function also improved significantly in the group treated with fish oil when compared with dogs receiving a placebo. These findings suggest that heart failure patients with cachexia may benefit from the alterations of cytokine production brought about by omega-3 fatty acid supplementation or other methods.

Taurine Supplementation (Page 546)

Cats and dogs with myocardial failure may benefit from taurine supplementation to their regular food or use of foods that contain high levels of taurine. Animals with documented taurine deficiency are more likely to respond favorably to taurine supplementation. In dogs, the association between taurine deficiency and dilated cardiomyopathy is strongest in American cocker spaniels and golden retrievers. Cats should receive 250- to 500-mg taurine per os daily, whereas dogs should receive 500- to 1,000-mg taurine per os three times daily. Some foods formulated for nutritional management of cardiovascular disease already contain high levels of taurine (**Tables 18-5** and **18-6**). Animals eating these foods usually do not need additional taurine supplementation. **Table 18-7** lists levels of taurine found in natural food sources.

Taurine supplementation of feline foods can be discontinued within 12 to 16 weeks if: 1) clinical signs of heart failure have resolved, 2) echocardiographic values are near normal and 3) the cat will eat a food known to support normal whole blood taurine concentrations. The length of time needed for taurine supplementation of canine foods is currently unknown.

Carnitine Supplementation (Page 546)

The recommended oral dosage for dogs with myocardial carnitine deficiency is 50 to 100 mg L-carnitine/BW_{kg} three times daily. Dogs weighing 25 to 40 kg are most often affected and should receive 2 g of L-carnitine mixed with food three times daily. This high oral dosage will elevate plasma carnitine concentration 10 to 20 times above usual pretreatment values. These high plasma carnitine levels will usually, but not always, increase myocardial carnitine concentrations into the normal range. The cost of this level of L-carnitine supplementation is approximately $100 to $200 (U.S.) per month for a large-breed dog. Carnitine is usually available in human health food stores.

Table 18-7. Taurine concentrations (mg/kg dry weight) in selected natural food sources.

Sources	Taurine concentrations
Beef muscle, uncooked	1,200
Chicken muscle, uncooked	1,100
Cod fish, uncooked	1,000
Lamb muscle, uncooked	1,600
Mouse carcass	7,000
Pork muscle, uncooked	1,600
Tuna, canned	2,500

Dogs that respond dramatically to carnitine therapy do so in a predictable manner. Owners often report generalized improvement in clinical signs within one to four weeks and echocardiographic improvement is noted after eight to 12 weeks of supplementation. Improvement may continue for about six to eight months, at which time patients often reach a plateau and though they appear clinically normal they have depressed ventricular function as determined by echocardiography.

Determine a Feeding Method (Page 547)

The method of feeding is often not altered in the nutritional management of cardiovascular disease. If a new food is fed, the amount to feed can be determined from the product label or other supporting materials. The food dosage may need to be changed if the caloric density of the new food differs from that of the previous food. The food dosage is usually divided into two or more meals per day. The food dosage and feeding method should be altered if the animal's body weight and condition are not optimal.

For clinical nutrition to be effective, there needs to be good compliance. Enabling compliance includes limiting access to other foods and knowing who feeds the animal. If the animal comes from a household with multiple pets, it should be determined whether the pet with cardiovascular disease has access to other pets' food. Access to other food (table food, other pets' food, etc.) may contribute to cardiovascular disease and thus should be denied.

REASSESSMENT (Pages 547-549)

Nutrient-Drug Interactions (Pages 547-548)

Most patients with cardiovascular disease in which nutritional management is used also receive drug therapy. In the past, drug-drug interactions received considerable attention but few investigators evaluated or dis-

HEART

cussed how nutrient levels might affect drug availability and pharmacokinetics and vice versa (See Chapter 27.). Because many cardiovascular patients are treated with a combination of veterinary therapeutic foods and drugs, potential food-drug or nutrient-drug interactions are important.

Diuretics (Pages 547-548)

Diuretics continue to be a pharmacologic mainstay of acute therapy for heart failure. Sodium restriction, ACE inhibition, venodilating drugs and diuretics represent the major available methods for preload reduction. Sodium chloride restriction is a key component of CHF treatment even with the use of diuretics. Well-controlled studies have demonstrated that loop diuretics such as furosemide given once a day fail to achieve a negative sodium balance in people with high sodium intake. Although there is an impressive natriuresis for several hours after furosemide administration, a compensatory increase in sodium reabsorption during the next 24 hours exactly matches the earlier losses. Thus, it is essential to limit sodium intake to ensure negative sodium balance.

Blood volume contraction and circulatory impairment are potential complications of aggressive diuretic therapy. These complications can exacerbate pre-existing renal disease, alter excretion of drugs dependent on renal elimination and reduce cardiac output by reducing cardiac filling pressures. Reduced levels of sodium in the food have been implicated, but have not been proven to contribute to volume depletion from excessive diuresis. Fractional excretion of sodium in the urine actually decreases in normal dogs fed a sodium-restricted food. The influence of diuretics on sodium and chloride balance in dogs with heart disease and failure fed sodium- and chloride-restricted foods has not been evaluated.

Furosemide contributes to hypokalemia and hypomagnesemia because of increased urinary loss of potassium and magnesium. The role of magnesium and potassium in the development of cardiac dysrhythmias has not received attention beyond the recognition that digitalis toxicosis appears to be much more dysrhythmogenic in hypomagnesemic and hypokalemic patients. Hypomagnesemia may potentiate cardiac dysrhythmias caused by catecholamine release and is also associated with increased vascular reactivity.

Conflicting reports have been published about the serum electrolyte and magnesium concentrations of dogs with CHF. A study of 113 dogs with CHF identified only four dogs with hypomagnesemia. Three of the four hypomagnesemic dogs received combined therapy with a commercial sodium-restricted veterinary therapeutic food, furosemide and either hydralazine or enalapril. In another study, furosemide-treated dogs with heart failure had significantly lower serum magnesium and potassium values than did age-matched healthy controls. A third study showed no significant differences in serum magnesium concentrations between clinically normal dogs, dogs with heart failure before any treatment, heart-failure dogs treated only with furosemide and heart-failure dogs treated with furosemide and digoxin. The feeding history was not included in the last two studies; therefore, specific food-diuretic interactions could not be interpreted. Normal dogs treated with a commercial sodium-restricted veterinary therapeutic food and furosemide for four weeks had no significant change in serum potassium concentrations.

Several studies have shown that the RAA system is not activated in human patients with moderate heart failure in the absence of diuretic therapy. The major increase in plasma renin activity and plasma aldosterone concentration occurs with the introduction of diuretic drugs into the treatment regimen rather than as a result of the disease process itself. Furosemide apparently stimulates renin release by inhibiting chloride transport in the ascending limb of the loop of Henle, even if blood volume contraction is prevented. Treatment of normal geriatric dogs with moderate doses of furosemide profoundly stimulates the RAA system, irrespective of the sodium level in the food. Use of furosemide with either hydralazine or enalapril also stimulates the RAA system in dogs with heart failure due to acquired mitral valve regurgitation.

Although diuretics will remain important first-line drugs for management of acute cardiogenic pulmonary edema, findings in people suggest that diuretics continue to stimulate the RAA system and may play a pivotal role in the progressive self-perpetuating cycle of heart failure. Veterinary cardiologists are now questioning the use of diuretic monotherapy early in the management of symptomatic heart failure. Diuretics should be reserved for managing more advanced heart failure in patients already receiving sodium-chloride-restricted foods, ACE inhibitors, digoxin or combination therapy. Feeding patients foods without excess sodium chloride may allow lower dosages of diuretics to be used for control of the clinical signs of CHF.

The sodium, chloride, potassium and magnesium levels in commercial pet foods and veterinary therapeutic foods for dogs and cats with cardiovascular disease and heart failure vary tremendously (**Tables 18-4** through **18-6**). The levels of minerals in the food should be considered when using concurrent diuretic therapy.

Long-term furosemide therapy may be associated with clinically significant thiamin deficiency, due to excessive urinary loss of thiamin, and may contribute to impaired cardiac performance in patients with CHF. Animals receiving long-term diuretic therapy should be given supplements containing thiamin and other water-soluble vitamins or be fed a commercial food with increased concentrations of these vitamins. Veterinary therapeutic foods for patients with cardiac and renal disease are often formulated with higher levels of water-soluble vitamins to offset excessive urinary losses.

HEART

ACE Inhibitors (Page 548)

Captopril, enalapril and lisinopril, all ACE inhibitors, have emerged as drugs of choice for treating many dogs and cats with CHF. Inhibition of angiotensin II results in vascular dilatation and decreased circulating plasma aldosterone concentrations. Angiotensin II and aldosterone play important roles in the maintenance of vascular volume and potassium balance. Both increase the reabsorption of sodium and chloride, and aldosterone promotes the excretion of potassium.

The use of ACE inhibitors in human patients with severe renal insufficiency or in patients given potassium supplements may increase the risk for hyperkalemia. In a study, more than half the dogs with CHF developed mild serum potassium elevations when treated with a commercial sodium-restricted veterinary therapeutic food, furosemide and captopril. Another study confirmed that heart-failure dogs treated with furosemide, digoxin and an ACE inhibitor had significantly higher mean serum potassium concentrations when compared with clinically normal dogs, dogs with heart failure before any treatment, heart-failure dogs treated only with furosemide and heart-failure dogs treated with furosemide and digoxin. Mild elevations in serum potassium concentrations have also been observed in dogs treated with enalapril. In another study, serum potassium concentration decreased in a subset of heart-failure dogs treated with ACE inhibitors and furosemide, although the specific feeding history was not reported.

When mild hyperkalemia occurs in people with heart failure, reducing oral potassium intake and discontinuing potassium-sparing diuretics is recommended. Although clinically significant hyperkalemia (serum potassium ≥ 6.5 mEq/l) is uncommon, the use of ACE inhibitors in dogs with CHF or renal insufficiency fed commercial or veterinary therapeutic foods with high potassium content may increase the risk for hyperkalemia.

Functional renal insufficiency occurs in up to one-third of human patients with severe CHF treated with sodium chloride restriction, ACE inhibitors and diuretics. This decline in renal function has been attributed to loss of angiotensin II-mediated systemic and intrarenal vasoconstrictor effects, which maintain renal perfusion pressure and glomerular filtration rate in low-output heart failure. Functional renal insufficiency appears to be alleviated in human patients when efforts are made to replenish total body stores of sodium by reducing the diuretic dosage and liberalizing sodium intake. Renal insufficiency is a potential complication of ACE inhibitor therapy in dogs with CHF, but the role of sodium restriction is unknown. Four of 10 heart-failure dogs treated with captopril, furosemide and a sodium-restricted veterinary therapeutic food developed azotemia during the first five weeks of treatment; one of these dogs developed clinical signs of uremia. Two of the dogs that developed severe azotemia had isosthenuria on the initial urinalysis, which suggested some degree of pre-existing renal insufficiency. Azotemia is a more frequent complication when canine heart failure is treated with furosemide and enalapril rather than with furosemide alone.

Reducing the ACE inhibitor dose (usually by half), reducing the diuretic dose (usually by half), increasing the sodium intake to the next level (**Table 18-4**) or combining these tactics is used to treat drug-induced azotemia in heart-failure patients.

Cardiac Glycosides *(Page 548)*

Absorption of the cardiac glycosides is influenced by the formulation of the drug and its administration in relation to meals (See Chapter 27.). Because administering digoxin or digitoxin with food may result in up to a 50% reduction in serum concentrations, these drugs are best given between meals. The body condition of the animal can also influence the pharmacokinetics of these drugs. Digoxin is minimally distributed in adipose tissue; therefore, the dosage of the drug should be based on lean body weight even for obese patients. Digitoxin is more lipid soluble than digoxin; therefore, its dosage need not be adjusted for overweight animals.

The dosage of digoxin for cats is influenced by concurrent drug and nutritional therapy. Reduce the digoxin dose by one-third if the cat is receiving concomitant furosemide, aspirin and a sodium-restricted veterinary therapeutic food.

Metabolic derangements associated with increased risk of digoxin toxicosis include hypokalemia, hypomagnesemia, hypercalcemia, renal insufficiency, hypothyroidism and obesity. Serum electrolyte and magnesium concentrations should be measured and corrections made before starting cardiac glycoside therapy.

Overall Reassessment Strategy (Page 549)

In general, the survival of patients with heart failure is related to the degree of myocardial failure, whereas their clinical signs are related more to CHF and its compensatory mechanisms. The overall objectives of treatment for chronic heart failure, as for almost any cardiovascular disease, are threefold: 1) prevention (prevent myocardial damage, prevent recurrence of heart failure), 2) relief of clinical signs (eliminate edema and fluid retention, increase exercise capacity, reduce fatigue and respiratory compromise) and 3) improvement of prognosis (reduce mortality).

Dogs and cats with suspected cardiovascular disease should undergo a routine serum biochemistry profile and urinalysis before any nutritional or drug therapy is initiated. Dogs and cats with heart failure and evidence of preexisting renal disease, including isosthenuria, may be at increased risk for developing azotemia during combined food-drug therapy. There

HEART

are no universal recommendations for controlling: levels of sodium, chloride and potassium; fluid intake; ACE inhibition and diuretic administration for animals with cardiovascular disease. Rather, each patient must be monitored frequently (weekly for the first four to six weeks). Reassessment should include: 1) measurement of body weight, 2) assessment of body condition, 3) determination of serum electrolyte and magnesium concentrations and 4) evaluation of renal function.

Clinical cases that illustrate and reinforce the nutritional concepts presented in this chapter can be found in Small Animal Clinical Nutrition, 4th ed., pp 552-562.

CHAPTER 19

Renal Disease

For a review of the unabridged chapter, see Allen TA, Polzin DJ, Adams LG. Renal Disease. In: Hand MS, Thatcher CD, Remillard RL, et al, eds. Small Animal Clinical Nutrition, 4th ed. Topeka, KS: Mark Morris Institute, 2000; 563-604.

CLINICAL IMPORTANCE (Pages 564-567)*

Nutritional management of patients with chronic renal disease includes measures to reduce signs of uremia and to slow progression to the uremic phase. There is little controversy regarding nutritional management of chronic renal failure (CRF) when overt signs are present. However, there is controversy regarding the impact of nutritional management on slowing progression. Thus, in a sense, the question is when to initiate nutritional management. Because detection of chronic renal disease in its early phases is difficult and there appears to be no harm in avoiding nutrient excess during earlier phases, nutritional management should be instituted as soon as renal dysfunction is demonstrated. Verified loss of urine concentrating ability or significant renal proteinuria, even in the absence of azotemia, reflects marked renal damage and signals the need for nutritional management.

ASSESSMENT (Pages 567-589)

Assess the Animal (Pages 567-588)
History (Page 567)
Table 19-1 lists the historical information important to obtain for patients with suspected renal disease or failure. Historical findings in patients with CRF usually include polyuria/polydipsia (less frequent in cats than dogs), anorexia, lethargy, vomiting, weight loss, nocturia, constipation, diarrhea, acute blindness (because of hypertension) and seizures or coma (late). Cats may also have ptyalism and muscle weakness with cervical ventriflexion (hypokalemic myopathy).

Physical Examination (Page 567)
Physical examination of a patient suspected to have renal failure should include the factors listed in Table 19-2. Dehydration (70%) and

*Page numbers in headings refer to Small Animal Clinical Nutrition, 4th ed., where additional information may be found.

Key Nutritional Factors–Renal Disease (Page 565)

Factors	Associated conditions	Dietary recommendations
Water	Dehydration Electrolyte imbalances Blood volume contraction Renal hypoperfusion	Offer water free choice at all times Parenteral fluid therapy if dehydration, blood volume contraction and/or renal hypoperfusion are clinically significant
Phosphorus	Hyperphosphatemia Hyperparathyroidism	Avoid excess dietary phosphorus Restrict phosphorus to 0.15 to 0.3% dry matter (DM) in dogs and 0.4 to 0.6% DM in cats
Protein	Uremia Metabolic acidosis Glomerular capillary hypertension Glomerular hyperfiltration Proteinuria Hypoproteinemia (protein-losing nephropathy)	Avoid excess dietary protein Restrict protein to ≤15% DM in dogs and ≤30% DM in cats Lower levels may be necessary for dogs with advanced CRF
Sodium/ chloride	Systemic hypertension Increased renal oxygen consumption	Avoid excess dietary sodium Restrict sodium to <0.25% DM in dogs and <0.35% DM in cats Chloride levels are typically 1.5 times sodium levels
Acid load	Renal ammoniagenesis Tubulointerstitial changes	Avoid excess dietary acid Maintain plasma bicarbonate levels between 17 and 22 mEq/l Consider oral alkalinization therapy if serum bicarbonate levels decrease below this target range

(Continued on next page.)

Key Nutritional Factors–Renal Disease (Page 565) (Continued.)

Factors	Associated conditions	Dietary recommendations
Energy	Cachexia	Ensure adequate energy intake in the form of non-protein calories The daily energy requirement (DER) for most canine renal patients is 1.1 to 1.6 x resting energy requirement (RER) The DER for most feline renal failure patients is 1.1 to 1.4 x RER Adjust food dose as needed to maintain optimal body weight and condition
Fatty acids	Hyperlipidemia Progressive renal failure	For dogs, feed a food with an n-6/n-3 ratio of <3:1 See text for information about n-6/n-3 ratios in foods for cats
Potassium	Hypokalemia (especially cats)	Provide oral potassium supplementation (3 to 5 mEq or mmol/BW_{kg}/day) Change to a food with a higher potassium concentration Dogs: 0.3 to 0.5% DM potassium Cats: 0.8 to 1.2% DM potassium
	Hyperkalemia	Discontinue potassium supplementation Switch from potassium-sparing diuretics to loop or thiazide diuretics Change to a food with a lower potassium concentration

Table 19-1. Historical questions for owners of pets suspected of having renal dysfunction.

1. Age?
2. Breed? Family history of renal disease?
3. Duration of the present illness?
4. Previous illness, injury, anesthesia, surgery or drug administration?
5. History of renal disease? Previous laboratory evaluation of renal function?
6. Exposure to toxins?
7. Polyuria or polydipsia?
8. Urination: Frequency? Urinary incontinence? Change in color, odor or volume? "Accidents" in the house?
9. General signs potentially referable to renal failure: Anorexia? Vomiting? Diarrhea? Weight loss? Are these signs increasing, decreasing or unchanged?

an underweight body condition (58%) were the most common abnormal physical examination findings in a clinical series of feline renal failure patients. An abnormally large kidney was detected by palpation in 25% of cases and an abnormally small kidney in 16% of cases in this series. Gingivitis, halitosis and oral ulcers were only occasionally reported.

Laboratory and Other Clinical Data (Pages 568-571)

Clinically, disorders of one of five major categories of renal function are recognized: 1) glomerular filtration, 2) membrane permselectivity, 3) urine concentration, 4) tubular resorption and 5) endocrine function. See Small Animal Clinical Nutrition, 4th ed., pp 568-571 for an in-depth discussion of the five major categories of renal function. Functionally, renal diseases can be generalized or specific and involve only one function (e.g., tubular resorption in Fanconi syndrome). Each major function can be evaluated diagnostically. **Table 19-3** lists the minimum diagnostic database for most patients with suspected renal disease.

Risk Factors (Pages 571-572)

The presence of risk factors associated with kidney disease should increase the index of suspicion of renal disease. If risk factors are noted, additional laboratory tests may be warranted to rule in or rule out renal disease.

BREED/FAMILY Familial nephropathy has been reported to occur in cocker spaniels, Norwegian elkhounds, Lhasa apsos, Shih Tzus, Doberman pinschers, standard poodles, soft-coated wheaten terriers, bull terriers, Samoyeds, Bernese mountain dogs, rottweilers, beagles, golden retrievers, and chow chows. Hereditary nephropathy has been reported to occur in Samoyeds, English cocker spaniels and bull terriers. Renal failure was rec-

Table 19-2. Elements of the physical examination that should be emphasized in patients with suspected renal failure.

1. Temperature, pulse, heart and respiratory rate
2. Assessment of hydration status
3. Body weight and body condition score
4. Oral examination: Mucosal ulcers? Pallor? Necrosis or discoloration of tongue?
5. Cardiovascular system: Abnormal heart sounds? Increased tortuosity of superficial veins? Systemic blood pressure (direct or indirect measurement) abnormalities? Pulse rate and character?
6. Kidneys: Both palpable? Size? Shape? Position? Surface contours? Pain? Bilaterally symmetrical?
7. Urinary bladder: Size? Position? Shape? Pain? Thickness of wall? Intraluminal masses? Grating sensation?
8. Genitourinary tract (urethra, prostate, penis, prepuce, vulva): Shape? Position? Pain? Discharge?
9. Fundus: Retinal detachment? Hemorrhage? Increased tortuosity of arteries? Retinal edema? Lipemia retinalis?
10. Skeleton: Rubber jaw?
11. Cervical region: Thyroid masses (cats)?

Table 19-3. Minimum diagnostic database for patients with suspected renal disease.

1. Urinalysis, including microscopic examination of urine sediment
2. Serum biochemistry profile
3. Automated cell count, differential white cell count, if indicated
4. Urine culture, especially if bacteriuria, pyuria or hematuria is present
5. Abdominal radiography
6. Blood pressure measurement
7. Renal ultrasonography
8. Urine protein-creatinine ratio, if significant proteinuria is evident by dipstick analysis
9. Intravenous urography, if indicated
10. Renal biopsy, if indicated

ognized more than twice as often in Maine coon, Abyssinian, Siamese, Russian blue and Burmese feline breeds.

INFECTIONS It is possible that urinary tract infection (UTI) superimposed on other processes can accelerate the rate of renal injury. Consequently, whenever UTI is diagnosed, prompt treatment with renal-sparing antimicrobials is indicated.

DRUG THERAPY A substantial, but as yet undetermined proportion of renal disease in dogs and cats may be due to drug-induced nephrotoxicity. **Table 19-4** lists drugs that can cause nephrotoxicity.

Table 19-4. Drugs that cause nephrotoxicity.

Analgesics	**Antimicrobials**
Acetaminophen	Aminoglycosides
Aspirin	Sulfonamides
Ibuprofen	Tetracycline
Phenylbutazone	**Cardiovascular drugs**
Anesthetics	Captopril
Enflurane	**Chemotherapeutic drugs**
Methoxyflurane	Cisplatin
Antifungals	Daunorubicin
Amphotericin B	Methotrexate
	Immunosuppressive agents
	Penicillamine

ADVANCING AGE Dogs frequently develop renal dysfunction with advancing age. Hyperthyroidism and CRF occur commonly in older cats. Hyperthyroidism may also more directly contribute to the development of chronic renal disease in older cats. Systemic hypertension is present in most cats with hyperthyroidism.

ISCHEMIA Ischemic events that could lead to renal damage include: 1) shock (i.e., septic, hypotensive, hemorrhagic and hypovolemic), 2) decreased cardiac output (i.e., congestive heart failure [CHF] and severe dysrhythmias), 3) hypotension due to anesthesia or blood loss, 4) renal vessel thrombosis, 5) disseminated intravascular coagulation and 6) decreased renal prostaglandin formation.

DIABETES MELLITUS Diabetic nephropathy is not frequently recognized in dogs and cats. The lower prevalence of diabetic nephropathy in pets compared with people may reflect the much shorter survival of animals with diabetes mellitus.

Etiopathogenesis (Pages 573-582)
Several etiopathogenic mechanisms may be involved in naturally occurring renal disease. The specific mechanisms for progressive renal injury are unknown. Possible mechanisms include a variety of compensatory or adaptive responses. Examples include glomerular capillary hypertension, glomerular hyperfiltration, increased renal ammoniagenesis, increased renal oxygen consumption, secondary renal hyperparathyroidism and compensatory renal growth (hypertrophy). In addition, sequelae of CRF (e.g., hypertension, metabolic acidosis and tubulointerstitial injury) may contribute to progression. Changes in lipid metabolism, coagulation and normal renal aging may also contribute to progression. Some of these mechanisms are initially adaptive when renal function declines, but they may ultimately lead to progressive renal injury. These etiopathogenic mechanisms are not mutual-

ly exclusive and in some instances may act synergistically. See Small Animal Clinical Nutrition, 4th ed., pp 573-582 for an in-depth discussion of etio-pathogenic mechanisms in naturally occurring renal disease.

Key Nutritional Factors (Pages 582-587)

Dietary therapy is the mainstay of conservative medical management of canine and feline CRF. Dietary management prevents or mitigates the clinical signs of uremia, minimizes signs of mineral and electrolyte imbalance and may slow progression of CRF. Dietary therapy with foods that contain appropriate levels of key nutritional factors has been shown to increase longevity and improve the quality of life of pets with chronic renal failure.

WATER Healthy animals are in water balance. Water balance occurs when the sum of water intake and metabolic water produced equals output.

Renal disease causes a progressive decline in urine concentrating capacity. The maximum urine osmolality approaches that of plasma (300 mOsm/kg) in renal insufficiency. If total solute excretion remains normal, but the maximal achievable urine osmolality decreases, obligatory water loss will occur to eliminate the osmolar load. This obligatory water loss may lead to the development of polyuria.

Because concentrating urine predisposes an individual to renal injury, reducing the need to concentrate urine may be beneficial. Dehydration, volume depletion, renal hypoperfusion and dietary salt intake stimulate urine concentration. Avoiding dehydration and hypoperfusion reduces the work of concentrating urine and maintains intrarenal protective mechanisms such as prostaglandin and dopamine production. Reducing urinary solute transport activity may protect against renal injury. CRF patients should have unlimited access to fresh water for free-choice consumption.

PHOSPHORUS The objectives of mineral nutriture in patients with decreased renal function are to: 1) maintain plasma concentrations of phosphorus and ionized calcium within reference limits, 2) prevent or reduce secondary hyperparathyroidism and 3) prevent or reverse renal osteodystrophy (i.e., soft-tissue mineralization and skeletal disease).

Avoiding excess dietary phosphorus reduces renal pathology in cats and improves survival of dogs with experimentally induced renal failure.

An intake of 0.15 to 0.3% dry matter (DM) phosphorus is recommended for most dogs with CRF and 0.4 to 0.6% DM phosphorus for most cats with CRF.

PROTEIN One rationale for manipulation of diet in managing renal failure is the observation that many of the clinical and metabolic disturbances associated with uremia are direct results of accumulated waste products derived from protein catabolism. Excessive dietary protein is catabolized to urea and other nitrogenous compounds normally excreted by the kidneys. Decreased renal function leads to accumulation of these compounds. Endogenous proteins will be degraded if amino acid intake is insufficient to maintain nitrogen balance. One goal of nutritional therapy is to achieve nitrogen balance and limit the accumulation of waste products by proportionally decreasing protein intake as renal function declines. Dietary protein restriction also potentially limits the progression of renal insufficiency in dogs and cats. Avoidance of excess dietary protein can also improve acid-base balance.

A veterinary therapeutic food with 11 to 12% of calories as high biologic value protein (14.5 to 15% DM protein) is recommended for most dogs with CRF and 19 to 21% of calories as high biologic value protein (28 to 30% DM protein) for most cats with CRF. Even lower levels may be necessary for dogs with advanced CRF. The nutritional objective of feeding lower protein levels is to obviate uremic signs. Patients should be monitored for signs of protein insufficiency.

FATTY ACIDS Several studies have addressed the effect of fat and fatty acids on progression of renal failure. Potential mechanisms include alterations in renal eicosanoid production and plasma lipid concentration. Thus, dietary fat could influence progression of renal failure by changing platelet aggregation, fibrinolytic activity, immunologic responses, lipid peroxidation, blood pressure and renal hemodynamics.

Optimal n-6 to n-3 fatty acid ratios in foods for dogs with CRF are <3:1. Cats are unable to elongate 18-carbon fatty acids; therefore, only 20-carbon fatty acids and longer are important when calculating n-6 to n-3 fatty acid ratios for this species.

ENERGY Energy requirements of patients with renal failure are presumed to be similar to those of normal animals. Energy intake tends to decline as renal function declines because of progressive anorexia. Dietary protein and energy are interrelated. When considering the level of dietary protein intake, it is important to assess the adequacy of energy intake. Numerous factors (e.g., gender, environment, activity) influence the energy requirement. The starting point for estimating daily energy requirement (DER) for an individual patient is to calculate the resting energy requirement (RER) and multiply this number by a factor that varies based on the severity of chronic metabolic disease. The formula for calculating RER in kcal/day is $70 \times (BW_{kg})^{0.75}$. The recommended range for most canine patients with CRF is 1.1 to 1.6 x RER. The range for most feline patients with CRF is 1.1 to 1.4 x RER. Food dose should be adjusted to maintain optimal body weight and condition.

POTASSIUM Normally, the majority of ingested potassium is excreted in the urine with the remainder excreted in the feces. In CRF, the proportion excreted in the feces increases. Potassium excretion per nephron also increases and the amount of potassium excreted approaches and may even exceed the filtered load of potassium. In CRF, aldosterone and the rate of fluid flow in the distal nephron mediate the increased potassium secretion in the distal tubule. In general, these mechanisms maintain plasma potassium concentration within the reference limits until renal function is severely reduced. It is possible that low aldosterone and renin levels could produce hyperkalemia earlier in the course of CRF. Plasma potassium concentrations might also increase because of redistribution of potassium between intracellular and extracellular compartments as acidosis develops. Intracellular potassium leaves the cells and is replaced by hydrogen and sodium.

The secretory rate for potassium may be maximal at very low glomerular filtration rates (GFR) to maintain homeostasis. Thus with advanced renal failure, oliguria, sudden increases in potassium intake and sudden worsening of metabolic acidosis or catabolism can produce severe hyperkalemia. Severe hyperkalemia can develop in advanced renal failure despite the fact that total body potassium might be markedly reduced.

Cats with renal failure appear to be particularly predisposed to disorders in potassium homeostasis. Decreased dietary intake due to anorexia or vomiting and increased urinary losses due to polyuria can contribute to hypokalemia in renal failure.

Potassium depletion leads to functional and morphologic changes in the kidneys of dogs and cats. Functional changes include reduction in GFR and reduced urine concentrating ability.

The potassium requirement for cats is proportional to the protein content of the food. Using purified diets, 0.3% potassium was required for growth in kittens fed a 33% protein diet; however, 0.5% potassium was required with a 68% protein diet. Acidifying foods and chronic metabolic acidosis may contribute to hypokalemia.

An intake of 0.08 to 0.12 g of potassium/100 kcal metabolizable energy (ME) (0.30 to 0.50% DM potassium) is recommended for most dogs with CRF. In cats, the objective is to maintain serum potassium concentrations above 4.0 mEq/l. This is generally achieved by feeding a food that contains approximately 0.18 g of potassium/100 kcal ME (0.80 to 1.20% DM potassium). Oral supplementation with potassium gluconate should be considered if diet alone does not maintain serum potassium concentration above 4.0 mEq/l.

Potassium supplementation is indicated whenever hypokalemia is documented. Oral administration is safest and is the preferred route unless a critical emergency exists or if oral administration is impossible or contraindicated. Oral potassium gluconate appears to be tolerated well. The initial recommended dose for potassium gluconate is 2 to 6 mEq/cat/day, depending on the size of the cat and severity of clinical signs. The potassium gluconate dose should be adjusted based on clinical response and serial analyses of serum potassium concentration. During the initial treatment, serum potassium concentration should be checked every two to four days. Later, serum potassium level should be checked every two to four weeks. Routine potassium supplementation of all cats with renal disease, regardless of serum potassium concentration, is not recommended. Potassium supplementation does not have a significant effect on muscle potassium levels.

ACID LOAD Approximately 80% of renal failure cats in a retrospective clinical series had metabolic acidosis based on decreased levels of plasma bicarbonate and blood pH. In human patients, severe acidemia due to renal failure generally does not occur until GFR is roughly less than 20% of normal. Plasma bicarbonate levels are frequently decreased when GFR is above 20% of normal. Even at equivalent levels of GFR, the degree of acidosis is highly variable. Variability is due to differences in the nature of the underlying renal disease, diet, extracellular fluid volume status, potassium status and degree of respiratory compensation.

Dietary acid is derived from several sources. Sulfuric acid is formed when sulfur-containing amino acids (i.e., methionine and cysteine) are oxidized to sulfate. In general, animal-source proteins are higher in sulfur-containing amino acids than are plant-source proteins. Exogenous and endogenous proteins are equally important; foods deficient in calories that result in protein catabolism increase hydrogen ion production. Urinary urea production and total urinary hydrogen ion excretion are directly proportional.

Organic acids are produced from intermediary metabolism products formed from partial oxidation of carbohydrates, fats, proteins and nucleic acids. The organic acids generated contribute to net acid production when conjugate bases are excreted in the urine as negatively charged organic anions. If complete oxidation of organic acids occurs, the hydrogen ions are reclaimed and eliminated as water with carbon dioxide.

Phosphoric acid can be ingested in the food or it can be produced endogenously. Phosphoric acid is used in cat foods as a palatability enhancer, either separately or as a component of topically applied animal digests. Phosphoric acid can be derived from hydrolysis of phosphate esters in protein and nucleic acid providing ingredients, if they are not neutralized by mineral cations (e.g., sodium, potassium and magnesium). The contribution of dietary phosphate to acid production depends on the type of protein ingested. Some proteins generate phosphoric acid, whereas others generate only neutral phosphate salts.

Hydrochloric acid is generated when positively charged cationic amino acids (e.g., lysine and arginine) are broken down into neutral products.

Mineral salts vary in their effect on urinary pH. Differences in the absorption of the cation and anion portion of the salt are important. Intestinal absorption of calcium and magnesium is relatively low; however, absorption of accompanying anions can be high and influence urinary pH. Nonmetabolizable anions (e.g., chloride, phosphate and sulfate) absorbed in excess of their accompanying cations are acidifying, whereas oxides and carbonates are alkalinizing.

In general, protein metabolism is the major source of hydrogen ions. Consequently, avoiding excess dietary protein and decreasing endogenous protein catabolism for energy contribute markedly to the maintenance of acid-base balance.

The plasma bicarbonate concentration should be maintained between 17 and 22 mEq/l. If the plasma bicarbonate concentration is not within this target range two to four weeks after changing to a food that avoids excess acid load, additional restriction of acid load should be considered. Finally, if dietary change alone does not achieve the targeted bicarbonate level, alkali therapy should be considered.

SODIUM/CHLORIDE As renal function deteriorates, fractional sodium excretion increases to maintain external sodium balance and preserve extracellular fluid volume. When dietary sodium intake changes, the fractional excretion of sodium must change markedly to maintain sodium balance. Patients with decreased renal function can only vary sodium excretion over a limited range, which narrows progressively as GFR declines. Thus, renal failure patients cannot tolerate excessively high or low dietary sodium intake levels. If excessive sodium is ingested, sodium retention with expansion of extracellular fluid volume can occur and produce or worsen pre-existing hypertension, fluid overload and edema. If sodium intake is inadequate, negative sodium balance develops with resultant declines in extracellular fluid volume, plasma volume and GFR.

The incidence of hypertension in dogs with renal disease has been estimated to be between 58 and 93%. This wide range may be due to differences in types of renal disease, method of measurement or variable criteria for hypertension. The mechanism for hypertension in renal parenchymal disease is unknown. It has been postulated that reduced intrarenal blood flow activates the renin-angiotensin-aldosterone system, which leads to chronic expansion of the extracellular fluid and elevations in blood pressure. Other possible mechanisms include secondary renal hyperparathyroidism and reduced levels of renal vasodilators such as prostaglandins. The fact that blood pressure in dogs with renal failure decreases when a sodium-restricted food is fed supports chronic volume expansion as a causative mechanism.

The kidney has a dual role in hypertension. Renal disease may cause hypertension, and the kidneys may suffer the consequences of uncontrolled hypertension. The mechanism by which hypertension damages the kidney is not completely understood.

A number of recent studies have examined the interaction of dietary sodium with other ions, including chloride. The full expression of sodium chloride-sensitive hypertension in people is dependent on the concomitant administration of both sodium and chloride. In experimental models using rodents with sodium chloride-sensitive hypertension and in clinical studies with small numbers of hypertensive people, blood pressure or blood volume was not increased by a high dietary sodium intake provided with anions other than chloride. Furthermore, high chloride intake without sodium has less effect on blood pressure than does sodium chloride intake. The failure of nonchloride sodium salts to produce hypertension or hypervolemia may be related to their failure to expand plasma volume because the renal tubular signal for renin release is responsive to renal tubular chloride. Chloride may also act as a direct renal vasoconstrictor. These findings suggest that both sodium and chloride are nutrients of concern in patients with hypertension and CRF.

A sodium intake between 0.1 and 0.25% DM is recommended for most dogs with CRF. The recommended daily intake for most cats with CRF is 10 to 40 mg/BW$_{kg}$/24 hours or 0.2 to 0.35% DM. Although the chloride requirement of dogs and cats has not been established, a chloride level at least 1.5 times the sodium requirement is often recommended. Some animals may have obligatory urinary sodium losses; therefore, abruptly changing these animals to a sodium-restricted food may result in dangerous contraction of the extracellular fluid volume.

Other Nutrients of Concern *(Pages 587-588)*

VITAMIN A Hypervitaminosis A is a common finding in human renal failure patients. Serum creatinine concentrations correlate positively with plasma vitamin A concentrations. The major factor causing high plasma vitamin A concentrations is increased serum concentrations of retinol-binding protein in renal failure. It is unknown whether similar changes occur in dogs or cats with renal failure. Exact data are not available to support a recommendation for vitamin A levels for patients with CRF.

VITAMIN D Calcitriol (1,25-dihydroxyvitamin D) plays an important role in the pathogenesis of secondary renal hyperparathyroidism. Animals with severe CRF have decreased circulating levels of 1,25-dihydroxyvitamin D because of decreased synthesis by the kidney.

Avoiding excess dietary phosphorus and using phosphate binders reduce the inhibitory effects of hyperphosphatemia on renal 1-α-hydroxylase activity, thereby increasing calcitriol production by tubular cells. Oral administration of very low doses of calcitriol (1.7 to 3.4 ng/BW_{kg}) reportedly normalizes serum parathyroid hormone levels and slows progression of naturally occurring renal disease in dogs and cats.

B VITAMINS There is limited information concerning vitamin nutrition in animals with CRF. Animals with renal failure are at risk for B-vitamin deficiency because of decreased appetite, vomiting, diarrhea and polyuria. Human renal failure patients apparently are especially prone to pyridoxine and folate deficiency. Thiamin and niacin deficiency may contribute to anorexia associated with renal failure. Empirical administration of vitamins seems appropriate in anorectic renal failure patients. However, care must be taken not to give excessive amounts of fat-soluble vitamins.

CALCIUM Hypercalcemia can hasten the progression of renal disease and cause calcification of blood vessels and soft tissues. Calcium-phosphorus ratios in foods for chronic renal failure patients should typically be 1:1 to 1.2:1.

TRACE MINERALS Presumably, CRF alters metabolism of trace minerals. For example, nutrients such as copper and zinc that are highly bound to protein may be lost with severe proteinuria. Aluminum may accumulate in renal failure patients treated with aluminum-containing phosphate binders. Aluminum toxicity can cause metabolic bone disease, encephalopathy and anemia. However, exact data are not available to support making a routine recommendation for trace mineral levels for patients with CRF.

SOLUBLE FIBER Soluble fiber causes bacterial proliferation in the large intestine. Bacterial growth requires a source of nitrogen. Although dietary protein provides some nitrogen, blood urea is the largest and most available source of nitrogen for bacterial protein synthesis in the large intestine. Urea is the major end product of protein catabolism in mammals. When blood urea diffuses into the large bowel, it is broken down by bacterial ureases and used for bacterial protein synthesis. These bacterial proteins are excreted in the feces. The net effect is increased fecal urea excretion, reduced serum urea nitrogen concentration and reduced urinary urea excretion.

Assess the Food (Pages 588-589)

The food for animals with renal disease should be evaluated for all the key nutritional factors mentioned above. Levels of the nutrients of concern (i.e., phosphorus, protein, fatty acids [n-6/n-3 ratio]), energy, potassium, sodium and chloride) and other specific food factors (i.e., acid load) in the current food should be compared with the levels appropriate for renal failure patients. (See the Key Nutritional Factors–Renal Disease table and text above.) A new food should be selected if discrepancies exist between the recommended levels of key nutritional factors and those in the current food.

The protein content of the food can be estimated from the information provided by the guaranteed analysis or typical analysis on the commercial pet food label. Some commercial pet food labels also include information about the energy density of the food, although this is not required and is actually prohibited in some countries. However, the caloric density can be predicted from the fat content, which is listed on the pet food label. Levels of phosphorus, sodium, chloride, potassium, n-3 fatty acids and n-6/n-3 fatty acid ratios are not required on the guaranteed or typical analysis of pet food labels and are usually not listed. Levels of these nutrients and the expected acid load that is supplied by the food must be obtained by contacting the manufacturer or consulting published product information. Levels of phosphorus, protein, sodium, chloride and potassium in commercial pet foods intended for normal animals greatly exceed minimum requirements and are often excessive for renal failure patients. (See Appendix 2 and Small Animal Clinical Nutrition, 4th ed., pp 1073-1083.)

Other sources of nutrients include commercial treats and snacks for pets and human foods offered as snacks or part of the regular diet. Commercial pet treats and processed human foods are often high in sodium, phosphorus and other minerals. (See Appendix 3 and Small Animal Clinical Nutrition, 4th ed., pp 1084-1091.)

The caloric intake should be determined for patients with obesity or cachexia. A diary maintained by the client is helpful in documenting what types and quantities of foods and supplements are being offered and eaten by the animal. This caloric intake can be compared with the number of calories usually needed to maintain the animal's ideal body weight and condition.

Assess the Feeding Method (Page 589)

Changing the feeding method in the management of CRF may not be necessary, especially in patients with uncomplicated renal disease or early renal failure. It is important, however, to verify that an appropriate feeding method is being used. Items to consider include access to water, amount fed, how food is offered, access to other foods and who feeds the animal. The amount fed is probably appropriate if the animal has a normal body condition score (BCS 3/5). Animals with uremia and other signs of systemic disease may be partially or completely anorectic and require alternate feeding methods.

KIDNEY

FEEDING PLAN (Pages 590-592)

Anorexia, vomiting and diarrhea may be prominent in patients with moderate to severe renal disease and evidence of systemic illness (uremia). These patients should receive aggressive fluid and electrolyte therapy in an attempt to ameliorate the azotemia, uremia, electrolyte abnormalities and acidosis before initiating a traditional feeding plan.

Select a Food (Pages 590-592)

Typical pet foods contain excessive amounts of phosphorus, protein, sodium, chloride and acid load for patients with renal failure. **Tables 19-5** through **19-8** list nutrient profiles of selected veterinary therapeutic foods commonly used in patients with renal failure. These tables also provide average nutrient levels found in other types of foods.

Avoiding Excess Dietary Protein (Pages 590-591)

Although it has not been clearly established at what point during progressive renal insufficiency protein restriction should be initiated, it seems logical to feed a food with reduced protein levels when renal dysfunction is documented. In practice, documentation might consist of persistent renal proteinuria, loss of urine concentrating ability or azotemia. Protein restriction is more likely to be accepted by the patient and owner if it is initiated before the onset of severe gastrointestinal signs and anorexia.

Although uncommon, the potential problem of protein deficiency due to prolonged dietary protein restriction should be considered. The protein status of patients should be periodically evaluated. (See Reassessment below.) The ratio of serum urea nitrogen to serum creatinine can be used to crudely assess the adequacy of protein restriction. This ratio will be increased if excess protein is being consumed. However, increased tissue catabolism and degradation of endogenous protein can also increase serum urea nitrogen levels. Inadequate energy intake, metabolic acidosis, concurrent infections and corticosteroid administration may also alter the ratio. Protein deficiency may be manifested by worsening anemia, hypoalbuminemia, loss of muscle mass and/or a dry, unthrifty coat.

A common mistake is to insist that an owner feed only a protein-restricted food even if caloric intake is inadequate. Although avoiding excess dietary protein and minerals is important in renal failure patients, offering only such a food should not be imposed to the detriment of overall nutrient intake. Changing to a different commercial food or homemade food may be a more beneficial solution for some patients. Appetite may be cyclical in patients with advanced renal failure, both in respect to overall appetite and food preferences. A dedicated owner is often required and a trial and error approach must be used with different foods and feeding methods.

Table 19-5. Levels of selected nutrients in typical moist commercial dog foods (average analyses) and moist foods for dogs with renal disease.* See the Key Nutritional Factors—Renal Disease table for information about energy.

	Prot	Phos	N-6/ N-3	Na**	K	Acid load***
Recommended levels	**≤15**	**0.15-0.3**	**<3:1**	**<0.25**	**0.3-0.5**	**Reduced**
Grocery brand foods (avg. 31 products)†	41.7	1.39	na	0.87	1.15	na
Specialty brand foods (avg. 40 products)†	31.8	0.92	na	0.52	0.88	na
Hill's Prescription Diet Canine g/d	18.9	0.41	3.9:1	0.22	0.78	Normal
Hill's Prescription Diet Canine k/d	14.7	0.22	2.2:1	0.18	0.40	Reduced
Hill's Prescription Diet Canine u/d	11.5	0.14	3.8:1	0.25	0.39	Reduced
Hill's Prescription Diet Canine w/d	18.2	0.54	12.0:1	0.27	0.62	na
Leo Specific Renil CKW	18.8	0.38	na	0.15	0.94	na
Leo Specific Uremil CUW	13.1	0.35	na	0.14	0.96	na
Medi-Cal Canine Reduced Protein	17.3	0.33	na	0.24	na	na
Purina CNM Canine NF-Formula	16.5	0.30	7.0:1	0.24	0.72	na
Select Care Canine Modified Formula	16.8	0.35	na	0.24	0.96	na
Waltham/Pedigree Low Protein	17.4	0.18	na	0.36	0.30	na
Waltham/Pedigree Medium Protein	27.6	0.19	na	0.31	0.77	na

Key: Prot = protein, Phos = phosphorus, n-6/n-3 = n-6 fatty acid to n-3 fatty acid ratio, Na = sodium, K = potassium, na = information not available from manufacturer.
*This list represents products with the largest market share and for which published information is available.
Manufacturers' published values; expressed as % dry matter.
**Chloride is also a key nutritional factor; chloride levels are typically 1.5 times sodium levels.
***Maintain plasma bicarbonate levels between 17 and 22 mEq/l.
†Averages for grocery and specialty foods were obtained from Appendix 2.

Control of Mineral Imbalances (Pages 591-592)

The keys to prevention and treatment of renal secondary hyper-parathyroidism and renal osteodystrophy are: 1) avoiding excess dietary phosphorus, 2) using phosphate binders and 3) providing supplemental calcium. Controlling hyperphosphatemia can reduce secondary renal

Table 19-6. Levels of selected nutrients in typical dry commercial dog foods (average analyses) and dry foods for dogs with renal disease.* See the Key Nutritional Factors—Renal Disease table for information about energy.

	Prot	Phos	N-6/ N-3	Na**	K	Acid load***
Recommended levels	≤15	0.15-0.3	<3:1	<0.25	0.3-0.5	Reduced
Grocery brand foods (avg. 32 products)†	25.0	1.02	na	0.42	0.64	na
Specialty brand foods (avg. 93 products)†	28.0	1.00	na	0.41	0.72	na
Hill's Prescription Diet Canine g/d	18.7	0.41	3.5:1	0.18	0.63	Normal
Hill's Prescription Diet Canine k/d	14.5	0.24	1.9:1	0.23	0.66	Reduced
Hill's Prescription Diet Canine u/d	9.4	0.19	4.3:1	0.24	0.63	Reduced
Hill's Prescription Diet Canine w/d	18.9	0.58	11.5:1	0.22	0.68	na
Iams Eukanuba Kidney Formula-Early Stage	20.6	0.43	5.0:1	0.52	0.70	Reduced
Iams Eukanuba Kidney Formula-Advanced	15.1	0.25	5.0:1	0.50	0.58	Reduced
Leo Specific Renil CKD	14.7	0.33	na	0.14	0.99	na
Medi-Cal Canine Reduced Protein	14.9	0.27	na	0.28	na	na
Purina CNM Canine NF-Formula	15.9	0.29	9.4:1	0.22	0.86	na
Select Care Canine Modified Formula	14.4	0.34	na	0.28	0.88	na
Waltham/Pedigree Low Phos Low Protein	17.8	0.18	na	0.26	0.78	na
Waltham/Pedigree Low Phos Med Protein	24.4	0.19	na	0.28	0.62	na

Key: Prot = protein, Phos = phosphorus, n-6/n-3 = n-6 fatty acid to n-3 fatty acid ratio, Na = sodium, K = potassium, na = information not available from manufacturer.
*This list represents products with the largest market share and for which published information is available.
Manufacturers' published values; expressed as % dry matter.
**Chloride is also a key nutritional factor; chloride levels are typically 1.5 times sodium levels.
***Maintain plasma bicarbonate levels between 17 and 22 mEq/l.
†Averages for grocery and specialty foods were obtained from Appendix 2.

hyperparathyroidism. The nutritional goal is to normalize fasting serum phosphorus concentrations. Limiting dietary phosphate intake and intestinal absorption of phosphate controls hyperphosphatemia.

Table 19-7. Levels of selected nutrients in typical moist commercial cat foods (average analyses) and moist foods for cats with renal disease.[*] See the Key Nutritional Factors—Renal Disease table for information about energy.

	Prot	Phos	Na[**]	K	Acid load[***]
Recommended levels	≤30	0.4-0.6	<0.35	0.8-1.2	Reduced
Grocery brand foods (avg. 31 products)[†]	51.5	1.54	0.90	1.04	na
Specialty brand foods (avg. 36 products)[†]	45.8	0.97	0.51	0.92	na
Hill's Prescription Diet Feline g/d	35.1	0.54	0.74	0.29	Normal
Hill's Prescription Diet Feline k/d	29.5	0.39	1.05	0.32	Reduced
Leo Specific Renil FUW	34.9	0.60	0.28	0.83	na
Medi-Cal Feline Reduced Protein	35.3	0.59	0.27	na	na
Purina CNM Feline NF-Formula	31.1	0.52	0.16	0.96	Normal
Select Care Modified Formula	35.0	0.49	0.23	1.00	na
Waltham/Whiskas Low Phos Low Protein	34.5	0.39	0.33	0.99	na

Key: Prot = protein, Phos = phosphorus, Na = sodium, K = potassium, na = information not available from manufacturer.
*This list represents products with the largest market share and for which published information is available.
Manufacturers' published values; expressed as % dry matter.
**Chloride is also a key nutritional factor; chloride levels are typically 1.5 times sodium levels.
***Maintain plasma bicarbonate levels between 17 and 22 mEq/l.
†Averages for grocery and specialty foods were obtained from Appendix 2.

Reducing dietary intake alone may be sufficient to control hyperphosphatemia in mild to moderate CRF. Typical commercial dog foods contain 1 to 2% DM phosphorus. Veterinary therapeutic foods designed for canine renal failure patients contain 0.11 to 0.30% DM phosphorus. Typical commercial cat foods contain about 1 to 1.5% DM phosphorus. Veterinary therapeutic foods designed for feline renal failure patients contain approximately 0.40 to 0.60% DM phosphorus. Because protein-providing ingredients in pet foods tend to be high in minerals, avoiding excess dietary protein usually limits dietary phosphate intake. High-phosphorus human foods (e.g., milk, milk products, cheese, fish, beef liver, chocolate, nuts and legumes) should be avoided or limited.

Oral administration of phosphate binders with each meal is indicated if hyperphosphatemia persists despite feeding a food that avoids excess phos-

Table 19-8. Levels of selected nutrients in typical dry commercial cat foods (average analyses) and dry foods for cats with renal disease.* See the Key Nutritional Factors—Renal Disease table for information about energy.

	Prot	Phos	Na**	K	Acid load***
Recommended levels	≤30	0.4-0.6	<0.35	0.8-1.2	Reduced
Grocery brand foods (avg. 25 products)†	35.1	1.18	0.42	0.76	na
Specialty brand foods (avg. 43 products)†	35.3	0.95	0.46	0.69	na
Hill's Prescription Diet Feline g/d	33.4	0.55	0.34	0.75	Normal
Hill's Prescription Diet Feline k/d	28.2	0.46	0.25	0.76	Reduced
Iams Eukanuba Multi-Stage Renal	30.4	0.54	0.49	0.71	na
Purina CNM Feline NF-Formula	30.8	0.41	0.20	0.88	Normal
Select Care Modified Formula	28.3	0.52	0.27	0.92	na
Waltham/Whiskas Low Phos Low Protein	26.1	0.34	0.22	0.89	na

Key: Prot = protein, Phos = phosphorus, Na = sodium, K = potassium, na = information not available from manufacturer.
*This list represents products with the largest market share and for which published information is available.
Manufacturers' published values; expressed as % dry matter.
**Chloride is also a key nutritional factor; chloride levels are typically 1.5 times sodium levels.
***Maintain plasma bicarbonate levels between 17 and 22 mEq/l.
†Averages for grocery and specialty foods were obtained from Appendix 2.

phate. Initially, serum phosphorus concentrations should be monitored every two to three weeks until levels stabilize within the laboratory reference range.

Serum phosphorus concentrations tend to decrease slowly at first. Administering phosphate binders without reducing dietary intake of phosphorus is ineffective. Aluminum carbonate administered at doses between 1,500 and 2,500 mg does not consistently correct hyperphosphatemia in dogs consuming foods containing more than 1% DM phosphorus. See Small Animal Clinical Nutrition, 4th ed., pp 591-592 for more information about controlling mineral imbalances in CRF patients.

Prevention/Treatment of Metabolic Acidosis (Page 592)

Metabolic acidosis occurs when the balance between the addition of hydrogen ions to body fluids and their excretion by the kidney is disrupted. Ingestion of dietary acids can add hydrogen ions to body fluids. Failure to excrete hydrogen ions in the required quantities can occur with decreased kidney function. In a retrospective clinical series of feline renal failure, 80% of cats reviewed were acidemic. Veterinary therapeutic foods used in patients with renal failure should be formulated to reduce the dietary acid load (**Tables 19-5** through **19-8**).

Determine a Feeding Method (Pages 592-593)

Changing to a new food should take place gradually over several days to weeks. A gradual transition improves acceptance and decreases the likelihood of problems in those animals that cannot rapidly adjust urinary sodium levels because of their renal dysfunction. Dogs with CRF usually tolerate a dietary change over seven to 10 days, whereas cats with renal failure may need one to four weeks or longer to make a successful transition to a new food.

Occasionally, some patients may resist a change to a commercial food with reduced protein and mineral levels. This may occur because of: 1) advanced illness associated with renal failure, 2) established feeding habits of older patients and their owners, 3) anorexia associated with concurrent drug administration and 4) the "all or nothing" approach to feeding, rather than transitioning to the new food over several days to weeks. Changing the eating habits of most dogs and cats is relatively easy, but changing the feeding habits and preconceptions of some pet owners and veterinarians is often much more difficult (e.g., the idea that low-protein food is always unpalatable).

When switching to a veterinary therapeutic food, the patient should usually be maintained on a familiar form of food (e.g., moist, dry or combination). These foods can be made more palatable by warming them and adding water or flavoring agents such as tuna juice, clam juice, chicken broth, low-sodium soup, garlic, brewer's yeast or sweeteners (dogs only) such as honey or syrup.

The food dose should be changed if the energy density of the new food differs from the energy density of the previous food. The total daily food dose is usually divided into two or more meals per day. The food dose and feeding method should be altered if body condition is not normal (i.e., BCS < or >3/5).

Owner compliance and pet acceptance of the food must be adequate for clinical nutrition to be effective. Knowing who feeds the patient is important for compliance, and limiting the patient's access to other foods improves acceptability. It is usually necessary to limit a patient's access to another animal's food (e.g., a dog having access to cat food). Uremic patients with anorexia and nausea should be offered small quantities of food several times per day. Feeding location and presentation are important. Timid animals should be fed in a quiet place. Cats should be fed away from loud, persistent barking. Food should be offered in wide bowls to avoid stimulation of tactile whiskers. Placing small quantities of palatable food in a patient's mouth or on its paws to stimulate licking or swallowing (i.e., hand feeding) may facilitate eating. Licking and swallowing in turn may activate neural and hormonal responses associated with normal feeding cues. Pieces of dry veterinary therapeutic foods (**Tables 19-6** and **19-8**) can be offered as treats.

Food aversion is possible if a nauseated animal is force-fed or if a painful or unpleasant experience is associated with feeding. Disagreeable medications (e.g., phosphate binders) should not be mixed with veterinary therapeutic foods. Managing underlying abnormalities in fluid, electrolyte and acid-base balance will help minimize nausea and vomiting. Pharmacologic agents (e.g., cimetidine, ranitidine, metoclopramide, sucralfate) can be used to limit uremic gastritis, nausea and vomiting. Veterinary therapeutic foods intended for long-term management of renal failure should not be offered during periods of nausea and vomiting to prevent possible food aversions. Consider using an appropriate, alternative food temporarily during hospitalization for dogs and cats with uremic crises.

Recombinant human erythropoietin (r-HuEPO, Epogen, Amgen Inc., Thousand Oaks, CA, USA) has been used successfully to improve clinical well being of cats with renal failure. The most consistent and conspicuous change in renal failure cats treated with r-HuEPO is an increased appetite that may precede improvement in hematocrit values and other clinical parameters. Some cats will only eat their veterinary therapeutic food readily after treatment with r-HuEPO. This finding reinforces the necessity to manage anemia as a part of the comprehensive approach to CRF management.

REASSESSMENT (Pages 593-594)

Hospitalized patients with renal failure should be monitored for the following abnormalities that could be acutely life-threatening: 1) severe dehydration, 2) severe metabolic acidosis, 3) severe hypokalemia or hyperkalemia and 4) severe hypocalcemia. Other abnormalities that are less likely to be life threatening in the short-term include: 1) severe azotemia, 2) severe hyperphosphatemia, 3) anemia and 4) hypercalcemia.

Frequency of reassessment depends on the stage of renal failure. Patients with azotemia should be rechecked every two to three months and uremic patients should be rechecked as often as every two to four weeks. Parameters included in the reassessment are listed in **Table 19-9**.

A food that avoids excess sodium chloride is recommended for hypertensive dogs and cats. A few animals with renal dysfunction may adapt slowly to changes in dietary sodium chloride intake. Therefore, the amount of the new food with reduced sodium chloride levels should be increased gradually over several days (dogs) to weeks (cats) while the

Table 19-9. Reassessment methods for patients with chronic renal failure.

Physical examination
Abdominal palpation (size and contour of kidneys, presence of ascites)
Blood pressure measurement
Body condition/muscle mass
Body weight
Coat quality
Fundic examination (retinal hemorrhage, detached retina suggests hypertension)
Hydration status
Oral examination (uremic ulcers, ammoniacal odor)
Laboratory evaluation
Serum albumin
Serum creatinine
Serum electrolytes (calcium, potassium, chloride, sodium, magnesium)
Serum urea nitrogen
Total serum carbon dioxide or venous blood gases (blood pH, bicarbonate, base excess) to evaluate acid-base status
Urinalysis
   Microscopic sediment exam (pyuria or bacteriuria may indicate UTI)
   Urine specific gravity (crude index of tubulointerstitial function)
   Dipstick (protein evaluation is a screening test for glomerular involvement)
   pH (very crude index of acid-base status)
Imaging
Abdominal radiographs (assess kidney shape and size, reference L_2 vertebra)
Intravenous urogram (assess obstruction due to nephroliths, pyelonephritis, etc.)
Ultrasound (assess kidney and prostate gland, determine presence of hydronephrosis or hydroureter)

amount of the old food is decreased. In general, if blood pressure does not normalize, the next step is to administer a diuretic. However, care should be exercised not to cause dehydration, which may further decrease renal function. Antihypertensive drugs should be instituted only if it is possible to serially monitor blood pressure. Indirect methods (i.e., oscillometric and ultrasonic) are adequate for clinical assessment of blood pressure in dogs and cats. All modifications to the antihypertensive regimen should be based on measured changes in blood pressure after a minimum of 14 days.

Nutrient-Drug Interactions (Pages 593-594)

In general, catabolic drugs (e.g., glucocorticoids, tetracyclines and antineoplastic agents) should be avoided in patients with renal failure. If the animal is volume-depleted, nonsteroidal antiinflammatory agents should also be avoided. Urine acidifiers (e.g., ammonium chloride and methionine) will exacerbate metabolic acidosis. Similarly, urinary antiseptics (e.g., methenamine and nalidixic acid) should be avoided.

Aminoglycoside (Gentamicin) Toxicity (Page 594)

Feeding a food containing 27% DM protein to dogs before and during administration of toxic levels of gentamicin reduces nephrotoxicosis. In studies, the renoprotective effect was due to more rapid clearance of gentamicin and a larger volume of distribution in the 27% DM protein group than in the 14% or 9.4% DM protein groups. Therefore, feeding foods with moderate protein levels before and during gentamicin administration may decrease the risk of nephrotoxicity in patients without pre-existing renal dysfunction.

Erythropoietin-Iron Interaction (Page 594)

The dose of erythropoietin needed to achieve an adequate hematocrit response in adult people with chronic renal insufficiency may depend on the severity of secondary hyperparathyroidism and the extent of bone marrow fibrosis. Repletion of iron stores is necessary for the optimal hematologic response to erythropoietin. Commercial pet foods are usually replete with iron but homemade foods may need to be supplemented. An adequate dialysis prescription is required in human patients for optimal response to exogenous erythropoietin.

Angiotensin-Converting Enzyme Inhibitors (Page 594)

Angiotensin-converting enzyme (ACE) inhibitors block angiotensin II production, which results in dilatation of efferent arterioles with a subsequent decline in glomerular transcapillary pressure. In dogs with unilateral nephrectomies and experimentally induced diabetes mellitus, ACE inhibition with lisinopril reduced glomerular transcapillary hydraulic pressure, glomerular cell hypertrophy and proteinuria.

Treatment of Samoyeds affected by familial x-linked nephritis with enalapril improved renal function and prolonged survival. The treated dogs survived a mean of 261 days, whereas the untreated control dogs survived 197 days.

Treatment with ACE inhibitors probably decreases proteinuria and slows progression of renal disease by several mechanisms. ACE inhibitors may decrease glomerular capillary hypertension, which may be a major cause of the progressive decline in renal function observed in naturally occurring renal disease. The use of ACE inhibitors in people, rats and dogs has been associated with a decreased decline of GFR, decreased glomerular cell hypertrophy and decreased glomerulosclerosis. Enalapril prevents the loss of glomerular heparin sulfate that can occur with glomerular disease. Heparin sulfate contributes to the negative charge that is a component of the permselectivity of the glomerular membrane. ACE inhibitors may also decrease proteinuria by decreasing the size of glomerular capillary cell pores. Reducing proteinuria may have a renoprotective effect. Long-term enalapril treatment in aging mice decreases renal interstitial fibrosis.

ACE inhibitors should be used with caution in dogs with renal failure. These drugs can cause an abrupt decline in renal function if systemic hypotension develops or if angiotensin II is critical for the maintenance of glomerular pressures (e.g., in dogs with concurrent congestive renal failure).

Heart Failure and Renal Function (Page 594)

Therapy for CHF in dogs frequently includes the use of sodium chloride-restricted foods and diuretics to reduce preload and venous congestion and vasodilators to reduce preload, afterload and venous congestion. Renal insufficiency has been demonstrated in up to one-third of human patients with severe CHF receiving a salt-restricted diet, ACE inhibitors and furosemide. Decreased renal function, in this situation, has been attributed to loss of angiotensin II-mediated vasoconstriction, which maintains renal perfusion and glomerular filtration when cardiac output is reduced. ACE inhibition in conjunction with decreased renal function and potassium supplementation has been incriminated as a cause of hyperkalemia.

Electrolyte concentrations and renal function were evaluated in normal dogs and dogs with CHF receiving captopril, furosemide and a sodium-restricted food. Four of 10 dogs with CHF treated with captopril, furosemide, sodium-restricted food and digoxin became azotemic at some time during the study period. Dogs with evidence of pre-existing renal dysfunction (e.g., isosthenuria) appear to be at increased risk for developing azotemia. Dogs receiving a combination of diuretics, vasodilators, cardiac glycosides and sodium-restricted foods should be monitored carefully.

Clinical cases that illustrate and reinforce the nutritional concepts presented in this chapter can be found in Small Animal Clinical Nutrition, 4th ed., pp 598-604.

CHAPTER 20

Canine Urolithiasis

For a review of the unabridged chapter, see Osborne CA, Bartges JW, Lulich JP, et al. Canine Urolithiasis. In: Hand MS, Thatcher CD, Remillard RL, et al, eds. Small Animal Clinical Nutrition, 4th ed. Topeka, KS: Mark Morris Institute, 2000; 605-688.

CLINICAL IMPORTANCE (Pages 605-608)*

Clinical signs of urolithiasis may be the first indication of underlying systemic disorders, or defects in the structure or function of the urinary tract (**Table 20-1**). Uroliths may pass through various parts of the excretory pathway of the urinary tract, they may dissolve, they may become inactive or they may continue to form and grow. If uroliths associated with clinical signs are allowed to remain untreated, they may result in more serious sequelae. Uroliths frequently recur if risk factors associated with their formation are not corrected, despite urolith removal by voiding, dissolution protocols or surgery.

Urolithiasis should not be viewed as a single disease, but rather as a sequela of one or more underlying abnormalities. The fact that urolith formation is often erratic and unpredictable indicates that several interrelated complex physiologic and pathologic factors are involved. Therefore, detection of uroliths is only the beginning of the diagnostic process. Determination of urolith composition narrows etiologic possibilities. Knowledge of the patient's diet and serum and urine concentrations of calculogenic minerals, crystallization promoters, crystallization inhibitors and their interactions aids in the diagnosis, treatment and prevention of urolithiasis.

ASSESS THE ANIMAL (Pages 613-625)

See Small Animal Clinical Nutrition, 4th ed., pp 608-612 for a discussion of factors that influence urolith formation and growth.

History and Physical Examination (Page 613)

The history of dogs with urolithiasis is dependent on: 1) anatomic location(s) of uroliths, 2) duration of uroliths in specific location(s), 3) physical characteristics of uroliths (size, shape, number), 4) secondary urinary

*Page numbers in headings refer to Small Animal Clinical Nutrition, 4th ed., where additional information may be found.

Key Nutritional Factors–Canine Urolithiasis (Page 607)

Factors	Dietary recommendations
Ammonium urate/other purine uroliths (dissolution and prevention)	
Water	Water intake should be encouraged to achieve urine specific gravity <1.020
Protein	Avoid excess dietary protein Restrict dietary protein to 10 to 18% dry matter (DM)
Urinary pH	Feed a food that maintains an alkaline urine (urinary pH = 7.0 to 7.5)
Calcium oxalate uroliths (prevention)	
Water	Water intake should be encouraged to achieve urine specific gravity <1.020
Protein	Avoid excess dietary protein Restrict dietary protein to 10 to 18% DM
Calcium	Avoid excess dietary calcium Restrict dietary calcium to 0.3 to 0.6% DM
Sodium	Avoid excess dietary sodium Restrict dietary sodium to <0.3% DM
Magnesium	Avoid excess or deficient dietary magnesium (0.04 to 0.15% DM)
Oxalate	Avoid foods high in oxalates (See **Table 20-15**.)
Vitamin D	Avoid excess dietary vitamin D Restrict dietary vitamin D to 500 to 2,500 IU/kg Avoid using supplements that contain vitamin D
Urinary pH	Feed a food that maintains a urinary pH of 6.8 to 7.2
Calcium phosphate uroliths (prevention)	
Water	Water intake should be encouraged to achieve urine specific gravity <1.020
Calcium	Avoid excess dietary calcium Restrict dietary calcium to 0.3 to 0.6% DM
Protein	Avoid excess dietary protein Restrict dietary protein to 10 to 18% DM
Phosphorus	Avoid excess dietary phosphorus Restrict dietary phosphorus to <0.6% DM
Sodium	Avoid excess dietary sodium Restrict dietary sodium to <0.3% DM
Vitamin D	Avoid excess dietary vitamin D Restrict dietary vitamin D to 500 to 2,500 IU/kg

Key Nutritional Factors–Canine Urolithiasis (Page 607) (Continued.)

Factors	Dietary recommendations
Cystine uroliths (dissolution and prevention)	
Water	Water intake should be encouraged to achieve urine specific gravity <1.020
Protein	Avoid excess dietary protein
	Restrict dietary protein to 10 to 18% DM
Urinary pH	Feed a food that maintains an alkaline urine (urinary pH = 7.1 to 7.7)
Struvite uroliths	
Water	Water intake should be encouraged to achieve urine specific gravity <1.020
Protein	Avoid excess dietary protein
	Dissolution: restrict dietary protein to <8% DM
	Prevention: restrict dietary protein to <25% DM
Phosphorus	Avoid excess dietary phosphorus
	Dissolution: restrict dietary phosphorus to 0.1% DM
	Prevention: restrict dietary phosphorus to <0.6% DM
Magnesium	Avoid excess dietary magnesium
	Dissolution: restrict dietary magnesium to 0.02% DM
	Prevention: restrict dietary magnesium to 0.04 to 0.1% DM
Urinary pH	Feed a food that maintains an acid urine
	Dissolution: urinary pH = 5.9 to 6.1
	Prevention: urinary pH = 6.2 to 6.4

K-9 UROLITH

tract infection (UTI) and virulence of infecting organism(s) and 5) presence of concomitant diseases in the urinary tract and other body systems. After a diagnosis of urolithiasis has been confirmed, the history and physical examination should focus on detection of any underlying illness that may predispose the dog to urolith formation.

A dietary history should also be obtained for all patients with urolithiasis, with the objective of identifying risk factors that predispose the patient to specific mineral types. Likewise, owners should be questioned about vitamin-mineral supplements, previous illnesses and medications that may predispose the patient to various types of uroliths.

Signs typical of lower urinary tract disease include dysuria, pollakiuria, hematuria, urge incontinence, paradoxical incontinence and voiding small uroliths during micturition. Signs of uremia may occur if urine flow has

Table 20-1. Clinical importance of urolithiasis.

First evidence of an underlying systemic disorder
Hypercalcemia
  Calcium oxalate uroliths
  Calcium phosphate uroliths
Cushing's syndrome
  Calcium oxalate uroliths
  Calcium phosphate uroliths
  Struvite uroliths
Defects in purine metabolism
  Portal vascular anomalies
    Ammonium urate uroliths
  Enzyme defects
    Xanthine uroliths
First evidence of an underlying urinary tract disorder
Renal tubular transport defect
  Cystinuria
    Cystine uroliths
  Renal tubular acidosis
    Calcium oxalate uroliths
    Calcium phosphate uroliths
Defect in local host defenses against urease-producing microbes
  Struvite uroliths
Foreign bodies in urinary tract
  Suture material
    Usually struvite uroliths
  Catheters
    Usually struvite uroliths
Sequelae to urolithiasis
Dysuria, pollakiuria, urge incontinence
Secondary microbial urinary tract infection
Partial or total obstruction to urine outflow
  Bacterial urinary tract infection that may progress
  Impaired renal function and postrenal azotemia
  Rupture of the outflow tract
    Uroperitoneum
    Inflammation of tissues adjacent to various portions of the urinary tract
Formation of inflammatory bladder polyps

been obstructed for a sufficient period, or if there is extravasation of urine into the peritoneal cavity due to rupture of the excretory pathways.

Signs of upper tract disease include painless hematuria, polyuria if sufficient nephrons have impaired function and abdominal pain if there is overdistention of the renal pelvis with urine due to outflow obstruction (**Table 20-2**). Many patients with uroliths have no clinical signs. Absence of signs is especially common in patients with nephroliths.

If gross hematuria is present, determining when during the process of micturition it is most severe may be of value in localizing its source. If hematuria occurs throughout micturition, lesions (including uroliths) may be present in the kidneys, ureters, urinary bladder, prostate gland and/or urethra. If hematuria occurs primarily at the end of micturition, lesions of the ventral bladder wall or intermittent renal hematuria should be suspected. If hematuria occurs at the beginning or is independent of micturition, lesions in the urethra or genital tract should be suspected.

Digital palpation of the entire urethra, including evaluation by rectal examination, may reveal urethroliths or uroliths lodged in the bladder neck. A firm, nonyielding mass may be palpated in the urinary bladder if a solitary urolith is present; a grating sensation confined to the bladder may be detected if multiple uroliths are present. It may be impossible to palpate small or solitary urocystoliths if the bladder wall is contracted and/or thickened due to inflammation. Likewise, it may be impossible to palpate uroliths in a distended or overdistended bladder. In this situation, the bladder should be repalpated after urine has been eliminated by voiding, manual compression of the bladder, cystocentesis or catheterization. One should suspect urethral uroliths when urethral catheters cannot be advanced into the bladder. However, inability to advance a catheter through the urethra may also be associated with urethral strictures or space-occupying lesions that partially or totally occlude the urethral lumen.

In the absence of infection or outflow obstruction, abnormalities are usually not associated with renoliths unless bilateral renoliths are associated with sufficient renal damage to cause uremia. If infection or obstruction is present, there may be pain in the area of the kidneys and/or palpable enlargement of the affected kidney(s). Concomitant bacterial pyelonephritis may be associated with polysystemic signs due to sepsis.

Diagnostic Studies (Pages 613-625)
Urinalysis (Pages 613-616)

Results of urinalysis are usually characterized by abnormalities typical of inflammation (pyuria, proteinuria, hematuria and increased numbers of epithelial cells), which may or may not be associated with infection. Whereas urease-producing microbes (staphylococci, *Proteus* spp, ureaplasmas) may cause infection-induced struvite uroliths to form, opportunistic bacteria that are not calculogenic (e.g., *Escherichia coli* and streptococci) may colonize the urinary tract as a result of urolith-induced alterations in local host defenses. Quantitative urine culture of all patients with uroliths is recommended because knowledge of bacterial type is important in predicting the mineral composition of uroliths, and in selecting an appropriate antimicrobial agent for treatment.

Table 20-2. Clinical signs of uroliths that may be associated with urinary system dysfunction.

Urethroliths
Asymptomatic
Dysuria, pollakiuria and urge incontinence
Gross hematuria
Palpable urethral uroliths
Spontaneous voiding of small uroliths
Partial or complete urine outflow obstruction
   Overflow incontinence
   Anuria
   Palpation of an overdistended and painful urinary bladder
   Urinary bladder rupture, abdominal distention and abdominal pain
   Signs of postrenal azotemia (anorexia, depression, vomiting and diarrhea)
Signs associated with concurrent urocystoliths, ureteroliths and/or renoliths

Urocystoliths
Asymptomatic
Dysuria, pollakiuria and urge incontinence
Gross hematuria
Palpable bladder uroliths
Palpably thickened urinary bladder wall
Partial or complete urine outflow obstruction of bladder neck
   (See Urethroliths.)
Other signs associated with concurrent urethroliths, ureteroliths and/or
   renoliths

Ureteroliths
Asymptomatic
Gross hematuria
Constant abdominal pain
Unilateral or bilateral urine outflow obstruction
   Palpably enlarged kidney(s)
   Signs of postrenal azotemia (See Urethroliths.)
May have other signs associated with concurrent urethroliths, urocystoliths
   and/or renoliths

Renoliths
Asymptomatic
Gross hematuria
Constant abdominal pain
Signs of systemic illness if generalized renal infection is present (anorexia,
   depression, fever and polyuria)
Palpably enlarged kidney(s)
Signs of postrenal azotemia (See Urethroliths.)
Other signs associated with concurrent urethroliths, urocystoliths and/or
   ureteroliths

The pH of urine obtained from patients with uroliths is variable; however, it may become persistently alkaline if secondary infection with urease-producing bacteria occurs. The significance of a single urinary pH measurement should be interpreted with appropriate caution because there are significant fluctuations throughout the day, especially with respect to the time(s), amount and types of food consumption. In general, magnesium ammonium phosphate and calcium phosphate uroliths are associated with alkaline urine, whereas ammonium urate, sodium urate, uric acid, calcium oxalate, cystine and silica uroliths are associated with acidic urine.

The advent of effective medical protocols to dissolve and prevent uroliths in dogs and cats has resulted in renewed interest in detection and interpretation of crystalluria. Evaluation of urine crystals may aid in: 1) detection of disorders predisposing animals to urolith formation, 2) estimation of the mineral composition of uroliths and 3) evaluation of the effectiveness of medical protocols initiated to dissolve or prevent uroliths.

Crystals form only in urine that is or recently has been supersaturated with crystallogenic substances. Therefore, crystalluria represents a risk factor for urolithiasis. However, detection of urine crystals is not synonymous with urolithiasis and clinical signs associated with uroliths. Nor are urine crystals irrefutable evidence of a urolith-forming tendency. For example, crystalluria that occurs in individuals with anatomically and functionally normal urinary tracts is usually harmless because the crystals are eliminated before they aggregate or grow to sufficient size to interfere with normal urinary function. In addition, crystals that form after elimination or removal of urine from the patient often are of no clinical importance. Identification of crystals that have formed in vitro does not justify therapy.

On the other hand, detection of some types of crystals (e.g., cystine and ammonium urate) in clinically asymptomatic patients, frequent detection of large aggregates of crystals (e.g., calcium oxalate or magnesium ammonium phosphate) in apparently normal individuals or detection of any form of crystals in fresh urine collected from patients with confirmed urolithiasis may be of diagnostic, prognostic and therapeutic importance. Large crystals and aggregates of crystals are more likely to be retained in the urinary tract, and therefore may be of greater clinical significance than small or single crystals.

Although there is not a direct relationship between crystalluria and urolithiasis, detection of crystals in urine is proof that the urine sample is oversaturated with crystallogenic substances. However, oversaturation may occur as a result of in vitro events in addition to or instead of in vivo events. Therefore, care must be used not to overinterpret the significance of crystalluria. In vivo variables that influence crystalluria include: 1) the concentration of crystallogenic substances in urine (which in turn is influenced by their rate of excretion and the volume of water in which they are excreted), 2) urinary pH, 3) the solubility of crystallogenic substances and 4) excretion of diagnostic agents (e.g., radiopaque contrast media) and medications (e.g., sulfonamides).

In vitro variables that influence crystalluria include: 1) temperature, 2) evaporation, 3) urinary pH and 4) the technique of specimen preparation (e.g., centrifugation vs. non-centrifugation and volume of urine examined) and preservation. When knowledge of in vivo urine crystal type is especially important, fresh, warm specimens should be serially examined. The number, size and structure of crystals should be evaluated, as well as their tendency to aggregate.

Urinary pH influences the formation and persistence of several types of crystals. Therefore, it is often useful to consider pH when interpreting crystalluria. Different crystals tend to form and persist in certain urinary pH ranges, although there are exceptions.

Refrigeration is an excellent method to preserve many physical, chemical and morphologic properties of urine sediment.

Crystalluria may also be influenced by diet, including water intake. Dietary influence on crystalluria is of diagnostic importance because urine crystal formation that occurs while patients are consuming hospital foods may be dissimilar to urine crystal formation that occurs when patients are consuming foods fed at home.

Microscopic evaluation of urine crystals should not be used as the sole criterion to predict the mineral composition of macroliths in patients with confirmed urolithiasis.

Only quantitative analysis can provide definitive information about the mineral composition of the entire urolith. However, interpretation of crystalluria in light of other clinical findings often allows the clinician to tentatively identify the mineral composition of uroliths, especially their outermost layers. Subsequent reduction or elimination of crystals by therapy provides a useful index of the efficacy of medical protocols designed to dissolve or prevent uroliths.

Radiography and Ultrasonography *(Page 616)*

The primary objective of radiographic or ultrasonographic evaluation of patients suspected of having uroliths is to determine the site(s), number, density and shape of uroliths. However, the size and number of uroliths are not a reliable index of the probable efficacy of therapy. After urolithiasis has been confirmed, radiographic or ultrasonographic evaluation also aids in detection of predisposing abnormalities.

Very small uroliths (<3 mm in diameter) may not be visualized by survey radiography or ultrasonography. Uroliths greater than 1 mm in diameter can usually be detected by double-contrast cystography provided excessive contrast medium is not used.

Hematology and Serum Chemistry (Page 617)

Hemograms of dogs with uroliths are usually normal unless there is concomitant generalized infection of the kidneys or prostate gland associated with leukocytosis. Microcytosis, anemia, target cells and leukocytosis have occasionally been associated with portal vascular anomalies in dogs with and without urate uroliths.

Serum chemistry values are usually normal in patients with infection-induced magnesium ammonium phosphate, cystine and silica uroliths unless obstruction of urine outflow or generalized renal infection leads to changes characteristic of renal failure. Although most patients with calcium oxalate and calcium phosphate uroliths are normocalcemic, some are hypercalcemic.

A variety of biochemical alterations may exist in patients with urate urolithiasis. The following changes may be observed in patients with urate uroliths due to congenital or acquired hepatic disorders: 1) decreased urea nitrogen concentrations, 2) decreased total protein and albumin concentrations, 3) abnormal bile acid concentrations, 4) increased concentrations of total bilirubin and fasting blood ammonia and 5) increased serum alanine aminotransferase and serum alkaline phosphatase enzyme activities. Dogs with portal vascular anomalies typically have reduced hepatic functional mass and altered portal blood flow evidenced by abnormally elevated bile acid concentrations, prolonged sulfobromophthalein (BSP) retention times and abnormal ammonia tolerance tests.

Urine Chemistry (Page 617)

Detection of the underlying causes of specific types of urolithiasis is often linked to evaluation of the biochemical composition of urine. For best results, at least one and preferably two consecutive 24-hour urine samples should be collected because determination of fractional excretion of many metabolites in "spot" urine samples does not accurately reflect 24-hour metabolite excretion.

Laboratory results may be markedly affected by changes in foods fed in a home environment vs. different foods fed in a hospital environment. For example, urinary excretion of potentially calculogenic metabolites while animals consume foods fed in the hospital may be different from those excreted by animals eating at home. To determine the influence of home-fed foods on laboratory test results, consider asking clients to provide home-fed foods for use during periods of diagnostic hospitalization.

Urolith Analysis (Pages 617-625)

Small uroliths in the urinary bladder or urethra are commonly voided during micturition by female dogs and occasionally by male dogs. Uroliths with a smooth surface (e.g., those composed of ammonium urate or calcium oxalate monohydrate) are more likely to pass through the urethra than uroliths with a rough surface (e.g., those composed of calcium oxalate dihydrate or silica). Commercially manufactured tropical fish nets designed for household aquariums facilitate retrieval of uroliths during voiding. Urocystoliths may also be obtained by catheter or voiding urohydropropulsion. For more information, see Small Animal Clinical Nutrition, 4th ed., pp 617-620.

COLLECTION AND QUANTITATIVE ANALYSIS OF URINE CRYSTALS
If available data do not indicate the probable mineral composition of uroliths and if uroliths cannot be retrieved with the aid of a urethral catheter, consider preparing a large pellet of urine crystals by centrifugation of urine in a conical-tip centrifuge tube. The quantity of crystalline sediment available for analysis may be increased by repeatedly removing the supernatant after centrifugation, adding additional uncentrifuged urine to the tube containing sediment and again centrifuging the preparation. If the conditions that caused urolith formation are still present, evaluation of the pellet formed from crystalline sediment by quantitative methods designed for urolith analysis may provide meaningful information about the mineral composition of a patient's uroliths. However, crystals identified by this method may only reflect the outer portions of compound uroliths.

QUANTITATIVE ANALYSIS OF UROLITHS The location, number, size, shape, color and consistency of uroliths removed from the urinary tract should be recorded. All uroliths should be saved in a container (preferably a sterile one) and submitted for analysis. Do not give uroliths to owners before analysis. If multiple uroliths are present, one may be placed into a container of 10% buffered formalin for demineralization and microscopic examination. However, formalin should not be used to preserve uroliths for mineral analysis because formalin may alter the results.

Because many uroliths contain two or more mineral components, it is important to examine representative portions. The mineral composition of crystalline nuclei may be identical or different from outer layers of uroliths. Uroliths should not be broken before submission for analysis because the central core may be distorted or lost.

Routine analysis of uroliths by qualitative methods of chemical analysis is not recommended. The major disadvantage of this procedure is that only some of the chemical radicals and ions can be detected. In contrast to chemical methods of analysis, physical methods (quantitative methods such as x-ray diffractometry) have proved to be far superior in identification of crystalline substances.

UROLITH CULTURE Bacterial culture of the interior of uroliths is indicated if: 1) urine obtained from the patient has not been previously cultured, 2) culture of urine obtained from patients suspected of having struvite uroliths yields no growth or 3) the patient has a UTI with bacteria that do not produce urease. Bacteria harbored inside uroliths are not always the same as those present in urine. Bacteria detected within uroliths probably represent those present at the time they were formed, and may serve as a source of recurrent UTI. Bacteria may remain viable within uroliths for long periods.

GUESSTIMATION OF UROLITH COMPOSITION Formulation of effective medical protocols for urolith dissolution is dependent on knowledge of the mineral composition of uroliths. Because a variety of different types of uroliths and nephroliths occur in dogs, the veterinarian should use a protocol that facilitates determination of the mineral composition of uroliths based on probability (**Table 20-3**).

Attempts to induce dissolution of uroliths may be hampered if the uroliths are heterogeneous in composition. This has not been a significant problem in dogs with uroliths composed primarily of magnesium ammonium phosphate with small quantities of calcium apatite because the solubility characteristics of the two minerals are similar. Veterinarians have encountered difficulty in dissolving uroliths composed primarily of struvite with an outer shell composed primarily of calcium apatite. Difficulty will also be encountered in attempting to induce complete dissolution of a urolith with a nucleus of calcium oxalate, calcium phosphate, ammonium urate or silica and a shell of struvite because the solubility characteristics of this combination of minerals are dissimilar. These phenomena should be considered if medical therapy seems to be ineffective after initially reducing the size of uroliths. For more information see Principles of Urolith Treatment and Prevention, Small Animal Clinical Nutrition, 4th ed., pp 622-623.

AMMONIUM URATE AND OTHER PURINE UROLITHS
(Pages 625-636)

Prevalence and Mineral Composition (Pages 625-626)

Purine uroliths (ammonium urate, sodium urate, calcium urate, uric acid and xanthine) accounted for 8% of all canine uroliths and 12% of all canine nephroliths analyzed at the Minnesota Urolith Center from 1981 to 1997. All dogs with xanthine uroliths had a history of treatment with allopurinol. Information about breed, gender and age predisposition, serum and urine profiles, crystal appearance and urate urolith radiographic density and con-

Table 20-3. Predicting mineral composition of common canine uroliths.

Mineral types	Predictors								
	Urinary pH	Crystal appearance	Urine culture	Radio-graphic density	Radio-graphic contour	Serum abnormalities	Breed predis-position	Gender predis-position	Common ages
Magnesium ammonium phosphate	Neutral to alkaline	Three- to eight-sided colorless prisms	Urease-producing bacteria (staphyl-ococci, *Proteus* spp, *Ureaplasma* spp)	1+ to 4+ (sometimes laminated)	Smooth, round or faceted May assume shape of renal pelvis, ureter, bladder or urethra	None	Miniature schnauzer, miniature poodle, bichon frise, cocker spaniel	Female (>80%)	2 to 9 years
Calcium oxalate	Acidic to neutral	Colorless envelope or octahedral shape (dihy-drate salt) Spindles or dumbbell shape (monohy-drate salt)	Negative	2+ to 4+	Rough or spiculated (dihydrate salt) Small, smooth, round (mono-hydrate salt) Sometimes jackstone	Usually normocalcemic, occasionally hypercalcemic	Miniature schnauzer, standard schnauzer, Lhasa apso, Yorkshire terrier, miniature poodle, Shi Tzu, bichon frise	Males (>70%)	5 to 12 years
Urate	Acidic to neutral	Yellow-brown amorphous shapes (ammonium urate)	Negative	0 to 2+	Smooth (occasionally irregular), round or oval	Low serum urea nitrogen and albumin values in dogs with hepatic porto-systemic shunts	Dalmatian, English bulldog, miniature schnauzer, Yorkshire terrier, Shi Tzu	Males (>90%)	1 to 5 years

(Continued on next page.)

K-9 UROLITH

Table 20-3. Predicting mineral composition of common canine uroliths (Continued.).

Mineral types	Predictors									
	Urinary pH	Crystal appearance	Urine culture	Radio-graphic density	Radio-graphic contour	Serum abnormalities	Breed predis-position	Gender predis-position	Common ages
Calcium phosphate	Alkaline to neutral (brushite forms in acidic urine)	Amorphous or long thin prisms	Negative	2+ to 4+	Smooth or irregular, round or faceted	Occasionally hypercalcemic	Yorkshire terrier, miniature schnauzer, Shi Tzu	Males (>55%)	<1 year, 6 to 10 years
Cystine	Acidic to neutral	Flat colorless hexagonal plates	Negative	1+ to 2+	Smooth (occasionally irregular), round or oval	None	English bulldog, dachshund, basset hound, Newfoundland	Males (>98%)	1 to 7 years
Silica	Acidic to neutral	None observed	Negative	2+ to 3+	Round center with radial spoke-like projections (jackstone)	None	German shepherd, golden retriever, Labrador retriever, miniature schnauzer, cavalier King Charles spaniel	Males (>95%)	3 to 10 years

Table 20-4. Common characteristics of canine purine uroliths.

Chemical names	Formulas
Ammonium acid urate	$C_5H_3N_4O_3NH_4•H_2O$
Sodium acid urate	$C_5H_3N_4O_3Na•H_2O$
Uric acid	$C_5H_4N_4O_3•2H_2O$
Xanthine	$C_5H_4N_4O_2$

Variations in mineral composition
Ammonium acid urate only
Sodium acid urate only
Uric acid only
Xanthine only
Ammonium urate mixed with variable quantities of sodium urate, magnesium ammonium phosphate and/or calcium oxalate
Sodium and calcium oxalate
Xanthine and uric acid

Physical characteristics
Color: Light or dark brown, brown-green
Shape: Variable. Usually round or ovoid in urinary bladder, may assume shape of renal pelvis (funnel shaped), may assume jackstone appearance. Usually smooth, occasionally irregular or rough.
Nuclei: Nuclei and concentric laminations are common.
Density: Usually dense and brittle. Radiographically, purine uroliths have marginal radiodensity compared with soft tissue. Some may be radiolucent.
Number: Single or multiple
Location: May be located in kidneys, ureters, urinary bladder (most common) and/or urethra.
Size: Usually small (1 mm to 1 cm in diameter), occasionally large (more than 1 cm)

Prevalence
Approximately 7 to 8% of all canine uroliths. Approximately 12% of canine nephroliths.
May be recurrent

Characteristics of affected canine patients
In Dalmatians, most common in males
Mean age at diagnosis is four years (range <1 to >17 years).
Most commonly observed in Dalmatians, miniature schnauzers, Yorkshire terriers, Shih Tzus and English bulldogs.

tour are listed in **Table 20-3**. Common characteristics of canine purine uroliths are listed in **Table 20-4**.

Etiopathogenesis and Risk Factors (Pages 626-629)

Uric acid is one of several biodegradation products of purine nucleotide metabolism. Uroliths composed of uric acid (anhydrous uric acid, uric acid dihydrate, sodium urate, ammonium urate) or xanthine form because urine is oversaturated with these substances. Ammonium urate (also known as ammonium acid urate and ammonium biurate) is the monobasic ammonium salt of uric acid. It is the most common naturally occurring purine urolith form observed in dogs.

Uric Acid, Sodium Urate, Ammonium Urate *(Page 626)*

Risk factors for urate lithogenesis in dogs include: 1) increased renal excretion and urine concentration of uric acid, 2) increased renal excretion, renal production or microbial urease production of ammonium ions, 3) low urinary pH and 4) presence of promoters or absence of inhibitors of urate urolith formation. Genetic factors may be important because urate uroliths are common in certain breeds of dogs. For example, Dalmatian dogs have an inherent predisposition to forming urate uroliths. Dietary components may promote urate urolith formation in predisposed dogs because dietary purines may be digested, absorbed, incorporated into the body's purine pool and eventually excreted in the urine (**Tables 20-5** and **20-6**). Thus, metabolism of dietary purines may result in oversaturation of urine with urate calculogenic substances. In studies of normal dogs, consumption of high-protein foods was associated with greater urinary uric acid excretion and increased urine saturation with uric acid, sodium urate and ammonium urate, when compared with consumption of low-protein foods.

Xanthine *(Page 626)*

Xanthine is a product of purine metabolism and is converted to uric acid by the enzyme xanthine oxidase. Hereditary xanthinuria is a rarely recognized disorder of people characterized by a deficiency of xanthine oxidase. As a consequence, abnormal quantities of xanthine are excreted in urine as a major end product of purine metabolism. Because xanthine is the least soluble of the purines naturally excreted in urine, xanthinuria may be associated with formation of uroliths. The most common cause of xanthinuria is treatment with allopurinol. However, naturally occurring xanthinuria and xanthine urolithiasis have been reported to occur in cavalier King Charles spaniels.

Dalmatian Dogs *(Pages 626-627)*

Dalmatian dogs are predisposed to urate uroliths because their ability to oxidize uric acid to allantoin is intermediate between that of people and most non-Dalmatian dogs. Compared with non-Dalmatians, Dalmatians convert uric acid to allantoin at a reduced rate.

The proximal renal tubules of Dalmatians reabsorb less uric acid than those of non-Dalmatian dogs; a small amount is secreted by the distal tubules. In non-Dalmatian dogs, 98 to 100% of the uric acid in the glomerular filtrate is reabsorbed by the proximal tubules and returned to the liver for further metabolism.

Table 20-5. Some potential risk factors for canine purine uroliths.

Diet	Urine	Metabolic	Drugs
High purine content (See **Table 20-6**.)	Hyperuricuria	Males	Urine acidifiers
	Hyperammonuria	Breed	Salicylates
Acidifying potential	Acidic pH	Dalmatians	Chemothera-peutic agents (especially 6-mercapto-purine)
	Urine concentration	English bulldogs	
Low moisture content	Urine retention	Miniature schnauzers	
Ascorbic acid?	Urease-producing microbiuria	Yorkshire terriers	
	Increased promoters?	Shih Tzus	
	Decreased inhibitors?	Hyperuricemia	
		Hyperammonuria	
		Hepatic dysfunction	
		Neoplasia with rapid cell destruction	

Table 20-6. Purine content of selected foods.

Foods to avoid (High purine concentration)	Foods to use sparingly (Moderately high purine concentration)	Foods that can be fed (Negligible purine concentration)
Anchovies	Asparagus	Breads (whole grain cereal products)
Brains	Cauliflower	Butter/fats
Clams	Fish[*]	Cheese
Goose	Legumes (beans/peas)	Eggs
Gravies	Lentils	Fruits/fruit juices
Heart	Meats	Gelatin
Kidney	Mushrooms	Milk
Liver	Spinach	Nuts
Mackerel		Refined cereals
Meat extracts including bouillon		Vegetable soups
Mussels		Cream soups
Oysters		Vegetables[**]
Salmon		
Sardines		
Scallops		
Shrimp		
Sweetbreads		
Tuna		
Yeast (baker's/ brewer's)		

[*]Except those listed in the first column.
[**]Except those listed in the second column.

K-9 UROLITH

The definitive mechanism of urate urolith formation in Dalmatian dogs is unknown. Increased urinary excretion of uric acid is a risk factor rather than a primary cause. Urate uroliths occur more commonly in males than females; the average age of dogs when uroliths are diagnosed is 4.5 years. Although all Dalmatian dogs excrete relatively high quantities of uric acid in their urine, apparently only a small percentage form urate uroliths.

Although urate uroliths commonly affect Dalmatian dogs, not all uroliths formed by Dalmatians are composed of ammonium urate. For example, of 2,020 uroliths formed by Dalmatian dogs, 93% were composed of purines (ammonium urate, sodium urate, uric acid and xanthine), 3% were of mixed composition, 1% were struvite, 1% were calcium oxalate, 2% were compound uroliths and less than 1% were cystine.

Non-Dalmatian Dogs *(Pages 627-628)*

Comparatively little is known about urate lithogenesis in non-Dalmatian dogs that do not have portal vascular anomalies. Urate urolithiasis reportedly affects many breeds of dogs. Although urate uroliths are commonly encountered in Dalmatian dogs, approximately 30 to 60% of all canine urate uroliths analyzed by quantitative methods are found in other breeds. Other non-Dalmatian breeds that appear to have a significantly higher incidence of urate urolithiasis based on quantitative urolith analyses are listed in **Table 20-3**.

Urate uroliths from non-Dalmatian dogs are recognized most frequently in males. Urate uroliths have been detected from dogs of all ages; however, they are most frequently detected in dogs three to six years of age.

Regardless of cause, severe hepatic dysfunction may predispose dogs to urate lithogenesis, especially ammonium urate uroliths. Observations and evidence derived from experimental models suggest that prolonged consumption of foods with markedly restricted levels of protein may be associated with formation of urate uroliths in dogs. Biochemical and histologic evaluation of these dogs suggests that long-term consumption of foods severely restricted in protein may induce hepatocellular dysfunction and concomitant hyperuricemia. Hepatic cirrhosis has also been associated with urate uroliths in dogs and other species. However, cirrhosis, foods with severely restricted protein levels and other causes of hepatic dysfunction have been uncommon causes of ammonium urate urolithiasis.

Dogs with Portal Vascular Anomalies *(Page 628)*

A high incidence of ammonium urate uroliths has been observed in dogs with portal vascular anomalies. These uroliths occur in males and females and usually have been detected before dogs reach three years of age.

Direct communication between the portal and systemic vasculature shunts blood around the liver, resulting in severe hepatic atrophy and diminished hepatic function. Hepatic dysfunction in turn is associated with reduced hepatic conversion of uric acid to allantoin, and reduced conversion of ammonia to urea. The predisposition of dogs with portal vascular anomalies to urate urolithiasis is probably associated with concomitant hyperuricemia, hyperammonemia, hyperuricuria and hyperammonuria. However, not all dogs with portal systemic anomalies develop concurrent ammonium urate urolithiasis.

Dietary Risk Factors *(Pages 628-629)*

Concentrations of calculogenic substances in urine are dependent on urine volume. Because commercial dry foods are associated with production of less volume of urine compared with moist foods, consumption of dry foods may be considered a risk factor for urate urolith formation.

Dalmatian dogs consuming foods containing more than 20% protein (dry matter), and having protein sources that are high in purines and purine precursors (**Table 20-6**) are at increased risk for urate lithogenesis. However, purine uroliths have been reported to form in some dogs consuming lesser amounts of dietary purines; therefore, other factors are apparently involved. In dogs with portosystemic shunts, the degree of urine saturation with purines is probably related, at least in part, to the degree of vascular shunting in addition to the dietary protein consumption. Because urate uroliths associated with portal vascular anomalies are often diagnosed in dogs less than one year of age, it is probable that they were consuming foods with increased protein content.

Urine acidity is a risk factor for urate lithogenesis because the solubility of most purines, especially ammonium urate, is pH-dependent. Therefore, consumption of foods that promote aciduria (e.g., high-protein foods or those with other acidifying ingredients) may be a risk factor. Overconsumption of purines or purine precursors increases the risk of urate urolith formation. Therefore, dietary purines should be restricted in animals at risk for forming purine uroliths.

Biologic Behavior *(Page 629)*

Purine uroliths have the potential to undergo spontaneous dissolution, remain active (grow) or become inactive (remain unchanged).

Recurrence of urate uroliths may be influenced by several factors including: 1) persistence of underlying causes, 2) incomplete removal of all uroliths from the urinary tract at the time of surgery, 3) persistence or recurrence of UTIs with urease-producing bacteria or 4) failure to comply with therapeutic or prophylactic recommendations. Frequent recurrence of urate uroliths is not surprising considering the persistence of disorders associated with urate urolithiasis.

A relatively high incidence of recurrence following surgical removal is a unique characteristic of urate urolithiasis in Dalmatian and non-Dalmatian dogs. In several studies using qualitative methods of urolith analysis, recurrence was reported in 33 to 50% of dogs with urate uroliths. In these dogs, uroliths generally recurred within one year after diagnosis and treatment. Recurrence of urate urolithiasis in non-Dalmatian dogs with portal vascular anomalies also appears to be similar. In dogs, recurrence of urolithiasis with uroliths composed of minerals other than those present during the initial episode is uncommon. However, uroliths predominantly composed of minerals other than ammonium urate, sodium urate or uric acid may form in canine patients originally affected with urate uroliths.

K-9 UROLITH

Feeding Plan and Treatment (Pages 629-633)

Current recommendations for medical dissolution of canine ammonium urate uroliths include a combination of: 1) calculolytic foods, 2) administration of xanthine oxidase inhibitors (i.e., allopurinol), 3) alkalinization of urine, 4) eradication or control of UTIs and 5) formation of an increased quantity of less concentrated urine (**Table 20-7**).

Calculolytic Foods (Page 630)

The goal of dietary modification for patients with uric acid or ammonium urate uroliths is to reduce urine concentration of uric acid, ammonium ions and hydrogen ions (**Tables 20-7** and **20-8**). Use a purine-restricted, alkalinizing food that does not contain supplemental sodium (See the Key Nutritional Factors–Canine Urolithiasis table.). Consumption of this food by healthy and urate urolith forming dogs results in marked reductions in urinary uric acid and ammonia excretion.

Xanthine Oxidase Inhibitors (Pages 630-631)

Allopurinol is a synthetic isomer of hypoxanthine. It rapidly binds to and inhibits the action of xanthine oxidase and thereby decreases production of uric acid by inhibiting the conversion of hypoxanthine to xanthine and xanthine to uric acid. The result is a reduction in serum and urine uric acid concentration within approximately two days and a concomitant but lesser increase in the serum concentrations of hypoxanthine and xanthine.

The dosage of allopurinol for dissolution of ammonium urate uroliths in dogs is 15 mg/BW_{kg} q12h. According to the manufacturer, the drug has been given to normal dogs at this dosage for one year without causing significant abnormalities. When owners supplement a diet with foods containing purine precursors, a layer of xanthine can form around ammonium urate uroliths. Therefore, to minimize xanthine formation, allopurinol should only be administered to animals consuming purine-restricted foods.

Table 20-7. Summary of recommendations for medical dissolution and prevention of canine ammonium acid urate uroliths.

1. Perform appropriate diagnostic studies, including complete urinalysis, quantitative urine culture and diagnostic radiography. Determine precise location, size and number of uroliths. The size and number of uroliths are not a reliable index of probable therapeutic efficacy.
2. If uroliths are available, determine their mineral composition. If unavailable, determine their composition by evaluating appropriate clinical data.
3. Consider surgical correction if uroliths obstruct urine outflow. Small urocystoliths may be removed by voiding urohydropropulsion.
4. Determine baseline pretreatment serum uric acid concentrations and (if possible) 24-hour excretion of urine uric acid.
5. Initiate therapy with a low-purine calculolytic food (Prescription Diet Canine u/d*). Other foods or supplements should not be fed to the patient. Reduction in serum urea nitrogen concentration (usually <10 mg/dl) suggests compliance with dietary recommendations.
6. Initiate therapy with allopurinol at a dosage of 30 mg/BW_{kg}/day divided into two equal subdoses (azotemic patients require a lesser dose). Xanthine uroliths may form if foods containing excessive purines are fed or if excessive allopurinol is given.
7. If necessary, administer sodium bicarbonate or potassium citrate orally to eliminate aciduria. Strive for a urinary pH of approximately 7.
8. If necessary, eradicate or control urinary tract infections (UTI) with appropriate antimicrobial agents. Maintain antimicrobial therapy during and for an appropriate period after urate urolith dissolution.
9. Devise a protocol to monitor efficacy of therapy.
   a. Try to avoid diagnostic follow-up studies that require urinary tract catheterization. If they are required, give appropriate peri-catheterization antimicrobial agents to prevent iatrogenic UTI.
   b. Perform serial urinalyses. Determination of urinary pH and specific gravity and microscopic examination of sediment for urate crystals are especially important. Remember, crystals formed in urine stored at room or refrigeration temperatures may represent in vitro artifacts.
   c. Serially evaluate serum uric acid concentrations and (if possible) fractional excretion of urine uric acid.
   d. Evaluate the location(s), number, size, density and shape of uroliths at monthly intervals. Intravenous urography or ultrasonography may be used for radiolucent uroliths located in the kidneys, ureters or urinary bladder. Retrograde contrast urethrocystography may be required for radiolucent uroliths in the bladder and urethra.
   e. If necessary, perform quantitative urine cultures. They are especially important in patients that are infected before therapy and in patients that are catheterized during therapy.
10. Continue the calculolytic food, allopurinol and alkalinizing therapy for approximately one month following the disappearance of uroliths as detected by radiography.

(Continued on next page.)

Table 20-7. Summary of recommendations for medical dissolution and prevention of canine ammonium acid urate uroliths (Continued.).

11. Prevention. Urate uroliths are highly recurrent. Preventive therapy should be directed at minimizing urine concentrations of ammonia and uric acid. This may be achieved by feeding a food low in protein that also promotes an alkaline urine. The effectiveness of dietary management for the prevention of ammonium urate uroliths in dogs with portosystemic shunts is unknown. The long-term use of allopurinol is discouraged because of the potential for development of xanthine uroliths.

*Hill's Pet Nutrition, Inc., Topeka, KS, USA.

K-9 UROLITH

Table 20-8. Levels of key nutritional factors in selected foods for dissolution and prevention of ammonium urate and other purine uroliths in dogs.*

Foods	Protein (% DM)	Purines	Urinary pH**
Recommended levels	**10-18**	**Reduced**	**7.0-7.5**
Hill's Prescription Diet Canine u/d, moist	11.5	Reduced	7.1-7.7
Hill's Prescription Diet Canine u/d, dry	9.3	Reduced	7.1-7.7
Purina CNM NF-Formula, dry	15.9	Reduced	na

Key: na = information not available from manufacturer, DM = dry matter.
*Manufacturers' published values. This list represents products with the largest market share for which published information is available.
**Protocols for measuring urinary pH may vary.

Alkalinization of Urine (Page 631)

Because ammonium ions and hydrogen ions appear to precipitate urates in dog urine, oral administration of alkalinizing agents (i.e., sodium bicarbonate or potassium citrate) may be of value in preventing acid metabolites from increasing renal tubular production of ammonia.

Dosage of urine alkalinizing agents should be individualized for each patient. Preliminary dosages of sodium bicarbonate vary from approximately 25 to 50 mg/BW_{kg} q12h depending on the status of the patient and pretreatment urinary pH values. Alternatively, potassium citrate in wax matrix tablets (Urocit-K, Mission Pharmacal, San Antonio, TX, USA) or as a liquid (Polycitra-K, Willen Drug Co., Baltimore, MD, USA) (40 to 75 mg/BW_{kg} q12h) may be given. Because sodium may combine with uric acid to form sodium urate, potassium citrate may be preferable to sodium bicarbonate as a urine alkalinizing agent. Divided doses should be administered to maintain a consistently nonacidic environment in the urinary tract. The food used for urate dissolution (Prescription Diet Canine u/d, Hill's Pet Nutrition, Inc., Topeka, KS, USA) is formulated to contain potassium citrate, and its consumption typically results in alkaluria in dogs.

The goal of treatment with urine alkalinizing agents or the urate calculolytic food is to maintain a urinary pH of approximately 7.0. Higher values (>7.5) should be avoided until it is determined whether or not they provide a significant risk factor for formation of calcium phosphate uroliths. Deposition of a layer of calcium phosphate crystals around existing urate uroliths may impede urolith dissolution. Owners may monitor urinary pH with pH paper or handheld "pocket" pH meters.

Eradication or Control of UTIs *(Pages 631-632)*

Clinical studies indicate that UTIs in dogs with ammonium urate uroliths usually occur as a consequence of altered local host defenses. These alterations may be caused by urolith-induced trauma to the urothelium, or they may occur as a consequence of catheterization or other invasive diagnostic procedures. Efforts should be made to prevent, eradicate or control infections because they may cause problems of equal or greater severity as the uroliths.

Appropriate antimicrobial agents selected on the basis of susceptibility or minimum inhibitory concentration tests should be used at therapeutic dosages. The fact that diuresis reduces the urine concentration of the antimicrobial agent should be considered when formulating antimicrobial dosages.

Augmenting Urine Volume *(Page 632)*

Augmenting urine volume with the goal of decreasing urine uric acid and ammonium concentrations and enhancing urine flow through the excretory pathway appears to be a logical recommendation. Because the calculolytic food used for urate urolith dissolution impairs urine-concentrating capacity by decreasing renal medullary urea concentration, additional diuretic agents are unnecessary. Because excessive sodium excretion may cause hypercalciuria, excessive dietary sodium should be avoided, particularly if the urinary pH is high. This event may in turn cause calcium phosphate crystals to form. (See Calcium Phosphate Urolithiasis section.)

Dogs Without Portal Vascular Anomalies *(Page 632)*

At the University of Minnesota, 25 dogs with ammonium urate uroliths were treated with dietary and allopurinol therapy. Complete dissolution occurred in nine dogs (36%), partial dissolution occurred in eight dogs (32%) and no dissolution occurred in eight dogs (32%). A similar dissolution protocol in seven dogs with sodium urate uroliths resulted in complete dissolution in two dogs (29%), partial dissolution in three dogs (42%) and no dissolution in two dogs (29%). Inability to dissolve urate uroliths was usually associated with formation of xanthine. In some dogs with partial urolith dissolution, the remaining uroliths were completely retrieved using voiding urohydropropulsion or catheter-assisted retrieval. The mean time of urate urolith dissolution in 11 dogs was 3.5 months (median one month, range one to 18 months). Using the above protocol, a nephrolith presumed to be composed of urate was dissolved in nine months in a six-year-old, neutered female English bulldog.

Dogs with Portal Vascular Anomalies (Pages 632-633)

It is logical to hypothesize that elimination of hyperuricuria and reduction of urine ammonium concentration following surgical correction of anomalous shunts would result in spontaneous dissolution of uroliths composed primarily of ammonium urate. Occasionally, success has been reported in medically dissolving urate uroliths in dogs with portal vascular anomalies. For example, there is a report of dissolution of a urolith presumed to be composed of ammonium urate in a two-year-old female miniature schnauzer with a portal vascular anomaly. The dog was consuming a veterinary therapeutic food designed for treatment of renal failure (Prescription Diet Canine k/d, Hill's Pet Nutrition, Inc., Topeka, KS, USA). The mechanisms involved were presumably decreased production of ammonium ions from urea and reduced formation of uric acid from dietary protein.

Likewise, a renolith in the right renal pelvis of a seven-year-old female malamute with a portal vascular anomaly disappeared while the dog consumed Prescription Diet Canine k/d.

Additional clinical studies are needed to evaluate the relative value of calculolytic foods, allopurinol and/or alkalinization of urine in dissolving ammonium urate uroliths in dogs with portal vascular anomalies. The efficacy of allopurinol may be altered in such dogs because biotransformation of this drug, which has a very short half-life, to oxypurinol, which has a longer half-life, requires adequate hepatic function.

Immature Dogs with Urate Uroliths (Page 633)

Providing safe and effective therapy for urate uroliths in immature dogs presents a challenge. Growing dogs usually consume greater quantities of protein and, thus purines than adult dogs. The safety and efficacy of calculolytic foods in young dogs with urate uroliths are unknown. Adding non-purine-containing protein to the calculolytic food may be effective; however, no studies have yet been performed to confirm this hypothesis. Also, the metabolism of allopurinol in young animals has not been evaluated. Therefore, surgical removal of large uroliths remains the option with the most predictable short-term outcome.

However, the dry formulation of Prescription Diet Canine u/d can be modified for growing dogs (**Table 20-9**). The safety and efficacy of this modified food in young dogs with urate or other uroliths is unknown. Growing dogs should be monitored closely when fed these recipes.

Surgical and Medical Combinations (Page 633)

There are several situations in which a combination of surgical removal of urate uroliths followed by medical dissolution protocols might be beneficial. One involves the inability to remove all uroliths by surgery. This occasionally occurs because ammonium urate uroliths are frequently multiple and

Table 20-9. Modified recipes for growing dogs based on the dry formulation of Prescription Diet Canine u/d.*

Recipe A

1 cup dry Prescription Diet Canine u/d
1 tsp dicalcium phosphate
1 cup cottage cheese
Multivitamin-mineral supplement for dogs

Nutrient levels	Dry matter (%)
Protein	30.5
Fat	19.5
Calcium	1.0
Phosphorus	1.0
Magnesium	0.02
Sodium	0.6
Potassium	0.5

Recipe B

1 cup dry Prescription Diet Canine u/d
3/4 tsp dicalcium phosphate
2 cooked eggs
Multivitamin-mineral supplement for dogs

Nutrient levels	Dry matter (%)
Protein	17.6
Fat	27.1
Calcium	1.1
Phosphorus	1.0
Magnesium	0.02
Sodium	0.4
Potassium	0.6

*Hill's Pet Nutrition, Inc., Topeka, KS, USA.

small. The fact that they may be radiolucent creates an additional problem by interfering with their radiographic detection immediately after surgery.

In some patients, immediate surgery may be required to remove uroliths obstructing the renal pelvis, ureter(s) or urethra. Initiation of medical dissolution protocols may prove advantageous if such patients have multiple uroliths in several locations, and if circumstances preclude their surgical removal at the time the obstructing urolith is removed.

Certain patients with portal vascular anomalies and urate uroliths may also benefit from a combination of surgical and medical urolith dissolution protocols. Techniques have been devised to correct some types of intrahepatic and extrahepatic shunts in dogs. However, the condition of the patient and factors related to anesthesia and surgery may preclude urolith removal at the time the anomalous portal vessels are corrected. In this situation, postsurgical medical therapy designed to dissolve uroliths should be considered. Also, some types of portal vascular anomalies are not amenable to surgical correction. If the uroliths are causing unacceptable signs of urinary tract disease, they should be surgically removed and postsurgical preventive measures should be initiated. Voiding urohydropropulsion may be used to remove small urocystoliths (**Table 20-10**).

Reassessment (Pages 634-635)

Ammonium urate urocystoliths have a propensity to move into the urethra of dogs. This finding may be related to their small size, round to ovoid shape and smooth surface. If small enough, they readily pass through the urethra. However, they often become lodged behind the os penis of male dogs. Owners should be informed of this likelihood and given a written summary of associated clinical findings. Urethroliths causing clinical signs may be easily returned to the bladder lumen by urohydropropulsion. The physical characteristics that promote their passage into the urethra also facilitate their removal from the urethra.

When attempting medical dissolution of urate uroliths, advise owners to only feed low-purine foods. Consumption of a high-purine diet by dogs while receiving allopurinol may result in formation of a xanthine shell around urate uroliths or formation of xanthine uroliths. Xanthine uroliths may not dissolve. However, spontaneous dissolution of xanthine shells and underlying uroliths may occur by discontinuing allopurinol and continuing the low-purine food. Alternatively, dissolution of urate uroliths may occur as a result of a combination of feeding a low-purine food and administering allopurinol at a lower dose to dogs that previously formed xanthine shells.

Because allopurinol and its metabolites are excreted from the body primarily in urine, the drug should be used with appropriate caution in patients with renal dysfunction.

The size of the uroliths should be periodically monitored by survey and (if necessary) double-contrast radiography or ultrasonography (**Table 20-11**).

To prevent iatrogenic UTIs, appropriate prophylactic antibiotics should be administered around the time of urinary catheterization, if retrograde, double-contrast urethrocystography is used to monitor dissolution of radiolucent urethrocystoliths. Excretory urography or ultrasonography should be used to monitor dissolution or recurrence of urate nephroliths.

Urinary pH should be monitored at appropriate intervals (**Table 20-11**). Periodic evaluation of urine sediment for crystalluria should also be considered. Ammonium urate crystals should not form in fresh urine if therapy has been effective in promoting formation of urine that is undersaturated with ammonium ions and uric acid. Periodic evaluation of serum urea nitrogen concentration, serum uric acid concentration and (if possible) urine uric acid concentration is recommended. Reduction of serum urea nitrogen concentration below pretreatment values (usually <10 mg/dl in

Table 20-10. Voiding urohydropropulsion: A nonsurgical technique for removing small urocystoliths.[*]

1. Perform appropriate diagnostic studies, including complete urinalysis, quantitative urine culture and diagnostic radiography. Determine the location, size, surface contour and number of urocystoliths.
2. Anesthetize the patient, if needed.
3. If the urinary bladder is not distended with urine, moderately distend it with a physiologic solution (e.g., saline, Ringer's, etc.) injected through a transurethral catheter. To prevent overdistention, palpate the bladder per abdomen during infusion. Remove the catheter.
4. Position the patient such that the vertebral spine is approximately vertical.
5. Gently agitate the urinary bladder, with the objective of promoting gravitational movement of urocystoliths into the bladder neck.
6. Induce voiding by manually expressing the urinary bladder. Use steady digital pressure rather than an intermittent squeezing motion.
7. Collect urine and uroliths in a cup. Compare urolith number and size to those detected by radiography and submit them for quantitative analysis.
8. If needed, repeat Steps 3 through 7 until the number of uroliths detected by radiography are removed or until uroliths are no longer voided.
9. Perform double-contrast cystography to ensure that no uroliths remain in the urinary bladder. Repeat voiding urohydropropulsion if small urocystoliths remain.
10. Administer prophylactic antimicrobials for three to five days, or longer if needed.
11. Monitor the patient for adverse complications (i.e., hematuria, dysuria, bacterial urinary tract infection and urethral obstruction with uroliths).
12. Formulate appropriate recommendations to minimize urolith recurrence or to manage uroliths remaining in the urinary tract on the basis of quantitative mineral analysis of voided urocystoliths.

*Also see Figure 20-5, Small Animal Clinical Nutrition, 4th ed., p 619.

previously nonazotemic patients), reduction of urine specific gravity (usually <1.020) and an increase in urinary pH (usually >7.0) indicate owner and patient compliance with dietary therapy (**Table 20-11**). Reductions in serum and urine uric acid concentrations also indicate compliance with recommendations for dietary and allopurinol therapy.

There is no rigid time interval after which response to dissolution therapy is unlikely. The fact that current medical protocols are not designed to induce dissolution of urolith matrix may be a factor that influences dissolution rate. The time required to induce dissolution of nine episodes of urate urolithiasis in a clinical study ranged from four to 40 weeks (mean 14.2 weeks). Reevaluation of the diagnosis and/or alternate methods of management should be considered if uroliths enlarge during therapy or do not begin to decrease in size after approximately eight weeks of appropriate medical therapy (**Table 20-12**). If it is difficult to completely dis-

Table 20-11. Expected changes associated with medical therapy of ammonium urate uroliths.

Factors	Pre-therapy	During therapy	Prevention therapy
Polyuria	±	1+ to 3+	1+ to 3+
Pollakiuria	0 to 4+	↑ then ↓	0
Hematuria	0 to 4+	↓	0
Urine specific gravity	Variable	1.004 to 1.015	1.004 to 1.015
Urinary pH	<7.0	>7.0	>7.0
Pyuria	0 to 4+	↓	0
Urate crystals	0 to 4+	0	Variable
Bacteriuria	0 to 4+	0	0
Bacterial culture of urine	0 to 4+	0	0
Urea nitrogen (mg/dl)	Variable	≤15	≤15
Urolith size/number	Small to large	↓	0

K-9 UROLITH

solve urate uroliths by creating urine that is undersaturated with uric acid and ammonium ions, consider that: 1) the wrong mineral component was identified, 2) the nucleus of the urolith is of different mineral composition than the outer portions of the urolith, 3) a xanthine shell or xanthine uroliths have formed or 4) the owner or patient is not complying with therapeutic recommendations.

Prevention of Urate Urolithiasis (Pages 635-636)
Dalmatian Dogs (Page 635)

Prophylactic therapy should be considered for urate-forming Dalmatian dogs because of the high risk for recurrent urate uroliths. As a first choice, foods that are restricted in purines and that promote formation of less concentrated alkaline urine should be considered (**Table 20-8**). In one study of naturally occurring ammonium urate urocystoliths in Dalmatian dogs, a low-protein nonacidifying moist food (Prescription Diet Canine u/d) reduced urolith recurrence by 50% compared with an adult moist maintenance food. If dry foods are fed, water should be added with the goal of maintaining a urine specific gravity less than approximately 1.025.

If urate crystalluria or hyperuricuria persists, serial urinary pH measurements are indicated to ensure appropriate alkalinization. If necessary, urine alkalinizing agents may be added to the protocol. If difficulties persist, low doses of allopurinol (approximately 10 to 20 mg/BW_{kg}/day) may be given cautiously. Prolonged administration of high doses (30 mg/BW_{kg}/day) of allopurinol may result in formation of xanthine uroliths. The risk of xanthine urolithiasis is enhanced if dietary purines are not restricted during allopurinol therapy. Therefore, appropriate caution in long-term administration of this drug is indicated.

Table 20-12. Managing urate uroliths refractory to complete dissolution.

Causes	Identification	Therapeutic goals
Client and patient factors		
Inadequate dietary compliance	Question owner Persistent urate crystalluria Urea nitrogen >10-17 mg/dl Urine specific gravity >1.010-1.020 Urinary pH <7.0-7.5 during treatment with Prescription Diet Canine u/d* (use lower values for canned food)	Emphasize need to exclusively feed dissolution food
Inadequate allopurinol administration	Question owner Count remaining pills	Emphasize need to administer allopurinol Determine if owner is capable and willing to administer medication Demonstrate a variety of methods to administer medication
Clinician factors		
Incorrect prediction of mineral type	Analysis of retrieved urolith	Alter therapy based on identification of mineral type
Excessive allopurinol administration	Xanthine urolith formation	Reduce allopurinol administration in conjunction with appropriate dietary therapy to minimize purine consumption Clinically active uroliths may require surgical removal Remove small uroliths by voiding urohydropropulsion

(Continued on next page.)

Table 20-12. Managing urate uroliths refractory to complete dissolution (Continued.).

Causes	Identification	Therapeutic goals
Disease factors		
Xanthine urolith formation	Analysis of retrieved urolith Allopurinol administration without concomitant reduction in dietary protein consumption Excessive allopurinol dose	Clinically active uroliths may require surgical removal Remove small uroliths by voiding urohydropropulsion
Inadequate hepatic function	Suspect hepatic portosystemic shunts in breeds other than Dalmatians and English bulldogs Elevated postprandial serum bile acid concentration Microhepatica	Clinically active uroliths may require surgical removal Remove small uroliths by voiding urohydropropulsion Repair vascular anomaly
Compound urolith	Radiographic density of nucleus and outer layer(s) of urolith is different Analysis of retrieved urolith	Alter therapy based on identification of a new mineral type Uroliths not causing clinical signs should be monitored for potentially adverse consequences (obstruction, urinary tract infection, etc.) Clinically active uroliths may require surgical removal Remove small uroliths by voiding urohydropropulsion

K-9 UROLITH

*Hill's Pet Nutrition, Inc., Topeka, KS, USA.

Non-Dalmatian Dogs (Page 636)

Preventive measures for Dalmatian dogs should also be used for non-Dalmatian dogs.

There have been few studies of the biologic behavior of ammonium urate uroliths in dogs and cats with portal vascular anomalies. It is logical to hypothesize that elimination of hyperuricuria and reduction of urine ammonium concentration following surgical correction of anomalous shunts would result in spontaneous dissolution of uroliths composed primarily of ammonium urate.

Additional clinical studies are needed to evaluate the relative value of calculolytic foods, allopurinol and/or alkalinization of urine in dissolving ammonium urate uroliths in dogs and cats with portal vascular anomalies. The likelihood of adverse side effects or further deterioration in hepatic function following administration of allopurinol to dogs with portal vascular anomalies has not been determined.

Because tetracycline exacerbates hepatic and renal dysfunction in dogs with experimentally produced portal vascular anomalies, it should not be routinely used to treat UTIs in dogs with naturally occurring portal vascular anomalies.

CALCIUM OXALATE UROLITHS (Pages 636-644)

Prevalence and Mineral Composition (Page 636)

Calcium oxalate accounted for 31% of all canine uroliths and 38% of all canine nephroliths analyzed at the Minnesota Urolith Center from 1981 to 1997. Information about breed, gender, and age predisposition, serum and urine profiles, crystal appearance and calcium oxalate urolith radiographic density and contour are listed in **Table 20-3**. **Table 20-13** describes characteristics of canine calcium oxalate uroliths.

Etiopathogenesis and Risk Factors (Pages 636-637)

In order for uroliths to form, urine must be supersaturated with respect to that crystal system. Therefore, increasing the urine concentration of calcium or oxalic acid promotes calcium oxalate crystal formation.

Dietary Risk Factors (Pages 637-638)

Dietary ingredients that promote hypercalciuria or hyperoxaluria represent nutritional risk factors for calcium oxalate urolith formation (**Tables 20-14** and **20-15**). Therefore, reduction of dietary calcium and oxalate appears to be a logical therapeutic goal; however, it is not necessarily harmless. Reducing consumption of only one of these substances (e.g., calcium) may increase the availability of the other (e.g., oxalic acid) for intestinal absorption and subsequent urinary excretion. In general, reduction in dietary calcium should be accompanied by an appropriate reduction in dietary oxalate.

Dogs with calcium oxalate urolithiasis frequently consume human food. Calcium oxalate is the most common urolith type recognized in people liv-

Table 20-13. Common characteristics of canine calcium oxalate uroliths.

Chemical names	Formulas	Crystal names
Calcium oxalate monohydrate	$CaC_2O_4 \bullet H_2O$	Whewellite
Calcium oxalate dihydrate	$CaC_2O_4 \bullet 2H_2O$	Weddellite

Variations in mineral composition
Calcium oxalate monohydrate only
Calcium oxalate dihydrate only
Combinations of calcium oxalate monohydrate and dihydrate
Calcium oxalate (monohydrate and/or dihydrate) mixed with variable quantities of calcium phosphate. Variable quantities of struvite or ammonium acid urate may also be present.
Calcium oxalate (monohydrate and/or dihydrate) nucleus surrounded by other minerals especially infection-induced struvite

Physical characteristics
Color: Calcium oxalate monohydrate uroliths are usually tan or brown. Calcium oxalate dihydrate uroliths are usually white or cream-colored. Surfaces may be red to black if uroliths are coated with blood.
Shape: Variable. Calcium oxalate monohydrate uroliths are usually round or elliptical and have a smooth, polished surface. On occasion, they may develop a jackstone or mulberry shape. Calcium oxalate dihydrate uroliths and mixed calcium oxalate monohydrate/calcium oxalate dihydrate uroliths are usually round to ovoid and have an irregular surface caused by protrusion of sharp-edged crystals. On occasion, they may develop a jackstone shape.
Nuclei: Radial striations and concentric laminations may occur.
Density: Very dense and brittle. Survey radiographs reveal that calcium-containing uroliths are radiodense compared with soft tissue.
Number: Single or multiple
Location: May be located in renal pelves, ureters, urinary bladder (most common) and/or urethra.
Size: Sub-visual to several centimeters

Prevalence
Approximately 30% of all canine uroliths. More than 38% of canine nephroliths. May be recurrent (more than 50% recur by three years after removal)

Characteristics of affected canine patients
More common in males (57%) than females (43%)
Mean age at diagnosis is about eight years (range <1 to >25 years)
Most commonly observed in miniature schnauzers, miniature poodles, Lhasa apsos, Yorkshire terriers, Shih Tzus and bichon frises

ing in developed countries. As people feed their dogs the same dietary proportions and ingredients they feed themselves, it is logical to assume that dogs would be exposed to the same nutritional risk factors for urolith formation (**Tables 20-15**).

Certain dietary excesses and deficiencies have also been recognized as potential risk factors. Excessive administration of vitamin D, sodium or magnesium promotes hypercalciuria.

Table 20-14. Some potential risk factors for canine calcium oxalate uroliths.

Diet	Urine	Metabolic	Drugs
Acidifying potential	Hypercalciuria	Chronic meta-	Urine
High protein	Hyperoxaluria	bolic acidosis	acidifiers
content	Hypocitraturia?	Males	Furosemide
High sodium	Hypomagnesuria?	Breed	Glucocorti-
content	Hyperuricuria?	Miniature	coids
Excessive calcium	Increased crystal	schnauzers	Sodium
content	promoters	Miniature	chloride
Excessive restriction	Decreased crystal	poodles	Vitamin D
of calcium	inhibitors	Lhasa apsos	Ascorbic acid
Low moisture	Urine concentration	Yorkshire	
content	Urine retention	terriers	
Excessive restriction		Shih Tzus	
of phosphorus		Bichon frises	
Excessive magnesium		Older age	
content		Hypercalcemia	
Excessive magnesium		Glucocorticoid	
restriction		excess	
Excessive vitamin D		Hypophospha-	
content		temia	
Excessive vitamin C		Osteolysis?	
content			
Deficient pyridoxine?			
High oxalate content			

Other Risk Factors (*Page 638*)

Certain clinical conditions also represent potential risk factors for calcium oxalate urolith formation. Hyperparathyroidism, hyperadrenocorticism, hypervitaminosis D, paraneoplastic hypercalcemia and furosemide administration promote hypercalciuria. Intestinal resection, hereditary hyperoxaluria and excessive ascorbic acid administration promote hyperoxaluria. See **Table 20-14** and Small Animal Clinical Nutrition, 4th ed., pp 638-640 for a discussion of these risk factors.

Biologic Behavior (Page 640)

Calcium oxalate uroliths may be voided in the urine or become lodged in any portion of the urinary tract. Uroliths that remain in the urinary tract may continue to grow or become inactive (no further growth). Not all persistent uroliths are associated with clinical signs. Most calcium oxalate uroliths are not associated with UTI. Uroliths composed of the dihydrate salt of calcium oxalate are less likely to cause complete urinary obstruction because their irregular contour prevents the urolith from forming a continuous seal within the lumen of the urethra. However, if uroliths remain in the urinary tract, dysuria, UTI, partial or total urinary obstruction and polyp formation are potential sequelae. Spontaneous dissolution of calcium oxalate uroliths in dogs is unlikely.

In a retrospective clinical survey of 438 dogs surgically treated for urolithiasis at the Animal Medical Center in New York, 111 patients had 155 known recurrences. Recurrence was observed in 25% of dogs with calcium oxalate uroliths. At the University of Minnesota, results of a retrospective study indicated that the rate of recurrence of calcium oxalate uroliths increased with the length of time that dogs were evaluated: 3% recurred after three months, 9% after six months, 36% after one year, 42% after two years and 48% after three years. Owner and patient compliance with therapy and persistence of factors responsible for urolith initiation at the time of urolith eradication influence the tendency for uroliths to recur.

Assess the Food (Page 640)

Patients with calcium oxalate uroliths have usually consumed a variety of foods. Dogs consuming dry commercial foods may be at greater risk for urolithiasis than dogs consuming moist foods because dry foods are often associated with higher urine concentrations of calcium and oxalic acid and more concentrated urine. Likewise, foods with high protein content also contribute to hypercalciuria and hyperoxaluria.

Feeding foods designed to dissolve struvite uroliths provide some benefits, but also some risks to patients with calcium oxalate uroliths (See the Key Nutritional Factors–Canine Urolithiasis table and **Table 20-14**). The lower protein content and potential to enhance formation of less concencentrated urine help reduce calcium and oxalate concentrations in urine. Although formation of acidic urine is desirable for management of struvite uroliths, foods that promote acidic urine promote hypercalciuria and hypocitraturia. Therefore, consumption of foods that result in formation of acidic urine enhances the risk of calcium oxalate urolithiasis in susceptible dogs.

Feeding Plan and Treatment (Pages 640-643)

Though struvite, urate and cystine uroliths dissolve when urine is no longer supersaturated with calculogenic substances, calcium oxalate uroliths in dogs do not. Surgery is the time-honored method to remove calcium oxalate uroliths from the urinary tract; however, complete surgical removal of all visible uroliths may be difficult due to their small size and irregular contour. Small urocystoliths may be aspirated through a transurethral catheter or removed by voiding urohydropropulsion (**Table 20-10**). In some patients, however, calcium oxalate uroliths are clinically silent, obviating the need for intervention. For those patients in which surgery is not indicated, the clinical status of uroliths should be periodically assessed by urinalyses, renal function tests and radiography or ultrasonography.

Table 20-15. Selected human foods to limit or avoid feeding to dogs with calcium oxalate uroliths.

Food categories	Moderate-/high-calcium foods
Meats	Bologna (M)
	Herring (M)
	Oysters (M)
	Salmon (H)
	Sardines (H)
Vegetables	Baked beans (M)
	Broccoli (H)
	Collards (H)
	Lima beans (M)
	Spinach (M)
	Tofu (soybean curd) (M)
Milk and dairy products	Cheese (H)
	Ice cream (H)
	Milk (H)
	Yogurt (H)
Breads, grains, nuts	Brazil nuts (M)
Miscellaneous	Cocoa (M)
	Hot chocolate (M)

Food categories	Moderate-/high-oxalate foods
Meats	Sardines (M)
Vegetables	Asparagus (M)
	Broccoli (M)
	Celery (H)
	Corn (M)
	Cucumber (H)
	Eggplant (H)
	Green beans (H)
	Green peppers (H)
	Lettuce (M)
	Spinach (H)
	Summer squash (H)
	Sweet potatoes (H)
	Tomatoes (M)
Fruits	Apples (H)
	Apricots (H)
	Cherries (M)
	Citrus peel (H)
	Oranges (M)
	Peaches (M)
	Pears (M)
	Pineapple (M)
	Tangerine (H)

(Continued on next page.)

Table 20-15. Selected human foods to limit or avoid feeding to dogs with calcium oxalate uroliths (Continued.).

Food categories	Moderate-/high-oxalate foods
Breads, grains, nuts	Cornbread (M)
	Fruitcake (H)
	Grits (H)
	Peanuts (H)
	Pecans (H)
	Soybeans (H)
	Wheat germ (H)
Miscellaneous	Beer (H)
	Chocolate (H)
	Cocoa (H)
	Coffee (M)
	Tea (H)
	Tomato soup (H)
	Vegetable soup (H)

Key: M = moderate; feed in limited amounts. H = high; avoid feeding.

K-9 UROLITH

In general, medical therapy should be implemented in stepwise fashion, with the initial goal of reducing the urine concentration of calculogenic substances (**Table 20-16**). Medications that have the potential to induce unwanted, sustained, detrimental alterations in the composition of metabolites should be reserved for patients with active or frequently recurring calcium oxalate uroliths. Caution should be used to ensure that side effects of treatment are not more detrimental than the effects of uroliths. The cause of hypercalcemia (e.g., primary hyperparathyroidism) should be corrected in patients with hypercalcemia and resorptive hypercalciuria. An attempt should be made to identify risk factors for urolith formation in patients with normal serum calcium concentrations (**Table 20-14**). Amelioration or control of the consequences of risk factors (e.g., urine oversaturation with calculogenic minerals) should minimize urolith growth and recurrence.

Select a Food (Pages 641-643)

The goals for preventing the recurrence of calcium oxalate uroliths include: 1) reducing calcium concentration in urine, 2) reducing oxalic acid concentration in urine, 3) promoting high concentration and activity of inhibitors of calcium oxalate crystal growth and aggregation in urine and 4) reducing urine concentration (See the Key Nutritional Factors–Canine Urolithiasis table.).

Reduction of dietary calcium appears to be a logical therapeutic goal because intestinal hyperabsorption of calcium has been identified as one mechanism promoting hypercalciuria in dogs with calcium oxalate uroliths. However, reducing consumption of calcium may increase the availability of

Table 20-16. Recommendations for the management of calcium oxalate urolithiasis in dogs.

1. Obtain baseline data (postsurgical radiography, complete urinalysis, serum concentrations of calcium, urea nitrogen and creatinine) to evaluate effectiveness of surgery, renal function and calcium homeostasis.
2. If the dog is hypercalcemic, correct underlying cause.
3. If the dog is normocalcemic, consider foods with reduced calcium, oxalate, sodium and protein that do not promote formation of acidic urine. Ideally foods should contain additional water and citrate and have adequate phosphorus and magnesium. Avoid vitamins C and D. Such foods include Prescription Diet Canine u/d* and Canine w/d.*
4. Reevaluate patient in two to four weeks to verify dietary compliance (urine specific gravity and pH and serum urea nitrogen concentration) and amelioration of crystalluria (urine sediment examination).
5. Consider additional potassium citrate if calcium oxalate crystals and aciduria persist.
6. Reevaluate patient in two to four weeks to verify dietary compliance (urine specific gravity and pH and serum urea nitrogen concentration) and amelioration of crystalluria (urine sediment examination). Consider vitamin B_6 supplementation (2 to 4 mg/BW_{kg} every 24 to 48 hours) if calcium oxalate crystalluria persists.
7. Again, reevaluate patient in two to four weeks to verify dietary compliance and amelioration of crystalluria. Consider administration of hydrochlorothiazide (2 mg/BW_{kg} every 24 to 48 hours) if calcium oxalate crystalluria persists. Adverse effects of hydrochlorothiazide administration include dehydration, hypokalemia and hypercalcemia.
8. After three to six months, reevaluate patient to verify dietary compliance and amelioration of crystalluria. Check for urolith recurrence by abdominal radiography. If no uroliths are present, continue current therapy and reevaluate in three to six months. If uroliths have recurred, consider voiding urohydropropulsion. If unsuccessful and clinical signs referable to urocystoliths are persistent, consider surgery. Continue therapy to minimize urolith growth if clinical signs are not present.

*Hill's Pet Nutrition, Inc., Topeka, KS, USA.

oxalic acid for intestinal absorption and subsequent urinary excretion. As in the urinary bladder, calcium and oxalic acid in the intestinal lumen form an insoluble compound, thereby preventing the absorption of one another. A reduction in dietary calcium should be accompanied by an appropriate reduction in dietary oxalic acid (**Tables 20-15** and **20-17**).

Consumption of high levels of sodium may augment renal excretion of calcium. Daily urinary calcium excretion of normal dogs consuming foods with 0.8% sodium (dry weight analysis) was comparable to calcium excretion observed in dogs with calcium oxalate uroliths. Therefore, moderate dietary restriction of sodium (<0.3% sodium dry matter basis) is recommended for active calcium oxalate urolith formers.

Table 20-17. Selected human foods with minimal calcium or oxalate content.

Food categories	Low-calcium foods	Low-oxalate foods
Meats and eggs	Eggs Poultry	Beef Eggs Fish and shellfish* Lamb Pork Poultry
Vegetables		Cabbage Cauliflower Mushrooms Peas, green Radishes Potatoes, white
Milk and dairy products		Cheese* Milk* Yogurt*

Fruits		Apple
		Avocado
		Banana
		Bing cherries
		Grapefruit
		Grapes, green
		Mangos
		Melons
		  Cantaloupe
		  Casaba
		  Honeydew
		  Watermelon
		Plums

Breads, grains, nuts	Almonds Macaroni Pretzels Rice Spaghetti Walnuts	Bread, white Macaroni Noodles Rice Spaghetti
Miscellaneous	Popcorn	Jellies Preserves Soups with allowed ingredients

*Low in oxalate, but not low in calcium content.

Studies involving laboratory animals, dogs and people suggest that dietary phosphorus should not be restricted in patients with calcium oxalate urolithiasis because reduction in dietary phosphorus is often associated with augmentation of intestinal calcium absorption and hypercalciuria. If calcium oxalate urolithiasis is associated with hypophosphatemia and normal serum calcium concentration, oral phosphorus supplementation should be considered. However, caution must be used because excessive dietary phosphorus may predispose patients to formation of calcium phosphate uroliths.

Although supplemental dietary magnesium contributes to formation of magnesium ammonium phosphate uroliths in some species (cats and ruminants), urine magnesium apparently impairs formation of calcium oxalate crystals. Therefore, supplemental magnesium has been used in people in an attempt to minimize recurrence of calcium oxalate uroliths. However, normal dogs given supplemental magnesium have shown a five-fold increase in urinary excretion of calcium. Therefore, pending further studies, dietary magnesium restriction or supplementation is not recommended for treatment of canine calcium oxalate uroliths.

Ingestion of foods that contain high quantities of animal protein may contribute to calcium oxalate urolithiasis by increasing urinary calcium excretion and decreasing urinary citrate excretion. Some of these consequences result from obligatory acid excretion associated with protein metabolism. Hypercalciuria occurs in normal dogs fed high-protein foods (40% dry weight analysis). Therefore, excessive dietary protein consumption should be avoided in dogs with active calcium oxalate urolithiasis.

A food that avoids excess protein, calcium, oxalate and sodium (See the Key Nutritional Factors–Canine Urolithiasis table and **Table 20-18**.) may be considered to minimize recurrence of active calcium oxalate uroliths in dogs. Dry foods are not generally recommended because they tend to result in formation of concentrated urine. Ideally, foods should not be restricted or supplemented with phosphorus or magnesium. Excessive levels of vitamin D (which promotes intestinal absorption of calcium) and ascorbic acid (a precursor of oxalate) should also be avoided. Vitamin D levels in foods for dogs with calcium oxalate uroliths should be between 500 and 2,500 IU/kg.

A deficiency of pyridoxine should be avoided because vitamin B_6 promotes endogenous production of oxalate. Vitamin B_6 (pyridoxine) increases the transamination of glyoxylate, an important precursor of oxalic acid, to glycine. Experimentally induced pyridoxine deficiency resulted in renal precipitation of calcium oxalate and hyperoxaluria in kittens. Commercial foods routinely fortified with vitamin supplements would not be deficient in pyridoxine or other vitamins. A homemade food might be deficient in pyridoxine if a multivitamin supplement is not added. The ability of supplemental pyridoxine (above nutritional requirements) to reduce urinary oxalic acid excretion in dogs is unknown and not recommended.

A urinary pH target range of 6.8 to 7.2 may be beneficial for dogs at increased risk for calcium oxalate urolith formation.

Table 20-18. Levels of key nutritional factors in selected foods for prevention of calcium oxalate uroliths in dogs.*

Foods	Protein	Calcium	Magnesium	Sodium	Vitamin D	Urinary pH**
Recommended levels	**10-18**	**0.3-0.6**	**0.04-0.15**	**<0.3**	**500-2,500**	**7.0-7.5**
Hill's Prescription Diet Canine k/d, moist	14.7	0.83	0.13	0.18	2,248	6.8-7.2
Hill's Prescription Diet Canine k/d, dry	14.5	0.80	0.09	0.23	1,897	6.8-7.2
Hill's Prescription Diet Canine u/d, moist	11.5	0.29	0.03	0.24	2,263	7.1-7.7
Hill's Prescription Diet Canine u/d, dry	9.4	0.38	0.04	0.24	1,959	7.1-7.7
Purina CNM NF-Formula, moist	16.5	0.50	0.08	0.24	na	na
Purina CNM NF-Formula, dry	15.9	0.76	0.07	0.22	na	na
Select Care Canine Modified Formula, moist	16.8	0.61	0.07	0.18	na	mean 7.0
Select Care Canine Modified Formula, dry	13.7	0.85	0.10	0.22	na	mean 7.0
Waltham S/O Lower Urinary Tract Support, moist	25.2	1.50	0.10	2.23	na	5.5-6.1
Waltham S/O Lower Urinary Tract Support, dry	16.0	0.80	0.07	1.23	na	5.5-6.1

Key: na = information not available from manufacturer.
*Manufacturers' published values. Nutrients expressed as % dry matter; vitamin D expressed as IU/kg. This list represents products with the largest market share for which published information is available. Foods containing high levels of oxalates should be avoided (**Table 20-15**).
**Protocols for measuring urinary pH may vary.

K-9 UROLITH

Citric Acid (Page 643)

Citric acid forms soluble salts with calcium thereby inhibiting calcium oxalate crystal formation. Citric acid is also beneficial because it is metabolized to bicarbonate and promotes formation of alkaline urine. In dogs, chronic metabolic acidosis inhibits renal tubular reabsorption of calcium, whereas metabolic alkalosis enhances tubular reabsorption of calcium. Potassium citrate is preferred to sodium bicarbonate as an alkalinizing agent because oral administration of sodium enhances urinary calcium excretion. If persistent aciduria or hypocitraturia is recognized in dogs (mean urinary citrate excretion of 33 normal beagles was 2.57 ± 2.31 mg/BW_{kg}/day), therapy with wax matrix tablets of potassium citrate (Urocit-K) should be considered. A liquid product (Polycitra-K) works well for small dogs. A dose of 40 to 75 mg/BW_{kg} q12h is recommended. Because Prescription Diet Canine u/d already contains adequate quantities of potassium citrate, additional potassium citrate is often not needed.

Thiazide Diuretics (Page 643)

Thiazide diuretics have been recommended to reduce recurrence of calcium-containing uroliths in people because of their ability to reduce urinary calcium excretion. There is a beneficial reduction in urinary calcium excretion in dogs with calcium oxalate urolithiasis following administration of hydrochlorothiazide (2 to 4 mg/BW_{kg} q12h) for two weeks. However, a reduction in urinary calcium excretion was not detected following chlorothiazide administration (20 to 65 mg/BW_{kg} q12h) to clinically healthy beagles. Thiazide diuretic administration should be accompanied by appropriate clinical and laboratory monitoring for early detection of adverse effects (dehydration, hypokalemia, hypercalcemia).

Reassessment (Pages 643-644)

The goal of therapy is to prevent calcium oxalate urolith recurrence. However, this expectation may be unrealistic because the primary causes responsible for urolith formation are multifactorial and incompletely understood. With the information and techniques currently available, however, veterinarians can minimize urolith recurrence and prevent future surgical urolith removal with careful and planned monitoring and intervention.

Therapy should be instituted in a stepwise fashion (**Table 20-16**). Dietary and pharmacologic management should result in formation of less concentrated urine without calcium oxalate crystalluria (**Table 20-19**). Strive to achieve urine specific gravity values less than 1.020. After this is achieved, a urinalysis and survey lateral abdominal radiograph should be performed every two to four months. Dietary and pharmacologic changes can be made if crystalluria or concentrated urine persists (**Table 20-20**). By following these recommendations, recurrent urocystoliths detected by radiography

Table 20-19. Expected changes associated with medical therapy to minimize recurrence of calcium oxalate uroliths.

Factors	Pre-therapy	Prevention therapy
Polyuria	±	Variable
Pollakiuria	0 to 4+	0
Hematuria	0 to 4+	0
Urine specific gravity	Variable	1.004-1.015
Urinary pH	<7.0	>7.0
Pyuria	0 to 4+	0
Calcium oxalate crystals	0 to 4+	0
Bacteriuria	0 to 4+	0
Bacterial culture of urine	0 to 4+	0
Urea nitrogen (mg/dl)	>15	<15
Urolith size and number	Small to large	0

should be small enough to remove by voiding urohydropropulsion. After a rate of urolith recurrence has been established, the frequency of evaluation can be modified such that predicted recurrences can be diagnosed and managed accordingly.

CALCIUM PHOSPHATE UROLITHS (Pages 644-649)

Prevalence and Mineral Composition (Pages 644-645)

Calcium phosphate accounted for 0.6% of all canine uroliths and 2.3% of canine nephroliths analyzed at the Minnesota Urolith Center from 1981 to 1997. Information about breed, gender and age predisposition, serum and urine profiles, crystal appearance and calcium phosphate urolith radiographic density and contour are listed in **Table 20-21**.

Etiopathogenesis and Risk Factors (Pages 645-647)
Solubility of Calcium Phosphates in Urine (Pages 645-646)

The solubility of calcium phosphates in urine is dependent on: 1) urinary pH, 2) urine calcium ion concentration, 3) total urine inorganic phosphate concentration, 4) urine concentration of inhibitors of calcium crystallization and 5) urine concentration of potentiators of crystallization. Factors that decrease calcium phosphate solubility predispose patients to urolith formation. For a discussion of these factors, see Small Animal Clinical Nutrition, 4th ed., pp 645-647.

Disorders Associated with Formation of Calcium Phosphate Uroliths (Pages 646-647)

Several medical conditions are associated with calcium phosphate uroliths; for more information, see **Table 20-22** and Small Animal Clinical Nutrition, 4th ed., pp 646-647.

Table 20-20. Managing highly recurrent calcium oxalate uroliths.

Causes	Identification	Therapeutic goals
Client and patient factors		
Inadequate dietary compliance	Question owner Persistent calcium oxalate crystalluria Urea nitrogen >10-15 mg/dl Urine specific gravity >1.010-1.020 Urinary pH <7.0-7.5 during treatment with Prescription Diet Canine u/d* (use lower values for the canned food)	Emphasize need to feed dissolution food exclusively
Administration of vitamin-mineral supplements	Question owner	Discontinue vitamin-mineral supplements containing calcium and vitamins C and D
Clinician factors		
Incomplete surgical removal of uroliths	Postsurgical radiography revealing uroliths Persistence of clinical signs after cystotomy or recurrence of clinical signs soon after cystotomy (within one to three months)	Uroliths not causing clinical signs should be monitored for potentially adverse consequences (obstruction, urinary tract infection [UTI], etc.) Clinically active uroliths may require surgical removal Remove small uroliths by voiding urohydropropulsion
Inappropriate food choice	Persistent calcium oxalate crystalluria	Choose foods with reduced levels of calcium, oxalate, protein and sodium that do not promote formation of acidic urine Consider adding potassium citrate if aciduria persists

(Continued on next page.)

Table 20-20. Managing highly recurrent calcium oxalate uroliths (Continued.).

Causes	Identification	Therapeutic goals
Clinician factors		
Inadequate monitoring	Postsurgical radiography to verify complete urolith removal was not performed	Perform postsurgical radiography to evaluate success of surgery
	Urinalysis or urine sediment examinations were not performed within three to six months of initiation of therapy	Perform complete urinalysis within one to three months of initiation of therapy
		Once stable, urinalysis should be performed every four to six months
		Perform survey lateral abdominal radiography every four to six months to assess recurrence
Corticosteroid administration	Corticosteroids were prescribed to manage other disease conditions	If possible, discontinue corticosteroid administration
Disease factors		
Hypercalcemia	Elevated serum calcium concentration	Identify and, if possible, eliminate underlying cause for hypercalcemia (hyperparathyroidism, neoplasia, hypervitaminosis D, etc.)
Recurrence of uroliths despite appropriate management	Lateral radiograph of abdomen	Uroliths not causing clinical signs should be monitored for potentially adverse consequences (obstruction, UTI, etc.)
		Clinically active uroliths may require surgical removal
		Remove small uroliths by voiding urohydropropulsion

Assess the Food (Pages 647-648)

Dogs with calcium phosphate uroliths have consumed a variety of foods. Dogs consuming dry commercial foods may be at greater risk for urolith formation than dogs consuming moist foods because dry foods tend to be associated with formation of more concentrated urine. Foods with high protein content contribute to hypercalciuria and hyperphos-

Table 20-21. Common characteristics of canine calcium phosphate uroliths.

Chemical names	Formulas	Crystal names
β-tricalcium phosphate (calcium orthophosphate)	β-Ca$_2$(PO$_4$)$_2$	Whitlockite
Carbonate apatite	Ca$_{10}$(PO$_4$CO$_3$OH)$_6$(OH)$_2$	Carbonate apatite
Calcium hydrogen phosphate dihydrate	CaHPO$_4$•2H$_2$O	Brushite
Calcium phosphate	Ca$_{10}$(PO$_4$)$_6$(OH)$_2$	Hydroxyapatite or calcium apatite

Variations in mineral composition
Calcium apatite only
Brushite only
Calcium apatite mixed with calcium oxalate
Brushite mixed with calcium oxalate
The carbonate apatite form of calcium phosphate is most commonly detected as a minor component of infection-induced struvite

Physical characteristics
Color: Calcium phosphate uroliths are usually cream or tan. Blood clots mineralized with calcium oxalate are black.
Shape: Variable. With the exception of brushite, calcium phosphate uroliths do not have a characteristic shape. Brushite uroliths are typically round and smooth.
Nuclei: Brushite uroliths are often laminated.
Density: Generally dense and brittle, sometimes chalklike. Mineralized blood clots may be softer. All forms of calcium phosphate are radiodense compared to soft tissue.
Number: Single or multiple
Location: Kidneys, ureters, urinary bladder (most common) and/or urethra
Size: Variable, with smaller sizes more common

Prevalence
Approximately 1% of all canine uroliths. Approximately 2% of canine nephroliths.

Characteristics of affected canine patients
No gender prevalence for calcium apatite. Brushite is more common in males. Mean age at diagnosis is seven years (range <1 to >16 years)

phaturia. Foods with restricted quantities of protein tend to reduce urine concentrating capacity and urinary phosphorus and calcium excretion. Foods with higher quantities of sodium or vitamin D may promote hypercalciuria (See the Key Nutritional Factors–Canine Urolithiasis table and **Table 20-23**). Foods with higher levels of phosphorus tend to augment hyperphosphaturia. However, excessive restriction of dietary phosphorus may enhance the availability of dietary calcium for intestinal absorption, and also enhance production of 1,25-vitamin D by the kidneys, thereby promoting hypercalciuria. Although these trends serve as guidelines, the optimal level of phosphorus for dogs with calcium phosphate urolithiasis

Table 20-22. Disorders that may predispose dogs to formation of calcium phosphate uroliths.

Primary hyperparathyroidism
Other hypercalcemic disorders
   Neoplasia
   Vitamin D intoxication
   Excess calcium intake
   Thyrotoxicosis
   Hyperadrenocorticism
   Immobilization
Distal renal tubular acidosis
Normocalcemic hypercalciuria
   Intestinal hyperabsorption
   Renal leak

Table 20-23. Some potential risk factors for canine calcium phosphate uroliths.

Diet	Urine	Metabolic
Alkalinizing potential	Alkaline pH	Hypercalcemia
High calcium content	Hypercalciuria	Distal renal tubular
High sodium content	High-phosphate ion	acidosis
High phosphorus content?	concentration	
Low-moisture content	Increased concentration of promoters	
Excessive vitamin D content	Decreased concentration of inhibitors	
Others?	Hypocitraturia	
	Hypomagnesuria	
	Blood clots	
	Urine concentration	
	Urine retention	

has yet to be determined.

Most forms of calcium phosphate are least soluble in alkaline urine. Formation of calcium phosphate is enhanced because alkaline urine favors dissociation of monobasic phosphate ($H_2PO_4^-$) to dibasic phosphate (HPO_4^{2-}) and phosphate ions (PO_4^{3-}). Although formation of acidic urine may enhance the solubility of calcium phosphate (except brushite), acidosis promotes hypercalciuria and hypocitraturia.

Feeding Plan and Treatment (Pages 648-649)

Surgery remains the most reliable way to remove active calcium phosphate uroliths from the urinary tract. However, surgery may be unnecessary for clinically inactive calcium phosphate uroliths. Voiding urohydropropulsion may be used to remove small urocystoliths (**Table 20-10**).

Table 20-24. Levels of key nutritional factors in selected foods for prevention of calcium phosphate uroliths in dogs.*

Foods	Protein	Phosphorus	Calcium	Sodium	Vitamin D
Recommended levels	10-18	<0.6	0.3-0.6	<0.3	500-2,500
Hill's Prescription Diet Canine u/d, moist	11.5	0.14	0.29	0.24	2,263
Hill's Prescription Diet Canine u/d, dry	9.4	0.19	0.38	0.24	1,959

*Nutrients expressed as % dry matter; vitamin D expressed as IU/kg.

For prevention of urolith recurrence, it seems reasonable to recommend trial therapy with foods lower in certain nutrient levels (i.e., foods formulated to avoid excessive protein, sodium, calcium and vitamin D may be of benefit) compared with the food consumed at the time the urolith formed (**Table 20-24**). Encouraging water consumption is likely to be of benefit; therefore, enhancing urine volume by feeding a moist food (and/or a protein-restricted food to reduce renal medullary urea concentration) may be helpful. Although understandably difficult in some patients, fluid intake should be encouraged throughout the day to promote a constantly high urine volume. Excessive restriction or supplementation of dietary phosphorus should probably be avoided. In people, some high-fiber foods reduce intestinal absorption and urinary excretion of calcium.

Reassessment (Page 649)
The likelihood of recurrence of calcium phosphate uroliths following removal is not well established. Therefore, patients should be periodically monitored by urinalysis, radiographic or ultrasonographic procedures and other hematologic and urologic laboratory tests, if indicated (**Table 20-25**). Small, recurrent urocystoliths may be removed by voiding urohydropropulsion or by aspiration through a urinary catheter. Medical therapy of patients with recurring calcium phosphate uroliths should be directed toward removing or minimizing risk factors that contribute to supersaturation of urine with calcium phosphate.

CYSTINE UROLITHIASIS (Pages 649-653)

Prevalence and Mineral Composition (Pages 649-650)
The prevalence of cystine uroliths in dogs varies with geographic location. The prevalence is 1 to 3% of the uroliths removed from dogs in the United States and as high as 39% in some European centers. Cystine accounted for 1.0% of all canine uroliths and 0.4% of canine nephroliths analyzed at the Minnesota Urolith Center from 1981 to 1997. Information about breed, gen-

Table 20-25. Summary of recommendations for management of canine calcium phosphate uroliths.

1. Surgery remains the most reliable way to remove active calcium phosphate uroliths from the urinary tract. However, surgery may be unnecessary for clinically inactive calcium phosphate uroliths. Small urocystoliths may be nonsurgically removed by voiding urohydropropulsion (**Table 20-10**) or by aspiration through a urinary catheter. Medical therapy of patients with recurrent calcium phosphate uroliths should then be directed toward removing or minimizing risk factors that contribute to supersaturation of urine with calcium phosphate.
2. Patients with hypercalcemia and primary hyperparathyroidism usually require surgery. Parathyroidectomy may result in dissolution of uroliths and generally prevents their recurrence.
3. Several different medical protocols have been reported to be of value in people with normocalcemic hypercalciuria. Ideally, the choice of therapy should be based on the cause of idiopathic hypercalciuria.
   a. There has been little clinical experience with the use of drugs in dogs and cats with calcium phosphate uroliths. However, medications that enhance calcium excretion (glucocorticoids, furosemide and those containing large quantities of sodium) should be avoided if possible.
   b. Foods designed to avoid excessive protein, sodium, calcium and vitamin D may be of benefit. Excessive restriction or supplementation of dietary phosphorus should probably be avoided. Enhancing urine volume by feeding a moist food (and/or a protein-restricted food to dogs to reduce renal medullary urea concentrations) and encouraging water consumption may be of benefit. Although understandably difficult to accomplish in some patients, fluid intake should be encouraged throughout the day to promote a constantly high urine volume. In people, some high-fiber diets reduce intestinal absorption and urinary excretion of calcium.

(Continued on next page.)

der and age predisposition, serum and urine profiles, crystal appearance and cystine urolith radiographic density and contour is listed in **Table 20-3**.

Pure cystine uroliths are usually multiple, ovoid and smooth. They are light yellow and vary from 0.5 mm to several cm in diameter.

Etiopathogenesis and Risk Factors (Page 650)

Cystinuria is an inborn error of metabolism characterized by abnormal transport of cystine (a nonessential sulfur-containing amino acid composed of two molecules of cysteine) and other amino acids by the renal tubules.

Cystine is normally present in low concentrations in plasma. Normally, circulating cystine is freely filtered at the glomerulus and most is actively reabsorbed in the proximal tubules. The solubility of cystine in urine is pH dependent. It is relatively insoluble in acidic urine, but becomes more soluble in alkaline urine.

Table 20-25. Summary of recommendations for management of canine calcium phosphate uroliths (Continued.).

---

    c. With the exception of brushite, calcium phosphates tend to be less soluble in alkaline urine. Whether or not patients with such mineral types would benefit from appropriate dosages of urine acidifiers is unknown. Acidification tends to enhance urine calcium excretion and is a risk factor for calcium oxalate urolith formation. Pending further studies, routine use of urine acidifiers for patients with calcium phosphate urolithiasis is not recommended.

4. Medical dissolution of calcium phosphate uroliths has not been attempted in dogs with distal renal tubular acidosis (RTA). Foods designed to dissolve struvite uroliths would generally not be expected to promote dissolution of calcium phosphate uroliths, in part because they may tend to promote acidemia and aciduria, thus potentially enhancing hypercalciuria and hypocitraturia. However, correction of hypercalciuria, hyperphosphaturia and hypocitraturia by alkalinization therapy with potassium citrate might promote dissolution of these uroliths in patients with complete or incomplete distal RTA. Long-term alkalinization therapy appears to be beneficial in preventing calcium phosphate urolith formation in people with distal RTA. Such therapy has been advocated for patients with complete or incomplete forms of distal RTA because it decreases urolith formation and nephrocalcinosis and increases urine citrate concentration. Oral administration of sodium chloride, long recommended for all forms of urolithiasis, may promote hypercalciuria and calcium phosphate urolith formation. Therefore, oral salt therapy is not recommended to promote diuresis in dogs with uroliths containing calcium salts.

Unlike normal dogs, some cystinuric dogs reabsorb a much smaller proportion of the amino acid from the glomerular filtrate.

The exact mechanism of cystine urolith formation is unknown. Because not all cystinuric dogs form uroliths, cystinuria is a predisposing factor rather than a primary cause of cystine urolith formation.

Biologic Behavior (Page 650)

Cystine uroliths are often not recognized in most affected dogs until after they reach maturity. The average age at detection in many breeds is approximately two to five years.

Compared with the situation in other breeds, the severity of the tubular transport defect for cystine in Newfoundlands appears to be more severe. This provides a plausible explanation for the earlier onset of cystine urolith formation in this breed and for the involvement of the kidneys in addition to the urinary bladder in female as well as male dogs.

Because cystinuria is an inherited defect, uroliths commonly recur in two to 12 months unless prophylactic therapy has been initiated. Recurrence in Newfoundlands appears to be more rapid than in other breeds.

Feeding Plan and Treatment (Pages 650-652)

Current recommendations for dissolution of cystine uroliths encompass reducing urine concentration of cystine and increasing the solubility of cystine in urine. This may be accomplished by various combinations of: 1) dietary modification, 2) administration of thiol-containing drugs and 3) alkalinization of urine, if necessary (See the Key Nutritional Factors–Canine Urolithiasis table and **Table 20-26**). Small cystine urocystoliths may be removed by voiding urohydropropulsion (**Table 20-10**).

Dietary Modification (Pages 650-651)

Reduction of dietary protein has the potential of minimizing formation of cystine uroliths. Pilot studies performed on cystinuric dogs at the University of Minnesota revealed a 20 to 25% reduction in urinary cystine excretion when subjects consumed a moist veterinary therapeutic food (Prescription Diet Canine u/d, **Table 20-27**) vs. when they received a moist canine adult maintenance food. Reducing the renal medullary urea concentration and the associated urine concentration is an important indirect effect.

Thiol-Containing Drugs (Page 651)

D-penicillamine (dimethylcysteine) is a nonmetabolizable degradation product of penicillin that may combine with cysteine to form cysteine-D-penicillamine disulfide. This disulfide exchange reaction is facilitated by an alkaline pH. The resulting compound has been reported to be 50 times more soluble than free cystine. Although D-penicillamine is effective in reducing urine cystine concentrations, drug-related adverse events limit its use. To minimize such adverse drug events use N-(2-mercaptopropionyl)-glycine (2-MPG) rather than D-penicillamine.

2-MPG decreases the concentration of cystine by a thiol-disulfide exchange reaction similar to that of D-penicillamine. Studies in dogs indicate that the drug is highly effective in reducing urinary cystine concentration and has less toxicity than D-penicillamine.

Oral administration of 2-MPG at a daily dosage of approximately 30 to 40 mg/BW_{kg} (divided in two equal doses) was effective in inducing dissolution of multiple cystine urocystoliths in nine of 17 dogs evaluated. Dissolution required two to four months of therapy.

Appropriate hematologic and biochemical evaluations should be performed during use of 2-MPG in dogs with a history of D-penicillamine hypersensitivity. A combination of a calculolytic food (**Table 20-27**) and 2-MPG therapy is more effective than either alone. Dissolution of 18 episodes of cystine urocystoliths affecting 14 dogs using combination dietary and drug therapy has been reported. The mean time required to dissolve the cystine uroliths was 78 days (range 11 to 211 days).

Table 20-26. Summary of recommendations for medical dissolution and prevention of canine cystine uroliths.

1. Perform appropriate diagnostic studies including complete urinalysis, quantitative urine culture and diagnostic radiography. Determine precise location, size and number of uroliths. The size and number of uroliths are not a reliable index of probable therapeutic efficacy.
2. If uroliths are available, determine their mineral composition. If they are unavailable, determine their composition by evaluation of appropriate clinical data.
3. Consider surgical correction if uroliths obstruct urine outflow and/or if correctable abnormalities predisposing the patient to recurrent urinary tract infection (UTI) are identified by radiography or other means. Small urocystoliths may be removed by voiding urohydropropulsion (**Table 20-10**).
4. Initiate therapy with a calculolytic food (Prescription Diet Canine u/d*). No other food or mineral supplements should be fed to the patient. Compliance with dietary recommendation is suggested by a reduction in urea nitrogen concentration (usually <10 mg/dl).
5. Initiate therapy with N-(2-mercaptopropionyl)-glycine (2-MPG)** at a daily dosage of approximately 30 mg/BW$_{kg}$, divided into two equal subdoses.
6. If necessary, administer potassium citrate orally to eliminate aciduria. Strive for a urinary pH of approximately 7.5.
7. If necessary, eradicate or control UTIs with appropriate antimicrobial agents.
   a. Devise a protocol for follow-up therapy.
      1. Try to avoid diagnostic follow-up studies that require urinary catheterization. If they are required, give appropriate peri-catheterization antimicrobial agents to prevent iatrogenic UTI.
      2. Perform serial urinalyses. Urinary pH, specific gravity and microscopic examination of sediment for crystals are especially important. Remember, crystals formed in urine stored at room or refrigeration temperatures may represent in vitro artifacts.
      3. Perform serial radiography monthly to evaluate urolith location(s), number, size, density and shape. Intravenous urography may be used to identify radiolucent uroliths in the kidneys, ureters and urinary bladder. Antegrade contrast cystourethrography may be required for radiolucent uroliths located in the bladder and urethra.
   b. Continue calculolytic food, 2-MPG and alkalinizing therapy for approximately one month after disappearance of uroliths as detected by radiography.
   c. Prevention. Feeding a low-protein food that promotes alkaline urine has been effective in preventing cystine urolith recurrence. If necessary, low doses of 2-MPG may also be given.

*Hill's Pet Nutrition, Inc., Topeka, KS, USA.
**Thiola, Mission Pharmacal, San Antonio, TX, USA.

Table 20-27. Levels of key nutritional factors in selected foods for dissolution and prevention of cystine uroliths in dogs.

Foods	Protein (% DM)	Urinary pH
Recommended levels	10-18	7.1-7.7
Hill's Prescription Diet Canine u/d, moist	11.5	7.1-7.7
Hill's Prescription Diet Canine u/d, dry	9.4	7.1-7.7

Alkalinization of Urine (Page 652)

The solubility of cystine is pH dependent. In dogs, the solubility of cystine at a urinary pH of 7.8 is approximately double that at a urinary pH of 5.0. If cystine uroliths fail to dissolve in dogs whose urinary pH does not become sufficiently alkaline as a result of dietary therapy, a sufficient quantity of potassium citrate or sodium bicarbonate should be given orally in divided doses to sustain a urinary pH of approximately 7.5. Potassium citrate may be preferable to sodium bicarbonate as a urine alkalinizing agent.

Reassessment (Page 652)

The goal of therapy is to promote cystine urolith dissolution. Careful and planned monitoring is necessary for consistent effectiveness (**Tables 20-28 and 20-29**). Dietary management should result in formation of less concentrated urine without cystine crystalluria; strive to achieve urinary specific gravity values less than 1.020. Orally administered potassium citrate may be considered if owners and patients are compliant with dietary therapy and the urinary pH remains acidic. A urinalysis and survey abdominal radiographs should be performed about every four weeks. Reduction in serum urea nitrogen concentration provides supportive evidence that the owner is complying with dietary recommendations.

Prevention (Pages 652-653)

Because cystinuria is an inherited metabolic defect, and because cystine uroliths recur in a high percentage of young to middle-aged dogs within two to 12 months after surgical removal, prophylactic therapy should be considered. Dietary therapy and if necessary, urine alkalinization may be initiated with the objective of minimizing cystine crystalluria and promoting a negative cyanide-nitroprusside test result. If necessary, 2-MPG may be added to the regimen in sufficient quantities to maintain a urine concentration of cystine less than approximately 200 mg/liter. If the dosage cannot be titrated by measuring urine cystine concentration, 2-MPG may be given at a dosage of 15 mg/BW$_{kg}$ q12h. Continuous therapy of urolith-free cystinuric dogs with 2-MPG has been effective in preventing formation of cystine uroliths in studies performed in Sweden and at the University of Minnesota.

K-9 UROLITH

Table 20-28. Expected changes associated with medical therapy of cystine uroliths.

Factors	Pre-therapy	During therapy	Prevention therapy
Polyuria	±	1+ to 3+	1+ to 3+
Pollakiuria	0 to 4+	↑ then ↓	0
Hematuria	0 to 4+	↓	0
Urine specific gravity	Variable	1.004-1.014	1.004-1.014
Urinary pH	<7.0	>7.0	>7.0
Pyuria	0 to 4+	↓	0
Cystine crystals	0 to 4+	0	Variable
Bacteriuria	0 to 4+	0	0
Bacterial culture of urine	0 to 4+	0	0
Urea nitrogen (mg/dl)	Variable	<15	≤15
Urolith size and number	Small to large	↓	0

Table 20-29. Managing cystine uroliths refractory to complete dissolution.

Causes	Identification	Therapeutic goals
Client and patient factors		
Inadequate dietary compliance	Question owner Persistent cystine crystalluria Urea nitrogen >10-17 mg/dl Urine specific gravity >1.010-1.020 Urinary pH <7.0-7.5 during treatment with Prescription Diet Canine u/d* (use lower values for the canned food)	Emphasize need to feed dissolution food exclusively
Inadequate 2-MPG** administration	Question owner Count remaining pills	Emphasize need to administer the full dose of medication Determine if owner is capable and willing to administer medication Demonstrate a variety of methods to administer medication

(Continued on next page.)

Table 20-29. Managing cystine uroliths refractory to complete dissolution (Continued.).

Causes	Identification	Therapeutic goals
Clinician factors		
Incorrect prediction of mineral type	Analysis of retrieved urolith	Alter therapy based on identification of mineral type
Inadequate 2-MPG dose for degree of diuresis	No change in urolith size after two months of appropriate therapy	Increase 2-MPG dose to 20 mg/BW$_{kg}$ q12hr
Disease factors		
Compound urolith	Radiographic density of nucleus and outer layer(s) of urolith is different Analysis of retrieved urolith	Alter therapy based on identification of new mineral type Uroliths not causing clinical signs should be monitored for potentially adverse consequences (obstruction, urinary tract infection, etc.) Clinically active uroliths may require surgical removal Remove small uroliths by voiding urohydropro-pulsion

*Hill's Pet Nutrition, Inc., Topeka, KS, USA.
**Thiola, Mission Pharmacal, San Antonio, TX, USA.

K-9 UROLITH

STRUVITE UROLITHIASIS (Pages 653-662)

Prevalence and Mineral Composition (Page 653)

The most common type of mineral encountered in uroliths from dogs is magnesium ammonium phosphate hexahydrate (MAP) or struvite. Struvite accounted for 50% of all canine uroliths and 33% of canine nephroliths analyzed at the Minnesota Urolith Center from 1981 to 1997. Information about breed, gender and age predisposition, serum and urine profiles, crystal appearance and struvite urolith radiographic density and contour is listed in **Table 20-3**. Common characteristics of canine struvite uroliths are listed in **Table 20-30**.

Etiopathogenesis and Risk Factors (Pages 653-656)
Infection-Induced Struvite Uroliths (Pages 653-655)

See Small Animal Clinical Nutrition, 4th ed., pp 653-654 for sequential steps in urolith formation.

Table 20-30. Common characteristics of canine struvite uroliths.

Chemical name	Formula	Crystal name
Magnesium ammonium phosphate hexahydrate	$MgNH_4PO_4\bullet6H_2O$	Struvite

Variations in mineral composition
Struvite only
Struvite mixed with lesser quantities of calcium apatite and/or ammonium acid urate
Nucleus of a different mineral surrounded by variable layers composed primarily of struvite. Small quantities of calcium apatite and/or ammonium acid urate also may be present.

Physical characteristics
Color: Struvite uroliths are usually white, cream or light brown. The surface of uroliths is commonly red because of concomitant hematuria and may be green due to bile pigments.
Shape: Variable. Solitary urocystoliths are commonly round or elliptic. Multiple urocystoliths may be any shape, but are often pyramidal. Rapidly growing uroliths with a large quantity of matrix may form a cast of the lumen (renal pelvis, ureter, bladder, urethra) in which they are formed.
Nuclei and laminations: Common in infection-induced uroliths
Density: Variable. Soft if they contain a large quantity of matrix. Dense and harder to cut if little matrix is present. A combination of hard and soft internal density may occur within the same urolith. Radiodense compared with nonskeletal tissue on survey radiographs. Degree of radiodensity is related to the quantity of matrix (inversely proportional) and other minerals, especially calcium apatite (more proportional).
Number: Single or multiple
Location: May be located in the kidney, ureter, urinary bladder and/or urethra. Most occur in the urinary bladder.
Size: Subvisual to a size limited by the capacity of the structure (kidney and urinary bladder) in which they form. Very large uroliths are often composed of struvite.

Predisposing factors
Urinary tract infections with urease-producing microbes in patients whose urine contains a large quantity of urea
Alkaline urinary pH
Unidentified factors

Characteristics of affected patients
Mean age: six years (<1 to >9 years). Especially common in miniature schnauzers, bichon frises, miniature poodles and cocker spaniels; however, any breed may be affected. More common in females (84%) than males (16%).

BACTERIAL UTIs Clinical and experimental studies of dogs have repeatedly demonstrated a close relationship between formation of struvite uroliths and UTIs caused by urease-producing bacteria.

Staphylococcus and *Proteus* spp are consistent and potent urease producers and have been commonly isolated from animals with infection-induced struvite uroliths. Staphylococci have been more commonly associated with struvite uroliths in dogs than *Proteus* spp.

The bacterial flora of urine may change after formation of struvite uroliths in dogs as a result of staphylococcal UTI. The change in bacterial flora may be associated with damage to local host defense mechanisms by uroliths, iatrogenic infection induced by urinary catheters or administration of antimicrobial agents.

A small percentage of dogs with struvite uroliths have sterile urine. In some of these cases, however, bacteria have been isolated from the inside of uroliths. This observation indicates that bacterial infection of the urinary tract may undergo spontaneous remission after initiating urolith formation in some patients. Bacteria that become trapped within struvite uroliths may remain viable for long periods. Several studies have revealed that calculogenic bacteria harbored within uroliths are protected from the destructive effects of antimicrobial agents in urine.

In contrast to struvite uroliths, bacterial infection of the urinary tract is not a consistent finding in dogs with nonstruvite uroliths (ammonium urate, calcium oxalate, cystine, silica, etc.). When infection does occur in association with these so-called metabolic uroliths, it appears to be a sequela rather than a predisposing cause of urolith formation.

UREAPLASMA UTIs Ureaplasmas differ from all other mycoplasmas by their production of urease and, therefore, their ability to hydrolyze urea. Urea is required for growth of these organisms.

DIET The quantity of dietary protein catabolized for energy influences formation and dissolution of infection-induced struvite uroliths. Consumption of dietary protein in quantities that exceed daily protein requirements for anabolism results in formation of urea from catabolism of amino acids. Hyperammonuria, hypercarbonaturia and alkaluria mediated by microbial urease are dependent on the quantity of urea (the substrate of urease) in urine (**Table 20-31**).

GENETICS The high incidence of struvite urolithiasis in some breeds of dogs such as miniature schnauzers suggests a familial tendency (**Table 20-31**).

Sterile Struvite Uroliths (Pages 655-656)

Clinical studies indicate that microbial urease is not involved in formation of struvite uroliths in some dogs. Several observations suggest that

K-9 UROLITH

Table 20-31. Some potential risk factors for canine infection-induced struvite uroliths.

Diet	Urine	Metabolic	Drugs
High protein content (source of urea)	Urease-positive UTI	Females	Glucocorticoid-associated bacterial UTI
Urine alkalinizing potential	High urea concentration	Breeds Miniature schnauzers	
High phosphorus content	Hyperammonuria	Miniature poodles	
High magnesium content	High-ionic phosphorus concentration	Bichon frises	
	High magnesium levels	Cocker spaniels	
	High pH	Hyperadreno-corticism associated with bacterial UTI	
	Urine retention		
	Concentration of urine and calculogenic substances		

Key: UTI = urinary tract infection.

dietary or metabolic factors may be involved in the genesis of sterile struvite uroliths.

Biologic Behavior (Pages 656-657)

Struvite uroliths can form within two to eight weeks after infection with urease-producing staphylococci. Struvite uroliths associated with UTI caused by staphylococci or *Proteus* spp have been detected in puppies as young as five weeks of age. Spontaneous dissolution of uroliths appears to be uncommon.

Uroliths located in the urinary bladder frequently pass into the urethra. In male dogs, they commonly lodge behind the os penis, but in female dogs they are frequently voided to the exterior. Small renoliths may pass into the ureters. The rapid rates at which struvite uroliths form and the potential they have to migrate to lower portions of the urinary tract are of clinical importance. If several days have elapsed between the date of diagnostic radiography or ultrasonography and the date of surgery scheduled to remove uroliths, the number and location of uroliths should be reevaluated by radiography.

Struvite uroliths have a tendency to recur after surgical removal or medical dissolution. The rate of recurrence following medical dissolution of canine struvite uroliths is less frequent than that associated with surgery. In addition, time elapsed between recurrent episodes is longer following medical dissolution.

Many episodes of multiple recurrences have been associated with lack of removal of all uroliths at the time of surgery (pseudorecurrence), and poor control of recurrent UTI with urease-producing microbes. With the advent of effective therapeutic and preventive antimicrobial protocols to control recurrent or persistent UTI, the frequency of recurrent infection-induced struvite urolithiasis in dogs has declined. The tendency for uroliths to recur after surgery may also be associated with persistence of an environment that favors initiation and growth of struvite at the time of removal.

Assess the Food (Page 657)
Infection-Induced Struvite Uroliths (Page 657)
Both urea and microbial urease are required for ammonia production, alkalinization, supersaturation and subsequent precipitation of struvite crystals. The majority of urea in urine originates from dietary protein, whereas the urease in vertebrates must be derived from microbes (some bacteria, some yeasts or ureaplasmas). Hyperammonuria and alkaluria mediated by microbial urease are dependent on the quantity of urea (the substrate of urease) in urine. In addition, foods high in protein are also high in phosphorus (See the Key Nutritional Factors–Canine Urolithiasis table and **Table 20-31**). The high concentration of urea and phosphorus normally present in urine of individuals consuming dietary protein in excess of daily requirement for protein anabolism makes urine an environment well suited to support the pathogenic effects of urease-producing microbes.

Sterile Struvite Uroliths (Page 657)
Foods high in magnesium and phosphorus would be expected to predispose susceptible dogs to sterile struvite urolith formation.

Feeding Plan and Treatment (Pages 657-659)
Current recommendations for medical dissolution of canine struvite uroliths include: 1) eradication or control of UTI (if present), 2) use of calculolytic foods and 3) administration of urease inhibitors (acetohydroxamic acid) to patients if struvite uroliths persist because of persistent UTI caused by urease-producing microbes (**Table 20-32**).

Infection-Induced Struvite Urocystoliths (Pages 657-659)
ERADICATION OR CONTROL OF UTIs Sterilization of urine appears to be an important objective in creating a state of struvite undersaturation that may prevent further growth of uroliths or that promotes their dissolution.

Appropriate antimicrobial agents selected on the basis of susceptibility or minimum inhibitory concentrations should be used at therapeutic dosages. The fact that diuresis reduces the urine concentration of antimicrobial agents should be considered when formulating antimicrobial dosages. Antimicrobial agents should be administered as long as uroliths can be identified by survey radiography. This recommendation is based on the fact that bacterial pathogens harbored inside uroliths may be protected from antimicrobial agents. Although the urine and surface of

Table 20-32. Summary of recommendations for medical dissolution of canine struvite uroliths.

1. Adult dogs with urinary tract infection (UTI)
   a. Perform appropriate diagnostic studies including complete urinalysis, quantitative urine culture and diagnostic radiography.
      Determine precise location, size and number of uroliths. The size/number of uroliths are not a reliable index of probable therapeutic efficacy.
   b. If uroliths are available, determine their mineral composition. If unavailable, determine their composition by evaluation of appropriate clinical data.
   c. Consider surgical correction if uroliths obstruct urine outflow and/or if correctable abnormalities predisposing the patient to recurrent UTI are identified by radiography or other means. Small urocystoliths may be removed by voiding urohydropropulsion (**Table 20-10**).
   d. Eradicate or control UTIs with appropriate antimicrobial agents. Maintain full-dose antimicrobial therapy during and for three to four weeks after urolith dissolution.
   e. Initiate therapy with calculolytic foods. No other food or mineral supplements should be fed to the patient. Compliance with dietary recommendations is suggested by a reduction in urea nitrogen concentration (usually <10 mg/dl).
   f. Devise a protocol to monitor efficacy of therapy.
      1. Try to avoid diagnostic follow-up studies that require urinary tract catheterization. If they are required, give appropriate peri-catheterization antimicrobial agents to prevent iatrogenic UTIs.
      2. Perform serial urinalyses. Determination of urinary pH and specific gravity and microscopic examination of sediment for crystals are especially important. Remember, crystals formed in urine stored at room or refrigeration temperatures may represent in vitro artifacts.
      3. Perform serial radiography monthly to evaluate urolith location(s), number, size, density and shape.
      4. If necessary, perform quantitative urine cultures. They are especially important in patients infected before therapy and in patients catheterized during therapy.
      5. Feed patients a calculolytic food for one month following disappearance of uroliths as detected by survey radiography.
      6. Consider alternative methods if uroliths increase in size during dietary management, or do not begin to decrease in size after four to eight weeks of appropriate medical management. Difficulty in inducing complete dissolution of uroliths by creating urine that is undersaturated with the suspected calculogenic crystalloids should prompt consideration that: a) the wrong mineral component was identified, b) the nucleus of the uroliths is of different mineral composition than other portions of the urolith and c) the owner or the patient is not complying with medical recommendations.

(Continued on next page.)

Table 20-32. Summary of recommendations for medical dissolution of canine struvite uroliths (Continued.).

---

g. Consider administration of acetohydroxamic acid (25 mg/BW$_{kg}$/day divided into two equal doses) to patients with persistent uroliths and persistent urease-producing microburia despite the use of antimicrobial agents and calculolytic foods.
2. Adult dogs with persistently sterile urine
   a. Follow the protocol described above, but do not administer antimicrobial agents or acetohydroxamic acid.
   b. Periodically culture urine specimens obtained by cystocentesis to detect secondary UTIs. Initiate antimicrobial therapy if a UTI develops.
3. Immature dogs
   a. Use caution when feeding protein-restricted foods to growing dogs.
   b. Short-term therapy with calculolytic foods has been effective in dissolving struvite urocystoliths. If initiated, monitor the patient for evidence of nutritional deficiencies (especially protein malnutrition).
   c. Acetohydroxamic acid has not been evaluated in growing dogs.
   d. Small urocystoliths may be removed by voiding urohydropropulsion (**Table 20-10**). Pending further studies, surgery remains the safest means of removing large uroliths from immature dogs.

---

uroliths may be sterilized following appropriate antimicrobial therapy, the original and secondary infecting organisms may remain viable below the surface of the urolith. Therefore, discontinuation of antimicrobial therapy may result in relapse of bacteriuria and infection.

Although use of antimicrobial agents alone may result in dissolution of struvite uroliths in some patients, experimental studies in rats and dogs and clinical studies in people indicate that this phenomenon represents the exception rather than the rule. In addition to the unpredictable response to this form of therapy, the time required to induce urolith dissolution with antimicrobial agents is usually measured in multiples of months rather than in multiples of weeks.

CALCULOLYTIC FOODS The goal of dietary modification for patients with struvite uroliths is to reduce urine concentration of urea (the substrate of urease), phosphorus and magnesium (See the Key Nutritional Factors–Canine Urolithiasis table and **Table 20-33**.). A commercial veterinary therapeutic food (Prescription Diet Canine s/d, Hill's Pet Nutrition, Inc., Topeka, KS, USA) was formulated that contained a reduced quantity of a high-quality protein and reduced quantities of phosphorus and magnesium. The food was supplemented with sodium chloride to stimulate thirst and induce compensatory polyuria. Reduction of hepatic production of urea from dietary protein reduced renal medullary urea concentration and further contributed to diuresis.

Table 20-33. Levels of key nutritional factors in selected foods for dissolution and prevention of struvite uroliths in dogs.*

Foods	Protein	Phosphorus	Magnesium	Urinary pH**
Recommended levels for dissolution	**<8**	**<0.1**	**<0.02**	**5.9-6.1**
Hill's Prescription Diet Canine s/d, moist	7.6	0.10	0.02	5.9-6.1
Recommended levels for prevention	**<25**	**<0.6**	**0.04-0.10**	**6.2-6.4**
Hill's Prescription Diet Canine c/d, moist	23.5	0.48	0.06	6.2-6.4
Hill's Prescription Diet Canine c/d, dry	21.5	0.54	0.11	6.2-6.4
Hill's Prescription Diet Canine w/d, moist	18.2	0.54	0.09	6.2-6.4
Hill's Prescription Diet Canine w/d, dry	18.9	0.58	0.10	6.2-6.4
Purina CNM DCO-Formula, dry	25.3	0.93	0.13	na
Purina CNM EN-Formula, dry	25.8	0.90	0.08	na
Select Care Canine Control Formula, moist	23.8	0.71	0.08	mean 6.5
Select Care Canine Control Formula, dry	21.2	0.71	0.08	mean 6.5
Waltham S/O Lower Urinary Tract Support, moist	25.2	1.32	0.10	5.5-6.1
Waltham S/O Lower Urinary Tract Support, dry	16.0	0.52	0.07	5.5-6.1

Key: na = information not available from manufacturer.
*Manufacturers' published values. Nutrients expressed as % dry matter. This list represents products with the largest market share for which published information is available.
**Protocols for measuring urinary pH may vary.

The efficacy of Prescription Diet Canine s/d in inducing dissolution of infected struvite uroliths has been confirmed by controlled experimental studies in dogs. The calculolytic food was highly effective in inducing dissolution of struvite uroliths in five of six dogs despite persistent infection with urease-producing bacteria. The uroliths underwent dissolution in about 3.5 months (range eight to 20 weeks). The urolith in the remaining dog decreased to less than one-half its pretreatment size at the termination of the study, six months following initiation of dietary therapy. UTIs persisted in these dogs until the uroliths dissolved, at which time they underwent remission in three dogs. In the corresponding control group fed a maintenance food (10% protein, 0.19% phosphorus and 0.06% magnesium, as fed basis), uroliths increased by a mean of 5.5 times their pretreatment size (range three to eight times). A urolith developed in the renal pelvis of one of these dogs. UTIs persisted in control dogs throughout the six-month study.

Consumption of Prescription Diet Canine s/d by dogs with induced staphylococcal UTI and struvite uroliths was associated with a marked reduction in the serum concentration of urea nitrogen and mild reductions in the serum concentrations of magnesium, phosphorus and albumin. A mild increase in the serum activity of hepatic alkaline phosphatase also was observed. These alterations in serum chemistry values were of no clinical consequence during six-month experimental studies or during clinical studies. However, they underscore the fact that Canine s/d is designed for short-term (weeks to months) dissolution therapy rather than long-term (months to years) prophylactic therapy. Appropriate reduction in concentrations of serum urea nitrogen may be used as one index of client and patient compliance with dietary therapy (**Table 20-34**).

UREASE INHIBITORS Experimental and clinical studies in dogs have revealed that administration of microbial urease inhibitors in pharmacologic doses is capable of inhibiting struvite urolith growth and promoting struvite urolith dissolution. Acetohydroxamic acid given orally to dogs at a dosage of 25 mg/BW_{kg} (divided into two daily subdoses) reduces urease activity, struvite crystalluria and urolith growth. By reducing the pathogenicity of staphylococci, acetohydroxamic acid may also result in less severe dysuria, bacteriuria, pyuria, hematuria and proteinuria.

Although higher dosages of acetohydroxamic acid may result in urolith dissolution, they are not recommended because they may cause a reversible hemolytic anemia and abnormalities in bilirubin metabolism. Likewise, acetohydroxamic acid should not be administered to pregnant dogs because it is teratogenic.

Acetohydroxamic acid may be added to the therapeutic regimen if infection-induced struvite uroliths do not dissolve after an appropriate therapeutic trial with diet modification and antimicrobial agents.

Infection-Induced Struvite Nephroliths (Page 659)

Nephroliths and ureteroliths causing outflow obstruction and marked impairment of renal function should be managed by surgical intervention or, if possible, by percutaneous nephropyelonephrostomy, especially if associated with concomitant bacterial infection. Medical therapy designed to induce urolith dissolution over several weeks is unlikely to be effective in patients with poorly functioning kidneys because uroliths must be completely surrounded by urine that is undersaturated with struvite for prolonged periods to be dissolved. Intermittent passage of urine through a partially obstructed kidney or ureter would logically preclude dissolution of struvite nephroliths or ureteroliths.

Table 20-34. Characteristic clinical findings before and after initiation of medical therapy to dissolve struvite uroliths in nonazotemic dogs.*

Factors	Pre-therapy	During therapy	After successful therapy**
Polyuria	±	1+ to 3+	Negative
Pollakiuria	1+ to 4+	Transient ↑; subsequent ↓	Negative
Gross hematuria	0 to 4+	↓ by 5 to 10 days	Negative
Abnormal urine odor	0 to 4+	↓ by 5 to 10 days	Negative
Small uroliths voided	±	Common in females	Negative
Urine specific gravity	Variable	1.004 to 1.014	Normal
Urinary pH	≥7	Decreased (usually acidic)	Variable
Urine protein	1+ to 4+	Decreased to absent	Negative
Urine RBC	1+ to 4+	Decreased to absent	Negative
Urine WBC	1+ to 4+	Decreased to absent	Negative
Struvite crystals	0 to 4+	Usually absent	Variable
Other crystals	Variable	May persist	May persist
Bacteriuria	0 to 4+	Decreased to absent	Negative
Quantitative bacterial urine culture	0 to 4+	Decreased to absent	Negative
Serum urea nitrogen (mg/dl)	>15	<10	Dependent on food
Serum creatinine	Normal	Normal	Normal
Serum alkaline phosphatase	Normal	↑ by 2 to 5 times	Normal
Serum albumin	Normal	↓ by 0.5 to 1 g/dl	Normal
Serum phosphorus	Normal	Slight decrease	Normal
Urolith size (radiographic)	Small to large	Progressive decrease	Negative
Hemogram	Normal	Normal	Normal

*For dogs with urinary tract infection, therapy consists of a calculolytic food and antimicrobial agents. For dogs without urinary tract infection, therapy consists of a calculolytic food.
**All forms of therapy withdrawn.

Dissolution of nephroliths presumed to be composed of infection-induced struvite in six dogs has been reported. The mean time required for dissolution was 184 days (range 67 to 300 days). Although the dogs had varying degrees of impaired capacity to concentrate urine as a result of pyelonephritis, none had primary renal azotemia at the time therapy was initiated with Prescription Diet Canine s/d and antimicrobial agents. This point is emphasized because dogs with moderate to severe primary renal failure require a greater quantity of protein for anabolism than normal.

Sterile Struvite Uroliths (Page 659)
Current recommendations include use of calculolytic foods (**Table 20-33**) and urine acidifiers (**Table 20-32**). Antibiotics and urease inhibitors are not required unless secondary UTI develops.

Controlled experimental and clinical studies have confirmed the efficacy of Prescription Diet Canine s/d in inducing sterile struvite urolith dissolution. The time required to induce dissolution of sterile struvite uroliths is usually less than that required for infection-induced struvite uroliths.

In a study of induced sterile struvite uroliths in dogs, consumption of Prescription Diet Canine s/d resulted in urolith dissolution in a mean of 3.3 weeks (range two to four weeks).

When the calculolytic food was fed to nine dogs with naturally occurring sterile uroliths presumed to be composed of struvite, uroliths dissolved in a mean of six weeks (range one to three months).

Preliminary studies indicate that protein restriction is not essential for dissolution of canine sterile struvite uroliths. Acidification of urine to approximately 6.0 has been effective in promoting sterile struvite urolith dissolution. However, dietary protein restriction has the advantage of contributing to obligatory polyuria by decreasing renal medullary urea concentration and thus enhancing the rate of sterile struvite urolith dissolution.

Struvite Uroliths in Immature Dogs *(Page 659)*

Struvite urocystoliths have been successfully dissolved in several immature dogs (**Table 20-32**).

Do not feed Prescription Diet Canine s/d to immature dogs for more than two weeks. If the food is used, serially monitor body weight, body condition, serum albumin concentration and packed cell volume for evidence of protein/calorie malnutrition. Appropriate adjustments in dietary management should be made if marked reductions in these variables are observed. The urocystoliths may be removed by voiding urohydropropulsion if they have been reduced enough to pass through a distended urethra.

Reassessment (Pages 659-661)

Because calculolytic foods stimulate thirst and promote diuresis, the magnitude of pollakiuria in dogs with urocystoliths may increase for a variable time following initiation of dietary therapy. Pollakiuria and the abnormal urine odor caused by bacterial degradation of urea usually subside as infection is controlled and uroliths decrease in size (**Table 20-34**).

The size of uroliths should be monitored by survey radiography or ultrasonography at monthly intervals. Survey radiography or ultrasonography is usually preferred to retrograde double-contrast radiography because use of catheters during retrograde radiographic studies may result in iatrogenic UTI. Alternatively, intravenous urography may be considered.

Periodic evaluation of urine sediment for crystalluria also may be considered. Struvite crystals should not form in fresh uncontaminated urine if therapy has been effective in promoting formation of urine that is undersaturated with magnesium ammonium phosphate.

UTIs may persist despite antimicrobial therapy in patients having infection-induced struvite uroliths and consuming the calculolytic food. In most patients, however, the magnitude of bacteriuria is markedly reduced (i.e., from >100,000 to 100 to 1,000 bacteria/ml of urine) and the associated inflammatory response progressively subsides. Although the urine is not sterile, reduction in bacterial colony counts by logarithmic magnitudes (e.g., from 10^6 to 10^4 colony forming units) has a marked effect in reducing the quantity of microbial urease in urine. Diet-induced diuresis should be considered when formulating dosages of antimicrobial agents that will achieve minimum inhibitory concentrations in urine. Concomitant use of calculolytic foods, antimicrobial agents and acetohydroxamic acid is the most effective method of inducing dissolution of uroliths when UTI complications persist.

Urine collected by cystocentesis should be quantitatively cultured during therapy and five to seven days after antimicrobial therapy is discontinued. Results of urine culture may not be the same as results obtained before therapy or from cultures of the interior of uroliths. Rapid recurrence of UTI caused by the same type of organism (relapse) or a different type of bacterial pathogen (reinfection) following withdrawal of antimicrobial therapy may indicate residual uroliths within the urinary tract or other abnormalities in local host defense mechanisms that predispose the patient to UTI and subsequent urolithiasis.

Because small uroliths may escape detection by survey radiography or ultrasonography, Prescription Diet Canine s/d and (if necessary) antimicrobial agents should be continued for at least one month after radiographic or ultrasonographic documentation of urolith dissolution. This protocol is likely to prevent rapid recurrence of radiographically detectable uroliths and bacterial UTI after therapy is discontinued.

Alternate methods of management should be considered if uroliths increase in size during therapy or if urolith size remains unchanged after approximately eight weeks of appropriate medical therapy. Small uroliths that become lodged in the urethra of male or female dogs during therapy may be readily returned to the urinary bladder lumen by retrograde urohydropropulsion. Complete obstruction of a ureter or renal pelvis with a urolith, especially with concomitant UTI, is an indication for surgical intervention.

Attempts to induce dissolution of struvite uroliths may be hampered if the uroliths are heterogeneous in composition (**Table 20-35**). This has not been a significant problem in dogs with uroliths composed primarily of magnesium ammonium phosphate with lesser quantities of calcium apatite because the solubility characteristics of the two minerals are similar. However, difficulty may be encountered in dissolving uroliths composed primarily of struvite with an outer shell composed primarily of calcium apatite. Difficulty will also be encountered in attempting to induce complete dissolution of a urolith with a nucleus of calcium oxalate or silica and a shell struvite because

Table 20-35. Managing magnesium ammonium phosphate uroliths refractory to complete dissolution.

Causes	Identification	Therapeutic goals
Client and patient factors		
Inadequate dietary compliance	Question owner Persistent struvite crystalluria Urea nitrogen >8-12 mg/dl Urine specific gravity >1.010-1.015 Urinary pH is alkaline during treatment with Prescription Diet Canine s/d*	Emphasize need to feed dissolution food exclusively
Inadequate antibiotic administration	Question owner Count remaining antibiotic pills	Emphasize need to administer the full dose of antibiotics Determine if owner is capable and willing to administer medication Demonstrate a variety of methods to administer medication
Clinician factors		
Incorrect prediction of mineral type	Analysis of retrieved urolith	Alter therapy based on identification of mineral type
Inappropriate antibiotic choice	Positive urine culture with poor susceptibility for chosen antibiotic	Choose antibiotics based on susceptibility testing
Inappropriate antibiotic dose for degree of diuresis	Positive quantitative urine culture with same bacterial species and same susceptibility; number of bacteria may be lower (See text.)	Administer antibiotic at the higher recommended dose or consider a higher dose than recommended
Premature discontinuation of antibiotic	Discontinuing antibiotic before complete urolith dissolution Positive urine culture with same bacterial species and the same susceptibility (See text.)	Prescribe full antibiotic dose for the entire period of urolith dissolution

K-9 UROLITH

(Continued on next page.)

Table 20-35. Managing magnesium ammonium phosphate uroliths refractory to complete dissolution (Continued.).

Causes	Identification	Therapeutic goals
Disease factors		
Change in bacterial susceptibility	Positive urine culture with susceptibility results different from those of previous culture	Choose antibiotic based on susceptibility testing
New bacterial infection	Positive urine culture identifying new bacterial species	Choose antibiotic effective against both bacteria
		Avoid procedures requiring urinary tract catheterization
Compound urolith	Radiographic density of nucleus and outer layer(s) of urolith is different	Alter therapy based on identification of new mineral type
	Analysis of retrieved urolith	Uroliths not causing clinical signs should be monitored for potentially adverse consequences (obstruction, urinary tract infection, etc.)
		Clinically active uroliths may require removal
		Remove small uroliths by voiding urohydropropulsion

*Hill's Pet Nutrition, Inc., Topeka, KS, USA.

the solubility characteristics of these minerals are dissimilar. This phenomenon should be considered if medical therapy seems to be ineffective after initially reducing the size of a urolith.

Precautions with Calculolytic Foods (Pages 660-661)

There are benefits and risks associated with feeding calculolytic foods. Not all patients are candidates for dietary medical management. Benefits and risks of such therapy should be considered and discussed with the client if the following problems coexist in dogs with struvite uroliths, or if risk factors for their development are present.

ABNORMAL FLUID RETENTION The food (Prescription Diet Canine s/d) designed to dissolve canine struvite uroliths is restricted in protein and supplemented with sodium chloride. Both could affect fluid balance. Therefore, the food should not be routinely fed to patients with concomitant diseases associated with positive fluid balance (e.g., heart failure, nephrotic syndrome) or hypertension.

AZOTEMIC PRIMARY RENAL FAILURE Complete obstruction of urine outflow caused by uroliths in patients with a concomitant UTI should be regarded as an emergency. In this situation, rapid spread of infection and associated damage to the urinary tract, especially the kidneys, are likely to induce septicemia and acute renal failure by a combination of obstruction and pyelonephritis. The risks and benefits associated with medical therapy to dissolve uroliths should not be considered until adequate urine flow has been restored.

Nonobstructing struvite nephroliths have been dissolved in patients with nonazotemic renal failure caused by ascending pyelonephritis. Nevertheless, protein-restricted calculolytic foods should be used with caution in patients with azotemic primary renal failure. Such foods may induce protein malnutrition if given for prolonged periods to dogs with moderate azotemic primary renal failure.

To minimize adverse drug reactions/events, adjustments in doses and maintenance intervals of drugs excreted primarily by the kidneys should be considered in patients with azotemic primary renal failure.

PATIENTS AT RISK FOR PANCREATITIS Approximately one in 250 dogs seen in private veterinary practice is affected by pancreatitis (0.4%). There appears to be no relationship between pancreatitis and gender, but there is a significant relationship between the disease and age. The mean age of dogs with pancreatitis in private veterinary practice is eight years vs. 5.5 years for the general canine population. Breed is another strong risk factor for pancreatitis. For example, miniature schnauzers have a fivefold increase in risk for pancreatitis (i.e., about one in 50 miniature schnauzers can be expected to have pancreatitis.) Other breeds at increased risk include bichon frises, Yorkshire terriers, Chihuahuas, Jack Russell terriers, Japanese spaniels, Labrador retrievers, Maltese, and Shetland sheepdogs. (See Chapter 22.)

Investigators conducting an independent epidemiologic study recently surveyed veterinarians to ascertain the health of dogs fed Prescription Diet brand pet foods. This study disclosed an association between feeding Prescription Diet Canine s/d and acute pancreatitis. The risk of a dog developing pancreatitis when fed Canine s/d is comparable to that of a miniature schnauzer developing acute pancreatitis, or about one in 40 (i.e., about one in 40 dogs fed Canine s/d might develop pancreatitis).

The calculolytic food is relatively high in fat, which serves primarily as a source of calories. Because dietary fat is a risk factor for pancreatitis, the serum activity of pancreatic enzymes (amylase, lipase, trypsin-like immunoreactivity) should be monitored before initiating therapy in patients known to be at increased risk for pancreatitis. These tests should be repeated if signs of pancreatitis develop during therapy. Because abnormal increases in activity of these enzymes are not pathognomic for pancreatitis, other relevant findings should also be considered.

Female miniature schnauzers are at increased risk for infection-induced struvite uroliths and pancreatitis. Likewise, patients with hyperadrenocorticism are at increased risk for UTIs (which could include staphylococci) and pancreatitis. Although risk factors are not synonymous with cause and effect, clients should be informed of these associations and advised of how to respond to adverse events if they occur. They should be informed about adverse events that need medical attention and those that need medical attention only if they continue or are bothersome.

Prevention (Pages 661-662)
Infection-Induced Struvite Uroliths (Pages 661-662)

Eradication or control of UTIs due to urease-producing bacteria is the most important factor in preventing recurrence of most infection-induced struvite uroliths. If UTI persists or is recurrent, indefinite therapy is indicated with prophylactic dosages of antimicrobial agents eliminated in high concentration in urine. These may include amoxicillin, nitrofurantoin, enrofloxacin and trimethoprim-sulfadiazine; however, the final choice is best determined by the results of the most recent antimicrobial susceptibility test. In light of the effectiveness of foods in inducing dissolution of struvite uroliths, use of dietary modification (**Table 20-33**) to prevent recurrence of uroliths is logical and feasible. Long-term use of calculolytic foods with severely reduced protein levels is recommended only if patients develop frequently recurrent urolithiasis despite attempts to control infection, augment fluid intake and urine acidification.

Administration of 25 mg of acetohydroxamic acid/BW_{kg}/day to dogs with urinary bladder foreign bodies (zinc disks) and experimentally induced urease-positive staphylococcal UTIs was effective in preventing formation of and minimizing the growth rate of uroliths.

Sterile Struvite Uroliths (Page 662)

Sterile struvite uroliths have a greater tendency to recur than do infection-induced struvite uroliths in which the UTI has been eradicated or controlled. Administration of urine acidifiers should be considered if the urinary pH of patients with sterile struvite uroliths remains alkaline despite dietary therapy.

Uncontrollable risk factors (i.e., defective inhibitors of crystal formation and/or defective inhibitors of crystal aggregation) may be present in those situations in which dogs have documented occurrences of either calcium oxalate or calcium phosphate followed by struvite urolithiasis. If struvite urolithiasis is associated with urease-positive UTI, appropriate therapy should be devised to eradicate the UTI and prevent its recurrence. When considering dietary management (**Table 20-33**), emphasis should be placed on minimizing recurrence of calcium oxalate or calcium phosphate uroliths, because medical management cannot dissolve these types of uroliths. Should struvite uroliths recur, they often can be dissolved by dietary management and antimicrobial agents, if necessary. This strategy tends to minimize the need for repeated surgical intervention.

SILICA UROLITHIASIS (Pages 662-664)

Prevalence and Mineral Composition (Page 662)

Silica accounted for 0.9% of all canine uroliths and 0.5% of all canine nephroliths analyzed at the Minnesota Urolith Center from 1981 to 1997. Information about breed, gender and age predisposition, serum and urine profiles and silica urolith radiographic density and contour is listed in **Table 20-3**.

Etiopathogenesis and Risk Factors (Pages 662-663)

Relationship of Silica Uroliths to Diet (Pages 662-663)

Corn gluten feed, a byproduct of the wet milling and distilling process designed to separate shelled corn into various components, may be a source of silica in some pet foods. Corn gluten feed remains after extraction of starch, gluten and germ from shelled corn. Corn gluten feed contains about 40% protein and is found in some low-quality pet foods. Corn gluten feed is not the same as corn gluten meal. Corn gluten meal is found in many higher quality manufactured foods because it is readily digestible and a good source of protein (approximately 60%), vitamins, minerals and energy. Corn gluten meal is an unlikely source of the silica in uroliths. Another potential source of silica in foods is microfine silica, which is used in small quantities as an anti-caking agent in the pet food manufacturing process. Although a cause and effect relationship between microfine silica and silica urolithiasis is unlikely, it seems logical to avoid giving foods containing this ingredient to dogs with recurrent silica urolithiasis.

Assess the Food (Page 663)

Foods with large quantities of plant-derived ingredients are suspected to be risk factors for silica uroliths in susceptible dogs. Corn gluten feed, rice hulls and soybean hulls have been incriminated.

Concentrations of calculogenic substances in urine are dependent on urine volume. Because commercial dry foods are associated with production of a lesser volume of urine than moist foods, consumption of dry foods may also be considered a risk factor for silica urolith formation.

Feeding Plan and Treatment (Pages 663-664)

Effective medical protocols to induce dissolution of canine silica jackstones have not yet been developed. Calculolytic foods that do not contain large quantities of vegetable proteins and that induce diuresis may prevent further growth of silica uroliths. Voiding urohydropropulsion may be used to remove small urocystoliths (**Table 20-10**). Surgery is the only viable alternative to remove large silica uroliths.

Because initiating and perpetuating causes of silica urolithiasis are unknown, only nonspecific measures to reduce the degree of supersaturation of urine with calculogenic substances can be recommended for prevention. At this time, recommendations include change of diet, augmentation of urine volume and possibly altering urinary pH (**Table 20-36**).

Although the role of diet in the genesis of canine silica uroliths is speculative, it seems reasonable to recommend that the food of affected patients be changed, especially if the problem is recurrent. Although empirical, this maneuver is unlikely to be harmful and may be helpful. Based on the assumption that the primary source of excessive silica in foods is vegetable in origin, selection of a food with reduced quantities of plant ingredients is recommended.

For dogs with recurrent silica urolithiasis, increasing the volume of urine produced by increasing water consumption will increase the volume of urine in which calculogenic substances are dissolved or suspended. Moist foods rather than dry foods should be considered. Oral administration of sodium chloride has been a favored empirical method to induce diuresis in dogs with uroliths. However, the use of sodium chloride to promote diuresis in dogs that form silica uroliths is not recommended because of the unpredictable but marked occurrence of calcium oxalate in silica uroliths and because orally administered sodium chloride is associated with hypercalciuria.

Silica is less soluble in acidic than alkaline water and currently available information suggests that silica is less soluble in acidic than alkaline biologic environments. It is noteworthy that the urinary pH of eight uninfected dogs with silica uroliths was acidic to neutral at the time of diagnosis (mean 6.0, range 5.0 to 7.0). Whether or not alkalinization of urine is of benefit in increasing the solubility of silica or silicates in urine is unknown. Likewise, the effects of orally administered alkalinizing agents (e.g., sodium bicarbonate) on the absorbability of silica from the gastrointestinal tract have not been evaluated. Nonetheless, it seems prudent to avoid efforts to deliberately acidify the urine of dogs with recurrent silica uroliths.

Table 20-36. Summary of recommendations for prevention of canine silica uroliths.

1. Perform appropriate diagnostic studies including complete urinalysis, quantitative urine culture and diagnostic radiography. Determine precise location, size and number of uroliths.
2. If uroliths are available, determine their mineral composition. If unavailable, determine their composition by evaluation of appropriate clinical data.
3. Small urocystoliths may be removed by voiding urohydropropulsion (**Table 20-10**). Consider surgical removal of larger uroliths causing clinical disease.
4. To prevent further growth of existing silica uroliths or to prevent recurrence of silica uroliths after surgical removal:
   a. Avoid use of foods containing large quantities of plant proteins, and especially avoid those containing soybean hulls or corn gluten feed.
   b. Enhance diuresis by adding moisture to the food.
   c. Avoid efforts to deliberately acidify urine.
5. If necessary, eradicate or control urinary tract infections with appropriate antimicrobial agents.

K-9 UROLITH

Reassessment (Page 664)

The goal of therapy is to prevent silica urolith recurrence. Dietary management should minimize exposure to minerals predisposing to silica uroliths and result in formation of less concentrated urine; strive to achieve urine specific gravity values less than 1.020. After this is achieved, a urinalysis and lateral abdominal radiograph should be performed every three to four months. By following these recommendations, recurrent urocystoliths detected by radiography should be small enough to remove by voiding urohydropropulsion. After a rate of urolith recurrence has been established, the frequency of evaluation can be modified such that predicted recurrences can be diagnosed and managed accordingly.

COMPOUND UROLITHS (Page 664)

Compound uroliths (i.e., nucleus composed of one mineral type and shells of a different mineral type) occur in approximately 7% of the canine uroliths analyzed at the Minnesota Urolith Center. Examples include: 1) a nucleus of 100% calcium oxalate monohydrate surrounded by a shell of 80% magnesium ammonium phosphate and 20% calcium phosphate, 2) a nucleus composed of 95% magnesium ammonium phosphate and 5% calcium phosphate surrounded by a shell of 95% ammonium acid urate and 5% magnesium ammonium phosphate and 3) a nucleus composed of 95% silica and 5% calcium oxalate monohydrate surrounded by a shell of 100% calcium oxalate monohydrate.

Voiding urohydropropulsion may be used to remove small compound urocystoliths (**Table 20-10**). Surgery remains the only reliable method to remove large compound uroliths.

Because risk factors that predispose patients to precipitation (nucleation) of different minerals vary, the occurrence of compound uroliths poses a unique challenge in terms of preventing recurrence. In the absence of clinical evidence to the contrary, it seems logical to recommend management protocols designed primarily to minimize recurrence of minerals composing the nucleus (rather than those in shells) of compound uroliths. Follow-up studies designed to evaluate efficacy of preventive protocols should include complete urinalyses, radiography or ultrasonography and if available, evaluation of the urine concentration of calculogenic metabolites.

Clinical cases that illustrate and reinforce the nutritional concepts presented in this chapter can be found in Small Animal Clinical Nutrition, 4th ed., pp 670-688.

CHAPTER 21

Feline Lower Urinary Tract Disease

For a review of the unabridged chapter, see Allen TA, Kruger JM. Feline Lower Urinary Tract Disease. In: Hand MS, Thatcher CD, Remillard RL, et al, eds. Small Animal Clinical Nutrition, 4th ed. Topeka, KS: Mark Morris Institute, 2000; 689-723.

CLINICAL IMPORTANCE (Pages 689-690)*

FLUTD

Approximately 4.6% of all cats are affected by lower urinary tract disease. Urolithiasis, urinary tract infection (UTI), neoplasia, inflammation and congenital defects of the urinary tract cause signs of lower urinary tract disease. Although feline lower urinary tract disease (FLUTD) has diverse causes, including bacterial and viral infections and interstitial cystitis, this chapter will focus on struvite and calcium oxalate urolithiasis because these disorders are most amenable to nutritional therapy. Generally, idiopathic lower urinary tract disease resolves within seven to 10 days, regardless of the feeding plan.

ASSESSMENT (Pages 690-714)

Assess the Animal (Pages 690-713)
History (Pages 690-691)
The dietary history should include specific brands of food fed, the form (dry, moist, semi-moist or a combination), method of feeding (meal fed, free choice) and whether table food, supplements and treats are offered. Access to other food should be assessed (e.g., other pets in the household that eat different foods, access to food at other households, etc.). Trends in water consumption (i.e., increased, decreased, unchanged) should be ascertained and recorded.

The pet owner should be carefully questioned about: 1) the duration of clinical signs, 2) progression of clinical signs (same, better, worse), 3) whether the episode was the patient's first or a recurrence, 4) the interval between recurrences, 5) previous treatments (medical, surgical, nutritional), including doses of pharmaceutical agents prescribed and response to therapy, 6) presence of other illnesses, injuries or trauma (cur-

*Page numbers in headings refer to Small Animal Clinical Nutrition, 4th ed., where additional information may be found.

Key Nutritional Factors–Feline Lower Urinary Tract Disease
(Page 691)

Factors	Dietary recommendations
Struvite uroliths and urethral plugs	
Water	Promote water intake by using a moist food or other measures
Protein	Avoid excess dietary protein
	Dissolution and prevention: restrict protein to 30 to 45% DM*
Phosphorus	Avoid excess dietary phosphorus
	Dissolution: restrict dietary phosphorus to 0.5 to 0.8% DM
	Prevention: restrict dietary phosphorus to 0.5 to 0.9% DM
Magnesium	Avoid excess dietary magnesium
	Dissolution: restrict dietary magnesium to 0.04 to 0.06% DM
	Prevention: restrict dietary magnesium to 0.04 to 0.10% DM
Fat	Increased energy density reduces overall mineral intake
	Avoid obesity, a risk factor for FLUTD in some studies
	Dissolution and prevention: foods with 8 to 25% DM fat can routinely be fed, depending on the cat's body condition and the desired energy density of the food
Mean daily urinary pH	Use a food that maintains an acidic urine
	Dissolution: urinary pH = 5.9 to 6.1
	Prevention: urinary pH = 6.2 to 6.4
Calcium oxalate uroliths (prevention)	
Water	Promote water intake by using a moist food or other measures
Protein	Avoid excess dietary protein
	Restrict dietary protein to 30 to 45% DM
Fat	Increased energy density reduces overall mineral intake
	Avoid obesity, a risk factor for FLUTD in some studies
	Foods with 8 to 25% DM fat can routinely be fed, depending on the cat's body condition and the desired energy density of the food
Calcium	Avoid excess dietary calcium
	Restrict dietary calcium to 0.5 to 0.8% DM
Magnesium	Avoid excess or deficient dietary magnesium (0.04 to 0.10% DM)
Mean daily urinary pH	Use a food that maintains a urinary pH between 6.6 to 6.8

*DM = dry matter, FLUTD = feline lower urinary tract disease.

rent or previous), 7) presence of systemic signs (anorexia, vomiting, diarrhea, weight loss) and 8) presence of localizing signs such as licking at the prepuce or vulva and altered micturition or altered urine characteristics.

The pet owner's description of micturition is especially important. Questions should be directed to determine the presence or absence of the following parameters: 1) dysuria, 2) pollakiuria, 3) urinary incontinence, 4) micturition in unusual places, 5) hematuria and 6) uroliths or urethral plugs voided during micturition. The approximate urine volume and changes should be determined.

Pharmaceutical agents administered should be recorded as described above. Specific pharmaceutical agents may be risk factors for FLUTD. Corticosteroids and furosemide can predispose cats to hypercalcemia and hypercalciuria. Sulfadiazine-containing drugs may predispose cats to uroliths containing varying amounts of sulfadiazine. Allopurinol may predispose cats to xanthine uroliths or shells. Urinary acidifiers can cause metabolic acidosis and subsequent hypercalciuria and may predispose cats to calcium-containing uroliths.

Physical Examination (Page 691)

The urinary bladder should be palpated to evaluate its size, shape, surface contours and thickness of the bladder wall. Pain, masses and grating within the bladder lumen should be carefully noted. Most feline urocystoliths, however, cannot be detected by abdominal palpation. The penis and prepuce should be examined for urethral abnormalities. Rectal palpation should be performed to assess the size, position and shape of the urethra and any associated masses or pain. The kidneys should be evaluated for size, shape, surface contour and bilateral symmetry. The patient should be observed micturating, if possible (i.e., size of urine stream, dysuria, urine color).

Laboratory and Other Clinical Information (Pages 691-696)

The following diagnostic studies are often used to evaluate cats with lower urinary tract disease: 1) urinalyses, 2) quantitative urine culture(s), 3) survey abdominal radiography, 4) contrast urethrocystography if urethral plugs or uroliths are noted, 5) intravenous urography or contrast cystography if cystoliths are noted, 6) intravenous urography if ureteral or renal uroliths are noted, 7) ultrasonography of kidneys and urinary bladder, 8) complete blood cell counts, 9) serum biochemistry profiles and 10) cystoscopy (**Table 21-1**).

URINALYSIS Urine sediment findings in cats with urolithiasis usually indicate inflammation (i.e., pyuria, proteinuria, hematuria and increased numbers of epithelial cells). Microscopic examination of urine sediment

Table 21-1. Summary of diagnostic plans—feline lower urinary tract disease.

Clinical findings	Rule outs	Uncomplicated min. database	Recurrent min. database
Hematuria with dysuria	Idiopathic disease Urolithiasis	Urinalysis Survey radiography	Bacterial urine culture Double-contrast cystography Serum biochemistry profile Ultrasonography Cystoscopy
Hematuria without dysuria	Renal hemorrhage Lower urinary tract hemorrhage Coagulopathy Catheter or cystocentesis related	Voided urinalysis Survey radiography Complete blood count Bleeding time Prothrombin time	Intravenous urography Platelet count Cystoscopy Serum biochemistry profile
Urethral obstruction	Urethral plug (matrix + crystals) Urolithiasis	Urinalysis Urine culture Survey radiography Urea nitrogen Creatinine Potassium	Urethrocystography Ultrasonography Serum biochemistry profile

helps: 1) detect conditions that may predispose cats to urolith or urethral plug formation, 2) infer mineral composition of uroliths or urethral plugs and 3) evaluate response to treatment or preventive measures. Hematuria and proteinuria are typical urinalysis findings in cats with idiopathic lower urinary tract disease.

Crystals only form when urine is supersaturated with crystallogenic materials. Therefore, crystalluria is a risk factor for formation of uroliths and urethral plugs. However, crystalluria is not, in itself, pathognomonic for uroliths or urethral plugs. Urine crystals can be an artifact due to external factors that influence crystal formation, such as temperature, evaporation, urinary pH and method of sediment specimen preparation. Conversely, urolithiasis is possible without associated crystalluria.

Because crystals formed after voiding are not significant, examination of sediment from fresh, warm urine specimens is recommended. Diagnostic agents (e.g., radiographic contrast agents) and drugs (e.g., sulfonamides) can produce urinary crystals. Several factors influence the number of crystals, including the volume of urine centrifuged, centrifugation speed and the volume of sediment resuspended and transferred to the microscope slide. Consequently, it is difficult to attach clinical significance to the number of crystals observed. In addition to evaluating crystal type, the sediment should be evaluated for tendencies for crystals to aggregate. Detection of large aggregates of struvite or calcium oxalate crystals is an important finding when monitoring the efficacy of preventive measures.

URINE CULTURE Infection with urease-producing staphylococci and *Proteus* spp may be associated with formation of struvite uroliths (**Table 21-2**). Urinary pH is persistently alkaline with UTI due to urease-producing microorganisms. Infection can also result from bacterial colonization of the urinary tract due to urolith-induced changes in host defense mechanisms. Thirty percent of feline patients with urocystoliths have positive urine cultures. Although UTIs are uncommon in young cats, they become a significant cause of urinary tract disease in cats 10 years of age and older. Microorganisms may remain viable within uroliths. Culture of urine may be negative or yield the same or different organisms than cultures from uroliths. Urine cultures are negative in cats with idiopathic lower urinary tract disease and feline interstitial cystitis.

URINARY pH Urinary pH influences formation of several crystal types. Although there are exceptions, certain crystal types tend to form and persist at certain urinary pH ranges (**Table 21-2**). In general, struvite uroliths are associated with an alkaline urinary pH and calcium oxalate uroliths are associated with an acidic urinary pH.

pH values measured with multi-test reagent strips and test tapes are only accurate to within 0.5 pH units. Indicator squares on reagent strips should be compared with the manufacturer's color standards only in well-illuminated areas.

Urinary pH varies throughout the day due to the influence of food, time of eating, method of feeding and amount of food consumed. Consequently, it is difficult to interpret a single urinary pH value, especially if time of eating and food are unknown.

RADIOGRAPHY The rationale for radiography is to confirm the diagnosis of urolithiasis. Radiography can determine the size, shape, location and number of uroliths. Survey radiography or ultrasonography may fail to detect small uroliths (i.e., <3 mm in diameter). Uroliths greater than 1 mm in diameter can usually be detected with double-contrast cystography, if care is taken not to infuse too much contrast medium.

The relative radiodensity of uroliths can be used to make a rough guess of mineral composition (**Table 21-2**). Calcium oxalate and struvite uroliths are usually radiodense. Radiographic shape, contour and size can also be used as an inexact predictor of mineral composition. Struvite uroliths can be

Table 21-2. Checklist of factors that suggests probable mineral composition of feline uroliths.

Urinary pH

Struvite and calcium apatite uroliths, usually alkaline. Sterile struvite uroliths may be observed with urinary pH values of ≥6.5.

Ammonium urate uroliths, acidic to neutral.*

Cystine uroliths, acidic.*

Calcium oxalate uroliths, often acidic to neutral.*

Identification of crystals in uncontaminated fresh urine sediment, preferably at body temperature

Type of bacteria, if any, isolated from urine

Urease from bacteria, especially staphylococci and less frequently *Proteus* spp, may be associated with struvite uroliths.

Urinary tract infections (UTIs) often are absent in patients with calcium oxalate, cystine or ammonium urate uroliths.

Calcium oxalate, cystine or ammonium urate uroliths may predispose to UTIs. If infections are caused by urease-producing bacteria, struvite may precipitate around metabolic uroliths.

Radiographic density and physical characteristics of uroliths

Struvite, 1+ to 4+ radiopacity, uroliths are rough or smooth, round or faceted, sometimes disk-shaped.

Calcium oxalate, 3+ to 4+ radiopacity, uroliths are rough or smooth, usually small, occasionally jackstone shaped.

Calcium phosphate, 4+ radiopacity, uroliths are smooth or rough, round or faceted.

Cystine, 0 to 2+ radiopacity, uroliths are smooth, small.

Ammonium urate/uric acid, 0 to 2+ radiopacity, uroliths are smooth, occasionally irregular.

Serum biochemistry evaluation

Hypercalcemia may be associated with calcium-containing uroliths.

Hyperuricemia may be associated with uric acid or urate uroliths.

Hyperchloremia, hypokalemia and acidemia may be associated with distal renal tubular acidosis and calcium phosphate or struvite uroliths.

Urine chemistry evaluation

Patient should be consuming a standard diagnostic food or the food consumed when uroliths formed.

Excessive quantities of one or more minerals contained in the urolith are expected. The concentration of crystallization inhibitors may be decreased.

Breed of cat and history of uroliths in patient's ancestors or littermates

Drugs

Corticosteroids and furosemide predispose to hypercalciuria.

Allopurinol predisposes to xanthine uroliths.

Drugs containing sulfadiazine predispose to formation of uroliths containing varying quantities of sulfadiazine.

Quantitative analysis of uroliths voided during micturition or collected via catheter technique

---

*Concomitant infection with urease-producing microbes may result in alkaline urine.

smooth or rough, round or faceted. Calcium oxalate dihydrate uroliths are usually small, rough and round to oval. Calcium oxalate monohydrate uroliths are usually small, smooth and round. Occasionally, calcium oxalate monohydrate uroliths have a jackstone appearance. The size and number of urocystoliths does not predict whether medical dissolution will be successful.

Feline nephroliths are more commonly composed of calcium salts than struvite. Only approximately 5% of nephroliths are struvite. Nephroliths must be differentiated from dystrophic or metastatic calcification of renal parenchyma, calcified mesenteric lymph nodes and ingesta or medications in the intestinal tract.

SERUM BIOCHEMISTRY PROFILE Serum biochemistry profiles are useful in cases of recurrent FLUTD (**Table 21-2**). Approximately one-third of cats with calcium oxalate uroliths are hypercalcemic. In evaluating cats with concurrent calcium oxalate uroliths and hypercalcemia, the veterinarian should rule out potential causes of persistent hypercalcemia such as hyperparathyroidism, malignancy-associated hypercalcemia and hypervitaminosis D. Presumably, persistent hypercalcemia increases the risk of forming calcium-containing uroliths by increasing the excretion of calcium in urine. It is also possible that the processes involved with formation of calcium-containing uroliths and hypercalcemia are unrelated.

Acidemia, as evidenced by decreased total carbon dioxide, is also common in patients with calcium oxalate uroliths. Metabolic acidosis may contribute to calcium urolith formation. Normal serum concentrations do not rule out increased urinary concentrations of calculogenic substances. Increased serum concentrations of calculogenic substances (e.g., calcium) may provide clues about the underlying etiology of uroliths.

UROLITH ANALYSIS Because recommendations for urolith dissolution and prevention are mineral composition specific, it is important to analyze uroliths whenever possible. In addition to surgical removal, several less invasive techniques for obtaining uroliths should always be considered. These methods include retrieval with a urinary catheter, voiding urohydropropulsion and voiding into an empty or plastic bead-filled litter box.

Uroliths can be analyzed qualitatively or quantitatively. Qualitative analysis uses spot tests to identify radicals and ions; however, these tests do not reveal the proportion of mineral types and do not detect certain mineral crystals such as silica and drug crystals such as sulfadiazine. Uroliths, therefore, should be analyzed quantitatively. See Small Animal Clinical Nutrition, 4th ed., p 694 for a list of laboratories that analyze uroliths quantitatively.

Different minerals may be deposited in layers or mixed throughout the urolith. Although one mineral type predominates, the composition of uroliths is frequently mixed. Thus, sampling and reporting results from different parts of the urolith become important when considering urolith dissolution and prevention.

Risk Factors *(Page 696)*

A review of data from the Veterinary Medical Data Base, Purdue University, West Lafayette, IN, collected between 1980 and 1990 revealed that lower urinary tract disease is more prevalent in cats one to 10 years of age than in cats less than one year of age or more than 10 years of age. The mineral type of urocystoliths and urethroliths in kittens is usually struvite, associated with infection with urease-producing microorganisms. In general, metabolic uroliths (e.g., calcium oxalate) are rarely observed in kittens (**Figure 21-1**). The risk of calcium oxalate urolith formation increases with age.

Potential risk factors for formation of sterile struvite uroliths include mineral and moisture content and energy density of the food, method of feeding, urine concentration and retention of urine (**Table 21-3**).

Burmese, Persian and Himalayan breeds are at increased risk for developing calcium oxalate uroliths, but at reduced risk for developing struvite uroliths.

Nutritional risk factors associated with calcium oxalate urolith formation in cats include urinary acidifiers, acidifying foods, limited food variety, free-choice feeding and dry foods. By far, the strongest association in one epidemiologic study was urinary acidifiers. Nephroliths are more likely to be composed of calcium oxalate than struvite.

Etiopathogenesis *(Pages 696-708)*

The clinical signs of hematuria, dysuria, pollakiuria and/or urethral obstruction are the hallmarks of FLUTD. It is well established that these signs may result from a number of different etiologies affecting the lower urinary tract (**Table 21-4**).

The causes of FLUTD may act alone or in combination. The appropriate descriptive term should be applied if a specific cause has been identified. If clinical signs are present and a specific cause is not identified after appropriate evaluation, the preferred term is idiopathic FLUTD.

Between January 1980 and June 1993, 221,477 cats were examined at 23 veterinary colleges and findings recorded in the Veterinary Medical Data Base, Purdue University. Of these 221,477 cats, 15,349 (6.9%) were diagnosed with some type of lower urinary tract disease. Within this category, the five most common recorded diagnoses were feline urologic syndrome (34.3%), cystitis (29.8%), urethral obstruction (21.9%), urethral uroliths (7.6%) and urocystoliths (4.7%). Other less common diagnoses included urinary incontinence, bacterial cystitis, urethral stricture, urinary bladder diverticula and neoplasia.

The real prevalence of the various etiologies of FLUTD is unknown and difficult to discern from a retrospective analysis of a large multicenter data-

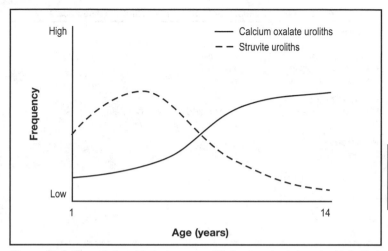

Figure 21-1. The relationship between urolith type and age in cats. Note that struvite urolithiasis occurs more frequently in young cats, whereas calcium oxalate urolithiasis occurs more frequently in older cats.

Table 21-3. Some risk factors reported to occur in cats with lower urinary tract disease.

Factors	Comments
Age	Uncommon in cats younger than one year. Most common between ages one and 10 years, with peak between two and six years.
Gender	Urethral obstruction occurs most commonly in males. Males and females have a similar risk for nonobstructive forms of the disease.
Neuter status	Increased risk of disease in neutered males and females, regardless of age at neutering.
Food	Consumption of an increased proportion of dry food in the daily ration is associated with increased risk of disease.
Feeding frequency	Increased frequency of feeding is associated with an increased risk of disease, regardless of food.
Excessive weight	Obesity is associated with increased risk of disease.
Water consumption	Decreased daily water consumption is associated with increased risk of disease.
Sedentary lifestyle	Inactive cats are at increased risk of disease.
Spring or winter season	Seasonal variation has been implicated as a risk factor by some investigators, but not by others.
Indoor lifestyle	Cats using indoor litter boxes for micturition and defecation have increased risk of disease.

FLUTD

Table 21-4. Summary of causes of lower urinary tract disease in cats.

Uroliths

Struvite	Uric acid
Calcium oxalate	Cystine
Calcium phosphate	Xanthine
Ammonium urate	Matrix

Urethral plugs

Struvite crystals only	Matrix and struvite crystals
Matrix only	Matrix and other crystals

Infections

Bacterial	Viral?
Fungal	Others?
Parasitic	

Other causes

Interstitial cystitis	Neurogenic
Anatomic abnormalities	Traumatic
Neoplastic	Iatrogenic

base. The largest and most detailed prospective, clinical series characterizing FLUTD was performed from 1982 to 1985 at the University of Minnesota. **Table 21-5** presents updated data from that study. Three groups of untreated cats were selected for this series: unobstructed female cats with hematuria, dysuria or both; unobstructed male cats with hematuria, dysuria or both; and male cats with urethral obstruction. Uroliths were detected in 23% (32 of 141) of cats in this study. The majority of the uroliths were struvite. An additional 21% (30 of 141) had urethral plugs. Struvite was the most commonly identified mineral component in these plugs. Struvite crystalluria was also more common in cats with urethral plugs than in cats with other causes of FLUTD and in control cats. Thus, it can be argued that struvite urinary precipitates were involved in roughly 44% of the cases in this series. Seventy-seven (55%) of these cases were diagnosed with idiopathic FLUTD by exclusion of known causes of hematuria, dysuria and urethral obstruction.

URETHRAL PLUGS Urethral plugs are typically white or tan unless blood clots are present. Plugs are often cylindrical but sometimes are shapeless. Because plugs contain large amounts of matrix, they tend to be soft, compressible and friable. The diameter of cylindrical plugs approximates the diameter of the urethra and their length varies from a few mm to several centimeters. Plugs can be single or multiple.

Urethral plugs and urethral uroliths are physically different and may be due to different pathogenic mechanisms. Urethral plugs are typically composed of large amounts (>50%) of matrix mixed with smaller amounts of crystalline minerals (**Table 21-6**). However, on occasion, plugs can be

Table 21-5. Types of disorders in 143 cats with hematuria and dysuria.

Disorders	No. cats	Percent
Idiopathic conditions	77	53.8
Urethral plugs	32	22.4
Uroliths	30	21.0
Uroliths and bacterial UTIs	2	1.4
Bacterial UTIs	2	1.4
Total	143	100.0

Key: UTI = urinary tract infection.

Table 21-6. Mineral composition of 1,050 feline urethral plugs analyzed by quantitative methods.

Predominant mineral types	Percent
Struvite	76
Matrix	16
Mixed	3.6
Calcium phosphate	2.1
Calcium oxalates	1.4
Ammonium acid urate	0.6

FLUTD

composed almost completely of matrix, blood cells, inflammatory cells and sloughed tissue.

It has been hypothesized that formation of matrix-crystalline urethral plugs requires two simultaneous but unrelated events. One event is the formation of matrix that might be due to bacterial or viral UTI or some other inflammatory process (e.g., feline idiopathic cystitis). The other event is the formation of crystalline precipitates. If matrix forms without concomitant crystals, the noncrystalline gel is voided; however, nonobstructive dysuria and hematuria result. In the presence of crystals, a more rigid plug forms that may obstruct the urethra. The mineral composition of crystals can serve as the basis for preventive efforts.

UROLITHS Uroliths consist of small amounts of matrix and macroscopic crystalline mineral concretions. Urolithiasis is a multifaceted process that begins with the formation of microcrystals in urine and ends with the formation of mature uroliths somewhere in the urinary tract. The most common urolith mineral types found in cats are struvite (magnesium ammonium phosphate) and calcium oxalate. Usually one mineral type predominates, but the composition may be mixed. Different mineral types may be dispersed throughout the urolith or may be organized into separate, discrete bands or layers.

The numbers in **Table 21-7** represent cumulative totals over several years. The cumulative numbers tend to mask an apparent trend that was first recognized in the late 1980s. The proportion of struvite uroliths sub-

Table 21-7. Mineral composition of 9,481 feline uroliths analyzed by quantitative methods.

Predominant mineral types	Percent
Struvite	48
Calcium oxalates	40
Ammonium acid urate	6.1
Mixed/compound	3.6
Matrix	1.4
Calcium phosphate	0.8
Cystine	0.3
Xanthine	0.1

mitted to specialized urolith analysis laboratories has decreased and the proportion of calcium oxalate uroliths has increased. Although the prevalence of calcium oxalate uroliths submitted for analysis appears to be increasing, several relevant questions remain unanswered: 1) How has the apparent aging feline population affected the relative frequency of different urolith types? 2) Is the apparent increase in calcium oxalate uroliths relative because fewer struvite uroliths are being submitted for analysis due to the availability of struvite dissolution protocols? 3) Are veterinarians better at diagnosing calcium oxalate uroliths? 4) Why has the prevalence of calcium oxalate uroliths also increased in dogs and people?

Formation of Uroliths The initiation and growth of crystals involves chemical precipitation of dissolved ions or molecules from urine that has become supersaturated with these components. To understand this from a physiochemical point of view, it is useful to describe how the degree of supersaturation or undersaturation of urine influences the probability that a crystal will form or dissolve.

In its simplest form, the initiation and growth of uroliths, and perhaps urethral plugs, involves chemical precipitation of dissolved ions or molecules from a solution that has become supersaturated with respect to those components. From a physiochemical perspective, the degree of saturation or undersaturation of urine influences the probability that precipitates will form, or if already present, dissolve. Relatively simple diagrams depict the states of saturation of any solution (**Figure 21-2**). These diagrams provide the framework for understanding the concept of how nutritional management influences the probability of urolith formation or dissolution. Units and numerical values are not included in these diagrams because they differ for each of the urolith components; however, the general features apply to all crystalline materials. A more detailed description of this process is found in Small Animal Clinical Nutrition, 4th ed., pp 608-612.

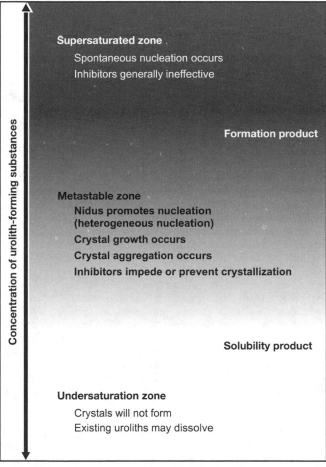

FLUTD

Figure 21-2. Increasing concentrations of urolith-forming substances result in metastable and supersaturated urine. Crystal growth and aggregation may occur in the metastable zone. Presence of a nidus promotes nucleation (heterogeneous nucleation) and subsequent crystal formation. Inhibitors in the metastable zone may impede or prevent crystallization. Spontaneous nucleation occurs when concentrations of urolith-forming substances increase to the point of supersaturation. At this point, inhibitors of crystal formation are generally ineffective.

The solubility product is the concentration product at which dissolved and crystalline components are in equilibrium with each other at a given set of conditions. At concentrations below the solubility product (under-saturation zone) it is impossible for crystals to form under any circumstances. Crystals added to such a solution would dissolve. Crystals would grow if added to a solution with a concentration greater than the solubility product. The formation product is the concentration at which crystals will begin to precipitate at a significant rate in the absence of preformed crystalline material.

The solubility product is constant for a pure crystalline material. The formation product is much more difficult to demonstrate experimentally. Thus, this area is illustrated by a shaded band rather than a line (**Figure 21-2**). Strictly speaking, the ionic activities not concentrations of the species govern the solubility principles described here. Ionic activities are influenced by the presence of other ions in solution (ionic strength) and by the presence of other species that form complex ions, thereby reducing their "free" concentrations in solution.

The metastable zone is of most interest from a clinical perspective (**Figure 21-2**). In this concentration range: 1) crystal growth will occur, 2) crystal aggregation will occur and 3) inhibitors will impede or prevent crystallization. This so-called metastable (unstable) region corresponds to the urinary concentration of crystal-forming substances found in normal people and many urolith-forming human patients. Risk factor reduction and nutritional interventions may be most beneficial with urine in this region. A precarious balance exists between crystal formation and inhibition in the metastable zone. Anatomic defects within the urinary tract that allow for stasis of metastable urine will lead to formation and growth of crystals. Urine containing microscopic impurities will facilitate crystal formation and growth. This process is called heterogeneous nucleation. Crystal formation is much less likely in urine without impurities (homogeneous solution).

Struvite Uroliths Struvite uroliths form as a result of supersaturation of urine with magnesium ammonium phosphate. This supersaturation can occur in the presence of infection with a urease-producing organism. However, approximately 70% of struvite uroliths in cats form in sterile urine. Urinary magnesium levels are related to dietary intake (**Figure 21-3**).

Calcium Oxalate Uroliths Urinary oxalate is a more important determinant of calcium oxalate supersaturation than calcium because small increases in oxalate excretion profoundly influence the activity product ratio (APR). Oxalate is a metabolic byproduct of glycine usage and forms a number of complexes and salts in solution. Oxalic acid, a relatively

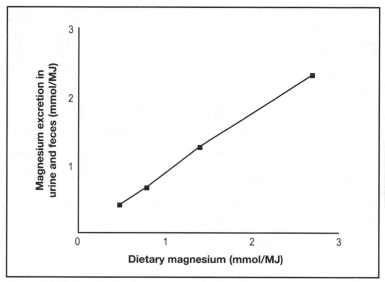

Figure 21-3. Relationship between dietary magnesium and magnesium excretion in urine and feces of cats. Note the direct linear relationship between dietary magnesium and magnesium excretion.

strong organic acid, is a simple two-carbon dicarboxylic acid. The calcium salt is relatively insoluble and pH does not influence solubility over the physiologic range.

Oxalate is absorbed throughout the large and small bowel by facilitated diffusion using an anion exchange mechanism. The calcium salt of oxalate is just as insoluble in the luminal content of the intestinal tract as in other complex solutions. Consequently, dietary calcium is an important determinant of oxalate availability and thus absorption. Because of analytical problems, no data are available about the oxalate content of pet food ingredients and finished pet foods.

Human foods relatively high in oxalate include rhubarb, spinach, peanuts, pecans, wheat germ, peppers, beets, okra, strawberries, chocolate and tea. Cats are unlikely to eat excessive amounts of these foods.

The major metabolic sources of oxalate are ascorbic acid and glyoxylate. Under normal circumstances, the pathways for metabolism of ascorbic acid to oxalate are saturated. Thus, a reasonable increase in dietary intake does not increase urinary oxalate excretion. Megadoses of ascorbic acid increase oxalate production and excretion; however, the clinical significance of this increase is unknown.

Table 21-8. Crystal inhibitors found in urine.

Small molecules
Citrate
Pyrophosphate

Macromolecules
Nephrocalcin
Crystal matrix protein
Tamm-Horsfall protein (inhibitor or promoter)
Osteopontin (uropontin)

The oxidation of glyoxylate accounts for the remainder of daily oxalate production. A large part of the metabolic pool of glyoxylate is transaminated to glycine by the enzyme alanine glyoxylate aminotransferase, which requires pyridoxine as a cofactor. Pyridoxine deficiency leads to increased oxalate production and urinary excretion in cats. Documented cases of clinical pyridoxine deficiency, however, are extremely rare.

Urinary excretion is the primary route of oxalate elimination from the extracellular pool. Circulating oxalate is freely filtered by the glomeruli and there is net tubular secretion of oxalate.

Inhibitors Urine contains substances that modify and inhibit nucleation, growth and aggregation of crystals (**Table 21-8**). These substances include small molecules such as citrate and pyrophosphate and macromolecules, such as mucoproteins, glycoproteins, glycosaminoglycans and proteoglycans.

Some recurrent urolith formers may have defective inhibitory substances (**Table 21-8**). For example, some highly recurrent calcium urolith formers excrete Tamm-Horsfall mucoprotein, which compared with normal Tamm-Horsfall mucoprotein, self-aggregates at low pH and high sodium concentrations and is less effective at inhibiting calcium oxalate monohydrate aggregation.

The urine of most human beings is continuously saturated with calcium oxalate, yet only 4% of the human population will form calcium oxalate uroliths during their lifetime. This underscores the importance of nucleation, crystal growth and crystal aggregation in urine. Patients with a history of urolithiasis should particularly avoid saturation conditions that favor specific urolith types.

Relative Supersaturation vs. Activity Product Ratios Analytical data indicate that urine is often supersaturated with respect to most common urolith components. Thus, the question is not why a specific animal formed a urolith, but rather why doesn't every animal form uroliths? Inhibitors in urine probably explain the less than predicted prevalence.

Because we don't know how to manipulate or change inhibitors, the current therapeutic strategy is to reduce risk factors by decreasing the degree of supersaturation. One way to express urinary saturation is to determine the relative supersaturation (RSS) of a urolith type such as calcium oxalate or struvite. RSS is determined by measuring the concentration of a number of urinary analytes, including sodium, calcium, oxalate, magnesium, and potassium. The concentrations of these analytes are entered into a computer program that calculates the saturation of the urolith elements compared with a standard human urine sample. RSS has limitations because it is highly dependent on urine volume and involves comparison with standard values for human urine.

A better technique for predicting the likelihood of crystal formation is the APR. The APR for calcium oxalate is the mathematical product of the activity of calcium and the activity of oxalate. Activity is different than simple concentration of the substance of interest in an aqueous solution. Activity refers to the ionic activity, which is influenced by the concentration of the substance of interest, other substances in urine and factors such as pH and temperature. Like RSS, APR involves measurement of a number of analytes in urine such as sodium, potassium, calcium, oxalate and magnesium. Values are also entered into a computer program. But unlike RSS, the APR technique requires incubation of seed crystals such as calcium oxalate in an aliquot of urine. After incubation, the urinary analytes are measured again and the postincubation activity product is determined. Dividing the pre-incubation activity product by the postincubation activity product yields the APR. An APR less than 1 indicates that crystals dissolved during incubation. The APR provides a better indication of risk of crystal formation than RSS because APR considers the influence of unmeasured inhibitors and promoters and is not unduly influenced by urine volume. APRs can be used to quantitatively evaluate the influence of nutrients, complete foods and drugs on the risk of crystal formation.

Key Nutritional Factors *(Pages 708-712)*

In its simplest form, the initiation and growth of urinary precipitates involve a chemical precipitation of dissolved ions or molecules from a solution that is supersaturated with precipitate-forming components. The degree of supersaturation or undersaturation of urine with regard to these components influences the probability of precipitates forming. In turn, the urinary concentration of these components is influenced by nutritional factors. The urinary precipitates of primary concern in FLUTD are struvite (magnesium ammonium phosphate) and calcium oxalate. Some nutritional factors (e.g., urinary pH) have opposing effects on the probability of struvite and calcium oxalate formation. Consequently, the goal is to achieve balance in reducing the likelihood of struvite and calcium oxalate formation.

WATER The percent of water in cat foods varies. In general, dry foods contain less than 10% water and moist foods contain more than 72% water. The volume of water consumed each day by a cat in a thermoneutral environment depends to a large extent on the composition and quantity of food ingested. The volume of water drunk increases as the moisture content of the food decreases; however, cats fed dry foods usually have lower total water intake than those fed moist foods.

The solute load of the food also influences water consumption; urea is a major contributor to the renal solute load. Increasing the protein content of food increases the solute load; therefore, foods with higher protein content are associated with higher water intake. The metabolism of energy substrates yields endogenous water. The daily volume of endogenously produced water is small (approximately 10 to 15%) compared with the total daily water intake. Fats provide the most water per gram and carbohydrates provide the most water per calorie. When glycogen is oxidized, additional water is released during glycogen mobilization. The amount of water generated differs slightly depending on the source of fat, chain length and degree of saturation. Proteins vary more than fats in the range of endogenous water produced. Because they lead to greater water intake and subsequent higher urine volume, moist foods are recommended in the management of calcium oxalate and struvite urolithiasis. Salt should be used cautiously as a strategy to increase water consumption in patients with FLUTD, because increased sodium levels may exacerbate hypertension and subclinical renal disease. Sodium levels in foods for patients with FLUTD should generally not exceed 0.6% DM for long-term maintenance.

URINARY pH The kidneys eliminate the acid that is produced each day by metabolism and food. Therefore, to define the "normal" urinary pH, it is necessary to consider the "normal" or habitual diet. The diet of feral house cats has been well studied. On a volume basis, the gastric content of feral cats is approximately 90% small mammals, such as mice and rats. The average urinary pH is approximately 6.3 when cats are fed a normal diet of rat carcasses.

Blood pH is normally maintained within a narrow range by extracellular and intracellular buffering processes and respiratory and renal regulatory mechanisms. Chemical buffering and respiratory compensation protect against abrupt deviations in body pH; however, the kidneys provide longer-term defense against acid and alkali deviations. This process occurs continuously as endogenous acids are generated. The kidneys must conserve bicarbonate in the glomerular filtrate and regenerate bicarbonate degraded by the reaction with metabolic acids to maintain normal plasma bicarbonate levels. The kidneys can increase the amount of net acid excretion in urine and generate bicarbonate in response to exogenous acid loads. Urinary organic anion excretion parallels changes in urinary bicarbonate. Urinary organic anions (e.g., citrate and α-ketoglutarate) increase in response to an alkali load and decrease in response to an acid load.

Normally, the kidneys synthesize urinary ammonia and thus ammonium almost exclusively. Ammonia is derived from glutamine and other amino acids in the proximal tubule cells. Adaptive changes in renal ammonia production occur in response to changes in systemic acid-base balance. More ammonia is produced and urinary ammonium ion concentration is increased with chronic metabolic acidosis. The kidneys excrete hydrogen ions (H^+) in the form of titratable acid and ammonium ions. The acidification process is an integrated function that occurs at several nephron segments. The proximal convoluted tubules reabsorb most of the filtered bicarbonate and add ammonia to the tubule fluid. Residual bicarbonate is titrated with filtered buffers (conjugate bases) to form titratable acid in the cortical collecting tubules. There is also a parallel bicarbonate secretory mechanism. The medullary collecting ducts decrease urinary pH and trap ammonia as ammonium. Ammonia exists in two forms in aqueous solution, nonionized ammonia (NH_3) and the monovalent cation ammonium (NH_4^+). Nonionized ammonia is lipid soluble and readily traverses cell membranes in the direction determined by the concentration gradient. At the pH of renal tubular fluid, hydrogen ions avidly combine with ammonia to form ammonium. The ionized form is not lipid soluble; consequently, the ammonium is trapped within the tubular lumen and is excreted in the form of neutral salts (i.e., NH_4Cl).

Bicarbonate resorption in the proximal tubules is affected primarily by neutral Na^+/H^+ exchange, which is driven by the lumen-to-cell sodium gradient. Hydrogen ions secreted into the lumen of the proximal convoluted tubule combine with filtered bicarbonate to form carbonic acid. Carbonic acid, catalyzed by membrane-bound carbonic anhydrase, is broken down to carbon dioxide and water. At equilibrium, dissolved carbon dioxide equilibrates with the gas phase (i.e., the respiratory gas in the pulmonary alveoli). Dissolved carbon dioxide diffuses into the cell and combines with hydroxyl ions. Newly formed bicarbonate and reabsorbed sodium are added to the blood in the process. The net result is the excretion of hydrogen ions and the replenishment of bicarbonate. If a short-term (three to five days) acid load is encountered, the kidneys respond by increasing titratable acid and ammonium excretion.

Physiologic limits for urinary pH in cats are 8.5 to 5.5. Reduction in urinary pH from the upper to lower end of this range changes the ratio of NH_4^+/NH_3 from 3.4:1 to 3,400:1. Thus, acidifying foods increase the urinary concentration of ammonium ions, one component of struvite. However, this increase is more than offset by an even greater decrease in anionic phosphate concentration, another component of struvite, as the urinary pH decreases from 8.5 to 5.5.

Table 21-9. Urine acidifying and alkalinizing pet food ingredients.

Protein acidifying ingredients	
Poultry meal	Corn gluten meal
Other acidifying ingredients	
Ammonium chloride*	dl-methionine
Calcium chloride	Phosphoric acid
Calcium sulfate	
Alkalinizing ingredients	
Calcium carbonate	Magnesium oxide
Potassium citrate	

*Not approved in the United States as a food additive.

The effect of a food on urinary pH is the net effect of its constituent nutrients. Dietary acid is derived from several nutrients (**Table 21-9**). Sulfuric acid is formed when sulfur-containing amino acid residues of proteins (e.g., methionine and cysteine) are oxidized to sulfate. In general, animal-source protein ingredients contain more sulfur-containing amino acid residues than do plant-source proteins. The quantity of sulfuric acid generated is reflected by the amount of sulfate in the urine.

Phosphorus has strong effects on acid-base balance, depending on its chemical form. Inorganic phosphorus can be ingested as phosphoric acid, monobasic and dibasic or anionic phosphate. Phosphoric acid is used in cat foods to enhance palatability, either separately or as a component of topically applied animal digests. Phosphoric acid has a strong acidifying effect. Monobasic phosphate is also acidifying, dibasic phosphate has little effect on urinary pH and anionic phosphate is alkalizing. At the physiologic pH of 7.4, inorganic phosphorus from phospholipids, phosphoproteins and nucleic acids generates dibasic and monobasic phosphate at a ratio of 1:4, thus its acidifying effect is nearly as strong as that of monobasic phosphate.

The effect of other inorganic cations depends on their pKa and to some extent on their absorption from the intestine in relation to their accompanying cations. Mineral salts vary in their effect on urinary pH and thus are potential acid or base sources. Oxides and carbonates are alkalizing. Differences in absorption of the cation and anion portion of a salt are important. Intestinal absorption of calcium and magnesium is relatively low; however, absorption of accompanying anions can be high and influences urinary pH. Nonmetabolizable anions (e.g., chloride, phosphate and sulfate) absorbed in excess of their accompanying cations are acidifying. For example, ammonium chloride, calcium chloride and calcium sulfate decrease urinary pH, and magnesium oxide and calcium carbonate increase urinary pH.

Carbohydrates, fat and the sulfur- and phosphorus-free residues of protein have little effect on acid-base balance if they are fully oxidized to water and carbon dioxide. However, organic acids produced from intermediary metabolism products formed from partial oxidation of carbohydrates or fats (e.g., lactate during anaerobic exercise or β-hydroxybutyrate during ketosis) have a marked acidifying effect. The organic acids generated contribute to net acid production when conjugate bases are excreted in the urine as negatively charged organic anions. The net amount of hydrogen ion contributed by organic acids can be estimated by measuring the amount of organic anions in the urine. Healthy cats fed a balanced diet are unlikely to produce considerable amounts of organic acids from carbohydrate, fat or sulfur- and phosphorus-free protein residues; therefore, it is unlikely that their acid-base-balance is affected by these nutrients.

Organic acids (e.g., ascorbic acid, lactic acid or citric acid) can be fully oxidized in intermediary metabolism. Consequently, they have little effect on urinary pH unless they are excreted in the urine. Most organic acids are rather weak; therefore, marked renal excretion is necessary to affect urinary pH. For example, ascorbic acid in doses less than $1,000 \, mg/BW_{kg}$ do not significantly affect urinary pH of cats fed a balanced diet. By comparison, the conjugated bases of organic acids may change acid-base balance. They take up hydrogen ions during oxidation to water and carbon dioxide. Thus, calcium and magnesium salts containing conjugated bases of organic acids can be as strongly alkalizing as the corresponding carbonates or oxides.

Ammonium and positively charged cationic amino acids (e.g., lysine and arginine) release hydrogen ions when they are transformed into neutral products. Hydrochloric acid is generated if their accompanying anion is chloride. The acidifying effect is obvious.

Urinary pH is not a direct function of acid excretion. Urinary buffer systems can modify the effect of acid intake on urinary pH. Small changes in the acid load of the diet lead to much greater changes in urinary pH when the phosphorus intake of cats is decreased.

Although decreasing urinary pH theoretically increases urinary ammonium concentration, the same change in urinary pH decreases anionic phosphate levels in urine. Thus as urine becomes more acidic, precipitation of struvite becomes less likely. The risk reduction for struvite crystal formation becomes clinically important as the mean daily urinary pH declines to about 6.4.

For most cats less than about seven years of age, the ideal mean daily urinary pH is between 6.1 and 6.4. Urinary pH less than 6.1 may reflect systemic acidosis that may be accompanied by hypercalciuria. Urinary pH values above 6.4 favor struvite precipitation; therefore, mean daily urinary pH values for struvite dissolution are between 5.9 and 6.1. Values between 6.2 and 6.4 are appropriate for struvite prevention. A slightly higher urinary pH target (i.e., 6.6 to 6.8) may be beneficial for cats at

Table 21-10. Recommended levels of selected nutrients in commercial foods used for calcium oxalate urolith prevention and struvite urolith prevention and dissolution in cats.*

Nutrients	Calcium oxalate prevention	Struvite prevention	Struvite dissolution
Calcium	0.5-0.8	–	–
Phosphorus	0.5-0.7	0.5-0.9	0.5-0.8
Magnesium	0.04-0.10	0.04-0.10	0.04-0.06

*Nutrients expressed as percent dry matter.

increased risk for calcium oxalate urolith formation because of their age and/or other risk factors.

MAGNESIUM For struvite precipitates to form, the urine must be supersaturated with magnesium, ammonium and anionic phosphate. Avoiding excess dietary magnesium intake can reduce the urinary concentration of magnesium. Excess magnesium is present in some commercial cat foods because they contain ingredients high in magnesium (e.g., high-ash meat and bone, fish and poultry meals). Foods recommended for prevention of struvite precipitation should meet the nutritional allowance for magnesium but avoid excess.

Magnesium is an inhibitor of calcium oxalate crystal formation in vitro; therefore, low urinary magnesium concentrations have been suggested as a potential risk factor in the formation of calcium-containing uroliths. At physiologic concentrations of magnesium, in vitro studies demonstrate that magnesium decreases the rate of nucleation and growth of oxalate crystals.

The proposed mechanisms of action of magnesium supplementation include increased urinary pH, increased urinary excretion of citrate and formation of magnesium oxalate complexes in urine. Magnesium oxalate is more soluble than calcium oxalate. The high excretion of magnesium in urine and the formation of magnesium oxalate in theory reduce the concentration of oxalate available for precipitation as calcium oxalate. **Table 21-10** lists magnesium recommendations in foods for calcium oxalate prevention, struvite prevention and struvite dissolution.

PHOSPHORUS Varying dietary phosphorus levels can alter urinary phosphate concentrations in cats, thereby influencing the likelihood of urinary struvite precipitates. Urinary phosphate can exist in several states. Anionic phosphate (PO_4^{3-}) is the important form in precipitation and dissolution of struvite. Urinary concentration of anionic phosphate is reversibly influenced by pH as explained above. Thus as urine becomes more acidic, anionic phosphate is converted to monobasic and dibasic phosphate, thereby reducing the concentration of anionic phosphate available for incorporation in struvite precipitates. As the urine becomes more alkaline, the reaction proceeds in the opposite direction and the concentration of anionic phosphate increases.

In cats, increased levels of dietary phosphorus decrease urinary excretion of calcium. The inhibitory effect of dietary phosphorus on calcium and magnesium absorption can be explained by formation of insoluble calcium-magnesium-phosphate complexes in the intestinal lumen. Formation of these complexes decreases the concentration of soluble calcium and magnesium, thereby decreasing the availability of these minerals for absorption.

The intake of magnesium and calcium influences urinary phosphate concentration. Rats fed a very low-phosphorus food (0.07% dry matter [DM]) had marked hypercalciuria. Feeding rats this very low level of phosphorus for one week resulted in urine highly supersaturated with calcium oxalate and containing large amounts of calcium oxalate crystals. Adult maintenance cat foods contain at least 0.5% DM phosphorus. **Table 21-10** lists phosphorus recommendations in foods for calcium oxalate prevention, struvite prevention and struvite dissolution.

CALCIUM Excess dietary calcium should be avoided to prevent recurrence of calcium oxalate uroliths. The most important sources of excess calcium are commercial foods and mineral supplements containing high levels of calcium. High intake of dietary calcium may lead to hypercalciuria and urolith formation in patients with intestinal hyperabsorption of calcium. However, severe calcium restriction can cause negative calcium balance and may contribute to hyperoxaluria.

Intestinal absorption of calcium occurs primarily in the duodenum. Transport of calcium across the gut is vitamin D-dependent and saturable. Calcium availability varies markedly and may be influenced by nondietary and dietary factors. Generally, calcium absorption from the gut is inversely proportional to dietary intake (i.e., absorption is high from low-calcium foods and low from high-calcium foods). Other dietary factors such as vitamin D, sucrose, fructose, glucose, xylose, dietary fiber, oxalic acid, phytic acid, protein and phosphorus reportedly affect calcium availability. The role of individual dietary factors in dogs and cats consuming mixed meals is difficult to assess.

Table 21-11 lists calcium-rich foods that should be avoided in homemade diets and in treats. In addition to foods naturally high in calcium, a number of different foods and beverages (e.g., breads and breakfast cereals) are fortified with calcium. The amount of calcium added to these foods is listed on the product label. Another potential unrecognized source of excess dietary calcium is vitamin-mineral supplements, especially specific calcium supplements. A wide variety of calcium supplements are available over the counter. These supplements differ in the amount of elemental calcium provided. Calcium carbonate, for example,

Table 21-11. Calcium-rich foods that should be avoided in cats at risk for calcium oxalate urolithiasis.*

Food items	Serving sizes	Calcium (mg)
Low-fat yogurt	1 cup (8 oz)	415
Whole milk	1 cup (8 oz)	291
Cheese	1 oz	200-270
Ice cream or ice milk	1 cup (8 oz)	176
Cottage cheese, creamed	1 cup (8 oz)	136
Broccoli, cooked	1 large stalk	88

*Mineral supplements and some commercial cat foods contain much more calcium than these foods, emphasizing the need for a thorough, complete dietary history.

contains 40% calcium (by weight), whereas calcium lactate and calcium gluconate contain 13 and 9% calcium, respectively. There is little information available about the relative availability of calcium from different supplements. Calcium supplements differ not only in their calcium content but also in solubility.

Table 21-10 lists calcium recommendations in foods for calcium oxalate prevention.

PROTEIN A high-protein diet reportedly increases urinary calcium excretion in dogs. The 24-hour urinary calcium excretion almost doubles when dogs are fed a food containing 31% DM protein compared with a food containing 10% DM protein. The type of protein, duration of protein intake and phosphorus intake influence the effect of protein on calcium.

Animal proteins are rich in sulfur-containing amino acids. Sulfur-containing amino acids are metabolized to sulfate and thus may reduce urinary pH and increase urinary calcium and uric acid concentrations. The hypocalciuric effect of phosphorus may vary depending on the form of phosphorus. Excess dietary protein should be avoided by feeding a food that contains 30 to 45% DM protein.

FAT Higher fat foods can be advantageous because the increased energy density reduces overall mineral intake, including magnesium. Compared with protein and carbohydrate, fat provides the highest metabolic water contribution. Alternatively, foods with increased fat content will contribute to obesity, if they are not fed in a controlled manner. Obesity has been identified in some studies as a risk factor for FLUTD. Foods with 8 to 25% DM fat can routinely be fed, depending on the cat's body condition and the desired energy density of the food.

Other Food Factors (Page 713)
OXALATE Excessive dietary intake of oxalate is unlikely in cats because foods high in oxalate are not likely to be major components of the overall

diet. The following foods are considered high-oxalate foods: sugar (in large quantities), chocolate, beets, beet greens, spinach, Swiss chard, rhubarb, whole soybeans, leeks, citrus pulp, wheat germ, sweet potatoes, raspberries, beans, carrots, celery, oranges, squash, corn and potatoes. These ingredients are not commonly used in cat foods.

VITAMIN D Calcium absorption occurs primarily in the duodenum, and transport is vitamin D-dependent. The metabolically active form of vitamin D (1,25-dihydroxycholecalciferol) is produced when the precursor undergoes hydroxylation in the proximal tubular cells of the kidneys. The primary actions of the active form of vitamin D are: 1) stimulation of intestinal calcium absorption, 2) inhibition of parathyroid hormone (PTH) synthesis, 3) inhibition of more active vitamin D synthesis in the kidneys and 4) facilitation of osteoclastic bone resorption. PTH, hypophosphatemia, growth hormone, estrogen and prolactin increase renal synthesis of 1,25-dihydroxycholecalciferol. Hyperphosphatemia, hypercalcemia and damage to the kidneys decrease renal synthesis of 1,25-dihydroxycholecalciferol. The Association of American Feed Control Officials (AAFCO) dietary allowance for vitamin D for adult cats is 500 to 10,000 IU/kg of food. However, cats at risk for calcium oxalate urolithiasis and those with hypercalcemia associated with calcium oxalate urolithiasis should be fed foods that do not exceed 2,000 IU/kg of food. Dietary supplements containing vitamin D should not be fed to at-risk cats.

POTASSIUM Transient negative potassium balance may occur in adult cats receiving long-term dietary acidification (i.e., for struvite urolith prevention) with phosphoric acid and ammonium chloride.

Potassium affects ammonia synthesis. Acute increases in potassium concentration suppress ammonia synthesis. Chronic potassium depletion stimulates ammonia synthesis at the same site as chronic metabolic acidosis. Acidifying foods, therefore, should have potassium levels in excess of the AAFCO minimum allowance of 0.6% (DMB).

SODIUM In people, increased dietary sodium levels promote hypercalciuria. However, this finding hasn't been proven in cats. Salt should be used cautiously as a strategy to increase water consumption in patients with FLUTD, because increased sodium levels may exacerbate hypertension and subclinical renal disease. Sodium levels in foods for patients with FLUTD should generally not exceed 0.6% DM for long-term maintenance.

FIBER Certain dietary fiber sources bind calcium in the small intestine and may reduce the amount of calcium absorbed from the gut. Some clinicians advocate adding fiber to the current food or using a commercial fiber-enhanced food for cats with persistent hypercalcemia and/or recurrent calcium oxalate urolithiasis. Anecdotal reports suggest that moderate levels of crude fiber (9 to 12% DMB) may be useful in these patients.

Less Common Urolith Types (Page 707)

Less common mineral types include ammonium acid urate, uric acid, xanthine, calcium phosphate and cystine (**Tables 21-7** and **21-12**). In general, metabolic uroliths (e.g., ammonium urate and cystine) are rarely observed in immature cats. Ammonium urate uroliths in cats with portosystemic shunts are an important exception to this generalization about metabolic uroliths.

Likewise, these mineral types make up a small percentage of the mineral in feline urethral plugs (**Table 21-6**). However, detection of certain crystals (e.g., cystine and ammonium urate), even in patients without clinical signs, suggests an important underlying metabolic defect.

In general, ammonium urate and cystine uroliths are associated with an acidic urinary pH. Cystine and ammonium urate uroliths are less dense than struvite and calcium-containing uroliths. Cystine uroliths are small, smooth and round to oval. Ammonium urate and uric acid uroliths are smooth, but occasionally irregular and round to oval (**Table 21-2**).

Assess the Food (Pages 713-714)

Food for cats with lower urinary tract disease should be evaluated for all the key nutritional factors described above. Levels of magnesium, calcium and phosphorus and information about the urinary pH are not required on the guaranteed or typical analysis of pet food labels and usually are not listed. Levels of these nutrients and urinary pH produced by feeding a particular food must be obtained by contacting the manufacturer or consulting published product information (**Tables 21-13** and **21-14**).

Other sources of key nutritional factors include treats, supplements and human food offered as treats or part of the pet's daily intake. Commercial cat treats and processed human foods may have very high levels of minerals, such as phosphorus.

Product claims (e.g., low magnesium) may help in evaluating a food. In the United States, a food with a "low magnesium" claim contains a maximum of 0.12% DM magnesium or 25 mg magnesium/100 kcal metabolizable energy.

The United States Food and Drug Administration, Center for Veterinary Medicine (FDA-CVM) has issued guidelines for pet food companies to establish the following claims: "reduces urinary pH," "low magnesium" and "improves urinary tract health." The FDA-CVM suggests that submissions requesting permission to make "reduces urinary pH" or "improves urinary tract health" claims include supportive utility data demonstrating efficacy (i.e., the ability of the food to produce an appropriately acidic urine compared with a nonacidifying control food) and safety data.

A food with a label claim such as "helps maintain urinary tract health" is low in magnesium and produces appropriately acidic urine. The FDA-

Table 21-12. Key nutritional factors–uncommon feline urolith types.

Nutritional factors	Dietary recommendations
Calcium phosphate uroliths (prevention)	
Water	Promote water intake by using a moist food or other measures
Calcium	Avoid excess dietary calcium
	Restrict dietary calcium to 0.6 to 0.8% dry matter (DM)
Phosphorus	Avoid excess dietary phosphorus
	Restrict dietary phosphorus to <0.8% DM
Sodium	Avoid excess dietary sodium
	Restrict dietary sodium to <0.3% DM
Vitamin D	Avoid excess dietary vitamin D
	Restrict dietary vitamin D to ≤2,000 IU/kg food
Purine uroliths (prevention)	
Water	Promote water intake by using a moist food or other measures
Protein	Avoid excess dietary protein
	Use protein sources with low purine content (i.e., those with few cell nuclei), such as milk and egg
	Avoid proteins with many cell nuclei such as brain, liver and yeast
	Restrict dietary protein to 28 to 30% DM
Urinary pH	Use a food that helps maintain a more alkaline urine (6.6 to 6.8)

CVM has promulgated guidelines for protocols to support urinary tract health claims. The guidelines focus on prevention of struvite urinary precipitates, but do not address the question of calcium oxalate precipitate formation.

Assess the Feeding Method (Page 714)

The feeding method can influence urinary pH. Ingestion of food stimulates secretion of gastric acid and a temporary net acid loss from the body. This alkalization of urine is referred to as the postprandial alkaline tide. Specifically, secretion of bicarbonate into the blood by gastric parietal cells causes the alkaline tide. This alkali load produces a transient bicarbonaturia that increases urinary pH unless offset by absorption of acidifying ingredients. When offered food free choice, most cats will eat small amounts every few hours, resulting in a smaller but more prolonged alkaline tide than with meal feeding. The smaller alkaline urinary pH

Table 21-13. Levels of key nutritional factors in commercial foods for struvite urolith dissolution and prevention in cats.*

	Protein	Fat	Phosphorus	Magnesium	Urinary pH**
Struvite urolith dissolution					
Recommended levels	**30-45**	**8-25**	**0.5-0.8**	**0.04-0.06**	**5.9-6.1**
Hill's Prescription Diet Feline s/d, moist	41.7	33.8	0.52	0.04	5.9-6.1
Hill's Prescription Diet Feline s/d, dry	34.3	26.3	0.77	0.06	5.9-6.1
Leo Specific Struvil FSW, moist	48.7	32.5	0.52	0.05	<6.3
Leo Specific Struvil FSD, dry	33.7	28.8	0.49	0.05	<6.3
Struvite urolith prevention					
Recommended levels	**30-45**	**8-25**	**0.5-0.9**	**0.04-0.10**	**6.2-6.4**
Hill's Prescription Diet Feline c/d-s, moist	43.3	21.7	0.50	0.06	6.2-6.4
Hill's Prescription Diet Feline c/d-s, dry	34.6	16.3	0.67	0.05	6.2-6.4
Hill's Prescription Diet Feline w/d, moist	41.3	16.7	0.60	0.08	6.2-6.4
Hill's Prescription Diet Feline w/d, dry	38.8	9.3	0.85	0.08	6.2-6.4
Iams Eukanuba Urinary Formula Low pH/S, moist	46.2	29.9	0.92	0.08	5.9-6.3
Iams Eukanuba Urinary Formula Low pH/S, dry	36.2	18.2	0.94	0.09	5.9-6.3
Leo Specific Precal FCW, foil pack	57.6	31.6	0.67	0.04	<6.4
Leo Specific Precal FCD, dry	33.9	25.1	0.62	0.05	<6.4
Medi-Cal Feline Preventive Formula, moist	46.7	31.1	1.06	0.05	Acid
Medi-Cal Feline Preventive Formula, dry	33.4	22.9	0.71	0.07	Acid
Purina CNM UR-Formula, moist	41.4	16.5	0.82	0.05	Mean 6.1
Purina CNM UR-Formula, dry	35.4	11.6	0.84	0.08	Mean 6.1
Select Care Control Formula, moist	46.5	30.4	0.96	0.09	6.0-6.8
Select Care Control Formula, dry	34.2	23.2	0.77	0.08	6.0-6.7
Waltham S/O Control pHormula/pH Control, moist	41.1	44.2	1.16	0.10	6.3-6.5
Waltham S/O Control pHormula/pH Control, dry	36.5	19.7	0.88	0.06	6.3-6.5
Whiskas Feline Low pH Control Diet, moist	44.8	39.3	0.65	0.08	Mean 6.3
Whiskas Feline Low pH Control Diet, dry	38.3	18.9	1.03	0.08	Mean 6.1

*This list represents products with the largest market share and for which published information is available. Manufacturers' published values; nutrients expressed as % dry matter.
**Protocols for measuring urinary pH may vary.

Table 21-14. Levels of key nutritional factors in commercial foods for calcium oxalate urolith prevention in cats.*

Recommended levels	Protein 30-45	Fat 8-25	Phosphorus 0.5-0.7	Calcium 0.5-0.8	Magnesium 0.04-0.10	Urinary pH** 6.6-6.8
Hill's Prescription Diet Feline c/d-oxl, moist	41.6	19.7	0.55	0.67	0.09	6.6-6.8
Hill's Prescription Diet Feline c/d-oxl, dry	34.2	16.7	0.66	0.80	0.08	6.6-6.8
Hill's Prescription Diet Feline g/d, moist	35.1	19.4	0.54	0.66	0.08	6.4-6.6
Hill's Prescription Diet Feline g/d, dry	33.4	18.9	0.55	0.51	0.05	6.4-6.6
Hill's Prescription Diet Feline k/d, moist	29.5	26.7	0.39	0.63	0.04	6.6-6.9
Hill's Prescription Diet Feline k/d, dry	28.2	22.3	0.46	0.76	0.04	6.6-6.9
Iams Eukanuba Urinary Formula Moderate pH/O, moist	46.2	29.9	0.87	1.17	0.13	6.3-6.9
Iams Eukanuba Urinary Formula Moderate pH/O, dry	35.6	18.1	0.96	1.11	0.09	6.3-6.9
Select Care Control Formula, moist	46.5	30.4	0.96	1.04	0.09	6.0-6.8
Select Care Control Formula, dry	34.2	23.2	0.77	0.89	0.08	6.0-6.7
Waltham S/O Control pHormula/pH Control, moist	41.1	44.2	1.16	0.95	0.10	6.3-6.5
Waltham S/O Control pHormula/pH Control, dry	36.5	19.7	0.88	0.70	0.06	6.3-6.5

*This list represents products with the largest market share and for which published information is available. Manufacturers' published values; nutrients expressed as % dry matter.
**Protocols for measuring urinary pH may vary.

FLUTD

excursions observed with free-choice feeding reduce the likelihood of struvite precipitate formation (**Figure 21-4**). However, free-choice feeding may be associated with obesity, which in turn is a risk factor for FLUTD.

FEEDING PLAN (Pages 714-715)

By definition, idiopathic cystitis is not amenable to nutritional therapy; therefore, the focus of this section is on nutritional management of urinary precipitates. However, idiopathic cystitis represents a risk factor for formation of urethral plugs. Feeding strategies designed to minimize struvite crystalluria may be appropriate for male cats at risk for plug formation.

Select a Food (Pages 714-715)

After a presumptive diagnosis of struvite urolithiasis has been made, a decision regarding surgical or medical therapy with a struvite calculolytic food is required. Treatment with a struvite calculolytic food is contraindicated in growing kittens, reproducing queens and patients with renal failure, metabolic acidosis, hypokalemia, hypertension or congestive heart failure. Surgery may be contraindicated if patient factors increase the risk of general anesthesia or adverse surgical outcomes. Because cystotomy does not necessarily remove all uroliths and because of high recurrence rates, multiple surgeries may be required. The size and the number of uroliths influence the rate of dissolution, but do not determine ultimate success. In theory, medical dissolution may slow resolution of UTI or produce obstructive uropathy. Fortunately, these adverse sequelae are rare.

If a patient has a history of forming both struvite and calcium oxalate uroliths, the food should be selected to reduce the likelihood of calcium oxalate formation (i.e., less acidifying). The rationale for favoring formation of struvite over calcium oxalate is that struvite uroliths usually can be dissolved medically whereas calcium oxalate uroliths generally require surgical removal. Clients should be encouraged to feed only foods with key nutritional factors appropriate for the cat's urolith type. Other foods, treats and supplements may alter the key nutritional factors and increase the risk of recurrence. Clients who own cats that form calcium oxalate crystals and uroliths should be cautioned about grocery brand foods with urinary tract health claims. Such foods are formulated for healthy cats to avoid struvite crystals and uroliths. Feeding such foods may increase the risk for development of calcium oxalate uroliths in at-risk cats.

The Key Nutritional Factors–Feline Lower Urinary Tract Disease table and **Tables 21-13** and **21-14** summarize the key nutritional factors for cats with struvite and calcium oxalate crystals and uroliths. Nutrient content

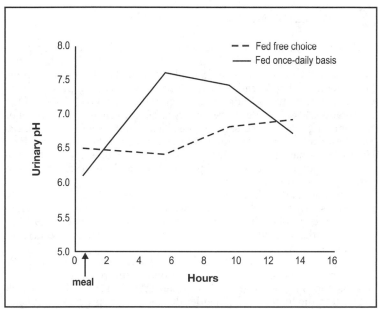

Figure 21-4. Mean urinary pH of cats fed a commercial food either free choice or once daily. Note how once-daily feeding results in a significant increase in urinary pH (i.e., postprandial alkaline tide).

of foods for affected cats should be compared with information in these tables to determine if discrepancies exist between the ideal profile and the product under consideration. In many cases, it will be necessary to contact the manufacturer or review published information to determine the levels of some nutrients. A new food should be selected if discrepancies exist between the patient's current food and the Key Nutritional Factors table. **Tables 21-13** and **21-14** list selected foods marketed for cats with specific urinary tract problems.

Determine a Feeding Method (Page 715)

The feeding method selected should consider concurrent problems such as obesity. Although free-choice feeding minimizes the postprandial alkaline tide, it may lead to excessive consumption of calories and contribute to obesity. If moist food is used to increase water consumption and thus reduce the urinary concentration of crystal-forming elements, free-choice feeding is not feasible.

REASSESSMENT (Pages 715-718)

Urinary pH measurements and urinalyses, including microscopic sediment examination, should be conducted periodically. The time of day markedly influences urinary pH. Samples obtained early in the morning, before food is offered, tend to be more acidic, whereas samples obtained within several hours of eating tend to be more alkaline (because of the postprandial alkaline tide). When interpreting the urinary pH of a spot or random sample, it is necessary to consider when the sample was collected relative to the time of eating. When evaluating the effect of a food change on urinary pH it is necessary to standardize time of collection relative to the time of eating. See **Figures 21-5** and **21-6** for algorithms to assist in monitoring patients with FLUTD.

Recurrence of uroliths appears to be quite variable. In general, it is possible for struvite uroliths to recur within weeks to months. Calcium oxalate, calcium phosphate and ammonium urate uroliths tend to recur within months rather than weeks. In two retrospective studies, recurrence rates ranged from 11 to 37%.

Interpretation of recurrence and interval until recurrence should be based on a number of factors. Were all uroliths removed from the urinary tract at the time of surgery? Did nonabsorbable suture materials left exposed in the lumen of the bladder during surgery provide a nidus for precipitation of crystalline material? What diagnostic methods were used to detect recurrence? How often was the patient evaluated for recurrence? Did the owner comply with recommendations to decrease the likelihood of recurrence? Has infection with a urease-producing microorganism persisted or recurred? Has an underlying anatomic defect gone uncorrected?

Recurrence rates for calcium oxalate uroliths in human patients without any form of preventive management are 10% at one year, 33% at five years and 50% at 10 years. Current preventive management in people consists of increased fluid intake and avoidance of foods rich in calcium and oxalate. Other measures include treatment with thiazide diuretics, pyridoxine, potassium citrate, orthophosphate and allopurinol.

Medical protocols for dissolution of calcium oxalate uroliths in cats are not available. Therefore, there is a need for an integrated nutritional, medical and surgical approach. If urocystoliths are small enough to pass through the urethra, an alternative to cystotomy is voiding urohydropropulsion. Because the urethra is smaller in males than in females, voiding urohydropropulsion is generally more successful in female cats. Voiding urohydropropulsion is usually ineffective in cats with uroliths lodged in the urethra. Excessive manual compression of the bladder in cats with UTI can cause vesicoureteral reflux of infected urine.

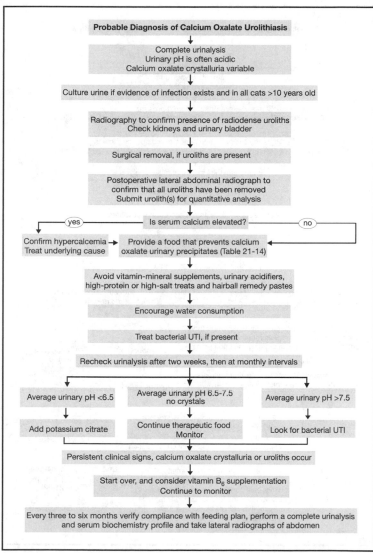

Figure 21-5. Algorithm for management and reassessment of cats with calcium oxalate urolithiasis.

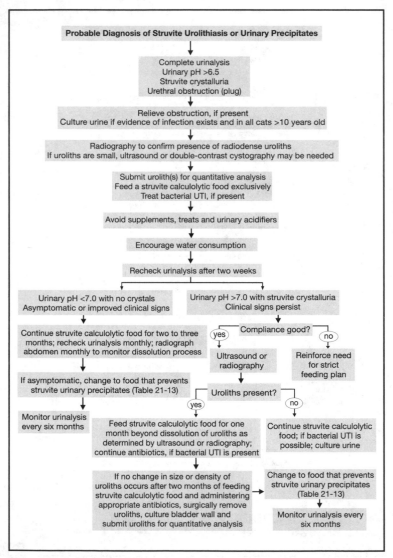

Probable Diagnosis of Struvite Urolithiasis or Urinary Precipitates

⬇

Complete urinalysis
Urinary pH >6.5
Struvite crystalluria
Urethral obstruction (plug)

⬇

Relieve obstruction, if present
Culture urine if evidence of infection exists and in all cats >10 years old

⬇

Radiography to confirm presence of radiodense uroliths
If uroliths are small, ultrasound or double-contrast cystography may be needed

⬇

Submit urolith(s) for quantitative analysis
Feed a struvite calculolytic food exclusively
Treat bacterial UTI, if present

⬇

Avoid supplements, treats and urinary acidifiers

⬇

Encourage water consumption

⬇

Recheck urinalysis after two weeks

Urinary pH <7.0 with no crystals
Asymptomatic or improved clinical signs

Urinary pH >7.0 with struvite crystalluria
Clinical signs persist

Continue struvite calculolytic food for two to three months; recheck urinalysis monthly; radiograph abdomen monthly to monitor dissolution process

Compliance good?
yes / no

Ultrasound or radiography

Reinforce need for strict feeding plan

If asymptomatic, change to food that prevents struvite urinary precipitates (Table 21-13)

Uroliths present?
yes / no

Monitor urinalysis every six months

Feed struvite calculolytic food for one month beyond dissolution of uroliths as determined by ultrasound or radiography; continue antibiotics, if bacterial UTI is present

Continue struvite calculolytic food; if bacterial UTI is possible; culture urine

If no change in size or density of uroliths occurs after two months of feeding struvite calculolytic food and administering appropriate antibiotics, surgically remove uroliths, culture bladder wall and submit uroliths for quantitative analysis

Change to food that prevents struvite urinary precipitates (Table 21-13)

Monitor urinalysis every six months

Figure 21-6. Algorithm for management and reassessment of cats with struvite urolithiasis.

In selected situations, periodic monitoring of venous blood gases for evidence of chronic metabolic acidosis is recommended. These situations include reduced renal function or the concurrent use of an acidifying food and a urinary acidifier. Recently, a handheld analyzer (I-STAT, Sensor Devices, Inc., Waukesha, WI, USA) developed for human bedside testing was introduced to the veterinary market. This portable analyzer uses a test cartridge to measure blood gases on whole blood specimens of about 0.1 ml.

Drug-Nutrient Interactions (Pages 715-718)

The use of urinary acidifiers in conjunction with an acidifying food is not recommended. With continued acid loading, renal excretion of acid increases but does not quite equal the acid intake, resulting in a positive acid balance (i.e., metabolic acidosis). Overacidification leading to metabolic acidosis is more likely to occur in young cats and cats with renal insufficiency. Supplemental acidifiers may be considered if nonacidifying foods are used or if the food is ineffective. The lack of convincing evidence that urinary acidification alone prevents recurrence of idiopathic lower urinary tract disease and the many adverse effects of overacidification preclude the widespread use of urinary acidifiers for nonobstructive lower urinary tract disease in cats.

Although thiazide diuretics may be beneficial in minimizing urinary calcium excretion in people and dogs, no data have been provided to prove their efficacy in cats with calcium oxalate uroliths. Because thiazide diuretics may be associated with adverse effects such as dehydration, hypokalemia and hypercalcemia, their routine use in cats is not recommended, pending further investigation. Drugs that promote hypercalciuria should also be avoided in patients with calcium oxalate uroliths. This precludes routine administration of furosemide or glucocorticoids to these patients.

Investigators have detected sulfadiazine and/or its metabolites in uroliths and a urethral plug from cats with lower urinary tract disease. Factors that predispose cats to precipitation of sulfonamides and/or their metabolites in the urinary tract include administration of high doses of these drugs for prolonged periods, acidic urine and highly concentrated urine. Based on these findings, sulfadiazine should not be used to empirically treat lower urinary tract signs, especially in cats known to: 1) have uroliths, 2) be at increased risk for metabolic uroliths and 3) form acidic and/or highly concentrated urine.

Clinical cases that illustrate and reinforce the nutritional concepts presented in this chapter can be found in Small Animal Clinical Nutrition, 4th ed., pp 720-723.

CHAPTER 22

Gastrointestinal and Exocrine Pancreatic Disease

For a review of the unabridged chapter, see Davenport DJ, Remillard RL, Simpson KW, et al. Gastrointestinal and Exocrine Pancreatic Disease. In: Hand MS, Thatcher CD, Remillard RL, et al, eds. Small Animal Clinical Nutrition, 4th ed. Topeka, KS: Mark Morris Institute, 2000; 725-810.

INTRODUCTION (Pages 725-728)*

Altering food ingredients, nutrient profiles, food form and feeding method can be powerful tools in managing gastrointestinal (GI) and exocrine pancreatic diseases (**Table 22-1**). Drug therapy instituted without concomitant dietary therapy often yields less than desirable results in the long-term management of most GI diseases. Occasionally, foods or ingredients may function as diagnostic tools in evaluating patients with GI and pancreatic disorders.

A multitude of factors, including trophic hormones, adequate blood flow, neurologic input and nutrient composition of digesta, are involved in maintaining intestinal integrity (mass and function). Nutrients and ingredients can positively or negatively affect the bowel (**Table 22-2**). Malnutrition profoundly affects the bowel; effects include decreased pancreatic enzyme production and secretion, intestinal mucosal atrophy and reduced gastric emptying rates. The resultant diarrhea and malassimilation further exacerbate the malnutritive state. In addition, starvation can markedly affect bowel immune response and mucosal integrity.

The outline for this chapter is as follows: sections include assessment and feeding plans for: 1) oral, pharyngeal and esophageal disorders, 2) gastric disorders, 3) small intestinal disorders, 4) colonic and rectal disorders and 5) exocrine pancreatic disorders (**Table 22-1**).

Vomiting and diarrhea, in particular, have a myriad of causes and feeding plans vary according to the underlying condition. When a specific cause of acute vomiting or diarrhea is unknown, then feeding plans outlined for acute gastroenteritis are most appropriate. When a specific cause of chronic small bowel diarrhea is not identified, then feeding plans outlined for exocrine pancreatic insufficiency are most appropriate. Finally, when a specific cause of chronic large bowel diarrhea is not identified, then feeding plans outlined for colitis are most appropriate.

*Page numbers in headings refer to Small Animal Clinical Nutrition, 4th ed., where additional information may be found.

Key Nutritional Factors–GI and Exocrine Pancreatic Disease
(Page 727)

Table 22-6 lists key nutritional factors for dogs with swallowing disorders.
Table 22-9 lists key nutritional factors for patients with gastritis and/or gastroduodenal ulceration.
Table 22-15 lists key nutritional factors for patients with gastric motility/ emptying disorders.
Table 22-23 lists key nutritional factors for patients with acute enteritis.
Table 22-29 lists key nutritional factors for patients with inflammatory bowel disease.
Table 22-35 lists key nutritional factors for patients with lymphangiectasia/ protein-losing enteropathy.
Table 22-36 lists key nutritional factors for patients with small intestinal bacterial overgrowth.
Table 22-42 lists key nutritional factors for patients with acute or chronic colitis.
Table 22-48 lists key nutritional factors for patients with exocrine pancreatic insufficiency.

GI/PANCREAS

Each patient should be seen as an individual variant of the norm; therefore, multiple dietary manipulations may be needed for any one patient. Because of the diverse nature of GI disorders and exocrine pancreatic conditions, a number of food types may be appropriate. Nutrient profiles should be considered as starting points on a continuum of possible nutrient concentrations that can be adjusted for each patient as needed. The reference point should be the current food that the owner feeds. Changes include increases or decreases, usually in 5 to 10% increments, of nutrient concentrations relative to the previous food.

Food Types Useful in the Management of GI and Exocrine Pancreatic Disease (Pages 729-730)
Gastrointestinal Foods (Page 729)
Several commercial veterinary therapeutic foods have been specially formulated for managing GI disease in dogs and cats. Typically, these products are highly digestible and have consistent ingredient and nutrient profiles.

The term highly digestible is not defined in a regulatory sense. However, highly digestible has generally been reserved for products with protein digestibility ≥87% and fat and soluble carbohydrate digestibilities ≥90%. The average digestibility coefficients for popular commercial foods are 78 to 81%, 77 to 85% and 69 to 79% for crude protein, crude fat and

Table 22-1. Gastrointestinal and exocrine pancreatic diseases amenable to dietary management.

Mouth	
Inflammatory disorders (stomatitis, radiation-induced mucositis)	Physical abnormalities (trauma, neoplasia, congenital malformations)
Pharynx and esophagus	
Inflammatory disorders (esophagitis)	Obstructive disorders (vascular ring anomalies, strictures, neoplasia)
Motility disorders (cricopharyngeal achalasia, megaesophagus)	
Stomach	
Gastric dilatation/gastric dilatation-volvulus	Gastritis
	Gastroduodenal ulceration
Gastric motility/emptying disorders	
Small intestine	
Acute enteritis	Protein-losing enteropathy
Inflammatory bowel disease	Short bowel syndrome
Intestinal neoplasia	Small intestinal bacterial overgrowth
Lymphangiectasia	Wheat-sensitive enteropathy
Large intestine	
Colitis	Irritable bowel syndrome
Constipation	Perianal fistula
Flatulence	
Pancreas	
Exocrine pancreatic insufficiency	Pancreatitis

Table 22-2. Potential dietary influences on the gastrointestinal tract.

Food may alter:	
Absorption	Microflora
Cellular turnover rate	Motility
Luminal ammonia concentration	Secretory rate
Luminal volatile fatty acid content	Villous height
Food may be a source of:	
Chemical/bacterial toxins	Dietary antigens
Food may correct:	
	Nutritional deficiencies

carbohydrate (nitrogen-free extract [NFE]), respectively. Commercial veterinary therapeutic foods formulated for GI disease usually contain highly refined meat and carbohydrate sources to increase digestibility.

Carbohydrates make up the largest nonwater fraction of foods formulated for managing GI diseases. Carbohydrate digestibility of pet foods is influenced by source and processing. Dogs digest most properly cooked starches very well, including starch components in corn, rice, barley and wheat. Other starches, including potato and tapioca, are less digestible, particularly when inadequately cooked. Although cats also efficiently digest carbohydrates, some clinicians feel that cats with small bowel disorders are less tolerant of dietary carbohydrate than dogs with similar causes of malassimilation.

Recent research has identified a link between particle size and carbohydrate digestibility of moist foods. As a result, carbohydrate ingredients (e.g., rice, corn, etc.) should be chopped or ground before they are incorporated into moist foods. These findings probably do not apply to extruded dry products because as part of the dry food manufacturing process, carbohydrate sources are almost always chopped or ground. In fact, other studies have demonstrated almost complete ileal carbohydrate digestibility in dogs consuming extruded grains.

The requirements for many macro- and microminerals in the face of GI disease are not well understood. However, sodium, potassium and B-vitamin losses are expected with vomiting and diarrhea. Therefore, foods formulated for managing GI diseases should contain sodium, potassium and B vitamins in excess of maintenance allowances. Patients with fat malabsorption are at risk for developing fat-soluble vitamin deficiencies. Highly digestible foods formulated for feeding steatorrheic patients should, therefore, be fortified with fat-soluble vitamins.

It is unusual for GI foods to contain crude fiber levels greater than 5% dry matter (DM) because fiber reduces dry matter digestibility and decreases pancreatic enzymatic activity in vitro. More recently, manufacturers of some highly digestible commercial veterinary therapeutic foods have added small amounts (<5% DM) of soluble or mixed fibers because short-chain fatty acids produced by intestinal microbial fermentation of fiber may positively affect the intestinal mucosa.

Veterinarians recommend GI foods most often for managing acute gastroenteritis or malassimilation associated with small bowel disease or exocrine pancreatic insufficiency. The utility of highly digestible foods has been demonstrated through anecdotal reports and by the use of such foods in clinical trials involving animals with spontaneous and experimental exocrine pancreatic insufficiency. Some gastroenterologists also recommend these foods for patients with certain colonic disorders to reduce exposure of the colonic mucosa to ingesta. This therapeutic strategy has been suggested for management of inflammatory colitides and constipation.

Fiber-Enhanced Foods (Page 729)

Commercial veterinary therapeutic foods contain varying levels and sources of fiber. Based on the combined knowledge obtained from research in people, ongoing research in dogs and cats and clinical experience, fiber is beneficial in managing many large and some small bowel diseases.

GI/PANCREAS

Soluble fibers (e.g., pectins and gums) increase the viscosity of intestinal contents, which delays gastric emptying and slows small bowel transit time. Viscosity markedly affects the extent of intraluminal mixing of digesta and digestive enzymes, which can shift sites of absorption and subsequently the rate of nutrients entering the bloodstream. Bacteria in the colon ferment soluble fiber to short-chain fatty acids, including acetic, propionic and butyric acids. Colonocytes apparently use butyrate, whereas propionic and acetic acids are absorbed. Short-chain fatty acids are nutritive to the colonic mucosa and foster normal colonic flora while discouraging pathogenic flora. These properties result in an acidic colonic pH and increased colonic bacterial numbers, colonic mucosal mass and fecal dry matter and water content. Soluble fiber may bind and decrease macronutrient absorption and decrease protein digestibility. Certain fiber types, especially gels and gums, may be of benefit in GI disease because they bind toxins and irritating bile acids. This binding effect prevents these substances from further damaging the intestinal mucosal surface.

Insoluble fiber is primarily composed of cellulose and structural polysaccharides that are relatively resistant to digestion and that ferment slowly, increase intestinal residue and normalize intestinal transit time. These fibers have little or no effect on gastric emptying, mineral absorption or colonic microflora unless fed in high concentrations (>20% DM).

One of the most profound effects of fiber on the GI tract is the normalization of gut motility, particularly in the stomach, proximal small bowel and colon. This effect appears to be greatest for insoluble fibers such as cellulose. In general, increasing the insoluble fiber content of the food resolves or modulates most cases of colitis. There are several plausible mechanisms by which insoluble fiber controls large bowel diarrhea. Undigested residues absorb water and increase bacterial mass, which increases fecal bulk. Fecal bulk provides physical intraluminal stimulation to reestablish neuromuscular-endocrine coordinations and normalize intestinal transit times. Fecal bulk increases intestinal residue, which absorbs toxins and offending agents.

Restricted- and Moderate-Fat Foods (Pages 729-730)

In general, dietary fat is more digestible than soluble carbohydrate and protein and provides 2.25 times more calories by weight. Average fat digestibility in commercial dog food is approximately 90%. Average fat digestibility of commercial cat foods ranges from 74 to 91%. Patients with GI or pancreatic disease may not tolerate high-fat foods (>25% DM), which may contribute to diarrhea and steatorrhea. Foods containing moderate amounts of fat (12 to 15% DM for dogs and 15 to 22% DM for cats) are generally tolerated and have sufficient caloric density for most patients. Commercial veterinary therapeutic foods containing less than 10% DM fat need to be fed in larger volumes to meet the patient's caloric requirement. Some patients may not tolerate this volume of food.

Restricted-fat foods are often recommended for patients with gastroenteritis in which the complex process of fat digestion and absorption may be disrupted. Unabsorbed fat in the bowel lumen may cause secretory diarrhea. Dietary fat should be reduced when fat maldigestion or malabsorption is present due to exocrine pancreatic insufficiency or reduced bowel surface area. The latter occurs in short bowel syndrome and other conditions in which inflammation, infectious agents, neoplasia or surgery markedly reduces the intestinal villus surface area. For example, intestinal malabsorption of fat is seriously impaired in primary and secondary lymphangiectasia. Fat restriction is also useful in small intestinal bacterial overgrowth in which many of the side effects of the condition can be ameliorated by removing the inciting cause of the secretory diarrhea.

Elimination Foods (Page 730)
Elimination foods are most often recommended for patients with GI signs due to suspected food intolerance or food hypersensitivity. Chapter 14 discusses adverse food reactions and elimination foods in more detail.

Gluten- and Gliadin-Free Foods (Page 730)
Several potential antigens are found in flour when cereal grains are processed. One polypeptide, gliadin, is found in wheat, barley, rye, buckwheat and oat flours. Gliadin is responsible for gluten-sensitive enteropathies in people and dogs. Homologous gliadin polypeptides are not present in whole grains and flours produced from rice and corn.

In people, gluten-induced enteropathy or celiac disease is an important malabsorptive disorder. An analogous condition, termed wheat-sensitive enteropathy, has been identified in Irish setter dogs and is suspected to affect dogs of other breeds as well. Affected animals develop small bowel diarrhea due to malabsorption secondary to villous atrophy. Gluten- and gliadin-free foods are most commonly recommended for managing dogs suspected of having wheat-sensitive enteropathy. In most cases, withdrawal of the offending gliadin antigen from the diet results in resolution of the villous atrophy and clinical signs.

Monomeric Foods (Page 730)
Monomeric foods are water-soluble, liquid foods containing nutrients in their simplest absorbable form. Mixtures of di- and tripeptides and/or individual amino acids most commonly provide nitrogen. Fats are present as triglycerides or as fatty acids. Carbohydrates are generally present as mono- or disaccharides. Minerals and vitamins are present to meet human

requirements. These foods minimize GI and pancreatic secretions and allow nutrient usage with minimal requirements for digestion. In addition, the small size of nitrogen sources (i.e., amino acids, dipeptides and tripeptides) in monomeric products ensures delivery of a truly "hypoallergenic" food.

Monomeric foods should be considered for patients with severe malabsorption or short bowel syndrome and in initial refeeding of patients with acute pancreatitis. In addition, these foods may provide "bowel rest" for patients with severe inflammatory bowel disease. Monomeric foods are often unpalatable and are not well accepted by dogs or cats. Thus, these foods are usually administered for several days via indwelling feeding tubes. **Table 12-6** lists selected monomeric foods.

Performing Dietary Trials in Patients with GI Disease (Page 728)

Nutritional therapies are extremely useful for treating GI disease in dogs and cats. Several commercial and homemade foods are available to practitioners and pet owners for this purpose. Unfortunately, there is no historical or clinical finding that will predict the success of a specific food type. Therefore, selection of the most appropriate food for an individual patient is often based on results of a dietary trial.

Dietary trials are easily performed in most clinical and home settings. Oral food consumption is preferred for managing GI diseases, except in those rare situations in which the animal is intolerant of enteral feeding.

After the veterinarian identifies those foods to be included in the trial, selection of the initial test food is often based on clinical experience and the animal's nutritional history. In general, foods that have been used unsuccessfully in the past to manage the patient should be avoided. Typically, highly digestible GI or elimination foods are good first choices for patients with gastric or small intestinal disorders. Fiber-enhanced foods are often the initial selection when large bowel signs predominate. No other foods, supplements, table foods or treats should be offered during the dietary trial. Dietary trials are most useful if continued for at least seven to 10 days. In certain settings (e.g., adverse reactions to food), trials lasting two to 12 weeks may be necessary to determine efficacy. (See Chapter 14.) Successful dietary trials are marked by partial or complete resolution of clinical signs.

ORAL DISORDERS (Pages 728-732)

Clinical Importance (Pages 728-731)

The oral cavity is susceptible to a number of acquired and congenital disorders. These conditions, however, are relatively uncommon compared with the prevalence of dental disease. Chapter 16 discusses dental disease in detail. Inflammatory lesions and physical abnormalities such as neoplasia, trauma and congenital malformations (e.g., cleft palate) are among the common conditions affecting the oral cavity.

Acquired inflammatory lesions of the oral cavity and tongue are uncommon in dogs and cats. These conditions include eosinophilic granuloma complex, lymphoplasmacytic stomatitis, labial granuloma and mucositis due to radiation therapy of the head and neck. Infectious oral disorders (e.g., candidiasis or fusospirochetal infections) are very rare and usually occur in immunocompromised animals. Oral neoplasia is the fourth most common cancer of dogs and cats. Trauma to the oral cavity may arise from fights among animals, falls (high-rise syndrome), motor vehicle accidents, chemical and electrical burns and penetrating foreign bodies. Oral congenital anomalies such as cleft palate are uncommon but may have nutritional causes (e.g., copper deficiency in pregnant queens) or profound consequences due to malnutrition and secondary aspiration pneumonia in growing animals.

Assessment (Pages 731-732)
Assess the Animal (Pages 731-732)
HISTORY AND PHYSICAL EXAMINATION Dogs and cats with oral disease have variable clinical signs depending on the type and location of the lesions. Animals may exhibit dysphagia or pain associated with eating. Owners may report excessive salivation, oral hemorrhage, halitosis and reluctance to eat. In some cases, careful questioning of the owner will reveal ingestion of foreign bodies or caustic materials or a history of trauma. Puppies and kittens with congenital anomalies such as cleft palate may be presented to veterinarians for ineffectual suckling, poor weight gain and coughing or gagging following attempts at nursing.

Sedation may be required to facilitate examination of the oropharynx and tongue. Various conditions may present with specific signs. Congenital defects may be noted in the soft or hard palate. Epulides originate from periodontal stroma and are most commonly located in the gingiva near the incisor teeth and appear as pedunculated or smooth, nonulcerated masses. Odontogenic tumors (e.g., ameloblastoma and odontoma) are typically expansile, slow-growing odontogenic masses that often form in the incisor region. Malignant tumors (e.g., squamous cell carcinoma, malignant melanoma and fibrosarcoma) grow rapidly and are characterized by early invasion of the gingiva and bone. Pets with suspected oral or tonsillar tumors should be carefully evaluated for peripheral lymphadenopathy.

Raised, erythremic cobblestone-like lesions at the glossopalatine arches characterize lymphoplasmacytic stomatitis in cats, whereas feline eosinophilic granuloma complex manifests as ulcers, plaques and granulomas on the maxillary lips, tongue and palate. In dogs, inflammatory lesions are most often present on the tongue or palatine and labial mucosa.

Head trauma in pets often results in mandibular symphyseal fractures, maxillary fractures, displaced teeth and separation of the hard palate. These injuries may result in reluctance or inability to eat.

Chemical, electrical and thermal burns are characterized by ulceration and necrosis of affected tissues. Animals with oral burns may suffer life-threatening consequences such as pulmonary edema or cardiogenic shock.

LABORATORY AND OTHER CLINICAL INFORMATION Laboratory values are often unremarkable in animals with oral disease and generally reflect underlying conditions when present. Leukocytosis and a polyclonal hyperglobulinemia are frequent findings in cats with lymphoplasmacytic stomatitis. Radiography is often of value in cases with suspected trauma to assess the extent of bony injury. Radiography is invaluable for tumor staging in patients with oral neoplasia. Generally, both skull and thoracic films are evaluated. In addition, thoracic films allow assessment of aspiration pneumonia in young animals with cleft palate. Diagnosis of lesions within the oral cavity often requires biopsy and histopathologic examination.

RISK FACTORS Age and breed are risk factors for several oral disorders. Young animals are more likely to present with congenital and traumatic lesions, whereas older dogs and cats are more likely to suffer from oral neoplasia and inflammatory disorders. Animals undergoing radiation therapy of the head and neck for cancer are susceptible to radiation-induced mucositis. See Small Animal Clinical Nutrition, 4th ed., p 732 for information about feeding patients undergoing radiation therapy. In addition, certain breeds are predisposed to various oral disorders (**Table 22-3**).

ETIOPATHOGENESIS Pets with oral disease often exhibit dysphagia or reluctance to eat resulting in malnutrition. Often this nutritional state is compounded by inflammatory, traumatic or neoplastic processes. The etiology of oral inflammatory lesions such as lymphoplasmacytic stomatitis and faucitis, and eosinophilic granuloma complex is unknown. Lymphoplasmacytic stomatitis in cats has been theorized to be an aberrant immunologic response to antigenic stimuli. Various bacterial, viral, periodontal, dietary and immune factors have been implicated. There is a strong association between this disorder and infection with feline immunodeficiency virus (FIV) or calicivirus. Approximately 50% of cats with FIV infection and 60% of cats with feline calicivirus infection have chronic oral disease.

KEY NUTRITIONAL FACTORS/Water Dehydration is a frequent problem in dogs and cats with oral disorders that interfere with consumption

Table 22-3. Breed-associated oral disorders.

Disorders	Breeds
Cleft palate	Brachycephalic dogs and cats
Epulides	Boxer
Gingivitis/stomatitis	Maltese dog
	Siberian husky
Lymphoplasmacytic stomatitis	Abyssinian cat
	Burmese cat
	Himalayan cat
	Maltese cat
	Persian cat
	Siamese cat
Neoplasia	Cocker spaniel
	German shepherd dog
	German shorthaired pointer
	Golden retriever
	Weimaraner

of water. Whenever possible, fluid balance should be maintained via oral consumption of fluids. However, parenteral fluid administration is often needed for dehydrated patients and those unable or unwilling to drink adequate amounts of water.

KEY NUTRITIONAL FACTORS/Energy A relatively high fat concentration is helpful in meeting the patient's caloric requirement in a small volume of food relative to lower fat foods. Foods with energy densities in excess of 4.5 kcal/g (18.8 kJ/g) DM for dogs and 5 kcal/g (20.9 kJ/g) DM for cats are recommended.

KEY NUTRITIONAL FACTORS/Food Form The veterinarian or owner should experiment with foods of differing consistency. Often liquid foods or slurries made from moist pet food and water are more readily accepted. A dilute consistency is often associated with less discomfort, and is less likely to accumulate in oral lesions or adhere to surgical sites within the oral cavity.

Assess the Food and Feeding Method (Page 732)

The energy density of the patient's food should be compared with the levels described above. Underweight patients may need a nutrient profile similar to that found in a growth or recovery-type formula to regain normal body condition. In addition, the food should be suitable for any other conditions present that are amenable to dietary management.

Because the feeding method is often altered in patients with oral disease, a thorough assessment should include verification of the feeding method currently being used. Items to consider include feeding frequency, amount fed, how the food is offered, access to other food sources including table food and who feeds the animal. All of this information should have been gathered when the history of the animal was obtained.

Feeding Plan (Page 732)

The goals of dietary management for patients with oral disease are to provide adequate nutrition while minimizing discomfort to the pet and enhancing resolution of the oral lesions.

Select a Food (Page 732)

Foods should be energy dense (≥4.5 kcal/g [18.8 kJ/g] DM for dogs and ≥5 kcal/g [20.9 kJ/g] DM for cats) and in a form that is easily ingested and less irritating to the oral mucosa (e.g., moist foods, gruels or slurries; some critical care and growth-type foods are appropriate). Animals with extensive oral injuries or inflammation of the oral cavity may benefit from foods designed for assisted feeding or recovery. (See Chapter 12.) Animals with oral neoplasia may benefit from foods specifically formulated for patients with cancer. (See Chapter 25.)

Determine a Feeding Method (Page 732)

Animals with oral disease should initially be fed several small meals daily if they are able and willing to consume food voluntarily. After each meal, the oral cavity should be flushed with water to remove particulate matter adhered to the oral mucous membranes. In many cases, tube-feeding methods are preferred until oral discomfort is reduced and voluntary food consumption resumes. (See Chapter 12 and Appendix 4 for more information about tube feeding.)

Reassessment (Page 732)

Body condition scores and hydration status should be evaluated to determine adequacy of food and water consumption. Assisted feeding should be instituted if oral feeding is not adequate to maintain body weight and condition (Chapter 12).

PHARYNGEAL AND ESOPHAGEAL DISORDERS
(Pages 733-738)

Clinical Importance (Page 733)

Compared with vomiting and diarrhea, swallowing disorders are relatively uncommon in dogs and cats. However, these conditions are often profoundly debilitating due to malnutrition (i.e., lack of adequate food intake) and recurrent pulmonary infections resulting from aspiration. Pharyngeal and esophageal disorders most commonly encountered include: 1) motility disorders (e.g., cricopharyngeal achalasia, megaesophagus), 2) inflammatory disorders (e.g., esophagitis, gastroesophageal reflux) and 3) obstructive lesions (e.g., vascular ring anomalies, stricture and foreign bodies).

Assessment (Pages 733-736)
Assess the Animal (Pages 733-736)
HISTORY AND PHYSICAL EXAMINATION Congenital pharyngeal and esophageal disorders are typically diagnosed in young animals soon after weaning. Rarely, dogs with congenital malformations of the aortic arches, also known as vascular ring anomalies, may present with late-onset regurgitation as adults. Acquired pharyngeal and esophageal disease can affect dogs and cats of any age.

Owners of dogs with dysphagia due to pharyngeal disease typically report coughing or gagging as the dog chews and swallows food. The hallmark of esophageal disorders is regurgitation. Additional clinical signs include ptyalism, gurgling esophageal noises, halitosis and apparent pain on swallowing. The frequency of regurgitation is variable. Owners may report immediate postprandial regurgitation of undigested food, water or saliva or describe signs manifested several hours after feeding. Affected dogs and cats often have a voracious appetite despite regurgitation. Dyspnea, coughing, exercise intolerance and syncope may be referable to severe respiratory compromise associated with aspiration pneumonia.

Esophageal disorders are frequently associated with neuromuscular diseases and endocrinopathies. Owners may describe weakness or incoordination and many dogs have a history of recurrent skin and coat problems associated with hypothyroidism.

Poor body condition is often evident (body condition score [BCS] 1/5 or 2/5). Body condition should be monitored closely during reassessment and the BCS should be recorded. Young animals with congenital megaesophagus, vascular ring anomalies or cricopharyngeal achalasia are often stunted in comparison to littermates.

Auscultatory findings often indicate secondary aspiration pneumonia and may include crackles and prominent bronchovesicular sounds. Dogs with active respiratory disease may be febrile and manifest a mucopurulent nasal discharge.

A complete neurologic examination should be performed on an adult dog with a swallowing disorder because acquired megaesophagus is often associated with neuromuscular disorders. Signs of lower motor neuron disease may provide evidence of a generalized polymyopathy, polyneuropathy or neuromuscular junctionopathy.

GI/PANCREAS

LABORATORY AND OTHER CLINICAL INFORMATION A complete blood count may provide evidence of aspiration pneumonia and some sense of the severity of infection. In chronically affected patients, serum protein and albumin concentrations may provide an indication of nutritional status. Additionally, other serum biochemical abnormalities may provide evidence for an underlying disorder (e.g., hypoadrenocorticism, hypothyroidism).

Radiography is a vital diagnostic aid for evaluating dogs and cats with suspected swallowing and esophageal disorders. Survey films may provide definitive information in cases of megaesophagus and esophageal foreign bodies. Radiographic findings in dogs and cats with megaesophagus include a dilated, air-filled esophagus. In the case of vascular ring anomalies, characteristic esophageal dilatation proximal to the heart base can be identified. Thoracic radiography also allows the clinician to assess the patient for aspiration pneumonia. Additionally, thoracic films may reveal a cranial thoracic mass. Thymoma and thymic lymphosarcoma have been associated with secondary acquired megaesophagus and generalized inflammatory myopathies.

The esophagram offers additional diagnostic information, especially in cases of obstructive lesions, esophagitis and esophageal hypomotility without megaesophagus. When coupled with fluoroscopy, an esophagram allows sensitive evaluation of the swallow reflex and esophageal motility.

Esophagoscopy is a valuable tool for evaluating dogs and cats with suspected obstructive, neoplastic or inflammatory lesions of the esophagus and pharynx. This tool allows visualization of the entire area and collection of tissue specimens for microbiologic and histopathologic examination, if indicated. Additionally, in cases of esophageal foreign bodies or strictures, the flexible endoscope can provide definitive treatment of the lesion. Foreign bodies can be retrieved using a variety of forceps, whereas esophageal strictures are best managed with endoscopic bougienage, balloon dilatation or both procedures.

Acquired megaesophagus can occur secondary to several neuromuscular disorders such as myasthenia gravis, dysautonomia, hypothyroidism, hypoadrenocorticism, systemic lupus erythematosus and other causes of generalized myopathy or neuropathy.

RISK FACTORS Swallowing disorders are exceedingly rare in cats. In dogs, risk factors for swallowing disorders are primarily breed and age related. Several breeds appear to be predisposed to the development of congenital disorders such as cricopharyngeal achalasia, congenital megaesophagus and vascular ring anomalies (**Table 22-4**). The condition, however, occurs more often in large-breed dogs.

Table 22-4. Breed-associated disorders of the pharynx and esophagus.

Conditions	Breeds
Cricopharyngeal achalasia	Cocker spaniel
Congenital megaesophagus	Chinese Shar-Pei
	Fox terrier
	German shepherd dog
	Great Dane
	Irish setter
	Labrador retriever
	Miniature schnauzer
	Newfoundland
	Siamese cat
	Wire-haired terrier
Idiopathic acquired megaesophagus	German shepherd dog
	Golden retriever
	Great Dane
	Irish setter
Vascular ring anomalies	Boston terrier
	English bulldog
	German shepherd dog
	Irish setter
	Labrador retriever
	Poodle

GI/PANCREAS

There does not seem to be any gender predisposition for idiopathic acquired swallowing disorders. Middle-aged to older dogs are more likely to develop myasthenia gravis and other neuromuscular disorders resulting in esophageal disease. Nearly 90% of dogs with focal or generalized myasthenia gravis develop megaesophagus. In addition, those breeds predisposed to endocrinopathies (e.g., hypothyroidism or hypoadrenocorticism) are at risk for development of megaesophagus as a manifestation of their disease. Also, it has been reported that dogs with laryngeal paralysis, another neuromuscular condition, are also at risk for development of megaesophagus. In certain areas (e.g., northeastern United States), exposure to lead has been linked to cases of secondary acquired megaesophagus.

ETIOPATHOGENESIS Pharyngeal and esophageal disorders can generally be attributed to one of three basic pathophysiologic mechanisms: aberrant motility, obstructive lesions or inflammatory degenerative conditions (**Table 22-5**).

ETIOPATHOGENESIS/Aberrant Motility Cricopharyngeal achalasia is characterized by asynchrony of the swallowing reflex. In this condition, the cricopharyngeal muscle fails to relax in coordination with pharyngeal muscle contractions, thus preventing passage of a food bolus from the oropharynx to the esophagus.

Table 22-5. Mechanisms of pharyngeal and esophageal disorders.

Mechanisms	Disorders
Aberrant motility	Congenital megaesophagus
	Cricopharyngeal achalasia
	Dysautonomia
	Endocrinopathies (hypothyroidism, hypoadrenocorticism)
	Idiopathic megaesophagus
	Infectious diseases (canine distemper)
	Myasthenia gravis
	Paraneoplastic syndromes (lymphosarcoma, thymoma)
	Polymyopathies
	Polyneuropathies
	Secondary megaesophagus
	Toxin ingestion (lead)
Inflammatory conditions	Foreign body esophagitis
	Pharyngitis
	Reflux esophagitis
Obstructive lesions	Foreign bodies
	Neoplasia
	Spirocerca lupi granulomas
	Strictures
	Vascular ring anomalies

Megaesophagus is a disorder within the body of the esophagus. Dogs with idiopathic megaesophagus have a defect in their afferent neural pathway. These findings have clinical implications because they suggest that foods containing more bulk or prepared in larger boluses may have the ability to stimulate esophageal motility in mildly affected animals.

ETIOPATHOGENESIS/Obstructive Lesions Persistent right aortic arch is the most common vascular ring anomaly recognized in dogs and cats. This anomaly results in constriction of the esophagus at the level of the heart base by the right fourth aortic arch and the ligamentum arteriosum. Esophageal dilatation develops proximal to the vascular ring, leading to regurgitation. Esophageal motility defects may persist if the obstructive lesion is not surgically corrected soon after detection.

Esophageal obstruction due to stricture formation may occur as a consequence of recurrent or severe esophageal injury. Strictures occur most commonly due to esophageal foreign bodies or as sequelae to gastroesophageal reflux.

ETIOPATHOGENESIS/Inflammation/Degeneration Esophagitis arises most often as a consequence of gastroesophageal reflux or foreign

Table 22-6. Key nutritional factors for dogs with swallowing disorders.*

Disorders	Energy (kcal/g)	Energy (kJ/g)	Fat (%)	Protein (%)
Esophagitis/gastroesophageal reflux	<4.0	<16.7	<15	≥25
Motility disorders	≥4.5	≥18.8	≥25	≥25
Obstructive disorders	≥4.5	≥18.8	≥25	≥25

*Nutrients expressed on a dry matter basis.

body ingestion. The gastroesophageal junction (GEJ) serves as a barrier preventing reflux of gastric contents including pepsin and hydrochloric acid into the lumen of the esophagus. Postprandially, GEJ pressure increases in response to neural and hormonal stimuli. Certain GI hormones, including gastrin, pancreatic polypeptide, motilin and substance P increase GEJ pressure, whereas others (i.e., secretin, cholecystokinin) reduce GEJ pressure. Dietary influences on GEJ pressure are presumably mediated via GI hormone release. High-protein meals increase GEJ pressure through gastrin release, whereas high-fat foods reduce GEJ pressure via cholecystokinin release.

Certain sedative agents, including acepromazine, xylazine and diazepam reduce GEJ pressure and may predispose an animal to reflux esophagitis following anesthetic episodes. Iatrogenic esophagitis may occur as a sequela to nasoesophageal intubation when the feeding tube crosses the GEJ resulting in incompetence of the sphincter. Occasionally, consumption of irritative substances such as strong acids or alkalis may cause serious esophagitis.

KEY NUTRITIONAL FACTORS Key nutritional factors for patients with swallowing disorders are listed in **Table 22-6** and discussed below. Animals with swallowing disorders are often debilitated and growth of very young animals is often stunted. In addition to the key nutritional factors discussed here, other nutritional factors may be important depending on the lifestage and body condition of the animal.

KEY NUTRITIONAL FACTORS/Energy and Fat In patients with motility and obstructive disorders, a relatively high fat concentration is helpful in meeting the patient's caloric requirement in a small volume of food relative to lower fat foods. Foods with greater than 25% DM fat and energy densities in excess of 4.8 kcal/g (20.1 kJ/g) DM are recommended. However, a lower fat content (<15% DM) is a better option for cases of esophagitis with gastric reflux. High dietary fat delays gastric emptying and reduces lower esophageal sphincter pressure, which promotes reflux of food and gastric secretions into the esophagus.

KEY NUTRITIONAL FACTORS/Protein Protein is required in amounts adequate for tissue repair and to support growth in young animals.

Additionally, dietary protein may play an important role in reducing episodes of gastroesophageal reflux because protein stimulates an increase in gastroesophageal sphincter pressure. This effect is linked to dietary protein's stimulatory effect on gastrin and gastric acid secretion. By increasing the lower esophageal sphincter pressure, episodes of gastroesophageal reflux are decreased, thus limiting the potential for further esophageal injury or aspiration pneumonia. For these reasons, dietary protein content should exceed 25% DM.

KEY NUTRITIONAL FACTORS/Food Form Foods of differing consistency should be used to determine the best texture for individual patients. A liquid or gruel consistency is usually best for animals with cricopharyngeal achalasia, esophageal obstructive lesions and/or esophagitis. Esophageal performance may improve in animals with megaesophagus when the swallowing reflex is maximally stimulated by the texture of dry foods or when moist foods are formed into large boluses. Dry food or boluses of moist food may act as a stimulus (secondary peristalsis) to any remaining normal esophageal tissue and, therefore, are the form of choice. Gruels or liquids may not stimulate secondary peristalsis, thereby increasing the risk of aspiration pneumonia.

Assess the Food and Feeding Method (Page 736)

The nutrient profile of the patient's current food should be compared with the key nutritional factors described above (**Table 22-6**). Increasing the energy, fat and protein concentration relative to the current food is often necessary for patients with poor body condition.

Patients with swallowing disorders often require specialized feeding methods because the current feeding protocol of one to three meals per day fed in a bowl on the floor is rarely appropriate. The key tools of nutritional management in these cases are a change in food form and a change in the feeding method.

Feeding Plan (Pages 736-738)

The goals of dietary management for patients with megaesophagus are to minimize regurgitation, avoid secondary aspiration pneumonia and to provide adequate nutrition to regain or maintain proper body weight and condition.

Select a Food (Pages 736-737)

OBSTRUCTIVE LESIONS AND ABERRANT MOTILITY Feeding a high-calorie (≥4.5 kcal/g [≥18.8 kJ/g] DM), high-fat (≥25% DM), high-protein (≥25% DM) balanced growth or recuperative food is appropriate for most patients with megaesophagus, cricopharyngeal achalasia or obstructive lesions (See Chapter 12 and Appendix 2.). The food consistency that best promotes flow through the esophagus to the stomach is determined in each case by trial and error. However, gruels often work well, which necessitates using foods with high water content (>80%). Highly digestible (soluble carbohydrate/fat digestibility ≥90%; protein digestibility ≥87%), calorically dense moist foods are made with ingredients that blenderize easily with water. For example, meat ingredients containing connective tissue and bone do not blenderize as easily as skeletal muscle and organ protein sources. Therefore, using nutrient-dense products made from highly digestible ingredients is more likely to meet the nutrient requirements of the patient in the smallest volume possible. Recommending larger cans of calorically dense cat food can help reduce the volume and cost of feeding a large dog.

INFLAMMATORY CONDITIONS Foods with lower levels of dietary fat (<15% DM) are recommended for managing patients with esophagitis and gastroesophageal reflux. Higher dietary fat levels may precipitate gastroesophageal reflux by delaying gastric emptying and reducing lower esophageal sphincter pressure. Increased dietary protein (≥25% DM) enhances lower esophageal sphincter tone.

Determine a Feeding Method (*Pages 737-738*)
Small-volume, frequent meals are recommended when feeding animals with swallowing disorders. Gruel-type foods are often necessary because the liquid form is more amenable to gravity fill of the stomach. Feeding a high-calorie food to a patient in an upright position and maintaining this position for 20 to 30 minutes after feeding provides ample time for gravitational flow of the food through the esophagus to the stomach. Upright feeding can be accomplished by several methods. The most common technique is to elevate the food bowl so that the dog or cat has to sit down or stand on its hind legs to eat. Pets can be trained to eat on stairs or from a counter or stool. Alternatively, small dogs and cats can be cradled in an upright position in the owner's arms while eating. Large dogs have been trained to lie in sternal recumbency on an inclined board for the required period of time. Several companies manufacture devices to facilitate upright feeding.

In some animals, upright feeding is not adequate to control regurgitation or is impractical because of the pet's temperament or the owner's schedule. In those cases, placement of a gastrostomy or enterostomy tube is recommended to bypass the esophagus entirely. Nasoesophageal, nasogastric and esophagostomy tubes are not appropriate in this situation. Patients with ongoing signs of malnutrition at presentation should receive a large-bore gastrostomy feeding tube, if possible, and immediate alimentation via the tube until adequate oral intake can be achieved. Gastrostomy tubes have been used successfully for long periods to maintain the nutritional status of dogs with megaesophagus. A permanent button-type gastrostomy tube should be considered in cases in which owners are willing to feed their pet long term via gastrostomy tube. Some clinicians prefer feeding via enterostomy tube because of the potential for gastroesophageal reflux and recurrent aspiration. Readers are referred to Chapter 12 and Appendix 4 for additional details regarding these assisted-feeding techniques. Owners should be made aware that regurgitation will not completely cease even if all food and water is administered through the gastrostomy tube. Many animals will continue to regurgitate fluid (i.e., salivary secretions). However, the likelihood of aspiration pneumonia is reduced greatly.

Pharyngeal and esophageal tissues heal slowly and are susceptible to secondary bacterial infections. Therefore, surgeons have traditionally recommended withholding oral feedings of regular pet foods for three to four days for patients with inflammation, trauma or surgery to these tissues. Patients with no history or evidence of malnutrition may be safely held off food for two to three days if necessary, but should receive nutrition by the fourth day. Dietary goals in these patients are to provide adequate nutrition to the patient using foods that minimize irritation and trauma to sensitive pharyngeal and esophageal tissues.

Concurrent Therapy (Page 738)

The feeding plan is often used in conjunction with other therapeutic modalities including surgery (cricopharyngeal myotomy, esophageal stricture, vascular ring anomaly, esophageal foreign bodies), bougienage (esophageal stricture), endoscopy (foreign body removal) and drugs (antibiotics, prokinetic agents, corticosteroids, antacids, H_2-receptor blockers, mucosal protective agents).

Reassessment (Page 738)

Nutritional reassessment of patients with swallowing disorders includes: 1) monitoring changes in body weight and condition, 2) evaluating owner compliance in delivering the daily food dosage to the patient, 3) determining the extent of ongoing dysphagia or regurgitation and 4) monitoring resolution of other concurrent disease processes (e.g., pneumonia, myopathies, endocrinopathies). Daily food dosage should be adjusted as indicated by changes in the patient's body weight and condition.

GASTRIC DISORDERS (Pages 738-750)

Vomiting is the hallmark of gastric disorders in dogs and cats. Vomiting may be acute or chronic with a long list of possible etiologies.

Dietary goals are to meet the nutritional requirements of the patient with foods that minimize gastric irritation, promote gastric emptying and normalize gastric motility. In most vomiting cases of less than 48 hours' duration, withholding water for 24 hours and food for 24 to 48 hours generally controls the episode. The patient's regular food should then be gradually reintroduced in small frequent meals over two to three days. Episodes of acute vomiting that occur for longer than three days and cases of chronic vomiting (i.e., persisting longer than 21 days) with signs of malnutrition require more intensive nutritional and medical management.

Hairballs (Page 740)

Hairballs occur commonly in cats due to their normal grooming behavior and sharp barbs on the tongue that enhance hair ingestion. Cats with longer, thicker coats and those with fastidious grooming behavior usually have more problems with hairballs. Swallowed hair initially accumulates as loose aggregates or more compacted, soft aggregates mixed with mucus. Hairballs are periodically regurgitated from the oropharynx or esophagus or vomited from the stomach, or they pass into the intestinal tract where they are voided in the feces. Owners observe periodic gagging, retching and regurgitation or vomiting of hair and mucus (usually not containing food or bile). Hairballs are usually tubular.

Although hairballs do not usually cause significant clinical disease, their associated clinical signs are a nuisance for many cat owners. Various laxatives, lubricants, treats and foods are available for routine management of these problems. Several commercial foods are available to help reduce the frequency with which cats vomit hairballs. Most of these foods have increased amounts of dietary fiber. Kibble size is another important feature of foods designed to reduce vomiting associated with hairballs. Radiographic studies indicate that a larger kibble size is associated with an increased tendency for hairballs to exit the stomach and be eliminated in the feces, thereby reducing the frequency of vomiting. Laxatives and lubricants should be used intermittently because large daily doses may interfere with normal digestion and nutrient absorption.

Frequent regurgitation or vomiting of hairballs (every day) with or without diarrhea, weight loss, anorexia or abdominal pain usually indicates an underlying problem (e.g., gastric motility defect or lymphoplasmacytic enteritis). Cats with severe or frequent clinical signs should be evaluated more extensively with diagnostics including hematology, serum biochemistry profiles, radiography and upper gastrointestinal endoscopy.

Gastritis and Gastroduodenal Ulceration (Pages 738-743)
Clinical Importance (Page 738)

Gastritis is one of the most common causes of vomiting in dogs and cats. The prevalence of gastritis in the pet population is unknown but is

Table 22-7. Potential causes of gastritis and/or gastroduodenal ulceration.

Adverse reactions to food	
Food allergy (hypersensitivity)	Food intolerance
Dietary indiscretion	
Chemicals	Gluttony
Foreign bodies	Heavy metal toxicosis
Garbage toxicosis	Plants
Drug administration	
Corticosteroids	Nonsteroidal antiinflammatory agents
Idiopathic gastritis	
Infectious agents	
Fungi	Spiral bacteria
Parasites	
Inflammatory bowel disease	
Neoplasia	
Gastrinoma	Primary gastric neoplasia
Mastocytosis	
Reduced gastric blood flow	
Disseminated intravascular coagulopathy	Sepsis
	Shock
Neurologic disorders	
Reflux gastritis	
Systemic disease	
Hypoadrenocorticism	Renal disease
Liver disease	

thought to be high because many different insults can result in gastric mucosal inflammation (**Table 22-7**). In one survey, 9% of research beagles had histologic evidence of gastritis in the absence of clinical signs.

The prevalence of GI ulcer disease in dogs and cats is low compared with the prevalence reported in people. The infrequent diagnosis of GI ulceration is possibly due to the absence of clinical signs in many cases. In experimental studies involving dogs, extensive gastroduodenal ulceration existed with only mild clinical signs.

Assessment (Pages 738-742)

ASSESS THE ANIMAL/History and Physical Examination The patient history is often adequate to provide a presumptive diagnosis of gastritis. Owners should be questioned closely about the potential for toxin exposure (e.g., lead, arsenic) and foreign body ingestion (e.g., bones, coins, garbage) by the animal. A history of nonsteroidal antiinflammatory drug (NSAID) administration provides a presumptive diagnosis of drug-induced GI erosions or ulcerations. The veterinarian should query the owner specifically about the use of over-the-counter agents, such as aspirin and ibuprofen.

Vomiting is the most common presenting complaint for dogs with acute or chronic gastritis. Typically, owners report intermittent vomiting of food or bile-stained fluid. Fresh or digested blood appearing as "coffee grounds" may be present in the vomitus. Associated signs may include diarrhea, abdominal pain and melena.

Physical examination is often unremarkable in dogs and cats with gastritis or GI ulcerations. Reduced skin turgor and tacky mucous membranes indicate dehydration. In longstanding cases, weight loss and poor body condition may be noted. Pallor and weakness may be present in patients with significant GI blood loss. Other findings may reflect the underlying cause of gastritis (e.g., cutaneous masses or hepatosplenomegaly associated with mastocytosis).

ASSESS THE ANIMAL/Laboratory and Other Clinical Information
Routine hematology, serum biochemistry profiles and urinalyses help rule out metabolic causes of gastritis. These tests readily identify renal disease, hepatopathies and hypoadrenocorticism. The hematocrit and hemogram are useful in assessing severity and chronicity of gastric disease. Inflammatory leukograms may be identified in animals with neoplasia, perforated GI ulcers, inflammatory bowel disease and phycomycosis. Eosinophilia may indicate parasitism or eosinophilic gastritis. In cats, extreme eosinophilia is suggestive of the hypereosinophilic syndrome or systemic mastocytosis. Identification of circulating mast cells is generally diagnostic for mast cell tumors.

Fecal examinations for parasites and occult blood are important screening tests. Parasites are an unlikely cause of gastritis but should be considered. Gastric parasites such *Ollulanus tricuspis* or *Physaloptera* spp are more readily identified in vomitus and gastric juice. The accuracy of fecal occult blood testing has been confirmed in dogs consuming dry foods. Moist meat-based foods often yield false-positive results. Both the modified guaiac and orthotoluidine tests are sensitive and specific for detecting occult blood in feces.

Imaging modalities such as survey and contrast radiography and ultrasonography offer noninvasive diagnostic techniques for evaluating pets with gastritis or GI ulceration. Survey radiography may be useful in the diagnosis of radiopaque foreign bodies. Abnormalities in renal size and/or shape may suggest renal insufficiency as the cause of gastritis. Hepatosplenomegaly in cats is suggestive of systemic mastocytosis or alimentary lymphosarcoma. Free air in the abdomen is diagnostic for viscus rupture associated with a perforated GI ulcer and indicates the need for immediate exploratory surgery.

Contrast radiographic examinations may be useful. Iodinated contrast agents should be used if GI perforation is suspected. Otherwise, barium sulfate is the contrast agent of choice for GI studies because of its superior ability to coat the GI mucosa.

Endoscopic examination is the most sensitive test for detection of gastritis and gastroduodenal ulcerative disease. Gastric fluid can be collected for para-

Table 22-8. Breed-associated gastric disorders.

Disorders	Breeds
Chronic hypertrophic gastritis	Basenji
	Drentse patrijshond
	Lundehund
Chronic hypertrophic pyloric gastropathy	Lhaso apso
	Maltese dog
	Pekingese
	Shih Tzu
Gastric dilatation-volvulus	Basset hound
	Doberman pinscher
	Gordon setter
	Great Dane
	Irish setter
	Saint Bernard
	Weimaraner
Hemorrhagic gastroenteritis	Dachshund
	Miniature schnauzer
	Toy poodle
Pyloric stenosis	Boston terrier
	Boxer
	Siamese cat

sitic and microbiologic examination. Endoscopic evaluation allows for the identification of mucosal and submucosal hemorrhages, erosions and ulcers. Most important, endoscopic procedures allow collection of multiple gastric and duodenal biopsy specimens.

ASSESS THE ANIMAL/Risk Factors Older animals are more likely to be suffering from metabolic and neoplastic causes of gastritis. Older dogs receiving NSAIDs for management of osteoarthritis are at risk for gastritis and gastroduodenal ulceration. Younger dogs and cats and unsupervised pets are more likely to suffer from gastritis secondary to foreign bodies and dietary indiscretion. A number of breed-associated causes of gastritis have been recognized (**Table 22-8**).

ASSESS THE ANIMAL/Etiopathogenesis/*Acute Gastritis and Gastroduodenal Ulceration* Several metabolic disorders have been associated with the development of acute gastritis and GI ulceration. Uremia may result in diffuse GI tract hemorrhage. GI erosions and ulcers are thought to result from the effects of uremic toxins on the gut mucosa. Additionally, increased circulating concentrations of gastrin have been identified in patients with uremia. Hypergastrinemia promotes hyperacidity.

Liver disease is a common cause of GI ulcerations, which may be manifested as hematemesis. The pathogenesis of mucosal ulceration associated with hepatopathies is multifactorial and associated coagulopathies may worsen clinical manifestations. Potential mechanisms include altered gastric blood flow due to portal hypertension, delayed epithelial turnover, gastric hyperacidity and hypergastrinemia.

Experimentally induced and spontaneous gastritis and gastroduodenal ulcerations have been reported to occur in dogs in conjunction with the use of a variety of NSAIDs including aspirin, indomethacin, naproxen, ibuprofen, phenylbutazone, flunixin meglumine, piroxicam, sulindac and meclofenamic acid. The ulcerogenicity of NSAIDs is attributed to inhibition of the enzyme cyclooxygenase in the prostaglandin synthesis pathway, resulting in the loss of the gastric protective effects of prostacyclin and prostaglandin E.

Stress ulcerations are poorly defined entities in veterinary patients. However, gastroduodenal ulcerations have been noted in companion animals in conjunction with severe burns, heat stroke, multiple trauma, head injuries and spinal cord disorders. In addition, hypovolemic shock and sepsis may be complicated by development of GI ulcers.

Gastrin-producing pancreatic tumors, histamine-producing tumors (e.g., mast cell tumors, basophilic leukemia) and a pancreatic polypeptide-producing pancreatic tumor have been associated with gastric or duodenal ulceration in dogs and cats. Persistent gastric hyperacidity stimulated by gastrin, histamine or pancreatic polypeptide possibly induced ulcers in these animals.

ASSESS THE ANIMAL/Etiopathogenesis/*Chronic Gastritis* The etiopathogenesis of chronic gastritis in dogs and cats is not fully understood. In some cases, an underlying etiology such as parasitism or a metabolic disorder (e.g., uremia, liver disease) can be identified. In most cases, however, an immune-mediated response is hypothesized to be responsible for inflammatory infiltrates within the gastric mucosa. Experimentally, chronic gastritis can be produced in dogs via mucosal irritants, systemic administration of gastric juices or prenatal thymectomy. Each of these disturbs oral tolerance. Chronic idiopathic gastritis is probably a subset of the inflammatory bowel disease syndrome or may arise as an adverse reaction to food antigens. Readers are referred to the Small Intestinal Disorders section below and Chapter 14 for a more complete discussion of the pathogenesis of inflammatory bowel disease and adverse reactions to food.

Once present, gastric inflammation interferes with gastric motility and gastric reservoir function leading to vomiting. Nutrients including proteins are lost through the inflamed mucosal surface.

Table 22-9. Key nutritional factors for patients with gastritis and/or gastroduodenal ulceration.*

Factors	Recommended levels
Chloride	0.50 to 1.30%
Fat	<15% for dogs
	<22% for cats
Potassium	0.80 to 1.10%
Protein**	Limit dietary protein to one or two sources
	Use protein sources to which animal has not been exposed previously or feed a protein hydrolysate (See Chapter 14.)
	16 to 26% for dogs
	30 to 45% for cats
	Protein digestibility ≥87%
Sodium	0.35 to 0.50%

*Nutrients expressed on a dry matter basis.
**Protein is a key nutritional factor for patients with chronic idiopathic gastritis.

ASSESS THE ANIMAL/Key Nutritional Factors Key nutritional factors for patients with gastritis and/or gastroduodenal ulceration are listed in **Table 22-9** and discussed in detail below.

ASSESS THE ANIMAL/Key Nutritional Factors/*Water* Water is the most important nutrient for patients with acute vomiting because of the potential for life-threatening dehydration due to excessive fluid loss and inability of the patient to replace those losses. Moderate to severe dehydration should be corrected with appropriate parenteral fluid therapy rather than using the oral route.

ASSESS THE ANIMAL/Key Nutritional Factors/*Minerals* Gastric and intestinal secretions differ from extracellular fluids in electrolyte composition, so their loss can result in systemic electrolyte abnormalities. Dogs and cats with vomiting and diarrhea may have low, normal or high serum sodium, potassium and chloride concentrations. The derangement that predominates in a particular animal depends on the severity of the disease, nutritional status, site of the disease process, etc. For these reasons, serum electrolyte concentrations are helpful in tailoring the fluid therapy and nutritional management of these patients. Mild hypokalemia, hypochloremia and either hypernatremia or hyponatremia are the electrolyte abnormalities most commonly associated with acute vomiting and diarrhea.

Total body depletion of potassium is a predictable consequence of severe or chronic GI disease because the potassium concentration of both gastric and intestinal secretions is high. Hypokalemia in association with GI disease will be particularly profound if losses are not matched by sufficient intake of potassium.

Electrolyte disorders should be corrected initially with appropriate parenteral fluid and electrolyte therapy. Foods for patients with acute gastroenteritis should contain levels of sodium, chloride and potassium above the minimum allowances for normal dogs and cats. Recommended levels of these nutrients are 0.35 to 0.5% DM sodium, 0.50 to 1.3% DM chloride and 0.80 to 1.1% DM potassium.

ASSESS THE ANIMAL/Key Nutritional Factors/*Protein* Because dietary antigens are suspected to play a role in chronic gastritis, some authors recommend the use of "hypoallergenic" or elimination foods for patients with chronic idiopathic gastritis. In certain cases, elimination foods may be used successfully without pharmacologic intervention. Mild to moderate chronic gastritis may respond to dietary management alone. Ideal elimination foods should: 1) avoid protein excess 2) have high protein digestibility ($\geq$87%) and 3) contain a limited number of novel protein sources to which the animal has never been exposed. Elimination foods are discussed in more detail in Chapter 14.

ASSESS THE ANIMAL/Key Nutritional Factors/*Fat* Solids and liquids higher in caloric density and fat levels are emptied more slowly from the stomach than similar foods with lower caloric density and fat levels. Fat in the duodenum stimulates the release of cholecystokinin, which delays gastric emptying. Foods with less than 15% DM fat are probably appropriate for dogs with gastritis and those with less than 22% DM fat for cats.

ASSESS THE ANIMAL/Other Nutritional Factors/*Vitamins and Minerals* Iron, copper and B-vitamin supplementation may be of benefit in patients with gastroduodenal ulceration and GI blood loss. The use of hematinics is indicated in patients with nonregenerative, microcytic/hypochromic anemias attributable to iron deficiency. Hematinics are probably not necessary in most animals that receive blood transfusions.

ASSESS THE ANIMAL/Other Nutritional Factors/*Acid Load* Alkalemia should be expected if the vomiting patient loses hydrogen and chloride ions in excess of sodium and bicarbonate. Hypochloremia perpetuates the alkalosis by increasing renal bicarbonate reabsorption. Mild alkalemia is common, but profound alkalemia is more likely to occur with pyloric or upper duodenal obstruction rather than with acute gastritis.

Acidemia may occur in vomiting patients if the vomited gastric fluid is relatively low in hydrogen and chloride ion content (e.g., during fasting) or if there is concurrent loss of intestinal sodium and bicarbonate. Severe acid-base disorders are best corrected with appropriate parenteral fluid and electrolyte therapy. Foods for patients with acute vomiting and diar-

GI/PANCREAS

rhea should avoid excess dietary acid load. Foods that normally produce alkaline urine are less likely to be associated with acidosis.

ASSESS THE FOOD AND FEEDING METHOD Levels of the key nutritional factors should be evaluated in foods currently fed to patients with gastritis and compared with recommended levels (**Table 22-9**). Information from this aspect of assessment is essential for making any changes to foods currently provided. Changing to a more appropriate food is indicated if key nutritional factors in the current food do not match recommended levels.

A thorough assessment should include verification of the feeding method currently used. Items to consider include feeding frequency, amount fed, how the food is offered, access to other food and who feeds the animal. All of this information should have been gathered when the history of the animal was obtained. If the animal has a normal BCS (3/5), the amount of food previously fed (energy basis) was probably appropriate.

Feeding Plan (Pages 742-743)
The first objective in managing acute or chronic gastritis should be to correct dehydration and electrolyte and acid-base imbalances, if present. The dietary goals are to provide a food that meets the patient's nutrient requirements, allows normalization of gastric motility and function and controls vomiting. In most cases of acute vomiting, initial fasting for 24 to 48 hours, with parenteral fluid administration, either reduces or resolves vomiting by simply removing the effects of undigested food and the offending agents from the GI tract. Chronic vomiting cases generally require a more detailed diagnostic and therapeutic (i.e., medical and nutritional) approach.

SELECT A FOOD Food selection for vomiting patients should be based on the key nutritional factors described in **Table 22-9**. In general, veterinary therapeutic foods formulated for the management of GI disease (**Tables 22-10** and **22-11**) or adverse reactions to food (**Tables 14-3**, **14-4** and **14-5**) contain appropriate levels of key nutritional factors.

Bland foods are often recommended for veterinary patients with gastritis. The term bland is not a useful recommendation to pet owners; instead specific ingredients or nutrients to avoid should be clearly stated.

DETERMINE A FEEDING METHOD Two feeding methods have been described for patients with acute gastric disorders. The more classic feeding method for patients with acute gastritis begins by discontinuing oral intake of food and water for 24 to 48 hours. After this period, animals should be offered small amounts of water or ice cubes every few hours. If water is well tolerated, small amounts of food can be offered several times (i.e., six to

Table 22-10. Key nutritional factors in selected low-residue commercial foods for dogs with gastritis and/or gastroduodenal ulceration.*

Products	Protein**	Protein digestibility	Fat	Potassium	Sodium	Chloride
Recommended levels	16-26	≥87	<15	0.8-1.1	0.35-0.5	0.5-1.3
Moist canine products						
Hill's Prescription Diet Canine i/d	25.9	86	14.0	0.92	0.44	1.23
Leo Specific Digest CIW	24.4	na	14.7	na	na	na
Iams Eukanuba Low-Residue Adult/Canine	36.3	na	22.0	1.00	0.75	1.00
Medi-Cal Canine Gastro Formula	22.1	na	12.3	na	na	na
Purina CNM Canine EN-Formula	30.5	85	13.8	0.61	0.37	0.78
Select Care Canine Sensitive Formula	22.1	na	12.3	na	na	na
Waltham/Pedigree Canine Low Fat	33.8	na	6.9	na	na	na
Dry canine products						
Hill's Prescription Diet Canine i/d	26.4	92	13.6	0.92	0.46	1.10
Iams Eukanuba Low-Residue Adult/Canine	25.4	na	10.6	0.89	0.46	1.13
Iams Eukanuba Low-Residue Puppy	33.3	na	22.3	0.78	0.42	0.85
Leo Specific Digest CID	27.5	na	14.3	na	na	na
Medi-Cal Canine Gastro Formula	23.4	na	11.7	na	na	na
Purina CNM Canine EN-Formula	25.8	87	11.7	0.59	0.36	0.87
Select Care Canine Sensitive Formula	23.4	na	11.7	na	na	na
Waltham/Pedigree Canine Low Fat	24.1	na	5.4	na	na	na

Key: na = information not available from manufacturer.
*This list represents products with the largest market share and for which published information is available. Manufacturers' published values. Nutrients expressed as % dry matter.
Dietary protein may need to be limited to one or two sources that the patient has not been exposed to previously. **Tables 14-3 and **14-5** contain foods with these characteristics.

GI/PANCREAS

Table 22-11. Key nutritional factors in selected low-residue commercial foods for cats with gastritis and/or gastroduodenal ulceration.*

Products	Protein**	Protein digestibility	Fat	Potassium	Sodium	Chloride
Recommended levels	30-45	≥87	<22	0.8-1.1	0.35-0.5	0.5-1.3
Moist/semi-moist feline products						
Hill's Prescription Diet Feline i/d	40.6	91	20.1	1.08	0.36	1.04
Iams Eukanuba Low-Residue Adult/Feline	41.6	na	16.9	na	na	na
Leo Specific Dermil FDW	34.7	na	34.7	na	na	na
Medi-Cal Feline HYPOallergenic/ Gastro Formula	35.5	na	25.6	na	na	na
Purina CNM Feline EN-Formula	41.9	90	17.0	0.80	0.32	0.55
Waltham/Whiskas Feline Selected Protein	38.2	na	29.0	na	na	na
Dry feline products						
Hill's Prescription Diet Feline i/d	39.9	94	20.1	0.98	0.36	1.03
Iams Eukanuba Low-Residue Adult/Feline	34.8	na	15.4	0.98	0.32	0.88
Medi-Cal Feline HYPOallergenic/ Gastro Formula	29.8	na	13.6	na	na	na
Select Care Feline Neutral Formula	27.7	na	13.6	na	na	na
Waltham/Whiskas Feline Selected Protein	33.3	na	12.2	na	na	na

Key: na = information not available from manufacturer.
*This list represents products with the largest market share and for which published information is available. Manufacturers' published values. Nutrients expressed as % dry matter.
Dietary protein may need to be limited to one or two sources that the patient has not been exposed to previously. **Tables 14-4 and **14-5** contain foods with these characteristics.

Table 22-12. Pharmacologic agents useful in managing gastritis, gastroduodenal ulceration and gastric motility/emptying disorders.

Antacids	Give as needed
Antiemetic agents	
Chlorpromazine	0.2 to 0.5 mg/BW$_{kg}$, PO, SC, IM every six to eight hours
Prochlorpromazine	0.5 mg/BW$_{kg}$, PO, SC, IM every six to eight hours
Antihistamines	
Diphenhydramine	2 to 4 mg/BW$_{kg}$, PO
Dimenhydrinate	25 to 50 mg PO per dog or 12.5 mg PO per cat
Antiprostaglandin agents	
Misoprostol	1 to 3 µg/BW$_{kg}$, PO every eight to 12 hours (dogs)
H$_2$-receptor blockers	
Cimetidine	5 to 10 mg/BW$_{kg}$, PO, SC every six to eight hours
Ranitidine	1 to 4 mg/BW$_{kg}$, PO, SC, IV every eight to 12 hours
Prokinetic agents	
Metoclopramide	0.2 to 0.4 mg/BW$_{kg}$, IM, SC every eight hours or 1.0 mg/BW$_{kg}$/day as a constant IV infusion
Cisapride	0.25 to 0.5 mg/BW$_{kg}$, PO every eight hours
Proton pump inhibitors	
Omeprazole	0.7 to 2.0 mg/BW$_{kg}$, PO every 24 hours (dogs)
Sucralfate	1 g/25 BW$_{kg}$, PO every six to eight hours

Key: BW = body weight, PO = per os, SC = subcutaneously, IM = intramuscularly, IV = intravenously.

eight times) a day. If the pet eats food without vomiting, the amount fed is gradually increased over three to four days until the animal is receiving its estimated daily energy requirement (DER) in two to three meals per day. Food should be withdrawn and offered again after a few hours if the animal begins to vomit during this period.

In some cases, persistent vomiting may complicate refeeding. In such cases, metoclopramide or antiemetic agents are recommended (**Table 22-12**). Rarely, some animals may require parenteral feeding. (See Chapter 12.)

The second approach, feeding through vomiting, has been a successful alternative to "nothing per os" (NPO) therapy in some vomiting patients. Pregnant women suffering hyperemesis reported feeling less nausea and preferred the placement of a nasogastric tube with slow frequent self feeding of small liquid meals to eating small regular meals or NPO therapy. A possible explanation for the continued vomiting is that the normal motility pattern throughout the length of the bowel cannot be reestablished without strong intraluminal stimulation. In fact, vomiting and mucosal atrophy probably perpetuate bowel dysfunction. Feeding restarts normal patterns of motility beginning in the esophagus and food probably reestablishes motility patterns as it passes down the bowel. The physical presence of food and nutrients serves as mechanical and chemical stimuli to the intestine, which in turn releases endogenous and hormonal secretions to reestablish normal bowel motility and function.

Simply refeeding dogs (orally) and cats (via nasoesophageal tube) with protracted vomiting (i.e., lasting more than seven days) may stop vomiting without antiemetic drugs. Feedings are continued even though the patient may vomit. Most cases of protracted vomiting cease within 24 hours of administering liquid food. These patients are then offered small frequent meals of a highly digestible, moderate-fat food 24 hours after the last episode of vomiting (**Tables 22-10** and **22-11**).

Concurrent Therapy (Page 743)
Nutritional management is often used in conjunction with other therapeutic modalities including parenteral fluids, antacids, H_2-receptor antagonists, cytoprotective drugs, prostaglandin E_2 analogues, antibiotics and anthelmintics (**Table 22-12**).

Reassessment (Page 743)
Nutritional reassessment of patients with gastritis or gastroduodenal ulcers includes monitoring changes in body weight and condition and determining the extent of vomiting. Daily food dosage should be adjusted as indicated by changes in body weight and condition.

If vomiting persists in the face of appropriate medical and nutritional therapy, different foods should be tried. If anemia was identified as a problem in pets with GI ulcers, reassessment of the hemogram is recommended to ensure adequate repletion of iron and copper. In addition, frequent monitoring for fecal occult blood loss is recommended.

Gastric Dilatation/Gastric Dilatation-Volvulus (Pages 743-747)
Clinical Importance (Page 743)
Gastric dilatation (GD) is distention of the stomach with a mixture of air, food and fluid. GD often occurs intermittently, usually in young dogs, particularly as a result of overeating or some other dietary indiscretion. Gastric dilatation-volvulus (GDV) is characterized by rotation of the stomach on its mesenteric axis, entrapping gastric contents and compromising vascular supply to the stomach, spleen and pancreas. Acute GDV is a

medicosurgical emergency with high morbidity and mortality. Rarely, chronic, intermittent GDV may occur associated with a partial (i.e., <90 degree) rotation of the stomach.

GDV most commonly affects large-breed, deep-chested dogs and has been estimated to affect 40,000 to 60,000 dogs annually. GDV accounted for 3.4% of deaths of military dogs based on necropsy findings. A review of data from the Veterinary Medical Data Base, Purdue University, West Lafayette, IN, suggests a 1,500% increase in the frequency of GDV from 1964 to 1974 within cases presented to veterinary teaching hospitals.

Assessment (Pages 743-746)

ASSESS THE ANIMAL/History and Physical Examination Clinical signs of GD include nausea, belching and vomiting. Conversely, there may be no effort to vomit, but instead lethargy, reluctance to move and grunting sounds with respiratory effort. The onset of GDV is usually acute and often occurs at night or in the early morning. Owners often report some precipitating stressful event. Boarding, hospitalization, travel and participation in shows have been associated with GDV. Affected dogs exhibit restlessness, progressive abdominal distention with tympany, abdominal pain, hypersalivation and repeated, nonproductive attempts to vomit. Occasionally, owners will find affected dogs dead or in shock.

Chronic GDV is a rare manifestation of the syndrome. Dogs present with intermittent, progressive signs including vomiting, inappetence and weight loss. Periods of illness are interspersed with periods of normalcy. If untreated, these dogs often progress to acute GDV.

The most prominent sign of GD and GDV is abdominal distention. In some dogs, concurrent splenomegaly may be identified by abdominal palpation. Clinical manifestations of cardiovascular shock, including tachycardia, delayed capillary refill time, pallor and weak pulses may be present.

ASSESS THE ANIMAL/Laboratory and Other Clinical Information Laboratory assessment of patients with GDV or GD should include a complete blood count, serum biochemistry profile, urinalysis and blood gas analysis. The complete blood count often reflects stress and can provide early evidence for disseminated intravascular coagulopathy, if thrombocytopenia is present. If faced with thrombocytopenia, a complete coagulation panel is recommended before surgery.

Hypokalemia is common in patients with GDV and should be managed with intravenous potassium supplementation because hypokalemia can potentiate cardiac dysrhythmias. Metabolic acidosis, metabolic alkalosis, respiratory acidosis and mixed acid-base disorders have been reported to occur in dogs with GDV. Routine use of alkalinizing fluids and sodium bicarbonate, therefore, is not recommended.

Table 22-13. Risk factors for canine gastric dilatation-volvulus.

Eating only one meal per day
Excluding moist food, table food and treats from the diet
Exclusive feeding of one food type
Exercising more than two hours per day
Fearful, nervous or aggressive temperament
Feeding food with a mean particle size <5 mm
Increased chest or abdominal depth:width ratio
Increased adult weight, based on breed standards
Increasing age
Large or giant breed status
  Great Dane, weimaraner, Saint Bernard, Gordon setter, Irish setter,
  standard poodle, basset hound, Doberman pinscher, Old English sheep-
  dog, German shorthaired pointer
Lean body condition (body condition score ≤2/5)
Male gender
Purebred status
Rapid eating
Stressful events (boarding in kennel or travel)

Radiography is critical to the diagnosis of GD and GDV. Dorsoventral and right lateral views should be evaluated to distinguish simple GD from GDV. In most cases, gastric rotation is clockwise (i.e., with the dog in a dorsoventral position) and ranges from 90 to 360 degrees.

Other significant findings may include splenomegaly and free abdominal air, which indicates gastric rupture.

Electrocardiograph recordings should be monitored in patients with GDV pre- and postoperatively because cardiac dysrhythmias occur in approximately half of patients. The distended, malpositioned stomach compresses the caudal vena cava and portal vein resulting in cardiovascular compromise. Reduction in venous return and cardiac output leads to myocardial ischemia and cardiovascular shock. Cardiac dysrhythmias, gastric necrosis and multiple organ ischemia are potential consequences if gastric decompression is not performed expeditiously. Generally, dysrhythmias are ventricular in origin and can be life-threatening.

ASSESS THE ANIMAL/Risk Factors A number of predisposing and precipitating risk factors have been demonstrated through epidemiologic studies (**Table 22-13**).

ASSESS THE ANIMAL/Etiopathogenesis A single cause of GDV will probably not be found. GDV is more likely a condition that arises because of the interaction of two or more risk factors. The gastric distention manifested in GDV is associated with an as yet uncharacterized functional or mechanical gastric outflow obstruction. This obstruction results in loss of the normal means for removing air from the stomach (i.e., eructation, vomiting and gastroduodenal flow). In some dogs, gastric volvulus apparently develops as a consequence of gastric distention, but, in others, gastric volvulus may precede the dilatation.

Gas in the stomach of dogs with GDV is primarily atmospheric air, which differs greatly in composition from the gas produced by bacterial fermentation. For that reason, aerophagia is believed to be the primary source of gastric gas in dogs with GDV.

Normally, swallowed air is eructated and does not accumulate in excessive quantities. It has been hypothesized that dogs with GDV have defective eructation mechanisms. In one study, esophageal motility abnormalities were observed in 60% of dogs with GDV. It is possible that such abnormalities are linked to defective eructation complicated by the anatomic relationship of the stomach and esophagus in deep-chested, large dogs, which also may interfere with effective eructation of air. Aerophagia increases with rapid food consumption, excitement, stress and exercise; thus, controlling these factors is recommended in high-risk dogs.

ASSESS THE ANIMAL/Key Nutritional Factors There are no known nutrients of particular concern for dogs with an increased risk for GD or GDV. There are, however, several prudent feeding management recommendations for owners of high-risk, large-breed dogs. These will be discussed below. The key nutritional factors for postoperative patients are similar to those for patients with acute gastritis. Readers should refer to that section for appropriate recommendations.

ASSESS THE FOOD AND FEEDING METHOD Because no specific key nutritional factors have been established for GD and GDV, foods appropriate for the patient's current lifestage and activity level should be provided. Key nutritional factors for normal dogs are found in Chapter 9. If key nutritional factors in the current food do not match the recommended levels described in Chapter 9, then changing to a more appropriate food is indicated.

Because feeding methods are often altered in postoperative patients and patients at risk for GD and GDV, a thorough assessment should include verification of the feeding method currently being used. Items to consider include feeding frequency, amount fed, how the food is offered, access to other food, relationship of feeding to exercise and who feeds the animal. All of this information should have been gathered when the history of the animal was obtained. If the animal has a normal BCS (3/5), the amount of food it was fed previously (energy basis) was probably appropriate.

GI/PANCREAS

Feeding Plan (Page 746)

Without early diagnosis and appropriate treatment, GDV is usually fatal. Initial management includes cardiovascular stabilization (treatment of shock and cardiac dysrhythmias), gastric decompression (orogastric intubation, gastric trocarization), surgery (gastric repositioning and permanent gastropexy) and appropriate postsurgical care. The feeding plan is implemented as part of a prevention strategy or after rapid, aggressive emergency management.

SELECT A FOOD For preventive purposes, foods should be chosen that are appropriate for the dog's lifestage and activity level. (See Chapter 9.) In the postoperative period, foods should be used that provide levels of the key nutritional factors outlined for acute gastritis (**Tables 22-9** and **22-10**).

DETERMINE A FEEDING METHOD It would appear prudent to recommend feeding a dog at risk for GDV two to three times per day in an environment that decreases competitive eating. If the dog typically eats too fast, placing large balls or rocks in the food bowl or feeding the dog from a muffin tin may slow consumption of food and decrease aerophagia. Feeding a mixture of moist and dry food appears to reduce the risk of GDV. Alternatively, feeding foods with particle sizes greater than 30 mm is also thought to reduce the risk of GDV. Although no definitive link between exercise and GDV has been found, limiting exercise within three to four hours of eating (normal gastric emptying time) is prudent. Avoiding vigorous exercise one hour before and two hours after feeding is recommended.

Postoperative patients are best fed small meals frequently. Judicious use of antiemetics and/or metoclopramide in conjunction with continuous feeding may allow adequate caloric intake by patients with persistent vomiting. If tube gastrostomy was chosen as the method of permanent gastropexy, this indwelling catheter should be used for feeding. (See Chapter 12.)

Reassessment (Pages 746-747)

Postoperative patients should be monitored closely for cardiac dysrhythmias, coagulopathies, surgical dehiscence, electrolyte and acid-base abnormalities and infections. After the patient is discharged, the owner should monitor its appetite, activity level and attitude. Rechecks should include body weight and body condition assessment. Food dosages should be adjusted to maintain the dog at ideal body condition. The ultimate marker of success in GDV patients is the prevention of recurrent disease. Rarely, GD will develop in dogs that have had a gastropexy. Any

Table 22-14. Potential causes of gastric emptying disorders in dogs and cats.

Functional obstruction (primary motility defects)	
Gastric ulcers	Infectious gastroenteritis
Idiopathic asynchronous motility	Postoperative ileus
Idiopathic hypomotility	
Functional obstruction (secondary motility defects)	
Drug therapy	Electrolyte disturbances
Anticholinergics	Hypercalcemia
Narcotic analgesics	Hypocalcemia
Inflammation	Hypokalemia
Acute pancreatitis	Hypomagnesemia
Peritonitis	Metabolic disorders
	Diabetes mellitus
	Hepatic encephalopathy
Mechanical obstruction	
Congenital or acquired antral	Gastric or duodenal granulomatous
pyloric hypertrophy	lesions
Extraluminal compression	Gastric or duodenal neoplasia or
Gastric or duodenal foreign bodies	polyps

episode of dilatation and precipitating factors should be reported and evaluated.

Gastric Motility/Emptying Disorders (Pages 747-750)
Clinical Importance (Page 747)
Gastric motility disorders arise from conditions that directly or indirectly disrupt three of the basic functions of the stomach: 1) storage of ingesta, 2) mixing and dispersion of food particles and 3) timely expulsion of gastric contents into the duodenum. **Table 22-14** outlines a number of primary and secondary causes of gastroparesis reported to occur in dogs and cats. The importance of these disorders in the general pet population is unknown, but primary gastric motility disorders are probably rare.

Assessment (Pages 747-749)
ASSESS THE ANIMAL/History and Physical Examination Delayed gastric emptying due to any cause results in vomiting. Owners may report vomiting of undigested or partially digested food more than 12 hours after the pet eats. The onset of clinical signs may be gradual in acquired cases of chronic hypertrophic pyloric gastropathy or acute in the case of foreign body ingestion. Clinical signs may have been present since weaning in dogs and cats with congenital pyloric stenosis.

Weight loss and poor body condition are often present in chronic cases. Other manifestations may include intermittent gastric bloating, nausea, partial or complete inappetence and belching. Occasionally, patients will present with unrelenting or projectile vomiting; complete gastric outflow obstruction should be suspected in such cases.

Physical examination findings are often unremarkable beyond evidence of weight loss. Body condition should be assessed and used as a reassessment tool. Gastric distention and tympany may be evident in some cases. Patients with unrelenting vomiting may present with dehydration, depression and malaise. In rare cases, severe electrolyte abnormalities resulting from persistent vomiting may manifest as weakness.

ASSESS THE ANIMAL/Laboratory and Other Clinical Information Hematologic and serologic findings in animals with gastroparesis or gastric obstruction are nonspecific and may be more reflective of an underlying disorder. Chronic, persistent vomiting may precipitate dehydration and electrolyte (hypokalemia, hypochloremia) and acid-base abnormalities. Prerenal azotemia is common. Hypochloremic metabolic alkalosis with paradoxical aciduria may be present in dogs and cats with complete pyloric outflow obstruction.

Survey abdominal radiographs are often of benefit when evaluating dogs and cats with gastric motility disorders. Typical findings include a stomach distended by fluid, air or food. The presence of food in the stomach 12 to 18 hours after the last meal is evidence of an emptying disorder. Occasionally, gastric wall thickening may be recognized on survey radiographs. Rarely, extraluminal masses causing pyloric obstruction may be identified.

GI contrast studies confirm delayed gastric emptying. If liquid contrast media (i.e., barium sulfate) remains in the stomach for more than four hours in dogs or 30 minutes in cats, gastroparesis or mechanical obstruction should be suspected. Liquid contrast media, however, is not representative of a typical meal. For that reason, feeding barium mixed with food or administering radiopaque particles (BIPS, Med-ID, Grand Rapids, MI, USA) mixed with food more completely assesses gastric function. GI contrast studies may also identify thickened gastric walls, intraluminal foreign bodies and extraluminal masses.

Gastric emptying disorders may first be suspected at the time of upper GI endoscopy. Food in the stomach after a 12- to 18-hour fast is good evidence of the condition. In some cases, endoscopic findings may be diagnostic. Chronic hypertrophic pyloric gastropathy, for example, has a typical endoscopic appearance, including hyperplastic mucosal folds surrounding the pylorus, protuberance of the pylorus and polyps. In the case of antropyloric or proximal duodenal foreign bodies, endoscopy can be both diagnostic and curative.

Ultrasonography may be useful in the evaluation of pyloric masses and extraluminal sources of pyloric compression.

ASSESS THE ANIMAL/Risk Factors Several breeds are associated with gastric motility disorders (**Table 22-8**). Congenital pyloric stenosis is most often

Table 22-15. Key nutritional factors for patients with gastric motility/emptying disorders.*

Factors	Recommended levels
Fat	<15% for dogs
	<22% for cats
Fiber	Avoid foods with gel-forming fiber sources
Food form	Liquid or semi-liquid consistency
Food temperature	Offer foods between room and body temperature
Minerals	Appropriate for current lifestage

*Nutrients expressed on a dry matter basis.

encountered in brachycephalic dogs and Siamese cats. Chronic hypertrophic pyloric gastropathy usually affects small purebred middle-aged dogs. Young animals are more at risk for gastric foreign bodies, whereas older pets are more likely to have neoplastic lesions that may obstruct gastric outflow.

ASSESS THE ANIMAL/Etiopathogenesis Gastric motility disorders may arise from functional or mechanical obstruction of gastric outflow. Functional disorders of gastric emptying arise from abnormal or asynchronous gastric motility. Myenteric neuronal or gastric smooth muscle function or antropyloroduodenal coordination may be impaired.

A number of benign and malignant anatomic lesions of the stomach and proximal duodenum may result in mechanical gastric outflow obstruction (**Table 22-14**). The most common of these is chronic hypertrophic pyloric gastropathy, which refers to an acquired hypertrophic mucosal or muscular lesion of the pyloric antrum. In addition, congenital pyloric stenosis occurs in young dogs and cats as a consequence of benign muscular hypertrophy of the pylorus. Certain gastric and proximal duodenal neoplasms and granulomatous conditions (e.g., pythiosis, eosinophilic gastritis) can also result in pyloric obstruction.

ASSESS THE ANIMAL/Key Nutritional Factors Key nutritional factors for patients with gastric motility/emptying disorders are listed in **Table 22-15** and are discussed in detail below.

ASSESS THE ANIMAL/Key Nutritional Factors/*Water* Dehydration is a common problem in animals with persistent vomiting. Dehydration should be corrected with appropriate parenteral fluid therapy. Thereafter, water should be available free choice. Water should be offered at a temperature between room and body temperature. Colder water should be avoided because cold water and food delay gastric emptying.

ASSESS THE ANIMAL/Key Nutritional Factors/*Minerals* Abnormalities in serum electrolyte concentrations, especially potassium, sodium and chloride, are common in animals with chronic vomiting and can adversely affect gastric motility and emptying. Initial abnormalities should be corrected with appropriate parenteral fluid therapy. Thereafter, the food should contain mineral levels appropriate for the animal's lifestage.

ASSESS THE ANIMAL/Key Nutritional Factors/*Fat* Solids and liquids higher in caloric density and fat levels are emptied more slowly than similar foods with lower caloric density and fat. Fat in the duodenum stimulates release of cholecystokinin, which delays gastric emptying. Foods with less than 15% (dogs) or less than 22% (cats) DM fat are probably appropriate for patients with gastric emptying or motility disorders.

ASSESS THE ANIMAL/Key Nutritional Factors/*Fiber* Many grocery brand moist foods contain gelling agents such as gums or hydrocolloids to enhance the aesthetic characteristics of the food. Foods containing gel-forming soluble fibers should be avoided in patients with gastric emptying and motility disorders because they increase the viscosity of ingesta and slow gastric emptying.

ASSESS THE ANIMAL/Key Nutritional Factors/*Food Form and Temperature* Liquids are emptied from the stomach more quickly than solids due to lower digesta osmolality. Water is emptied most quickly, whereas liquids containing nutrients are emptied more slowly. High-osmolality fluids are emptied more slowly than dilute fluids. Solids and liquids high in fat are the slowest to be emptied from the stomach.

The ideal food for patients with gastric emptying disorders has a liquid or semi-liquid consistency. Cold meals slow gastric emptying. Therefore, food should be offered between room and body temperature. Refrigerated or frozen foods should be warmed before being fed.

ASSESS THE FOOD AND FEEDING METHOD The food form and levels of fat and minerals should be assessed in the current food and compared with the recommendations outlined in the key nutritional factors section (**Table 22-15**). Most importantly, the food should be complete and balanced for the current lifestage of the animal. The food may not need to be altered for patients with mild disease or few clinical signs.

Patients with gastric motility disorders often require specialized feeding methods; the current feeding protocol is rarely appropriate. A thorough assessment should include verification of the feeding method currently used. Items to consider include feeding frequency, amount fed, how the food is offered, access to other food, relationship of feeding to exercise and who feeds the animal. All of this information should have been gathered when the history of the animal was obtained. If the animal has a normal BCS (3/5), the amount of food fed previously was probably appropriate.

Table 22-16. Fat levels in selected moist commercial foods for dogs with gastric motility/emptying disorders.*

Products	Fat (%)
Recommended levels	**<15**
Hill's Prescription Diet Canine i/d	14.0
Leo Specific Digest CIW	14.7
Iams Eukanuba Low-Residue Adult/Canine	22.0
Medi-Cal Canine Gastro Formula	12.3
Purina CNM Canine EN-Formula	13.8
Select Care Canine Sensitive Formula	12.3
Waltham/Pedigree Canine Low Fat	6.9

*This list represents products with the largest market share and for which published information is available. Manufacturers' published values. Fat content expressed on a dry matter basis.

The food should have mineral levels appropriate for the patient's current lifestage. (See Chapter 9.) Foods with gel-forming fiber sources should be avoided. Foods with liquid or semi-liquid consistency are preferred. Foods should be offered between room and body temperature.

Feeding Plan (Pages 749-750)

Dehydration, electrolyte and acid-base abnormalities and gastric outflow obstruction should be corrected with appropriate fluid therapy and surgical intervention before the feeding plan is initiated. For dogs or cats with functional gastric motility disorders, several prokinetic agents are available (**Table 22-12**) and should be considered if dietary management alone is insufficient to control clinical signs.

SELECT A FOOD Liquid or semi-liquid foods with restricted levels of fat may promote gastric emptying and should be used initially in patients with gastric motility or emptying disorders. A liquid or semi-liquid consistency can be obtained by adding water to a moist veterinary therapeutic food (**Tables 22-16** and **22-17**). Alternatively, liquid enteral products containing restricted levels of fat may be used (**Tables 12-5** and **12-6**).

Feeding foods with lower fat levels means that larger or more frequent meals are required to meet the animal's DER. Larger meals may promote more vomiting. Therefore, each animal should be managed individually and the optimal fat level determined according to the patient's ability to tolerate meal size and maintain optimal body condition.

DETERMINE A FEEDING METHOD Foods should be offered at a temperature between room and body temperature. Frequent small meals (at least three per day) are preferred. In some cases of complete pyloric outflow obstruction, parenteral nutritional support may be necessary to meet the animal's needs before surgical alleviation of the obstruction. This is in-

Table 22-17. Fat levels in selected moist and semi-moist commercial foods for cats with gastric motility/emptying disorders.*

Products	Fat (%)
Recommended levels	<22
Hill's Prescription Diet Feline i/d	20.1
Iams Eukanuba Low-Residue Adult/Feline	16.9
Leo Specific Dermil FDW	34.7
Medi-Cal Feline HYPOallergenic/Gastro Formula	25.6
Purina CNM Feline EN-Formula	17.0
Waltham/Whiskas Feline Selected Protein	29.0

*This list represents products with the largest market share and for which published information is available. Manufacturers' published values. Fat content expressed on a dry matter basis.
The food should have mineral levels appropriate for the patient's current lifestage. (See Chapter 11.) Foods with gel-forming fiber sources should be avoided. Foods with liquid or semi-liquid consistency are preferred. Foods should be offered between room and body temperature.

dicated when the patient's body condition is poor (BCS 1/5 or 2/5) and the patient is deemed at increased risk for postsurgical complications.

Most patients can be fed using a feeding method similar to that used for normal pets if normal gastric function is restored after surgery. The best feeding method will need to be individualized for each patient and determined by trial and error based on remaining gastric function.

Reassessment (Page 750)

Body weight and condition should be assessed every two to four weeks. Presence or absence of vomiting should be documented. If vomiting continues, the food or feeding pattern should be altered. Dividing the daily food intake into additional meals may also increase GI tolerance. Use of prokinetic agents (e.g., metoclopramide, cisapride) should be considered if vomiting persists despite implementation of these therapeutic strategies.

Gradual attempts to normalize the feeding regimen can be made if the animal is doing well on the recommended therapy. Feeding more solid foods and larger, less frequent meals are more convenient for the pet owner.

The prognosis for dogs and cats with gastric motility disorders varies with the underlying cause. Mechanical obstructions can often be managed effectively through surgical or endoscopic (e.g., foreign body retrieval) means, resulting in an excellent prognosis. Occasionally, dogs and cats with longstanding gastric outflow obstruction with gastric distention may have residual gastric motility abnormalities. These animals may benefit from use of prokinetic agents.

Table 22-18. Potential causes of acute small bowel diarrhea in dogs and cats.

Dietary	
Dietary indiscretion	Garbage toxicity
Foreign bodies	
Infectious agents	
Bacteria	Parasites
Campylobacter spp	Helminths (roundworms,
Clostridium spp	hookworms)
Escherichia coli	Protozoa (*Giardia lamblia*,
Salmonella spp	coccidia)
Staphylococcus spp	
Yersinia spp	
Rickettsia	Viruses
Salmon poisoning	Canine distemper
	Coronavirus
	Panleukopenia
	Parvovirus
	Rotavirus
Miscellaneous	Hemorrhagic gastroenteritis
Toxin or drug induced	
Chemotherapeutic agents	Laxatives (magnesium oxide,
Digoxin	lactulose)
Heavy metals	Nonsteroidal antiinflammatory drugs

SMALL INTESTINAL DISORDERS (Pages 750-772)

Disorders of the small intestine are frequently encountered in veterinary practice. A number of acute and chronic enteropathies are recognized (**Tables 22-18 and 22-19**) and must be distinguished from diseases of other organ systems resulting in GI signs. Typical clinical manifestations of small intestinal disease include diarrhea, weight loss, poor body condition, vomiting, borborygmus and flatulence.

Diarrhea is defined as a change in the frequency, consistency or volume of bowel movements and stools. Diarrhea is the most common manifestation of small intestinal disease. The diarrhea associated with small intestinal conditions differs from that typically associated with large intestinal disorders (**Table 22-20**).

Mechanisms of Diarrhea (Pages 750-751)

An understanding of normal gut physiology and the common pathophysiologic mechanisms responsible for diarrhea in companion animals allows for a rational approach to evaluation and treatment of patients with small intes-

(margin tab) GI/PANCREAS

Table 22-19. Potential causes of chronic small bowel diarrhea in dogs and cats.

Dietary	
Food allergy (hypersensitivity)	Lactose intolerance
Infectious agents	
Bacteria	Fungi
Campylobacter spp	Histoplasmosis
Salmonellosis	Pythiosis
Small intestinal bacterial overgrowth	
Parasites	
Helminths (roundworms, hookworms)	
Protozoa (coccidia, *Giardia lamblia*)	
Inflammatory bowel disease	
Eosinophilic gastroenteritis	Lymphoplasmacytic enteritis
Lymphocytic enteritis	Suppurative gastroenteritis
Miscellaneous	Juvenile diarrhea of cats
Neoplasia	
APUD cell tumors	Mast cell tumor
Lymphosarcoma	

Key: APUD = amine precursor uptake and decarboxylation.

Table 22-20. Characteristics of small and large bowel diarrhea.

Characteristics	Small bowel	Large bowel
Blood in feces	Melena	Hematochezia
Fecal quality	Loose, watery, "cow-pie"	Loose to semi-formed, "jelly-like"
Fecal volume	Large quantities	Small quantities
Frequency of defecation	Normal to slightly increased	Increased
Malaise	May be present	Rare
Mucus in feces	Usually absent	Usually present
Steatorrhea	May be present	Absent
Tenesmus	Absent	Usually present
Urgency	Absent	Usually present
Vomiting	May be present	Uncommon
Weight loss	May be present	Rare

tinal disorders. There are four major mechanisms for diarrhea: 1) osmotic, 2) altered mucosal permeability, 3) abnormal motility and 4) secretory.

Osmotic Diarrhea (Page 750)

Osmotic diarrhea, also referred to as diarrhea of malabsorption, is the most common cause of diarrhea in dogs and cats. Osmotic diarrhea may occur in conjunction with other pathophysiologic processes. The presence of unabsorbed nutrients (solutes) in the bowel results in passive diffusion of water into the gut lumen. (See the sidebar "Disaccharide Intolerance" in Small Animal Clinical Nutrition, 4th ed., p 753) This process continues until the osmolality of the intestinal chyme is approximately that of plasma. Osmotic diarrhea may occur as a result of maldigestion, malabsorption, administration of osmotic laxatives and overeating. Clinical manifestations of osmotic diarrhea include passage of large volumes of fluid and/ or soft stools. Stools may appear greasy if steatorrhea is present. The diarrhea usually resolves following a 24- to 36-hour fast.

Diarrhea Due to Altered Mucosal Permeability *(Page 750)*

Altered mucosal permeability (i.e., exudative diarrhea) is another common cause of diarrhea in dogs and cats. The large or small bowel may be affected. The intestinal permeability barrier is composed of epithelial tight junctions, mucosal lymphatics and capillaries and the local immune system. Failure of any one of these components can result in diarrhea. GI diseases that result in erosions, ulcerations and mucosal inflammation or infiltration are potential causes of gut permeability changes and diarrhea. Diarrhea associated with increased gut permeability may present as a protein-losing enteropathy (i.e., hypoproteinemia, hypoalbuminemia, weight loss). Fresh and/or melenic blood may be present in the stool. Fecal examination may reveal inflammatory cells. Often, these diarrheas do not completely resolve if food is withheld.

Diarrhea Due to Abnormal GI Motility *(Pages 750-751)*

Diarrhea may be associated with deranged GI motility. It is often difficult to determine whether abnormal GI motility is a primary entity or a secondary consequence of another disorder. In general, deranged GI motility is not a common cause of small bowel diarrhea in dogs and cats. The most common motility derangement is rapid intestinal transit associated with a decreased frequency of rhythmic segmental contractions, also termed ileus. The reduction in segmental contractions results in a "pipe" effect with little resistance to ingesta flow. Ileus may occur in conjunction with infiltrative diseases, severe abdominal pain, parvoviral enteritis or may develop postoperatively. In many cases, iatrogenic ileus complicates the management of animals treated inappropriately with anticholinergic agents. Increased frequency of peristaltic contractions is probably not an important cause of diarrhea in companion animals. However, it may play a role in the irritable bowel syndrome. (See text below.) A reduction in peristaltic or interdigestive motility may result in small intestinal bacterial overgrowth. Response to dietary manipulation is variable.

Secretory Diarrhea *(Page 751)*

Secretory diarrhea is relatively uncommon in companion animals vs. people (cholera is the prototypical example) and food animal species.

Table 22-21. Clinical signs associated with life-threatening acute gastroenteritis.

Abdominal pain	Fecal leukocytes
Dehydration	Fever
Depression	Melena or hematochezia

Crypt epithelial cells produce intestinal fluid, whereas enterocytes lining the villous tips are responsible for absorption. Normally, absorption exceeds intestinal secretion. Most secretagogue effects are mediated via a second messenger (e.g., cyclic AMP, cyclic GMP, calmodulin). Secretagogues include GI hormones, bacterial enterotoxins, certain pharmacologic agents, deconjugated bile acids and hydroxy fatty acids. Clinical manifestations of secretory diarrhea are often extreme. Patients have large volumes of fluid diarrhea and often become dehydrated rapidly. Generally, fasting is not successful in alleviating clinical signs.

Acute Enteritis (Pages 751-757)
Clinical Importance (Page 751)
Acute gastroenteritis is one of the most common illnesses of dogs and cats. A number of infectious, toxic and dietary factors can trigger the sudden onset of vomiting and diarrhea (**Tables 22-7** and **22-18**). This section addresses the diagnosis and management of veterinary patients with an acute onset of diarrhea.

Assessment (Pages 751-756)
ASSESS THE ANIMAL/History and Physical Examination Patients are usually presented for the sudden onset of diarrhea, vomiting or both signs. In many cases, the owner will report that the pet is depressed and has a poor appetite. The number and character of the defecations should be assessed. Large fluid stools are typical of small bowel disorders. Melenic or hemorrhagic stools may indicate a potentially life-threatening disorder (**Table 22-21**).

Careful attention should be paid to the dietary history. Diet-induced diarrhea is relatively common; therefore, a recent change to a moist high-fat or meat-based food may be the source of the animal's diarrhea. Often, it is possible to elicit a history of dietary indiscretion, feeding table foods over a holiday, or access to garbage, carrion or abrasive materials. Cats that hunt birds may have been exposed to *Salmonella* spp and dogs eating raw salmon are at risk for salmon poisoning.

Feeding uncooked meat in homemade foods and racing greyhound rations has been linked to bacterial enteritis. Greyhound diets often contain raw ground beef, which has been identified as source of salmonellosis and colibacillosis. Incorporation of raw poultry in foods has been linked to campylobacteriosis and salmonellosis. (See Chapter 7.)

Other husbandry issues are also important. Records of vaccinations and anthelmintic treatments should be scrutinized. Questions should be asked about the health of other pets and people in the household. A positive answer to inquiries increases the likelihood that an infectious organism is involved.

Often, affected dogs and cats are depressed and dehydrated. Typically, the diarrhea is most consistent with small bowel disease (**Table 22-20**). Occasionally, animals may present with signs suggestive of both small and large bowel involvement. Abdominal discomfort may be recognized on palpation. Patients should be carefully evaluated for evidence of septic shock. Animals exhibiting systemic signs of illness such as fever and congested mucous membranes in addition to GI signs should be treated more aggressively.

ASSESS THE ANIMAL/Laboratory and Other Clinical Information Because there are many potential causes of acute gastroenteritis, achieving a definitive diagnosis can be difficult. It is more important to determine whether the animal's condition is self-limiting or if it is potentially life-threatening. This decision, based on historical and physical findings, is critical. **Table 22-21** lists factors that suggest a potentially life-threatening condition. Cases of a serious nature should be pursued aggressively with the use of hematology, serum biochemistry profiles, urinalyses and fecal examinations for parasites and other infectious pathogens. Abdominal films or GI contrast radiographs are recommended to rule out obstruction. Self-limiting cases are usually approached more conservatively. Diagnostics are often limited to assessment of hydration status (i.e., packed cell volume, total protein concentration and body weight) and thorough examination of feces for evidence of parasites and bacterial pathogens (e.g., spores of *Clostridium* spp).

ASSESS THE ANIMAL/Risk Factors Risk factors for acute gastroenteritis include age, breed, immune status and environment. Young animals are more susceptible to a variety of infectious pathogens including parasites, viruses and bacteria. Several canine breeds (e.g., Chinese Shar-Pei, German shepherd dog, beagle) may have IgA deficiency; therefore, these dogs may be more susceptible to development of a number of GI conditions, including giardiasis and small intestinal bacterial overgrowth (**Table 22-22**). Likewise, immunocompromised animals are at risk for contracting viral and bacterial enteritides. Several conditions including cancer, diabetes mellitus, feline leukemia virus and FIV infections may result in deranged immune function.

Environment also plays an important role in exposure to pathogens. Dogs and cats kept in unsanitary or overcrowded conditions are much more likely to develop infectious enteropathies. In addition, animals kept in poorly controlled environments have higher risk for exposure to high-

Table 22-22. Breed-associated small intestinal disorders.

Eosinophilic gastroenteritis	German shepherd dog Irish setter
Hemorrhagic gastroenteritis	Dachshund Miniature schnauzer
Immunoproliferative small intestinal disease	Basenji Ludenhund
Intestinal adenocarcinoma	Siamese cat
Lymphoplasmacytic enteritis	German shepherd dog Chinese Shar-Pei Soft-coated wheaten terrier Domestic shorthaired cat
Parvoviral enteritis	Doberman pinscher Rottweiler Labrador retriever (black)
Small intestinal bacterial overgrowth	German shepherd dog Beagle
Lymphangiectasia*	Yorkshire terrier Golden retriever Dachshund Basenji (IPSID) Ludenhund (IPSID)
Wheat-sensitive enteropathy	Irish setter

Key: IPSID = immunoproliferative small intestinal disease.
*Soft-coated wheaten terriers may be affected by a protein-losing enteropathy that may occur in conjunction with a protein-losing nephropathy.

fat table foods, garbage and toxins. Dogs in particular indulge in indiscriminate eating. Consumption of rotten garbage, decomposing carrion or abrasive materials (e.g., hair, bones, rocks, plastic, aluminum foil) can result in severe enteritis. Poor husbandry practices including inadequate parasite control, overcrowding and poor vaccination measures also put pets at risk for acute gastroenteritis.

ASSESS THE ANIMAL/Etiopathogenesis In acute gastroenteritis, diarrhea may occur as a result of any or all of the four mechanisms of diarrhea described above. Many viral organisms and cancer chemotherapeutic agents destroy intestinal villi. Consequently, diarrhea may occur due to altered gut permeability and/or osmotic mechanisms. Additionally, ileus may arise as a consequence of abdominal pain in pets with parvoviral enteritis. Finally, bacterial pathogens may elaborate enterotoxins that serve as potent secretogogues.

Table 22-23. Key nutritional factors for patients with acute enteritis.*

Factors	Recommended levels
Chloride	0.50 to 1.30%
Crude fiber	0.5 to 15% (see text)
Digestibility	≥87% for protein and ≥90% for fat and soluble carbohydrate
Fat	12 to 15% for dogs
	15 to 22% for cats
Potassium	0.80 to 1.10%
Protein	Suitable for lifestage
Sodium	0.35 to 0.50%

*Nutrients expressed on a dry matter basis.

Small bowel atrophy begins within days in the absence of luminal stimulation. Atrophy, the small intestinal response to disuse, occurs in several species with simple stomachs, including cats and dogs. The hallmarks of small bowel atrophy are decreased villus height with an overall reduced absorptive surface area and brush border enzyme activity.

Food in the lumen of the small bowel stimulates intestinal integrity (mass and function) by several mechanisms. Ingested nutrients present mechanical and chemical stimuli to the intestine, increasing intestinal secretory and endocrine activity. The type and amount of ingested nutrients mechanically alter the mucosal cell mass by affecting the rate of stem cell division and the rate of mucosal cell renewal. Gastric, duodenal and pancreatobiliary secretions, which normally accompany eating, digestion and absorption, promote mucosal structure and function. Refeeding the atrophied small bowel should be done bearing in mind the altered function of the small bowel.

ASSESS THE ANIMAL/Key Nutritional Factors Key nutritional factors for patients with acute enteritis are listed in **Table 22-23** and are discussed in detail below.

ASSESS THE ANIMAL/Key Nutritional Factors/*Water* Water is the most important nutrient for patients with acute vomiting and diarrhea because of the potential for life-threatening dehydration due to excessive fluid loss and inability of the patient to replace those losses. Moderate to severe dehydration should be corrected with appropriate parenteral fluid therapy rather than using the oral route.

Oral fluid therapy is typically reserved for patients with minor fluid deficits or to supply maintenance fluid requirements. Oral rehydration solutions have been commonly used in people and production animals with acute diarrhea. Oral rehydration solutions have also been advocated for use in dogs and cats. Oral rehydration solutions contain glucose,

GI/PANCREAS

amino acids and electrolytes in addition to water. The physiologic basis for these solutions is the coupled transport of sodium and glucose and other actively transported small organic molecules. The maximum uptake of water and electrolytes occurs when the ratio of carbohydrate to sodium approaches 1:1. Oral rehydration solutions (**Table 22-24**) are useful as an alternate fluid source, provided the patient readily consumes them.

ASSESS THE ANIMAL/Key Nutritional Factors/*Minerals* The electrolyte composition of gastric and intestinal secretions differs from that of extracellular fluids; therefore, loss of gastric and intestinal secretions may result in systemic electrolyte abnormalities. Dogs and cats with vomiting and diarrhea may have low, normal or high serum sodium, potassium and chloride concentrations. The derangement that predominates in a particular animal depends on the severity of the disease, nutritional status, site of the disease process, etc. For these reasons, serum electrolyte concentrations are helpful in tailoring the fluid therapy and nutritional management of these patients. Mild hypokalemia, hypochloremia and either hypernatremia or hyponatremia are the electrolyte abnormalities most commonly associated with acute vomiting and diarrhea.

Depletion of total body potassium is a predictable consequence of severe or chronic GI disease because the potassium concentration of both gastric and intestinal secretions is high. Hypokalemia in association with GI disease will be particularly profound if losses are not matched by sufficient intake of dietary potassium.

Electrolyte disorders should be corrected initially with appropriate parenteral fluid and electrolyte therapy. Foods for patients with acute gastroenteritis should contain levels of sodium, chloride and potassium above the minimum allowances for normal dogs and cats. Recommended levels of these nutrients are 0.35 to 0.50% DM sodium, 0.50 to 1.30% DM chloride and 0.80 to 1.10% DM potassium.

ASSESS THE ANIMAL/Key Nutritional Factors/*Fat* In comparison to processes involved with other macronutrients, fat digestion and absorption are relatively complex and may be disrupted in patients with GI disease. Ingestion of a fatty meal decreases gastroesophageal tone, slows gastric emptying and is a potent stimulus for pancreatic secretion.

On the other hand, dietary fat is a concentrated source of calories; higher fat foods allow smaller amounts of food to be ingested to meet the patient's DER. This is an important consideration in many patients because limiting the amount of food entering the GI tract helps control clinical signs. Fat also improves the palatability of food, which is important in patients with nausea.

For these reasons, foods for patients with acute gastroenteritis and many other GI diseases should contain moderate amounts of fat. Recom-

Table 22-24. Selected commercial oral rehydration solutions available for use in dogs and cats.

Products (manufacturers)	Nutrient content (mEq/l)							ME (kcal)	Comments
	Na	K	Cl	Mg	Ca	P	Citrate		
Electramine (Vitae Inc.)	69.8	15.4	69.7	–	–	–	–	–	Contains glycine
Ritrol (Nutramax Labs)	90	20	–	tr	tr	0	30	165	Rice-based WHO formula, mOsm/l = 270, contains glutamine
Pedigree/Whiskas Electrolyte Instant Fluid (Waltham)	40	20	40	–	0	0	9	197	Contains glycine, maltodextrins
Pedialyte Solution (Ross Laboratories)	45	20	35	–	–	–	30	3	–
Ricelyte Oral (Mead Johnson)	50	25	45	–	–	–	34	4.2	–
Resol Solution (Wyeth-Ayerst)	50	20	50	4	4	5	34	2.5	–
Rehydralyte Solution (Ross)	75	20	65	–	–	–	30	3	–
Biolyte (Pharmacia/Upjohn)	134	22.8	75.8	6.6	–	–	–	–	–

Key: mEq/l = milliequivalents per liter, ME = metabolizable energy, WHO = World Health Organization, tr = trace, Na = sodium, K = potassium, Cl = chloride, Mg = magnesium, Ca = calcium, P = phosphorus.

GI/PANCREAS

mended dietary fat levels are 12 to 15% on a dry matter basis (DMB) for dogs and 15 to 22% DMB for cats.

ASSESS THE ANIMAL/Key Nutritional Factors/*Fiber* Although dietary fiber primarily affects the large bowel of dogs and cats, fiber can also affect gastric, small intestinal and pancreatic structure and function. Beneficial effects of dietary fiber include: 1) modifying gastric emptying, 2) normalizing intestinal motility and intestinal transport rate, 3) buffering toxins in the GI lumen, 4) binding or holding excess water, 5) supporting growth of normal GI microflora, 6) buffering gastric acid and 7) altering viscosity of GI luminal contents. Dietary fiber also adds nondigestible bulk and decreases the dry matter digestibility of the food.

Various types and levels of dietary fiber have been advocated for patients with acute gastroenteritis. The traditional approach is to recommend very low-fiber foods (<1% DM crude fiber) to enhance dry matter digestibility and provide "low residue" in the GI tract. However, two other approaches are commonly used: 1) small amounts (0.5 to 5% DM total dietary fiber) of a mixed (i.e., soluble/insoluble) fiber type may be used in conjunction with a highly digestible food, or 2) foods containing moderate levels (10 to 15% DM crude fiber) of insoluble fiber may be fed. Each of these strategies can be successful in managing selected patients with acute gastroenteritis and other GI disorders.

ASSESS THE ANIMAL/Key Nutritional Factors/*Digestibility* The term "highly digestible" is not defined in a regulatory sense. However, the term has generally been reserved for products with protein digestibility ≥87%, and fat and soluble carbohydrate digestibility ≥90%. The average digestibility coefficients for popular commercial dog and cat foods are 78 to 81%, 77 to 85% and 69 to 79% for crude protein, crude fat and NFE (carbohydrate), respectively. Commercial veterinary therapeutic foods formulated for patients with GI disease (**Tables 22-10** and **22-11**) usually contain meat and carbohydrate sources that have been highly refined to increase digestibility. Meat ingredients in many veterinary therapeutic foods are usually composed of muscle and organ sources rather than meat and bone meals. Typical meat/animal source ingredients in commercial GI foods include egg, cottage cheese, chicken and ground beef.

Carbohydrates make up the largest non-water fraction (i.e., 60 to 80% DM) of commercial and homemade foods formulated for managing patients with GI diseases. In pet foods, carbohydrate digestibility is influenced by source and processing. Dogs digest most properly cooked starches very well including corn, rice, barley and wheat. Other starches (i.e., potato and tapioca) are less digestible, especially when inadequately cooked. Cats, despite their obligate carnivorous nature, also efficiently digest carbohydrates.

Recent work has identified a link between particle size and carbohydrate digestibility in moist foods. These findings support chopping or grinding carbohydrate ingredients (e.g., rice, corn, etc.) before they are incorporated into moist foods. These findings probably do not apply to extruded dry foods because, as part of the dry food manufacturing process, carbohydrate sources are almost always chopped or ground. In fact, other studies have demonstrated almost complete ileal carbohydrate digestibility in normal dogs consuming extruded grains.

In general, dietary fat is more digestible than soluble carbohydrates and protein. The digestibility of fat in average commercial dog foods is approximately 90%. The average digestibility of fat in commercial cat foods ranges from 74 to 91%.

Digestibility of protein, soluble carbohydrates and fat in foods for patients with acute GI disease should be high because normal digestion and absorption of nutrients are often compromised. Moderate amounts of fiber decrease the dry matter digestibility of the overall food; however, digestibility of the nonfiber nutrients is usually unaffected.

ASSESS THE ANIMAL/Other Nutritional Factors/*Glutamine* Glutamine is considered a conditionally essential nutrient for pets with severe GI disorders. As the preferred energy substrate for enterocytes, glutamine is necessary for maintaining gut mucosal integrity. Commercial and homemade pet foods containing meat ingredients provide glutamine. Unfortunately, an analytical method for determining glutamine levels in foods is not widely available, making selection of foods based on glutamine content impossible. Glutamine intake can be increased by orally administering a 2% solution of glutamine in water; 0.5 g of glutamine/BW_{kg} should be provided daily. Many pets will readily consume a glutamine solution, or these solutions can be administered by dose syringe or indwelling feeding tubes. Such dosing regimens have been used for treating dogs with parvoviral enteritis. Alternatively, commercial liquid or moist homogenized enteral foods enhanced with glutamine may be offered (**Tables 12-5** and **12-6**).

ASSESS THE ANIMAL/Other Nutritional Factors/*Acid Load* Acidemia is common in pets with diarrhea because fluid secreted in the caudal small intestine and large intestine contains bicarbonate concentrations higher than those in plasma and sodium in excess of chloride ions. The acidosis is compounded in some patients by development of hypovolemia (i.e., severe dehydration). Severe acid-base disorders are best corrected with appropriate parenteral fluid therapy. Foods for patients with acute vomiting and diarrhea should avoid excess dietary acid load and preferably contain buffering salts (e.g., potassium gluconate and calcium carbonate). Ideally, foods that normally produce a urinary pH greater than 6.8 should be selected.

ASSESS THE FOOD AND FEEDING METHOD Levels of the key nutritional factors should be evaluated in foods currently fed to patients with acute gastroenteritis and compared with recommended levels (**Table 22-23**). Information from this aspect of assessment is essential for making any changes to foods currently provided. Changing to a more appropriate food is indicated if key nutritional factors in the current food do not match recommended levels.

A thorough assessment should include verification of the feeding method currently used. Items to consider include feeding frequency, amount fed, how the food is offered, access to other food and who feeds the animal. All of this information should have been gathered when the history of the animal was obtained. If the animal has a normal BCS (3/5), the amount of food previously fed (energy basis) was probably appropriate.

Feeding Plan (Pages 756-757)

The first objective in managing acute gastroenteritis should be to correct dehydration and electrolyte, glucose and acid-base imbalances, if present. The dietary goals are to provide a food that meets the patient's nutrient requirements and allows normalization of intestinal motility and function. Medical therapy may include antibiotics, NSAIDs (e.g., flunixin meglumine), anti-endotoxin sera and anthelmintics.

SELECT A FOOD There are several plausible dietary strategies for managing small bowel diarrhea and they may be attempted in any order. The traditional approach is to first feed a highly digestible, low-residue food with moderate levels of fat (12 to 15% DMB for dogs, 15 to 22% DMB for cats). Small amounts (≤5%) of soluble or mixed fiber sources may be included in such foods. Including fiber at these levels does not usually impair digestibility or increase fecal volume. This approach can be accomplished by feeding a variety of commercial veterinary therapeutic foods formulated for GI disease (**Tables 22-25** and **22-26**) or homemade foods (**Table 6-9**). Foods for puppies and kittens with GI disease should also meet the nutritional requirements for growth.

Fiber has several physiologic characteristics that are beneficial in managing small bowel diarrhea. Moderate amounts of dietary fiber (10 to 15 % DMB) add nondigestible bulk, which buffers toxins, holds excess water and, perhaps more important, provides intraluminal stimuli to reestablish the coordinated actions of hormones, neurons, smooth muscles, enzyme delivery, digestion and absorption. Fiber normalizes transit time through the small bowel, which means fiber slows a hypermotile state, but also

Table 22-25. Key nutritional factors in selected highly digestible commercial foods for dogs with acute enteritis.* (See **Table 22-27** if more fiber is desired.)

Products	Fat	Digestibility			Fiber	Potassium	Sodium	Chloride
		Prot ≥87	Fat ≥90	CHO ≥90				
Recommended levels	12-15				0.5-5.0	0.8-1.1	0.35-0.5	0.5-1.3
Moist canine products								
Hill's Prescription Diet Canine i/d	14.0	Y	Y	Y	0.7	0.92	0.44	1.23
Leo Specific Digest CIW	14.7	na	na	na	4.2	na	na	na
Iams Eukanuba Low-Residue Adult/Canine	22.0	na	na	na	2.2	1.00	0.75	1.00
Medi-Cal Canine Gastro Formula	12.3	na	na	na	1.7	na	na	na
Purina CNM Canine EN-Formula	13.8	na	na	na	0.9	0.61	0.37	0.78
Select Care Canine Sensitive Formula	12.3	na	na	na	1.7	na	na	na
Waltham/Pedigree Canine Low Fat	6.9	na	na	na	2.4	na	na	na
Dry canine products								
Hill's Prescription Diet Canine i/d	13.6	Y	Y	Y	1.2	0.92	0.46	1.10
Iams Eukanuba Low-Residue Adult/Canine	10.6	na	na	na	2.1	0.89	0.46	1.13
Iams Eukanuba Low-Residue Puppy	22.3	na	na	na	1.9	0.78	0.42	0.85
Leo Specific Digest CID	14.3	na	na	na	1.1	na	na	na
Medi-Cal Canine Gastro Formula	11.7	na	Y	na	2.0	na	na	na
Purina CNM Canine EN-Formula	11.7	Y		Y	1.1	0.59	0.36	0.87
Select Care Canine Sensitive Formula	11.7	na	na	na	2.0	na	na	na
Waltham/Pedigree Canine Low Fat	5.4	na	na	na	0.4	na	na	na

Key: na = information not available from manufacturer, Prot = protein, CHO = soluble carbohydrate, Y = yes.
*This list represents products with the largest market share and for which published information is available.
Manufacturers' published values. Nutrients expressed as % dry matter.

Table 22-26. Key nutritional factors in selected highly digestible commercial foods for cats with acute enteritis.*
(See Table 22-28 if more fiber is desired.)

Products	Fat 15-22	Digestibility Prot ≥87	Fat ≥90	CHO ≥90	Fiber 0.5-5.0	Potassium 0.8-1.1	Sodium 0.35-0.5	Chloride 0.5-1.3
Recommended levels	15-22	≥87	≥90	≥90	0.5-5.0	0.8-1.1	0.35-0.5	0.5-1.3
Moist feline products								
Hill's Prescription Diet Feline i/d	20.1	Y	Y	Y	1.6	1.08	0.36	1.04
Iams Eukanuba Low-Residue Adult/Feline	16.9	na	na	na	2.2	na	na	na
Leo Specific Dermil FDW	34.7	na	na	na	1.5	na	na	na
Medi-Cal Feline HYPOallergenic/ Gastro Formula	25.6	na	na	na	1.65	na	na	na
Purina CNM Feline EN-Formula	17.0	Y	Y	Y	1.1	0.80	0.32	0.55
Waltham/Whiskas Feline Selected Protein	29.0	na	na	na	2.9	na	na	na
Dry feline products								
Hill's Prescription Diet Feline i/d	20.1	Y	Y	Y	1.3	0.98	0.36	1.03
Iams Eukanuba Low-Residue Adult/Feline	15.4	na	na	na	2.2	0.98	0.32	0.88
Medi-Cal Feline HYPOallergenic/ Gastro Formula	13.6	na	na	na	1.6	na	na	na
Select Care Feline Neutral Formula	13.6	na	na	na	1.6	na	na	na
Waltham/Whiskas Feline Selected Protein	12.2	na	na	na	3.3	na	na	na

Key: na = information not available from manufacturer, Prot = protein, CHO = soluble carbohydrate, Y = yes.
*This list represents products with the largest market share and for which published information is available.
Manufacturers' published values. Nutrients expressed as % dry matter.

Table 22-27. Key nutritional factors in selected fiber-enhanced commercial foods for dogs with acute enteritis.*
(See **Table 22-25** if less fiber is desired.)

Products Recommended levels	Fat 12-15	Fiber 10-15	Potassium 0.8-1.1	Sodium 0.35-0.5	Chloride 0.5-1.3
Moist canine products					
Hill's Prescription Diet Canine r/d	8.4	21.8	0.83	0.25	0.50
Hill's Prescription Diet Canine w/d	13.0	12.2	0.62	0.27	0.73
Leo Specific CRW	8.9	17.8	na	na	na
Medi-Cal Canine Fibre Formula	7.9	15.9	na	na	na
Medi-Cal Canine Wt. Control/Geriatric	10.8	6.1	na	na	na
Purina CNM Canine OM-Formula	8.4	19.2	1.06	0.28	0.51
Select Care Canine Hifactor Formula	9.1	15.0	na	na	na
Waltham/Pedigree Canine High Fiber	8.0	10.3	na	na	na
Dry canine products					
Hill's Prescription Diet Canine r/d	8.5	23.4	0.82	0.30	0.43
Hill's Prescription Diet Canine w/d	8.7	16.9	0.68	0.22	0.54
Leo Specific CRD	5.6	8.9	na	na	na
Medi-Cal Canine Fibre Formula	9.8	15.8	na	na	na
Medi-Cal Canine Wt. Control/Geriatric	8.5	5.9	na	na	na
Purina CNM Canine DCO-Formula	12.4	7.6	0.70	0.34	0.82
Purina CNM Canine OM-Formula	6.7	10.7	0.82	0.21	0.32
Select Care Canine Hifactor Formula	10.6	14.3	na	na	na
Waltham/Pedigree Canine High Fiber	8.2	4.9	1.09	0.33	1.09

Key: na = information not available from manufacturer.
*This list represents products with the largest market share and for which published information is available.
Manufacturers' published values. Nutrients expressed as % dry matter.

Table 22-28. Key nutritional factors in selected fiber-enhanced commercial foods for cats with acute enteritis.*
(See **Table 22-26** if less fiber is desired.)

Products	Fat 15-22	Fiber 10-15	Potassium 0.8-1.1	Sodium 0.35-0.5	Chloride 0.5-1.3
Recommended levels	15-22	10-15	0.8-1.1	0.35-0.5	0.5-1.3
Moist feline products					
Hill's Prescription Diet Feline r/d	8.9	17.4	0.72	0.30	0.76
Hill's Prescription Diet Feline w/d	16.7	10.7	0.87	0.44	1.07
Leo Specific FRW	9.3	18.7	na	na	na
Medi-Cal Feline Fibre Formula	14.4	8.6	na	na	na
Medi-Cal Feline Weight Control	17.7	4.7	na	na	na
Select Care Feline Hifactor Formula	16.1	7.2	na	na	na
Dry feline products					
Hill's Prescription Diet Feline r/d	9.1	13.7	0.71	0.30	0.67
Hill's Prescription Diet Feline w/d	9.3	7.9	0.74	0.27	0.67
Leo Specific FRD	7.1	19.0	na	na	na
Medi-Cal Feline Fibre Formula	13.8	7.3	na	na	na
Medi-Cal Feline Weight Control	12.3	4.7	na	na	na
Purina CNM Feline OM-Formula	8.5	8.0	0.69	0.26	0.88
Select Care Feline Hifactor Formula	12.9	5.0	na	na	na
Select Care Feline Weight Formula	12.2	3.6	na	na	na

Key: na = information not available from manufacturer.
*This list represents products with the largest market share and for which published information is available.
Manufacturers' published values. Nutrients expressed as % dry matter.

improves a hypomotile state to reestablish normal peristaltic action. **Tables 22-27** and **22-28** list selected fiber-enhanced commercial foods.

In cases of protracted small bowel disuse (i.e., more than seven days), a third strategy may be used. This strategy involves initially feeding small amounts of a monomeric liquid food containing maltodextrins and glutamine (**Table 12-6**). This type of initial feeding may ease the transition to other foods. Feeding puppies recovering from parvoviral enteritis a monomeric liquid food containing maltodextrins and glutamine reduces nausea and vomiting, and subsequently eases the transition to feeding other commercial veterinary therapeutic foods.

DETERMINE A FEEDING METHOD Withholding all oral intake of food and water for 24 to 48 hours is the first step in the feeding method for patients with acute gastroenteritis. After this period, animals should be offered small amounts of water or ice cubes every few hours. If water is well tolerated, small amounts of food can be offered several times (i.e., six to eight times) a day. If the pet can eat food without episodes of diarrhea or vomiting, the amount fed can be increased over three to four days until the animal is receiving its estimated DER in two to three meals per day. During this period, if the animal begins to vomit, food should be withdrawn and offered again after several hours. As discussed above, monomeric liquid foods can also be offered.

Persistent vomiting may complicate refeeding in some cases of parvoviral enteritis; some puppies develop gastroparesis and may require prokinetic drugs to facilitate feeding. In such cases, intravenous infusion of metoclopramide (1.0 mg/BW_{kg}/day) is recommended. Alternatively, metoclopramide can be administered to well-hydrated patients subcutaneously or intramuscularly at a dose of 0.5 mg/BW_{kg} q8h. Some animals may require parenteral feeding. (See Chapter 12.)

Reassessment (Page 757)

The prognosis for recovery in most cases of acute gastroenteritis is good. Body weight should be recorded daily until recovery is complete. Changes in body weight from day to day usually reflect changes in hydration status rather than loss or gain of body tissue. Further diagnostic testing is warranted if vomiting or severe diarrhea persists.

Dogs and cats presenting with multiple or recurrent episodes of small bowel diarrhea require further diagnostic workup and, most probably, a combination of dietary and medical therapies. Parasitic causes, however, should be ruled out or treated empirically before pursuing further diagnostics. The diagnostic approach to patients with chronic small bowel diarrhea is beyond the scope of this book; readers are referred to internal medicine and gastroenterology texts for more information.

GI/PANCREAS

Inflammatory Bowel Disease (Pages 757-763)

Clinical Importance (Pages 757-758)

The term inflammatory bowel disease (IBD) refers to a group of chronic, idiopathic GI disorders. Each is characterized by inflammatory infiltrates within the lamina propria of the GI tract. Currently, IBD is considered the commonest cause of chronic diarrhea and vomiting in dogs and cats. The generic term IBD encompasses lymphoplasmacytic enteritis, lymphocytic gastroenterocolitis, eosinophilic gastroenterocolitis, segmental granulomatous enterocolitis, suppurative enterocolitis and histiocytic colitis. Specific types are categorized on the basis of the type of inflammatory cells found in the lamina propria. The lymphoplasmacytic form is probably the most common type of IBD.

The severity of the condition varies from relatively mild clinical signs to life-threatening protein-losing enteropathies. In particular, the basenji and Ludenhund breeds may present with a very severe variant that has been termed immunoproliferative small intestinal disease.

Inflammatory infiltrates may involve the stomach, small bowel and colon. In cats, the stomach and small bowel are affected most often. In dogs, IBD is common in the small and large bowel. In many cases, multiple segments of the bowel are involved and clinical signs may be mixed, reflecting the broad distribution of mucosal lesions.

Assessment (Pages 758-762)

ASSESS THE ANIMAL/History and Physical Examination The most common clinical signs in dogs and cats with IBD are chronic vomiting, diarrhea and weight loss. The predominant GI sign varies with the portion or portions of bowel affected. Vomiting tends to be the predominant clinical sign when the stomach and proximal duodenum are affected. Loose, fluid or steatorrheic stools are most common when the small intestine is involved. Diarrhea marked by tenesmus, mucus and small scanty stools is noted with colonic lesions. Clinical signs may be intermittent or persistent. Clinical signs tend to increase in frequency and intensity as IBD progresses. The presence of systemic signs is also variable. Some animals present with a history of depression, malaise and inappetence. Others are alert and active at the time they are examined.

The frequency and character of the vomitus and stools are important. At times, vomiting will be temporally related to food intake and the vomitus will contain food particles. In other cases, animals may vomit only fluid or froth. Vomiting hairballs is a typical finding in cats and suggests that the stomach is affected. Owners should be questioned closely about the appearance of the vomited material. Dark black or coffee grounds material suggests gastric ulceration. The diarrhea may be small or large bowel in origin. The color of the stools should be assessed to determine the presence of GI bleeding.

Physical examination findings in dogs and cats with IBD are variable. Many animals have no abnormalities. Others present only with evidence of weight loss and poor body condition. Weight loss may be severe in longstanding cases. Mild to moderate peripheral lymphadenopathy may be detected in rare cases of IBD. This finding is most often recognized in cats with eosinophilic gastroenteritis and hypereosinophilic syndrome, which is characterized by multisystemic eosinophilic infiltrates.

On occasion, thickened loops of bowel may be detected by abdominal palpation. This finding is more easily detected in cats. A segmental thickening of bowel is consistent with eosinophilic gastroenteritis in cats or granulomatous enteritis in dogs. This finding should also be distinguished from intestinal intussusceptions, foreign bodies, histoplasmosis and neoplastic lesions. Occasionally, animals with IBD present with abdominal pain, which is suggestive of gastroduodenal ulceration.

Evidence of hemorrhage or hypoproteinemia may be noted in very severe cases. A vitamin K-dependent coagulopathy may develop in animals with marked steatorrhea. At times, IBD may result in a protein-losing enteropathy. When severe, hypoalbuminemia and external manifestations of hypoproteinemia (i.e., pitting edema, ascites) may be present. Surprisingly, some animals with protein-losing enteropathy may present with only mild or no diarrhea.

ASSESS THE ANIMAL/Laboratory and Other Clinical Information Laboratory findings in patients with IBD are often nonspecific. Hematologic findings are variable and may include blood loss anemia, anemia of chronic disease and/or eosinophilia. In cats with eosinophilic gastroenteritis and hypereosinophilic syndrome, eosinophil counts may exceed 100,000/μl. Patients with chronic diarrhea should be assessed with serum biochemistry profiles and urinalyses to determine the systemic effects of the GI disorder and to rule out concurrent disease. Electrolyte abnormalities, including hypokalemia, may be identified. Hypoproteinemia and hypoalbuminemia may be recognized in severe cases with protein-losing enteropathy. Prerenal azotemia may be present in dehydrated patients. In cats, IBD may be associated with pancreatitis and hepatitis. In such cases, increased hepatic enzyme activities, hyperamylasemia, hyperlipasemia and hyperbilirubinemia may be noted. IBD is often associated with a protein-losing nephropathy in soft-coated wheaten terriers. Varying degrees of azotemia and proteinuria are common in these cases.

Fecal examinations are very important in the evaluation of patients with chronic diarrhea. Multiple fecal examinations using concentration techniques are necessary to rule out parasitism. Qualitative and quantitative fecal fat examinations assess fat absorption capacity. Marked steatorrhea usually indicates severe infiltrative disease.

GI/PANCREAS

Radiographic findings in IBD are usually nonspecific and nondiagnostic. Occasionally, thickened bowel loops are detected.

Endoscopic abnormalities in IBD include mucosal granularity, hyperemia, friability and inability to visualize colonic submucosal blood vessels. Multiple biopsy specimens should be collected from several bowel segments even if the endoscopic appearance is normal because histologic changes may be present despite a normal appearance.

The definitive diagnosis of IBD is based on histopathologic examination of biopsy specimens collected by endoscopic or surgical techniques. Interpretation of histologic changes can be difficult when the lesions are mild or suggest lymphosarcoma. The latter finding is a serious concern in cases of lymphoplasmacytic enteritis and lymphocytic enteritis.

ASSESS THE ANIMAL/Risk Factors There does not appear to be an age or gender predisposition for any of the forms of IBD. The condition usually arises in adult dogs and cats, but has been diagnosed in puppies and kittens (i.e., less than six months of age). A genetic influence has been recognized in veterinary medicine (**Table 22-22**).

The environment may also play an important role in IBD. Animals maintained in overcrowded, contaminated quarters are at risk for development of parasitic infections, viral and bacterial enteritis and small intestinal bacterial overgrowth, which have been hypothesized to play a role in the pathogenesis of IBD. The role of parasites in the pathogenesis of IBD is poorly understood; however, occult parasitism has been suggested as a cause for these disorders. For example, in German shepherd dogs, visceral larval migrans has been linked to eosinophilic gastroenteritis. In cats, feline infectious peritonitis has been associated with granulomatous and suppurative enterocolitis. In addition, small intestinal bacterial overgrowth has been reported in association with lymphoplasmacytic infiltrates and enteritis.

ASSESS THE ANIMAL/Etiopathogenesis The pathophysiology of IBD is not completely understood. The disorder is undoubtedly immune-mediated, yet the pathogenesis of the various forms of IBD is poorly defined. The fundamental pathway for the development of IBD involves hypersensitivity.

Mucosal inflammatory infiltrates are responsible for the clinical manifestations of IBD. Mucosal inflammation disrupts normal absorptive processes resulting in malabsorption and osmotic diarrhea. Altered gut permeability can result in leakage of fluid, protein and blood into the gut lumen. Malabsorbed fats, carbohydrates and bile acids result in secretory diarrhea. Inflammatory mediators may also directly trigger intestinal secretion and mucus production by goblet cells. Mucosal inflammatory infiltrates may

Table 22-29. Key nutritional factors for patients with inflammatory bowel disease.*

Factors	Recommended levels
Crude fiber	0.5 to 15% (see text)
Digestibility	≥87% for protein and ≥90% for fat and soluble carbohydrate
Fat	12 to 15% for dogs
	15 to 22% for cats
Potassium	0.80 to 1.10%
Protein	Limit dietary protein to one or two sources
	Use protein sources to which animal has not been exposed previously or feed a protein hydrolysate (See Chapter 14.)
	16 to 26% for dogs
	30 to 45% for cats

*Nutrients expressed on a dry matter basis.

alter intestinal and colonic motility patterns, a mechanism attributed to the influence of prostaglandins and leukotrienes on smooth muscle. Inflammation of the proximal bowel (stomach and small bowel) may stimulate visceral afferent receptors that trigger vomiting. Delayed gastric emptying associated with gastroparesis or ileus may exacerbate vomiting.

ASSESS THE ANIMAL/Key Nutritional Factors Key nutritional factors for patients with IBD are listed in **Table 22-29** and discussed in more detail below.

ASSESS THE ANIMAL/Key Nutritional Factors/*Water* Dehydration is a frequent problem in patients with IBD. Reduced water consumption is often aggravated by fluid losses from vomiting and/or diarrhea. Whenever possible, fluid balance should be maintained via oral consumption of fluids. However, dehydrated patients and those with persistent vomiting often need parenteral fluid administration.

ASSESS THE ANIMAL/Key Nutritional Factors/*Minerals* Serum electrolyte concentrations should be assessed regularly to allow early detection of abnormalities as vomiting and diarrhea persist. Hypokalemia is particularly common in patients with IBD. Thus, foods containing 0.85 to 1.1% DM potassium are recommended for dogs and cats with IBD. Initially, potassium levels should be restored with intravenous potassium supplementation.

ASSESS THE ANIMAL/Key Nutritional Factors/*Fat* High-fat foods may contribute to osmotic diarrhea and GI protein losses, which complicate IBD.

Thus, it is often advantageous to initially provide a food with moderate fat levels (12 to 15% DMB for dogs and 15 to 22% DMB for cats). Foods with higher fat levels can be offered if the patient tolerates these nutrient levels.

There appears to be a difference in how dogs and cats are able to tolerate dietary fat in the face of GI disease. Normal cats can tolerate much higher concentrations of dietary fat than dogs. Anecdotal information suggests that foods with increased fat content may actually benefit cats with small bowel disease. Controlled evaluations are needed to confirm these observations.

ASSESS THE ANIMAL/Key Nutritional Factors/*Protein* Protein malnutrition may occur in dogs and cats with IBD due to fecal losses. Recommended dietary protein intake will be discussed in more detail in the following section about protein-losing enteropathies. Protein should be provided at levels sufficient for the appropriate lifestage for patients without excessive GI protein loss (**Table 22-29**). High biologic value, highly digestible (≥87%) protein sources should be used.

Because dietary antigens are suspected to play a role in the pathogenesis of IBD, some authors recommend the use of "hypoallergenic" or elimination foods. In some cases, elimination foods may be used successfully without pharmacologic intervention. Mild to moderate feline lymphoplasmacytic enteritis, canine lymphoplasmacytic colitis and canine eosinophilic gastroenteritis are more likely to respond to dietary management alone. A protein hydrolysate-based elimination diet has been used successfully in refractory canine IBD cases. Chapter 14 discusses elimination foods and protein hydrolysates in more detail.

ASSESS THE ANIMAL/Key Nutritional Factors/*Fiber* A number of substrates including beet pulp, soy fiber, inulin and fructooligosaccharides (FOS) have been demonstrated by in vitro fermentation to produce volatile fatty acids that may be beneficial in IBD involving the distal small intestine and colon. These fibers are usually incorporated at rates of 0.5 to 5% DM in commercial products. Moderate levels of dietary fiber (10 to 15% DMB) may also be included as another strategy. (See Select a Food below.)

ASSESS THE ANIMAL/Key Nutritional Factors/*Digestibility* Feeding highly digestible (**Table 22-29**) foods provides several advantages in the management of dogs and cats with IBD. Nutrients from low-residue foods are more completely absorbed in the proximal gut. Furthermore, low-residue foods are associated with: 1) reduced osmotic diarrhea due to fat and carbohydrate malabsorption, 2) reduced production of intestinal gas due to carbohydrate malabsorption and 3) decreased antigen loads because smaller amounts of protein are absorbed intact. Ideal foods for IBD patients are free of lactose to avoid the complication of lactose intolerance.

Table 22-30. Potential causes of zinc deficiency in patients with inflammatory bowel disease.

Decreased absorption	
Intestinal inflammation	Surgical resection of distal duodenum
Supplemental iron and/or copper	
Inadequate dietary intake	
Anorexia	Parenteral nutrition
High fiber or phytate intake	
Increased losses	
Chronic blood loss	Increased metabolism
Increased requirements	
Growth	Pregnancy
Lactation	Wound healing

ASSESS THE ANIMAL/Other Nutritional Factors/*Vitamins* Adequate intake of water-soluble and fat-soluble vitamins is critical for patients with IBD. In many cases, the limited stores of water-soluble vitamins have been depleted by diarrheic losses and the large fluid flux through the animal. Thiamin deficiency, in particular, occurs commonly and can profoundly affect appetite. Loss of fat-soluble vitamins can be significant in patients with steatorrhea (e.g., vitamin K-deficient coagulopathies may occur in patients with IBD). Initially, parenteral administration of vitamins may be necessary. Dietary intake is often sufficient when the disease responds to treatment and fat absorption is reestablished.

ASSESS THE ANIMAL/Other Nutritional Factors/*Zinc* Zinc deficiency is well recognized in people as a complication of IBD. The small intestine is the primary site of zinc homeostasis and there are several potential mechanisms for zinc deficiency in IBD (**Table 22-30**). Dietary zinc intake should be assessed if dogs and cats with IBD have poor coat quality or dermatitis.

ASSESS THE ANIMAL/Other Nutritional Factors/*N-3 Fatty Acids* N-3 fatty acids (omega-3 fatty acids) derived from fish oil or other sources have been hypothesized to have a beneficial effect in controlling mucosal inflammation in IBD. There is some clinical evidence that dietary supplementation with these fatty acids can modulate the generation and biologic activity of inflammatory mediators. Although use of n-3 fatty acids warrants further consideration in veterinary medicine, there is no well-established effective dose for dogs and cats. A reasonable starting dose is approximately 175 mg (range 50 to 300 mg) n-3 fatty acids/BW_{kg}/day. (See Chapter 26 for additional information about the use of n-3 fatty acids in inflammatory disorders.)

ASSESS THE FOOD AND FEEDING METHOD Levels of key nutritional factors should be evaluated in foods currently being fed to patients with

GI/PANCREAS

IBD and compared with recommended levels (**Table 22-29**). Information from this aspect of assessment is essential for making any changes to foods currently provided. Changing to a more appropriate food is indicated if key nutritional factors in the current food do not match recommended levels.

A thorough assessment should include verification of the feeding method currently being used. Items to consider include feeding frequency, amount fed, how the food is offered, access to other food and who feeds the animal. All of this information should have been gathered when the history of the animal was obtained. If the animal has an ideal BCS (3/5), the amount of food previously fed (energy basis) was probably appropriate. If the animal has poor body condition (BCS 1/5 or 2/5), the amount of food previously fed may have been inappropriate or significant malassimilation may be occurring due to IBD.

Feeding Plan *(Pages 762-763)*

The justification for nutritional management of IBD is twofold. First, dietary factors may contribute to the initiation or perpetuation of the disease. Second, malabsorption and malnutrition are common sequelae to IBD.

Dietary intervention should be aimed at controlling clinical signs while providing adequate nutrients to meet requirements and compensate for ongoing losses through the GI tract. Some dogs and cats with IBD may require only dietary manipulation. In other cases, dietary therapy is better used in concert with pharmacologic agents. Antibiotics (tylosin, tetracycline, metronidazole), anthelmintics (fenbendazole) and immunosuppressive agents (corticosteroids, azathioprine, cyclophosphamide) are often used for managing IBD.

SELECT A FOOD Food selection should focus on foods that reduce intestinal irritation/inflammation and normalize intestinal motility. Three types of foods may be useful in managing diarrhea associated with IBD: 1) highly digestible, low-residue foods formulated for GI disease, 2) fiber-enhanced foods and 3) elimination foods. Unfortunately, no physical examination finding, laboratory test result or historical fact will dictate which method will be successful in any one patient. Dietary trials are often needed to determine which food type works best.

The most commonly used strategy is to feed a highly digestible, low-residue GI food with moderate levels of fat (i.e., 12 to 15% DMB for dogs, 15 to 22% DMB for cats). This can be accomplished by feeding a variety of commercial veterinary therapeutic foods (**Tables 22-31** and **22-32**) or homemade foods (**Table 6-9**).

A second approach is to increase dietary fiber content to normalize intestinal motility, water balance and microflora. Fiber has several physi-

Table 22-31. Key nutritional factors in selected highly digestible commercial foods for dogs with inflammatory bowel disease.* (See **Table 22-33** if more fiber is desired; see **Tables 14-3** and **14-5** if foods with novel protein sources or protein hydrolysates are desired.)

Products	Protein	Fat	Digestibility			Fiber	Potassium
			Prot ≥87	Fat ≥90	CHO ≥90		
Recommended levels	**16-26**	**12-15**	**≥87**	**≥90**	**≥90**	**0.5-5.0**	**0.80-1.1**
Moist canine products							
Hill's Prescription Diet Canine i/d	25.9	14.0	Y	Y	Y	0.7	0.92
Leo Specific Digest CIW	24.4	14.7	na	na	na	4.2	na
Iams Eukanuba Low-Residue Adult/Canine	36.3	22.0	na	na	na	2.2	1.00
Medi-Cal Canine Gastro Formula	22.1	12.3	na	na	na	1.7	na
Purina CNM Canine EN-Formula	30.5	13.8	na	na	na	0.9	0.61
Select Care Canine Sensitive Formula	22.1	12.3	na	na	na	1.7	na
Waltham/Pedigree Canine Low Fat	33.8	6.9	na	na	na	2.4	na
Dry canine products							
Hill's Prescription Diet Canine i/d	26.4	13.6	Y	Y	Y	1.2	0.92
Iams Eukanuba Low-Residue Adult/Canine	25.4	10.6	na	na	na	2.1	0.89
Iams Eukanuba Low-Residue Puppy	33.3	22.3	na	na	na	1.9	0.78
Leo Specific Digest CID	27.5	14.3	na	na	na	1.1	na
Medi-Cal Canine Gastro Formula	23.4	11.7	na	na	na	2.0	na
Purina CNM Canine EN-Formula	25.8	11.7	Y	Y	Y	1.1	0.59
Select Care Canine Sensitive Formula	23.4	11.7	na	na	na	1.1	na
Waltham/Pedigree Canine Low Fat	24.1	5.4	na	na	na	0.4	na

Key: na = information not available from manufacturer, Prot = protein, CHO = soluble carbohydrate, Y = yes.
*This list represents products with the largest market share and for which published information is available.
Manufacturers' published values. Nutrients expressed as % dry matter.

GI/PANCREAS

Table 22-32. Key nutritional factors in selected highly digestible commercial foods for cats with inflammatory bowel disease.* (See **Table 22-34** if more fiber is desired; see **Tables 14-4** and **14-5** if foods with novel protein sources or protein hydrolysates are desired.)

Products	Protein	Fat	Digestibility			Fiber	Potassium
			Prot ≥87	Fat ≥90	CHO ≥90		
Recommended levels	30-45	15-22				0.5-5.0	0.80-1.1
Moist feline products							
Hill's Prescription Diet Feline i/d	40.6	20.1	Y	Y	Y	1.6	1.08
Iams Eukanuba Low-Residue Adult/Feline	41.6	16.9	na	na	na	2.2	na
Leo Specific Dermil FDW	34.7	34.7	na	na	na	1.5	na
Medi-Cal Feline HYPOallergenic/Gastro Formula	35.5	25.6	na	na	na	1.65	na
Purina CNM Feline EN-Formula	41.9	17.0	Y	Y	Y	1.1	0.80
Waltham/Whiskas Feline Selected Protein	38.2	29.0	na	na	na	2.9	na
Dry feline products							
Hill's Prescription Diet Feline i/d	39.9	20.1	Y	Y	Y	1.3	0.98
Iams Eukanuba Low-Residue Adult/Feline	34.8	15.4	na	na	na	2.2	0.98
Medi-Cal Feline HYPOallergenic/Gastro Formula	29.8	13.6	na	na	na	1.6	na
Select Care Feline Neutral Formula	27.7	13.6	na	na	na	1.6	na
Waltham/Whiskas Feline Selected Protein	33.3	12.2	na	na	na	3.3	na

Key: na = information not available from manufacturer, Prot = protein, CHO = soluble carbohydrate, Y = yes.
*This list represents products with the largest market share and for which published information is available.
Manufacturers' published values. Nutrients expressed as % dry matter.

Table 22-33. Key nutritional factors in selected fiber-enhanced commercial foods for dogs with inflammatory bowel disease.* (See **Table 22-31** if less fiber is desired; see **Tables 14-3** and **14-5** if foods with novel protein sources or protein hydrolysates are desired.)

Products Recommended levels	Protein 16-26	Fat 12-15	Fiber 10-15	Potassium 0.80-1.1
Moist canine products				
Hill's Prescription Diet Canine r/d	25.5	8.4	21.8	0.83
Hill's Prescription Diet Canine w/d	18.2	13.0	12.2	0.62
Leo Specific CRW	31.6	8.9	17.8	na
Medi-Cal Canine Fibre Formula	24.7	7.9	15.9	na
Medi-Cal Canine Weight Control/Geriatric	21.8	10.8	6.1	na
Purina CNM Canine OM-Formula	44.1	8.4	19.2	1.06
Select Care Canine Hifactor Formula	24.8	9.1	15.0	na
Waltham/Pedigree Canine High Fiber	29.8	8.0	10.3	na
Dry canine products				
Hill's Prescription Diet Canine r/d	24.7	8.5	23.4	0.82
Hill's Prescription Diet Canine w/d	18.9	8.7	16.9	0.68
Leo Specific CRD	24.4	5.6	8.9	na
Medi-Cal Canine Fibre Formula	23.7	9.8	15.8	na
Medi-Cal Canine Weight Control/Geriatric	20.0	8.5	5.9	na
Purina CNM Canine DCO-Formula	25.3	12.4	7.6	0.70
Purina CNM Canine OM-Formula	32.0	6.7	10.7	0.82
Select Care Canine Hifactor Formula	25.1	10.6	14.3	na
Waltham/Pedigree Canine High Fiber	21.9	8.2	4.9	1.09

Key: na = information not available from manufacturer.
*This list represents products with the largest market share and for which published information is available.
Manufacturers' published values. Nutrients expressed as % dry matter.

GI/PANCREAS

Table 22-34. Key nutritional factors in selected fiber-enhanced commercial foods for cats with inflammatory bowel disease.* (See **Table 22-32** if less fiber is desired; see **Tables 14-4** and **14-5** if foods with novel protein sources or protein hydrolysates are desired.)

Products	Protein 30-45	Fat 15-22	Fiber 10-15	Potassium 0.80-1.1
Recommended levels				
Moist feline products				
Hill's Prescription Diet Feline r/d	36.0	8.9	17.4	0.72
Hill's Prescription Diet Feline w/d	41.3	16.7	10.7	0.87
Leo Specific FRW	42.1	9.3	18.7	na
Medi-Cal Feline Fibre Formula	40.2	14.4	8.6	na
Medi-Cal Feline Weight Control	40.2	17.7	4.7	na
Select Care Feline Hifactor Formula	34.1	16.1	7.2	na
Dry feline products				
Hill's Prescription Diet Feline r/d	37.1	9.1	13.7	0.71
Hill's Prescription Diet Feline w/d	38.8	9.3	7.9	0.74
Leo Specific FRD	37.0	7.1	19.0	na
Medi-Cal Feline Fibre Formula	34.1	13.8	7.3	na
Medi-Cal Feline Weight Control	34.4	12.3	4.7	na
Purina CNM Feline OM-Formula	38.4	8.5	8.0	0.69
Select Care Feline Hifactor Formula	38.2	12.9	5.0	na
Select Care Feline Weight Formula	36.2	12.2	3.6	na

Key: na = information not available from manufacturer.
*This list represents products with the largest market share and for which published information is available.
Manufacturers' published values. Nutrients expressed as % dry matter.

ologic characteristics that are beneficial in managing small bowel diarrhea. Moderate levels of dietary fiber (10 to 15% DMB) add nondigestible bulk, which buffers toxins, holds excess water and, perhaps more important, provides intraluminal stimuli to reestablish the coordinated actions of hormones, neurons, smooth muscles, enzyme delivery, digestion and absorption. Fiber normalizes transit time through the small bowel, which means fiber slows a hypermotile state, but also improves a hypomotile state to reestablish normal peristaltic action. **Tables 22-33** and **22-34** list selected fiber-enhanced commercial foods.

The third dietary option in IBD cases is the use of an elimination food with a limited number of highly digestible, novel protein sources (**Tables 14-3** and **14-4**) or a protein hydrolysate (**Table 14-5**). Commercial veterinary therapeutic foods or homemade foods that contain novel protein sources often combine lamb, rabbit, venison, duck or fish with a highly digestible or novel carbohydrate source. All other possible dietary sources of protein and carbohydrate should be eliminated including treats, snacks, table foods, vitamin-mineral supplements and chewable/flavored medications. Clinical signs should abate within the first three weeks of strict dietary management (e.g., feeding only the novel ingredient food). After signs abate, owners may add individual specific ingredients previously fed in an effort to identify the allergen. Clinical GI signs may recur within 12 hours after the offending ingredient is fed.

DETERMINE A FEEDING METHOD Initially the IBD patient should be fed multiple small meals per day as indicated by animal acceptance and tolerance for the food. Meal size can be increased and meal frequency can be reduced as tolerated by the patient after the clinical signs have been successfully managed for several weeks.

Reassessment (Page 763)

Regaining or maintaining optimal body weight and condition, normal activity level, a positive attitude and absence of clinical signs are measures of successful dietary and medical management. The feeding method and amount fed can be adjusted as needed to maintain body weight and condition.

The prognosis for IBD varies with the specific entity present, severity of the condition at the time of presentation and owner compliance. The hypereosinophilic form of eosinophilic gastroenteritis in cats and immunoproliferative enteropathy and histiocytic colitis in dogs may be refractory to treatment. Likewise, response to therapy may be poor when animals present late in the course of disease and with evidence of protein-losing enteropathy.

In most cases, judicious use of dietary and medical regimens controls the disease. Often, medical measures can be withdrawn after three to six months; thereafter, animals maintain remission with appropriate foods. In some cases, however, pharmacologic treatment may be required for the life of the animal.

GI/PANCREAS

The most common causes for failure to respond include noncompliance on the part of the owner and failure of the clinician to tailor a program incorporating both dietary and pharmacologic measures for each patient. On occasion, treatment failures occur because of misdiagnosis of alimentary lymphosarcoma or progression of IBD to lymphosarcoma. This progression has been reported to occur in dogs and cats.

Lymphangiectasia and Protein-Losing Enteropathies
(Pages 763-767)
Clinical Importance (Page 763)
Lymphangiectasia is a chronic enteropathy characterized by abnormalities of the intestinal lymphatic system. The condition may occur as a primary lymphatic defect or secondarily as a consequence of severe intestinal infiltrative disease (e.g., IBD, alimentary lymphosarcoma, immunoproliferative enteropathy, fungal enteritis). Lymphangiectasia is the most common cause of protein-losing enteropathy (PLE) in dogs and cats. However, PLE is a relatively rare manifestation of diarrheic disorders in dogs and cats.

Assessment (Pages 763-766)
ASSESS THE ANIMAL/History and Physical Examination Typically, signs of lymphangiectasia are insidious in onset and follow a waxing and waning course over several weeks to months before becoming flagrant. The clinical manifestations of lymphangiectasia are generally attributable to the loss of lymph constituents (i.e., albumin, lymphocytes, fat) or to underlying enteric disease. Many patients present with chronic intermittent diarrhea or vomiting; however, not all patients have GI signs. Progressive weight loss, often in the face of good appetite, is a consistent finding in longstanding cases. Excessive protein loss from leaky intestinal lymphatics results in hypoalbuminemia and loss of colloidal oncotic pressure. External manifestations of hypoalbuminemia may include pitting edema, ascites and pleural effusion. In some cases, chylous effusions of the abdomen, subcutis or thoracic cavity may occur in conjunction with primary or congenital lymphangiectasia. These findings probably represent multisystemic lymphatic defects.

Physical examination findings may be unremarkable in dogs and cats with PLE. Patients with severe hypoproteinemia may present with dyspnea and abdominal enlargement due to accumulation of fluid in the thoracic or abdominal cavities, respectively. Pitting edema of the limbs may be noted. Body condition assessment should be performed because many animals are underweight at the time of presentation.

ASSESS THE ANIMAL/Laboratory and Other Clinical Information A consistent pattern of laboratory results can be identified in many dogs with PLE. Panhypoproteinemia (i.e., hypoglobulinemia and hypoalbuminemia) and hypocholesterolemia are classic findings and reflect the loss of lymphatic constituents into the gut lumen. Occasionally, when PLE arises as a consequence of chronic inflammatory enteropathies, hypergammaglobulinemia will be present as well. In lymphangiectasia, lymphopenia is an important finding that can be used to differentiate this condition from other causes of PLE. Other common laboratory findings include the anemia of chronic disease, a stress leukogram and hypocalcemia. Hypocalcemia may be present due to malabsorption of calcium and/or vitamin D or to a decrease in the protein-bound calcium fraction. Symptomatic hypocalcemia is rare; therefore, ionized calcium levels should be assessed before initiating intravenous calcium supplementation.

Typical intestinal function tests (i.e., D-xylose absorption tests, fecal fat determinations, breath hydrogen tests) yield inconsistent results and are of little value for assessing patients with lymphangiectasia and other types of PLE.

Endoscopic examination of patients with suspected PLE or lymphangiectasia can be helpful. Mucosal granularity and glistening white patches, which indicate dilated lacteals, may be noted. Endoscopy also provides a noninvasive route for obtaining intestinal biopsy specimens. A definitive diagnosis of lymphangiectasia and other types of PLE is made through histologic demonstration of characteristic mucosal lesions. In lymphangiectasia, these lesions include dilated, chyle-engorged lacteals and submucosal lymphatics. Often, mucosal edema is present. In some cases, lipogranulomas may be identified adjacent to intestinal and mesenteric lymphatics.

The potential for surgical dehiscence should be considered before full-thickness intestinal biopsy specimens are collected from patients with PLE. Thus, samples obtained by endoscopy or per oral suction biopsy capsule are preferred. Full-thickness biopsy specimens should be obtained only if a diagnosis cannot be made based on results of endoscopic examination and evaluation of biopsy specimens collected by that procedure.

ASSESS THE ANIMAL/Risk Factors Several breeds appear to be at risk for development of lymphangiectasia (**Table 22-22**). PLE often occurs in conjunction with a protein-losing nephropathy in soft-coated wheaten terriers.

ASSESS THE ANIMAL/Etiopathogenesis Normally, plasma proteins are lost into the GI lumen daily. This loss is attributed to protein leakage at the time of villous tip extrusion. Typically, these plasma proteins are reassimilated through digestive and absorptive processes. Certain GI disorders can disturb protein balance. Intestinal protein loss can be accelerated when the mucosal barrier is disrupted or disorders interfere with lym-

Table 22-35. Key nutritional factors for patients with lymphangiectasia/protein-losing enteropathy.*

Factors	Recommended levels
Crude fiber	0.5 to 15% (see text)
Digestibility	≥87% for protein and ≥90% for fat and soluble carbohydrate
Fat	<10% for dogs
	<15% for cats
Protein	>25% for dogs
	>35% for cats

*Nutrients expressed on a dry matter basis.

phatic drainage. Altered intestinal lymphatic drainage results in reflux of protein-rich lymph into the gut lumen. When the intestinal mucosa is damaged, excess protein can be lost through exudation or hemorrhage. Hypoproteinemia develops in either case after protein losses exceed compensatory synthesis.

Intestinal lymphangiectasia can arise as a primary disorder of the lymphatic system or secondary to chronic IBD. Severe inflammatory infiltrates and lipogranulomas can obstruct lymphatic drainage. Normally, the intestinal lymphatics transport absorbed fats from enterocytes to the venous circulation via the thoracic duct. Lacteals become distended with chyle if lymphaticovenous flow is impaired. Overdistended lacteals rupture and release intestinal lymph (containing protein, lymphocytes, fat and cholesterol) into the lumen. In some patients with primary lymphangiectasia, the lymphatic defects are not limited to the GI tract. In these animals, abnormal lymph flow may result in chylothorax, chylous abdominal effusions and subcutaneous chyle accumulations.

ASSESS THE ANIMAL/Key Nutritional Factors Key nutritional factors for patients with lymphangiectasia and other types of PLE are listed in **Table 22-35** and discussed in detail below.

ASSESS THE ANIMAL/Key Nutritional Factors/*Fat* The key factor in dietary management of primary and secondary lymphangiectasia is controlling dietary fat. In most pet food products, long-chain triglycerides (LCT) compose approximately 90% of dietary fat. After digestion and lymphatic absorption, LCT provide a major stimulus for intestinal lymph flow. LCT are absorbed as chylomicrons and are transported from the mucosal epithelium via lacteals to the thoracic duct and into the systemic circulation. LCT absorption increases both lymph protein content and lymph flow two- to threefold for four to six hours postprandially. The protein content of lymph tends to increase with dietary fat content. Limiting fat intake (i.e., <10% DMB for dogs and <15% DMB for cats) minimizes lymph flow, reduces lacteal and lymphatic distention and minimizes protein loss.

Unfortunately, foods with dietary fat restriction have a lower caloric density. Many animals with PLE are cachectic. Animals fed low-fat foods must consume larger volumes of food to meet caloric needs. If the patient continues to lose weight, adding medium-chain triglycerides (MCT) may be necessary. MCT are water-soluble, do not require micellarization for absorption, are absorbed directly across enterocytes into the portal vasculature and do not affect lymph flow. However, MCT oil does not contain the essential fatty acids required by dogs and cats, has been linked to hepatic lipidosis in cats, may decrease diet palatability and unabsorbed portions cause diarrhea.

ASSESS THE ANIMAL/Key Nutritional Factors/*Protein* Foods fed to animals with PLE should contain high biologic value proteins to replace depleted tissue proteins. Protein content should be adequate to meet the needs of the patient and should be tailored to the species and age of the animal. In general, in excess of 25% DM protein is recommended for dogs and in excess of 35% DM protein for cats. Feeding high-protein or all-meat foods without other dietary alterations has not been successful. If severe IBD is the underlying cause of PLE, the use of a low-fat, elimination food containing reduced numbers of highly digestible, novel protein sources should also be considered. (See the Inflammatory Bowel Disease section.)

ASSESS THE ANIMAL/Key Nutritional Factors/*Fiber* Foods containing high levels of fiber (>15% DM) are not routinely recommended for the dietary management of intestinal lymphangiectasia. Fiber binds digestive enzymes and bile acids, decreases pancreatic secretion of lipase and reduces pancreatic enzyme activity. Fiber, through these mechanisms, decreases intraluminal fat digestion and micelle formation, which selectively inhibits long-chain fatty acid absorption. Therefore, fiber may play a secondary role in reducing long-chain fatty acid absorption and decreasing lymphatic flow and subsequent lymph fluid losses.

However, moderate- (10 to 15% DM) and high-fiber (>15%) foods may be detrimental in managing patients with lymphangiectasia. MCT may need to be added to foods with low energy density for some patients to maintain good body condition. Low-fat (<10% DMB), low-fiber (<5% crude fiber DMB) veterinary therapeutic foods are available and may be of value in the management of primary and acquired lymphangiectasia. (See **Tables 13-3** and **13-4**.)

ASSESS THE ANIMAL/Key Nutritional Factors/*Digestibility* Feeding highly digestible (**Table 22-35**) foods provides several advantages for managing lymphangiectasia in dogs and cats. Nutrients in low-residue foods are more completely absorbed in the proximal gut. Furthermore, low-residue foods are associated with: 1) reduced osmotic diarrhea due to fat and carbohydrate malabsorption, 2) reduced production of intestinal gas due to carbohydrate malabsorption and 3) decreased antigen loads because smaller amounts of protein are absorbed intact.

ASSESS THE ANIMAL/Other Nutritional Factors Vitamin and mineral supplementation is rarely necessary when feeding commercially prepared foods. Dogs and cats usually have body stores of vitamins A, D, E and K to last several months. However, parenteral supplementation with fat-soluble vitamins may be needed if marked steatorrhea persists. Fat-soluble vitamin supplementation is warranted in cases of long-term fat malabsorption. It is simple and cost effective to administer 1 ml of a vitamin A, D and E product, divided into two intramuscular sites, which should supply fat-soluble vitamins for approximately three months.

Patients with fat malabsorption fed foods containing higher levels of fat may have increased divalent cation losses (i.e., calcium, magnesium, zinc and copper) because of intraluminal saponification. Calcium supplementation is generally not needed because serum calcium levels usually increase in conjunction with serum albumin concentrations. However, intravenous calcium supplementation should be instituted if hypocalcemic tetany develops. Supplementation with other minerals should also be based on evidence of deficiency rather than given pro forma.

Normal animals fed homemade foods are typically at increased risk for vitamin and calcium deficiencies; animals with PLE fed homemade foods are at even greater risk, unless the owner is exceptionally diligent about providing vitamin-mineral supplements.

ASSESS THE FOOD AND FEEDING METHOD Levels of key nutritional factors in foods currently fed to patients with lymphangiectasia or PLE should be evaluated and compared with recommended levels (**Table 22-35**). Key nutritional factors include fat, protein, fiber and digestibility. Information from this aspect of assessment is essential for making any changes to foods currently provided. Changing to a more appropriate food is indicated if key nutritional factors in the current food do not match recommended levels.

Because the feeding method is often altered for patients with lymphangiectasia and PLE, a thorough assessment should include verification of the feeding method currently being used. Items to consider include feeding frequency, amount fed, how the food is offered, access to other food and who feeds the animal. All of this information should have been gathered when the history of the animal was obtained.

Feeding Plan (Pages 766-767)

The goal of therapy for patients with lymphangiectasia or PLE is to decrease the enteric loss of plasma protein. In some cases, dietary manipulation alone is adequate. In others, concurrent medical management is necessary.

SELECT A FOOD A food containing low fat, moderate fiber and high carbohydrate and protein should be fed for at least several weeks. In general, fat should be restricted to less than 10% DMB for dogs and less than 15% DMB for cats; however, some cases may require levels as low as 5 to 7% DM fat. Commercially prepared foods formulated for weight control (**Tables 13-3** and **13-4**) or home-prepared foods (**Table 6-6**) may be suitable. MCT oil (9 kcal/ml) should be added to the diet to increase caloric density if patients are unable to maintain optimal body weight and condition. This supplement, however, should be used with caution, introduced gradually and should not exceed 25% of the caloric requirement. Dietary protein levels should exceed 25% DMB for dogs and 35% DMB for cats. Some cases of PLE may require additional protein. Dogs may be fed a low-fat (<10% DMB) cat food that has a higher protein (>35% DMB) content and nutrient density than comparable dog food. Protein may also be added in the form of cooked egg whites. Egg whites contain protein of the highest biologic value and are a useful supplement for some animals with PLE.

DETERMINE A FEEDING METHOD Initially, patients with lymphangiectasia or PLE should be fed multiple small meals per day as indicated by animal acceptance and tolerance for the food. Meal size can be increased as tolerated by the patient after clinical signs have been successfully managed for several weeks. In longstanding cases in which the patient is hospitalized and in poor body condition, the patient should be given a parenteral solution containing calories, protein and essential micronutrients. (See Chapter 12.) Calories can also be easily administered peripherally to dogs and cats using an isomolar 20% lipid solution piggybacked with standard fluid therapy at volumes sufficient to meet the patient's resting energy requirement (RER).

Concurrent Therapy (Page 767)

Immunosuppressive therapy as described for IBD is indicated when lymphangiectasia or PLE occurs as a consequence of mucosal inflammatory infiltrates. In addition to quieting the underlying enteric lesions, corticosteroid therapy has the added advantage of controlling the inflammatory lesions of lymphangiectasia, lymphangitis and lipogranulomas.

When hypoalbuminemia is severe, plasma or dextran infusions may be necessary to restore colloidal oncotic pressure. In general, aggressive nutri-

GI/PANCREAS

tional support will be more successful than plasma transfusions in restoring normoalbuminemia. Plasma transfusions may, however, benefit those patients with hypercoagulability resulting from panhypoproteinemia; plasma serves as a rich source of coagulation factors and antithrombin III.

Reassessment *(Page 767)*

Initially, patients with PLE should be reassessed weekly following discharge from the hospital. Each reexamination should include assessment of body weight and condition. Biweekly assessment of serum albumin and calcium concentrations and lymphocyte counts are useful. In addition, serial radiography can be used to assess the resolution of abdominal or thoracic effusion.

If the patient's condition is improving, dietary therapy should continue until the underlying enteropathy is resolved. Failing that, dietary manipulation should continue for the lifetime of the pet. Over time, it may be possible to increase dietary fat intake; however, this should be done cautiously and only for patients having difficulty maintaining ideal body weight and those manifesting evidence of essential fatty acid deficiency.

Short Bowel Syndrome (Pages 767-770)

For information about nutritional management of short bowel syndrome, see Small Animal Clinical Nutrition, 4th ed., pp 767-770.

Small Intestinal Bacterial Overgrowth (Pages 770-772)
Clinical Importance *(Page 770)*

Small intestinal bacterial overgrowth (SIBOG), a diarrheic disorder characterized by excessive numbers of small intestinal bacteria, has received much attention in recent years. Although the incidence of SIBOG is unknown, some authors have suggested that it is present in as many as 50% of dogs with chronic small bowel diarrhea.

Assessment *(Pages 770-772)*
ASSESS THE ANIMAL/History and Physical Examination Affected dogs usually present with a history of weight loss and intermittent small bowel diarrhea. Borborygmus and flatulence are also common complaints. Physical examination findings are often unremarkable. Poor body condition (BCS 1/5 or 2/5) and unthriftiness may be present if the condition is longstanding.

ASSESS THE ANIMAL/Laboratory and Other Clinical Information The gold standard for diagnosing SIBOG is quantitative aerobic and anaerobic culture of undiluted duodenal juice. Samples can be collected via endoscopy or direct needle aspiration at surgery. In dogs and cats, the small intestine normally contains a relatively sparse bacterial flora compared with the densely populated oral cavity and large bowel. Healthy dogs and cats may contain small intestinal bacterial counts in excess of 10^5 CFU/ml. See Small Animal Clinical Nutrition, 4th ed., p 771 for information about other diagnostic tests for SIBOG.

Response to therapy with antibiotics should not be overlooked as an effective diagnostic tool. A therapeutic trial may be particularly useful in situations in which quantitative cultures are not possible.

ASSESS THE ANIMAL/Risk Factors A number of risk factors have been identified for SIBOG. German shepherd dogs appear to be predisposed to this enteropathy possibly because of IgA deficiency. Exocrine pancreatic insufficiency is also a predisposing factor for SIBOG, and SIBOG can complicate management of exocrine pancreatic insufficiency.

Kenneled dogs (especially beagles) may be more likely to have duodenal fluid bacterial counts in excess of 10^5 CFU/ml. Potential causes for abnormal bacterial counts in kennel-housed dogs include environment (i.e., cleanliness), coprophagia and breed-specific characteristics (e.g., IgA deficiency).

ASSESS THE ANIMAL/Etiopathogenesis SIBOG can develop any time normal host defenses are impaired. Loss of gastric acid secretion, normal intestinal peristalsis and interdigestive ("housekeeper") motility, the ileocolic valve or local IgA production can result in SIBOG.

ASSESS THE ANIMAL/Key Nutritional Factors/*Digestibility* Feeding highly digestible (fat and soluble carbohydrate digestibility ≥90; protein digestibility ≥87%.) foods provides several advantages for managing dogs with SIBOG. Nutrients from these low-residue foods are more completely absorbed in the proximal gut. Highly digestible foods are also associated with reduced osmotic diarrhea due to fat and carbohydrate malabsorption and reduced production of intestinal gas due to carbohydrate malabsorption. The ideal food for SIBOG patients is lactose free to avoid the complication of lactose intolerance due to loss of brush border disaccharidases.

ASSESS THE ANIMAL/Key Nutritional Factors/*Fat* High-fat foods may contribute to osmotic diarrhea and GI protein losses, which complicate SIBOG. Therefore, it is often advantageous to initially provide a food that contains moderate levels of fat (12 to 15% DMB for dogs and 15 to 22% DMB for cats). Higher fat and energy denser foods can be offered if the patient tolerates these nutrient levels.

GI/PANCREAS

Table 22-36. Key nutritional factors for patients with small intestinal bacterial overgrowth.*

Factors	Recommended levels
Digestibility	≥90% for fat and soluble carbohydrate and ≥87% for protein
Fat	12 to 15% for dogs
	15 to 22% for cats

*Nutrients expressed on a dry matter basis.

ASSESS THE ANIMAL/Other Nutritional Factors FOS have been proposed for use in managing dogs with SIBOG. These nondigestible sugars are thought to promote beneficial bacteria at the expense of bacterial pathogens. When FOS was fed (1.0% as fed) to a group of German shepherd dogs with asymptomatic SIBOG, total bacterial counts were reduced within the duodenum. However, this reduction was smaller than the change in bacterial numbers demonstrated within the same dogs at different sampling intervals. Therefore, the clinical utility of FOS and other oligosaccharides in the treatment of SIBOG remains unproven. For more information, see Small Animal Clinical Nutrition 4th ed., p 772.

ASSESS THE FOOD AND FEEDING METHOD Levels of key nutritional factors should be evaluated in foods currently fed to patients with SIBOG and compared with recommended levels (**Table 22-36**). Key nutritional factors include food digestibility and fat content. Information from this aspect of assessment is essential for making any changes to foods currently provided. Changing to a more appropriate food is indicated if key nutritional factors in the current food do not match recommended levels.

Because the feeding method is often altered in patients with SIBOG, a thorough assessment should include verification of the feeding method currently being used. Items to consider include feeding frequency, amount fed, how the food is offered, access to other food and who feeds the animal. All of this information should have been gathered when the history of the animal was obtained. If the animal has a normal BCS (3/5), the amount of food previously fed was probably appropriate.

Feeding Plan (Page 772)

The feeding plan is often used in conjunction with other medical therapy. Underlying causes of SIBOG (e.g., partial intestinal obstruction) should be identified and treated before specific medical and dietary therapy is instituted. Antibiotic therapy is usually required for effective management of SIBOG. Antibiotic selection should be based on culture and antimicrobial sensitivity testing of specific pathogens identified in duodenal aspirates. Tetracycline or tylosin should be used if no pathogen is isolated.

Table 22-37. Key nutritional factors in selected highly digestible commercial foods for dogs with small intestinal bacterial overgrowth.*

Products	Fat	Digestibility		
		Prot	Fat	CHO
Recommended levels	**12-15**	**≥87**	**≥90**	**≥90**
Moist canine products				
Hill's Prescription Diet Canine i/d	14.0	Y	Y	Y
Leo Specific Digest CIW	14.7	na	na	na
Iams Eukanuba Low-Residue Adult/Canine	22.0	na	na	na
Medi-Cal Canine Gastro Formula	12.3	na	na	na
Purina CNM Canine EN-Formula	13.8	na	na	na
Select Care Canine Sensitive Formula	12.3	na	na	na
Waltham/Pedigree Canine Low Fat	6.9	na	na	na
Dry canine products				
Hill's Prescription Diet Canine i/d	13.6	Y	Y	Y
Iams Eukanuba Low-Residue Adult/Canine	10.6	na	na	na
Iams Eukanuba Low-Residue Puppy	22.3	na	na	na
Leo Specific Digest CID	14.3	na	na	na
Medi-Cal Canine Gastro Formula	11.7	na	na	na
Purina CNM Canine EN-Formula	11.7	Y	Y	Y
Select Care Canine Sensitive Formula	11.7	na	na	na
Waltham/Pedigree Canine Low Fat	5.4	na	na	na

Key: na = information not available from manufacturer, Prot = protein, CHO = soluble carbohydrate, Y = yes.
*This list represents products with the largest market share and for which published information is available. Manufacturers' published values. Nutrients expressed as % dry matter.

SELECT A FOOD Commercial veterinary therapeutic foods that are highly digestible and designed for patients with GI disease are recommended for patients with SIBOG (**Tables 22-37** and **22-38**). Many of these foods contain moderate levels of dietary fat. Young growing dogs with SIBOG should receive a food that meets the optimal levels of key nutritional factors for growth. (See Chapter 9.)

DETERMINE A FEEDING METHOD Ideally, patients with SIBOG should be fed multiple small meals per day as indicated by animal acceptance and tolerance for the food. Meal size can be increased as tolerated by the patient after clinical signs have been successfully managed for several weeks.

Reassessment (Page 772)

Owners of affected animals should be questioned regarding frequency of diarrhea, borborygmi and flatus. Body weight and condition should be evaluated frequently to assess resolution of malabsorption. In general, SIBOG can be managed effectively with a combination of medical (e.g., antibiotics) and nutritional therapy.

Table 22-38. Key nutritional factors in selected highly digestible commercial foods for cats with small intestinal bacterial overgrowth.*

Products	Fat	Digestibility		
		Prot	Fat	CHO
Recommended levels	15-22	≥87	≥90	≥90
Moist/semi-moist feline products				
Hill's Prescription Diet Feline i/d	20.1	Y	Y	Y
Iams Eukanuba Low-Residue Adult/Feline	16.9	na	na	na
Leo Specific Dermil FDW	34.7	na	na	na
Medi-Cal Feline HYPOallergenic/Gastro Formula	25.6	na	na	na
Purina CNM Feline EN-Formula	17.0	Y	Y	Y
Waltham/Whiskas Feline Selected Protein	29.0	na	na	na
Dry feline products				
Hill's Prescription Diet Feline i/d	20.1	Y	Y	Y
Iams Eukanuba Low-Residue Adult/Feline	15.4	na	na	na
Medi-Cal Feline HYPOallergenic/Gastro Formula	13.6	na	na	na
Select Care Feline Neutral Formula	13.6	na	na	na
Waltham/Whiskas Feline Selected Protein	12.2	na	na	na

Key: na = information not available from manufacturer, Prot = protein, CHO = soluble carbohydrate, Y = yes.
*This list represents products with the largest market share and for which published information is available.
Manufacturers' published values. Nutrients expressed as % dry matter.

LARGE INTESTINAL DISORDERS (Pages 772-784)

Colitis (Pages 772-776)
Clinical Importance (Pages 772-773)

Colitis is a common disorder of dogs and cats. A number of infectious, toxic, inflammatory and dietary factors can trigger the sudden onset of large bowel diarrhea (**Tables 22-39** and **22-40**).

Currently, IBD is thought to be the most common cause of chronic large bowel diarrhea in dogs and cats. The generic term, IBD, encompasses lymphoplasmacytic enterocolitis, lymphocytic enterocolitis, eosinophilic enterocolitis, segmental granulomatous enterocolitis, suppurative enterocolitis and histiocytic colitis. Specific types are categorized based on the type of inflammatory cells found in the lamina propria. The lymphoplasmacytic form probably occurs most commonly. The severity of the condition varies from relatively mild clinical signs to life-threatening PLE. The boxer breed may present with an especially severe variant termed histiocytic or ulcerative colitis.

Assessment (Pages 773-775)
ASSESS THE ANIMAL/History and Physical Examination The most common clinical sign in dogs and cats with acute or chronic colitis is large

Table 22-39. Potential causes of acute large bowel diarrhea in dogs and cats.

Dietary	
Dietary indiscretion	Garbage toxicity
Foreign bodies	
Drugs	
Cyclophosphamide	Doxorubicin
Infectious agents	
Bacteria	Parasites
Campylobacter spp	*Giardia lamblia*
Clostridium spp	*Trichuris vulpis*
Salmonella spp	
Viruses	
Panleukopenia	
Parvovirus	
Miscellaneous	Hemorrhagic gastroenteritis

Table 22-40. Potential causes of chronic large bowel diarrhea in dogs and cats.

Infectious causes	
Parasitic	Bacteria
Giardia lamblia	*Campylobacter* spp
Trichuris vulpis	*Salmonella* spp
Viral	Fungal
Feline immunodeficiency virus	Histoplasmosis
Feline leukemia virus	Pythiosis
Inflammatory bowel disease	
Eosinophilic colitis	Lymphoplasmacytic colitis
Lymphocytic colitis	Suppurative colitis
Dietary (adverse reactions to food)	
Food allergy (hypersensitivity)	Food intolerance
Neoplasia	
Adenocarcinoma	Lymphosarcoma
Adenoma/polyps	Mast cell tumor

bowel diarrhea characterized by tenesmus, dyschezia, urgency and passage of mucus and blood (**Table 22-20**). Clinical signs may be intermittent or persistent. The clinical signs tend to increase in frequency and intensity as colitis progresses. The presence of systemic signs is also variable. Some animals present with a history of depression, malaise and inappetence; however, most are alert and active when examined. Hemorrhagic stools indicate a potentially life-threatening disorder (**Table 22-21**).

When evaluating colitis cases, careful attention should be paid to the dietary history. Diet-induced diarrhea is common; a recent change to a moist high-fat or meat-based food may be the source of the animal's diarrhea.

Often, it is possible to elicit a history of dietary indiscretion, feeding table foods over a holiday or access to garbage, carrion or abrasive materials.

Other husbandry issues are also important (e.g., Records of anthelmintic treatments should be scrutinized.). The likelihood that an infectious organism is involved is increased if other animals or people in the household are similarly affected.

Dogs and cats with acute colitis may be depressed and dehydrated and may exhibit pain on abdominal palpation. Patients should be carefully evaluated for evidence of septic shock. Those animals with systemic signs of illness (i.e., fever and congested mucous membranes) in addition to GI signs should be treated more aggressively.

Physical examination findings vary in dogs and cats with chronic colitis. Many animals have no abnormalities. Rarely, dogs and cats with colitis present with weight loss and poor body condition. In such cases, serious infiltrative colonic disorders (e.g., histoplasmosis, neoplasia or histiocytic colitis) should be suspected.

On occasion, thickened loops of bowel may be detected by abdominal palpation, especially in cats. Segmental thickening of bowel is consistent with eosinophilic gastroenterocolitis in cats and granulomatous enteritis in dogs. This finding should also be distinguished from intussusceptions, foreign bodies, histoplasmosis and neoplastic lesions.

ASSESS THE ANIMAL/Laboratory and Other Clinical Information Because there are many potential causes of acute colitis, achieving a definitive diagnosis can be difficult. In acute cases, it is most important to determine whether the animal's condition is self-limiting or potentially life-threatening. This determination, based on historical and physical findings is critical. Some factors suggest a potentially life-threatening condition (**Table 22-21**). Cases of a serious nature should be pursued aggressively with diagnostics (i.e., hematology, serum biochemistry profiles, urinalyses and fecal examinations for parasites and other infectious pathogens). Self-limiting cases are usually approached more conservatively. Diagnostics are often limited to assessing hydration status (i.e., packed cell volume, total protein concentration and body weight) and thorough examination of feces for parasites and bacterial pathogens (e.g., spores of *Clostridium* spp).

Laboratory findings in patients with chronic colitis are often nonspecific. Hematologic findings are variable and may include blood loss anemia, anemia of chronic disease and/or eosinophilia. Serum biochemistry profiles and urinalyses should be performed on samples from patients with chronic diarrhea to assess the systemic affect of the GI disorder and to rule out concurrent disease. Electrolyte abnormalities, including hypokalemia, may be identified. Hypoproteinemia and hypoalbuminemia may be recognized in severe cases of PLE. Dehydrated patients may have prerenal azotemia.

Table 22-41. Breed-associated colonic disorders.

Disorders	Breeds
Flatulence	Brachycephalic dogs and cats
Hemorrhagic gastroenteritis	Dachshund, miniature schnauzer, toy poodle
Irritable bowel syndrome	Working breeds, toy breeds
Ulcerative colitis	Boxer, French bulldog

Fecal examinations are very important in the evaluation of patients with chronic large bowel diarrhea. Multiple fecal parasite examinations using concentration techniques are necessary to rule out parasitism. Endoscopic abnormalities in chronic colitis may include mucosal granularity, hyperemia, increased friability and inability to visualize colonic submucosal blood vessels. Multiple biopsy specimens should be collected from multiple bowel segments. Even if these areas appear normal endoscopically, histologic changes may still be present. The definitive diagnosis of IBD is based on histopathologic examination of endoscopic or surgical biopsy specimens.

ASSESS THE ANIMAL/Risk Factors The risk factors for acute colitis include age, breed, immune status and environment. Young animals are more susceptible to a variety of infectious pathogens including parasites, viruses and bacteria. Likewise, immunocompromised animals are at risk for contracting viral and bacterial enteritides. Hospitalization and administration of cancer chemotherapeutic drugs are associated with nosocomial infection with *Clostridium* and *Campylobacter* species.

Environment also plays an important role in exposure to pathogens. Dogs and cats kept in unsanitary or overcrowded conditions are much more likely to develop infectious enteropathics. In addition, animals kept in poorly controlled environments have a higher risk for exposure to high-fat table foods, garbage and toxins. Dogs in particular eat indiscriminately. Consumption of rotten garbage, decomposing carrion and abrasive materials (e.g., hair, bones, rocks, plastic, aluminum foil) can result in severe colitis. Poor husbandry practices including inadequate parasite control and overcrowding also put pets at risk for acute colitis.

There does not appear to be an age or gender predisposition for any of the forms of IBD. Certain breeds appear to be at risk for specific colonic disorders (**Table 22-41**). For example, the boxer breed is linked to histiocytic colitis. Other breeds at risk for chronic inflammatory colonopathies include German shepherd dogs and French bulldogs.

ASSESS THE ANIMAL/Etiopathogenesis In acute colitis, diarrhea may occur as a result of altered gut permeability or osmotic mechanisms. In

Table 22-42. Key nutritional factors for patients with acute or chronic colitis.*

Factors	Recommended levels
Chloride	0.50 to 1.30%
Digestibility	≥87% for protein and ≥90% for fat and soluble carbohydrate
Fat	12 to 15% for dogs
	15 to 22% for cats
Fiber	0.5 to 15% (see text)
Potassium	0.80 to 1.10%
Protein	Consider elimination foods or protein hydrolysates
Sodium	0.35 to 0.50%

*Nutrients expressed on a dry matter basis.

addition, many of the bacterial pathogens elaborate enterotoxins that serve as potent secretogogues.

Histiocytic colitis, also termed ulcerative or boxer colitis, is characterized by infiltration of the lamina propria with PAS-positive histiocytes. Some authors have suggested that the presence of these macrophages indicates an infectious etiology. However, to date no organisms have been consistently identified in tissues from affected animals.

ASSESS THE ANIMAL/Key Nutritional Factors Key nutritional factors for patients with acute or chronic colitis are listed in **Table 22-42** and discussed in more detail below.

ASSESS THE ANIMAL/Key Nutritional Factors/*Water* Water is the most important nutrient in patients with acute large bowel diarrhea because of the potential for life-threatening dehydration due to excessive fluid losses and inability of the patient to replace those losses. Moderate to severe dehydration should be corrected with appropriate parenteral fluid therapy rather than using the oral route.

ASSESS THE ANIMAL/Key Nutritional Factors/*Minerals* Potassium depletion is a predictable consequence of severe and chronic enteric diseases because the potassium concentration of intestinal secretions is high. Hypokalemia in association with colitis will be particularly profound if losses are not matched by sufficient dietary intake of potassium.

Electrolyte disorders should be corrected initially with appropriate parenteral fluid and electrolyte therapy. Foods for patients with acute gastroenteritis should contain levels of sodium, chloride and potassium above the minimum allowances for normal dogs and cats. Recommended levels of these nutrients are 0.35 to 0.50% DM sodium, 0.50 to 1.30% DM chloride and 0.80 to 1.10% DM potassium.

ASSESS THE ANIMAL/Key Nutritional Factors/*Fat* Compared with the processes involved with other macronutrients, fat digestion and absorption are relatively complex and may be disrupted in patients with GI disease. The action of bacterial flora on unabsorbed fats in the colon resulting in hydroxy fatty acid production is an important cause of large bowel diarrhea. Thus, foods for patients with colitis and many other GI diseases often contain moderate amounts of fat (i.e., 12 to 15% DMB for dogs and 15 to 22% DMB for cats). However, dogs and cats digest fat very efficiently and the process is rarely disrupted except in malassimilative disorders. Therefore, colitis patients can be fed foods containing higher concentrations of fat when greater caloric density is required.

ASSESS THE ANIMAL/Key Nutritional Factors/*Protein* Protein should be provided at levels sufficient for the appropriate lifestage of colitis patients unless a protein-losing enteropathy is present. High biologic value, highly digestible (≥87%) protein sources are preferred.

Some authors recommend the use of elimination foods because of the suspected role of dietary antigens in the pathogenesis of chronic colitis. In some cases, elimination foods may be used successfully without pharmacologic intervention. Mild to moderate lymphoplasmacytic and eosinophilic colitis are the forms most likely to respond to dietary management. Chapter 14 discusses elimination foods and protein hydrolysates in more detail. The suspected pathogenesis of IBD involves an increase in gut permeability; therefore, the use of "sacrificial" dietary antigens has been suggested in the treatment of IBD. For more information, see Small Animal Clinical Nutrition, 4th ed., p 762.

ASSESS THE ANIMAL/Key Nutritional Factors/*Digestibility* Feeding highly digestible (**Table 22-42**) foods provides several advantages for managing dogs and cats with longstanding inflammatory colitis. Nutrients from low-residue foods are more completely absorbed from the proximal gut. Low-residue foods are associated with: 1) reduced osmotic diarrhea due to fat and carbohydrate malabsorption, 2) reduced production of intestinal gas due to carbohydrate malabsorption and 3) decreased antigen loads because smaller amounts of protein are absorbed intact.

ASSESS THE ANIMAL/Key Nutritional Factors/*Fiber* Dietary fiber predominantly affects the large bowel of dogs and cats. Beneficial effects of dietary fiber include: 1) normalizing colonic motility and transit time, 2) buffering toxins in the GI lumen, 3) binding or holding excess water, 4) supporting growth of normal GI microflora, 5) providing fuel for colonocytes and 6) altering viscosity of GI luminal contents.

Various types and levels of dietary fiber have been advocated for use in patients with colitis. Some veterinarians recommend very low-fiber foods

(<1% DM crude fiber) to enhance dry matter digestibility and reduce quantities of ingesta presented to the colon. Small amounts (1 to 5% DM crude fiber) of a mixed (i.e., soluble/insoluble) fiber type can also be used in conjunction with a highly digestible food. Other authors have had success using moderate levels (10 to 15% DM crude fiber) to high levels (>15% DM crude fiber) of insoluble fiber. All three strategies have been used successfully in managing patients with colitis.

Some authors have suggested that feeding insoluble or slowly fermentable fibers is detrimental to the management of colonopathies; these suggestions are based on the results of a small, uncontrolled feeding trial comparing cellulose-containing foods with foods containing beet pulp. However, larger, controlled trials incorporating pre-study histopathology and electron microscopic examination of tissues have not identified any negative effects of slowly fermentable fiber on the colon. In fact, many clinicians select foods enhanced with insoluble fiber as their first food option in the management of acute and chronic colitis.

Several substrates including beet pulp, soy fiber, inulin and FOS have been demonstrated by in vitro fermentation to produce volatile fatty acids that may be beneficial in inflammatory colonopathies. Manufacturers of commercial products usually incorporate these fibers at 1 to 5% DMB.

ASSESS THE ANIMAL/Other Nutritional Factors/*Acid Load* Acidemia is common in patients with acute large bowel diarrhea because fluid secreted in the caudal small intestine and large intestine contains bicarbonate concentrations higher than those in plasma and sodium in excess of chloride ions. Hypovolemia (i.e., severe dehydration) compounds the acidosis in some patients. Severe acid-base disorders are best corrected with appropriate parenteral fluid therapy. Foods for patients with colitis should normally produce an alkaline urinary pH. These foods preferably contain buffering salts such as potassium gluconate or calcium carbonate.

ASSESS THE ANIMAL/Other Nutritional Factors/*N-3 Fatty Acids* N-3 fatty acids derived from fish oil or other sources may have a beneficial effect in controlling mucosal inflammation in patients with chronic inflammatory colitis. There is some clinical evidence that dietary n-3 fatty acid supplementation may modulate the generation and biologic activity of inflammatory mediators. See Chapter 26 and the earlier section about IBD for additional information.

ASSESS THE ANIMAL/Other Nutritional Factors/*Vitamins* Folic acid supplementation is recommended for patients receiving long-term sulfasalazine therapy.

ASSESS THE FOOD AND FEEDING METHOD Levels of key nutritional factors in foods currently fed to patients with colitis should be evaluated and compared with recommended levels (**Table 22-42**). Information from this aspect of assessment is essential for making any changes to foods currently provided. Changing to a more appropriate food is indicated if key nutritional factors in the current food do not match recommended levels.

A thorough assessment should include verification of the feeding method currently being used. Items to consider include feeding frequency, amount fed, how the food is offered, access to other food and who feeds the animal. All of this information should have been gathered when the history of the animal was obtained. In cases in which colitis is caused by exposure to garbage or inappropriate amounts or types of foods, avoiding foods other than the pet's regular food is recommended and will often prevent further occurrences. If the animal has a normal BCS (3/5), the amount of food previously fed (energy basis) was probably appropriate.

Feeding Plan (*Pages 775-776*)

Initially, the objectives for managing acute colitis should be to correct dehydration and electrolyte, glucose and acid-base imbalances, if present. Medical therapy may include antibiotics, anthelmintics or motility modifying agents (e.g., loperamide).

The dietary goal is to provide a food that meets the patient's nutrient requirements and allows normalization of colonic motility and function, and fecal water balance. In most cases of acute large bowel diarrhea, initial fasting for 24 to 48 hours, with access to water, either reduces or resolves the diarrhea by simply removing the effects of unabsorbed food and offending agents from the colon. Often, the patient's previous food can be gradually reintroduced over several days.

In chronic colitis, dietary intervention should be aimed at controlling clinical signs while providing adequate nutrients to meet requirements and compensate for ongoing losses through the GI tract. Optimal management of some dogs and cats with chronic colitis may require only dietary manipulation. In other cases, dietary therapy is better used in concert with pharmacologic agents. Antibiotics (metronidazole, tylosin), anthelmintics, antiinflammatory agents (sulfasalazine) and immunosuppressive agents (corticosteroids, azathioprine, cyclophosphamide) are often used to manage chronic colitis. Lifelong dietary therapy is often required to control clinical signs in longstanding colitis cases.

SELECT A FOOD Withholding food for one to two days and then reintroducing a homemade or commercial veterinary therapeutic GI-type or fiber-enhanced food is often palliative in managing acute colitis. After feeding the highly digestible or fiber-enhanced food for another three to four days, the pet's regular food may be reintroduced over another three-day period. Further workup is recommended if colitis recurs when the regular food is reintroduced.

Three types of food can be used to manage chronic colitis and they may be attempted in any order: 1) highly digestible, low-residue foods formulated for GI disease, 2) fiber-enhanced foods and 3) elimination foods. There is no physical examination finding, laboratory test or historical fact to predict which method will be successful in any one patient. Dietary trials are often needed to determine which food type works best for individual patients.

In chronic colitis, one option is to feed a highly digestible, low-residue food with moderate levels of fat and minerals. A variety of low-residue commercial veterinary therapeutic foods are available (**Tables 22-43** and **22-44**); alternatively, homemade foods can be prepared (**Table 6-9**). Foods for puppies and kittens should also meet the nutritional requirements for growth.

Another approach in chronic colitis is to increase dietary fiber content to normalize intestinal motility, water balance and microflora. Fiber has several physiologic characteristics that aid in managing large bowel diarrhea. Moderate levels of dietary fiber (10 to 15% DMB) add nondigestible bulk that buffers toxins, holds excess water and, perhaps more important, provides intraluminal stimuli to reestablish the coordinated actions of hormones, neurons, smooth muscles, enzyme delivery, digestion and absorption. Fiber normalizes transit time through the large bowel. **Tables 22-45** and **22-46** list selected fiber-enhanced commercial foods.

A third option in chronic colitis is to use an elimination food with a limited number of highly digestible, novel protein sources (**Tables 14-3** and **14-4**) or a protein hydrolysate (**Table 14-5**). Commercial veterinary therapeutic foods and homemade foods that contain novel protein sources are often formulated from lamb, rabbit, venison, duck or fish and a highly digestible or unusual carbohydrate source. All other possible dietary sources of protein and carbohydrate should be eliminated including treats, snacks, table foods, vitamin-mineral supplements and chewable/flavored medications. (See Chapter 14.)

DETERMINE A FEEDING METHOD The feeding method for patients with acute colitis begins by withholding all oral intake of food for 24 to 48 hours. After this period, animals should be offered small amounts of food several times (i.e., six to eight times) a day. If the pet tolerates food without a recurrence of diarrhea, the amount fed can be increased over three to four days until the animal is receiving its estimated DER in two to three meals per day.

Initially, chronic colitis patients should be fed multiple small meals per day as indicated by animal acceptance and tolerance for the food. Meal

Table 22-43. Key nutritional factors in selected low-residue commercial foods for dogs with acute or chronic colitis.* (See **Table 22-45** if more fiber is desired; see **Tables 14-3** and **14-5** if foods with novel protein sources or protein hydrolysates are desired.)

Products Recommended levels	Fat 12-15	Fiber 0.5-5.0	Potassium 0.8-1.1	Sodium 0.35-0.5	Chloride 0.5-1.3
Moist canine products					
Hill's Prescription Diet Canine i/d	14.0	0.7	0.92	0.44	1.23
Leo Specific Digest CIW	14.7	4.2	na	na	na
Iams Eukanuba Low-Residue Adult/Canine	22.0	2.2	1.00	0.75	1.00
Medi-Cal Canine Gastro Formula	12.3	1.7	na	na	na
Purina CNM Canine EN-Formula	13.8	0.9	0.61	0.37	0.78
Select Care Canine Sensitive Formula	12.3	1.7	na	na	na
Waltham/Pedigree Canine Low Fat	6.9	2.4	na	na	na
Dry canine products					
Hill's Prescription Diet Canine i/d	13.6	1.2	0.92	0.46	1.10
Iams Eukanuba Low-Residue Adult/Canine	10.6	2.1	0.89	0.46	1.13
Iams Eukanuba Low-Residue Puppy	22.3	1.9	0.78	0.42	0.85
Leo Specific Digest CID	14.3	1.1	na	na	na
Medi-Cal Canine Gastro Formula	11.7	2.0	na	na	na
Purina CNM Canine EN-Formula	11.7	1.1	0.59	0.36	0.87
Select Care Canine Sensitive Formula	11.7	2.0	na	na	na
Waltham/Pedigree Canine Low Fat	5.4	0.4	na	na	na

Key: na = information not available from manufacturer.
*This list represents products with the largest market share and for which published information is available.
Manufacturers' published values. Nutrients expressed as % dry matter.

GI/PANCREAS

Table 22-44. Key nutritional factors in selected low-residue commercial foods for cats with acute or chronic colitis.*
(See **Table 22-46** if more fiber is desired; see **Tables 14-4** and **14-5** if foods with novel protein sources or protein hydrolysates are desired.)

Products	Fat	Fiber	Potassium	Sodium	Chloride
Recommended levels	15-22	0.5-5.0	0.8-1.1	0.35-0.5	0.5-1.3
Moist/semi-moist feline products					
Hill's Prescription Diet Feline i/d	20.1	1.6	1.08	0.36	1.04
Iams Eukanuba Low-Residue Adult/Feline	16.9	2.2	na	na	na
Leo Specific Dermil FDW	34.7	1.5	na	na	na
Medi-Cal Feline HYPOallergenic/ Gastro Formula	25.6	1.65	na	na	na
Purina CNM Feline EN-Formula	17.0	1.1	0.80	0.32	0.55
Waltham/Whiskas Feline Selected Protein	29.0	2.9	na	na	na
Dry feline products					
Hill's Prescription Diet Feline i/d	20.1	1.3	0.98	0.36	1.03
Iams Eukanuba Low-Residue Adult/Feline	15.4	2.2	0.98	0.32	0.88
Medi-Cal Feline HYPOallergenic/ Gastro Formula	13.6	1.6	na	na	na
Select Care Feline Neutral Formula	13.6	1.6	na	na	na
Waltham/Whiskas Feline Selected Protein	12.2	3.3	na	na	na

Key: na = information not available from manufacturer.
*This list represents products with the largest market share and for which published information is available.
Manufacturers' published values. Nutrients expressed as % dry matter.

Table 22-45. Key nutritional factors in selected increased fiber commercial foods for dogs with acute or chronic colitis.* (See **Table 22-43** if less fiber is desired; see **Tables 14-3** and **14-5** if foods with novel protein sources or protein hydrolysates are desired.)

Products	Fat 12-15	Fiber 10-15	Potassium 0.8-1.1	Sodium 0.35-0.5	Chloride 0.5-1.3
Recommended levels					
Moist canine products					
Hill's Prescription Diet Canine r/d	8.4	21.8	0.83	0.25	0.50
Hill's Prescription Diet Canine w/d	13.0	12.2	0.62	0.27	0.73
Leo Specific CRW	8.9	17.8	na	na	na
Medi-Cal Canine Fibre Formula	7.9	15.9	na	na	na
Medi-Cal Canine Weight Control/Geriatric	10.8	6.1	na	na	na
Purina CNM Canine OM-Formula	8.4	19.2	1.06	0.28	0.51
Select Care Canine Hifactor Formula	9.1	15.0	na	na	na
Waltham/Pedigree Canine High Fiber	8.0	10.3	na	na	na
Dry canine products					
Hill's Prescription Diet Canine r/d	8.5	23.4	0.82	0.30	0.43
Hill's Prescription Diet Canine w/d	8.7	16.9	0.68	0.22	0.54
Leo Specific CRD	5.6	8.9	na	na	na
Medi-Cal Canine Fibre Formula	9.8	15.8	na	na	na
Medi-Cal Canine Weight Control/Geriatric	8.5	5.9	na	na	na
Purina CNM Canine DCO-Formula	12.4	7.6	0.70	0.34	0.82
Purina CNM Canine OM-Formula	6.7	10.7	0.82	0.21	0.32
Select Care Canine Hifactor Formula	10.6	14.3	na	na	na
Waltham/Pedigree Canine High Fiber	8.2	4.9	1.09	0.33	1.09

Key: na = information not available from manufacturer.
*This list represents products with the largest market share and for which published information is available.
Manufacturers' published values. Nutrients expressed as % dry matter.

GI/PANCREAS

Table 22-46. Key nutritional factors in selected increased fiber commercial foods for cats with acute or chronic colitis.* (See Table 22-44 if less fiber is desired; see Tables 14-4 and 14-5 if foods with novel protein sources or protein hydrolysates are desired.)

Products	Fat 15-22	Fiber 10-15	Potassium 0.8-1.1	Sodium 0.35-0.5	Chloride 0.5-1.3
Recommended levels					
Moist/semi-moist feline products					
Hill's Prescription Diet Feline r/d	8.9	17.4	0.72	0.30	0.76
Hill's Prescription Diet Feline w/d	16.7	10.7	0.87	0.44	1.07
Leo Specific FRW	9.3	18.7	na	na	na
Medi-Cal Feline Fibre Formula	14.4	8.6	na	na	na
Medi-Cal Feline Weight Control	17.7	4.7	na	na	na
Select Care Feline Hifactor Formula	16.1	7.2	na	na	na
Dry feline products					
Hill's Prescription Diet Feline r/d	9.1	13.7	0.71	0.30	0.67
Hill's Prescription Diet Feline w/d	9.3	7.9	0.74	0.27	0.67
Leo Specific FRD	7.1	19.0	na	na	na
Medi-Cal Feline Fibre Formula	13.8	7.3	na	na	na
Medi-Cal Feline Weight Control	12.3	4.7	na	na	na
Purina CNM Feline OM-Formula	8.5	8.0	0.69	0.26	0.88
Select Care Feline Hifactor Formula	12.9	5.0	na	na	na
Select Care Feline Weight Formula	12.2	3.6	na	na	na

Key: na = information not available from manufacturer.
*This list represents products with the largest market share and for which published information is available.
Manufacturers' published values. Nutrients expressed as % dry matter.

size can be increased and meal frequency can be decreased as tolerated by the patient after clinical signs have been successfully managed for several weeks.

Reassessment (Page 776)

The prognosis for recovery in most cases of acute colitis is good. Bouts of acute colitis often resolve within two to four days with conservative medical and nutritional management. Body weight should be recorded daily until recovery is complete. Changes in body weight from day to day usually reflect changes in hydration status rather than loss or gain of lean or adipose tissue. Further diagnostic testing is warranted if vomiting or severe diarrhea persists.

Weekly recordings of body weight and condition and stool evaluations are useful in assessing patients with chronic colitis. Regaining or maintaining optimal body weight and condition, normal activity level, a positive attitude and absence of clinical signs are measures of successful dietary/medical management. The feeding method and amount fed can be adjusted as needed to maintain body weight and condition. Additional medical therapies should be considered if dietary therapy alone fails to improve stool quality and maintain body weight.

Dogs and cats presenting with multiple or recurrent episodes of large bowel diarrhea require further diagnostic workup and, most probably, a combination of dietary and medical therapies; however, parasitic causes (e.g., whipworms) should be ruled out or treated empirically before pursuing further diagnostics.

Irritable Bowel Syndrome in Dogs (Pages 776-778)

Clinical Importance (Page 776)

Irritable bowel syndrome (IBS) is a poorly defined functional bowel disorder thought to be caused by GI dysmotility. IBS, also called spastic colitis or nervous colon, accounts for 5 to 17% of large bowel disorders in dogs. IBS has not been recognized in cats.

Assessment (Pages 776-777)

ASSESS THE ANIMAL/History and Physical Examination Dogs with IBS have chronic, intermittent bouts of diarrhea that are predominantly large bowel in character. Frequent, small-volume, fluid stools containing mucus are reported. Occasionally, explosive bouts of diarrhea and flatus may occur, often in association with abdominal pain. Dyschezia, tenesmus and, rarely, hematochezia may be seen. Some dogs have abdominal pain that is relieved by eating, eructation or defecation. Borborygmus, belching and flatus are frequent complaints in IBS. Typically, signs are variable and may change from bout to bout.

In many cases, GI signs can be linked to identifiable stressors. A thorough history may elicit such stress-causing variables as showing, work, owner anxiety, boarding in a kennel and changes in the home environment (e.g., new spouse, child, pet, house or apartment).

Generally, dogs with IBS are in good physical condition and do not exhibit the weight loss or poor body condition often associated with organic GI disorders. Affected dogs may exhibit discomfort during abdominal palpation if examined during an acute episode of GI distress. Rectal examination may reveal mucoid stools.

ASSESS THE ANIMAL/Laboratory and Other Clinical Information Results of routine laboratory tests are usually within normal limits. The diagnosis of IBS is applied to those dogs with the clinical signs and history described above in which other, more common organic causes have been ruled out. Radiography and colonoscopy are rarely useful in the diagnosis of IBS; findings are usually within normal limits.

ASSESS THE ANIMAL/Risk Factors As discussed above, psychological and physical stress appears to play an important role in IBS. Dogs with nervous, excitable temperaments and behavioral disorders such as separation anxiety seem predisposed to IBS. Working and toy-breed dogs are more commonly affected.

ASSESS THE ANIMAL/Etiopathogenesis The etiology of IBS is not clearly defined. The relationship of stress to the myoelectric and motility abnormalities present in IBS is not completely understood. However, psychological stress can trigger hypermotility.

ASSESS THE ANIMAL/Key Nutritional Factors/*Fiber* Reports suggest that many patients with IBS improve clinically when the fiber content of the diet is increased. Increasing dietary fiber alters fecal water content, colonic motility and intestinal transit rate, all of which may benefit patients with IBS. Either fiber sources can be added to typical foods or the patient can be fed fiber-enhanced veterinary therapeutic foods (**Table 22-45**) to increase dietary fiber intake. High-fiber breakfast cereals can be used to increase the patient's fiber intake. Fermentable fibers such as psyllium husks (e.g., Metamucil, Proctor & Gamble, Cincinnati, OH, USA) can be added to a typical or GI food at a rate of 1 to 6 tsp per meal. Soluble fiber typically improves stool quality, but does not alter the underlying pathophysiologic and motility abnormalities thought to be involved in IBS. Commercial foods containing small amounts of soluble fiber (1 to 5%) or moderate amounts of insoluble fiber (i.e., 10 to 15%) have been used successfully.

ASSESS THE FOOD AND FEEDING METHOD Levels of crude fiber should be evaluated in foods currently fed to patients with IBS. Information from this aspect of assessment is essential for making any changes to foods currently provided. Changing to a more appropriate food is indicated if levels of crude fiber in the food currently provided do not match recommended levels.

Because the feeding method may be altered in patients with IBS, a thorough assessment should include verification of the feeding method currently being used. Items to consider include feeding frequency, amount fed, how the food is offered, access to other food and who feeds the animal. All of this information should have been gathered when the history of the animal was obtained. If the animal has a normal BCS (3/5), the amount of food previously fed (energy basis) was probably appropriate.

Feeding Plan (Pages 777-778)

The feeding plan should be used in conjunction with efforts to decrease the frequency of events that seem to trigger the problem. Reasonable efforts should be made to identify and eliminate specific stressors. Psychotropic and GI antispasmodic agents may be beneficial in some cases. The feeding plan alone may not eliminate the problem but it may reduce the frequency and severity of clinical signs.

SELECT A FOOD IBS is managed most successfully by feeding a food with increased levels of insoluble fiber. Most commercial grocery and specialty brand dog foods contain less than 5% DM crude fiber. Additional fiber can be added to the diet by using fiber supplements or commercial fiber-enhanced foods. It is prudent to increase the fiber concentration in increments of 5% per week until clinical signs improve or resolve. High levels of fiber added to regular pet foods may make the food unpalatable and unbalanced. Alternatively, there are many balanced commercial pet food products offering a wide variety of fiber combinations and concentrations (**Table 22-45**). Some clinicians successfully manage IBS by gradually adding psyllium husk fiber to highly digestible GI foods. Recommended doses range from 1 to 6 tsp of psyllium husk fiber per meal.

DETERMINE A FEEDING METHOD Foods other than those determined to control the clinical signs should be strictly avoided for dogs in which recurring bouts are initiated by diet changes or exposure to garbage or table foods. Feeding once or twice a day is usually sufficient. Feeding three to four meals per day may be necessary in some cases to minimize the amount of ingesta passing into the large bowel at one time.

GI/PANCREAS

Reassessment *(Page 778)*

Regular body weight and condition assessment and stool evaluations are useful for monitoring patients with IBS. Well-compensated patients should be evaluated immediately if a change or decline in condition is noted. Maintaining optimal body weight and condition, normal activity level, a positive attitude and absence of clinical signs are measures of successful dietary and medical management. The feeding method and amount fed can be adjusted as needed to maintain body weight and condition. Additional medical therapy should be considered if dietary therapy alone is not sufficient to improve stool quality and maintain body weight and condition.

Constipation/Megacolon (Pages 778-782)
Clinical Importance (Page 778)

Classically, the term constipation is applied to those patients that pass stools infrequently or exhibit tenesmus in association with defecation. Because it is difficult to obtain accurate information about the defecation habits of pets, constipation is difficult to define in veterinary patients. Constipation is a clinical sign, not a disease; multiple disorders can result in constipation.

Obstipation is severe constipation that requires medical therapy for relief. Megacolon refers to dilatation of the colon and is usually seen as a consequence of severe chronic constipation or neurogenic disorders. Feline idiopathic megacolon is thought to result from longstanding constipation and/or an underlying innervation defect. A similar condition occurs rarely in dogs. Although relatively uncommon, idiopathic megacolon is a frustrating, recurring problem that often results in euthanasia of the affected animal.

Assessment (Pages 778-780)
ASSESS THE ANIMAL/History and Physical Examination Animals with megacolon are usually examined for constipation or obstipation. Affected animals exhibit tenesmus, dyschezia and abdominal pain. Chronically constipated animals often exhibit weight loss, inappetence, vomiting, depression and a poor coat.

Owners of constipated pets should be questioned about medications their pet is receiving. A number of commonly used drugs are associated with constipation (**Table 22-47**).

Depression and dehydration may be noted at physical examination. Abdominal palpation often reveals colonic distention and dry, hard feces. Digital rectal examination may also reveal dry, hard feces. Most constipated pets do not require diagnostic evaluation beyond a careful history and physical examination and the appropriate exclusion of systemic and GI causes.

Table 22-47. Drugs associated with constipation.

Antacids	Bismuth subsalicylate
Anticholinergics	Diuretics
Anticonvulsants (phenytoin)	Hematinics
Antidepressants	Opiates
Barium sulfate	Sucralfate

ASSESS THE ANIMAL/Risk Factors Besides drugs (**Table 22-47**), dietary indiscretion is frequently associated with constipation in dogs. Consumption of bones, rocks and clay may trigger an episode. Although not generally prone to dietary indiscretion, cats may develop constipation as a consequence of trichobezoar formation.

Perineal and perianal disorders (e.g., perineal hernias, perianal fistulas and anal sacculitis) often predispose pets to constipation because of the pain associated with defecation. Suppression of defecation results in increased fecal retention time, increased water absorption and inspissated feces. Orthopedic disorders may have a similar effect if the animal experiences pain when it assumes the defecation stance. Improperly healed pelvic fractures may reduce the size of the pelvic inlet and obstruct the colon externally. For information about dietary management of perianal fistulas in dogs, see Small Animal Clinical Nutrition, 4th ed., p 779.

ASSESS THE ANIMAL/Etiopathogenesis The colon is innervated by the parasympathetic nervous system and intrinsic myenteric and submucosal plexuses. Destruction or damage to either pathway results in reduced colonic motility and potentiates constipation.

Dehydration and electrolyte imbalances may induce constipation. Dehydration enhances colonic water absorption and leaves a dry, hard fecal mass. Electrolyte disturbances (e.g., hyponatremia, hypokalemia, hypocalcemia and hypercalcemia) may alter colonic muscular activity resulting in constipation.

Constipation may result from mechanical obstruction caused by intraluminal or extraluminal masses, rectal strictures and narrow pelvic outlets (e.g., improperly healed pelvic fractures). Additionally, a number of neurologic disorders may result in reduced colonic motility. These include cauda equina syndrome, dysautonomia (Key-Gaskell syndrome) and diabetic or hypothyroid polyneuropathy.

ASSESS THE ANIMAL/Key Nutritional Factors/*Water* Water is a key nutrient that is often overlooked in constipated animals. Maintaining normal hydration status is important in managing patients with chronic constipation. Methods should be used to encourage water intake. These include ensuring multiple bowls of potable water are available in prominent locations in the pet's environment, feeding moist rather than dry forms of foods, adding small amounts of flavoring substances such as bouillon or broth to water sources and offering ice cubes as treats or snacks.

ASSESS THE ANIMAL/Key Nutritional Factors/*Fiber* Many patients with constipation improve clinically when the fiber content of the diet is increased. Increasing dietary fiber alters fecal water content, colonic motility and intestinal transit rate, all of which may benefit patients with constipation. Nonfermentable (i.e., insoluble), moderately fermentable and highly fermentable (i.e., soluble) fiber sources have been advocated for management of constipation. Fiber sources can be added to typical foods or the patient can be fed fiber-enhanced veterinary therapeutic foods (**Tables 22-45** and **22-46**) to increase dietary fiber intake.

A number of gel-forming fibers have been recommended as an aid in managing constipation in domestic animals. These fibers, whether added or incorporated into food, swell to form emollient gels and facilitate passage of fecal matter. Pumpkin and psyllium husk fiber are not typically incorporated into commercial pet foods but can be used as supplements.

Flatulence and abdominal cramping are potential side effects to using these fermentable, bulk-forming fibers. These side effects can be reduced by a gradual transition to fiber supplementation, slowly increasing the level of added fiber until efficacy is achieved with minimal side effects. Such fibers should be added at no more than 5% of the total diet because soluble fibers can significantly reduce the availability of minerals, including zinc, calcium, iron and phosphorus.

Occasionally, an animal may develop megacolon, constipation or flatulence while being fed moderate- (10 to 15% DM crude fiber) or high-fiber (>15% DM crude fiber) foods. The prudent recommendation in such cases is to decrease the fiber content by 5% DMB, reassess the patient and then decrease the fiber content again, if necessary. This situation occurs more commonly in older overweight and obese cats consuming dry, high-fiber (>20% DM crude fiber), low-calorie foods for weight control than in cats being treated for constipation with fiber.

ASSESS THE ANIMAL/Key Nutritional Factors/*Digestibility* In situations in which colonic motility patterns are completely abolished (e.g., severe megacolon in cats), fiber-enhanced foods are no longer able to stimulate colonic motility and may actually contribute to obstipation. In this situation, changing to a food with high digestibility (fat and soluble carbohydrate digestibility ≥90%; protein digestibility ≥87%) will provide adequate calories with a marked reduction in fecal mass. These types of food can markedly reduce the burden of home management (i.e., administering stool softeners and enemas) for pet owners. In such cases, fecal production is reduced to such an extent that owners can generally remove feces via cleansing enemas once or twice weekly. In many cases, this food transition is made as the owner considers the surgical option of subtotal colectomy.

ASSESS THE FOOD AND FEEDING METHOD Levels of crude fiber and appropriate digestibility should be evaluated in foods currently fed to patients with constipation. Information from this aspect of assessment is essential for making any changes to foods currently provided. Changing to a more appropriate food is indicated if levels of crude fiber or digestibility of the current food do not match recommended levels.

Because the feeding method may be altered in patients with constipation, a thorough assessment should include verification of the feeding method currently being used. Items to consider include feeding frequency, amount fed, how the food is offered, access to other food and who feeds the animal. All of this information should have been gathered when the history of the animal was obtained. If the animal has a normal BCS (3/5), the amount of food previously fed (energy basis) was probably appropriate.

Feeding Plan (Pages 780-781)

Initial management of chronic constipation includes owner education, encouraging increased water intake, dietary changes and judicious use of laxatives and enemas. Obstipation often requires multiple cleansing enemas with or without mechanical removal of impacted feces before dietary changes are instituted.

Surgery may be necessary to remove the affected portion of the bowel in cases of idiopathic megacolon. In most cases, subtotal colectomy with ileorectal or cecocolic-rectal anastomosis is the treatment of choice.

SELECT A FOOD Constipation is often managed successfully by feeding a food with higher insoluble fiber levels. Most commercial grocery and specialty brand dog foods contain less than 5% DM crude fiber. Additional fiber can be added to the diet by using fiber supplements or feeding commercial fiber-enhanced foods. It is prudent to increase the fiber concentration in increments of 5% per week until clinical signs improve or resolve.

Almost any fiber concentration can be achieved by combining foods that have different levels of fiber. The calculation is done using the Pearson square (See Chapter 1.). When doing the calculation, be sure to use the same method of expressing the foods' fiber content (i.e., as fed or DMB). An example is shown in **Figure 22-1**.

High levels of fiber added as a supplement to regular pet foods may make the food unpalatable and unbalanced. Alternatively, there are many

1. Assume desired fiber level = 4% DMB.
2. Fiber levels of selected veterinary therapeutic foods (from **Tables 22-32** and **22-34**):
   Prescription Diet Feline i/d (moist) = 1.6% crude fiber DMB
   Prescription Diet Feline w/d (moist) = 10.7% crude fiber DMB
3. Pearson square calculation:

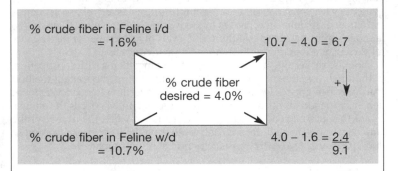

4. The final step converts fractions to percentages by dividing the numerator of the fractions by the denominator and multiplying by 100.

   Feline i/d: (6.7 ÷ 9.1) x 100 = 74%
   Feline w/d: (2.4 ÷ 9.1) x 100 = 26%
   Combining 3/4 Feline i/d with 1/4 Feline w/d (both moist) will provide a food with approximate 4% crude fiber DMB.

Figure 22-1. Example of combining selected veterinary therapeutic foods to provide a specific fiber level.

balanced commercial pet food products offering a wide variety of fiber combinations and concentrations (**Tables 22-45** and **22-46**).

Patients with severe megacolon and some patients with constipation may need foods with high dry matter digestibility and high energy density. Homemade foods (See Chapter 6.) or commercial veterinary therapeutic foods for stress/recovery (**Tables 12-5** and **12-7**) may be fed to affected patients.

DETERMINE A FEEDING METHOD In some cases, smaller more frequent meals may aid colonic motility patterns. Dogs should be walked immediately after feeding; both mild exercise and the gastrocolic reflex will often result in defecation during the immediate postprandial period. Feeding three to four meals per day also minimizes the amount of ingesta entering the large bowel at one time. As discussed above, water intake should be encouraged and multiple bowls of potable water should be available at all times.

DIET INTERACTIONS WITH MEDICAL THERAPY Medical therapies for mild to moderate constipation include enemas, stool softeners, laxatives and colonic motility modifiers (e.g., cisapride [Propulsid, Janssen Pharmaceutica, Inc., Titusville, NJ, USA] 0.25 mg/BW$_{kg}$, t.i.d. to q.i.d.). A number of poorly absorbed carbohydrates may prove useful as laxatives. These sugars, including polyethylene glycol, sorbitol, lactulose and lactitol are hydrolyzed to fatty acids by the colonic microflora. The metabolites of these sugars exert osmotic pressure and draw fluid into the colon lumen. Additionally, laxative therapy may occasionally be needed to promote fecal hydration and lubrication.

Reassessment (Pages 781-782)
Body weight and condition assessments and stool evaluations are useful for monitoring patients with constipation. Well-compensated patients should be evaluated immediately if a change or decline in condition is noted. Regaining or maintaining optimal body weight and condition, normal activity level, a positive attitude and absence of clinical signs are measures of successful dietary management. The feeding method and amount fed can be adjusted as needed to maintain body weight and condition. Additional medical therapies should be considered if dietary therapy alone is not sufficient to improve stool quality and maintain body weight. Although treatment is highly case specific, many cases can eventually be managed with diet alone after initial medical therapies are tapered.

Flatulence (Pages 782-784)
Clinical Importance (Page 782)
Excessive flatus is a chronic objectionable problem that occurs often in dogs and less commonly in cats. Although belching and borborygmus are rarely chief complaints of pet owners, routine questioning may elicit their presence. Flatus, belching and borborygmus occur in normal pets but often develop as a consequence of small intestinal or colonic disorders. At times, flatus is the primary reason pet owners seek veterinary advice.

Assessment (Pages 782-784)
ASSESS THE ANIMAL Pet owners often describe an increase in frequency of belching, flatus or an objectionable odor associated with flatus. At times, it may be possible to elicit a history of dietary change or dietary indiscretion in association with flatus. Occasionally, belching and flatus develop in conjunction with other GI signs including weight loss, diarrhea and steatorrhea. This type of history is very suggestive of an underlying small intestinal disorder.

In most cases, physical examination findings are unremarkable in dogs and cats with flatulence. Animals may be in poor body condition if objectionable flatus occurs secondary to an underlying GI condition.

GI/PANCREAS

Laboratory testing is usually not indicated. However, further evaluation is in order if concomitant GI signs are present. Readers are referred to earlier discussions of small and large bowel disorders for more information.

ASSESS THE ANIMAL/Risk Factors Excessive aerophagia is a risk factor for flatulence and is seen with brachycephalic, working and sporting canine breeds and aggressive and competitive eating behaviors. Dietary indiscretion and ingestion of certain pet food ingredients may be risk factors for certain individuals.

ASSESS THE ANIMAL/Etiopathogenesis Gas in the GI tract is normal and may be derived from three sources: air swallowing, intraluminal production (i.e., bacterial fermentation) and diffusion from the blood.

Swallowed air is thought to contribute the most to gas in the digestive tract. This may be the cause of flatus commonly seen in many brachycephalic breeds. Vigorous exercise and rapid and competitive eating situations may exacerbate aerophagia.

A large amount of gas is formed from bacterial fermentation of poorly digestible carbohydrates and fibers in the colon. Fiber-containing foods contribute to flatus indirectly through reduced dry matter digestibility. Many fibers used in pet foods are fermented by colonic microflora and may contribute to flatus directly. Foods that contain large amounts of nonabsorbable oligosaccharides (e.g., raffinose, stachyose and verbascose) are likely to produce large amounts of intestinal gas. Dogs and cats lack the digestive enzymes needed to split these sugars into absorbable monosaccharides. Therefore, *Clostridium* spp and other bacteria in the colon ferment these sugars producing hydrogen and carbon dioxide. Soybeans, beans and peas contain large quantities of nonabsorbable oligosaccharides. Diseases that cause maldigestion or malabsorption are often associated with excessive flatus because excessive amounts of malassimilated substrates are delivered to the colon where bacterial fermentation occurs. Flatus may be present in animals with lactose intolerance.

As much as 99% of flatus is composed of odorless gases (i.e., nitrogen, oxygen, carbon dioxide, hydrogen and methane). The residual 1% is composed of odoriferous gases including ammonia, hydrogen sulfide, indole, skatole, volatile amines and short-chain fatty acids. These gases contribute the objectionable odors associated with flatus. Spoiled food and many dietary substances including onions, spices and high-protein ingredients increase production of odoriferous gases.

ASSESS THE ANIMAL/Key Nutritional Factors/*Digestibility* Digestibility, especially of the carbohydrate fraction of food, is an important nutritional factor in patients with excessive flatulence. Foods with high digestibility are recommended for patients with objectionable flatus.

ASSESS THE ANIMAL/Key Nutritional Factors/*Food Ingredients* Certain protein, carbohydrate and fiber ingredients or levels may affect flatus production in individual animals.

Changing the sources of protein or carbohydrate in the food may benefit some animals. For example, changing from a dry food that contains corn, chicken meal and soybean meal to a dry food that contains lamb meal, rice and barley may be helpful. Fiber-enhanced foods may contribute to excessive flatus in some patients. Changing to a food that contains reduced amounts of fiber may benefit some pets. The lactose content of food and treats (e.g., cheese, ice cream, milk) may be a factor in adult dogs and cats, especially those with lactase deficiency. A series of dietary trials is often successful in finding a food that reduces flatulence in individual pets.

ASSESS THE FOOD AND FEEDING METHOD Obtaining a thorough dietary history is of paramount importance in evaluating animals with flatulence. Specific foods, major food ingredients, treats, supplements and opportunities for dietary indiscretion should be evaluated.

A thorough assessment should include verification of the feeding method currently being used. Items to consider include feeding frequency, amount fed, how the food is offered, access to other food, relationship of feeding to exercise and who feeds the animal. All of this information should have been gathered when the history of the animal was obtained. If the animal has a normal BCS (3/5), the amount of food previously fed (energy basis) was probably appropriate.

Feeding Plan (Page 784)

Dietary management of flatulence is primarily concerned with decreasing intestinal gas production by bacterial fermentation of undigested food. Changes in the feeding plan can be used in conjunction with other therapy. Recently, commercial products have been introduced that claim to reduce flatulence. These products contain α-galactosidase and reduce flatulence by improving digestion of nonabsorbable carbohydrates. Anecdotal reports suggest that these products may be beneficial in some dogs.

SELECT A FOOD In general, animals with excessive flatulence will benefit from highly digestible foods (fat and soluble carbohydrate digestibility ≥90%; protein digestibility ≥87%). Foods that are high in fiber should be avoided. Changing to a food that does not contain soybean meal or that contains different protein and carbohydrate sources benefits some animals. Feeding a food with rice as the sole or predominant carbohydrate source may be beneficial. In addition, vegetarian-based foods containing strongly flavored, sulfur-containing vegetables should be avoided. In some cases, reducing dietary protein content alleviates flatulence. In most cases, vitamin-mineral supplements should be avoided because these products may increase intestinal microbial activity. Avoid foods or treats containing lactose (e.g., cheese, milk, ice cream).

The initial recommendation is to feed a highly digestible food in small frequent meals. This protocol will reduce food residues available for bacterial fermentation in the large intestine and should reduce gas production.

DETERMINE A FEEDING METHOD Reducing aerophagia is important in the control of flatulence in dogs, especially in brachycephalic breeds. Feed several small meals daily in an effort to discourage rapid eating and gulping of air. Feeding in a quiet, isolated location will eliminate competitive eating and reduce aerophagia.

Reassessment (Page 784)
Patients should be evaluated for evidence of malassimilation if the methods outlined above are not successful in reducing or controlling flatulence. Relapses in animals that have been controlled often indicate dietary indiscretion.

The prognosis for control of flatulence is good in most cases. However, owners should be educated about normal intestinal gas production and should not expect complete cessation of flatulence.

PANCREATIC DISORDERS (Pages 784-792)

Exocrine Pancreatic Insufficiency (Pages 784-787)
Clinical Importance (Page 784)
Malassimilation is failure of nutrients to pass across the intestinal wall in quantities sufficient to maintain body weight and condition. Malassimilation is divided pathophysiologically into maldigestive and malabsorptive diseases. Malabsorption occurs with diseases that alter the structure and function of the small intestinal mucosa including the lymphatics. (See Small Intestinal Disorders section above.) Maldigestion occurs with defects in intraluminal digestion and may result from gastric, pancreatic or biliary dysfunction. Exocrine pancreatic insufficiency (EPI) refers to a partial or complete deficiency of pancreatic enzymes and is the most common cause of maldigestion in dogs. Occurring most commonly in young dogs as a congenital anomaly, EPI may also develop as a sequela to acute and chronic pancreatitis. EPI is rare in cats but has been reported in both the juvenile and acquired forms.

Assessment (Pages 784-786)
ASSESS THE ANIMAL/History and Physical Examination Affected dogs and cats have a history of chronic small bowel diarrhea, weight loss and

failure to thrive. Pets with EPI defecate frequently (six to 10 bowel movements per day) and stools are typically voluminous, greasy, foul smelling and pale in color. When stained with Sudan III and examined microscopically, fat droplets are readily identified in such feces. Polyphagia, borborygmus, pica and coprophagia are often reported. Vomiting and polydipsia occur less commonly.

Affected dogs and cats generally have a normal appearance except for poor body condition (BCS 1/5 or 2/5) and poor coat quality. Animals with pancreatic atrophy will be stunted in comparison to unaffected littermates or breed standards. Severely affected patients may have hemorrhages due to a vitamin K-deficient coagulopathy.

ASSESS THE ANIMAL/Laboratory and Other Clinical Information A presumptive diagnosis of EPI is often based on the signalment and patient history. Definitive diagnosis is achieved by radioimmunoassay of serum trypsin-like immunoreactivity (TLI). Low fasting TLI values (<2.5 µg/l) are indicative of EPI. TLI is a measure of serum trypsin and trypsinogen concentrations. Trypsinogen leaks out of pancreatic acini in trace amounts in healthy animals (normal canine serum TLI values = 5.0 to 35.0 µg/l; normal feline serum TLI values = 17.0 to 50.0 µg/l). In EPI, pancreatic acinar atrophy and fibrosis result in reduced serum TLI values.

ASSESS THE ANIMAL/Risk Factors EPI due to pancreatic acinar atrophy is most common in young, large-breed dogs. German shepherd dogs and rough-coated collies appear to have a genetic predisposition to pancreatic acinar atrophy; however, any breed can be affected. Siamese cats also appear to be at risk.

Acquired EPI may occur as a consequence of severe or recurrent pancreatic inflammation and resultant fibrosis. Thus, risk factors for acquired EPI are the same as for pancreatitis.

ASSESS THE ANIMAL/Etiopathogenesis Juvenile EPI results from atrophy of pancreatic acinar tissue rather than from congenital hypoplasia. This atrophy is idiopathic with minimal evidence of inflammation. One report suggests that histopathologic evidence of atrophy is present before the onset of clinical signs, which usually develop when patients are six to 18 months old. Clinical signs do not develop until 85 to 90% of functional exocrine tissue is lost. In the juvenile form of EPI, endocrine function is usually normal and diabetes mellitus does not develop. In rare cases, EPI and diabetes mellitus may occur concurrently in young dogs and cats.

The acquired form of EPI arises as a consequence of the inflammation and fibrosis of endstage chronic pancreatitis. Diabetes mellitus may devel-

GI/PANCREAS

Table 22-48. Key nutritional factors for patients with exocrine pancreatic insufficiency.*

Factors	Recommended levels
Digestibility	≥87% for protein and ≥90% for fat and soluble carbohydrate
Fat	12 to 15% for dogs
	15 to 22% for cats
Fiber	<2%

*Nutrients expressed on a dry matter basis.

op concurrently because pancreatic islet cells are similarly affected. EPI occurs rarely as a consequence of pancreatic adenocarcinoma or chole-cystoduodenostomy.

Several mechanisms are responsible for the severe nutrient malassimilation that occurs in EPI. Most important, the deficiency of pancreatic enzymes results in a failure of intraluminal digestion and inability of the patient to effectively use nutrients. In addition, intestinal mucosal enzyme activity is impaired in both experimental and naturally occurring EPI. Impaired mucosal enzyme function results in abnormal sugar, amino acid and fatty acid transport. The cause for the intestinal mucosal abnormality is unknown but is thought to result from the absence of trophic pancreatic secretions and concurrent SIBOG.

Dogs with EPI commonly have SIBOG because they lack the antibacterial factors present in pancreatic secretions and have changes in immunity secondary to malnutrition. In addition, many German shepherd dogs with EPI also have IgA deficiency. Bacterial overgrowth contributes to malnutrition in EPI by destroying exposed brush border enzymes and consuming unabsorbed intraluminal nutrients. In addition, bacterial hydroxylation of fatty acids may exacerbate fat malabsorption and contribute to osmotic and secretory diarrhea.

Diarrhea in EPI is usually characterized as osmotic. Distal ileal and colonic microflora ferment undigested sugars and fats, releasing osmotically active particles. These particles drive an efflux of fluid into the gut lumen, which overwhelms the colonic capacity for water reabsorption. Additionally, hydroxy fatty acids formed from bacterial metabolism of undigested fats can trigger secretory diarrhea.

ASSESS THE ANIMAL/Key Nutritional Factors Key nutritional factors for patients with EPI are listed in **Table 22-48** and discussed in detail below.

ASSESS THE ANIMAL/Key Nutritional Factors/*Digestibility* The primary nutritional factor in the management of EPI is food digestibility. Pancreatic enzyme preparations are often added to highly digestible foods (fat and soluble carbohydrate digestibility ≥90%; protein digestibility ≥87%) to maximize digestibility. In one study, the combination of a highly digestible commercial veterinary therapeutic food plus pancreatic enzymes provided more metabolizable energy to dogs with EPI than a grocery brand food with pancreatic enzyme supplementation. Further studies in naturally occurring EPI cases also demonstrated the benefits of feeding highly digestible foods.

Highly digestible veterinary therapeutic foods contain meat and carbohydrate sources that have been highly refined to increase digestibility. Typical ingredients in such commercial foods include eggs, cottage cheese and muscle and organ meats. Carbohydrates in highly digestible foods are primarily starches of corn, rice, barley and wheat, which are well digested if properly cooked.

ASSESS THE ANIMAL/Key Nutritional Factors/*Fat* Steatorrhea is the most prominent clinical sign in patients with EPI. As discussed above, feeding a highly digestible food in conjunction with pancreatic enzyme supplementation is more effective than simply decreasing the fat content of the current food. Overall fat digestion of a highly digestible food with added pancreatic enzymes can exceed 70%. The addition of MCT to the diet may increase caloric intake; however, MCT decreases diet palatability, which may decrease caloric intake and thus be counterproductive. Addition of MCT is unnecessary in most cases.

ASSESS THE ANIMAL/Key Nutritional Factors/*Fiber* Foods for patients with EPI should contain very little fiber (<2% DMB). Decreasing the fiber content from 4% to less than 1% in a study of people with EPI decreased both fecal weight and fat excretion by one-third and reduced bloating and flatus.

ASSESS THE ANIMAL/Other Nutritional Factors/*Vitamins* Micronutrients should be considered in the dietary management of patients with malassimilation. In EPI, the lack of pancreatic lipase results in failed solubilization and absorption of fat-soluble vitamins. The fat-soluble vitamins A and D may be initially administered intramuscularly (1 ml divided into two intramuscular sites every three months), if GI fat absorption remains impaired. Vitamin E supplementation is beneficial when serum concentrations are very low. Clinically, vitamin K deficiency has been described. Severe hemorrhage may occur when vitamin K stores are depleted because of the vitamin's pivotal role in the post-translational carboxylation of coagulation factors. Parenteral supplementation of vitamin K_1 is recommended (5 to 20 mg, q12h) if coagulopathies are detected in dogs or cats with EPI.

Folate and cobalamin are also of concern. Dogs and cats with EPI have low serum cobalamin concentrations. Several mechanisms may play a role in the development of cobalamin deficiency in EPI. The absence of pancreatic bicarbonate secretion may reduce the intestinal luminal pH and the affinity of cobalamin for intrinsic factor. In addition, it is suspected that a pancreatic rather than a gastric intrinsic factor may be necessary for ileal absorption of cobalamin in dogs. Finally, when SIBOG is present, the proximal gut microflora may consume dietary cobalamin before it can be absorbed. If serum levels of cobalamin are low, parenteral supplementation is recommended (100 to 250 μg, intramuscularly or subcutaneously, every seven days for several weeks) until serum cobalamin concentration is normalized.

Serum folate levels are elevated in most dogs with EPI probably due to SIBOG and bacterial elaboration of folate. Serum folate concentration may be decreased, however, in dogs with EPI and concurrent enteropathies involving the ileum. In such cases, parenteral supplementation of folate is recommended until the ileal pathology is resolved. Folate deficiency inhibits pancreatic exocrine function in rats.

ASSESS THE FOOD AND FEEDING METHOD Levels of key nutritional factors should be evaluated in foods currently fed to patients with EPI and compared with recommended levels (**Table 22-48**). Key nutritional factors include food digestibility, fat and fiber. Information from this aspect of assessment is essential for making any changes to foods currently provided. Changing to a more appropriate food is indicated if key nutritional factors in the food currently provided do not match recommended levels.

Because the feeding method is often altered in patients with EPI, a thorough assessment should include verification of the feeding method currently being used. Items to consider include feeding frequency, amount fed, how the food is offered, access to other food and who feeds the animal. All of this information should have been gathered when the history of the animal was obtained.

Patients with EPI usually should be fed multiple small meals per day with pancreatic enzyme supplementation to improve digestibility.

Feeding Plan (Pages 786-787)
Dietary management is an essential component in the medical management of patients with maldigestive diseases. Dietary intake should meet the patient's nutrient needs in a form that promotes nutrient absorption. The organs of the GI tract have very large reserve capacities and the small intestine has a very large and efficient absorptive area. About 90% of the pancreas must be dysfunctional before clinical signs of mal-

Table 22-49. Key nutritional factors in selected highly digestible commercial foods for dogs with exocrine pancreatic insufficiency.*

Products	Fat	Fiber
Recommended levels	12-15	<2
Moist canine products		
Hill's Prescription Diet Canine i/d	14.0	0.7
Leo Specific Digest CIW	14.7	4.2
Iams Eukanuba Low-Residue Adult/Canine	22.0	2.2
Medi-Cal Canine Gastro Formula	12.3	1.7
Purina CNM Canine EN-Formula	13.8	0.9
Select Care Canine Sensitive Formula	12.3	1.7
Waltham/Pedigree Canine Low Fat	6.9	2.4
Dry canine products		
Hill's Prescription Diet Canine i/d	13.6	1.2
Iams Eukanuba Low-Residue Adult/Canine	10.6	2.1
Iams Eukanuba Low-Residue Puppy	22.3	1.9
Leo Specific Digest CID	14.3	1.1
Medi-Cal Canine Gastro Formula	11.7	2.0
Purina CNM Canine EN-Formula	11.7	1.1
Select Care Canine Sensitive Formula	11.7	2.0
Waltham/Pedigree Canine Low Fat	5.4	0.4

*This list represents products with the largest market share and for which published information is available.
Manufacturers' published values.
Nutrients expressed as % dry matter.

digestion are seen. Consequently, patients with clinical signs of maldigestion have very little digestive capacity remaining.

SELECT A FOOD Commercial veterinary therapeutic foods that are highly digestible and designed for patients with GI disease are listed in **Tables 22-49** and **22-50**. Foods from these tables that match the key nutritional factors for patients with EPI are recommended. Feeding recommended foods to patients with EPI often allows smaller amounts of pancreatic enzyme preparations to be used, which results in significant cost savings for pet owners, especially those with large-breed dogs. Foods for young, growing dogs with EPI should also meet the optimal levels of key nutritional factors for growth. (See Chapter 9.)

DETERMINE A FEEDING METHOD Patients presenting with signs of malnutrition due to chronic maldigestion should be given parenteral nutritional support during the diagnostic workup. Parenteral nutrition in the management of these patients is primarily supportive, may be essential in the initial stages of case management and improves the patient's attitude and disposition. Parenteral nutrition also improves caloric, nitrogen and micronutrient balances in veterinary patients, thereby decreasing risks asso-

Table 22-50. Key nutritional factors in selected highly digestible commercial foods for cats with exocrine pancreatic insufficiency.*

Products	Fat	Fiber
Recommended levels	15-22	<2
Moist/semi-moist feline products		
Hill's Prescription Diet Feline i/d	20.1	1.6
Iams Eukanuba Low-Residue Adult/Feline	16.9	2.2
Leo Specific Dermil FDW	34.7	1.5
Medi-Cal Feline HYPOallergenic/Gastro Formula	25.6	1.65
Purina CNM Feline EN-Formula	17.0	1.1
Waltham/Whiskas Feline Selected Protein	29.0	2.9
Dry feline products		
Hill's Prescription Diet Feline i/d	20.1	1.3
Iams Eukanuba Low-Residue Adult/Feline	15.4	2.2
Medi-Cal Feline HYPOallergenic/Gastro Formula	13.6	1.6
Select Care Feline Neutral Formula	13.6	1.6
Waltham/Whiskas Feline Selected Protein	12.2	3.3

*This list represents products with the largest market share and for which published information is available.
Manufacturers' published values.
Nutrients expressed as % dry matter.

ciated with diagnostic procedures including exploratory surgery. Continued administration of parenteral nutrition (more than three days) is necessary in debilitated patients as a supportive procedure until nutrients can be adequately absorbed. Parenteral nutrition can be performed at most practices in a manner similar to other fluid therapies. (See Chapter 12.)

At home, feeding two to three times daily helps prevent dietary overload and osmotic diarrhea. Underweight patients should be fed in excess of DER until ideal body weight and condition (BCS 3/5) are reached. Even after patients reach ideal body weight, it may be necessary to offer DER plus 20% to allow for the persistent degree of malabsorption in patients with EPI. Pancreatic enzymes should be added immediately before feeding. (See below.)

DIET-DRUG INTERACTIONS In addition to dietary management, effective treatment of EPI requires oral administration of pancreatic enzymes. Most often, pancreatic enzymes are supplied as dried, powdered extracts of bovine or porcine pancreas (**Table 22-51**). Lipase activity of pancreatic enzyme preparations varies markedly. As a general rule, the more expensive preparations are the more effective lipase preparations. If available, raw pancreas can be fed successfully. Raw pancreas can be frozen for several months without losing enzyme activity. Dogs should receive 100 g of freshly thawed, chopped pancreas per meal. Cats should receive 30 to 90 g of freshly thawed, chopped pancreas per meal. Tablets, capsules and enteric-coated preparations are not recommended.

Table 22-51. Enzyme preparations used in patients with exocrine pancreatic insufficiency.*

Products (manufacturers)	Lipase	Protease	Amylase	Formulation
Viokase-V Powder (Fort Dodge)	57,000	285,000	428,000	Powder
Viokase-V Tablets (Fort Dodge)	9,000	57,000	64,000	Tablets
Viokase Powder (Robins)	16,800	70,000	70,000	Powder
Pancrezyme Powder (Daniels Pharmaceuticals)	61,000	330,000	440,000	Powder
Pancrezyme Tablets (Daniels Pharmaceuticals)	9,000	57,000	64,000	Tablets
Pancrease MT4 Capsules (McNeil)	4,000	12,000	12,000	Enteric-coated microtablets
Pancrease MT10 Capsules (McNeil)	10,000	30,000	30,000	Enteric-coated microtablets
Pancrease Capsules (McNeil)	4,000	25,000	20,000	Enteric-coated microspheres
Pancreatic Plus Powder (Butler)	71,400	388,000	460,000	Powder
Pancreatic Plus Tablets (Butler)	9,000	57,000	64,000	Tablets
Pancrelipase Capsules (Geneva)	4,000	25,000	20,000	Enteric-coated pellets

*Enzymatic contents (IU) per capsule, tablet or tsp of powder (2.8 g).

GI/PANCREAS

Pancreatic enzyme supplementation for dogs should be initiated at a dose of 2 tsp of powdered pancreatic extract (or crushed tablets) per 20 BW_{kg} at each meal. For cats, a starting dose of 1 tsp should be administered with each meal. Enzymes should be mixed with food immediately before the meal is fed. Owners may be able to decrease the dose of pancreatic enzymes based on their pet's response. Most dogs require at least 1 tsp of enzymes per meal.

Antacids or H_2-receptor blockers have been recommended in the therapeutic regimen to reduce gastric acid-induced destruction of orally administered enzymes. This practice, however, is costly and does not increase efficacy of pancreatic enzyme supplementation. Concurrent oral administration of sodium bicarbonate or bile salts and pre-incubation of the meal with pancreatic enzymes are also unnecessary. In one study, adding digestive enzymes to food 20 to 30 minutes before feeding did not improve the response to dietary management.

Oral antibiotics may be necessary to resolve clinical signs in dogs and cats with concurrent SIBOG. Tetracycline (20 mg/BW_{kg}, per os, t.i.d. for 21 days) is most often recommended for this purpose; however, metronidazole (10 to 20 mg/BW_{kg}, per os, every 24 hours for seven to 14 days) may be more effective if SIBOG with anaerobic organisms is suspected.

Concurrent diabetes mellitus in EPI cases must be managed with insulin. Unfortunately, the fiber-enhanced foods often recommended for diabetic pets are contraindicated for those with EPI. Dietary management of patients with concurrent diabetes mellitus and EPI often requires a modified profile of key nutritional factors. In many cases, foods containing 10 to 15% DM fat, 50 to 55% DM complex, soluble carbohydrate and 5 to 10% total dietary fiber can be used.

Reassessment (Page 787)

Clinical signs usually resolve within two to three days with proper dietary therapy, and weight gain is evident by five to 10 days. Successfully managed canine cases of EPI are recognized by weight gain (0.5 to 1 kg/week) and improved body condition and stool consistency. The food and enzyme dose should be reevaluated if less satisfactory results are obtained. Often, the initial dose of enzymes is inadequate and must be increased. Every effort should be made to rule out concurrent small bowel disease (e.g., eosinophilic gastroenteritis, lymphoplasmacytic enteritis, SIBOG) when clinical response is unsatisfactory.

Well-compensated patients should be evaluated immediately if a change or decline in condition is noted. Feeding more food than expected may be necessary to compensate for decreased digestibility and to maintain optimal body weight and condition. Regaining or maintaining optimal body weight and condition, normal activity level, a positive attitude and absence of clinical signs are measures of successful dietary management.

Acute and Chronic Pancreatitis (Pages 787-792)

Clinical Importance (Page 787)

Pancreatitis has been recognized as a clinical entity in dogs for more than a century. In dogs, acute pancreatitis is an important differential diagnosis for vomiting and abdominal pain. Because of difficulties in diagnosis, pancreatitis is a less common diagnosis in cats. However, based on recent clinical reports, recognition of feline pancreatitis is apparently increasing.

Assessment (Pages 787-790)

ASSESS THE ANIMAL/History and Physical Examination Dogs with pancreatitis most often present with acute vomiting. Vomiting may be sporadic and mild or very severe. Other clinical signs include abdominal pain, depression, anorexia, fever and diarrhea. Icterus and pale stools may be reported if pancreatic inflammation and edema are severe enough to result in common bile duct obstruction. If present, diarrhea is usually of large bowel origin because the transverse colon passes dorsal to the pancreas and is susceptible to local inflammation at that site.

An episode of dietary indiscretion has often occurred during the 24 hours before the onset of vomiting. The owner commonly relates that the pet consumed high-fat table food. Occasionally, the onset of clinical signs is preceded by administration of drugs associated with pancreatitis. Corticosteroids, in particular, have been linked to pancreatitis in dogs.

Cats with pancreatitis have highly variable clinical signs. In some cats, the disease may mimic the typical canine presentation (i.e., acute vomiting, lethargy, anorexia, diarrhea and abdominal pain). In others, a more indolent, smoldering course occurs, resulting in a mild chronic illness. In some cats, pancreatitis may be linked to diabetes mellitus and clinical signs may include polydipsia/polyuria and weight loss. In others, hepatic lipidosis or cholangiohepatitis may occur concurrently.

Depression, fever and dehydration may be the most prominent physical examination findings. Abdominal palpation may elicit splinting and discomfort that can be localized to the right cranial quadrant. Icterus, shock and coagulopathies may be detected in severe cases.

Clinical manifestations are variable in chronic pancreatitis. Weight loss and poor body condition may be the only signs noted in cats. An abnormal thickening or hardness of the falciform fat pad may be palpated in some cats, suggesting saponification and fat necrosis.

ASSESS THE ANIMAL/Laboratory and Other Clinical Information The laboratory diagnosis of acute and chronic pancreatitis can be very frustrating. Diagnosis is hampered by the poor specificity of available laboratory tests and the inaccessibility of tissue for cytologic or histopathologic examination. Serum amylase and lipase activities are the most commonly used laboratory tests for the diagnosis of pancreatitis in dogs and cats. Unfortunately, these tests are not very specific because they are influenced by a number of other disease conditions (e.g., renal failure, dehydration, hyperlipidemia). In addition, the short half-life of amylase and lipase often precludes their use as diagnostic aids unless the patient is presented promptly after the onset of clinical signs. If present, hyperamylasemia and hyperlipasemia should be considered supportive of a diagnosis of pancreatitis, if azotemia and hyperlipidemia are not present.

Serum TLI concentration has been suggested as a diagnostic aid for evaluating dogs and cats with suspected pancreatitis. Because TLI is specifically pancreatic in origin, high serum TLI concentrations may prove a more reliable indicator of clinical pancreatitis than high amylase or lipase activities.

A database consisting of a complete blood cell count, serum biochemistry profile and urinalysis should be collected for any dog or cat suspected to have pancreatitis to rule out other potential causes for the clinical signs. Additionally, these tests may aid in the diagnosis of concurrent medical conditions such as diabetes mellitus, hepatic lipidosis, interstitial nephritis and cholangiohepatitis. An inflammatory leukogram is typically identified in patients with pancreatitis. A degenerative left shift may indi-

GI/PANCREAS

Table 22-52. Risk factors for pancreatitis in dogs and cats.

Breed	
Briard	Miniature schnauzer
Himalayan and Siamese cats	Sheltie
Diet	
Dietary indiscretion	High-fat, low-protein foods
Drug administration	
Azathioprine	L-asparaginase
Corticosteroids	Organophosphate insecticides (cats)
Fasting hyperlipidemia	
Gender	
Castrated males	Spayed females
Hepatobiliary disease	Feline suppurative cholangiohepatitis
Hypercalcemia	
Hyperparathyroidism	Intravenous calcium infusion
Increasing age	
Intervertebral disk disease	
Ischemia or reperfusion	Postgastric dilatation-volvulus
Obesity	

cate severe necrotic pancreatitis. If thrombocytopenia is noted on the hemogram, a complete coagulation screen should be performed to rule out disseminated intravascular coagulation.

Imaging can be useful for diagnosing pancreatitis in dogs and cats. Findings consistent with pancreatic inflammation on survey abdominal radiographs may include haziness and widening of the gastroduodenal angle in the right cranial quadrant. Often, segmental gas distention of the proximal duodenum is noted. Ultrasonography may reveal fluid-filled cysts or abscesses within the pancreatic parenchyma and can be used to guide needle aspiration of pancreatic masses.

ASSESS THE ANIMAL/Risk Factors Several risk factors have been associated with pancreatitis in dogs and cats (**Table 22-52**). Most animals with these risk factors, however, do not develop pancreatitis.

An association has been made between hyperlipidemia and acute pancreatitis in dogs and people, which has led to speculation that disturbances in lipid metabolism may be involved. The exact relationship is not known in dogs or cats and information is often extracted from human cases. In people and pets, hyperlipidemia is thought to precede and cause the development of pancreatitis; however, it can also be evident during and after such episodes.

The incidence of hyperlipidemia in dogs or cats with pancreatitis is generally thought to be high. In a retrospective study of fatal acute pancreatitis in dogs, 26% of patients were hyperlipidemic. However, experimentally induced pancreatitis in dogs has not resulted in lipemia or hypertriglyceridemia. Hypertriglyceridemia is present in some but not all naturally occurring cases of canine pancreatitis as determined by serum lipid and electrophoretic patterns. Several pet breeds are predisposed to pancreatitis (e.g., miniature schnauzers, briards, Shetland sheepdogs, Siamese cats) because they are commonly hypertriglyceridemic. (See Chapter 24.)

Feeding a high-fat (>20% DMB) food, treats or table food has often been associated with the onset of acute pancreatitis. Experimentally, feeding high-fat, low-protein foods was associated with the development of pancreatitis and hepatic lipidic changes in dogs. The most widely repeated explanation for the association between hypertriglyceridemia and acute pancreatitis is that hydrolysis of serum triglycerides by lipase within the pancreatic microvasculature releases free fatty acids locally. Free fatty acids cause microthrombi and/or bind with calcium to cause further capillary damage, which, in turn, releases more pancreatic lipase. Consumption of calorically dense, high-fat foods also contributes to obesity in pets. Obesity is considered a risk factor for pancreatitis. In a recent report, 43% of dogs with acute pancreatitis were overweight or obese.

Pancreatitis has been associated with hypercalcemia in several dogs with hyperparathyroidism and in a dog receiving a calcium infusion. Experimentally, elevated ionized calcium concentrations can induce pancreatitis in cats. The pathophysiologic mechanism for pancreatitis in association with hypercalcemia has not been elucidated.

Reports of drug-induced pancreatitis in companion animals are rare. Anecdotal reports suggest that corticosteroids are the most common drug associated with pancreatitis in dogs. Pancreatitis is common in dogs with hyperadrenocorticism and in dogs receiving corticosteroids for management of intervertebral disk disease. Experimentally, corticosteroids increase the sensitivity of dispersed acinar cells to cholecystokinin and stimulate proliferation of the pancreatic ductular epithelium. Evidence is lacking for a role of corticosteroids in the development of pancreatitis in cats.

Ischemia and reperfusion injury have been linked to acute pancreatitis. Hypovolemic shock, GDV and abdominal trauma have been reported to precede acute pancreatitis. In addition, abdominal surgery marked by inept manipulation of the pancreas can result in pancreatitis.

ASSESS THE ANIMAL/Etiopathogenesis Acute pancreatitis is the sudden onset of inflammation of the pancreatic acinar tissue. Typically, the primary histopathologic lesion is edema. After resolution, there is usually no residual pancreatic lesion. However, in more severe cases, the pancreatic lesion may become hemorrhagic or may progress to necrosis. Mortality is high in acute necrotizing pancreatitis. Acute edematous or hemorrhagic pancreatitis may occur as a singular or recurrent event in dogs and cats.

Pancreatitis occurs as a consequence of intracellular pancreatic acinar enzymatic activation and resultant autodigestion of the pancreas. In the

Table 22-53. Protection against pancreatic autodigestion.

Enterokinase, produced by the duodenal brush border, is required for activation of proenzymes.
Pancreatic enzymes are synthesized as proenzymes (zymogens).
Serum protease inhibitors bind free trypsin.
The pancreas secretes a trypsin inhibitor in pancreatic juice that binds free trypsin.
Zymogens and lysosomal enzymes are stored in different intracytoplasmic membranes.

normal pancreas, safeguards ensure that harmful pancreatic enzymes are not activated until they reach the intestinal lumen (**Table 22-53**). Pancreatic enzymes are synthesized in endoplasmic reticuli, modified in Golgi apparatuses and stored in zymogen granules within acinar cells. Evidence suggests that intracellular pancreatic enzyme activation occurs as a result of abnormal zymogen activation. Normally, zymogens and lysosomes are segregated intracellularly. In pancreatitis, lysosomes containing proteases fuse with zymogen granules. The lysosomal contents (e.g., proteases such as cathepsin B) activate trypsinogen. In addition, the acidic environment of lysosomes interferes with self-regulating trypsin inhibitors stored with pancreatic enzymes in zymogen granules.

Cholecystokinin and acetylcholine are widely recognized as the principal physiologic mediators of pancreatic enzyme secretion. Normally, these substances initiate fusion of zymogen granules with the acinar cell membrane. Hyperstimulation of the pancreas with supraphysiologic doses of cholecystokinin appears to cause pancreatitis in experimental animals by interfering with the intracellular movement of zymogens resulting in fusion of zymogens and lysosomes. The lysosomal enzyme cathepsin B is then thought to activate trypsinogen and precipitate pancreatitis. Pancreatic duct obstruction also appears to facilitate fusion of zymogens and lysosomal enzymes. Foods that are high in fat and protein (particularly the amino acid arginine) stimulate production and release of cholecystokinin, gastrin and secretin. Organophosphate insecticides and intravenous calcium infusions are hypothesized to cause hyperstimulation via cholecystokinin.

Bile acid and enteric reflux into the pancreatic duct can also activate pancreatic enzymes interstitially (i.e., within the pancreatic duct system and interstitium). This mechanism is thought to be involved in the development of pancreatitis in cats in conjunction with suppurative cholangiohepatitis and enteritis. The anatomic configuration of the common bile duct and pancreatic duct in cats facilitates this mechanism. Experimentally, free fatty acids generated by the action of lipase on triglycerides damage acinar cell membranes, releasing lecithin, which causes marked necrosis of acinar cells when converted to lysolecithin by phospholipase A_2.

Regardless of the initiating cause, active pancreatic enzymes (trypsin, phospholipase, collagenase and elastase) and inflammatory mediators are released into the pancreatic tissues and blood vessels. These factors apparently activate coagulation, fibrinolytic, kinin and complement cascades. Circulating defense mechanisms include α_1-antitrypsin and α_2-macroglobulin, which bind to active enzymes to contain local damage and prevent systemic damage. After these defenses are overwhelmed, increased pancreatic permeability leads to fluid loss into the pancreas and the abdomen, a decline in pancreatic blood flow and an increase in the local concentrations of pancreatic enzymes and inflammatory mediators. Large numbers of leukocytes migrate to the inflamed pancreas and serve as a source of free radicals, inflammatory mediators and enzymes. This vicious, self-perpetuating cycle may ultimately lead to thrombosis of pancreatic blood vessels and pancreatic necrosis. Systemic complications may develop, including hypovolemic shock and disseminated intravascular coagulation.

Chronic pancreatitis is less commonly recognized in companion animals vs. people in whom alcohol can serve as a constant stimulus for smoldering acinar inflammation. Some authors suggest that chronic mild interstitial pancreatitis is the most common form of pancreatitis recognized in cats. Histopathologic examination of tissues from dogs and cats with chronic pancreatitis reveals irreversible fibrotic changes resulting from the persistent inflammatory condition. Both chronic and recurrent pancreatitis may result in acquired EPI.

ASSESS THE ANIMAL/Key Nutritional Factors/*Water* Water is the most important nutrient in patients with acute vomiting because of the potential for life-threatening dehydration due to excessive fluid loss and inability of patients to replace those losses. Moderate to severe dehydration should be corrected with appropriate parenteral fluid therapy rather than using the oral route.

ASSESS THE ANIMAL/Key Nutritional Factors/*Protein* Free amino acids (i.e., phenylalanine, tryptophan and valine) in the duodenum are a strong stimulus for pancreatic secretion, in fact, more so than fat. Excess dietary protein should be avoided; however, adequate protein should be provided for recovery and tissue repair. Protein levels (DMB) of 15 to 30% for dogs and 30 to 45% for cats are appropriate.

ASSESS THE ANIMAL/Key Nutritional Factors/*Fat* Obese and hypertriglyceridemic patients recovering from pancreatitis should receive low-fat foods (<10% DMB). Other patients can be fed moderate-fat foods (12 to 15% DMB). The most clinically relevant form of hypertriglyceridemia in veterinary patients is hyperchylomicronemia because triacylglycerides make up 84 to 89% of the lipids in chylomicrons. (See Chapter 24.) Plasma chylomicrons are derived from two sources. Large (12-carbon) triglycerides are present for a few hours after ingestion of dietary fat, whereas smaller triglycerides, secreted from the liver, are always present and independent of dietary fat.

ASSESS THE FOOD AND FEEDING METHOD Levels of protein and fat should be evaluated in foods currently fed to patients with pancreatitis and compared with recommended levels. Information from this aspect of assessment is essential for making any changes to foods currently provided. Changing to a more appropriate food is indicated if protein and fat levels in the food currently fed do not match recommended levels.

Because the feeding method is often altered in patients with pancreatitis, a thorough assessment should include verification of the feeding method currently being used. Items to consider include feeding frequency, amount fed, how the food is offered, access to other food sources including table food and garbage and who feeds the animal. All of this information should have been gathered when the history of the animal was obtained. In cases in which acute pancreatitis is associated with eating garbage or other inappropriate foods (most often during a holiday), strict avoidance of foods other than the pet's regular food is recommended.

Feeding Plan (Pages 790-791)

The goals of dietary management of patients with pancreatitis are to decrease stimuli to pancreatic secretion (and thus prevent pancreatic autodigestion) and still provide adequate nutrient levels to support tissue repair and recovery. Acute hemorrhagic or necrotizing pancreatitis should be considered a medical emergency. Initially, appropriate parenteral fluid therapy should be provided to correct dehydration and electrolyte and acid-base disturbances. Oral food intake stimulates pancreatic secretions by several mechanisms. Likewise, the physical presence of food in the stomach stimulates gastrin, which in turn stimulates pancreatic secretion. Therefore, NPO therapy is the initial treatment of choice.

Therapy used in conjunction with the feeding plan includes plasma transfusions, nasogastric suctioning of gastric secretions and air, anticholinergic agents, somatostatin analogues (octreotide), antibiotics and surgical exploration of the abdomen for extirpation or drainage of pancreatic abscesses or pseudocysts.

SELECT A FOOD Small amounts of water, ice cubes, oral rehydration solutions or monomeric foods can be offered after vomiting and abdominal discomfort subside. Monomeric foods are water-soluble, liquid foods containing nutrients in their simplest absorbable form. Thus, nutrients in these foods minimally stimulate pancreatic secretion. Some monomeric products also contain glutamine to stimulate enterocyte hyperplasia after several days of NPO therapy, which may have induced intestinal mucosal atrophy. In general, 1 to 2 ml/BW_{kg} q.i.d. are well tolerated and rarely induce vomiting.

If liquids are well tolerated for one to two days, solid food may be slowly reintroduced. Highly digestible, commercial veterinary therapeutic foods designed for patients with GI disease are often used initially (**Tables 22-31** and **22-32**). These foods also contain moderate levels of protein and fat. If vomiting recurs, NPO therapy should be reinstituted and feeding attempted again after 12 to 24 hours. A veterinary therapeutic food formulated for patients with GI diseases should be fed for another seven to 10 days before reintroducing the patient's regular food, if it is to be used at all.

Low-fat (<10% DM fat) foods are often used if obesity or hyperlipidemia was a contributing factor (**Tables 13-3** and **13-4**). High-fat commercial foods (>20% DM fat), table foods and snacks should be avoided. It may be necessary to remind clients of this around the holiday season, when many owners fall prey to the desire to share family meals with pets.

DETERMINE A FEEDING METHOD Discontinuing oral intake of food and water (NPO) is the cornerstone of initial therapy for acute pancreatitis. Factors that would normally stimulate pancreatic secretions (GI distention and hormone release [gastrin, secretin, cholecystokinin]) are reduced when food and water are withheld. Most patients respond within two to four days. After vomiting and abdominal discomfort resolve or lessen in severity, liquids and food can be reintroduced gradually over several days. Normal feeding methods can be reintroduced after several days without clinical signs, unless dietary indiscretion or inappropriate foods or feeding methods initially contributed to the problem.

After three days of the NPO protocol, patients with severe pancreatitis should receive enteral or parenteral nutritional support. The method deemed most desirable is the least invasive, supports the patient nutritionally and minimally stimulates pancreatic secretions.

Protracted cases of pancreatitis often require parenteral nutrition to meet the patient's energy, protein, electrolyte and B-vitamin requirements while minimizing pancreatic secretions. (See Chapter 12.) Parenteral administration of nutritional solutions (including lipid) has been associated with pancreatic atrophy; so pancreatic stimulation is minimal or nonexistent. Intravenous administration of nutrients to support pancreatic patients through a five- to 14-day course of vomiting is possible, safe and economical in most practices. (See Chapter 12.) Parenteral solutions may be of particular benefit in managing pancreatitis in cats, especially when complicated by hepatic disorders, IBD or interstitial nephritis.

Selection of parenteral solutions for feeding patients with pancreatitis is controversial because of the association between hyperlipidemia and pancreatitis. Some authors suggest that selection of parenteral solutions be based on amino acid and dextrose content only, whereas others advocate the use of lipid solutions in the admixture if the patient is not hyperlipidemic. Administering glucose as the sole nonprotein energy source perpetuates hyperglycemia and increases the risk of hepatic steatosis. Lipids in total nutrient admixtures (See Chapter 12.) have been used successfully in dogs and cats with pancreatitis. Lipid emulsions administered intravenously are synthetic 0.5-μm chylomicrons that appear to be well used by rats, people and dogs with pancreatitis.

Serum triglyceride levels should be assessed before lipids are administered intravenously. Although isolated cases of pancreatitis in people have been linked to lipid infusion, these cases are considered rare and were complicated by concurrent diseases such as alcoholism and IBD.

Lipid administration is well tolerated, most likely because the liver would be using endogenous fat stores if lipid were not supplied exogenously.

Placing a jejunostomy tube and bypassing the stomach and duodenum should also be considered in prolonged cases of pancreatitis. These tubes are best placed when patients must undergo general anesthesia and abdominal surgery for other reasons.

Jejunal feedings in dogs stimulate pancreatic secretion no more than parenteral feedings. In veterinary patients, however, a practical technique for nasojejunal feeding has not been developed; thus, jejunal feeding requires abdominal surgery. For that reason, some clinicians prefer parenteral feeding of patients with chronic, refractory pancreatitis.

Monomeric liquid foods infused directly into the duodenum of dogs stimulate some pancreatic output, whereas oral administration of the same monomeric foods stimulates a greater volume of pancreatic secretion. If jejunal tube feeding is selected, a liquid food supplemented with glutamine to maintain intestinal integrity that minimally stimulates the pancreas and meets the patient's RER is most suitable. Directly infusing a readily absorbable monomeric liquid food (vs. a polymeric product) into the jejunum should also reduce pancreatic secretions because whole nutrients elicit a greater response from the pancreas than monomeric nutrient forms. Monomeric liquid foods may be infused into the jejunum by slow continuous gravity drip (1 to 2 ml/BW_{kg}/hour) or, preferably, by an enteral pump. This rate of enteral feeding meets the RER of most patients and precludes other forms of nutritional support until oral intake is possible. If patients tolerate this rate of administration, solid food in small frequent meals may be given for several days in addition to the liquid feedings. Liquid feedings may cease and the number of oral meals per day can be increased when solid food is well tolerated.

Reassessment *(Pages 791-792)*

Hospitalized patients with pancreatitis should be assessed frequently. Assessment of body weight and condition are recommended to ensure adequate hydration and caloric intake, if instituted. Electrolyte and acid-base status should be monitored to assess adequacy of therapy. Certain laboratory parameters (leukogram and serum concentrations of amylase, lipase, TLI and bilirubin) are helpful markers of progress. However, the patient's attitude, appetite and presence or absence of vomiting and abdominal pain are often the most important predictors of progress. In addition, it is imperative that serum be evaluated for triglyceride concentration initially and then monitored daily for lipemia. It is important to distinguish between lipemia from endogenous sources vs. exogenous fat emulsions when parenteral nutrition is administered.

Discharged patients should be reevaluated in a number of weeks. If a low-fat, high-fiber food was recommended to control obesity or hyperlipidemia, body weights should be recorded and serum triglyceride concentration determined (or samples inspected visually for lipemia) to assess compliance with the dietary management program. Regaining or maintaining optimal body weight and condition, normal activity level and absence of clinical signs are measures of successful dietary management.

GI/PANCREAS

Clinical cases that illustrate and reinforce the nutritional concepts presented in this chapter can be found in Small Animal Clinical Nutrition, 4th ed., pp 799-810.

CHAPTER 23

Hepatobiliary Disease

For a review of the unabridged chapter, see Roudebush P, Davenport DJ, Dimski DS. Hepatobiliary Disease. In: Hand MS, Thatcher CD, Remillard RL, et al, eds. Small Animal Clinical Nutrition, 4th ed. Topeka, KS: Mark Morris Institute, 2000; 811-847.

CLINICAL IMPORTANCE (Pages 811-814)*

The normal liver carries out an estimated 1,500 essential biochemical functions. In addition to its role in drug metabolism, removal of environmental and endogenous noxious substances and synthesis of important substances (e.g., albumin and blood clotting factors), the liver plays a key role in digestion and metabolism of foods/nutrients. The liver influences nutritional status through its elaboration of bile salts and central role in intermediary metabolism of protein (amino acids), carbohydrate, fat and vitamins. **Table 23-1** lists selected hepatic functions that influence nutrient digestion and metabolism.

The liver has tremendous storage capacity, functional reserve and regenerative capabilities. All of these functions protect the body from profound metabolic alterations. However, these same characteristics complicate the clinical recognition of serious liver disease. Consequently, hepatobiliary disease must be severe or associated with cholestasis before clinical signs and/or laboratory tests reveal or confirm its presence. As a result, the patient is often suffering profound metabolic alterations by the time an appropriate feeding plan is implemented.

Malnutrition is a common finding in patients with advanced hepatic disease. Potential causes of malnutrition in animals with hepatic disease include: 1) anorexia, nausea and vomiting, 2) impaired nutrient digestion, absorption and metabolism, 3) increased energy requirements and 4) accelerated protein catabolism with impaired protein synthesis.

Unlike most terminally differentiated cells, hepatocytes in adult liver retain the capacity to proliferate. After partial (70%) hepatectomy, compensatory hyperplasia begins within minutes of resection and is typically completed within two weeks in rats and in less than one month in people. The management of many hepatic diseases should thus be predicated on using this capacity to maximum advantage.

Nutritional management of hepatobiliary disease is usually directed at

*Page numbers in headings refer to Small Animal Clinical Nutrition, 4th ed., where additional information may be found.

Key Nutritional Factors–Hepatobiliary Disease (Page 813)

Table 23-6 lists key nutritional factors for cats with hepatic lipidosis.
Table 23-7 lists key nutritional factors for dogs with copper-associated hepatotoxicosis.
Table 23-9 lists key nutritional factors for patients with portosystemic vascular shunts.
Table 23-10 lists key nutritional factors for patients with chronic hepatitis and cirrhosis.
Table 23-11 lists key nutritional factors for patients with portal hypertension.

clinical manifestations of the disease rather than the specific cause itself. The goals of nutritional management for hepatobiliary disease include: 1) maintaining normal metabolic processes, 2) correcting electrolyte disturbances, 3) avoiding toxic byproduct accumulation and 4) providing substrates to support hepatocellular repair and regeneration.

A recent report did not include liver disease in the list of the 25 most common diagnoses made in dogs or cats seen in private veterinary practices in the United States. Although the combined prevalence of hepatobiliary disease is small compared with the prevalence of other gastrointestinal (GI) disorders, hepatobiliary disease has increased clinical significance because the liver plays a central role in maintaining normal metabolic homeostasis. Hepatobiliary disease, despite its relatively uncommon occurrence, causes significant morbidity and mortality if not diagnosed early and managed appropriately.

Table 23-2 summarizes hepatic diseases or lesions commonly recognized in dogs and cats.

LIVER

ASSESSMENT (Pages 814-831)

Assess the Animal (Pages 814-831)
History and Physical Examination (Pages 814-816)
Patients with acquired hepatobiliary disease usually display vague clinical signs early in the disease process (**Table 23-3**). However, jaundice appears within the first 72 hours in animals with major bile duct obstruction. Animals with congenital portovascular anomalies may also develop clinical signs associated with hepatic encephalopathy in the first several months of life. Demonstration of certain metabolic uroliths (i.e., ammonium urate and other purine uroliths) may also indicate underlying liver disease such as portosystemic vascular shunts. (See Chapter 20.)

Table 23-1. Major hepatobiliary functions related to nutrient digestion and metabolism.

Metabolic functions
Converts glucose to glycogen and triacylglycerides during absorptive state
Converts glycogen to glucose in postabsorptive period
Synthesizes glucose from glucogenic precursors such as glycerol and amino acids in postabsorptive period (gluconeogenesis)
Transforms amino acids (transamination and deamination), synthesizes nonessential amino acids as needed for metabolism
Synthesizes triacylglycerols and secretes them as lipoproteins
Synthesizes and releases cholesterol into blood
Forms ketones from degraded fatty acids during fasting
Synthesizes urea from degraded amino acids (sole site in body)
Synthesizes plasma albumin and fibrinogen
Biliary functions
Synthesizes bile salts from cholesterol, which are secreted into bile for lipid emulsification and absorption in the small intestine
Secretes a bicarbonate-rich solution to help neutralize acid in the duodenum
Secretes plasma cholesterol into bile
Conjugates and excretes bilirubin in bile
Detoxifies substances by biotransformation before biliary excretion
Excretes endogenous and foreign organic molecules in bile
Storage functions
Stores glucose as glycogen and triacylglycerides
Stores vitamins, particularly A but also D, E, K, B_{12} and to a lesser extent other B vitamins
Stores minerals such as iron, copper, manganese and zinc
Stores blood, especially with pressure increases in the hepatic vein or posterior vena cava
Endocrine functions
Activates (partial) vitamin D by dehydroxylation
Converts thyroxine to triiodothyronine
Secretes IGF-1 in response to growth hormone
Metabolizes (deactivates) and excretes hormones
Miscellaneous functions
Removes bacteria and food antigens that regularly cross the intestinal epithelial barrier (Kupffer cells of mononuclear-macrophage system in the sinusoids)

GI abnormalities common in patients with hepatobiliary disease include anorexia, vomiting and diarrhea. Anorexia and ptyalism (hypersalivation) are especially common in cats. Hematemesis suggests GI ulceration, a complication of hepatobiliary disease. The anorexia, GI disturbances and metabolic alterations associated with liver disease often contribute to chronic weight loss. Other common clinical signs of hepatobiliary disease include: 1) polydipsia and polyuria, 2) intermittent pyrexia,

Table 23-2. Hepatic diseases and lesions commonly recognized in dogs and cats.

Canine necropsy (%)*	Feline necropsy (%)**	Feline biopsy (%)**	Canine/feline biopsy***
Hepatitis (18)	Hepatitis (22.9)	Lipidosis (49)	Steroid hepatopathy
Metastatic neoplasia (13.9)	Nonspecific hepatopathy (13.5)	Inflammatory liver disease (26)	Neoplasia
Steroid-induced hepatopathy (11.8)	Metastatic neoplasia (12.8)	Neoplasia (10)	Hepatitis (dogs)
Passive congestion (9.1)	Lipidosis (11.0)	Vacuolar changes (4)	Cholangiohepatitis (cats)
Necrosis (8.1)	Passive congestion (10.7)	Portal vascular anomalies (3)	Lipidosis (cats)
Nonspecific hepatopathy (7.2)	Necrosis (9.9)	Toxic hepatopathy (2)	Focal necrosis
Portosystemic vascular shunt (5.7)	Atrophy (3.8)	Miscellaneous disorders (6)	Cholestasis
Vacuolar hepatopathy (5.6)	Portosystemic vascular shunt (2.5)		Cirrhosis
Fibrosis (4.1)	Miscellaneous disorders (12.9)		Passive congestion
Lipidosis (3.9)			
Primary neoplasia (3.8)			
Cirrhosis (2.3)			
Miscellaneous disorders (6.5)			

*Strombeck DR, Guilford WG, eds. Pathogenesis and incidence of hepatic disease. In: Small Animal Gastroenterology, 2nd ed. Davis, CA: Stonegate Publishing, 1990; 526-527.

**Armstrong PJ, Weiss DJ, Gagne JM. Inflammatory liver disease. In: August JR, ed. Consultations in Feline Internal Medicine 3. Philadelphia, PA: WB Saunders Co, 1997; 63-78.

***Descending order of prevalence. Richter KP. Diseases of the liver. In: Tams TR, ed. Handbook of Small Animal Gastroenterology. Philadelphia, PA: WB Saunders Co, 1996; 409.

LIVER

3) icterus of the sclera, mucous membranes and skin, 4) pigmented urine (bilirubinuria), 5) changes in abdominal configuration due to hepatomegaly and/or ascites, 6) stunted or small body stature and 7) excessive bleeding (i.e., hemorrhages of the skin and mucous membranes, melena, hematuria). Bleeding tendencies develop due to malabsorption of vitamin K in patients with extrahepatic bile duct obstruction or failure of procoagulant synthesis. Subclinical blood clotting abnormalities may become clinically evident during liver biopsy procedures or surgery.

Altered (delayed) drug metabolism may be the first evidence of liver disease recognized by the owner or veterinarian. Prolonged recovery from anesthesia or sedation is a common finding in dogs and cats with liver disease.

Neurobehavioral signs of hepatic encephalopathy develop in young animals with congenital portosystemic vascular anomalies and in animals with severe acquired liver insufficiency. This manifestation of liver disease is uncommon; however, typical signs include aggression (cats), aimless wandering, manic barking (dogs), ataxia, lethargy, episodic weakness, ptyalism (cats especially), altered consciousness (disorientation, stupor or rarely coma), head pressing, sudden blindness, circling, pacing and seizures. As with other metabolic encephalopathies, these signs may be episodic and often can be historically linked to meals, dietary changes, GI hemorrhage or some other causal event.

Alterations in liver and abdominal size are common in dogs and cats with hepatobiliary disorders, though these changes are not always detected on routine physical examination. The normal liver can be difficult to palpate in dogs and cats and the edges are normally sharp, not rounded. In cats, most acquired hepatic disorders are associated with hepatomegaly that can be detected by abdominal palpation. Hepatomegaly is also readily detected in dogs. On the other hand, reduced liver size is difficult to detect in both species. Hepatomegaly may be caused by passive venous congestion, diffuse inflammation, nodular hyperplasia and infiltration by fat, glycogen or neoplastic cells. Pain on palpation of the liver usually indicates acute liver disease but must be differentiated from pain arising from the pancreas, stomach or spleen. Abdominal enlargement associated with ascites usually develops slowly and insidiously. Small amounts of effusion may go undetected, whereas moderate to severe abdominal effusion may be readily noted.

Changes in fecal color and consistency are noted in some patients. Pale, tan or gray acholic feces may be observed when bile flow is obstructed (as with either intrahepatic or extrahepatic cholestasis). Feces become dark green or green-orange when large quantities of bilirubin pigments enter the GI tract as occurs with hemolytic or prehepatic jaundice. Significant upper GI bleeding results in melena.

Table 23-3. Clinical signs associated with hepatobiliary disease.

Early signs	Major bile duct occlusion	Severe hepatic insufficiency	Portosystemic vascular anomaly
Anorexia	Anorexia	Anorexia	Stunted body size
Vomiting	Vomiting/hematemesis	Vomiting/hematemesis	Abnormal behavior (lethargy)
Diarrhea/constipation	Diarrhea/constipation	Diarrhea/constipation	Diarrhea/constipation
Weight loss	Weight loss	Weight loss	Weight loss
Pyrexia	Pyrexia	Pyrexia	Pyrexia
No jaundice	Jaundice within 72 hours	Jaundice as disease advances	No jaundice
Polydipsia/polyuria	Polydipsia	Polydipsia/polyuria	Polydipsia/polyuria
Clear to yellow urine	Orange urine	Clear to orange urine	Clear urine
Iris normal color	Iris normal color	Iris normal color	Copper-colored iris (cats)
	Bleeding tendencies	Bruising/bleeding tendencies	Normal blood clotting
	Acholic (pale) feces	Brown to melenic feces	Brown feces
	Melenic feces if bleeding	Green feces	Melena
	Hepatomegaly (firm, rounded borders)	Hepatomegaly (cats)	Microhepatica
		Normal to microhepatica (dogs)	
	Ascites (if >6 weeks)	Ascites	Ascites (rare)
		Edema (rare in cats)	Edema does not occur
		Hepatic encephalopathy	Hepatic encephalopathy
		Urinary tract obstruction (uroliths)	Urinary tract obstruction (uroliths)
		Ptyalism (cats)	Enlarged kidneys
			Cryptorchid (dogs)

LIVER

Laboratory Evaluation (Pages 816-817)

It is beyond the scope of a nutrition textbook to discuss the plethora of laboratory tests and imaging techniques (i.e., radiography, nuclear medicine, ultrasound) used to detect and confirm hepatobiliary disease and their interpretation. However, routine tests that may help establish parameters for developing feeding and reassessment plans will be summarized.

Liver disease is often discovered during routine hematologic, urine and serum biochemistry screening tests. Hematologic changes may include anemia, abnormal erythrocyte morphology, reduced platelet numbers or function and detection of icteric or lipemic plasma. A regenerative anemia caused by blood loss due to GI hemorrhage and/or a bleeding diathesis may by present. More commonly, a nonregenerative anemia is found and is associated with chronic disease, chronic blood loss, malnutrition and reduced erythrocyte survival. Target cells, poikilocytes and spur cells, Heinz bodies (cats) and microcytosis are erythrocytic abnormalities seen in animals with liver disease. Erythrocyte microcytosis is associated with both acquired and congenital portosystemic vascular shunts in dogs. A recent study suggested that cats with acute or chronic cholangiohepatitis had higher segmented and band neutrophil counts than cats with lymphocytic portal hepatitis.

The liver is the primary site for synthesis, degradation and regulation of plasma proteins. Total protein concentration reflects overall protein balance, but does not provide as much information as albumin and globulin concentration measurements. Albumin has been used commonly as an indicator of liver function, but it is a nonspecific marker because its concentration reflects hepatic synthesis, rate of degradation, pathologic excretion (e.g., urine, GI tract, draining cutaneous lesions) and volume of distribution. Hyperglobulinemia is common in animals with acquired liver disease and may be great enough to mask hypoalbuminemia if only total serum protein concentration is evaluated.

Liver enzymes typically included in serum biochemistry profiles include alanine aminotransferase (ALT, formerly SGPT), alkaline phosphatase (ALP), gamma-glutamyl transferase (GGT) and aspartate aminotransferase (AST, formerly SGOT). Increased serum liver enzyme activity is common in small animal patients and not necessarily associated with significant liver disease. Enzyme activity can increase as a result of induction, reversible and irreversible changes in cellular membranes, hepatocellular injury and/or biliary injury. Increased liver enzyme activity lacks specificity and provides no indication of functional capabilities of the liver. Cats with chronic cholangiohepatitis have higher ALT activities and total bilirubin concentrations than cats with lymphocytic portal hepatitis. Mild increases in liver enzyme activity (i.e., one-and-one-half-fold to twofold normal) in an otherwise normal animal should be evaluated again in two to four weeks. If liver enzyme activity remains abnormal, liver function tests are indicated.

Fasting and postprandial serum bile acid determinations and ammonia tolerance testing are the liver function tests used most often in clinical practice. Liver function studies such as serum bile acid concentrations are used to: 1) identify occult liver disease, 2) assess liver function when there is increased liver enzyme activity, 3) determine whether a liver biopsy is warranted and 4) monitor response to therapy.

Normal blood coagulation depends on production of plasma coagulation factors by the liver. Blood coagulation tests should always be performed in patients with significant liver disease before liver biopsy or surgery. Even if there are no clinical signs of excessive bleeding, coagulation test results are frequently abnormal. In one study, plasma coagulation factor abnormalities occurred in more than half of dogs with naturally occurring hepatic disease.

Imaging the Liver (Pages 817-818)

Routine imaging of the hepatobiliary system includes abdominal radiography and ultrasonographic imaging. The most important features evident during radiographic assessment of the liver are alterations in hepatic size, position and shape and variation in density. Blunting or rounding of the liver margins suggests diffuse hepatomegaly. Irregular or bumpy liver margins indicate hepatic neoplasia, regenerative nodules, hepatic cysts or other focal lesions. Detection of gas in the common bile duct, gallbladder or hepatic ducts is significant and may indicate anaerobic infection, recent surgery, gastroenteritis or paralytic ileus. Radiodense mineralized lesions may represent choleliths (gallstones) or dystrophic mineralization within hepatic parenchyma as a sequela to various hepatic diseases.

Contrast radiography can be used to evaluate the liver and portal blood flow. Portograms are indicated in dogs and cats with suspected congenital or acquired portosystemic vascular shunts. Contrast radiography has been replaced, in most cases, by ultrasonography and nuclear scintigraphy.

Hepatic ultrasonography is useful for initial disease identification and then as a method for monitoring disease progression. Ultrasonography can be used to detect and differentiate focal and diffuse liver disorders. Ultrasonic examination of the hepatobiliary system should include systematic evaluation of the hepatic parenchyma, portal and hepatic veins, gallbladder and biliary system. Ultrasonography is highly operator-dependent and imaging expertise takes time to develop.

Liver Biopsy (Page 818)

Cytologic or histopathologic tissue examination is essential for definitive diagnosis of hepatobiliary disease. Exceptions include patients with

LIVER

congenital portosystemic vascular anomalies, which are confirmed with liver function tests, ultrasonography, portography and/or nuclear scintigraphy. Liver biopsy is an invasive procedure that must be carefully considered before implementation. Common options for securing liver tissue include ultrasonographic-guided needle biopsy, laparoscopic needle or pinch biopsy and celiotomy for wedge biopsy. If a needle procedure is used, a minimum of three and optimally five to seven samples should be collected. The advantage of fine-needle aspiration cytology is decreased risk. However, a representative sample may not be obtained with this technique. Liver tissue should be submitted for histopathologic and cytologic evaluation, aerobic and anaerobic bacterial cultures and copper quantification when copper toxicosis is suspected. Specific stains for collagen, lipid, copper, iron and infectious agents may be required.

Hepatic copper content can be determined using fresh or formalin-fixed liver tissue. Most laboratories need 1 g or less of tissue for analysis. The normal copper content of canine hepatic tissue is debated. Generally, canine hepatic copper concentrations of 400 to 1,000 mg/g dry weight (DW) or less are considered normal. Concentrations from 1,000 to 2,000 mg/g DW may be either a cause or an effect of chronic liver disease. Hepatic copper concentrations greater than 2,000 mg/g DW are often associated with copper toxicosis.

Risk Factors (Pages 818-819)

Although any dog breed can be affected by chronic hepatitis and cirrhosis, certain breeds are predisposed to these disorders. These include: 1) Bedlington terriers (copper-associated hepatotoxicosis), 2) West Highland white terriers, Skye terriers and Doberman pinschers (chronic hepatitis) and 3) American and European cocker spaniels, standard poodles, Labrador retrievers and Scottish terriers (idiopathic cirrhosis). German shepherd dogs are predisposed to idiopathic hepatic fibrosis. Purebred dogs are at increased risk for portosystemic vascular shunts, especially miniature schnauzers, Irish wolfhounds and Yorkshire terriers. Extrahepatic vascular shunts usually occur in cats and small-breed dogs, whereas large-breed dogs are more likely to have an intrahepatic vascular shunt. In general, inflammatory hepatopathies are more common in females, whereas congenital liver disease is more common in males.

The age at onset of clinical signs may be helpful in differentiating congenital from acquired liver disease. Dogs with congenital portosystemic vascular shunts usually develop clinical signs within the first six months of life; most dogs are less than two years old when congenital portosystemic shunts are diagnosed. However, congenital vascular shunts are not diagnosed in some dogs until they are five to 10 years old. Cats are generally older than dogs when diagnosed with shunts. Acquired portosys-

Table 23-4. Clinically relevant hepatotoxins for dogs and cats.

Drugs

Acetaminophen	Methimazole
Amoxicillin/clavulanic acid	Methotrexate
Carprofen	Methoxyflurane
Ciprofloxacin	Methyltestosterone
Diazepam	Oxibendazole
Diethylcarbamazine	Phenobarbital
Diethylcarbamazine-oxibendazole	Phenylbutazone
Glucocorticoids	Phenytoin
Griseofulvin	Primidone
Halothane	Sulfasalazine
Isoniazid	Tetracycline
Ketoconazole	Thiacetarsemide
Mebendazole	Trimethoprim-sulfa
Megestrol acetate	

Chemicals and biologic substances

Aflatoxin	Heavy metals
Blue-green algae	Pennyroyal oil
Cycad seeds	Phenols
Gossypol	

temic vascular shunts secondary to liver disease and portal hypertension occur in animals of any age.

Obese cats and those with prolonged anorexia (from any cause) are at increased risk for hepatic lipidosis. Unvaccinated animals are at risk for infectious viral hepatitis. Exposure to wildlife, livestock and asymptomatic carriers is a risk factor for leptospirosis. Bacterial hepatitis is associated with omphalitis (neonates), septicemia, peritonitis, pancreatitis, trauma and immunosuppressive disorders (secondary to diabetes mellitus, hyperadrenocorticism, etc.). Pancreatitis, extrahepatic bile duct obstruction and inflammatory bowel disease are risk factors for cholangiohepatitis in cats.

The liver is a target for a wide array of chemicals and biologic substances because of its metabolic and detoxifying functions. Animals less than 16 weeks old may have immature hepatic enzyme function for metabolism and excretion of potentially hepatotoxic drugs. **Table 23-4** lists drugs, chemicals and biologic substances that are clinically important causes of liver disease in dogs and cats.

Etiopathogenesis (Pages 819-826)

Metabolic Alterations in Hepatocellular Dysfunction (Pages 819-822)

Hepatocellular dysfunction is responsible for a number of metabolic disturbances that alter usage of various nutrients by the body (**Table 23-**

Table 23-5. Metabolic alterations in hepatic failure.

Alterations	Mechanisms
Hyperglucagonemia	Portosystemic shunting
	Impaired hepatic degradation
	Increased plasma aromatic amino acid levels
	Hyperammonemia
Hyperinsulinemia	Increased peripheral insulin resistance
	Decreased insulin to glucagon ratio
	Impaired hepatic degradation
Increased plasma epinephrine and cortisol levels	Impaired hepatic degradation
Decreased liver and muscle carbohydrate stores	Accelerated glycogenolysis
	Impaired glycogenesis
Increased gluconeogenesis	Hyperglucagonemia
Hyperglycemia (fasting and postprandial)	Portosystemic shunting
	Increased gluconeogenesis
	Decreased insulin-dependent glucose uptake
	Decreased insulin-hepatic glycolysis
Increased plasma aromatic amino acid levels	Decreased hepatic clearance and incorporation into proteins
	Increased release into circulation
Decreased plasma branched-chain amino acid levels	Hyperinsulinemia and excessive uptake
	Increased usage as an energy source
Increased plasma methionine, glutamine, asparagine and histidine levels	Decreased hepatic clearance

5). Changes in protein, carbohydrate and fat metabolism are particularly prominent in the fasting state. Attempts to correct these alterations by manipulating nutrient supply represent an important strategy in the management of patients with significant hepatic disease.

Impaired hepatic metabolism and storage may result in vitamin and mineral deficiencies. A combination of these metabolic and storage problems usually exists in patients with hepatic disease, and each problem should be considered before appropriate dietary therapy is begun.

CARBOHYDRATE ALTERATIONS The liver plays a key role in the usage of the major monosaccharides glucose, fructose and galactose. Glucose can be used for energy production or to synthesize other substrates (e.g., amino acids, fatty acids), or it can be stored as glycogen. Liver glycogen is readily mobilized when glucose is in demand. Hepatic glycogen can normally meet glucose needs (primarily for the brain) for 24 to 36 hours. Gluconeogenesis, the production of glucose from amino acids, glycerol or lactate, is carried out only in the liver and the renal cortex. Glycolysis is the pathway by which glucose can be metabolized anaerobically with production of ATP. Regulation of glycolysis in the liver is highly integrated with that of gluconeogenesis, lipogenesis, glycogen synthesis and glycogenolysis. (See Chapter 2.)

Fasting hypoglycemia is uncommon in patients with liver disease because euglycemia can be maintained with as little as one-fourth to one-third of normal liver parenchymal mass. However, hepatogenic hypoglycemia can occur in dogs with cirrhosis, congenital portosystemic vascular anomalies, fulminant hepatic failure, septicemia and extensive hepatic neoplasia.

The importance and causes of glucose intolerance in dogs and cats with liver disease are poorly documented. Hyperglycemia has been observed in some dogs with cirrhosis and portosystemic vascular shunts and in some cats with hepatic lipidosis and cholangitis or cholangiohepatitis.

PROTEIN AND AMINO ACID ALTERATIONS The liver synthesizes the majority of circulating plasma proteins. The most abundant is albumin, which represents 55 to 60% of the total plasma protein pool. Albumin serves as a binding and carrier protein for hormones, amino acids, steroids, vitamins, calcium and fatty acids, as well as exogenous compounds, drugs, toxins, etc. Albumin also helps maintain normal plasma oncotic pressure. The other proteins synthesized and secreted by the liver are usually glycosylated proteins (i.e., glycoproteins) that function in hemostasis, protease inhibition, transport and ligand binding. Hypoalbuminemia, edema, ascites and increased bruising/bleeding tendencies result from decreased plasma protein production due to liver disease.

Protein regulatory events in the liver include amino acid storage and deamination of amino acids for intermediary metabolism. Generally, the liver degrades essential amino acids (including the aromatic amino acids [AAA], but not the branched-chain amino acids [BCAA]) and some of the nonessential amino acids. In dogs and other omnivores, the activities of key degradative enzymes are typically down regulated when minimal dietary protein is eaten to ensure amino acid availability for protein synthesis. Then, the activities of these key metabolic enzymes rapidly increase when excess dietary protein is ingested. This down regulation does not occur in carnivores such as cats. (See Chapter 11.) Amino acids not required for protein synthesis are deaminated and oxidized or will be converted to carbohydrate and lipid. In this way, the liver plays an important role in energy balance and regulation of plasma concentrations of important amino acids. (See Chapter 2.)

The deamination of amino acids is linked to carbohydrate and lipid metabolism via a number of common intermediates. These intermediates (e.g., pyruvate, fumarate, succinyl-CoA, oxaloacetate and acetyl-CoA) are entry points for amino acid carbon skeletons into the tricarboxylic acid (TCA or Krebs) cycle after deamination. (See Chapter 2.) Intermediates are used primarily for energy production, gluconeogenesis and storage of excess dietary energy as triacylglycerides.

Alterations in nitrogen metabolism are one of the most prominent biochemical changes in chronic liver failure. Hyperammonemia is a common finding and probably results from a combination of factors including: 1) active amino acid deamination and gluconeogenesis, 2) bacterial degradation of protein in the gut, 3) impaired or inadequate ureagenesis and 4) inadequate delivery of ammonia to the liver because of portosystemic vascular shunting.

Plasma amino acid concentrations may be altered in patients with liver disease. Plasma amino acid concentrations differ depending on the type of hepatic failure present. In health, the AAA (i.e., tyrosine, phenylalanine and tryptophan) are efficiently extracted from the portal circulation and metabolized by the liver. Reduced liver function is associated with an increase in circulating levels of AAA because of continued mobilization of amino acids for gluconeogenesis and impaired hepatic AAA metabolism. The plasma concentrations of BCAA (i.e., leucine, isoleucine and valine) and most other amino acids metabolized in peripheral tissues are reduced because of an increased rate of usage by muscle and adipose tissue. The molar ratio between BCAA and the AAA (i.e., BCAA:AAA) in healthy dogs usually ranges between 3.0 to 4.0. This ratio is often reduced to 1.0 or less in dogs with portosystemic vascular anomalies and chronic hepatitis. Conversely, massive acute hepatic necrosis in dogs causes an increase in the plasma concentrations of all amino acids except arginine. Increased circulating catecholamine, insulin and glucagon concentrations are thought to contribute to the altered amino acid metabolism seen in patients with liver disease.

Alterations in plasma amino acid profiles may also play a role in the pathogenesis of hepatic encephalopathy.

PROTEIN AND AMINO ACID ALTERATIONS/Ammonia Metabolism and the Urea Cycle Animals have developed different metabolic approaches to the need to excrete excess ammonia. Mammals use the urea cycle and glutamine synthesis as an ammonia disposal mechanism.

Urea is synthesized in the liver via the urea cycle. In herbivores and omnivores, the urea cycle is controlled by the activities of constituent enzymes, which in turn are controlled by the substrates they act upon.

Additionally, during periods of normal protein intake, most enzymes involved in urea synthesis in non-carnivorous animals operate only at 20 to 50% capacity, allowing for adaptation to high- or low-protein foods. These mechanisms conserve nitrogen during periods of food deprivation, but slow the response time for ammonia detoxification after ingestion of a high-protein meal.

The amino acid intermediates used in the urea cycle (i.e., ornithine, citrulline and arginine) are formed within the cycle itself and are provided by dietary sources of amino acids. In non-carnivorous mammals, amino acids for the urea cycle can be synthesized via alternative pathways. Therefore, non-carnivorous animals can better adapt to foods containing protein of lower quality that may not contain all of the amino acids required for urea cycle function or foods that vary in protein content over time.

In contrast to non-carnivorous animals, carnivores (e.g., cats and ferrets) have not developed adaptive mechanisms to conserve nitrogen during periods of low protein intake. Only minimal changes in enzymatic activity are seen in cats fed either high- or low-protein foods. Thus, urea cycle enzymes act continuously, independent of dietary protein intake. Because enzymatic activity is constant, carnivores control the urea cycle via concentrations of urea cycle intermediates, which allows for rapid detoxification of ammonia.

Carnivores are also unable to synthesize ornithine from proline and glutamate. Therefore, ornithine for the urea cycle must be synthesized exclusively from arginine. Although a small amount of arginine can be synthesized from citrulline in the kidney, the high activity of hepatic arginase dictates that food primarily supply arginine for the urea cycle. Adult cats and ferrets develop hyperammonemia and hepatic encephalopathy when fed foods devoid of arginine.

Glutamine synthesis is the second primary mechanism by which mammals can metabolize excess ammonia. Hepatic glutamine synthetase is compartmentalized in a small area surrounding the centrilobular vein; thus, perivenous cells serve as "scavengers" for any ammonia that has not been converted to urea by the periportal hepatocytes. Approximately one-third of the total ammonia from portal blood is detoxified by glutamine synthesis, although this percentage varies depending on the acid-base status. Glutamine synthesis acts as a backup system for ammonia detoxification, allowing urea production to be decreased as required for acid-base regulation, while preventing hyperammonemia.

LIPID ALTERATIONS Lipid metabolic processes in the liver include: 1) fatty acid and triacylglyceride synthesis, 2) phospholipid and cholesterol synthesis, 3) lipoprotein metabolism and 4) bile salt synthesis. Fatty acids are synthesized in the liver from carbohydrate precursors by conversion of these precursors to acetyl-CoA. Fatty acids are generally stored in the liver as triacylglycerides. After hepatic glycogen stores are depleted, fatty acids are mobilized from adipose tissue and their rate of hepatic oxidation increases. The ketone bodies produced are an important energy source for peripheral tissues (i.e., brain, skeletal muscle) and serve to decrease the rate of glucose usage.

The liver is a site for β-oxidation of fatty acids, producing energy from fatty acid substrates. (See Chapter 2.) Carnitine functions to transport long-chain fatty acids across the inner mitochondrial membrane into the mitochondrial matrix for β-oxidation. The liver is also a major site of cholesterol synthesis from acetyl-CoA. Cholesterol is found throughout the body as a structural component of cell membranes, a substrate for synthesis of steroid hormones and is important in the liver as the precursor for bile acid synthesis. The liver secretes lipoprotein particles and is an essential organ for their uptake and metabolism.

The composition of plasma lipids and lipoproteins is altered in patients with liver disease. These abnormalities are associated with changes in lipoprotein and cholesterol synthesis, lecithin-cholesterol acyltransferase deficiency, defective lipolysis, abnormal recognition and uptake of lipoproteins by the liver and regurgitation of biliary lipids into plasma. Obstructive jaundice may lead to hypercholesterolemia and hypertriglyceridemia. Hypocholesterolemia has been recognized in animals with portosystemic vascular anomalies and acquired hepatic insufficiency. Hypotriglyceridemia has been recognized in dogs with portosystemic shunts and hepatic necrosis. Little is known about changes in lipoprotein fractions in dogs and cats with liver disease.

VITAMIN AND MINERAL ALTERATIONS The liver serves as a storage reservoir for certain vitamins and minerals. Vitamin A can be stored in quantities sufficient for several months. The other fat-soluble vitamins (D, E and K) and vitamin B_{12} are also stored in the liver. The rest of the B vitamins are found in high concentration in hepatic tissue, but the liver is not generally considered as a storage reservoir. Iron from dietary sources and from erythrocyte degradation is sequestered in hepatic tissue. Copper, manganese, selenium and zinc are trace elements normally present in high concentrations in the liver. (See Chapter 2.)

In patients with significant liver disease, malabsorption and alterations in hepatic blood flow may decrease availability and liver concentrations of certain vitamins and minerals. An adequate supply of B-complex vitamins is essential for the liver to perform a myriad of metabolic activities.

Common Hepatobiliary Diseases (Pages 822-826)

FELINE HEPATIC LIPIDOSIS Feline hepatic lipidosis is a well-recognized syndrome characterized by accumulation of excess triacylglycerides in hepatocytes with resulting cholestasis and hepatic dysfunction. Many cats with idiopathic hepatic lipidosis are obese and often present with a history of prolonged anorexia after a stressful event. The biochemical mechanisms responsible for inducing hepatic lipidosis during fasting are not completely understood.

The prognosis for this life-threatening disorder has improved dramatically during the past several years as a result of long-term enteral feeding (i.e., three to eight weeks or longer). Resolution of hepatic lipidosis associated with pancreatitis, infection and the use of drugs depends on the success of treating the underlying disorder.

COPPER-ASSOCIATED HEPATOTOXICOSIS IN DOGS Bedlington terriers often develop copper storage disease and a subsequent hepatopathy. It is caused by an inherited autosomal recessive trait that results in impaired biliary excretion of copper. The frequency of the copper toxicosis gene in the Bedlington terrier breed has been estimated in England and the United States. These estimates suggest that about 25% of Bedlington terriers are affected and another 50% are carriers.

Homozygous recessive individuals invariably accumulate hepatotoxic levels of copper by two to four years of age. Without treatment, affected dogs develop liver disease and die, usually between three to seven years of age. It has become possible to distinguish affected, homozygous normal and carrier dogs in some Bedlington terrier pedigrees using DNA markers.

Hepatic mitochondria are important intracellular targets of hepatic copper toxicosis. Functional abnormalities of mitochondria associated with oxidative injury (i.e., lipid peroxidation) occur in people, rats and Bedlington terriers with copper-induced hepatic injury. Oxidative injury and abnormal hepatic mitochondrial respiration may be involved in the pathogenesis of copper toxicosis. This theory forms the basis for using vitamin E and other antioxidants as potential therapeutic agents.

Multifocal centrilobular hepatitis first appears in Bedlington terriers when hepatic copper concentrations exceed approximately 2,000 ppm (mg/g) DW. Hepatic copper levels greater than 3,000 ppm DW result in widespread hepatic necrosis in some dogs. Postnecrotic cirrhosis develops if the dog survives the episode of massive necrosis.

The role of copper in hepatic diseases observed in other dog breeds is less clear. This includes chronic hepatitis and cirrhosis seen in breeds such as West Highland white terriers, Skye terriers, Kerry blue terriers, cocker spaniels, Doberman pinschers and others.

The liver diseases in these dogs are distinct from copper toxicosis in Bedlington terriers in that hepatic copper concentrations are generally lower and do not increase with age. Further studies are needed to document the specific cause of elevated hepatic copper concentrations in non-Bedlington terrier dogs and the role, if any, of copper in the initiation and progression of hepatic injury in these breeds.

PORTOSYSTEMIC VASCULAR SHUNTS Portosystemic shunts are vascular communications between the portal and systemic venous systems. The communication usually occurs between the portal vein and caudal vena cava and allows access of portal blood to the systemic circulation without first passing through the liver. Congenital portosystemic vascular shunts are most common. They represent anomalous embryonal vessels that occur as single intrahepatic or extrahepatic shunts. Acquired portosystemic vascular shunts form in response to portal hypertension caused by fibrosis and chronic cirrhosis. Multiple extrahepatic shunts are typically seen.

Clinical signs of hepatic encephalopathy usually predominate as a result of inadequate hepatic clearance of enterically derived toxins and altered liver function.

Ammonium urate and other purine uroliths also occur in some animals because of high urinary excretion of ammonia and uric acid. (See Chapter 20.) Stunted growth or failure to gain weight may occur in young animals with congenital shunts.

PORTOSYSTEMIC VASCULAR SHUNTS/Hepatic Encephalopathy A number of conditions can trigger hepatic encephalopathy in animals with compensated liver disease, including azotemia, constipation, use of sedatives and anesthetics, GI bleeding, hypokalemia, alkalosis and high-protein meals.

Ammonia appears to play a role in hepatic encephalopathy, but its precise effect on cerebral function and importance in the pathogenesis remain undetermined. A number of other toxins may act synergistically to cause cerebral dysfunction in patients with hepatic encephalopathy including mercaptans (e.g., methanethiol), short-chain fatty acids, phenols, bile salts and other molecules.

Hyperammonemia and decreased serum urea nitrogen concentrations may also reflect decreased urea cycle function due to decreased hepatic perfusion.

Correction of impaired hepatic blood flow may improve urea cycle function. Most dogs and cats will have normal or markedly improved plasma ammonia concentrations subsequent to surgical ligation of a congenital shunt.

Profound alterations in the plasma concentrations of many amino acids occur in dogs with hepatic encephalopathy. However, the precise relationship of altered amino acid metabolism to the mechanism of cerebral dysfunction in patients with hepatic encephalopathy remains unknown. Plasma amino acid abnormalities also occur in dogs with chronic liver disease that do not have encephalopathic clinical signs.

Methionine may precipitate encephalopathic signs when fed or supplemented in high amounts to patients with portosystemic shunts. Levels of methionine found typically in commercial and homemade foods should not be harmful although excess supplementation (e.g., giving DL-methionine as a urinary acidifier) should be avoided.

Cats develop hepatic encephalopathy if fed foods deficient in arginine. Cats affected with hepatic lipidosis have low serum arginine concentrations. Most commercial cat foods and foods for stress and recovery contain animal-origin protein and thus are well fortified with this amino acid. Homemade vegetable-based foods and human enteral foods used in cats with encephalopathic clinical signs should be supplemented with arginine. Arginine levels in food should always be above the minimum dietary allowance for adult maintenance (>0.5% dry matter [DM] in dogs, >1.0% DM in cats). Dietary arginine levels of 1.2 to 2.0% DM and 1.5 to 2.0% DM seem appropriate for most dogs and cats with liver disease.

Although the precise neurochemical basis of hepatic encephalopathy has not been determined, its development is a severe complication of hepatobiliary disease and should be managed aggressively and appropriately. The encephalopathy is often fully reversible with amelioration of the underlying liver disease.

CHRONIC HEPATITIS AND CIRRHOSIS IN DOGS Chronic hepatitis (i.e., chronic active hepatitis) in dogs is a poorly defined group of clinicopathologic entities characterized by parenchymal necrosis, particularly piecemeal and/or bridging necrosis, with associated lymphoplasmacytic inflammation. Chronic hepatitis is a syndrome in dogs with many causes; it is not a specific disease entity. The presence of lymphoplasmacytic inflammation suggests an immune-mediated mechanism and autoantibodies have been recognized in dogs with chronic hepatitis. However, the target cell or structure of the immune reaction has not been identified. The insidious onset contributes to the poor understanding of the pathogenesis and the advanced stage of the disease when it is recognized in most patients.

Hepatic fibrosis is an accumulation of extracellular collagen and connective tissue within the liver. Fibrosis develops as a sequela to a single episode of massive hepatic necrosis or chronic hepatic parenchymal damage and inflammation. Hepatic cirrhosis is fibrosis with regenerative nodules. Fibrosis and regenerative nodules impair hepatic blood and bile flow, thus perpetuating hepatocellular injury.

Steatorrhea may occur in dogs with chronic hepatitis and cirrhosis, but is uncommon.

LIVER

CHOLANGITIS/CHOLANGIOHEPATITIS IN CATS Cholangitis (i.e., inflammation of the biliary ducts, especially the intrahepatic ducts) and cholangiohepatitis (i.e., inflammation of the biliary ducts and liver) are common feline liver diseases. Three histopathologic types are generally recognized: 1) suppurative, 2) lymphocytic and 3) lymphoplasmacytic. Bacterial infection (*Escherichia coli* and anaerobes are most common) occurs in many cases with suppurative inflammation, whereas immunologic mechanisms are probably involved in the lymphocytic and lymphoplasmacytic types. The suppurative form may precede the other two forms. The endpoint of these clinical entities is often cirrhosis.

Many cats with these conditions also have sludged or inspissated bile, which causes partial or complete biliary obstruction. Concurrent cholecystitis, pancreatitis, extrahepatic bile duct obstruction and inflammatory bowel disease are common in affected cats.

PORTAL HYPERTENSION Portal hypertension (i.e., a persistent increase in portal venous pressure) can be considered a "homeostatic" response to chronic fibrosis, cirrhosis and altered hepatic lobular architecture. Portal venous blood flow is gradually increased to maintain normal perfusion in hepatic lobules in which vascular resistance is increased due to fibrosis. Eventually, portal venous pressure may exceed systemic venous pressure and portosystemic shunts may develop. Shunting nutritionally depletes the liver and substrates are not delivered to the liver for degradation and metabolism; hepatic encephalopathy may result.

Increased vascular resistance within the liver also impairs lymphatic flow and results in ascites. Peritoneal fluid accumulation decreases intravascular volume and systemic venous pressure, further aggravating portal hypertension. Retention of sodium, chloride and water occurs because of pathophysiologic mechanisms similar to those in patients with congestive heart failure. (See Chapter 18.)

BILE DUCT OBSTRUCTION Extrahepatic bile duct obstruction is associated with several conditions. Cholestasis associated with occlusion of the major bile ducts leads to serious hepatobiliary injury within a few weeks. Obstructed bile flow and the resulting stagnation of bile acids and other compounds injure cell membranes and organelles. Bacterial cholecystitis may develop due to biliary reflux of intestinal bacteria or lymphohematogenous dissemination. Biliary injury is associated with cytokine-mediated inflammation and free radical injury. Long-term changes include biliary epithelial hyperplasia, cholangitis, multifocal parenchymal necrosis, fibrosis and cirrhosis. Coagulopathies associated with vitamin K deficiency may develop within three weeks.

Table 23-6. Key nutritional factors for cats with hepatic lipidosis.*

Factors	Recommended levels
Energy	≥4.4 kcal/g
	≥18.4 kJ/g
Fat	25 to 40%
Protein	30 to 45%
Potassium	0.8 to 1.0%
Carnitine	250 to 500 mg/day
Taurine	Feed foods with 2,500 to 5,000 ppm (0.25 to 0.5%)
	Supplement homemade foods and human enteral products with 250 to 500 mg/day
Arginine	1.5 to 2.0%

*Nutrients expressed on a dry matter basis.

Key Nutritional Factors (Pages 826-829)

Most nutritional recommendations for pets with various naturally occurring hepatobiliary diseases are based on understanding normal hepatic function, studies in animals with experimentally induced disease, results in human patients with comparable diseases and clinical experience. The wide range of hepatobiliary diseases and differing severity also mean that one nutrient profile will not be adequate for all patients.

Despite these challenges, general recommendations for key nutritional factors can be made that will benefit most patients with hepatobiliary disorders. The following section will discuss these key factors in more detail and outline specific recommendations for the most common hepatobiliary disorders.

Feline Hepatic Lipidosis (Pages 826-827)

Key nutritional factors for cats with hepatic lipidosis are listed in **Table 23-6** and discussed in more detail below.

ENERGY/FAT Provision of adequate daily energy intake is the cornerstone of successful medical management of cats with hepatic lipidosis. An adequate supply of energy is needed to: 1) prevent catabolism of amino acids for energy, 2) inhibit peripheral lipolysis and 3) avoid excess energy consumption, which will promote hepatic triacylglyceride accumulation. Cats with hepatic lipidosis are often fed commercial veterinary therapeutic products via assisted-feeding techniques. (See Chapter 12 and Appendix 4.) Foods containing 25 to 40% DM fat and energy densities equal to or in excess of 4.4 kcal/g (18.4 kJ/g) DM are well tolerated by most cats and result in clinical improvement when fed in appropriate amounts. The daily energy requirement (DER) for cats with hepatic lipi-

LIVER

dosis should be at least the resting energy requirement (RER) for ideal body weight when cats are managed in the hospital and 1.1 to 1.2 x RER when managed at home.

PROTEIN/AMINO ACIDS Protein and its constituent amino acids are important in cats with hepatic lipidosis. Cats are less efficient in sparing protein during fasting than other animals. As such, protein deficiency is thought to play a major role in the development of feline idiopathic hepatic lipidosis. In cats with hepatic lipidosis, signs of protein malnutrition include hypoalbuminemia, anemia, muscle wasting and negative nitrogen balance. Specific amino acids (e.g., methionine and arginine) become limiting during fasting in obese cats. Protein or amino acid deficiency may induce lipid accumulation in the liver by limiting lipoprotein synthesis needed for normal hepatic lipid metabolism and transport. Protein supplementation at only one-fourth of the daily requirement (22 g protein/day) significantly reduces lipid accumulation in the liver and promotes positive nitrogen balance during long-term fasting in obese cats.

Cats with hepatic lipidosis will usually tolerate moderate amounts of dietary protein unless they are suffering from concurrent hepatic encephalopathy, which is uncommon. Commercial veterinary therapeutic foods containing 30 to 45% DM protein are well tolerated by affected cats and have been used successfully in many cases. Foods for cats with hepatic lipidosis should contain 2,500 to 5,000 ppm taurine on a dry matter basis (DMB) and 1.5 to 2.0% arginine on a DMB.

POTASSIUM Hypokalemia may develop due to inadequate potassium intake, vomiting, polydipsia and polyuria, magnesium depletion and concurrent chronic renal failure. In one study, hypokalemia was present in 19 of 66 cats (29%) with severe hepatic lipidosis. Hypokalemia was significantly related to nonsurvival in this group of cats. Hypokalemia is dangerous because it may prolong anorexia and exacerbate expression of hepatic encephalopathy. Foods for cats with hepatic lipidosis should be potassium replete (0.8 to 1.0% DM potassium), or potassium supplementation (2 to 6 mEq potassium gluconate per day) should be considered.

CARNITINE Food and biosynthesis by the liver are the primary sources of carnitine for animals. Carnitine transports long-chain fatty acids across the inner mitochondrial membrane into the mitochondrial matrix for β-oxidation. Carnitine also removes potentially toxic acyl groups from cells and equilibrates ratios of free CoA/acetyl-CoA between the mitochondria and cytoplasm.

Obesity is a risk factor for feline hepatic lipidosis and several studies have investigated the relationship between carnitine, weight loss in obese cats and feline hepatic lipidosis. Mean concentrations of carnitine in plasma, liver and skeletal muscle are significantly greater in cats with idiopathic hepatic lipidosis than in control cats. These findings suggest that systemic carnitine deficiency does not contribute to the pathogenesis of feline idiopathic hepatic lipidosis. However, other studies have shown that feline foods supplemented with L-carnitine benefit obese cats undergoing rapid weight loss. Dietary L-carnitine supplementation protects obese cats from hepatic lipid accumulation during caloric restriction and rapid weight loss. Foods supplemented with L-carnitine can safely facilitate rapid weight loss in obese cats. Based on these studies, the use of L-carnitine supplements or L-carnitine-supplemented foods seems appropriate for obese cats undergoing weight reduction.

L-carnitine supplementation may also benefit cats with hepatic lipidosis. One author has recommended a dose of L-carnitine for cats with hepatic lipidosis of 250 to 500 mg per day. Others have found that lower doses (7 to 14 mg/BW_{kg}) also result in beneficial effects in weight loss, obesity prevention and in cats with hepatic lipidosis.

Copper-Associated Hepatotoxicosis in Dogs (Pages 827-828)

Key nutritional factors for dogs with copper-associated hepatotoxicosis are listed in **Table 23-7** and discussed in more detail below.

ENERGY/FAT Providing adequate daily energy intake is important in managing patients with copper toxicosis. An adequate supply of energy is needed to allow protein synthesis and prevent tissue catabolism that generates ammonia. The exact caloric needs of affected patients have not been determined, but they would be expected to be similar to those for other patients treated at home (1.4 to 1.8 x RER) or in the hospital (1.0 to 1.2 x RER). Patients with clinical evidence of hepatic encephalopathy should have energy partitioned between carbohydrates and fat as discussed under portosystemic vascular shunts. Foods for dogs with copper-associated hepatotoxicosis should contain 15 to 30 % DM fat and have a caloric density of ≥4.0 kcal/g (≥16.74 kJ/g).

PROTEIN Most dogs with copper toxicosis develop clinical manifestations of liver disease during adulthood (two to six years of age). Protein requirements have not been established for these dogs but would be expected to be similar to those for other adult dogs (15 to 30% DM). Patients with evidence of hepatic encephalopathy will often need more restricted dietary protein levels.

COPPER Avoiding excessive copper intake is important in those animals in which serious hepatic injury has not yet occurred. A minimum dietary copper requirement has been established as 2.9 ppm available copper

Table 23-7. Key nutritional factors for dogs with copper-associated hepatotoxicosis.*

Factors	Recommended levels
Energy	≥4.0 kcal/g
	≥16.74 kJ/g
Fat	15 to 30%
Protein	15 to 30%
	Restrict dietary protein if signs of hepatic encephalopathy develop
Copper	≤5 ppm
Zinc	Give supplemental zinc: 50 to 100 mg elemental zinc, per os, twice daily
Antioxidant vitamins	400 to 500 IU vitamin E, per os, daily
	500 to 1,000 mg vitamin C, per os, daily

*Nutrients expressed on a dry matter basis.

(DMB) for growth. A minimum dietary copper allowance of 7.3 ppm (DMB) for growth and adult maintenance has been established for typical dog foods. Studies have shown that Bedlington terriers achieve copper balance when consuming approximately 0.4 mg copper per day. Therefore, foods for dog with suspected or confirmed copper-associated hepatotoxicosis should contain no more than 5.0 ppm copper (DMB) from an available copper source.

Feeding selected commercial veterinary therapeutic foods or homemade foods can control copper intake. Dogs should not be fed supplements containing copper or table foods that have a high copper content (**Table 23-8**). Certain fiber sources and minerals in food inhibit copper absorption. The appropriate levels of these nutrients in foods for patients with copper toxicosis have not been investigated. Zinc supplementation is important for blocking copper absorption and is discussed below.

ZINC Treatment of hepatic copper toxicosis involves use of zinc and copper chelating agents such as D-penicillamine or trientine. Chelating agents bind to copper and increase its excretion in urine. Unlike chelating agents, zinc is thought to act by blocking intestinal absorption of copper. Animal and human studies have shown that zinc induces synthesis of intestinal metallothionein, which has greater affinity for copper than for zinc. In enterocytes, metallothionein acts as an intracellular ligand binding zinc, copper, mercury and cadmium to form mercaptides, thereby rendering them unavailable for systemic absorption. Thus, the metals are excreted in the feces with desquamated epithelial cells.

Zinc acetate has been recommended and used most often as a source of elemental zinc. Zinc sulfate, zinc methionine and zinc gluconate are also available. A loading dose of 100 mg elemental zinc per os twice daily is given for three months and the dose is then decreased to 50 mg twice

Table 23-8. Copper content of selected foods.

Foods with high to very high* copper content	
Cocoa	Mushrooms
Heart	Nuts
Kidney	Shellfish*
Legumes	Skeletal muscle (meat)
Liver*	

Foods low in copper	
Cheese	Rice
Cottage cheese	Tofu

daily. The zinc should not be given with meals unless nausea and vomiting occur. In those cases, zinc can be given with a small amount of food. Reduced hepatic copper concentrations, decreased hepatic enzyme activity and improved hepatic histologic features were noted after two years of zinc therapy in a small number of affected dogs.

ANTIOXIDANT VITAMINS Because lipid peroxidation has been implicated in the pathogenesis of copper toxicosis, use of supplemental vitamin E, vitamin C and other antioxidants may be beneficial. No specific dosages of vitamins E and C have been documented to be safe and efficacious in dogs with liver disease. However, 400 to 500 IU vitamin E and 500 to 1,000 mg vitamin C given per os daily have been recommended in dogs with inflammatory liver disease.

Portosystemic Vascular Shunts (Page 828)

Key nutritional factors for patients with portosystemic vascular shunts are listed in **Table 23-9** and discussed in more detail below.

ENERGY/FAT Providing adequate daily energy intake is important for managing patients with portosystemic vascular shunts. An adequate supply of energy is needed to allow protein synthesis and prevent tissue catabolism that generates ammonia. The exact caloric needs of these patients have not been determined but would be expected to be similar to those for other patients treated at home (1.4 to 1.8 x RER) or in the hospital (1.0 to 1.2 x RER). Young animals with congenital shunts may be stunted or underweight. DER calculations for these animals should be based on ideal weight rather than current body weight. Foods for patients with portosystemic vascular shunts should contain ≥4.0 kcal/g (≥16.74 kJ/g) DM.

The sources of energy in the food or foods may be important for patients with portosystemic shunts. Experimental studies suggest that feeding foods with a high carbohydrate component is advantageous. Pro-

Table 23-9. Key nutritional factors for patients with portosystemic vascular shunts.*

Factors	Recommended levels
Energy	≥4.0 kcal/g
	≥16.74 kJ/g
Fat	15 to 30% for dogs
	20 to 40% for cats
Protein	15 to 20% for dogs
	30 to 35% for cats

*Nutrients expressed on a dry matter basis.

viding at least 30 to 50% of dietary calories in the form of easily digested, complex soluble carbohydrate (e.g., corn, rice, wheat, barley) may help avert encephalopathic clinical signs.

Fat is an important source of calories. Only a minor decrease in fat digestibility (i.e., 92 to 85%) occurred in dogs with experimentally created portosystemic shunts. Other studies showed that dogs with experimental shunts tolerate foods containing 20 to 25% DM fat. Moderate dietary fat intake (15 to 30% DM fat for dogs and 20 to 40% DM fat for cats) seems appropriate for portosystemic shunt patients unless fat malabsorption (i.e., steatorrhea) is evident.

PROTEIN The protein requirement of patients with portosystemic vascular shunts has been roughly estimated from a nutritional study in adult dogs with surgically created shunts. This study showed that ingestion of 2.11 g crude protein/BW_{kg}/day with an 80% or greater availability was adequate to maintain body protein reserves without producing hepatic encephalopathy. In the absence of other data, this recommendation for dietary protein intake seems appropriate. This equates to approximately 14 to 16% protein calories (15 to 20% DM protein) for dogs and 25 to 30% protein calories (30 to 35% DM protein) for cats. These levels of dietary protein are approximately twice the minimum protein requirement for adult dogs and cats. Some animals with hepatic encephalopathy may need lower levels of dietary protein in conjunction with medical therapy to control behavioral signs.

In addition to the absolute amount of protein fed, the amino acid profile and digestibility are important for optimal protein usage. Amino acids from poor-quality protein sources will be deaminated and metabolized to a greater extent than amino acids from higher-quality protein sources and will exacerbate hyperammonemia. Poorly digested proteins may be degraded by intestinal bacteria and add to the body's ammonia burden.

The importance of the dietary protein source has been studied in human patients with hepatic encephalopathy and in several experimental studies in dogs with portosystemic vascular shunts. Vegetable and dairy protein sources have produced the best results in maintaining positive nitrogen balance with minimal encephalopathic signs in human patients with liver disease. Foods using soybean meal averted encephalopathic signs in dogs with experimentally created shunts. In addition, dairy products (especially cottage cheese) have been frequently recommended for use in homemade foods for dogs and cats with portosystemic shunts and chronic hepatic insufficiency. The amino acid composition of these protein sources is not significantly different from that of meat sources, suggesting that other food factors such as digestibility and levels of soluble carbohydrate and fermentable fiber are important. Fermentable carbohydrates increase microbial nitrogen fixation, reduce ammonia production and absorption and promote colonic evacuation.

The abnormal plasma amino acid profile in patients with hepatic disease can be improved by feeding a protein with an amino acid composition high in BCAA and low in AAA. However, a causal relationship between a deranged BCAA-AAA ratio in plasma and cerebrospinal fluid and hepatic encephalopathy remains unproved. The deranged BCAA-AAA ratio associated with portosystemic shunting correlates better with the severity of shunting and hepatic insufficiency than with the presence or absence of hepatic encephalopathy. Although formulating and using foods to minimize development of this adverse ratio have long been recommended, it is not clear whether the recommendation provides a benefit to people and dogs. In contrast, feeding a food with moderate protein restriction (rather than providing a theoretically optimal amino acid balance) prevents weight loss and development of neurologic signs in dogs with portosystemic shunts.

Chronic Hepatitis and Cirrhosis (Page 829)

Key nutritional factors for patients with chronic hepatitis and cirrhosis are listed in **Table 23-10** and discussed in more detail below.

ENERGY/FAT Providing adequate daily energy intake is important in managing patients with chronic hepatitis. An adequate supply of energy is needed to allow protein synthesis and prevent tissue catabolism that generates ammonia. The exact caloric needs of affected patients have not been determined, but would be expected to be similar to those for other patients treated at home (1.4 to 1.8 x RER) or in the hospital (1.0 to 1.2 x RER). Patients with clinical evidence of hepatic encephalopathy should have energy partitioned between carbohydrates and fat as discussed under portosystemic vascular shunts. Foods for dogs should contain 15 to 30% DM fat and have a caloric density of ≥4.0 kcal/g (≥16.74 kJ/g).

Table 23-10. Key nutritional factors for patients with chronic hepatitis and cirrhosis.*

Factors	Recommended levels
Energy	≥4.0 kcal/g
	≥16.74 kJ/g
Fat	15 to 30% for dogs
	20 to 40% for cats
Protein	15 to 30% for dogs
	30 to 45% for cats
Antioxidant vitamins	400 to 500 IU vitamin E, per os, daily
	500 to 1,000 mg vitamin C, per os, daily

*Nutrients expressed on a dry matter basis.

Foods for cats should contain 20 to 40% DM fat and have a caloric density of ≥4.0 kcal/g (≥16.74 kJ/g).

PROTEIN Protein and its constituent amino acids are important in patients with chronic hepatitis and/or cirrhosis. Hypoalbuminemia, which reflects depleted body stores and reduced protein synthesis, is a frequent and serious problem in patients with chronic liver disease. Protein plays a leading role in hepatic regeneration; therefore, patients with liver disease require adequate protein intake to remain anabolic and support regeneration of hepatocytes. On the other hand, dietary protein restriction may be important in patients with endstage cirrhosis, hyperammonemia and hepatic encephalopathy. Protein, or more accurately, nitrogen excess, is a major contributor to neurotoxic precursors formed when amino acids are metabolized to ammonia. For patients with liver disease, the goal is to provide adequate dietary protein to support hepatic regeneration while avoiding excess dietary protein that might contribute to hepatic encephalopathy.

Most dogs and cats with chronic hepatitis develop clinical problems during adulthood. Protein requirements have not been established for these animals, but would be expected to be similar to those for other adult dogs (15 to 30% DM) or cats (30 to 45% DM). Patients with evidence of hepatic encephalopathy will often need restricted dietary protein levels as discussed under portosystemic vascular shunts.

ANTIOXIDANT VITAMINS Because lipid peroxidation may be involved in the pathogenesis of some forms of acute liver injury and chronic hepatitis, use of supplemental vitamin E, vitamin C and other antioxidants may be beneficial. No specific dosages of vitamins E and C have been documented to be safe and effective for dogs with liver disease. However, 400 to 500 IU vitamin E and 500 to 1,000 mg vitamin C given per os daily have been recommended for dogs with inflammatory liver disease.

Table 23-11. Key nutritional factors for patients with portal hypertension.*

Factors	Recommended levels
Energy	≥4.0 kcal/g
	≥16.74 kJ/g
Fat	15 to 30% for dogs
	20 to 40% for cats
Protein	15 to 30% for dogs
	30 to 45% for cats
Sodium	0.10 to 0.25% for dogs
	0.20 to 0.35% for cats
Chloride	0.25 to 0.40% for dogs
	0.30 to 0.45% for cats

*Nutrients expressed on a dry matter basis.

Cholangitis/Cholangiohepatitis in Cats (Page 829)

Key nutritional factors for cats with cholangitis and cholangiohepatitis are similar to those outlined for cats with hepatic lipidosis (**Table 23-6**).

Portal Hypertension (Page 829)

Key nutritional factors for patients with portal hypertension are listed in **Table 23-11** and discussed in more detail below.

ENERGY/FAT/PROTEIN Portal hypertension usually occurs secondary to chronic hepatitis and cirrhosis. See the energy and protein recommendations in that section for more information and **Table 23-11** for specific dietary recommendations.

SODIUM AND CHLORIDE Excessive dietary sodium chloride should not be given to animals with ascites, portal hypertension and/or significant hypoalbuminemia. Sodium chloride restriction as recommended for patients with renal and cardiac failure is appropriate. Recommended dietary levels are 0.10 to 0.25% DM sodium for dogs and 0.20 to 0.35% DM sodium for cats. Optimal chloride levels have not been established, but are typically 1.5 times sodium levels.

Bile Duct Obstruction (Page 829)

Most patients with bile duct obstruction are candidates for exploratory celiotomy and corrective surgery. Parenteral or enteral assisted feeding is often used before and after surgery, while the patient recovers. (See Chapter 12.) Appropriate adult maintenance-type foods are generally indicated after recovery. Patients with concurrent pancreatitis, exocrine pancreatic insufficiency or inflammatory bowel disease may require a food with an altered nutrient profile. (See Chapter 22.)

LIVER

Other Nutritional Factors (Pages 829-831)

Depending on the hepatobiliary disease, some of the following factors may also be key nutritional factors. (See **Tables 23-6** and **23-7** and **Tables 23-9** through **23-11**.)

Fat (Pages 829-830)

The role of dietary fat in patients with hepatic disease has not been specifically determined. Dietary lipids are beneficial because they have a protein-sparing effect, reduce carbohydrate intolerance, augment fat-soluble vitamin absorption, enhance palatability and are an important source of energy and essential fatty acids.

A minor decrease in fat digestibility (i.e., from 92 to 85%) was found in dogs with experimentally created portosystemic vascular shunts. Clinically significant impaired fat digestion may occur in animals with biliary disease and/or cholestasis. A number of studies have shown that dogs with experimental lesions do well on foods containing 20 to 35% DM fat.

There seems to be no reason for routinely restricting dietary fat in dogs and cats with liver disease. Dietary fat levels of 15 to 30% DM for dogs and 20 to 40% DM for cats are probably appropriate for most patients with liver disease that do not have evidence of significant cholestasis or fat malassimilation (i.e., steatorrhea). One of two different situations may be occurring if steatorrhea is a problem in a patient with hepatobiliary disease. First, the patient may have concurrent disease that is contributing to fat malassimilation, such as exocrine pancreatic insufficiency. Second, the patient may have significant bile duct obstruction and may be a candidate for an exploratory celiotomy.

Increased dietary levels of n-3 fatty acids may benefit animals with inflammatory liver disease. See Chapter 26 for information about the use of fatty acids to modify inflammatory diseases.

Fiber (Page 830)

Foods with increased dietary fiber levels may benefit patients with hepatobiliary disease. Dietary fiber reduces the availability and production of nitrogenous wastes in the GI tract. Although highly digestible foods were previously advocated in an effort to maximize digestion and absorption and reduce colonic residues considered a major source of encephalopathic toxins, this practice is now not recommended. Increased amounts of fermentable fiber encourage nitrogen fixation by enteric bacteria, resulting in reduced quantities of nitrogenous substances available for absorption. Increased dietary fiber is also thought to bind noxious bile acids, endotoxins and other bacterial products. Dietary fiber is also useful in maintaining euglycemia (See Chapter 24.) and altering the pH of colonic contents. Commercial and homemade foods with low dietary fiber levels can be supplemented with psyllium husk fiber (1 tsp per 5 to 10 BW_{kg}, added to each meal).

Taurine *(Page 830)*

In dogs and cats, taurine is synthesized primarily in the liver and bile salts are exclusively conjugated with taurine. Compared to cats, dogs have a high capacity to synthesize taurine and dietary taurine is not essential in most instances. Food-induced bile salt excretion into the gut can result in significant loss of taurine, particularly when normal enterohepatic recycling is interrupted. In cats, taurine synthesis is limited and dietary taurine is essential. Therefore, assurance of adequate taurine nutriture is important in animals with enterohepatic circulation abnormalities and possibly in liver disease. In certain species, taurine also stimulates the synthesis and turnover of bile independent of its role as a bile acid conjugate. Taurine appears to aid in choleresis in dogs and possibly cats. This role may explain the observation that taurine can prevent cholestasis in certain models of liver disease. Most commercial cat foods and foods for stress and recovery are well fortified with taurine. However, homemade and human enteral foods fed to cats should be supplemented with taurine (250 to 500 mg/day). Providing a source of taurine is important for cats with hepatic lipidosis.

Iron *(Page 830)*

Iron deficiency may occur in patients with GI ulceration and hemorrhage associated with chronic hepatitis, portal hypertension or bile duct obstruction. Microcytosis, an erythrocyte abnormality associated with iron deficiency, also develops in dogs with portosystemic vascular shunts despite adequate iron stores. Iron supplementation is indicated when serum iron concentrations are low, hypochromia is recognized or gastroenteric bleeding or another source of chronic blood loss is diagnosed. Iron supplementation of homemade foods is usually necessary.

On the other hand, iron loading by hepatocytes and Kupffer cells has been recognized in some animals with inflammatory liver diseases. Iron is a potent catalyst of oxidative processes and iron-associated hepatic injury may involve lipid peroxidation of membranes and organelle damage. Foods for dogs with chronic hepatitis and those with secondary hemosiderosis documented by evaluation of liver biopsy specimens should avoid excessive iron levels. Iron levels of 80 to 140 ppm (DMB) meet the dietary allowance without providing excessive intake. Injectable or oral supplements containing iron should be avoided in these patients.

Zinc *(Page 830)*

Zinc deficiency could adversely affect multiple aspects of ammonia metabolism. Foods should contain more than 200 mg/kg DM zinc, or the food should be supplemented with zinc gluconate (3 mg/BW_{kg}/day) or zinc sulfate (2 mg/BW_{kg}/day) divided into three doses. Zinc supplementation in patients with copper-associated hepatotoxicosis was discussed previously.

Potassium (Page 830)

Hypokalemia may develop because of inadequate potassium intake, vomiting, hyperaldosteronism, polydipsia and polyuria, magnesium depletion and administration of loop diuretics (e.g., furosemide) for managing ascites. Hypokalemia is dangerous because it may exacerbate anorexia and expression of hepatic encephalopathy. Foods for dogs and cats with liver disease should be potassium replete (i.e., 0.8 to 1.0% DM potassium), or potassium supplementation should be considered.

Vitamins (Pages 830-831)

Vitamin deficiencies are common in patients with chronic hepatic disease. Deficient dietary intake and malabsorption are the principal causes for vitamin deficiency, although decreased storage, metabolic defects and increased requirements also may be involved.

Deficiency of water-soluble vitamins may occur due to inadequate intake, vomiting and urinary losses. Commercial pet foods usually contain sufficient quantities of water-soluble vitamins to meet the needs of most patients with liver disease. Supplementation with water-soluble vitamins is indicated in patients: 1) receiving aggressive diuretic therapy for ascites, 2) with profound polydipsia and polyuria, 3) with prolonged anorexia and 4) eating homemade foods.

Vitamin K becomes important in animals with chronic liver disease, those with prolonged cholestasis and those with evidence of excessive bleeding. Abnormal blood coagulation tests and excessive bleeding reflect impaired hepatic synthesis of clotting factors and/or a consumptive coagulopathy. Patients with chronic liver disease may be vitamin-K deficient or unable to convert vitamin K_1 to its active form. Vitamin K stores in the liver are limited and can be rapidly depleted when dietary sources are inadequate or lipid malabsorption is severe. Abnormal blood coagulation tests in many patients with liver disease will return to normal after parenteral administration of vitamin K_1 (1 to 5 mg/BW_{kg}/day, given intramuscularly or subcutaneously) for several days. This therapy is often given before liver biopsy or surgical procedures.

Assess the Food (Page 831)

After a diagnosis of hepatobiliary disease is made, the quantities of the key nutritional factors in the current food should be determined and compared to the levels listed in the appropriate key nutritional factors table (**Tables 23-6** and **23-7** and **Tables 23-9** through **23-11**). A more appropriate food should be selected if discrepancies exist.

Other important nutrients to consider in some patients with hepatobiliary disease include water-soluble vitamins, iron, copper, zinc, vitamin E, vitamin K and carnitine. The food should be balanced for the appropriate species and age, especially for cats and young animals with congenital hepatopathies. Nutrient quantities are especially important to consider in patients with hepatic encephalopathy.

Crude protein, crude fat and crude fiber quantities in the food can be estimated from the guaranteed or typical analysis on the product label. More detailed nutrient information must be obtained by contacting the manufacturer or consulting published information.

Assess the Feeding Method (Page 831)

It may not always be necessary to change the feeding method when managing a patient with hepatobiliary disease; however, a thorough assessment includes verification that an appropriate feeding method is being used. Items to consider include amount fed, how the food is offered, access to other food and who feeds the animal. All of this information should have been gathered when the patient history was obtained. If the animal has normal body condition (body condition score [BCS] 3/5), the amount of food previously fed (energy basis) was probably appropriate.

FEEDING PLAN (Pages 831-833)

The universal goals for dietary management of hepatobiliary disorders include maintaining metabolic balance while providing nutrients for healing and regeneration of damaged tissue. Other important objectives include: 1) correcting and preventing malnutrition, 2) reducing the need for hepatic "work," 3) avoiding production of hepatotoxic and neurotoxic compounds and 4) eliminating the underlying cause of hepatic disease. The goals of therapy in patients with hepatic encephalopathy also include: 1) recognizing and correcting precipitating causes of encephalopathy (e.g., GI bleeding and constipation) and 2) reducing intestinal production and absorption of neurotoxins.

Select a Food (Page 831)

A wide variety of foods are typically used or recommended for patients with hepatic disease. These include commercial veterinary therapeutic foods (**Tables 23-12** and **23-13**) and homemade foods (**Tables 6-5** and **6-8**). Commercial veterinary therapeutic foods designed for patients with

LIVER

renal disease were often recommended for patients with hepatobiliary disease. Recently, commercial veterinary therapeutic foods designed specifically for patients with hepatic disease have become available (**Tables 23-12** and **23-13**). Foods designed for assisted feeding and recovery are often used in cats with hepatic lipidosis or inflammatory liver disease (**Tables 12-5** and **12-7**).

Although the total protein content of some veterinary therapeutic foods is lower than that of regular commercial pet foods, protein quality and digestibility are usually high. Many of these foods should provide adequate protein to support hepatic function and hepatocyte repair and regeneration while avoiding higher protein levels that increase hepatic workload and might exacerbate hyperammonemia and hepatic encephalopathy. These foods are also balanced and contain appropriate amounts of other key nutritional factors. Short-term discontinuation of protein intake or further reduction of protein intake may be necessary in patients with severe liver failure and hepatic encephalopathy. Foods formulated for patients with renal failure may be appropriate in such instances (**Tables 19-5** through **19-8**).

Determine a Feeding Method (Pages 832-833)

Sick, anorectic and severely malnourished patients with hepatobiliary disease should be hospitalized to initiate supportive care and assisted-feeding techniques. Early tube feeding via nasogastric or gastrostomy tube remains the cornerstone of therapy for feline patients with hepatic lipidosis and all other anorectic patients with liver disease. See Chapter 12 and Appendix 4 for details about foods and enteral feeding techniques commonly used in dogs and cats. Patients that are eating enough food to meet their DER can usually be managed at home.

The DER of patients with hepatobiliary disease is similar to the DER for hospitalized animals and normal adult dogs and cats at home. Force feeding of moist food and appetite stimulants can be used to ensure caloric intake, but these strategies often fail to meet the pet's caloric requirements and frustrate the owner and pet. Appetite stimulants such as anabolic steroids and benzodiazepine derivatives should be used cautiously in patients with hepatic disease, because of the potential for hepatotoxicity.

Many animals may develop learned aversion to the foods they are offered if GI disturbances accompany liver disease. This is the classic scenario in cats with hepatic lipidosis. Cats that refuse to eat a food they associate with nausea may continue to avoid that food even after a complete recovery. Tube feeding is therefore preferable in cats with hepatic lipidosis and should be started immediately after a diagnosis is made. Such an approach is preferred to offering several commercial foods and possibly having the cat develop an aversion to them. The prognosis for feline hepatic lipidosis is influenced largely by the ability of the veterinarian or owner to aggressively meet the caloric requirements of the cat via enteral feeding.

Multiple daily feedings rather than one or two large meals may benefit patients with hepatobiliary disease. Multiple meals may minimize the release of free fatty acids from adipose tissue, improve digestibility and reduce the quantity of ingesta at any one time that enters the large intestine where bacterial fermentation occurs. Studies involving people with hepatic failure have shown that nitrogen balance can be improved if the food is divided into small, frequent meals, including a snack at bedtime. Nauseated patients may also better tolerate multiple small meals.

ADJUNCTIVE THERAPY
(Pages 833-834)

Dietary therapy is only beneficial when performed in conjunction with proper medical and surgical management of the specific hepatobiliary disease involved. Medical management often includes use of antimicrobials, diuretics, antiinflammatory agents, immunomodulators, nonabsorbable disaccharides and bile "altering" agents (**Table 23-14**). In acute hepatic failure, cor-

LIVER

Table 23-12. Levels of key nutritional factors in selected commercial products for dogs with hepatobiliary disease.*

	Protein (%)** 15-30	Arginine (%) 1.2-2.0	Fat (%) 15-30	Energy (kcal/g)*** ≥4.0	Potassium (%) 0.8-1.0	Sodium (%) 0.1-0.25	Chloride (%) 0.25-0.4	Copper (ppm) ≤5.0
Recommended levels								
Moist canine products								
Hill's Prescription Diet Canine k/d	14.9	0.87	26.1	4.8	0.40	0.18	0.47	7.3
Hill's Prescription Diet Canine l/d	18.0	1.40	24.1	4.3	0.90	0.20	0.70	4.5
Hill's Prescription Diet Canine u/d	11.5	0.65	27.2	5.1	0.39	0.25	0.43	3.2
Leo Specific Renil CKW	18.7	na	16.9	4.5	0.94	0.15	0.50	3.8
Leo Specific Uremil CUW	13.1	na	18.4	4.6	0.96	0.14	0.57	2.8
Medi-Cal Canine Reduced Protein	16.8	na	21.8	na	na	0.24	na	na
Pedigree Canine Hepatic Support	16.3	na	11.3	3.9	0.83	0.13	na	5.0
Purina CNM Canine NF-Formula	16.5	na	27.4	4.6	0.72	0.24	0.43	10.3
Select Care Canine Modified Formula	16.8	na	21.8	na	0.96	0.24	na	na
Waltham/Pedigree Canine Low Protein	17.4	na	30.8	5.1	0.30	0.36	na	26.6

(Continued on next page.)

Table 23-12. Levels of key nutritional factors in selected commercial products for dogs with hepatobiliary disease (Continued).*

	Protein (%)** 15-30	Arginine (%) 1.2-2.0	Fat (%) 15-30	Energy (kcal/g)*** ≥4.0	Potassium (%) 0.8-1.0	Sodium (%) 0.1-0.25	Chloride (%) 0.25-0.4	Copper (ppm) ≤5.0
Recommended levels								
Dry canine products								
Hill's Prescription Diet Canine k/d	14.5	0.98	19.0	4.3	0.66	0.23	0.69	15.9
Hill's Prescription Diet Canine l/d	17.8	1.24	24.1	4.8	0.91	0.21	0.78	4.9
Hill's Prescription Diet Canine u/d	9.4	0.61	20.5	4.4	0.63	0.24	0.44	3.5
Iams Eukanuba Kidney Formula-Early Stage	20.6	na	14.2	4.3	0.70	0.52	1.31	na
Iams Eukanuba Kidney Formula-Advanced Stage	15.1	na	14.8	4.6	0.58	0.50	1.41	na
Leo Specific Renil CKD	14.7	na	19.6	4.7	0.99	0.14	0.54	3.9
Medi-Cal Canine Reduced Protein	14.9	na	19.7	na	na	0.28	na	na
Pedigree Canine Hepatic Support Diet	17.8	na	12.2	4.2	0.90	0.13	na	6.0
Purina CNM Canine NF-Formula	15.9	na	15.7	4.4	0.86	0.22	0.57	12.1
Select Care Canine Modified Formula	14.4	na	19.7	4.2	0.88	0.28	na	na
Waltham/Pedigree Canine Low Protein	17.8	na	10.6	4.2	0.78	0.26	na	9.1

Key: na = information not available from manufacturer.
*This list represents products with the largest market share and for which published information is available. Manufacturers' published values. Nutrients expressed on a dry matter basis. Zinc and antioxidant vitamins are key nutritional factors for some patients with liver disease. See **Tables 23-7** and **23-10.**
**Patients with severe liver failure and hepatic encephalopathy may require less protein.
***To convert to kJ/g, multiply kcal/g by 4.184.

Table 23-13. Levels of key nutritional factors in selected commercial products for cats with hepatobiliary disease.*

	Protein (%)** 30-45	Arginine (%) 1.5-2.0	Fat (%) 20-40	Energy (kcal/g)*** ≥4.0	Potassium (%) 0.8-1.0	Sodium (%) 0.2-0.35	Chloride (%) 0.3-0.45	Taurine (%) 0.25-0.5
Recommended levels								
Moist feline products								
Hill's Prescription Diet Canine/Feline a/d	45.7	2.05	28.7	5.1	0.96	0.78	0.83	0.66
Hill's Prescription Diet Feline k/d	29.5	1.54	26.7	4.9	1.05	0.32	0.56	0.39
Hill's Prescription Diet Feline l/d	31.7	2.00	23.3	4.4	0.96	0.21	0.83	0.50
Iams Eukanuba Maximum-Calorie Canine & Feline	43.3	2.55	41.4	5.8	1.04	0.32	0.78	0.26
Leo Specific Renil FUW	34.9	na	39.7	5.6	0.83	0.28	0.44	0.28
Medi-Cal Feline Reduced Protein	36.4	na	45.5	na	na	na	na	na
Purina CNM Feline CV-Formula	42.5	na	26.8	4.9	1.33	0.20	1.09	0.31
Purina CNM Feline NF-Formula	31.1	na	29.5	5.2	0.96	0.16	0.45	0.45
Select Care Modified Formula	35.0	na	53.0	na	1.00	0.23	na	na
Waltham/Whiskas Feline Low Protein	34.5	na	51.0	5.8	0.99	0.33	na	na
Dry feline products								
Hill's Prescription Diet Feline k/d	28.2	1.39	22.3	4.4	0.76	0.25	0.61	0.16
Hill's Prescription Diet Feline l/d	31.8	1.98	23.4	4.5	0.91	0.28	0.68	0.51
Iams Eukanuba Maximum-Calorie Feline	44.2	na	29.6	5.4	0.91	0.54	1.06	0.22
Leo Specific Renil FKD	25.4	na	25.9	2.8	1.44	0.21	1.03	0.14
Purina CNM NF-Formula	30.8	na	12.9	4.3	0.88	0.20	0.64	0.18
Select Care Modified Formula	28.3	na	22.1	4.4	0.92	0.27	na	na
Waltham/Whiskas Feline Low Protein	26.1	na	22.2	na	0.89	0.22	na	na

Key: na = information not available from manufacturer.
*This list represents products with the largest market share and for which published information is available. Manufacturers' published values. Nutrients expressed on a dry matter basis. Carnitine and antioxidant vitamins are key nutritional factors for some patients with liver disease. See **Tables 23-6** and **23-10.**
**Patients with severe liver failure and hepatic encephalopathy may require less protein.
***To convert to kJ/g, multiply kcal by 4.184.

LIVER

rection of fluid and electrolyte imbalances and treatment of other complications such as metabolic acidosis, excessive bleeding, hypotension, hypoglycemia, cardiac dysfunction, renal failure, cerebral edema and infections take precedence over nutritional support. Surgical management includes ligation of portosystemic vascular shunts, correction of bile duct obstruction and removal of focal liver masses.

Nonabsorbable Disaccharides (Pages 833-834)

Administration of lactulose is considered one of the treatments of choice for hepatic encephalopathy. Lactulose is a synthetic disaccharide that is hydrolyzed by colonic bacteria principally to lactic and acetic acids. Lactulose probably exerts its beneficial effects by: 1) lowering colonic pH with subsequent trapping of ammonia, 2) inhibiting ammonia generation by colonic bacteria through a process known as catabolite repression, 3) increasing intestinal transit rate due to its cathartic properties and 4) suppressing bacterial and intestinal ammonia generation by providing a carbohydrate source. The dosage required to achieve these goals varies greatly, with a range of 2.5 to 25 ml, three times daily for dogs and 1.0 to 3.0 ml, three times daily for cats. The dosage should be reduced if watery diarrhea develops.

Lactulose also is highly effective when added to enema fluid (composed of 30% lactulose and 70% water) and given as a retention enema. Approximately 20 to 30 ml/BW$_{kg}$ are infused and retained in the colon for 20 to 30 minutes before evacuation. Lactulose requires intestinal bacteria for activation; neomycin and other antibiotics, however, inhibit bacterial growth. Despite this antagonism, lactulose and neomycin have been used simultaneously with additive or synergistic effect.

Copper Chelating Agents (Page 834)

Treatment of hepatic copper toxicosis is clearly indicated for Bedlington terriers with subclinical or clinical liver disease. Some investigators advocate treatment for other breeds of dogs with chronic hepatitis and cirrhosis in which copper accumulation is documented by liver histopathology and/or elevated hepatic copper concentrations (generally >1,000 to 2,000 ppm DW).

Treatment of hepatic copper toxicosis usually includes zinc and copper chelating agents such as D-penicillamine or trientine. Zinc blocks copper absorption, as discussed earlier under key nutritional factors. Chelating agents bind to copper and increase its excretion in urine. D-penicillamine, the copper chelating agent most frequently recommended for use in dogs, should be given at a dosage of 10 to 15 mg/BW$_{kg}$ twice daily, on an empty stomach. Vomiting is the most common side effect in dogs, but can be alleviated by giving the agent more frequently in reduced doses.

In a clinical trial, chelation results with trientine (10 to 15 mg/BW$_{kg}$, per os, twice daily) were comparable to those of D-penicillamine and

Table 23-14. General therapy for patients with hepatobiliary disease.

Fluid therapy	
Maintain hydration	Give appropriate parenteral fluid therapy
Prevent hypokalemia	Add KCl to maintenance fluids
	Use potassium-replete food or potassium supplement
Maintain acid-base balance	Avoid alkalosis in patients with hepatic encephalopathy
Prevent or control hypoglycemia	Add dextrose to parenteral fluids as needed
Nutritional support	
Maintain caloric intake	Ensure that DER is being met; if not, begin assisted feeding
Provide adequate vitamins/ minerals	Use complete and balanced food
	Add B vitamins to fluids or give as injection
Modify feeding plan to control complications	See specific complications below
Control hepatic encephalopathy	
Modify food and prevent formation and absorption of enteric toxins	Avoid excess dietary protein
	Use retention enemas q6h containing neomycin and lactulose or povidone iodine solution
	Give neomycin or metronidazole and lactulose orally
Control GI hemorrhage	Treat GI parasites, treat gastric ulcers, avoid drugs that exacerbate GI hemorrhage (aspirin, glucocorticoids)
Correct metabolic imbalances	See fluid therapy above
Avoid drugs or therapies that exacerbate hepatic encephalopathy	Do not administer sedatives, analgesics, anesthetics, diuretics, stored blood or methionine-containing products
Control seizures	Use appropriate anticonvulsant drugs (phenobarbital, potassium bromide)
Control infection	Give systemic antimicrobials (see below)
Control ascites and edema	
	Avoid excess dietary sodium chloride
	Administer diuretics (furosemide, spironolactone)
	Paracentesis for relief of dyspnea only
Control coagulation defects and anemia	
	Give vitamin K_1 parenterally
	Give fresh plasma or blood transfusion as needed
Control GI ulceration	
	Give H_2 blockers (cimetidine, ranitidine) or cytoprotective agents (sucralfate)

LIVER

(Continued on next page.)

Table 23-14. General therapy for patients with hepatobiliary disease (Continued.).

Control infection and endotoxemia	
	Give systemic antibiotics (penicillin, ampicillin, cephalosporins, aminoglycosides)
	Give intestinal antibiotics (neomycin)
	Give toxin binders (cholestyramine)
Manage cholestasis	Give bile "altering" or choleretic drugs (ursodiol)
	Surgically correct extrahepatic bile duct obstruction

Key: DER = daily energy requirement, GI = gastrointestinal.

fewer side effects were noted. Modification of 2,2,2-tetramine to 2,3,2-tetramine increases the potency as a copper chelating agent. Use of 2,3,2-tetramine reduced liver copper concentrations significantly in affected Bedlington terriers after 200 days of treatment at a dose of 15 mg/BW$_{kg}$. This drug is not commercially available but can be obtained from chemical supply companies in the form of N,N'-bis(2-aminoethyl)-1,3-propane-diamine and prepared as a salt for oral administration.

Long-term concurrent use of chelating agents and zinc may be counterproductive. However, during initial treatment of symptomatic human patients, physicians have recommended a combination of a chelating agent and zinc for four to six months, then switching to zinc alone for maintenance. This same approach can be used in canine patients.

Vitamins (Page 834)

Parenteral vitamin K$_1$ administration may benefit patients: 1) with chronic liver disease and prolonged cholestasis (i.e., more than two weeks), 2) with clinical evidence of increased bleeding tendencies (i.e., bruising, overt hemorrhage) and 3) undergoing surgical or liver biopsy procedures. Vitamin supplements that contain vitamin K should always be added to homemade foods.

Patients with chronic liver disease, those with polydipsia and polyuria, those receiving diuretics and those with anorexia lasting more than one week may benefit from parenteral administration of B vitamins. Levels of B-complex vitamins in commercial foods are usually adequate, but vitamin supplements that contain B vitamins should always be added to homemade foods.

Vitamin E and vitamin C can be given as antioxidant supplements for patients with inflammatory liver disease. Initial doses of 400 to 500 IU vitamin E and 500 to 1,000 mg vitamin C given per os daily have been recommended for dogs with inflammatory liver disease.

REASSESSMENT (Pages 834-835)

The owner and veterinarian should monitor the appetite, body weight and body condition of the patient, while observing the frequency and severity of GI disturbances (i.e., vomiting, diarrhea), icterus and neurobehavioral signs. One of the most important clinical findings is improvement in the animal's attitude and activity level. This finding is highly correlated with nutritional success. Serial laboratory evaluations (every few days to weeks) of serum liver enzyme activity, serum concentrations of bilirubin, bile acids and potassium and blood ammonia concentrations are also useful. Serial hepatic biopsy specimens (every few months) can be evaluated for hepatic copper concentrations and assessment of inflammatory hepatopathies. Body weight, abdominal configuration and ultrasonography can be used to monitor patients with ascites.

Enteral feeding tubes often can be removed from cats with hepatic lipidosis and/or cholangiohepatitis after several weeks or months of assisted feeding. Enteral tubes are usually removed after the cat has shown clinical improvement and has begun eating two-thirds to three-fourths of its normal DER on its own. Many patients can be fed foods for maintenance of normal adult animals after hepatobiliary disease is resolved. These include patients that have recovered from an acute hepatic insult or hepatic lipidosis and patients that have undergone successful repair of portosystemic shunts.

Vomiting is often a problem in patients with hepatobiliary disease, especially cats with hepatic lipidosis or hepatic encephalopathy. Small frequent meals, continuous tube feeding and antiemetics may be helpful. More aggressive medical treatment of hepatic encephalopathy may also be needed. Long-term administration of lactulose and/or neomycin for management of hepatic encephalopathy may lead to nutrient malabsorption secondary to altered intestinal transit time and suppressed bacterial flora activity.

LIVER

Clinical cases that illustrate and reinforce the nutritional concepts presented in this chapter can be found in Small Animal Clinical Nutrition, 4th ed., pp 838-847.

CHAPTER 24

Endocrine and Lipid Disorders

For a review of the unabridged chapter, see Zicker SC, Ford RB, Nelson RW, et al. Endocrine and Lipid Disorders. In: Hand MS, Thatcher CD, Remillard RL, et al, eds. Small Animal Clinical Nutrition, 4th ed. Topeka, KS: Mark Morris Institute, 2000; 849-885.

INTRODUCTION (Pages 849-851)*

The interactions of carbohydrate and lipid metabolism and their control by hormones comprise the basis for metabolic homeostasis in health and disease. Diagnosing and treating these disorders requires a working knowledge of cellular changes and the role of hormones in controlling nutrient flux. This chapter covers in sequence, diabetes mellitus, hyperthyroidism, hypothyroidism and lipid disorders.

DIABETES MELLITUS (Pages 851-860)

Diabetes mellitus describes an alteration in cellular transport and metabolism of glucose. Certain tissues (e.g., skeletal muscle, adipose tissue and cardiac muscle) depend on insulin stimuli for transport of glucose into their cytosol. In diabetes mellitus, these cells either never receive the signal or do not interpret it properly. Proposed pathophysiologic mechanisms include: 1) insufficient insulin release from the pancreas, 2) lack of insulin receptors (down regulation) and 3) inability of a normal number of insulin receptors to transduce the signal (**Table 24-1**).

Classification (Pages 851-852)

Diabetes mellitus has been classified historically by a variety of schemes. For purposes of this book, only two types will be considered: 1) insulin-dependent diabetes mellitus (IDDM or type I) and 2) non-insulin-dependent diabetes (NIDDM or type II).

In IDDM, beta cells of the pancreas are progressively destroyed and lose their ability to secrete insulin. Approximately 75% of the beta cells must be destroyed before insulin secretion is inadequate to maintain nor-

---

*Page numbers in headings refer to Small Animal Clinical Nutrition, 4th ed., where additional information may be found.

Key Nutritional Factors–Endocrine and Lipid Disorders
(Page 851)

> **Table 24-3** lists key nutritional factors for patients with diabetes mellitus.
> **Table 24-11** lists key nutritional factors for patients with hyperthyroidism.
> **Table 24-14** lists key nutritional factors for patients with hypothyroidism.
> **Table 24-16** lists key nutritional factors for patients with lipid disorders.

Table 24-1. Possible causes of diabetes mellitus in dogs and cats.

Concurrent illness (hyperadrenocorticism, acromegaly)	Infection
Drugs (glucocorticoids, progestins)	Islet amyloidosis
Genetics	Obesity
Immune-mediated insulitis	Pancreatitis

mal glucose tolerance. IDDM is characterized by insulinopenia and dependence on exogenous insulin administration for treatment.

Insulin resistance at the peripheral tissues and/or dysfunctional beta cells characterizes NIDDM. NIDDM has been referred to as a relative insulin deficiency because the amount of insulin actually secreted by the beta cells may be increased, decreased or normal. The concentration of glucose in serum is thus determined by the relative response of peripheral tissues to the secreted insulin, which is usually blunted. Patients with NIDDM may be misdiagnosed with IDDM depending on the severity of beta-cell dysfunction and peripheral insulin resistance. NIDDM patients may not be totally dependent on administration of exogenous insulin to maintain glucose homeostasis.

Assessment (Pages 852-858)
Assess the Animal (Pages 852-857)
HISTORY Dogs and cats with diabetes mellitus are usually examined because of polyuria, polydipsia, polyphagia, weight loss and diminished activity. Care should be taken to differentiate between polyphagia from underfeeding compared to polyphagia associated with disease (true polyphagia). Less commonly, complaints of blindness (dogs), rear-limb weakness (cats) and lethargy (dogs and cats) may be identified. If diabetic ketoacidosis (DKA) develops, affected animals may be presented for evaluation of vomiting, weakness, anorexia and coma. DKA may be precipitated by infection, severe stress, hypokalemia, hypomagnesemia, renal

ENDO/LIPID

Table 24-2. Differential diagnosis for hyperglycemia in dogs and cats.

Acromegaly	Hyperadrenocorticism
Diabetes mellitus	Hyperthyroidism (cats)
Diestrus (bitch)	Laboratory error
Drug therapy (glucocorticoids, progestogens, megestrol acetate)	Pancreatitis
	Pheochromocytoma
Exocrine pancreatic insufficiency	Renal insufficiency
Glucose-containing fluids	Stress

failure, drugs that decrease insulin secretion, drugs that cause insulin resistance or inadequate fluid intake. Subclinical diabetes mellitus may only become noticeable to owners when another disease process precipitates clinical signs associated with DKA. Concurrent disease is common in diabetic dogs and cats and may prompt diabetes mellitus to manifest itself or be a consequence of the diabetic state. Therefore, assessment is important in developing a management protocol.

PHYSICAL EXAMINATION Body condition scores (BCS) for diabetic dogs and cats range from emaciated (BCS 1/5) to obese (BCS 5/5) depending on the severity and duration of disease. Weight loss is probably more severe in animals with IDDM than with NIDDM. Weight loss, which becomes obvious with time, is a hallmark sign of diabetes mellitus. Other physical findings may include lethargy, unkempt coats (cats), hepatomegaly, cataracts (dogs), icterus (cats), rear-limb weakness (cats) and dehydration.

DIAGNOSTIC TESTING/Hemograms Results of complete blood counts are usually within normal ranges in uncomplicated cases of diabetes mellitus. An increase in packed cell volume may be present in dogs and cats with DKA due to decreased extracellular water attributable to osmotic diuresis. Leukocytosis or shifts of white cell morphology to more immature types may indicate an underlying infectious process that is confounding the diagnosis of uncomplicated diabetes mellitus.

DIAGNOSTIC TESTING/Serum Biochemistry Profiles The most consistent and requisite feature of diabetes mellitus is persistent fasting hyperglycemia and glucosuria in the absence of other disease processes. Other disease processes, physiologic states and drugs, however, may cause hyperglycemia (**Table 24-2**). A thorough assessment may help identify the underlying cause of hyperglycemia. Repetitive determination of serum glucose concentrations may be required in cats to differentiate diabetes mellitus from stress hyperglycemia. A diagnosis of DKA is established if ketonuria is present with systemic metabolic acidosis.

Other commonly identified abnormalities include increased serum concentrations of cholesterol and triglycerides. Increased serum concentrations of urea nitrogen and creatinine may be present when dehydration becomes severe enough to impair renal function (prerenal azotemia). Electrolyte and acid-base alterations are more common in animals with DKA and include: 1) hyponatremia, 2) hypokalemia, 3) hypocalcemia, 4) hypomagnesemia, 5) hypophosphatemia and 6) hypochloremia. A shift in acid-base balance towards metabolic acidosis with a compensatory respiratory alkalosis may occur.

Increased activity of alanine aminotransferase (ALT) in serum may be present in cases in which hepatic lipidosis has resulted in hepatocellular damage. Activity of serum alkaline phosphatase (ALP) may also be increased. Increased serum ALP activity is primarily associated with hepatomegaly and biliary stasis; however, pancreatic inflammation resulting in extrahepatic biliary obstruction may also be present. Less commonly, serum concentrations of bile acids and total bilirubin may be elevated.

Dogs and cats with diabetes mellitus may present with concurrent exocrine pancreatic insufficiency or pancreatitis. Increased activity of amylase and lipase in serum may indicate pancreatitis; however, the correlation of these two enzyme activities with pancreatitis is variable and is especially poor in cats. Other disease processes may also result in increased activity of these enzymes in serum. (See Chapter 22.)

DIAGNOSTIC TESTING/Other Biochemical Tests Determination of insulin concentration in serum is not routinely performed in suspected cases of diabetes mellitus. A reliable radioimmunoassay must be used when measuring serum insulin, especially in cats. Insulin exhibits variance in the primary amino acid sequence between species; therefore, the test methodology must be validated for each species. Serum insulin concentrations may be high, normal or low. Concentrations of insulin greater than 15 µU/ml may indicate the presence of functional beta cells. Conversely, concentrations of insulin less than 10 µU/ml do not preclude the possibility of functional beta cells. Serum trypsin-like immunoreactivity can be used to help differentiate pancreatitis (increased activity) from exocrine pancreatic insufficiency (decreased activity). (See Chapter 22.)

Serum thyroid-hormone concentrations are usually normal in diabetic dogs and cats. However, both hypothyroidism and hyperthyroidism may be associated with insulin resistance and can occur in conjunction with diabetes mellitus. As such, evaluation of thyroid function may be useful in patients with diabetes mellitus that is difficult to control with insulin and dietary intervention. Serum thyroxine (T_4) concentrations in sick dogs and cats must be interpreted carefully because T_4 concentrations may be falsely low in poorly regulated cases of diabetes mellitus. This alteration is presumed to be attributable to the euthyroid sick syndrome (A syndrome in which patients have a systemic illness in conjunction with subnormal levels of thyroid hormones. Thyroid function is actually normal in these patients.).

DIAGNOSTIC TESTING/Urinalyses Urine specific gravity is typically greater than 1.025 in diabetic dogs and cats. Urine specific gravity less than 1.015 should increase suspicion for concurrent disorders, such as renal insufficiency or hyperadrenocorticism. Glucosuria is a hallmark finding in untreated diabetic dogs and cats. Lack of glucosuria rules out overt diabetes mellitus as the cause of polyuria and polydipsia. Other common urinalysis findings include ketonuria, proteinuria and changes consistent with urinary tract infection (i.e., bacteriuria and pyuria). Proteinuria may result from either bacterial infection or glomerulosclerosis secondary to basement membrane damage from the primary disease process.

RISK FACTORS Most dogs diagnosed with diabetes mellitus have the IDDM classification type. IDDM may affect dogs of both genders and all ages and breeds, but is more common in the four- to 14-year-old age group, with females affected twice as often as males. Breeds apparently at higher risk include keeshonden, pulis, Cairn terriers, miniature pinschers and poodles. NIDDM is less common in dogs than cats but still accounts for approximately one in five cases.

IDDM and NIDDM may affect cats of any age, but are diagnosed more commonly in cats older than six years. In contrast to findings in dogs, diabetes mellitus occurs predominately in neutered male cats and is usually of the NIDDM type. There has been no breed predilection determined as of yet in cats; however, obesity increases the risk for NIDDM in cats by fourfold.

ETIOPATHOGENESIS/Insulin Physiology Insulin is produced in the beta cells of the endocrine pancreas and is released in response to increased concentrations of glucose in plasma. Active insulin is a dipeptide that is linked by disulfide bonds between cysteine amino acid side chains. Insulin is first synthesized as proinsulin in beta cells and is subsequently processed by a cleavage step that produces C-peptide and active insulin.

Active insulin released into the bloodstream normally interacts at target tissues via cell surface receptors specific for insulin. Most tissues have insulin receptors but some (e.g., skeletal and cardiac muscle and adipose tissue) depend more on insulin for the acquisition of glucose and amino acids than others, and are classified as insulin-dependent tissues. For example, brain tissue has insulin receptors, but is quite capable of transporting glucose intracellularly without the help of hormonal stimuli; therefore, it is considered an insulin-independent tissue.

Insulin receptors in liver and fat cells consist of two alpha and two beta units. The alpha unit provides the binding site for insulin and the beta unit is a transmembrane protein. Insulin receptors appear to transduce the signal by either receptor/insulin complex aggregation and internalization, or via a tyrosine kinase/autophosphorylation reaction that does not require internalization of the complex. The insulin signal then acts intracellularly to increase or decrease specific metabolic pathways such as glycolysis and gluconeogenesis. The insulin signal also increases uptake of glucose into insulin-dependent cells via translocation of the GLUT4 transport protein to the cell surface membrane from the cytosol. GLUT4 is one of five glucose transport proteins and is expressed only in insulin-dependent tissues.

ETIOPATHOGENESIS/Insulin-Dependent Diabetes Mellitus The etiology of IDDM in dogs and cats is not well understood but is probably multifactorial in origin. Genetics, immune-mediated disease, inflammatory conditions, infections, exogenous drugs, body condition, concurrent disease and pancreatic beta-cell degeneration have all been suggested as possible etiologies (**Table 24-1**). Immune-mediated insulitis may play a role in development of IDDM in dogs. In one study, beta-cell specific antibodies were identified in approximately 50% of diabetic dogs studied.

ETIOPATHOGENESIS/Non-Insulin-Dependent Diabetes Mellitus NIDDM is classically characterized by insulin resistance or beta-cell dysfunction. A number of models in other species have been proposed to account for the observed dysfunction including decreased GLUT4 expression or responsiveness in target tissues, leptin protein deficiency in mice, hepatic overproduction of glucose and many others. NIDDM in cats is often characterized by deposition of amylin in pancreatic beta islet cells. Whether amylin is a primary factor in beta-cell dysfunction or secondary to overstimulation of beta cells in response to peripheral insulin resistance is not known. In addition, prolonged elevation of plasma glucose in cats (i.e., glucose toxicity) causes hydropic degeneration of beta cells and decreased pancreatic insulin secretion. NIDDM in dogs and cats appears to be associated with obesity and possibly subsequent down regulation of peripheral insulin receptors.

KEY NUTRITIONAL FACTORS Key nutritional factors consist of nutrients of concern and other factors such as food type and digestibility (**Table 24-3**). This section emphasizes key nutritional factors that have significant variance in commercial foods and markedly affect management of diabetes mellitus. The degree to which any of these factors affects management of diabetes mellitus depends to a large degree on the efficacy of primary disease control through insulin or other pharmacologic treatment.

ENDO/LIPID

Table 24-3. Key nutritional factors for diabetic dogs and cats.*

Factors	Recommended levels

Soluble carbohydrate	Avoid simple sugars
	Provide foods with these levels of complex carbohydrates:
	    50 to 55% for dogs
	    20 to 40% for cats

Energy	For most animals with BCS between 2/5 and 4/5, feeding at the DER for ideal weight in conjunction with adequate control of diabetes mellitus will result in desired body weight:
	    Neutered dogs: 1.6 x RER
	    Intact dogs: 1.8 x RER
	    Neutered cats: 1.2 x RER
	    Intact cats: 1.4 x RER
	In obese animals, restrict energy to achieve weight loss

Fiber**	8 to 17%
Fat	<20%
Protein***	15 to 25% for dogs 28 to 45% for cats True protein digestibility >85%
Macrominerals	Ensure food meets AAFCO recommendations for adult maintenance to compensate for increased losses of magnesium, potassium, chloride, calcium and phosphorus
Food form	Avoid semi-moist foods

Key: BCS = body condition score, DER = daily energy requirement, RER = resting energy requirement, AAFCO = Association of American Feed Control Officials.
*Nutrients expressed on a % dry matter basis.
**Crude fiber underestimates the soluble and total dietary fiber content of foods.
***Animals with renal failure should receive protein levels at the low end of the range.

KEY NUTRITIONAL FACTORS/Water Increased water loss due to osmotic diuresis from glucose, and ketone bodies if DKA is present, must be compensated. Generally, a source of potable water is recommended in amounts sufficient to meet the increased water requirement. This is usually accomplished via free-choice access to water. Dehydrated patients and those with DKA may require parenteral fluid administration. Caution should be observed with type and rate of fluid replacement because of electrolyte perturbations. Rapid replacement of fluid loss with hypotonic solutions may lead to water intoxication and cerebral edema.

KEY NUTRITIONAL FACTORS/Energy Animals with diabetes mellitus display a classic clinical picture of polyphagia with weight loss. This dichotomy may be attributable to inappropriate hormonal signals that result in poor cellular use of glucose and amino acids with concomitant urinary loss of nutrients. Before making recommendations for daily energy requirement (DER), it is important to emphasize that clinical response of animals with diabetes mellitus to dietary manipulation depends on the level of control of the primary disease process and the presence or absence of concurrent disease. For example, if weight loss or weight gain is a continuing problem, it may be due to poorly controlled diabetes mellitus or concurrent disease such as thyroid disorders (dogs and cats), lymphoplasmacytic enteritis (cats) or hyperadrenocorticism (dogs), rather than inappropriate calculation of DER. Consistent reevaluation and owner education are important tools in adjusting food dose and managing diabetes mellitus.

The basal metabolic rate (BMR) may actually be decreased in animals with poorly controlled diabetes mellitus because of the euthyroid sick syndrome. Caution should therefore be taken to avoid over diagnosis of true hypothyroidism in light of the prevalence of euthyroid sick syndrome. Hyperthyroidism is rare in dogs but may occur in some cats with diabetes mellitus.

For most animals (BCS 2/5 to 4/5), feeding at the DER for ideal body weight in conjunction with adequate control of diabetes mellitus will achieve desired body weights. It is best to calculate a DER, as a multiple of resting energy requirement (RER), based on the standard formulas for normal animals. (See Chapter 1.) For neutered dogs, a factor of 1.6 x RER and for intact dogs, a factor of 1.8 x RER are good initial estimates of DER. For neutered cats, a factor of 1.2 x RER and for intact cats, a factor of 1.4 x RER are appropriate starting points (**Table 24-3**). All animals should be reevaluated regularly; food doses should be adjusted as indicated based on body condition.

For obese diabetic animals, a conservative weight-loss protocol may need to be instituted after medical problems are stabilized. Calculation of DER for ideal body weight is a good initial estimate for calculation of food doses. Frequent monitoring and readjustment should be the norm rather than the exception in weight-loss programs for animals with concurrent disease such as diabetes mellitus. Animals that are too lean may need to be fed a food with less than 10% crude fiber to increase food energy density to a level where body weight is increased.

KEY NUTRITIONAL FACTORS/Protein Diabetic animals may have increased loss of amino acids in urine attributable to inappropriate or inadequate hormonal signals and renal glomerulopathy. It is important to provide protein quantity and quality that will meet the requirements of diabetic animals in the face of increased aminoaciduria while avoiding excess protein content that may enhance renal damage or contribute to excessive insulin secretion. Protein should be approximately 15 to 25% of the food on a dry matter basis (DMB) for dogs and between 28 and 45% of the food on a DMB for cats, with a true protein digestibility greater than 85% (**Table 24-3**).

KEY NUTRITIONAL FACTORS/Soluble Carbohydrate Glucose is one of the most potent secretagogues of insulin in healthy subjects. Some details about the effect of carbohydrate composition have been investigated in management of diabetes mellitus in dogs and cats; however, absolute quantities have not been fully evaluated. Until more information is known, 50 to 55% soluble carbohydrates (DMB) for dogs and 20 to 40% soluble carbohydrate (DMB) for cats seem appropriate. The carbohydrate fraction of foods for diabetic patients should come from complex carbohydrates; simple sugars should be avoided (**Table 24-3**).

The use of fructose in foods for cats with diabetes mellitus should be avoided. Cats do not appear to metabolize fructose, which leads to fructose intolerance, polyuria and potential renal damage. Fructose may be found in commercial semi-moist foods, as a humectant in the form of sucrose, or high-fructose corn syrup. The potential effects of fructose in foods for dogs with diabetes mellitus have not been evaluated.

KEY NUTRITIONAL FACTORS/Fiber There is evidence to support the hypothesis that feeding foods with moderate amounts of fiber, substituted for starch, has a positive effect on glycemic control in dogs and cats (**Table 24-4**). Dogs eating foods with an insoluble fiber content of 10 to 15% cellulose (DMB) and more than 50% digestible carbohydrate have significantly better glycemic control than dogs fed the same food without insoluble fiber. Cats fed a food with 12% insoluble fiber also have increased glycemic control.

Addition of soluble fiber (13% DMB) to foods also increases glycemic control in dogs; however, coat and fecal quality may be altered clinically. Soluble fiber may be partially fermented to short-chain fatty acids and then used as energy by colonocytes or absorbed into the blood. These steps decrease the amount of carbohydrate in the bloodstream and do not require insulin for assimilation.

Although an ideal fiber content has not been established, it is evident that including approximately 8 to 17% (DMB) of insoluble or mixed insoluble and soluble dietary fiber aids nutritional management of IDDM and NIDDM in dogs and cats (**Table 24-3**). There appears to be no difference between soluble and insoluble fiber effects in dogs. Some soluble fibers and mixtures of soluble/insoluble fibers may decrease small intestinal digestion of certain nutrients without affecting total tract digestibility. Constipation may occur with use of either fiber type in the management of diabetes mellitus because hyperglycemia inhibits the gastrocolic

Table 24-4. Effect of feeding insoluble dietary fiber to dogs and cats with diabetes mellitus.

	Mean daily insulin dose (U/kg/day)	Mean fasting blood glucose (mg/dl)	Mean blood glucose/ 24 hrs (mg/dl)	Mean urine glucose excretion (g/24 hrs)	Mean glycosylated hemoglobin (%)
Dog food*					
Low-fiber food (1% DM)	1.9 ± 0.6	247 ± 99	246 ± 100	9.3 ± 14.0	6.9 ± 1.8
High-fiber food (13% DM)	1.7 ± 0.5	164 ± 69	184 ± 71	2.8 ± 3.3	5.9 ± 1.4
Cat food*					
Low-fiber food (1% DM)	1.2 ± 0.7	328 ± 153	285 ± 131	Not done	2.7 ± 0.8
High-fiber food (12% DM)	1.0 ± 0.6	191 ± 118	182 ± 99	Not done	2.1 ± 0.4

Key: DM = dry matter.
*By the parameters shown here, dogs and cats eating the higher fiber food had better glycemic control than comparable animals eating the low-fiber food.

response. In addition, increased fiber levels may trap water in the gastrointestinal (GI) tract; therefore, water balance may need to be more closely monitored in animals with poorly controlled diabetes mellitus fed foods with these fiber levels.

KEY NUTRITIONAL FACTORS/Food Type Semi-moist foods tend to have a hyperglycemic effect compared to dry foods because they contain increased levels of simple carbohydrates and other ingredients used as humectants (e.g., propylene glycol). Semi-moist foods should be avoided in dogs and cats with diabetes mellitus.

KEY NUTRITIONAL FACTORS/Fat Abnormalities of lipid metabolism are manifested in diabetic dogs and cats via increased serum concentrations of triglycerides, cholesterol or both. In addition, concurrent pancreatitis is common in dogs and cats with diabetes mellitus. Therefore, excess dietary fat should be avoided in diabetic dogs and cats (**Table 24-3**).

KEY NUTRITIONAL FACTORS/Macrominerals Increased urine output associated with diabetes mellitus increases obligatory loss of electrolytes such as sodium, potassium, chloride, calcium and phosphorus. Dogs and cats with diabetes mellitus may have whole body phosphorus deficits despite normal or high serum phosphorus concentrations. One-fourth of dogs and nearly half of cats with DKA have hypophosphatemia when initially examined. Diabetic ketoacidosis will also hasten loss of cations

ENDO/LIPID

(sodium, potassium, calcium, magnesium) because ketone bodies possess weak acid properties.

Diabetes mellitus may lead to depletion of body magnesium stores via osmotic diuresis, especially when hyperglycemia is poorly controlled. In addition, treatment of DKA may result in shifts of magnesium into intracellular compartments further decreasing serum magnesium concentrations. Magnesium is essential in glucose homeostasis at different levels including as: 1) a cofactor in enzymes of the glucose oxidation pathway, 2) a cofactor in glucose transport systems across plasma membranes, 3) a modulator of energy transfer from high-energy phosphate bonds and 4) having a possible role in the release of insulin. However, it is generally accepted that magnesium depletion is a result rather than a cause of diabetes mellitus; therefore, alleviation of signs of diabetes mellitus would not be expected with dietary repletion of magnesium. Generally, treatment of diabetes mellitus, which results in good glycemic control, corrects magnesium deficiency if the patient is fed a typical commercial pet food. In some cases of diabetes mellitus with severe magnesium depletion, supplemental magnesium may be required as part of the treatment regimen.

It is necessary to replace depleted electrolytes to achieve normal homeostasis. No studies have been performed to establish recommended levels of minerals in foods for animals with diabetes mellitus. Dogs and cats without renal impairment should be fed foods with adequate amounts of phosphorus to avoid and replace whole body phosphorus deficits. However, excess dietary phosphorus should be avoided in animals with renal impairment. Diabetic cats fed foods with low magnesium content should be monitored carefully to avoid magnesium depletion. In general, foods that meet Association of American Feed Control Officials (AAFCO) recommendations for adult maintenance should supply adequate amounts of macrominerals to compensate for the increased losses described above.

OTHER NUTRITIONAL FACTORS/Trace Minerals and Vitamins The role of zinc in diabetes mellitus is controversial; however, it may affect insulin release from the pancreas, glucose tolerance and insulin resistance through changes in insulin binding and activity. Zinc appears to have biphasic activity; low concentrations enhance insulin secretion and activity whereas higher levels reverse this effect. Whole body zinc stores are often low in patients with diabetes mellitus.

Chromium is an essential trace element and is thought to have a role in glucose homeostasis. Chromium has no known enzymatic cofactor function, but it may exist as a complex with nicotinic acid and amino acids to form a "glucose tolerance factor" that may aid insulin action. Chromium supplementation may improve glucose tolerance in malnourished subjects and subjects with poor glucose tolerance. At present, there is no reliable method to detect marginal chromium deficiency. Cats may display some GI side effects when supplemental chromium is administered.

Manganese deficiency has been associated with perturbations in insulin secretion and carbohydrate and lipid metabolism, including impaired glucose usage in laboratory animals; however, its importance in the etiopathogenesis of diabetes is controversial. Repletion of manganese in deficient animals restores normal glucose tolerance and improves insulin secretion. However, treatment of diabetic subjects with manganese supplements had no impact on glycemic control; therefore, it is inferred that manganese deficiency is not a major factor in the pathophysiology of diabetes mellitus.

Substantiation of trace mineral benefits in diabetic dogs and cats has been confounding. Improvement with supplementation appears to occur on a case-by-case basis. In general, until otherwise proven, providing a food with microminerals supplied according to AAFCO recommendations for the appropriate lifestage should suffice for most animals with diabetes mellitus.

Diabetes mellitus may increase or decrease vitamin balance. Conversely, vitamin status may affect the development and manifestations of diabetes mellitus. Much of the investigative work in this area is controversial and needs to be clarified. In general, foods that contain AAFCO recommended levels of vitamins for adult maintenance should meet most of the altered requirements induced by diabetes mellitus. In some cases of diabetes mellitus, it may be necessary to supplement the food with exogenous B vitamins.

Assess the Food *(Page 858)*

Levels of the key nutritional factors should be evaluated in foods currently being fed to diabetic animals. Key nutritional factors include the food form, amount of energy consumed and dietary levels of protein, soluble carbohydrate, fiber, fat and macrominerals. Amounts and levels of key nutritional factors should be compared to those established for diabetic dogs and cats (**Table 24-3**). Information from this aspect of assessment is essential for making any changes to foods currently provided. If key nutritional factors in the current food do not match the recommended levels, then changing to a more appropriate food is indicated. **Tables 24-5** and **24-6** list selected commercial foods often fed to patients with diabetes mellitus.

Assess the Feeding Method *(Page 858)*

It is imperative that feeding methods complement pharmaceutical treatment protocols. Insulin is usually administered in conjunction with meals; therefore, thorough evaluation is required to assess whether any modifications to the feeding plan may be necessary. This is especially true when considering concurrent disease and physiologic changes associated with aging or response to treatment.

ENDO/LIPID

Table 24-5. Selected commercial foods for dogs with diabetes mellitus.*

Products	Protein	Fat	Crude fiber**	Carbohydrate
Recommended levels	**15-25**	**<20**	**8-17**	**50-55**
Dry canine products				
Hill's Prescription Diet Canine w/d	18.9	8.7	16.9	51.0
Iams Eukanuba Glucose-Control	29.0	8.0	2.9	52.4
Purina CNM DCO-Formula	25.3	12.4	7.6	47.8
Select Care Canine Hifactor Formula	25.1	10.6	14.3	42.5
Waltham/Pedigree Canine High Fibre Diet	21.9	8.2	4.9	57.2
Moist canine products				
Hill's Prescription Diet Canine w/d	18.2	13.0	12.2	52.3
Select Care Canine Hifactor Formula	24.8	9.1	15.0	47.0
Waltham/Pedigree Canine High Fibre Diet	29.8	8.0	10.3	45.0

*The list represents products with the largest market share and for which published information is available. Manufacturers' published values. Nutrients expressed as % dry matter. See **Table 24-3** for information about macromineral and energy levels. Semi-moist foods should be avoided.
**Crude fiber underestimates the soluble and total dietary fiber content of foods.

Table 24-6. Selected commercial foods for cats with diabetes mellitus.*

Products	Protein	Fat	Crude fiber**	Carbohydrate
Recommended levels	**28-45**	**<20**	**8-17**	**20-40**
Dry feline products				
Hill's Prescription Diet Feline w/d	38.8	9.3	7.9	37.9
Purina CNM OM-Formula	38.4	8.5	8.0	37.6
Select Care Feline Hifactor Formula	38.2	12.9	5.0	38.3
Moist feline products				
Hill's Prescription Diet Feline w/d	41.3	16.7	10.7	24.6
Purina CNM DM-Formula	56.9	23.8	3.6	8.1
Select Care Feline Hifactor Formula	34.1	16.1	7.2	36.3

*The list represents products with the largest market share and for which published information is available. Manufacturers' published values. Nutrients expressed as % dry matter. See **Table 24-3** for information about macromineral and energy levels. Semi-moist foods should be avoided.
**Crude fiber underestimates the soluble and total dietary fiber content of foods.

Treatment and Feeding Plan (Pages 858-859)

Treatment for diabetes mellitus usually involves a combination of commonly available options. Treatment with injectable insulin or oral sulfonylurea agents has been the mainstay of pharmacologic intervention for uncomplicated diabetes mellitus. Nutritional intervention is the major non-pharmacologic treatment modality in diabetes mellitus. NIDDM in cats may resolve with appropriate dietary treatment and proper case management. Efficacy of dietary treatment depends to a great degree on the ability to manage the primary disease via medical treatment options.

Exercise should be constant from day to day because large variations in activity level may affect glycemic control. Food changes should be accompanied by concurrent monitoring to assess if glycemic control has been affected. In general, increasing fiber levels increases glycemic control and decreasing fiber content decreases glycemic control. Changes in food/fiber may result in the need to adjust the insulin dose by up to 20%. If the weight of the animal changes, in either planned or unplanned fashion, then reassessment should take place at least every two weeks because the insulin dosage may need to be readjusted with changes in body weight. It is also important to control concurrent disease processes via appropriate therapy.

Feeding Plan for IDDM (Pages 858-859)

Dietary modification, in conjunction with insulin therapy, is an effective adjunct in the control of diabetes mellitus in dogs and cats. The model for dietary management of diabetes mellitus is inferred from human nutrition and research in dogs and cats. In general, foods with decreased fat, moderate fiber, moderate protein and low levels of simple sugars have been recommended. Concerns have been raised about the composition of carbohydrate in cat foods because cats have a different capacity to metabolize carbohydrates than dogs.

The energy intake of diabetic animals must be assessed in relation to body condition. A number of diabetic animals may present in obese body condition, which necessitates implementation of a dietary plan to achieve weight loss. (See Chapter 13.) In other cases, feeding a food with moderate to high levels of fiber may pose problems for maintenance of current weight in thin patients. In such cases, feeding an increased quantity of food above the calculated maintenance requirement or slightly increasing the fat content of foods may prove useful. Increasing food dose may not result in increased weight gain if diabetes mellitus is poorly controlled.

Feeding must be coordinated with the time of exogenously administered insulin. This coordination of feeding and maximal insulin activity minimizes postprandial hyperglycemia and maximizes food usage. Ideally, several small meals given at regular intervals throughout the day with and following insulin administration result in minimal hyperglycemia. Generally for animals receiving insulin once daily, half of the caloric requirement should be fed at the time of exogenous insulin administration and the other half eight to 10 hours later. If insulin is given twice daily, half of the caloric requirement should be fed with each injection. It may be beneficial to offer a small snack between feedings. The food should be divided into four or more smaller meals for animals with poor glycemic control to aid in maintaining serum glucose concentrations within an acceptable range. Food should be offered just before insulin administration to avoid insulin-induced hypoglycemia. Commercial foods often used in patients with IDDM are listed in **Tables 24-5** and **24-6**.

Feeding Plan for NIDDM *(Page 859)*

The nutritional plan for NIDDM is similar to that for IDDM. Many cats with NIDDM are obese; therefore, caloric restriction may be a requisite part of dietary management. Care must be taken to avoid rapid weight loss that may predispose to hepatic lipidosis. Loss of 0.5 to 1% of initial body weight per week is considered safe. (See Chapter 13.) Hepatic lipidosis does not seem to be a weight-loss concern for dogs. Sulfonylurea agents may induce vomiting in cats; therefore, food may need to be offered a few hours before drug administration to ensure nutrient absorption.

Feeding Plan for Diabetic Ketoacidosis *(Page 859)*

Intensive care and intravenous fluid administration are not required if the animal is bright, alert and well hydrated. Administration of short- or intermediate-acting insulin can be initiated in conjunction with feeding recommendations similar to those for IDDM and NIDDM. Some DKA animals may require in-hospital intensive care. Goals are to correct dehydration, electrolyte disorders (hypokalemia, hypophosphatemia, hyponatremia, hypochloremia, hypomagnesemia), ketonuria and acidosis while initiating a feeding plan. Nutritional recommendations are similar to those for IDDM and NIDDM after the animal is stabilized.

Reassessment (Pages 859-860)

Clinical Signs *(Page 859)*

Response to treatment can be assessed through careful questioning of the owner of the diabetic pet. Favorable response to treatment may be indicated by decreased water intake, decreased urination, decreased food intake, achievement of weight goals and a generalized increased thriftiness. Continuation of polyuria, polydipsia, polyphagia and inability to achieve weight goals are unfavorable responses. Reassessment should take place every three to four months if the animal is stable and doing well. If the animal is symptomatic, reassessment should take place every one to two weeks until it is stable.

Body Weight and Condition *(Page 859)*

Achievement of weight goals can be measured through assessing body condition and weight. These measurements may also provide insight about the degree of glycemic control and the presence of other disease processes, especially in cases in which adjustments in food dose do not produce expected changes in body condition. Animals should be weighed once every two weeks and have body condition assessed at least once monthly. The owner should be encouraged to keep a chart of body weights and BCS. It may take several months to achieve weight-loss goals in obese animals. A loss of 10% body weight in already thin animals indicates a need for reassessment of the dietary and pharmacologic regimens.

Food Intake (Page 859)

Food intake, with maintenance of body weight, should decrease in animals with a favorable response to exogenous insulin administration. This response is caused by increased nutrient usage associated with hormonal treatment. If animals are anorectic or have depressed food intake, the relative palatability of the food may be poor and another food should be tried after ruling out medical causes. It is especially important to monitor food intake in cats because prolonged anorexia is a risk factor for hepatic lipidosis.

Urine Glucose and Ketones (Page 859)

Most owners can monitor urine glucose and ketones. A decrease in urinary ketone bodies and glucose signals a favorable response to treatment. In well-controlled diabetes mellitus, no ketone bodies should be present in the urine. Several urinalyses should be performed throughout the day to assess glycemic control. Ideally, urine should be free of glucose for the majority of the tests. Moderate amounts of glucose and any ketone bodies indicate a need to reassess insulin treatment or evaluate for concurrent disease.

Biochemistry Profiles (Page 859)

The biochemistry profile should return to normal with well-controlled diabetes and adequate nutritional intake. The primary exception is hyperglycemia that may or may not be present depending on when the blood sample is obtained in relation to insulin administration. Abnormalities of biochemical constituents in the face of controlled diabetes mellitus should be evaluated as separate disease entities.

Glycosylated Hemoglobin and Fructosamine (Page 860)

Hyperglycemia, which induces nonenzymatic glycosylation of body proteins, results in physiologic dysfunction in target proteins. Common complications include cataract formation in dogs, neuropathy in cats and basement membrane disease, nephropathy, weight loss, DKA and recurrent infections in both species. Increased monitoring should be performed to avoid these serious complications when situations with increased stress (e.g., holidays, boarding), unplanned food indiscretions and concurrent disease are present.

INSULIN-SECRETING TUMORS (Pages 860-863)

For information about the nutritional management of insulin-secreting tumors, see Small Animal Clinical Nutrition, 4th ed., pp 860-863.

HYPERTHYROIDISM (Pages 863-868)

Clinical Importance (Page 863)

Hyperthyroidism is primarily a disease of cats with the first clinical reports appearing in the late 1970s and early 1980s. Disease prevalence has been estimated at one in 300 from necropsy findings. The prevalence of the disease appears to be increasing; cases in one hospital increased from three per month to 22 per month over a 12-year period.

In dogs, hyperthyroidism is only found with thyroid neoplasia. However, most thyroid neoplasia in dogs tends to be nonfunctional compared to the functional type found in cats. About 10% of dogs with thyroid neoplasia exhibit hyperthyroidism attributable to functional thyroid tumors.

Assessment (Pages 863-868)

Assess the Animal (Pages 863-868)

The goals of animal assessment in patients are to: 1) establish the diagnosis of hyperthyroidism, 2) identify concurrent disease, 3) assess body condition for specific dietary recommendations and 4) identify inciting causes, if possible.

HISTORY AND PHYSICAL EXAMINATION Clinical signs of excessive thyroid activity are numerous and may be confused with other disease processes (**Tables 24-7** and **24-8**). The appearance of clinical signs may be insidious with some being more distinctive than others. A complete physical and laboratory examination is necessary to achieve a precise diagnosis. All body systems should be examined and evaluated individually and in the context of the overall disease process. Typically, hyperthyroidism is a disease of older cats with a mean age of 12 to 14 years, although cats as young as two years of age have been treated for hyperthyroidism. No gender or breed predilection has been established for hyperthyroid cats.

Dietary history of animals with hyperthyroidism usually includes polyphagia due to increased cellular metabolism. However, it is not uncommon to see decreased appetite following a prolonged period of polyphagia. Decreased appetite is usually associated with weakness, muscle wasting and severe weight loss.

Table 24-7. Clinical signs associated with hypothyroidism and hyperthyroidism.

Clinical signs	Hypothyroidism	Hyperthyroidism
Appetite	Normal to decreased	Increased to decreased
Behavior	Lethargy, mental dullness, inactivity, cold intolerance	Nervous, hyperactive to lethargic, excess vocalization, aggressive, heat intolerance
Coat	Dry/sparse (endocrine alopecia), seborrhea	Dry/greasy/patchy alopecia/unkempt
Eyes	Normal or corneal lipid deposits, corneal ulceration/uveitis	Normal
Heart rate/rhythm	Normal to decreased with possible dysrhythmias	Increased with possible dysrhythmias
Neck	Normal/mass	Normal/mass
Neuromuscular	Seizures, ataxia, circling, vestibular signs, weakness, knuckling, facial nerve paralysis	Weakness, tremors, ventriflexion of head, muscle wasting
Respiratory	Normal	Panting, respiratory distress, dysphonia (dogs)
Skin	Hyperpigmentation	Normal
Stools	Constipation to diarrhea	Bulky to diarrhea
Thirst	Normal to decreased	Increased
Urine	Normal	Excess urination
Vomiting	No	Possible
Weight	Normal to increased	Normal to decreased
Other	Reproductive dysfunction, poor growth	Reproductive dysfunction

Table 24-8. Differential diagnosis for hyperthyroidism.

Non-thyroid endocrine disease	
Acromegaly (rare)	Diabetes mellitus
Diabetes insipidus (rare)	Hyperadrenocorticism (rare)
Renal disease	
Heart disease	
Congestive cardiomyopathy	Idiopathic dysrhythmia
Hypertrophic cardiomyopathy	
Gastrointestinal disease	
Cancer	Inflammatory
Diffuse gastrointestinal disorders	Pancreatic exocrine insufficiency
Hepatopathy	
Cancer	Inflammatory
Pulmonary disease	

Hyperthyroid animals are typically underweight and often have muscle wasting. However, animals may be within normal weight ranges depending on how advanced the disease process is when recognized. The presence of concurrent diseases may influence these parameters and careful examination may reveal other abnormalities that may not be noted by owners.

LABORATORY AND OTHER DIAGNOSTIC TESTING The primary purpose of laboratory testing is to establish a diagnosis of hyperthyroidism and screen the animal for concurrent disease. Any number of abnormalities may be present in individual animals; however, clinical studies have enumerated common changes (**Table 24-9**). Specific diagnostics for thyroid dysfunction should be performed if thyroid disease is still consistent with and suspected from results of the initial screening.

LABORATORY AND OTHER DIAGNOSTIC TESTING/Serum Thyroid Concentrations Thyroid hormones in blood (total T_4 or triiodothyronine [T_3]) may exist either bound to carrier proteins or free in the water fraction of serum (free T_4 or T_3). Determination of total T_4 concentrations has proven more useful than total T_3 levels for differentiating hyperthyroidism in cats. Free T_4 concentrations (by dialysis) should be determined to substantiate disease in cases in which total T_4 values are nondiagnostic, but hyperthyroidism is still suspected.

Findings in dogs with thyroid neoplasms depend on the type of neoplasm and extent of proliferation or destruction of normal thyroid tissue within the gland. In dogs, measurement of total T_4 or T_3 is considered sufficient for diagnosis and has revealed that hyperthyroidism is present in about 10% of thyroid neoplasms. The remaining thyroid neoplasms in dogs cause either euthyroid (55%) or hypothyroid (35%) characteristics.

Table 24-9. Laboratory findings in animals with hyperthyroidism.

Laboratory tests	Cats	Dogs
Biochemical analysis	Increased ALT, ALP, creatinine, urea nitrogen, glucose, bilirubin and phosphate values	Mild increases in urea nitrogen, liver enzyme and possibly calcium values
Cardiac diagnostics	Tachycardia, PVCs, hypertrophic cardio-myopathy	Normal
Complete blood count	Erythrocytosis, leukocytosis, lymphopenia, eosinopenia, increased MCV	Mild, normocytic, normochromic, non-regenerative anemia
Imaging	Normal or cardiac/respiratory abnormalities	Normal or cardiac/respiratory abnormalities
Urinalysis	Increased or decreased specific gravity, glucosuria, signs of inflammation	Normal

Key: MCV = mean corpuscular volume, ALT = alanine aminotransferase, ALP = serum alkaline phosphatase, PVC = premature ventricular contraction.

LABORATORY AND OTHER DIAGNOSTIC TESTING/Thyroid-Hormone (T_3) Suppression Test This test is used to distinguish euthyroid from mildly hyperthyroid cats in cases in which T_4 and fT_4 (free T_4) values are nebulous. The T_3 suppression test is based on the theory that thyroid-stimulating hormone (TSH) is suppressed by increasing concentrations of thyroid hormones in normal cats. TSH secretion is suppressed and T_4 and T_3 concentrations are increased in hyperthyroid cats because of autonomous secretion. In this test, T_3 is administered orally, three times per day for seven treatments, while T_4 concentrations are determined before and after T_3 administration. T_4 concentrations will not be suppressed in hyperthyroid cats but will be suppressed in normal cats.

LABORATORY AND OTHER DIAGNOSTIC TESTING/Sodium Pertechnetate This test is used to identify functional thyroid tissue in dogs and cats. Radioactive sodium pertechnetate is infused intravenously and uptake by thyroid tissue is assessed by scintillation scan. Sodium pertechnetate uptake is useful for diagnosing unilateral compared to bilateral disease, extrathyroidal tissue, metastases and thyroid-tumor activity in cats with hyperthyroidism. In dogs, pertechnetate uptake may identify the extent of a thyroid tumor if the tumor takes up the radionuclide. However,

ENDO/LIPID

Table 24-10. Goitrogenic factors in foods and the environment.*

Nutrients or food types	
Cabbage (goitrin)	Millet
Canned foods	Rutabagas
Cassava (linamarin)	Sweet potatoes
Cyanides	Turnips
Excess iodine	Seaweed
Iodine deficiency	Various beans (including soybeans)
Environmental	
Polychlorinated biphenyls (fish-containing foods)	Polyphenols (fish-containing foods)
Pesticides	Propylthiouracil (drug)
Phthalates	Resorcinols (fish-containing foods)

*Epidemiologic associations and risk factors.

pertechnetate uptake does not correlate well with metastatic potential or the thyroid-functional status of thyroid tumors in dogs.

ETIOPATHOGENESIS For more information about normal thyroid function, see Small Animal Clinical Nutrition, 4th ed., pp 865-866.

RISK FACTORS/Feline Hyperthyroidism No definitive risk factors have been identified for the apparent increased prevalence of hyperthyroidism in cats. Whether the increased prevalence is real or not is subject to question. Several reasons have been suggested for the apparent increase including increased owner awareness, increased popularity of cats as house pets, increased longevity, environmental factors and increased diagnostic skills of veterinarians.

Nutritional factors have been suggested as potential inducing agents. Cats fed moist foods are apparently at significantly higher risk of hyperthyroidism than those fed dry or semi-moist foods. Most commercially prepared cat foods contain adequate amounts of iodine, with measured levels ranging from three to 100 times recommended amounts.

In addition, deficient or excessive iodine intake in homemade or poorly formulated foods may also be goitrogenic. Other goitrogenic compounds may be unknowingly included in the production process of commercial foods or in homemade foods (**Table 24-10**). Because several goitrogenic compounds are metabolized via hepatic glucuronidation, which is a limiting pathway in cats, they may contribute to hyperthyroidism. None of the above nutritional factors has been proven to be involved in the apparent increased prevalence of hyperthyroidism in cats.

Other non-nutritional risk factors suggested, but not proven, have included circulating thyroid stimulators similar to the immunoglobulins produced with Grave's disease in people. Growth factors (e.g., platelet-derived factors, epidermal growth factors and insulin-like growth factors) stimulate thyroid-cell proliferation and may affect the gland via autocrine or paracrine activity. Finally, it is possible that selection of an oncogene that promotes hyperthyroidism has occurred with the continuous in-breeding of domestic cats.

RISK FACTORS/Canine Hyperthyroidism The risk factors for development of hyperthyroidism in dogs are similar to those for cats. Iodine deficiency or excess, chronic stimulation of thyroid-hormone production, goitrogenic substances in the environment or food, ionizing radiation and gene abnormalities have all been suggested as possible etiologies. No definitive cause has yet been identified.

KEY NUTRITIONAL FACTORS Key nutritional factors for dogs and cats with hyperthyroidism are listed in **Table 24-11** and described in detail below.

KEY NUTRITIONAL FACTORS/Water Cats and dogs with hyperthyroidism often exhibit polyuria and polydipsia. Therefore, a readily available source of potable water is recommended for free-choice access.

KEY NUTRITIONAL FACTORS/Energy Uncompensated hyperthyroid animals are usually in an increased metabolic, energy-deficit state. Treatment of the primary disease usually results in equilibration of energy requirements to what is expected for age and physiologic status. Therefore, primary emphasis should be directed at regulation of the disease process rather than nutritional intervention. Provision of DER at the calculated ideal body weight of the animal should result in rapid return to normal body weight if primary disease processes are controlled (**Table 24-11**).

KEY NUTRITIONAL FACTORS/Protein Hyperthyroid animals are in a hypercatabolic state and may exhibit signs of protein wasting and deficiency. Increased protein intake may be needed during the recovery period to replenish body protein (**Table 24-11**). However, hyperthyroidism is frequently associated with renal failure, which should prompt a complete evaluation of renal function before feeding higher protein foods. (See Chapter 19.)

KEY NUTRITIONAL FACTORS/Fat Hyperthyroid animals may have decreased fat stores because they are in an increased metabolic state. Treatment of primary disease and use of a food that meets AAFCO nutrient allowances for the desired physiologic state should result in rapid normalization of body weight. If severe wasting of body mass has occurred, the fat content of foods may be increased to achieve higher energy density and enhance weight gain (**Table 24-11**).

KEY NUTRITIONAL FACTORS/Fiber Foods with increased fiber levels (8 to 23% dry matter [DM]) should be avoided in animals with poor body condition (**Table 24-11**).

ENDO/LIPID

Table 24-11. Key nutritional factors for hyperthyroid dogs and cats.*

Factors	Recommended levels
Energy	Feeding at the DER for ideal weight in conjunction with adequate control of hyperthyroidism will result in desired body weight: Neutered dogs: 1.6 x RER Intact dogs: 1.8 x RER Neutered cats: 1.2 x RER Intact cats: 1.4 x RER
Protein	In underweight animals, provide increased dietary protein. The following levels are adequate unless renal function is compromised. 15 to 25% for dogs 28 to 45% for cats True protein digestibility >85%
Fat	In underweight animals, provide increased dietary fat. Fat levels for normal animals are usually adequate until normal body condition is achieved: See Chapter 9 for dogs See Chapter 11 for cats
Fiber	Avoid fiber levels >8% in animals with poor body condition
Macrominerals	Ensure food meets AAFCO recommendations for adult maintenance to compensate for increased losses of magnesium, potassium, chloride, calcium and phosphorus
Trace minerals	Generally, foods that meet AAFCO minimum allowances for trace minerals are adequate; however, commercial products vary greatly in trace mineral content. It may be necessary to contact product manufacturers to determine iodine and selenium levels.

Key: DER = daily energy requirement, RER = resting energy requirement, AAFCO = Association of American Feed Control Officials.
*Nutrients expressed on a % dry matter basis.

KEY NUTRITIONAL FACTORS/Macrominerals Foods that meet AAFCO minimum nutrient allowances for phosphorus, potassium, sodium and calcium should suffice in most cases (**Table 24-11**). Decreased sodium chloride intake may benefit some cases in which hypertension and cardiac disease are primary problems. (See Chapter 18.)

KEY NUTRITIONAL FACTORS/Trace Minerals Iodine may be excessive or deficient in different states of thyroid disease. Iodine intake should be thoroughly evaluated to determine adequacy. Generally, foods that meet AAFCO minimum allowances for trace minerals are adequate; however, commercial products vary greatly in trace mineral content (**Table 24-11**).

Assess the Food (Page 868)

Information from assessing the food is essential for making any changes to foods currently fed. This information, in conjunction with assessment of the animal, leads to development of a rational, complete nutritional plan. After assessment of the animal and current food is completed, the following steps should be taken: 1) compare the current food's levels of key nutritional factors with recommended key nutritional factors and 2) identify discrepancies between key nutritional factors and current intake.

Assess the Feeding Method (Page 868)

It may not always be necessary to change the feeding method when managing animals with hyperthyroidism. However, a thorough evaluation includes verification that an appropriate feeding method is being used. Any deviations from ideal feeding methods should be identified and changes made as required.

Treatment and Feeding Plan (Page 868)

The success of nutritional management of hyperthyroidism depends to a great degree on the effectiveness of medical/surgical treatment for the primary disease. Three modes of treatment are generally accepted for hyperthyroidism in cats: 1) long-term antithyroid medication, 2) surgical thyroidectomy and 3) radioactive iodine.

Foods adequate in energy, protein, palatability and other factors outlined in **Table 24-11** are recommended for hyperthyroid animals. Commercial moist, semi-moist and dry foods that meet AAFCO minimum nutrient allowances for adult maintenance should suffice. (See Appendix 2.) Commercially prepared maintenance-type foods may be mixed with growth-type formulas to achieve higher protein and fat intakes, if needed, during the convalescent period. However, growth foods may add excessive sodium and phosphorus possibly complicating concurrent renal disease or primary cardiac disease. Known goitrogenic substances should be removed from the food and environment, if possible (**Table 24-10**).

Animals will usually return to normal body weight if provided energy at the calculated DER for ideal body weight. Feeding small amounts several times daily may be necessary during recovery. Two daily feedings are adequate after an animal resumes normal eating behavior.

Reassessment (Page 868)

Patient response to treatment may be assessed in a variety of ways. The simplest method is via owner observation of clinical signs, bimonthly body weight charting and monitoring food intake. Return to normal activity, size and appearance generally indicates successful response to treatment. No change in clinical signs, weight or activity necessitates reevaluation of the treatment and diagnosis.

HYPOTHYROIDISM (Pages 863-868)

Clinical Importance (Page 863)

Hypothyroidism is primarily a problem of dogs and rarely observed in cats. The most common cause of hypothyroidism in dogs is classified as primary (causes within the gland); however, extrathyroidal causes have been identified (**Table 24-12**). No overall incidence has been reported; however, hypothyroidism is considered the most commonly diagnosed endocrinopathy in dogs. Cases of hypothyroidism in cats are typically congenital or iatrogenic resulting from treatment of hyperthyroidism.

Assessment (Pages 863-868)

Assess the Animal (Pages 863-868)

The goals of animal assessment in patients with suspected thyroid disease are: 1) establish the diagnosis of hypothyroidism, 2) identify concurrent disease, 3) assess body condition for specific dietary recommendations and 4) identify inciting causes, if possible.

HISTORY AND PHYSICAL EXAMINATION Clinical signs of insufficient thyroid activity are numerous and may be confused with other disease processes (**Tables 24-7**). The appearance of clinical signs may be insidious; some signs may be more distinctive than others. A complete physical and laboratory examination is necessary to achieve a diagnosis. All body systems should be examined and evaluated individually and in the context of the overall disease process. Some canine breeds appear to be at increased risk for hypothyroidism; however, there is no documented gender predilection.

Hypothyroidism is usually associated with signs related to decreased cellular metabolism. Occasionally, iodine deficiency has been identified as a cause of hypothyroidism in dogs; however, most commercial dog foods contain adequate iodine. No other associations with dietary history have been reported as causes of hypothyroidism in dogs. However, selenium deficiency may influence thyroid status because deiodinase I and possibly deiodinase II are selenoproteins.

Animals with hypothyroidism often are overweight, in poor body condition and have poor skin and coat conditions. However, animals may be within normal weight ranges depending on how advanced the disease pro-

Table 24-12. Etiology of hypothyroidism in dogs.

Primary hypothyroidism

Lymphocytic thyroiditis	Iatrogenic
Idiopathic atrophy	Antithyroid medication
Follicular cell hyperplasia	Radioactive iodine treatment
(dyshormonogenesis)	Surgical removal
Neoplastic destruction	

Secondary hypothyroidism

Defective TSH molecule	Hypoplasia of pars distalis
Defective TSH-follicular cell receptor	Pituitary cyst
interaction	Pituitary destruction
Iatrogenic	Neoplasia
Drug therapy (glucocorticoids)	Pituitary thyrotropic cell suppression
Hypophysectomy	Euthyroid sick syndrome
Radiation therapy	Naturally acquired hyperadreno-
Pituitary malformation	corticism

Tertiary hypothyroidism

Congenital hypothalamic	Hemorrhage
malformation	Inflammation
Acquired destruction of	Neoplasia
hypothalamus	Defective TRH molecule
Abscess	Defective TRH-thyrotroph receptor
Granuloma	interaction

Congenital hypothyroidism

Circulating thyroid hormone	Ingestion of goitrogens
transport defects	Thyroid gland dysgenesis
Deficient dietary iodine intake	Aplasia
Dyshormonogenesis	Ectasia
(iodine organification defect)	Hypoplasia

Key: TRH = thyrotropin-releasing hormone, TSH = thyroid-stimulating hormone.

cess is when recognized. The presence of concurrent diseases may influence these parameters and careful examination may reveal other abnormalities that may not be noted by owners.

LABORATORY AND OTHER DIAGNOSTIC TESTING The primary purpose of laboratory testing is to establish a diagnosis of hypothyroidism and screen for concurrent disease. Several abnormalities may be present in individual animals; however, clinical studies have enumerated common changes (**Table 24-13**). Specific diagnostics for thyroid dysfunction should be performed if thyroid disease is still consistent with and suspected from results of the initial screening.

LABORATORY AND OTHER DIAGNOSTIC TESTING/Serum Thyroid Concentrations A diagnosis of hypothyroidism may be made through

Table 24-13. Laboratory findings in animals with hypothyroidism.

Biochemical analysis	Increased cholesterol, triglyceride, ALT (mild), ALP (mild) and CK (mild, variable) values
Cardiac diagnostics	Bradycardia, inverted T waves
Complete blood count	Normocytic, normochromic, nonregenerative anemia with leptocytes possible
Imaging	Normal/thyroid mass, metastatic lesions, thoracic or abdominal effusion
Urinalysis	Normal to nonspecific increase in white blood cells

Key: ALT = alanine aminotransferase, ALP = serum alkaline phosphatase, CK = creatine kinase.

thyroid-function testing, evaluation of clinical signs and response to trial thyroid-hormone administration. In theory, low concentrations of thyroid hormone should be associated with a diagnosis of hypothyroidism. However, hypothyroid dogs without neoplasia of the thyroid gland are usually not definitively diagnosed by determination of concentrations of total T_4 or T_3 in serum due to the considerable overlap of the concentration of these two hormones between euthyroid and hypothyroid animals. Measurement of free T_4 in serum has the same interpretative dilemma as total T_4 and is considered not as reliable. In addition, a small percentage of hypothyroid dogs have normal T_4 levels in the face of antithyroid hormone antibodies, which further clouds diagnostic accuracy of thyroid-hormone tests. Concentrations of these hormones should be interpreted in conjunction with clinical and physical signs to arrive at an educated assessment of thyroid status.

LABORATORY AND OTHER DIAGNOSTIC TESTING/Serum TSH Assays for canine TSH are available and appear quite promising in diagnosis of hypothyroidism. In addition, TSH testing in conjunction with baseline thyroid-hormone concentrations in blood appears to be a sensitive and specific method to diagnose canine hypothyroidism in preliminary studies.

ETIOPATHOGENESIS For more information about normal thyroid function, see Small Animal Clinical Nutrition, 4th ed., pp 865-866.

Hypothyroidism in dogs can be divided into primary, secondary, tertiary and congenital causes. Primary hypothyroidism is by far the most prevalent form, comprising approximately 95% of all cases. The two most common causes of primary hypothyroidism in dogs are lymphocytic thyroiditis and idiopathic follicular atrophy (**Table 24-12**).

RISK FACTORS The potential risk factors and causes of hypothyroidism in dogs are numerous. There appears to be a breed predilection with purebred medium- to large-breed dogs being more commonly represented. This finding may indicate a genetic predisposition. Iodine deficiency seems unlikely as a potential inciting agent because commercial dog foods usually contain adequate iodine. However, selenium metabolism and status may need to be investigated.

KEY NUTRITIONAL FACTORS Key nutritional factors for dogs with hypothyroidism are listed in **Table 24-14** and described in more detail below.

KEY NUTRITIONAL FACTORS/Energy Animals with untreated hypothyroidism usually have a decreased metabolic rate and use fewer calories than normal. Treatment of primary disease is most important to normalize metabolic processes and is a prerequisite to successful nutritional intervention, especially if weight loss is necessary. After primary disease treatment has been instituted, DER should be calculated based on ideal body weight, which will usually result in some normalization of weight (**Table 24-14**). If obesity is severe then a conservative weight-reduction program may be instituted. (See Chapter 13.) Failure of nutritional intervention to normalize weight most often results from inadequate primary disease control or lack of owner compliance.

KEY NUTRITIONAL FACTORS/Protein Hypothyroid animals do not have an increased need for protein and may even have a decreased need if not fully compensated by exogenous thyroid supplementation. In general, foods that meet AAFCO nutrient allowances for maintenance are adequate. Excess dietary protein should be avoided if renal disease is present (**Table 24-14**). (See Chapter 19.)

KEY NUTRITIONAL FACTORS/Fat Hypothyroid animals usually are in a state of decreased metabolism and therefore often have excessive body stores of fat. Animals with obesity, hyperlipidemia and hypercholesterolemia may benefit from restricted fat intake to near AAFCO minimum allowances (**Table 24-14**). (See Chapter 13 and the Lipid Disorders section of this chapter.)

KEY NUTRITIONAL FACTORS/Fiber Foods with increased fiber levels may be useful in managing obesity associated with hypothyroidism. In addition, increased fiber levels may help manage hyperlipidemia (**Table 24-14**).

KEY NUTRITIONAL FACTORS/Trace Minerals Iodine may be excessive or deficient in different states of thyroid disease. Iodine intake should be thoroughly evaluated to determine adequacy. In addition, animals with hypothyroidism may have decreased absorption of iron from the GI tract resulting in microcytic hypochromic anemia. Iron nutrition of these patients should

ENDO/LIPID

Table 24-14. Key nutritional factors for hypothyroid dogs and cats.*

Factors	Recommended levels
Energy	Feeding at the DER for ideal weight in conjunction with adequate control of hypothyroidism will result in desired body weight: Neutered dogs: 1.6 x RER Intact dogs: 1.8 x RER Neutered cats: 1.2 x RER Intact cats: 1.4 x RER In obese animals, restrict energy to achieve weight loss.
Protein	Foods that meet AAFCO nutrient allowances for maintenance are adequate. Excess dietary protein should be avoided if renal disease is present. 15 to 25% for dogs 28 to 45% for cats
Fat	Foods containing <17% fat are recommended for most patients. Animals with obesity, hyperlipidemia and hypercholesterolemia may benefit from restricted fat intake to near AAFCO minimum allowances. (See Chapter 13 and the Lipid Disorders section of this chapter.)
Fiber	Foods with increased fiber levels (8 to 23%) may be useful in managing obesity and hyperlipidemia associated with hypothyroidism.
Trace minerals	Generally, foods that meet AAFCO minimum allowances for trace minerals are adequate; however, commercial products vary greatly in trace mineral content. It may be necessary to contact product manufacturers to determine iodine, iron and selenium levels.

Key: DER = daily energy requirement, RER = resting energy requirement, AAFCO = Association of American Feed Control Officials.
*Nutrients expressed on a % dry matter basis.

receive special attention during the recovery phase of treatment. Selenium status may also be assessed. However, established reference values for selenium evaluation in dogs and cats have not been validated. Generally, foods that meet AAFCO minimum allowances for trace minerals are adequate; however, commercial products vary greatly in trace mineral content.

Assess the Food (Page 868)

Information from assessing the food is essential for making any changes to foods currently fed. This information, in conjunction with assessment of the animal, leads to development of a rational, complete nutritional plan. After assessment of the animal and current food is completed, the following steps should be taken: 1) compare the current food's levels of key nutritional factors with recommended key nutritional factors and 2) identify discrepancies between key nutritional factors and current intake.

Assess the Feeding Method *(Page 868)*

It may not always be necessary to change the feeding method when managing animals with hypothyroidism. However, a thorough evaluation includes verification that an appropriate feeding method is being used. Any deviations from ideal feeding methods should be identified and changes made as required.

Treatment and Feeding Plan (Page 868)

The success of nutritional management of hypothyroidism depends to a great degree on the effectiveness of medical treatment for the primary disease. The mainstay of treatment for primary hypothyroidism is administration of exogenous thyroid supplements (sodium levothyroxine). Secondary and tertiary hypothyroidism may require additional therapy (e.g., glucocorticoids or cobalt teletherapy).

Animals with hypothyroidism range from normal weight to obese. Successful treatment of primary disease processes will usually result in some weight loss when energy is provided at the calculated DER for ideal body weight. However, in some cases, a weight-loss program may need to be instituted. A food low in fat (<17% DM), higher in fiber (8 to 23% DM) and adequate in protein (15 to 25% DM for dogs, 28 to 45% DM for cats) is recommended. (See Chapter 13 for foods with these nutritional characteristics.) Special attention should be paid to mineral status because certain types of fiber may interfere with the mineral absorption, which may already be in low reserve. Known goitrogenic substances should be removed from the food and environment, if possible (**Table 24-10**).

The amount of food required/offered to attain adequate weight loss should follow the guidelines in Chapter 13. Numerous small meals may be necessary to alleviate owner concerns about providing inadequate energy. However, one to two daily feedings containing an appropriate quantity of energy are probably sufficient to achieve adequate results.

Reassessment (Page 868)

Patient response to treatment may be assessed in a variety of ways. The simplest method is via owner observation of clinical signs, bimonthly body weight charting and monitoring food intake. Return to normal activity, size and appearance generally indicates successful response to treatment. No change in clinical signs, weight or activity necessitates reevaluation of the treatment and diagnosis.

Evaluation of basal thyroid-hormone concentration in serum is another way of assessing response to treatment. Evaluation of thyroid-hormone concentrations may shed new information on owner compliance and response to treatment.

OTHER ENDOCRINE DISORDERS (Pages 868-869)

For information about the nutritional management of hypoadrenocorticism, hyperadrenocorticism, acromegaly and diabetes insipidus, see Small Animal Clinical Nutrition, 4th ed., pp 868-869.

DISORDERS OF LIPID METABOLISM (Pages 869-878)

Hyperlipidemia (also called hyperlipoproteinemia) refers to a disturbance of lipid metabolism that results in an elevated concentration of blood lipids, particularly triglycerides and/or cholesterol. In the fasted state, hyperlipidemia is an abnormal laboratory finding that represents either accelerated synthesis or retarded degradation of lipoproteins. Among dogs and cats, the most common, clinically important type of hyperlipidemia is characterized by an excess concentration of triglycerides in blood, a condition referred to as hypertriglyceridemia. The serum and plasma of affected animals typically appear milky white and turbid, or lipemic. In cases of extreme hypertriglyceridemia, the patient's serum can be so lipemic that it is opaque, or lactescent.

Clinical Importance (Page 869)

Hypercholesterolemia refers to an excess concentration of cholesterol in blood. Most of the circulating cholesterol in dogs and cats is carried on high-density lipoprotein (HDL), the smallest lipoprotein. Because HDL particles are small and do not refract light, patients with extreme cholesterol elevations will not have lipemic serum unless the triglyceride concentration is also elevated.

The clinical importance of hyperlipidemia in companion animal medicine centers around four facts: 1) lipemic serum may positively or negatively interfere with quantitative analyses of other serum analytes, 2) hyperlipidemia in fasted (>12 hours) dogs or cats is abnormal and should be addressed as a significant clinical finding, 3) hyperlipidemic patients are at risk for developing significant clinical illness, including acute pancreatitis and 4) specific dietary and/or drug intervention can eliminate or at least diminish the morbidity associated with hyperlipidemia.

Assessment (Pages 870-877)

Assess the Animal (Pages 870-877)

HISTORY AND PHYSICAL EXAMINATION Some hyperlipidemic dogs and cats do not manifest clinical signs but are considered to be at risk for developing overt signs in the future. Atherosclerosis is a rare manifestation of hyperlipidemia in dogs and cats as opposed to people.

HISTORY AND PHYSICAL EXAMINATION/Dogs with Hyperlipidemia
Table 24-15 lists the clinical signs associated with hypertriglyceridemia in dogs. The most common presenting complaints are vague and intermittent but usually center around vomiting and diarrhea. Accompanying signs include non-localizing abdominal discomfort and occasional pain, accompanied by a transient decrease in appetite. The owner may report that signs are episodic, lasting a few hours to a few days, and may resolve spontaneously with fasting. Abdominal distention is occasionally reported. There appears to be no gender predilection. Affected dogs are generally four years of age and older although younger dogs may be affected.

On physical examination, dogs may appear lethargic and may or may not manifest abdominal pain. Clinical signs and history are compatible with acute pancreatitis; however, abdominal radiographs, ultrasound and laboratory evidence supporting a diagnosis of pancreatitis are typically lacking. The term pseudopancreatitis has been suggested to describe the clinical manifestations associated with hypertriglyceridemia.

Lipemia retinalis, a condition characterized by the appearance of pale pink retinal arterioles and venules, is an incidental finding seen on funduscopic examination of lipemic dogs and cats. This condition does not affect vision. Laboratory analysis of affected animals will verify extreme hypertriglyceridemia, typically greater than 1,000 mg/dl.

Sustained hypertriglyceridemia is a principal risk factor among people and dogs for the development of acute pancreatitis. Dogs with acute abdominal pain and vomiting should be evaluated for hyperchylomicronemia at the time of presentation and during the recovery phase when food intake is restored.

Hypertriglyceridemia should also be considered in patients presenting with a history of seizures. A small number of patients, many of them miniature schnauzers, diagnosed with idiopathic epilepsy have elevated fasting triglyceride concentrations and lipemic serum. In some dogs, dietary therapy has successfully reduced blood triglyceride levels and eliminated seizures without concomitant use of anticonvulsant drugs. Interestingly, seizures associated with hyperlipidemia are not necessarily associated with other signs typically attributed to hyperlipidemia (e.g., vomiting and diarrhea).

Although owners of hypertriglyceridemic dogs rarely express concern about their pet's inactivity or lethargy at the time of initial presentation,

Table 24-15. Clinical signs and diseases associated with hypertriglyceridemia in dogs and cats.

Dogs

Abdominal discomfort*	Intermittent diarrhea*
Acute pancreatitis	Intermittent vomiting*
Behavioral (lethargy, inactivity)	Lipemia retinalis
Crystalline stromal dystrophy (especially cavalier King Charles spaniels)	Lipemic aqueous
	Lipid corneal dystrophy/arcus lipoides corneae
Cushing's syndrome	Seizures
Fasting lipemia (six to 12 hours)	

Cats

Cutaneous xanthomata	Horner's syndrome
Lipemia retinalis	Tibial nerve paralysis
Lipid keratopathy	Radial nerve paralysis
Peripheral nerve paralysis	Splenomegaly

*These clinical signs may occur concomitantly in the same patient. The collective term used to describe these signs is "pseudopancreatitis."

owners often remark that their pet's activity level increased as a result of lowering circulating triglyceride levels.

HISTORY AND PHYSICAL EXAMINATION/Cats with Hyperlipidemia
Clinical signs in hyperlipidemic cats are different than those in dogs (**Table 24-15**). The most common clinical finding in affected cats is cutaneous xanthoma, a painless, raised lesion caused by an accumulation of lipid-laden macrophages or foam cells in the skin. Xanthomas are most likely to occur over bony prominences and areas of skin subject to direct injury.

Xanthomata may also occur in other tissues such as liver, spleen, kidney, heart, skeletal muscle and intestines. Uniquely, xanthomata can form at the point where spinal nerves emerge through the vertebral foramina, the point at which nerves and vascular tissue are subject to mild injury associated with the movement of adjacent vertebrae. Peripheral neuropathy caused by neuronal xanthoma is characterized by motor paralysis. Signs vary depending on the specific nerves involved. Horner's syndrome, tibial nerve paralysis and radial nerve paralysis occur most often. In cases in which mixed motor and sensory nerves are affected, sensation to painful stimuli is retained.

Lipemia retinalis is more common in cats than dogs. Other ocular manifestations of hyperchylomicronemia in cats are uncommon but include iridocyclitis, arcus lipoides corneae and lipemic aqueous and lipid keratopathy. These lesions are thought to occur subsequent to existing ocular disease in lipemic cats.

LABORATORY EVALUATION Veterinarians assessing a dog or cat for hyperlipidemia should submit serum or plasma rather than whole blood.

Samples for cholesterol and triglyceride determinations can be refrigerated or frozen for several days without significant effect. For more information about the effects of excess serum triglycerides on laboratory results and how to handle lipemic samples, see Small Animal Clinical Nutrition, 4th ed., pp 871-872.

Visual inspection of the patient's serum provides valuable physical evidence about the presence or absence of an excessive concentration of triglycerides. In fasting patients (i.e., 24-hour fast or longer), lipemia or lactescent serum denotes hypertriglyceridemia and is usually associated with triglyceride concentrations in excess of 2,000 mg/dl (canine normal = 50 to 150 mg/dl, feline normal = 50 to 100 mg/dl). A diagnosis of hypertriglyceridemia should be based on laboratory determination of serum triglycerides in uncleared serum. By laboratory methods used in North America, serum triglyceride concentrations greater than 500 mg/dl are abnormal for fasted dogs and cats. Although a correlation has not been observed between triglyceride concentrations and the severity of clinical signs, dogs with a triglyceride concentration of 1,000 mg/dl or higher are at risk for developing clinical signs and, as such, are candidates for dietary intervention. Maintaining triglyceride levels less than 500 mg/dl in lipemic (familial) patients may be difficult with nutritional management alone. A more reasonable target range for dietary control is 500 to 1,000 mg/dl postprandially. Furthermore, clinical signs of hypertriglyceridemia appear to be uncommon in patients with postprandial triglyceride levels less than 1,000 mg/dl.

Significant hyperlipidemia characterized by lipemic serum and hypertriglyceridemia has been observed as an incidental finding in fasted adult dogs and cats. The absence of clinical signs at the time of presentation does not justify ignoring the significance of the lipemia. Because of the risks associated with hypertriglyceridemia, normal patients with persistent lipemia should be managed in the same manner as those presenting with clinical signs.

Hyperchylomicronemia is confirmed in fasted dogs with lipemic serum, hypertriglyceridemia (in uncleared serum) greater than 500 mg/dl and a positive chylomicron test. Clinical signs are neither a prerequisite for diagnosis nor for recommending therapeutic intervention. However, therapy in dogs that do not have associated signs is generally reserved for those having fasting hypertriglyceridemia on two consecutive samples two to four weeks apart.

Chylomicrons will normally appear in the serum of dogs and cats within 30 minutes to one hour after ingestion of a meal containing fat. This finding is associated with a transient (i.e., six to 12 hours) increase in serum triglycerides after which triglyceride levels rapidly return to baseline values. Physiologic hyperlipidemia is easily excluded from consideration if the patient is known to have fasted throughout the 12-hour period before

ENDO/LIPID

blood collection. In normal, postprandial animals, serum turbidity is associated with a modest elevation of serum triglycerides (from 150 to 400 mg/dl) that typically returns to normal within 10 hours.

LABORATORY EVALUATION/The Chylomicron Test Knowing that the patient has fasting lipemia provides immediate evidence of hypertriglyceridemia. The lipid disorder may be further characterized by performing a simple, in-hospital test for the presence of chylomicrons. The lipemic serum, separated from red cells, is refrigerated and allowed to stand undisturbed for six to 12 hours. Chylomicrons, if present, will float to the surface of the sample forming an opaque "cream layer" over a clear infranatant. This finding suggests a disorder of chylomicron metabolism, the most common form of hyperlipidemia in dogs. If the sample remains turbid, but doesn't form a cream layer, retention of very low-density lipoproteins (VLDL), rather than chylomicrons, is suggested. This finding also suggests that the hyperlipidemia is secondary to an underlying disorder. In some dogs, particularly poorly regulated diabetics, a cream layer may form over turbid, lipemic serum suggesting retention of chylomicrons and VLDL.

RISK FACTORS Familial (primary) hyperchylomicronemia has been reported as an autosomal recessive trait limited to certain lines of cats. The trait is thought to be present in mixed-breed cats throughout much of the world; therefore, clinically affected cats appear sporadically. Certain dog breeds, most notably miniature schnauzers, are also at increased risk of clinical illness associated with hypertriglyceridemia characterized by the inability to degrade chylomicrons. Though not definitively proven, a familial trait is thought to cause these disorders. Results of a limited survey of healthy adult dogs suggested that primary hypercholesterolemia might occur within some families of Doberman pinschers and rottweilers.

Secondary risk factors (i.e., particularly endocrine disorders, certain drugs and possibly certain diets leading to hyperlipidemia) occur but have not been well studied. For example, profound fasting hypertriglyceridemia occurs inconsistently in dogs with unregulated diabetes mellitus. Clinical signs associated with excess triglyceride concentrations typically include vomiting, diarrhea and abdominal discomfort. Approximately 30% of untreated hypothyroid dogs and from 25 to 30% of untreated dogs with pituitary-dependent hyperadrenocorticism have an excess serum cholesterol concentration. However, the relationship between clinical signs, if any, and the hyperlipidemia has not been established.

In some animals, drugs are known to either decrease lipoprotein degradation or increase lipoprotein production, thereby causing hyperlipidemia. For example, dogs receiving long-term phenobarbital therapy for regulation of idiopathic epilepsy may develop hypercholesterolemia. The clinical significance is unknown and may, in fact, be related to thyroid-hormone production or activity. Cats receiving megestrol acetate may secondarily develop diabetes mellitus, which may culminate in altered lipoprotein lipase activity and hyperchylomicronemia.

ETIOPATHOGENESIS For information about normal lipid metabolism, see Small Animal Clinical Nutrition, 4th ed., pp 873-874.

CLASSIFICATION OF HYPERLIPIDEMIC STATES Hyperlipidemic states can be classified as postprandial, familial or acquired. Familial hyperlipidemia, also called primary hyperlipidemia, refers to those defects in lipoprotein metabolism that are known or suspected to be inherited. Fasting lipemia is frequently recognized in miniature schnauzers and may be linked to a familial defect in chylomicron metabolism. Feline hyperchylomicronemia is the only hyperlipidemic state proven to be familial.

Acquired hyperlipidemia, also called secondary hyperlipidemia, refers to an excess concentration of lipid in blood resulting from an underlying disease in which normal lipoprotein metabolism is markedly altered. Several endocrine diseases alter lipid metabolism leading to secondary hyperlipidemia. For example, insulin-deficient states alter carbohydrate and lipid metabolism. Animals with insulin-dependent diabetes mellitus may have either hypertriglyceridemia or hypercholesterolemia. Hyperadrenocorticism, renal disease and hypothyroidism are variably associated with secondary hyperlipidemia.

In clinical practice, it is not unusual to encounter a patient with both primary and secondary hyperlipidemia. A miniature schnauzer presented with diabetes mellitus is likely to have extreme elevations in serum triglycerides and lactescent serum. From the clinician's perspective, hyperlipidemia, whether primary or secondary, can be associated with undesirable clinical effects. The ability to recognize the signs associated with hyperlipidemia and to make appropriate dietary and/or therapeutic recommendations becomes fundamental to the management of these cases.

CLASSIFICATION OF HYPERLIPIDEMIC STATES/Postprandial Hyperlipidemia Triglycerides are the predominant dietary fat present in pet food. Subsequent to consuming a meal, dogs and cats will experience transient, physiologic hyperlipidemia characterized by increased triglyceride concentration (circulating chylomicrons) and, depending on the amount of fat consumed, serum turbidity (lipemia). However, postprandial hyperlipidemia does not necessarily imply that a disorder of lipid or lipoprotein metabolism exists. In normal dogs and cats, postprandial hyperlipidemia normally persists from six to 12 hours after a meal. Even when a high-fat food is consumed, serum triglyceride levels are not expected to exceed

ENDO/LIPID

500 mg/dl in normal animals. In dogs and cats, hyperlipidemia associated with serum triglyceride levels greater than 1,000 mg/dl, whether fasted or not, is likely to result from an underlying disorder of lipid metabolism. Because chylomicrons carry only a fraction of circulating cholesterol, consumption of a meal has little impact on cholesterol during the six- to 12-hour postprandial period.

Postprandial hyperlipidemia, although physiologic, must be distinguished from intrinsic causes (primary or secondary). Confirming that a hyperlipidemic patient has fasted for 10 to 12 hours before collection of blood effectively excludes a recent meal as the cause for increased blood lipids and, therefore, justifies further evaluation in an attempt to determine the source of the hyperlipidemic state.

CLASSIFICATION OF HYPERLIPIDEMIC STATES/Canine Familial Hyperchylomicronemia Hypertriglyceridemia, particularly that associated with retention of chylomicrons, is the most prevalent lipid disorder recognized in dogs and cats and is associated with the greatest health risk. In dogs, the precise mechanism has not been elucidated; however, this disorder of lipoprotein metabolism is probably caused by either a lack of lipoprotein lipase activity or the absence of apo C-II. Several reports have been published suggesting that miniature schnauzers are predisposed to primary or familial hyperlipidemia. Although it is not definitively known that hyperlipidemia is an inherited disorder of miniature schnauzers, there appears to be a higher than expected prevalence of hypertriglyceridemia in the breed. Several other purebred and mixed-breed dogs have been identified as having fasting hyperchylomicronemia with significant clinical illness, but have no detectable underlying disease.

CLASSIFICATION OF HYPERLIPIDEMIC STATES/Canine Idiopathic Hypercholesterolemia Results of a limited survey of healthy, adult dogs suggested that primary hypercholesterolemia might occur within some families of Doberman pinschers and rottweilers. A relationship between the presence of peripheral corneal dystrophy, regarded by some ophthalmologists as containing cholesterol, and excess serum cholesterol concentration (>300 mg/dl) is of noteworthy interest. Lipoprotein profiles of affected dogs demonstrate elevations of LDL-cholesterol. To date, no studies have demonstrated whether or not administration of cholesterol-lowering drugs would either decrease the cholesterol concentration of hypercholesterolemic dogs or cause regression of the corneal dystrophy. Dietary management with a low-fat veterinary therapeutic food was successful in treatment of bilateral lipid keratopathy in one dog.

Occasionally, extreme elevations of cholesterol will be discovered incidentally in healthy, adult dogs with normal triglyceride values. The clinician is justified in evaluating the patient for evidence of an underlying disorder, such as diabetes mellitus or hyperadrenocorticism. However, in some dogs, hypercholesterolemia cannot be explained. Unless clear evidence of underlying disease exists, treatment specifically intended to lower serum cholesterol does not appear to be warranted.

CLASSIFICATION OF HYPERLIPIDEMIC STATES/Feline Inherited Hyperchylomicronemia In 1983, a primary, genetic disorder of young New Zealand cats was found to alter chylomicron metabolism. Cats that had inherited this disorder developed a form of hyperlipidemia similar to that reported to occur in miniature schnauzers.

CLASSIFICATION OF HYPERLIPIDEMIC STATES/Secondary Disorders of Lipid Metabolism Considering the prevalence of metabolic diseases that affect lipid metabolism, it is possible that secondary hyperlipidemia affects more animals than primary hyperlipidemia. Several endocrine diseases, as well as renal and hepatic diseases, variably alter lipoprotein metabolism resulting in either hypertriglyceridemia or hypercholesterolemia.

CLASSIFICATION OF HYPERLIPIDEMIC STATES/Secondary Disorders of Lipid Metabolism/*Diabetes Mellitus* Hyperlipidemia secondary to diabetes mellitus in dogs and cats may be characterized by hypertriglyceridemia and moderate hypercholesterolemia. In insulin-deficient states, clearance of chylomicrons is impaired due to insufficient activation of lipoprotein lipase in vascular endothelial cells by insulin. Examination of lipid profiles of diabetic dogs reveals lipemia, an increase in chylomicrons and VLDL and a corresponding increase in triglyceride concentration. In some diabetic dogs, excess serum cholesterol concentrations will be present independent of hypertriglyceridemia. In one study, diabetic dogs did not have cholesterol levels significantly different from those of a control population. LDL-cholesterol, on the other hand, was increased presumably as a result of increased LDL synthesis. The clinical significance of this finding is unknown.

Lipemia retinalis in dogs and cutaneous xanthomatosis in cats are associated clinical findings that may be apparent among insulin-dependent diabetics, particularly those with severe hypertriglyceridemia. The hyperlipidemia associated with diabetes mellitus usually improves or resolves as glycemic control is achieved. Diabetic dogs with excess serum triglyceride concentrations appear to be at risk for developing acute pancreatitis or pseudopancreatitis. Dietary fat restriction can be expected to lower the serum triglyceride concentration and may facilitate glycemic regulation in dogs receiving insulin.

CLASSIFICATION OF HYPERLIPIDEMIC STATES/Secondary Disorders of Lipid Metabolism/*Protein-Losing Nephropathy* Hyperlipidemia, characterized by increased serum cholesterol or triglyceride levels, may be detected in patients with proteinuria due to glomerulonephritis or amyloidosis. An inverse relationship between blood lipids/lipoproteins (elevated) and plasma albumin concentration (decreased) has been reported to occur in patients with nephrotic syndrome. The actual pathogenesis whereby the hyperlipidemia develops is complex and appears to be due to a combination of factors involving altered metabolism of lipoproteins. Hypercholesterolemia occurs inconsistently in dogs with heavy proteinuria. The lipoprotein profile of dogs and cats with nephrotic syndrome has not yet been characterized. The influence of hyperlipidemia on morbidity and mortality in nephrotic syndrome is unknown.

CLASSIFICATION OF HYPERLIPIDEMIC STATES/Secondary Disorders of Lipid Metabolism/*Hyperadrenocorticism* Hypercholesterolemia has been recognized in dogs with hyperadrenocorticism (Cushing's syndrome) without concomitant diabetes mellitus. Affected dogs have clear serum, increased plasma cholesterol and LDL-cholesterol levels, but no discrete clinical signs specifically attributable to the excess cholesterol. In a limited study of adult dogs confirmed to have hyperadrenocorticism, only 30% were hypercholesterolemic. There appears to be little diagnostic value to performing lipid determinations in dogs suspected of having endogenous cortisol excess. However, monitoring changes in a given patient's cholesterol profile may have prognostic value in dogs undergoing treatment.

CLASSIFICATION OF HYPERLIPIDEMIC STATES/Secondary Disorders of Lipid Metabolism/*Hypothyroidism* Hypercholesterolemia is present in up to two-thirds of hypothyroid dogs and is believed to result from impaired LDL clearance from the general circulation. It has been suggested that an absolute T_3 deficiency may lead to an increased hepatic cholesterol pool. In turn, LDL-receptor activity is down regulated preventing excess sterol accumulation in the liver. Atherosclerotic-type arterial lesions have occasionally been reported. This finding has led to the suggestion that cholesterol be included in an initial diagnostic screening for hypothyroidism. However, superior laboratory tests are available for evaluating thyroid disease in cats and dogs and should be considered before serum cholesterol evaluation. Therapy should be directed towards correcting the thyroid-hormone deficiency. Although hypothyroid people may experience decreased cholesterol levels after thyroid-replacement therapy is started, there is no apparent value in monitoring cholesterol in affected dogs.

KEY NUTRITIONAL FACTORS Chylomicrons are exclusively of dietary origin; therefore, the amount and type of dietary fat is of primary importance. Foods containing less than 12% fat (DMB) are most commonly rec-

Table 24-16. Key nutritional factors for dogs and cats with lipid disorders.*

Factors	Recommended levels
Fat	Foods containing <12% fat are recommended for most patients.
Fiber	Foods with increased fiber levels (8 to 23%) may be useful in managing secondary hyperlipidemia associated with diabetes mellitus or protein-losing nephropathy.

*Nutrients expressed on a % dry matter basis.

ommended. Other key nutritional factors (e.g., fiber) should be considered in patients with secondary hyperlipidemia and underlying diseases such as diabetes mellitus (See previous sections in this chapter.) and protein-losing nephropathy (**Table 24-16**). (See Chapter 19.)

Assess the Food (Page 877)

The food is the single most important element in the management of primary hyperlipidemia, particularly in hypertriglyceridemic patients. If the hyperlipidemic patient's current food contains more than 12% DM fat, then a dietary change should be considered. **Tables 24-17** and **24-18** list commercial foods that are commonly recommended for these patients. **Table 6-6** gives a low-fat homemade food recipe. Depending on their underlying disease, dogs and cats with secondary hyperlipidemia may benefit from lower fat foods that also meet recommendations for other key nutritional factors. As an example, protein is important in patients with protein-losing nephropathy and water, soluble carbohydrate and fiber are important in patients with diabetes mellitus.

Assess the Feeding Method (Page 877)

It may not always be necessary to change the feeding method when managing a patient with hyperlipidemia, but a thorough assessment includes verification that an appropriate feeding method is being used. Items to consider include amount fed, how the food is offered, access to other food and who feeds the animal. All of this information should have been gathered when the history of the animal was obtained. If the dog or cat has a normal BCS (3/5), the amount of food it was fed previously (energy basis) was probably appropriate. Dogs and cats with hyperlipidemia due to diabetes mellitus benefit from a feeding protocol that matches their insulin therapy.

Feeding Plan (Page 877)

Long-term dietary management of dogs and cats with lipemia caused by primary hypertriglyceridemia is indicated only after secondary causes of hypertriglyceridemia have been ruled out. The approach to treating any patient

Table 24-17. Levels of key nutritional factors in selected commercial products for dogs with lipid disorders.*

	Fat (% DM)	Crude fiber (% DM)
Recommended levels	<12	8-23
Dry canine products		
Hill's Prescription Diet Canine r/d	8.5	23.4
Hill's Prescription Diet Canine w/d	8.7	16.9
Hillís Science Diet Canine Maintenance Light	8.9	14.3
Iams Eukanuba Reduced Fat Adult Formula	10.5	1.9
Iams Eukanuba Restricted-Calorie	7.7	2.0
Iams Less Active	12.5	2.8
Leo Specific Fitness CRD	5.6	8.9
Medi-Cal Canine Fibre Formula	10.1	15.8
Medi-Cal Canine Weight Control/Geriatric	8.3	5.9
Purina CNM-OM Formula	6.7	10.7
Purina Fit & Trim	6.5	8.0
Purina O.N.E. Reduced Calorie Formula	9.6	3.0
Purina ProPlan Reduced Calorie Formula	9.4	2.8
Select Care Canine Hifactor Formula	10.6	14.3
Skippy Cycle Custom Fitness Lite	10.1	4.8
Waltham/Pedigree Calorie Control/Low Calorie	11.4	1.9
Moist canine products		
Hill's Prescription Diet Canine r/d	8.4	21.8
Hill's Prescription Diet Canine w/d	13.0	12.2
Hill's Science Diet Canine Maintenance Light	8.8	10.0
Iams Less Active (Beef, Liver & Rice Formula)	17.2	1.5
Leo Specific Fitness CRW (foil pack)	8.9	17.7
Medi-Cal Canine Fibre Formula	7.9	15.9
Medi-Cal Canine Weight Control	12.3	6.1
Purina CNM-OM Formula	8.4	19.2
Select Care Canine Hifactor Formula	9.1	15.0
Skippy Cycle Custom Fitness Lite	19.0	1.8
Waltham/Pedigree Calorie Control/ Low Calorie	16.9	2.1

*From manufacturers' published information.

with secondary hypertriglyceridemia is based on managing the underlying disease; an appropriate response to the medication should include resolution of the lipemia. Concurrent disorders may also influence the key nutritional factors and lead to other food and feeding method choices. The patient's BCS and body weight should be recorded before initiating therapy because these become important parameters to monitor during reassessment.

Select a Food (Page 877)

Because chylomicrons are exclusively of dietary origin, restriction of dietary fat is the first and most important line of therapy for dogs and cats

Table 24-18. Levels of key nutritional factors in selected commercial products for cats with lipid disorders.*

	Fat (% DM)	Crude fiber (% DM)
Recommended levels	<12	8-23
Dry feline products		
Hill's Prescription Diet Feline r/d	9.5	14.6
Hill's Prescription Diet Feline w/d	9.3	7.9
Hill's Science Diet Feline Maintenance Light	9.3	6.4
Iams Eukanuba Restricted Calorie	10.1	2.1
Iams Less Active Formula	13.1	2.4
Leo Specific Fitness FRD	7.2	18.7
Medi-Cal Feline Fibre Formula	13.8	7.3
Medi-Cal Feline Weight Control	12.3	3.8
Purina CNM-OM Formula	8.5	8.0
Purina ProPlan Reduced Calorie Formula	10.1	2.5
Select Care Feline Hifactor Formula	12.9	5.0
Select Care Feline Weight Formula	12.2	3.6
Waltham Calorie Control Food for Cats	7.7	4.8
Moist feline products		
Hill's Prescription Diet Feline r/d	8.9	17.4
Hill's Prescription Diet Feline w/d	16.7	10.7
Hill's Science Diet Feline Maintenance Light	12.5	10.0
Leo Specific Fitness CRW (foil pack)	9.3	18.7
Medi-Cal Feline Fibre Formula	17.7	5.5
Medi-Cal Feline Weight Control	14.4	8.6
Select Care Feline Hifactor Formula	16.1	7.2
Waltham/Whiskas Feline Calorie Control/Low Calorie	18.4	1.9

*From manufacturers' published information.

ENDO/LIPID

with primary hypertriglyceridemia. Several foods are available at reasonable cost to pet owners, an important consideration because many cases necessitate life-long nutritional therapy.

Foods containing less than 12% fat (DMB) should be used. **Tables 24-17 and 24-18** summarize the nutrient content of various commercial foods that avoid excess dietary fat and may be fed to hyperlipidemic dogs or cats. The foods listed in **Tables 24-17 and 24-18** represent appropriate recommendations for dogs and cats with hypercholesterolemia that have clinical signs associated with or caused by their hyperlipidemic state. Hypercholesterolemia without hypertriglyceridemia is not known to pose a significant health threat to affected dogs or cats. Because most dogs and cats with elevated cholesterol levels have a metabolic disease (e.g., diabetes mellitus or hypothyroidism), an attempt should be made to diagnose and treat the specific underlying disorder before recommending dietary therapy.

When recommending long-term dietary therapy, it is important to understand that the desired response to a particular food will vary from one animal to another.

Determine a Feeding Method (Page 877)

The method of feeding is often not altered in the nutritional management of lipid disorders. If a new food is fed, the amount to feed can be determined from the product label or other supporting materials. The food dosage may need to be changed if the fat level in the food is reduced, because the caloric density of the new food will probably differ from that of the previous food (i.e., the caloric density will usually be lower). If the patient's body weight and condition are optimal, the dosage of the new food should reflect the amount of kcal or kJ consumed by the animal previously. The food dosage is usually divided into two or more meals per day.

Good compliance is necessary for effective clinical nutrition. Enabling compliance includes limiting access to other foods and knowing who feeds the animal. If the dog or cat comes from a household with multiple pets, access to other pets' food should be denied (e.g., a dog with hyperlipidemia eating cat food).

Reassessment (Pages 877-878)

The effect of dietary therapy on hyperlipidemic patients is best determined three to four weeks after the feeding plan is initiated. Reassessment includes reviewing the client's assessment of the patient's response, documenting body weight and condition and evaluating the extent outward manifestations (i.e., ocular or cutaneous lesions) have resolved. Laboratory assessment involves: 1) collecting a blood sample from a fasted animal (10 to 12 hours), 2) evaluating the appearance of the serum for lipemia, 3) determining the triglyceride level in uncleared serum and 4) performing a chylomicron test. The veterinary health care team should assess the client's compliance with the outlined feeding plan. Feeding high-fat snacks and treats and access to other pet foods, even infrequently, can markedly increase circulating triglyceride levels in affected patients.

The goals of dietary therapy are to achieve a: 1) clear serum sample, 2) total triglyceride concentration less than 500 mg/dl and 3) negative chylomicron test. A reasonably acceptable goal is slight serum turbidity, a triglyceride concentration less than 1,000 mg/dl and an incomplete cream layer at the top of the sample. Unless weight loss is desired, the patient's body weight and BCS should be the same as it was before the feeding plan was initiated.

Some patients remain profoundly hyperlipidemic despite excellent owner compliance in feeding a fat-restricted food. The reason is still unknown. However, these animals should continue to receive a low-fat food; human table food should not be fed. The patient should be reassessed in one to two months. If a demonstrable reduction in fasting serum triglyceride concentrations hasn't occurred, drug therapy should be added to the dietary therapy. Drug therapy for patients with primary hypertriglyceridemia has included clofibrate, niacin, gemfibrozil and dietary supplementation with n-3 polyunsaturated fatty acids from fish oils. For more information, see Small Animal Clinical Nutrition, 4th ed., p 878.

Patients that lose a significant amount of weight (>1% of body weight/week) should receive gradually increasing amounts of the recommended low-fat food until desired weight can be maintained. Most dogs and cats with primary hyperlipidemia will experience a marked reduction in serum triglyceride and cholesterol concentrations if daily fat intake is decreased through exclusive feeding of a fat-restricted food. In addition, amelioration or elimination of clinical signs can be expected within two weeks (dogs with pseudopancreatitis) to as long as three months (cats with cutaneous xanthomata) after therapy with a low-fat food is implemented.

Clinical cases that illustrate and reinforce the nutritional concepts presented in this chapter can be found in Small Animal Clinical Nutrition, 4th ed., pp 881-885.

ENDO/LIPID

CHAPTER 25

Cancer

For a review of the unabridged chapter, see Ogilvie GK, Marks SL. Cancer. In: Hand MS, Thatcher CD, Remillard RL, et al, eds. Small Animal Clinical Nutrition, 4th ed. Topeka, KS: Mark Morris Institute, 2000; 887-905.

CLINICAL IMPORTANCE (Pages 887-888)*

Cancer is one of the most common causes of nonaccidental death of dogs and cats. One study documented the cause of death in a series of more than 2,000 necropsy cases. In that study, 45% of dogs that lived to 10 years or older died of cancer. Overall, 23% of pets examined at necropsy died of cancer.

Cats have the largest number of different retroviruses of any companion animal. Retroviral infections in cats produce a wide spectrum of diseases, including cancer. Retroviruses are considered the most common infectious cause of morbidity and mortality in the feline population. The overall prevalence (number of diagnosed cases/year) of feline leukemia virus infection in the United States is between 1 and 3%. The prevalence is less than 1% in single-cat households and as great as 30% in multi-cat households. The prevalence in sick cats is 11.5%.

The overall prevalence of cancer in pets also appears to be increasing. The prevalence is increasing for a variety of reasons, but is, in part, related to the fact that pets are living longer. Practicing veterinarians will more frequently diagnose and manage pets with cancer because of this increased prevalence.

Many pet owners have had cancer or will have a personal experience in which cancer affects themselves, a family member or an acquaintance. This fact suggests that veterinarians and health care teams should approach pets with cancer and their owners in a positive, compassionate and knowledgeable manner. These owners are also able to understand the importance of nutrition in cancer patients and how proper feeding can enhance the quality and length of life for pets with cancer.

Proper nutritional support can reduce or prevent toxicoses associated with cancer therapy and ameliorate the metabolic alterations induced by cancer. In addition, there is growing evidence that specific nutrients can be used to treat the malignant disease directly or indirectly. This chapter

*Page numbers in headings refer to Small Animal Clinical Nutrition, 4th ed., where additional information may be found.

Key Nutritional Factors–Cancer (Page 889)

Factors	Associated conditions	Dietary recommendations
Carbohydrate (NFE)	Altered carbohydrate metabolism Hyperlactatemia Hyperinsulinemia Glucose intolerance	Avoid excess dietary soluble carbohydrate NFE <25% of dry matter (DM) or <20% of metabolizable energy (ME) of the food
Protein	Altered protein metabolism Lower plasma amino acid concentrations Loss of lean muscle mass (cachexia) Altered immune function Impaired wound healing	Avoid protein deficiency Provide dietary protein in excess of adult requirements Dogs: protein = 30 to 45% of DM or 25 to 40% of ME of the food Cats: protein = 40 to 50% of DM or 35 to 45% of ME of the food
Fat	Altered lipid metabolism Altered blood lipid profiles Loss of body fat accumulation (cachexia) Reduced tumor cell use of fat	Provide a large proportion of energy from fat Fat = 25 to 40% of DM or 50 to 65% of ME of the food
Fatty acids	n-3 fatty acids inhibit tumorigenesis n-3 fatty acids influence cytokine production n-3 fatty acids reduce serum lactate concentrations	Provide foods with increased levels of n-3 fatty acids (>5.0% DM) Provide foods with an n-6:n-3 ratio <3.0
Arginine	Promotes wound healing Enhances immune function Inhibits tumorigenesis	Provide foods with arginine levels >2% (DM)

will focus on the nutritional management of patients with cancer rather than on cancer prevention. Much of the nutritional information presented was derived from canine studies and presumed to be accurate for cats, pending feline studies.

Table 25-1. Phases of clinical and metabolic alterations in cancer patients.

Phases	Clinical changes	Metabolic changes
Phase 1	Preclinical, silent phase No obvious clinical signs	Hyperlactatemia Hyperinsulinemia Altered blood amino acid profiles
Phase 2	Early clinical signs Anorexia Lethargy Mild weight loss More susceptible to side effects from chemotherapy, etc.	Similar metabolic changes
Phase 3	Cachexia Anorexia Lethargy More susceptible to side effects from chemotherapy, etc.	Similar changes but more profound
Phase 4	Recovery Remission	Metabolic changes may persist Changes secondary to surgery, chemotherapy or radiation therapy

ASSESSMENT (Pages 888-898)

Assess the Animal (Pages 888-897)
History and Physical Examination (Pages 888-889)

For convenience, metabolic and clinical alterations in cancer patients have been described in four phases (**Table 25-1**). The first phase is a preclinical "silent" phase in which patients do not exhibit clinical signs of disease. Most dogs and cats in the early stages of cancer often appear normal to their owners and veterinarians. As the underlying malignancy progresses, owners often state that their pet seems to be "slowing down" or aging more rapidly, or is less active and less willing to engage in normal activities. Despite normal clinical appearances, patients in Phase 1 have detectable metabolic changes such as hyperlactatemia, hyperinsulinemia and alterations in blood amino acid profiles.

The second phase is a clinical phase in which patients begin to exhibit anorexia, lethargy and early evidence of weight loss. These patients are more likely to exhibit side effects associated with chemotherapy, radiation therapy, immunotherapy and surgery. The third phase (cancer cachexia) is characterized by marked debilitation, weakness and biochemical evidence of negative nitrogen balance such as hypoalbuminemia. In this phase, cancer patients begin to lose body carbohydrate and fat stores.

Vomiting, diarrhea, weakness, lethargy and weight loss are clinical signs reported by owners of dogs and cats with endstage cancer.

A fourth phase (recovery or remission) occurs in those patients undergoing treatment with apparent elimination of their disease. Metabolic alterations persist in some patients despite elimination or control of the cancer via chemotherapy, radiation or surgery. In some individuals, the therapy itself may cause changes that affect the feeding plan. Animals may develop food aversions at any time because of treatment-induced alterations in taste and smell.

Clinical staging of cancer is performed by assessing the tumor size, depth of tumor invasion, presence of tumor in regional lymph nodes and by identifying tumors in distant sites. This information is used to stage tumors by the TNM system: T (tumor size and/or invasion), N (nodal involvement) and M (distant metastasis). Tumor staging may correlate with clinical behavior in certain types of cancer and, in the future, may help determine whether a tumor will respond to nutritional management. In general, body condition scoring is the most practical tool for monitoring the overall nutritional effects of cancer and cancer treatment in dogs and cats. (See Chapter 1.)

Laboratory and Other Clinical Information (Pages 889-890)

Laboratory evaluation of total lymphocyte count, hematocrit and serum albumin and urea nitrogen concentrations can be helpful to further evaluate nutritional status. The use of these parameters, however, is limited because hypoalbuminemia and lymphopenia have many causes unrelated to cancer. Albumin also has a relatively long half-life (eight days in normal dogs) and is slow to respond to changes in nutritional status. Body weight becomes an insensitive index in patients with severe intestinal malassimilation with marked hypoalbuminemia and ascites. Clearly, no single "gold standard" test exists for determining a cancer patient's nutritional status.

In certain tumors, grading the degree of malignancy histologically predicts biologic behavior. Tumor grading is somewhat subjective, but pathologists often evaluate several features to assign a grade, including: 1) degree of differentiation, 2) mitotic index, 3) degree of cellular or nuclear polymorphism, 4) invasiveness, 5) stromal reaction and 6) lymphoid response. Tumor grade may correlate with survival, metastatic rate, disease-free interval or with frequency or speed of local recurrence. Not only can a prognosis be determined based on tumor grade, but treatment may be modified to apply more aggressive therapies to higher grade tumors.

Nutritional Effects of Cancer (Pages 890-893)

Cancer cachexia, a common manifestation of a wide variety of malignancies in people, dogs and cats, is a complex syndrome that includes progressive weight loss that occurs even in the face of apparently adequate nutritional intake. People with cancer cachexia have decreased quality of life, decreased response to therapy and shortened survival time when compared with those with similar diseases but not exhibiting clinical or biochemical signs associated with this condition. Increasing evidence suggests that metabolic alterations are a significant problem in most canine and feline cancer patients.

The metabolic alterations associated with cancer occur before any overt clinical signs associated with cachexia are ever identified. The endstage of cancer cachexia is weight loss that is due not only to primary effects of the tumor, such as compression or infiltration of the alimentary tract, but also may be related to: 1) therapy (e.g., chemotherapy-induced anorexia, nausea or vomiting) or 2) alteration of metabolic pathways composing this paraneoplastic syndrome (**Figure 25-1**). Many tumor-bearing animals have altered metabolism, which necessitates special methods for delivering nutrients and specific types of fluid and nutrient support. (See Chapter 12.)

CARBOHYDRATE METABOLISM Carbohydrate metabolism is dramatically altered in dogs with cancer. Metabolic alterations are suspected to occur in cats but have not been proved. Altered metabolism occurs because tumors preferentially metabolize glucose for energy by anaerobic glycolysis, forming lactate as an endproduct. The host must then expend energy to convert lactate to glucose by the Cori cycle, resulting in a net energy gain by the tumor and a net energy loss by the host. Abnormalities have been documented in peripheral glucose disposal, hepatic gluconeogenesis, insulin effects and whole body glucose oxidation and turnover.

Altered carbohydrate metabolism occurs in dogs with lymphoma but without clinical evidence of cachexia. Following a 90-minute intravenous glucose tolerance test, serum lactate and insulin concentrations were significantly higher when compared with control values. The hyperlactatemia and hyperinsulinemia did not improve when these dogs achieved remission with doxorubicin chemotherapy. These metabolic alterations also occur in dogs with non-hematopoietic malignances (e.g., osteosarcoma, mammary adenocarcinoma and pulmonary bronchogenic adenocarcinoma). The hyperlactatemia and hyperinsulinemia associated with these cancers did not improve when a subset of dogs had their tumors completely surgically excised. The same metabolic alterations are suspected to occur in cats.

Investigators are beginning to understand the clinical significance of altered carbohydrate metabolism. In one study, researchers documented exacerbation of hyperlactatemia in dogs with lymphoma by infusing lactated Ringer's solution. In that study, blood lactate concentrations of relatively healthy, well-hydrated dogs with lymphoma were compared with

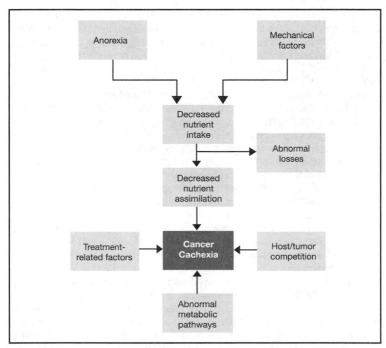

Figure 25-1. Mechanisms of cancer cachexia.

those of control dogs. Blood lactate concentrations in dogs with lymphoma were significantly elevated before, during and after lactated Ringer's solution was infused at a relatively modest rate (4.125 ml/BW_{kg}/hr). This lactated Ringer's-induced increase in lactate concentration may create an additional metabolic burden, requiring the host to convert lactate back to glucose, further exacerbating energy demands.

This finding may be even more important for septic, critically ill cancer patients that require more intensive fluid therapy. One could assume that glucose-containing fluids may increase hyperlactatemia as shown during glucose tolerance tests. Until further information is known about the effects of hyperlactatemia on critically ill animals with cancer, glucose- and lactate-containing fluids should generally be avoided.

Foods high in carbohydrate also appear to increase the total amount of lactate produced when fed to dogs with lymphoma. Mean blood glucose, lactate and insulin concentrations obtained during food tolerance testing were often higher in dogs fed a high-carbohydrate, low-fat food (9% fat,

58% carbohydrate on a dry matter basis [DMB]) compared to those fed a low-carbohydrate, high-fat food (37% fat, 14% carbohydrate on a DMB). However, although there was a positive initial response to chemotherapy, there was no difference in the duration of remission between the two groups.

PROTEIN METABOLISM Human cancer patients have decreased body muscle mass, decreased skeletal protein synthesis and altered nitrogen balance. These patients concurrently have increased skeletal protein breakdown, liver protein synthesis and whole body protein synthesis for tumor growth. Tumors often preferentially use amino acids for energy. The use of amino acids by tumors for energy becomes clinically significant when protein degradation exceeds intake. This imbalance can alter immune response, gastrointestinal (GI) function and surgical wound healing.

In one study, cancer-bearing dogs had significantly lower plasma concentrations of threonine, glutamine, glycine, valine, cystine and arginine and significantly higher concentrations of isoleucine and phenylalanine than did normal control dogs. The results were the same for dogs with different types of tumors. Alterations in plasma amino acid profiles do not normalize after tumors are surgically removed. This finding suggests that cancer induces long-lasting changes in protein metabolism that may be due to elaboration of specific cytokines such as tumor necrosis factor or toxhormone. It is hypothesized that these changes also occur in cats.

Arginine, a dibasic amino acid, is an essential amino acid for cats and is considered to be a conditionally essential amino acid for dogs. Arginine is synthesized endogenously in the kidney from gut-derived citrulline and is converted by the enzyme arginase into ornithine and urea. Arginine has potent secretagogue effects on several endocrine and neuroendocrine glands. Intravenous administration of arginine induces secretion of growth hormone, prolactin, insulin, glucagon, insulin-like growth factor-1, pancreatic polypeptide, somatostatin and catecholamines. Arginine, given in large doses, exerts numerous beneficial effects on the immune system, particularly on thymus-dependent and T-cell-dependent immune reactions. The exact mechanism whereby arginine stimulates T-cell function is unknown. In addition to its positive effects on immune function, arginine may also influence tumor growth, metastatic rate and survival time in patients with cancer.

LIPID METABOLISM Loss of body fat accounts for the majority of weight loss in patients with cancer cachexia, although protein loss also occurs. Thus, it is not surprising that animals and people with cancer have marked abnormalities in lipid metabolism. The decreased lipogenesis and increased lipolysis observed in people and rodents with cancer cachexia result in increased blood concentrations of free fatty acids, very low-density lipoproteins, triglycerides, plasma lipoproteins and hormone-dependent lipoprotein lipase activity, whereas concentrations of endothelial-derived lipoprotein lipase decrease.

Lipid profiles have been evaluated in dogs with lymphoma to determine if alterations similar to those reported in other species are present. In contrast to healthy control dogs, dogs with lymphoma had significantly higher concentrations of cholesterol associated with very low-density lipoprotein, total triglycerides as well as the triglyceride concentrations associated with very low-density lipoprotein (VLDL-TG), low-density lipoprotein and high-density lipoprotein (HDL-TG). Significantly lower concentrations of cholesterol associated with high-density lipoprotein (HDL-CH) were also noted. HDL-TG and VLDL-TG concentrations from dogs with lymphoma were significantly increased above pretreatment values after remission was lost. Additionally, dogs developed overt signs of cancer cachexia. These abnormalities did not normalize when clinical remission was obtained.

Abnormalities in lipid metabolism have been linked to a number of clinical problems including immunosuppression, which correlates with decreased survival in affected people.

In contrast to how tumor cells use carbohydrates and proteins, some tumor cells have difficulty using lipids as a fuel source, while host tissues continue to oxidize lipids for energy. This finding has led to the hypothesis that foods relatively high in fat may benefit animals with cancer compared with foods relatively high in carbohydrates.

Omega-3 (n-3) fatty acids inhibit tumorigenesis and cancer spread in animal models, and form the basis for work in prevention and treatment of cancer in people. Omega-3 fatty acids (eicosapentaenoic acid, docosahexaenoic acid) generally have an inhibitory effect on tumor growth. Metastases are enhanced by omega-6 (n-6) fatty acids (linoleic acid, γ-linolenic acid). In vivo studies have shown that eicosapentaenoic acid has selective tumoricidal action without harming normal cells. Omega-3 fatty acids reduce radiation-induced damage to pig skin. This effect appears to be specific for healthy nonmalignant cells.

Eicosapentaenoic acid decreases protein degradation without altering protein synthesis; the net effect is anticachectic. Foods with high n-3 fatty acid concentrations ameliorate endotoxin-induced lactic acidosis in guinea pigs. This finding may be of clinical importance because hyperlactatemia is a common problem in dogs with lymphoma. Administration of n-3 fatty acids reduced secretion of tumor necrosis factor alpha (TNF-α), interleukin-1β (IL-1β), interleukin-1α (IL-1α) and interleukin-2. This may be especially important because IL-1 and TNF are important mediators of cachexia and may act as tumor growth factors.

ENERGY EXPENDITURE Cancer cachexia may be partly due to negative energy balance secondary to decreased energy intake or altered energy expenditure. Many investigators have found alterations in basal metabolic rate (BMR) and resting energy requirement (RER) that were associated with altered carbohydrate, protein and lipid metabolism in human patients with cancer cachexia.

Because the thyroid gland and its constitutive hormones are intimately involved in the control of energy homeostasis, investigators have speculated that perturbations in thyroid function or thyroid hormone concentrations play a role in altering energy states in tumor-bearing individuals. In one study, researchers compared thyroid hormone concentrations in dogs with cancer (with and without chronic weight loss) with those in nontumor-bearing dogs (with and without chronic weight loss). Diminished serum concentrations of thyroxine (T_4), triiodothyronine (T_3) and free T_3 occurred proportional to the degree of weight loss, regardless of tumor-bearing status. Apparently, these reductions in hormone concentrations are related to the abnormal nutritional state or severity of illness rather than to a tumor-related phenomenon.

Several studies have evaluated energy expenditure in dogs with lymphoma and non-hematopoietic malignancies (e.g., carcinomas and sarcomas) and compared them to healthy, client-owned dogs. These studies found no significant differences in energy expenditure (and presumably caloric requirements) in dogs with a wide range of malignancies when compared with healthy, client-owned dogs. This finding suggests that, in general, dogs with cancer and no evidence of weight loss do not have energy requirements higher than those of apparently healthy dogs without cancer. Furthermore, these parameters do not change significantly in dogs with cancer when the tumor is removed surgically.

These findings suggest that daily energy requirements (DER) of animals with uncomplicated cancer are similar to those of normal animals. However, complications of cancer and cancer treatment, such as sepsis, may significantly alter the energy requirement in individual patients.

Nutritional Effects of Cancer Treatment (Pages 893-896)

Besides the effects of cancer itself, various modalities used to treat cancer (radiation, chemotherapy and surgery, used alone or in combination) may adversely affect a patient's nutritional status. The malnutrition that results from treatment assumes even more importance given that many cancer patients are already debilitated from their disease. Anticancer therapy may produce only mild disturbances, such as mucositis, or it may lead to severe, permanent problems, as in small bowel resection or in disabilities of mastication and swallowing after head and neck surgery or radiation. In general, nutritional problems should be anticipated, feeding tubes placed and animals fed earlier to lessen the toxic effects of treatment. (See Chapter 12.)

Table 25-2. Effects of surgery that may have nutritional implications in cancer patients.

Cancer sites	Surgical procedures	Possible nutritional problems
Head, neck, tongue	Mandibulectomy Maxillectomy Glossectomy	Difficulty prehending, chewing and swallowing food
Esophagus	Esophagectomy, with or without reconstruction	Dysphagia Regurgitation
Stomach	Gastrectomy, partial or complete	Altered gastric emptying Diarrhea
Small intestine	Resection	Malabsorption Diarrhea Intestinal obstruction
Large intestine	Colectomy, partial or complete	Fluid and electrolyte imbalances
Pancreas, liver	Pancreatectomy Cholecystectomy Cholecystoduodenostomy	Diabetes mellitus Maldigestion

SURGERY Nutritional problems that may develop depend on the surgical location and type of procedure performed (**Table 25-2**). Preliminary studies in dogs suggest that metabolic alterations associated with cancer persist even after tumors are removed surgically. In general, feeding tubes (gastrostomy, jejunostomy) should be placed at the time of surgery to avoid additional anesthesia and to allow early feeding.

Radical surgery of the head and neck may lead to significant malnutrition by altering a patient's ability to eat. Although some of these changes are temporary, many patients have permanent difficulty with chewing, swallowing and risk of aspiration. Proactive placement of gastrostomy tubes (See Appendix 4.) during head and neck surgery will facilitate enteral feeding during the immediate postoperative period. These tubes may be used for long-term enteral feeding of some cancer patients. (See Chapter 12.)

The nutritional sequelae of gastric and intestinal resection are directly related to the site and extent of resection and to the individual functions of the various segments. The ability of various segments of the small intestine to increase absorptive capabilities over a period of several months prevents major clinical problems after small bowel resection unless resection is massive. With massive resection, malabsorption (short bowel syndrome) becomes the primary nutritional problem. (See Small Animal Clinical Nutrition, 4th ed., pp 767-770.)

CHEMOTHERAPY Chemotherapeutic agents may contribute to malnutrition through a variety of direct and indirect mechanisms, including: 1)

CANCER

Table 25-3. Effects of chemotherapy that may have nutritional implications in cancer patients.

Stomatitis, glossitis, pharyngitis	Diarrhea
Alterations in smell or taste	Constipation
Decreased appetite	Nausea
Food aversions	Vomiting

anorexia, 2) nausea, 3) vomiting, 4) mucositis, 5) organ injury (toxicosis) and 6) food aversions (**Table 25-3**). These problems should be anticipated and feeding tubes placed before therapy. Early feeding lessens the adverse effects of therapy. Chemotherapeutic agents affect normal and malignant cells but have the greatest effect on rapidly proliferating cells such as epithelial cells of the GI tract. The degree to which GI function is affected depends on the chemotherapeutic agent, drug dosage, duration of treatment, rate of metabolism and the individual animal's susceptibility.

Small bowel villous damage is a major side effect of some chemotherapeutic agents and may be greatly intensified when radiation therapy is given concurrently. The rapid renewal rate of the alimentary tract epithelium usually means that clinical problems from drug-induced mucositis are short-lived.

Nausea and vomiting commonly accompany the administration of many anticancer drugs. Alterations in smell and taste are reported to occur in people and may occur in animals. Side effects experienced during chemotherapy make it difficult for some patients to consume adequate amounts of food.

Corticosteroids such as prednisone are used in chemotherapeutic protocols for some cancers, most notably lymphoma. High doses or prolonged therapy with corticosteroids causes profound polydipsia and polyuria and increased losses of water-soluble vitamins.

RADIATION Animals receiving treatment may have complications that affect food intake. The complications of radiation vary according to the region of the body radiated, dose, fractionation and associated antitumor therapy such as surgery or chemotherapy. Complications may develop acutely during radiation or become chronic and progress even after radiation therapy has been completed (**Table 25-4**).

Radiation therapy in animals is usually performed on five successive days per week with patients restrained by general anesthesia, which presents an opportunity to place a feeding tube. (See Appendix 4.) This treatment schedule requires careful planning of the feeding method to ensure that patients eat their required amount of food each day.

Unless nutritional intervention is provided, many patients lose weight during radiation therapy. Assisted feeding is indicated if oral feeding becomes impossible or food intake is inadequate. (See Chapter 12.)

Table 25-4. Effects of radiation therapy that may have nutritional implications in cancer patients.

Treatment areas	Acute effects	Chronic effects
Head and neck	Mucositis of mouth, tongue, esophagus	Dry mouth Dental disease Alterations in smell Alterations in taste
Thorax	Esophagitis	Esophageal fistula Esophageal stricture
Abdomen	Nausea, vomiting Enteritis, diarrhea Malabsorption	Intestinal obstruction Fistula formation Chronic enteritis

Risk Factors (*Pages 896-897*)

Numerous studies have outlined risk factors of certain nutrients and their relationship to the development of cancer. For example, decreased fiber and increased fat have been most commonly incriminated as causal factors for the development of a wide variety of malignant conditions of the GI tract, breast and urinary bladder in people.

To date, few data exist regarding the relationship between diet and cancer in pet animals. In one study, however, the risk of breast cancer among neutered dogs was significantly reduced in dogs that had been thin (body condition score [BCS] 2/5) at nine to 12 months of age. Results of this study suggest that nutritional factors resulting in altered body composition early in life may be important in canine breast cancer.

Key Nutritional Factors (*Page 897*)

Alterations in carbohydrate, lipid and protein metabolism precede obvious clinical disease and cachexia in cancer patients. These metabolic alterations may persist in patients with clinical remission or apparent recovery from their cancer. Key nutritional factors in cancer patients include soluble carbohydrate, fat, fatty acids, protein and a few specific amino acids, notably arginine.

SOLUBLE CARBOHYDRATES Carbohydrates may be poorly used because of peripheral insulin resistance. Feeding high levels may lead to hyperglycemia, glucosuria, hyperosmolarity, hepatic dysfunction and respiratory insufficiency. In addition, foods high in soluble carbohydrate may increase the total amount of lactate produced. Carbohydrates should supply no more than 20% and preferably 10% of the DER (soluble carbohydrate <25% dry matter [DM]).

FAT AND FATTY ACIDS A large proportion of the daily metabolizable energy (ME) requirement should come from dietary fat because tumor cells have difficulty using fats and loss of body fat accompanies cachexia. Dietary fat should provide 50 to 65% of the calories or 25 to 40% of the DM in the food. Increased levels of dietary n-3 fatty acids (>5.0% DM) may benefit cancer patients, according to studies that link n-3 fatty acids to tumor inhibition and immune enhancement.

PROTEIN AND AMINO ACIDS Dietary protein should exceed levels normally used for maintenance of adult animals because cancer patients have altered protein metabolism and often suffer loss of lean muscle mass (cachexia). Dietary protein should provide 25 to 40% of the ME (30 to 45% DM) in dogs and 35 to 45% of the ME (40 to 50% DM) in cats.

PROTEIN AND AMINO ACIDS/Arginine Adding arginine to parenteral solutions decreases tumor growth and metastatic rate in rodent cancer models. Increased dietary arginine, in conjunction with increased dietary n-3 fatty acid intake, influences clinical signs, quality of life and survival time in dogs with cancer. In a group of dogs receiving chemotherapy for lymphoma, food supplemented with arginine and n-3 fatty acids resulted in elevations in plasma arginine and eicosapentaenoic and docosahexaenoic acid levels. Plasma levels of arginine and n-3 fatty acids were positively correlated with survival time. Similarly, in dogs undergoing radiation therapy for nasal carcinomas, plasma levels of arginine, eicosapentaenoic and docosahexaenoic acid were positively correlated with quality of life and negatively correlated with inflammatory mediators and mucositis in irradiated areas. The minimum effective level of dietary arginine for cancer patients is unknown, but, based on work in other species, it is appropriate to provide more than 2% arginine (DMB).

Other Nutritional Factors (Pages 894-896)

Several vitamins, minerals and novel foods and ingredients have received considerable attention in cancer prevention and therapy.

AMINO ACIDS AND CANCER/Methionine and Asparagine Certain tumor cell lines require methionine for growth. Replacement of methionine with its precursor, homocysteine, locks these tumor cells into late S and G2 phases of the cell cycle. Because certain cancer chemotherapeutic agents are cell-cycle specific, the percentage of tumor cells sensitive to chemotherapy increases, improving the therapeutic index. Asparagine is essential for tumor cell growth in lymphoma. Treatment of dogs and cats with L-asparaginase has induced complete remissions in up to 80% of dogs and cats with lymphoma.

AMINO ACIDS AND CANCER/Tyrosine and Phenylalanine Tyrosine and phenylalanine restriction has been reported to suppress melanoma cell growth in tissue cultures and in rodent tumor models. The administration of tyrosine and phenylalanine increased the survival of melanoma tumor-bearing mice and increased the effectiveness of levodopa against melanoma. Studies are needed to sort out this discrepancy.

AMINO ACIDS AND CANCER/Glutamine Glutamine may have specific therapeutic value. Glutamine is an essential precursor for nucleotide biosynthesis and is the most important oxidative fuel for enterocytes. Supplementation of enteral preparations with glutamine is beneficial in several animal models of intestinal injury by improving intestinal morphometry, reducing bacterial translocation, enhancing local immunity and improving survival. Glutamine has only recently been recognized as a conditionally essential amino acid in certain pathophysiologic states. Glutamine is added to most human enteral formulas.

One study using a feline model of methotrexate-induced intestinal injury failed to demonstrate a beneficial role for glutamine supplementation to an amino acid-based purified food. Additional studies are warranted to determine whether glutamine supplementation of commercially available foods containing intact protein sources improves intestinal integrity during chemotherapy administration to dogs and cats with cancer.

AMINO ACIDS AND CANCER/Glycine Some amino acids may decrease the toxicity associated with chemotherapy. For example, glycine reduces cisplatin-induced nephrotoxicity.

VITAMINS AND CANCER Retinoids, β-carotene, vitamin C and vitamin E all appear to influence the growth and metastasis of cancer cells by a variety of mechanisms. Some of these mechanisms include selected receptor-mediated antiproliferative activities. These vitamins have been reported to bind their cytosolic receptors followed by translocation of the bound complex to the nucleus where the receptors mediate gene regulation. Other effects result from antioxidant, hormone-like and immunomodulator capabilities.

VITAMINS AND CANCER/Retinoids "Retinoids" refer to the entire group of naturally occurring and synthetic vitamin A derivatives, including retinol, retinal and retinoic acid. Retinoids appear to have the potential for regulating cancer cells either alone or in combination with other agents. Specific studies in human and veterinary medicine suggest that retinoids alone or with other agents can be effective for the treatment of certain types of malignancies. The synthetic retinoids, isotretinoin and etretinate, have been used successfully in some dogs with intracutaneous cornifying epitheliomas, other benign skin tumors, cutaneous lymphoma, solar-induced squamous cell carcinoma and associated preneoplastic lesions. The retinoids promote cellular differentiation and may enhance the susceptibility of neoplastic cells to chemotherapy and radiation therapy.

VITAMINS AND CANCER/Vitamin C Vitamin C (ascorbic acid) has been reported to inhibit nitrosation reactions and prevent chemical induction of cancers of the esophagus and stomach. Processed foods high in nitrates and nitrites, such as bacon and sausage, are often supplemented with vitamin C to reduce the carcinogenic capability of the resultant nitrosamines.

Ascorbic acid may be one therapeutic alternative for overcoming drug resistance in some cancer cells. Studies suggest that an ascorbic acid-sensitive mechanism may be involved in drug resistance to vincristine in certain cancer cell lines. Despite the extensive amount of vitamin C research, few direct data exist proving its efficacy in dogs and cats.

VITAMINS AND CANCER/Vitamin E Vitamin E (α-tocopherol) can also inhibit nitrosation reactions. Vitamin E has a broad capacity to inhibit mammary tumor carcinogenesis and colon carcinogenesis in rodents. Research indicates that vitamin E influences a variety of cell functions including free radical scavenging, which can prevent oxidative damage that leads to cell death.

In addition to its anticancer properties, vitamin E may potentially convey therapeutic efficacy against certain malignancies. Vitamin E has been reported to have antiproliferative activity, which involves the binding of the vitamin to salicylic receptors, followed by translocation to the nucleus where DNA binds on the domains of receptors that mediate gene regulatory events. Retrovirus-induced tumorigenesis involves transformation of normal cells into tumor cells. Evidence suggests that vitamin E may normalize the immune system by interacting with macrophages and T lymphocytes to inhibit retroviral-induced infections.

MINERALS AND CANCER Minerals that have been suggested as being important in patients with cancer include selenium, iron and zinc. Optimal levels of specific minerals for cancer prevention and treatment have not been established for pet animals.

MINERALS AND CANCER/Selenium Selenium has been one of the most heavily studied minerals associated with the development of cancer. Low serum selenium levels have been observed in human patients with GI cancer. In rodents, dietary supplementation with selenium inhibits colon, mammary gland and stomach carcinogenesis.

MINERALS AND CANCER/Iron Iron transferrin and ferritin have been linked to cancer risk and cancer cell growth. Lung, colon, bladder and esophageal cancer in people have been highly correlated with increased serum iron concentrations and increased transferrin saturation. Because many tumor cells require iron for growth, it has been suggested that the increased use of iron by the tumor depresses serum iron levels in human cancer patients. Mice with low levels of iron have slow tumor growth compared to those with normal iron levels.

MINERALS AND CANCER/Zinc In people, low levels of zinc in blood and diseased tissue have been observed in esophageal, pancreatic and bronchial cancer. Zinc deficiency appears to enhance carcinogenesis in laboratory animals.

Assess the Food (Page 897)

One of the first steps in evaluating the cancer patient with weight loss or anorexia is to obtain an accurate dietary history that clearly defines the patient's food, food intake and feeding methods. Levels of the key nutritional factors should be evaluated in foods currently being fed to dogs and cats with cancer. Key nutritional factors include dietary levels of protein, soluble carbohydrate, fat, fatty acids and arginine. Amounts and levels of key nutritional factors should be compared to those established for dogs and cats with cancer (See the Key Nutritional Factors–Cancer table.). Information from this aspect of assessment is essential for making any changes to foods currently provided. If key nutritional factors in the current food do not match the recommended levels, then changing to a more appropriate food is indicated. Evaluation of a food's guaranteed analysis can provide an estimate of the crude protein, crude fat and carbohydrate content of the food after the nutrients have been converted to a DM or energy basis. (See Chapter 1.) The manufacturer can also be contacted to obtain information about the energy density, protein, fat, fatty acid and soluble carbohydrate content of the food. **Tables 25-5** and **25-6** list selected commercial foods often fed to patients with cancer.

Assess the Feeding Method (Pages 897-898)

Careful assessment of the feeding method is important to determine whether the animal is receiving its caloric requirement and if it is able to prehend, masticate, swallow and assimilate its food. Calculation of the patient's DER, determination of the energy density and levels of key nutritional factors, careful measurement of the amount of food eaten by the animal and body condition scoring will help establish whether cancer patients with weight loss are actually receiving sufficient calories and nutrients.

RER is calculated for the current body weight and typical factors are used to establish an estimated DER. (See Chapter 1.) Hospitalized patients should eat enough food to at least meet their estimated RER. Patients at home should eat enough food to meet their estimated DER (RER x appropriate factor). Enteral or parenteral feeding techniques are indicated if hospitalized patients are unable or unwilling to eat their estimated RER. (See Chapter 12 and Appendix 4.) If hospitalized patients are eating their estimated RER and patients at home are eating their estimated DER, then frequent monitoring of body weight and body condition is indicated.

FEEDING PLAN (Pages 898-900)

Nutritional support of the cancer patient must be individualized. Nutritional therapy should be undertaken with the overall prognosis of the patient clearly in mind so that the aggressiveness of dietary intervention (e.g., supportive, adjunctive, definitive) can be appropriately adjusted. Owners of cancer patients should be educated about the integral part nutrition plays in the total management of their pet's disease. Dietary modification depends on the extent of disease, anorexia, nausea, weight loss and consequences of treatment.

Some underweight animals with cancer will stabilize at a less than optimal BCS (2/5 rather than 3/5). It may be difficult to achieve weight gain in these patients; therefore, the goal should change to maintaining this leaner body condition. (See Chapter 12, Accommodation.)

Select a Food (Page 898)

It is naive to believe that a single nutrient or set of nutrients will be effective for the treatment and prevention of all cancers. However, altered metabolism documented in cancer patients and clinical studies performed to date suggest that modifying nutrient intake may help reverse the adverse effects of cancer.

Several commercial veterinary therapeutic foods and moist specialty foods provide key nutrients in appropriate levels (**Tables 25-5** and **25-6**). However, only one commercial food (Prescription Diet Canine n/d, Hill's Pet Nutrition, Inc., Topeka, KS, USA) has been shown to improve the longevity and quality of life of canine patients with cancer. It is important to consider feeding cancer patients food with an appropriate nutrient profile. (See the Key Nutritional Factors–Cancer table.) Many pets eat commercial dry foods that are usually higher in soluble carbohydrates (>50% DM) and lower in fat (<20% DM) and protein (25 to 40% DM) than optimal.

Determine a Feeding Method (Pages 898-900)

The enteral route of feeding is the preferred route of nutritional support because it is easier, less expensive and more physiologic than parenteral administration. In addition, enteral feeding improves mucosal thick-

Table 25-5. Key nutritional factors in selected foods used in dogs with cancer.*

Products	Protein 30-45	Fat 25-40	Carbohydrate <25	n-3 fatty acids >5	Arginine >2
Recommended levels					
Hill's Prescription Diet Canine/Feline a/d, moist	45.7	28.7	16.5	2.6	2.04
Hill's Prescription Diet Canine n/d, moist	38.8	32.0	20.0	7.2	2.9
Hill's Science Diet Feline Maintenance Seafood Recipe, moist	45.1	25.4	20.1	0.82	2.83
Iams Eukanuba Maximum-Calorie/Canine, dry	40.1	29.0	22.7	0.9	na
Iams Eukanuba Maximum-Calorie/Feline, dry	44.2	29.6	19.1	0.93	na
Iams Eukanuba Maximum-Calorie/Canine & Feline, moist	43.3	41.1	7.6	0.78	2.6
Purina Feline CV-Formula, moist	42.5	26.8	23.1	na	na

Key: na = information not available from manufacturer.
*Nutrients expressed on a percent dry matter basis. Values obtained from manufacturers' published information.

Table 25-6. Key nutritional factors in selected foods used in cats with cancer.*

Products	Protein 40-50	Fat 25-40	Carbohydrate <25	n-3 fatty acids >5	Arginine >2
Recommended levels					
Hill's Prescription Diet Canine/Feline a/d, moist	45.7	28.7	16.5	2.6	2.04
Hill's Prescription Diet Feline p/d, moist	48.8	31.5	11.1	0.42	2.66
Iams Eukanuba Maximum-Calorie/Feline, dry	44.2	29.6	19.1	0.93	na
Iams Eukanuba Maximum-Calorie/Canine & Feline, moist	43.3	41.1	7.6	0.78	2.6
Purina Feline CV-Formula, moist	42.5	26.8	23.1	na	na
Select Care Feline Development Formula, moist	48.0	32.2	12.1	na	na

Key: na = information not available from manufacturer.
*Nutrients expressed on a percent dry matter basis. Values obtained from manufacturers' published information. The key nutritional factors recommended for cats are derived from canine studies and presumed to be accurate pending feline studies.

CANCER

ness, stimulates gut trophic hormones and stimulates IgA production. Enhancing the palatability of food is the simplest means of increasing voluntary intake. Warming a food may improve its aroma and mouth feel, thereby improving palatability.

If necessary, drug therapy can be attempted before offering food. Administration of a benzodiazepine derivative (diazepam or oxazepam) or cyproheptadine increases appetite transiently; however, these drugs are unreliable for ensuring adequate caloric intake. Benzodiazepine derivatives are contraindicated in patients with severely reduced hepatic function, especially when signs of hepatic encephalopathy are present. In addition, the appetite-stimulating properties of these agents appear to wane with time when used in sick animals. Megestrol acetate causes weight gain and increased appetite in people with cancer. The clinical benefit of this drug in veterinary patients remains to be determined. Controlled studies with human cancer patients have revealed that cyproheptadine, corticosteroids and nandrolone decanoate have little to no impact on improving food intake, body weight and clinical outcome. A deficiency of B vitamins is associated with anorexia and may occur in some cancer patients fed unbalanced homemade foods or patients that have decreased food intake.

Enteral-feeding techniques should be considered if these appetite-stimulating efforts fail or if long-term nutritional support (more than several days) is required. Enteral feeding via nasoesophageal intubation, esophagostomy tube or gastrostomy tube is the most reliable and efficient method for ensuring adequate alimentation. (See Chapter 12.) In many situations, it is best to proactively place an enteral feeding device during surgery or before radiation therapy is initiated. Examples include placing a gastrostomy tube in patients with oral tumor resections or before radiation treatment to the nose, oral cavity or neck.

Parenteral nutrition should be reserved for animals that are unable to assimilate nutrients or those with intractable vomiting.

REASSESSMENT (Page 900)

Reassessment of cancer patients should include monitoring the effects of: 1) cancer on the animal, 2) treatment and nutritional management on the tumor and 3) treatment and nutritional support on the patient. The overall effects of cancer, cancer treatment and nutritional management on the animal are best assessed by comparing the current body weight and BCS with previous assessments. The patient's appetite should be assessed and the daily caloric intake monitored closely. Improvements in body weight and condition are important indicators that overall treatment and feeding plans are adequate. Clinical staging of the cancer can be repeated to assess the specific tumor response to treatment. Appropriate modifications to the feeding plan should be made as the patient's status changes.

Clinical cases that illustrate and reinforce the nutritional concepts presented in this chapter can be found in Small Animal Clinical Nutrition, 4th ed., pp 902-905.

CHAPTER 26

Use of Fatty Acids in Inflammatory Disease

For a review of the unabridged chapter, see Schoenherr WD, Roudebush P, Swecker WS. Use of Fatty Acids in Inflammatory Disease. In: Hand MS, Thatcher CD, Remillard RL, et al, eds. Small Animal Clinical Nutrition, 4th ed. Topeka, KS: Mark Morris Institute, 2000; 907-921.

INTRODUCTION (Pages 907-908)*

Immunomodulation designates either suppression or augmentation of an immune response. Suppressing function of the immune system is important in cases of excessive inflammation or immune-mediated disease. Augmenting the immune response is helpful when increased resistance to disease is required. Nutrition plays an important role in modulation of an animal's immune system.

The majority of scientific literature about the interaction between nutrition and the immune system correlates the effects of nutrient deficiency and modulation of an immune response. These studies have evaluated deficiencies of protein, energy, the fat-soluble vitamins A, D and E, the B-complex vitamins, vitamin C and the minerals selenium, iron, zinc and copper and their relationship to immune dysfunction. Undoubtedly, the nutritional status of the animal plays an important role in resistance mechanisms against disease-causing organisms and may influence the outcome of disease in infected animals.

This chapter will concentrate on the role of dietary fatty acids in modulation of inflammatory disease of the skin and musculoskeletal system. These concepts may have broader application to modulation of inflammatory disease in other organ systems, as well.

THE ROLE OF POLYUNSATURATED FATTY ACIDS IN INFLAMMATION (Pages 909-911)

All mammals synthesize fatty acids de novo up to palmitic acid (16:0), which may be elongated to stearic acid (18:0) and converted into oleic acid (18:1). Plants, unlike mammals, can insert additional double bonds into oleic acid and produce the polyunsaturated fatty acid (PUFA) linole-

*Page numbers in headings refer to Small Animal Clinical Nutrition, 4th ed., where additional information may be found.

Key Nutritional Factors–Inflammatory Disease (Page 909)

Factors	Associated conditions	Dietary recommendations
Polyunsaturated fatty acids	Dermatitis Pruritus Arthritis Other inflammatory diseases	Provide additional dietary levels of specific n-6 (gamma-linolenic acid [GLA]) or n-3 (alpha-linolenic acid [ALA], eicosapentaenoic acid [EPA], docosahexaenoic acid [DHA]) fatty acids. GLA is provided in supplements containing borage oil, black currant oil or evening primrose oil. ALA is provided in supplements or foods with flax, flax oil or linseed oil. EPA and DHA are provided in supplements or foods with fishmeal or fish oil. Supplements or foods should initially provide 50 to 250 mg total n-3 fatty acids/BW_{kg}/day.

ic acid (LA, 18:2n-6) and α-linolenic acid (ALA, 18:3n-3). Both LA and ALA are considered essential fatty acids because animals cannot synthesize them from other series of fatty acids; thus, they must be supplied by the diet.

Dietary PUFA serve as substrates that may be metabolized to form important, biologically active compounds. To produce those metabolites, a number of cells contain a group of enzymes that desaturate, elongate and oxygenate fatty acids. All PUFA are categorized based on the position of the first double bond in the structure from the terminal end (**Figure 26-1**). The two most important PUFA series are the n-6 series (the first double bond is located at the sixth carbon atom) and the n-3 series (the first double bond is located at the third carbon atom).

In the n-6 series, linoleic acid can be desaturated to yield γ-linolenic acid (GLA, 18:3n-6), which is elongated to dihomo-γ-linolenic acid (DGLA, 20:3n-6) and ultimately desaturated again to produce arachidonic acid (AA, 20:4n-6) in the animal (**Figure 26-2**).

Many marine plants, especially algae, elongate chains and add double bonds to ALA to yield n-3 PUFA with 20 and 22 carbon atoms and five or

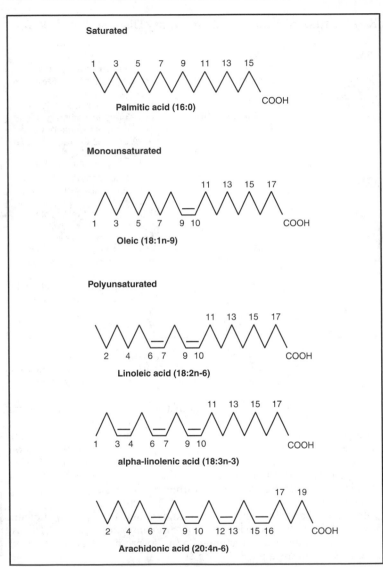

Figure 26-1. Chemistry and nomenclature of fatty acids.

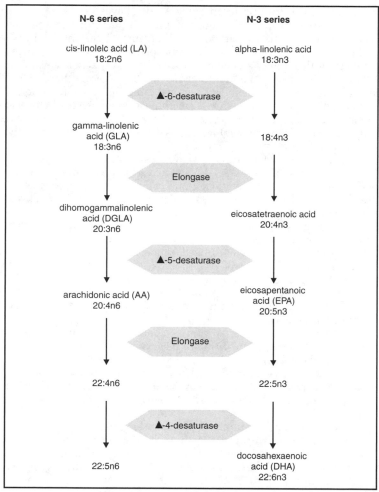

Figure 26-2. Metabolic transformation of two major unsaturated fatty acid families by desaturation and elongation.

six double bonds (**Figure 26-2**). Formation of these long-chain n-3 PUFA by marine algae and their transfer through the food chain to fish account for the abundance of eicosapentaenoic acid (EPA, 20:5n-3) and docosahexaenoic acid (DHA, 22:6n-3) in certain marine fish oils.

Eicosanoid Production from Fatty Acids (Pages 909-911)

Fatty acids have diverse functions in cells but their principal roles are as energy sources and as membrane constituents. In current theories of cell membrane structure, most of the phospholipid present in the membrane takes the form of a bimolecular sheet with fatty acid chains in the interior of the bilayer (the "fluid mosaic" model). Altering the composition of fatty acids consumed by the animal can change the proportion of different types of fatty acids in cell membranes. Modifying the fatty acid composition of the cell membrane changes membrane fluidity. Both the fatty acid composition of membrane phospholipids and the fluidity of the membrane affect membrane activities.

Certain membrane fatty acids also have specific roles in the regulation of cell functions. For example, AA, GLA and EPA act as precursors for the synthesis of eicosanoids, an important group of immunoregulatory molecules that function as local hormones and mediators of inflammation. Changing the type of fatty acids available to cells, including cells of the immune system (e.g., lymphocytes and monocytes), modifies the fatty acid composition of the membrane phospholipids of those cells. A change in cell function is apparent after the membrane fatty acid composition changes.

Eicosanoids are a family of 20-carbon oxygenated derivatives of GLA, AA and EPA, and include prostaglandins (PG) and thromboxanes (TX), which together are termed prostanoids, and leukotrienes (LT), lipoxins (LX), hydroperoxyeicosatetraenoic acids (HPETE) and hydroxyeicosatetraenoic acids (HETE).

Eicosanoid synthesis begins with metabolism of GLA, EPA or AA by one of two enzyme systems: 1) cyclooxygenase, which yields PG and TX (**Figure 26-3**) or 2) the 5-, 12- or 15-lipoxygenases, which yield LT, LX, HPETE and HETE (**Figure 26-4**). The amounts and types of eicosanoids synthesized are determined by the availability of the fatty acid precursor released from the membrane, and by the activities of cyclooxygenase and the lipoxygenases.

In most conditions the principal precursor for these compounds is AA, although GLA and EPA compete with AA for the same enzyme systems. The eicosanoids derived from AA appear to be more common in normal physiologic circumstances. This finding is explained by the higher content of AA than GLA or EPA in most membrane phospholipids, and a lower specificity of cyclooxygenase for GLA and EPA than for AA. The eicosanoids produced from AA are proinflammatory as compared to eicosanoids formed from GLA or EPA and may result in pathologic conditions when produced in excessive amounts.

The major products of the cyclooxygenase cascade formed from AA are the proinflammatory 2-series PG and TX, PGE_2 and TXA_2. Aspirin and non-steroidal antiinflammatory drugs (NSAIDs) primarily inhibit the cyclooxy-

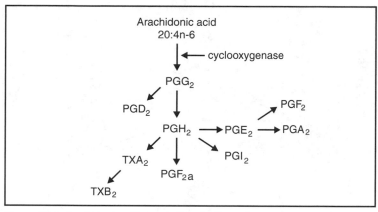

Figure 26-3. Arachidonic acid metabolism by the cyclooxygenase cascade.
Key: TX = thromboxane, PG = prostaglandin.

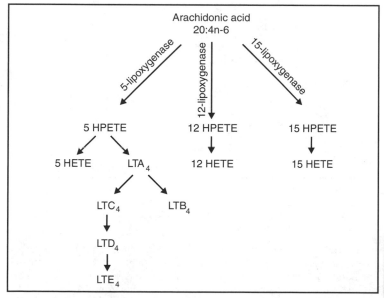

Figure 26-4. Arachidonic acid metabolism by the lipoxygenase pathway.
Key: HPETE = hydroperoxyeicosatetraenoic acid, HETE = hydroxyeicosatetraenoic
acid, LT = leukotriene

INFLAM DIS

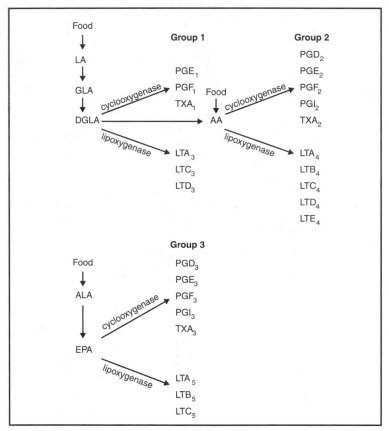

Figure 26-5. Polyunsaturated fatty acid metabolism by the lipoxygenase pathway. Key: LA = linoleic acid, GLA = γ-linolenic acid, DGLA = dihomo-γ-linolenic acid, ALA = α-linolenic acid, EPA = eicosapentaenoic acid, PG = prostaglandin, LT = leukotriene, TX = thromboxane.

genase cascade. The n-3 PUFA also competitively inhibit oxygenation of AA by cyclooxygenase. Ingestion of fish oils containing n-3 PUFA decreases membrane AA levels because n-3 PUFA replace AA in the substrate pool and decrease the capacity to synthesize eicosanoids from AA.

The products derived from AA through the lipoxygenase pathway are the proinflammatory 4-series LT: LTB_4, LTC_4, LTD_4, LTE_4, 12 HETE and 15 HETE (**Figure 26-5**). In contrast, eicosanoids derived from GLA and EPA

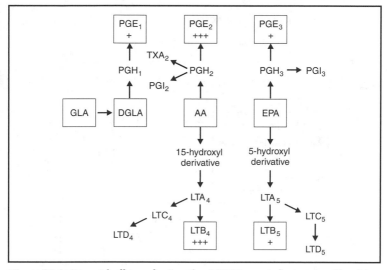

Figure 26-6. Potential effects of n-3 and n-6 PUFAs on inflammation. Key: LA = linoleic acid, GLA = γ-linolenic acid, DGLA = dihomo-γ-linolenic acid, AA = arachidonic acid, EPA = eicosapentaenoic acid, PG = prostaglandin, LT = leukotriene, TX = thromboxane, +++ = strongly proinflammatory, + = weakly proinflammatory.

promote minimal to no inflammatory activity. LTB_5 and 15 HETE are products derived from EPA via the lipoxygenase pathway that inhibit production of LTB_4.

EPA gives rise to the 3-series PG and TX and the 5-series LT (**Figure 26-5**). The eicosanoids produced from EPA do not always have the same biologic properties as the analogues produced from AA. For example, LTB_5 is less active than LTB_4 with regard to chemotactic and aggregatory properties in neutrophils and PGE_3 is a less potent inhibitor of lymphocyte proliferation than PGE_2 (**Figure 26-6**). DGLA also competes with AA for cyclooxygenase and, therefore, decreases production of cyclooxygenase products from AA and favors production of the 1-series PG and TX (**Figure 26-6**).

GLA is readily converted to DGLA (**Figure 26-2**) and increases the production of the 1-series PG and 15-hydroxy-DGLA. PGE_1 inhibits mobilization of AA from the membrane phospholipid stores restricting eicosanoid production from AA, and 15-hydroxy-DGLA reduces conversion of AA to its 5- and 12-lipoxygenase metabolites. GLA is an intermediate in the formation of AA from LA and increasing dietary GLA might lead to an increase in tissue levels of AA. The Δ-5 desaturation step, which converts DGLA to AA, is a rate-limiting step and is quite slow in animals. No change or a small

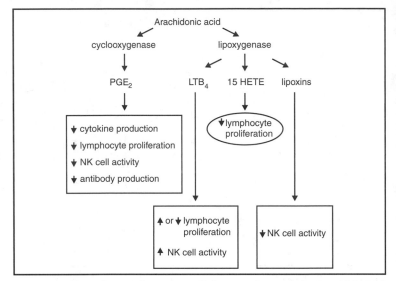

Figure 26-7. Regulation of immune cell function by arachidonic acid-derived eicosanoids. Key: PG = prostaglandin, LT = leukotriene, HETE = hydroxyeicosatetraenoic acid, NK = natural killer.

increase in tissue AA levels is found when GLA is fed; however, DGLA tissue levels increase in relation to dietary levels of GLA.

Modulation of Immune Response and Inflammation by Fatty Acids (Page 911)

Macrophages are the most significant source of eicosanoids (PG, TX, LT and HETE) because they possess cyclooxygenase and lipoxygenase enzymes, and are subject to regulatory effects of both enzyme systems. Macrophages modulate the intensity and duration of inflammatory and immune responses. Alteration of macrophage eicosanoid production modulates immune function.

AA is the major PUFA in membrane phospholipids of macrophages and lymphocytes. The response of tissue that is irritated or injured is inflammation, a mechanism by which tissue protects itself immunologically. In brief, phospholipases act on phospholipids of cell membranes to release fatty acids. AA, the fatty acid in greatest concentration, is released and converted into eicosanoids, which mediate inflammation.

Four AA-derived leukotrienes and one PG play a central role in the inflammatory process. LTB_4 stimulates neutrophil and eosinophil chemotaxis and increases vascular permeability. LTC_4, LTD_4 and LTE_4 encourage smooth muscle contraction and increase vascular permeability. PGE_2 inhibits T and B lymphocyte proliferation, reduces cytokine production and limits natural killer (NK) cell activity. Increased production of LT and PGE_2 has been reported in many chronic inflammatory diseases. **Figure 26-7** summarizes the effects of AA-derived eicosanoids on lymphocyte functions and cytokine production.

Consumption of fish oil with n-3 PUFA or oils high in the n-6 fatty acid GLA (e.g., evening primrose oil, borage oil and black currant oil) results in replacement of AA in the macrophage membrane with EPA or DGLA. The result is production of fewer AA-derived eicosanoids and more EPA- or GLA-derived eicosanoids, thereby reducing the immunologic response to an inflammatory episode. Consequently, changing the type of eicosanoid production and the subsequent alteration in cytokine production can reduce inflammation by eicosanoid-mediated effects. This premise is the basis for using EPA or GLA for treatment of chronic inflammatory conditions.

INFLAMMATORY SKIN DISEASE (Pages 911-914)

Clinical Importance (Pages 911-912)

Surveys and textbooks suggest that skin disorders are the most common reason for patient visits to veterinary hospitals. Surveys also indicate that 15 to 25% of all small animal practice activity is involved with the diagnosis and treatment of skin and coat problems.

The most commonly diagnosed canine skin disorders are allergy (flea-bite hypersensitivity, atopy), skin cancer, bacterial pyoderma, seborrhea, parasitic dermatoses, adverse reactions to food (food hypersensitivity or food intolerance), immune-mediated dermatoses and endocrine dermatoses. The most common feline skin disorders are abscesses, parasitic dermatoses, allergy (flea-bite hypersensitivity, atopy), papulocrustous (miliary) dermatitis, eosinophilic granuloma complex, fungal infections, adverse reactions to food, psychogenic dermatoses, seborrheic conditions, neoplastic tumors and immune-mediated dermatoses. These data confirm that inflammatory skin diseases, especially allergic dermatoses, are common and important clinical entities.

Assessment (Pages 912-918)
Assess the Animal (Pages 912-914)
HISTORY AND PHYSICAL EXAMINATION Numerous skin diseases have an inflammatory component. However, dietary fatty acid therapy has been used primarily in patients with allergic skin disease or patients with pruritus or papulocrustous dermatitis for which a specific cause has not been identified.

INFLAM DIS

Pruritus is the most common historical feature of allergic skin disease in dogs and cats. Clinical signs reportedly first occur in most dogs and cats with atopy between six months and three years of age. Lesions of canine atopy usually involve the muzzle, periocular region, pinnae and external ear canals, paws, axillae, groin and abdomen. Although the face and paws are most commonly involved, many animals will have generalized pruritus by the time they are examined. Chronic licking, rubbing, chewing or scratching can result in alopecia, lichenification, hyperpigmentation, scaling and excoriation. Other common lesions in atopic dogs include papules and erythematous macules, secondary superficial pyoderma, secondary *Malassezia* dermatitis, chronic otitis externa and seborrhea.

Atopic cats are most commonly described with symmetric alopecia, miliary dermatitis, eosinophilic plaques, indolent ulcer of the lip, pruritus of the head and neck with excoriations or generalized pruritus. Atopic cats are pruritic, but many are secretive and groom or traumatize themselves without the owner's knowledge.

Cats with miliary dermatitis have numerous small erythematous papules with adherent brownish crusts and various degrees of alopecia and pruritus. These lesions can usually be palpated over the dorsal lumbar and cervical regions long before they are visualized. Feline miliary dermatitis is most commonly a manifestation of flea allergy, but may occur with other ectoparasite infestations, dermatophytosis, bacterial folliculitis, adverse food reactions, atopy, drug eruptions and immune-mediated skin disease.

Canine flea-bite hypersensitivity is characterized by a pruritic, papular dermatitis. Flea bites induce an initial papule that may then form a crust. Chronic pruritus may lead to alopecia, lichenification, severe crusting and hyperpigmentation. Lesions are typically confined to the dorsal lumbosacral area, caudomedial thighs, ventral abdomen and flanks. Pyotraumatic dermatitis ("hot spots"), secondary bacterial pyoderma and secondary seborrhea are common in chronic cases. The presence of otitis externa, severe pedal pruritus or facial pruritus strongly suggests concurrent atopy or adverse food reaction.

There are numerous insects besides fleas and arachnids in the normal dog and cat environment that can stimulate hypersensitivity reactions. Blackfly, deerfly, horsefly, mosquito, red ant, black ant and tick bites may all contribute to allergic skin disease in dogs and cats. The primary clinical sign is pruritus, although erythematous maculopapular dermatitis may be present. Nodules and papules induced by mosquito bites are usually found on the bridge of the nose and pinnae of cats. Stable flies occasionally induce a granulomatous reaction, producing nodules or plaques and varying degrees of alopecia on the pinnae. Ticks may induce nodules due to granuloma formation at the site of attachment. Acute-onset nasal dermatitis has also been observed in dogs; pruritic papules and nodules are found on the bridge of the nose.

Adverse reactions to food mimic other allergic diseases. The clinical features and management of adverse food reactions are described in detail in Chapter 14.

LABORATORY AND OTHER CLINICAL INFORMATION Skin biopsy and histopathology can be used to confirm the presence of inflammatory skin disease. Chronic hyperplastic dermatitis is a common histopathologic reaction pattern seen in dogs with chronic allergy. The predominant types of inflammatory cells may suggest the specific allergic disease. However, many chronic dermatoses have similar histopathologic features, making specific diagnosis difficult.

Two methods of allergy testing are available to practitioners. Intradermal testing has been performed for many years. More recently, in vitro tests for detection of allergen-specific IgE have become commercially available.

Controversy continues over whether intradermal or in vitro testing is the better method for confirming a diagnosis of atopy and for selecting allergens for hyposensitization. Both tests have drawbacks. Furthermore, long-term studies are needed to evaluate responses of allergic animals to hyposensitization based on both types of testing. For more information, see Small Animal Clinical Nutrition, 4th ed., p 913.

RISK FACTORS Atopy is an inherited predisposition to the development of IgE antibodies to environmental allergens, resulting in allergic disease. Although the exact mode of inheritance is unknown, strong breed predilection and familial involvement in dogs indicate a genetically determined cause. Canine breeds reported to be predisposed to atopy include the Cairn terrier, West Highland white terrier, Scottish terrier, wire-haired fox terrier, Boston terrier, Sealyham terrier, Lhasa apso, Dalmatian, pug, Irish setter, English setter, golden retriever, Labrador retriever, boxer, miniature schnauzer, English bulldog, Bichon Frise, Chinese Shar-Pei, Shih Tzu, German shepherd dog, Belgian Tervuren, beauceron and cocker spaniel. However, canine atopy may be seen in any breed, including mixed breeds. Breed predisposition has not been reported for atopic cats.

Hypersensitivity requires environmental exposure to flea, other biting insect or arachnid allergens. Depending on the offending allergen, these cases may be seasonal in temperate climates; worse clinical signs occur during warm weather. The onset of clinical signs may be historically correlated with an increase in insect or arachnid numbers in the environment.

USE OF FATTY ACIDS AS ANTIPRURITIC AGENTS The inflammation and dermatitis associated with allergic skin disease may be partially caused

INFLAM DIS

by abnormal essential fatty acid metabolism and inappropriate eicosanoid synthesis. A unique feature of skin is that it lacks Δ-6- and Δ-5-desaturase enzyme activity, and thus is incapable of making AA from LA or EPA from ALA (**Figure 26-2**). Skin can elongate GLA to DGLA and EPA to DHA. Normal dogs metabolize dietary sources of ALA to EPA and DHA elsewhere in the body. These fatty acids are then incorporated into the skin.

DGLA, EPA and DHA in cutaneous cellular membranes may decrease inflammation through competition with AA for metabolic enzymes or because of the antiinflammatory nature of the eicosanoids produced. The rationale for specifically administering products high in GLA is that GLA can be incorporated into the skin, where it is rapidly elongated to DGLA. Because skin lacks desaturase enzymes, DGLA is not further metabolized to arachidonic acid. As a result, DGLA competes with AA for metabolic enzymes. Thus there is a decrease in AA-derived eicosanoids and an increase in the antiinflammatory eicosanoids PGE_1 via the cyclooxygenase cascade and 15-HETE via the lipoxygenase pathway.

ALA is an n-3 PUFA that is metabolized to EPA and DHA, and incorporated into the skin of normal dogs. Findings suggest that atopic dermatitis in human beings is associated with a deficiency of Δ-6-desaturase activity, which prevents the rapid conversion of ALA to EPA and DHA in atopic individuals. Comparable studies using atopic dogs and cats have not been published. However, subsets of atopic dogs may exist with different fatty acid metabolic capabilities.

The use of fatty acids for treating atopy and chronic pruritus has been extensively studied in dogs. Unfortunately, most of these studies have been uncontrolled, nonblinded clinical trials using low doses of fatty acids for short periods. In these studies, 0 to more than 75% of pruritic animals had degrees of clinical improvement. Three, well-controlled clinical studies using placebos and high doses of fatty acids for six weeks or more showed decreased pruritus in 0 to more than 50% of the patients. Dogs that did not have decreased pruritus still showed improvement in other clinical signs, including less erythroderma and skin edema. For more information, see Small Animal Clinical Nutrition, 4th ed., p 914.

The benefit of fatty acid supplementation is maximized in dogs if other contributing diseases such as adverse reactions to food, flea hypersensitivity, bacterial pyoderma and *Malassezia* dermatitis are controlled. Overall, it is probably safe to inform clients that up to 50% of dogs with allergic pruritus will improve with modification in fatty acid intake, if secondary bacterial and yeast infections are controlled. Synergistic effects have been documented between fatty acids and other antipruritic agents such as antihistamines and glucocorticoids.

The use of fatty acids for management of allergic skin disease and papulocrustous dermatitis in cats has been reported. More than 50% of allergic cats may improve, based on the results of uncontrolled, nonblinded clinical trials published to date.

Assess the Food (Page 916)
Assess the Feeding Method (Pages 916-917)
Feeding Plan (Pages 917-918)
Reassessment (Page 918)

Information about assessing the food and feeding method, developing a feeding plan and reassessing patients with allergic skin disease is presented below with comparable information for pets with arthritis.

ARTHRITIS (Pages 914-918)

Arthritis is classified into two broad categories for clinical diagnosis: 1) the degenerative types of arthritis, in which degradation of the articular cartilage is a prominent feature and 2) the inflammatory arthropathies, in which an obvious synovitis is the main pathologic feature.

Osteoarthritis is the most common form of arthritis recognized in people and most animal species. It generally is a slowly progressive disease, and is characterized by two main pathologic processes: degeneration of articular cartilage, with loss of proteoglycan and collagen, and proliferation of new bone. In addition, a variable inflammatory response occurs within the synovial membrane.

Inflammatory arthropathies are classified as infective, immune-based or crystal-induced. The most common category of the noninfective inflammatory arthritides in people and dogs is immune-based. Of these, rheumatoid arthritis is the most prevalent form. Rheumatoid arthritis is an immune-based arthritis characterized by chronic inflammation and synovitis associated with distortion of normal tissue architecture, degradation of the extracellular matrix and inadequate repair.

Clinical Importance (Page 915)

Arthritis (osteoarthritis) is the most common form of joint and musculoskeletal disease affecting dogs but is relatively uncommon in cats. Results of the National Companion Animal Study (40,000 dogs) revealed that arthritis was the eighth most common diagnosis in dogs seven to 10 years old and the fourth most common diagnosis in dogs more than 10 years old. Arthritis was clinically evident in 3% of all dogs over 10 years old. The prevalence of osteoarthritis in pets is likely to increase with increasing life spans.

Assessment (Pages 915-917)

Assess the Animal (Pages 915-916)

HISTORY AND PHYSICAL EXAMINATION Osteoarthritis is characterized by the gradual development of joint pain, stiffness and limitation of motion. Clinical signs may be exacerbated by exercise, obesity, long periods of recumbency and cold, damp weather. Dogs may have a history of joint trauma (fracture, luxation, ligament injury) or developmental disorders (patellar luxation, fragmented coronoid process, hip dysplasia, osteochondrosis, etc.).

The earliest sign noted by the owner is a loss of normal performance, often manifested as a pet's reluctance to jump or climb stairs. Temporary stiffness after rest is also an early sign of joint disease and is frequently present in osteoarthritis before the onset of overt lameness. Only in advanced cases of osteoarthritis is stiffness a problem for prolonged periods of time. Stiffness is invariably accompanied by obvious lameness. Other clinical signs of arthritis include reduced range of motion, crepitus, joint swelling, joint instability and pain.

LABORATORY AND OTHER CLINICAL INFORMATION Radiography remains the primary diagnostic modality for suspected cases of osteoarthritis.

Arthrocentesis and synovial fluid analysis can be used to support a diagnosis of degenerative joint disease. A slightly higher number of mononuclear cells is seen, generally fewer than 2,000 cells/ml. Large numbers of neutrophils are likely due to underlying immune-mediated or infectious arthritis. Synovial fluid can be submitted for bacterial culture and antimicrobial sensitivity testing. Biopsy of synovial tissue is helpful in ruling out other arthritides and neoplasia.

Coombs, antinuclear antibody (ANA) and rheumatoid factor tests help rule out immune-mediated disease whereas serum titers help rule out borreliosis, ehrlichiosis and other rickettsial infections.

RISK FACTORS Working dogs, athletic dogs and obese animals place more stress on their joints and are more likely to incur injury (fracture, luxation, ligament injury) and degenerative joint disease. Some dogs are at risk for developmental orthopedic diseases (hip dysplasia, osteochondrosis) that cause arthritis. (See Chapter 17.) Infective arthritis (bacterial, spirochetal, rickettsial, protozoal, fungal) and immune-based arthritides (rheumatoid, systemic lupus erythematosus, idiopathic polyarthritis) may mimic degenerative joint disease and carry their own set of risk factors.

USE OF FATTY ACIDS FOR ARTHRITIS In most inflammatory arthritides, prostaglandins are the active agent involved in the inflammation process. Diverting prostaglandin production from the inflammatory 2-series (PGE_2) to the less inflammatory 1- or 3-series (PGE_1 or PGE_3) may benefit patients (e.g., reduce inflammation and joint stiffness). Dietary manipulation of n-6 and n-3 fatty acids to abate production of certain prostaglandins in the inflammatory process has been evaluated in people with rheumatoid arthritis, but there has been little research reported on the use of these fatty acids in companion animals.

A consistent response to dietary fatty acid manipulation has been observed in studies of people with rheumatoid arthritis. Numerous well-designed clinical studies evaluating dietary supplementation with n-3 and n-6 PUFA have been conducted. The anticipated antiinflammatory effects were accompanied by reduced active synovitis in most trials. For more information, see Small Animal Clinical Nutrition, 4th ed., p 916.

In 16 of 20 trials, a positive response was found in those people or animals whose diet was supplemented with GLA, EPA or DHA. The degree of positive response varied based on the length of treatment, dosage, number of people or animals and severity of disease. The only study with negative results was an early rat study in which animals had an increased incidence of collagen-induced arthritis when fed a food supplemented with fish oil. Overall, arthritic animals treated with various fatty acid regimens improved clinically. For more information, see Small Animal Clinical Nutrition, 4th ed., p 916.

The most encouraging studies have been conducted in the last few years. In these long-term (six months to one year), double-blind, controlled studies, human patients were given various fish oil or EPA/DHA regimens in addition to physician-prescribed NSAIDs. In each trial, patients receiving the fish oil or EPA/DHA regimens were able to reduce or discontinue drug usage without experiencing pain or joint stiffness. Patient-assessed and physician-assessed pain scores were reduced for the fish oil or EPA/DHA treatment groups when compared with the control groups. The ability of patients to reduce or discontinue drug usage appeared to be related to fish oil or EPA/DHA dosage and length of treatment.

Many problems remain to be solved before these dietary supplements can be recommended as standard treatment for arthritides in dogs and cats. The only canine study was a compilation of observations of dog owners who perceived improvement in their pets' arthritic clinical signs when their dogs were treated with fatty acids for various dermatologic problems. In this preliminary study, 13 of 22 dogs with chronic, intercurrent hip arthritis had noticeable improvement in clinical signs of arthritis during treatment with a fatty acid supplement for a two-week period. Though encouraging, better-controlled studies in dogs with more objective measures of improvement need to be considered to validate the findings of this experiment and to confirm the results from the human tests.

Assess the Food *(Page 916)*

Patients with dermatitis or arthritis may benefit from changes in dietary fatty acid intake. The most common modification is to increase n-3 fatty acid

Table 26-1. Fatty acids found in pet food ingredients and supplements.

Fatty acids	Ingredient/supplement
Linoleic acid (n-6)	Vegetable oils (soy oil, corn oil, safflower oil, canola oil, etc.)
	Grains (corn, soybeans)
Gamma-linolenic acid (GLA, n-6)	Black currant oil
	Borage oil
	Evening primrose oil
Alpha-linolenic acid (ALA, n-3)	Flax
	Flax (linseed) oil
Eicosapentaenoic acid (EPA, n-3)	Fishmeal
	Cold water marine oils
Docosahexaenoic acid (DHA, n-3)	Fishmeal
	Cold water marine oils

intake and/or increase intake of GLA, an n-6 fatty acid. **Table 26-1** lists typical pet food ingredients and supplements with their associated fatty acids.

Changing the food, adding a supplement or doing both can modify fatty acid levels in the overall diet. Initially, the essential fatty acid levels in the current food should be assessed. Unfortunately, information about fatty acid concentrations in commercial pet foods is difficult to obtain. This information is not typically found in guaranteed or typical analysis statements on pet food labels and is often not published by the manufacturer. The manufacturer should be contacted directly to obtain information about fatty acid concentrations in specific products. **Table 26-2** contains information about fatty acid concentrations in selected commercial dog foods and in commercial fatty acid supplements. Comparable data for cat foods are lacking.

If the animal is given a supplement, the fatty acid concentrations in the supplement should also be determined. Most supplements marketed to improve skin and coat list the fatty acid concentrations on the product label or in published technical information. **Table 26-2** contains information about fatty acid concentrations in selected commercial fatty acid supplements. In many cases, fatty acid supplements contain much lower concentrations of fatty acids than concentrations already found in commercial pet food being consumed by the animal (**Table 26-2**). Thus, it may be more appropriate and convenient to change the animal's food to one with higher concentrations of appropriate fatty acids rather than adding a fatty acid supplement to the animal's current food. In some cases, changing the food and simultaneously adding a fatty acid supplement may be appropriate.

It is important to remember that the optimal concentrations and ratios of fatty acids have not been established for normal animals or dogs and cats with clinical disease. Trial and error with various food and supplement combinations may be needed in an individual animal to achieve the best clinical response.

Table 26-2. The total essential fatty acid intake for a 10-kg dog eating 600 kcal (2,510 kJ) per day of selected commercial foods or being given one of the selected supplements.

Foods	Food consumed (g)	Total n-6 consumed (mg)	Total n-3 consumed (mg)*
Hill's Prescription Diet Canine d/d Lamb & Rice, moist	431	4,398	864
Hill's Prescription Diet Canine d/d Rice & Duck, dry	164	4,992	492
Hill's Prescription Diet Canine d/d Rice & Egg, dry	154	5,670	570
Hill's Prescription Diet Canine d/d Rice & Salmon, dry	150	2,982	1,188
Hill's Prescription Diet Canine d/d Whitefish & Rice, moist	542	6,726	1,140
Hill's Prescription Diet Canine n/d, moist	380	2,772	8,088
Hill's Prescription Diet Canine z/d Low Allergen, dry	162	7,542	978
Hill's Prescription Diet Canine z/d ULTRA Allergen Free, dry	145	7,434	930
Hill's Prescription Diet Canine/Feline a/d, moist	500	6,714	3,048
Hill's Science Diet Adult Canine Maintenance, dry	162	5,574	810
Hill's Science Diet Canine Active, dry	130	6,150	678
Hill's Science Diet Adult Canine Maintenance Lamb Meal & Rice, dry	162	5,640	1,002
Hill's Science Diet Canine Senior, dry	163	4,866	1,692
Hill's Science Diet Light Canine Maintenance, dry	200	6,246	642
Hill's Science Diet Sensitive Skin Dog, dry	158	7,440	2,058
Iams Eukanuba Adult Maintenance Formula, dry	139	4,638	840
Iams Eukanuba Reduced Fat Adult Formula, dry	155	3,564	648
Iams Eukanuba Veterinary Diets Maximum-Calorie, dry	129	5,418	1,056
Iams Eukanuba Veterinary Diets Maximum-Calorie, moist	300	6,342	768
Iams Eukanuba Veterinary Diets Response Formula FP, dry	147	2,352	438
Iams Eukanuba Veterinary Diets Response Formula FP, moist	475	9,360	1,284
Iams Eukanuba Veterinary Diets Response Formula KO, dry	143	4,452	564
IVD Limited Ingredient Canine Duck & Potato, dry	188	2,280	450
NutroMax Special, dry	158	3,634	158
NutroMax, dry	140	5,460	280

(Continued on next page.)

INFLAM DIS

Table 26-2. The total essential fatty acid intake for a 10-kg dog eating 600 kcal (2,510 kJ) per day of selected commercial foods or being given one of the selected supplements (Continued.).

Foods	Food consumed (g)	Total n-6 consumed (mg)	Total n-3 consumed (mg)*
Purina Dog Chow, dry	158	3,160	158
Purina Veterinary Diets LA-Formula, dry	166	1,680	1,680
Supplements			
3V Caps Skin Formula for Large & Giant Breeds	1 capsule	0	417
3V Caps Skin Formula for Medium & Large Breeds	1 capsule	0	300
3V Caps Skin Formula for Small & Medium Breeds	1 capsule	0	171
3V Caps Skin Formula Liquid	1 ml	0	185
Dermage III Extra Strength	1 capsule	412	172
Dermage III Regular	1 capsule	310	63
DermCaps 100s	1 capsule	402	252
DermCaps ES	1 capsule	368	125
DermCaps ES Liquid	1 ml	375	130
DermCaps Liquid	1 ml	621	65
DermCaps Regular	1 capsule	402	42
EFA-Caps	1 capsule	10	75
EFA-Caps HP	1 capsule	30	160
EFA-Z Plus	1/4 oz	3,410	83
Pet-Derm OM Extra Strength	1 capsule	28	270
Pet-Derm OM Liquid	2 ml	28	270
Pet-Derm OM Regular	1 capsule	14	135

*Laboratory and clinical studies in a number of species have established a daily dosage for total n-3 fatty acids that seems to be a reasonable starting point in patients with inflammatory disease. An initial dose of 50 to 250 mg of total n-3 fatty acids/kg body weight/day seems to be effective in a large number of studies.

Other nutrients such as zinc, magnesium, biotin, pyridoxine, vitamin E and vitamin C are important cofactors in fatty acid metabolic pathways. Most commercial pet foods have adequate levels of these nutrients; routine supplementation would not be expected to improve clinical response. Many fatty acid supplements contain additional amounts of these cofactor nutrients.

Assess the Feeding Method *(Pages 916-917)*

Most feeding methods need not be changed for patients with arthritis or dermatitis. If the animal has a normal body condition score (3/5), the amount of food previously fed (energy basis) was probably appropriate.

Feeding Plan (Pages 917-918)

Select a Food or Supplement *(Pages 917-918)*

Dietary n-3 fatty acid concentrations can be increased by using a supplement (usually a cold water marine oil) or changing to food that contains flaxseed, fishmeal and/or fish oil as major ingredients. Dietary GLA concentrations can be increased by feeding a supplement containing evening primrose, borage or black currant oil. Most commercial pet foods already exceed the n-6 essential fatty acid requirement for linoleic acid by using vegetable oil and/or vegetable ingredients in their formula.

It is clear that dietary fatty acid levels well above those needed to avoid fatty acid deficiency benefit some animals with arthritis, allergic skin disease and chronic pruritus. What is less clear are answers to the following questions: 1) which fatty acid or combination of fatty acids is most effective, 2) what ratio of n-6 to n-3 fatty acids is optimal, 3) what absolute amount of n-6 and n-3 fatty acids is appropriate in normal animals and what amount is effective in animals with clinical disease, 4) what levels of other nutrients (vitamins, trace minerals) are needed to optimize fatty acid therapy and 5) what level of total dietary fat is needed to optimize fatty acid metabolism and clinical efficacy.

Although definitive answers to these questions are lacking, laboratory and clinical studies in a number of species have established a daily dosage for total n-3 fatty acids that seems to be a reasonable starting point in patients with inflammatory disease. An initial dose of 50 to 250 mg of total n-3 fatty acids (ALA, EPA and/or DHA)/BW_{kg}/day seemed to be effective in a large number of studies. This total dose can be supplied through a combination of appropriate foods and supplements (**Table 26-2**). Further studies using dogs and cats should help refine the most effective dose, type, ratio and time periods for fatty acid therapy.

The risks and side effects of high levels of dietary fatty acids are few. Soft feces, overt diarrhea, flatulence and oral malodor ("fishy breath") are most commonly noted at levels of fatty acid supplementation used in

INFLAM DIS

most patients. These risks and side effects are outweighed by the possibility that fatty acid supplements will allow practitioners to reduce or eliminate the use of NSAIDs in arthritis cases or discontinue corticosteroid therapy for pruritic dogs and cats. The veterinary community will need to confirm the benefits of fatty acid supplementation for inflammatory diseases in dogs and cats in well-designed, controlled studies.

Determine a Feeding Method (Page 918)

Other than supplementation, the method of feeding is often not altered in the nutritional management of dermatitis or arthritis. If a new food or supplement is fed, the amount to feed can be determined from the product label or other supporting materials. The food dosage may need to be changed if the caloric density of the new food differs from that of the previous food. The food dosage is usually divided into two or more meals per day. The food dosage and feeding method should be altered if the animal's body weight and condition are not optimal.

For clinical nutrition to be effective, there needs to be good client compliance. Enabling compliance includes limiting the patient's access to other foods and knowing who feeds the animal. If the animal comes from a multiple-pet household, it should be determined whether the pet with dermatitis or arthritis has access to the other pets' food.

Reassessment (Page 918)

Fatty acid-responsive diseases usually respond to dietary changes or supplementation over several weeks to months. After a dietary change or supplement has been started, the patient should be examined every four weeks for significant improvement in pruritus, skin erythema or lameness. Some patients may not respond for several months or may need concurrent therapy with antihistamines, topical agents (medicated shampoo), corticosteroids or nonsteroidal antiinflammatory agents.

Clinical cases that illustrate and reinforce the nutritional concepts presented in this chapter can be found in Small Animal Clinical Nutrition, 4th ed., pp 920-921.

CHAPTER 27

Dietary Effects on Drug Metabolism

For a review of the unabridged chapter, see Fettman MJ, Phillips RW. Dietary Effects on Drug Metabolism. In: Hand MS, Thatcher CD, Remillard RL, et al, eds. Small Animal Clinical Nutrition, 4th ed. Topeka, KS: Mark Morris Institute, 2000; 923-939.

INTRODUCTION (Page 923)*

When the effects of veterinary pharmaceuticals are evaluated and standardized doses are determined, researchers typically use relatively healthy, fasted animals that have been maintained on foods with acceptable nutrient balance. However, in clinical settings, animals receiving a drug often have variable food intake or specific nutrient imbalances, or they must be given a drug in conjunction with meal, or some combination of these factors may be in play. The patient's health status and the nutrient ingredient profile of the food being consumed may greatly affect drug absorption, distribution, metabolism, efficacy and toxicity (**Table 27-1**). Concurrent food intake also may markedly affect drug availability and pharmacokinetics (**Table 27-2**).

Veterinarians should be acquainted with the effects diet can have on drug metabolism in order to anticipate adjustments in the food or drug dose, properly time administration of drugs and allow for changes in the margin for error between efficacy and toxicity of pharmaceutical agents.

GENERAL TYPES OF FOOD AND DRUG INTERACTIONS
(Pages 923-924)

Food-drug interactions that occur as a result of the physical form or chemical properties of food may lead to drug binding, precipitation, inactivation or ionization, which alters gastrointestinal (GI) absorption. These interactions may occur in vitro after mixing the drug with food to make administration more convenient, to enhance palatability or to reduce GI irritation. Another concern is adsorption of drugs to synthetic surfaces of the equipment used for nutrient and drug administration (e.g., food containers, feeding syringes, tubing for assisted feeding). Physiochemical inter-

*Page numbers in headings refer to Small Animal Clinical Nutrition, 4th ed., where additional information may be found.

Table 27-1. Factors affecting the disposition of drugs that can be influenced by foods.

Absorption
GI transit time
GI luminal environment
Enterocyte function
Electrochemical gradient across the GI mucosa
pH gradient across the GI mucosa
Distribution
Drug-binding proteins
Blood cells that bind or metabolize drugs
Metabolism
Site of metabolism
   Organ
   Tissue
   Cell type
   Cell organelle
Biotransformation pathways
   Phase I oxidative vs. phase II conjugative pathways
Cofactors required for metabolism
   Vitamins
   Minerals
   Reducing agents
Non-nutrient enzyme inducers
   Phytochemicals
   Synthetic contaminants
   Preservatives
Excretion
Route of excretion
   Biliary
   Fecal
   Mammary
   Pulmonary
   Renal
   Salivary
   Sweat
Electrochemical gradient across mucosa of excretory organs
Rate of excretion

---

actions may occur in vivo, whereby drug absorption from the GI tract is decreased because of chelation by dietary fiber or minerals, or increased because of favorable changes in ionization or solvent partitioning.

Metabolic Interactions (Page 923)

Both nutrient and non-nutrient substances in foods can alter the metabolism of absorbed drugs. Malnutrition can alter the synthesis of plasma proteins, affecting drug distribution and pharmacokinetics. Individual dietary

Table 27-2. Various factors that can determine the effects of nutrients on drug absorption.

Factors	Examples
Physiochemical properties of drugs	Lipophilic or hydrophilic
Drug formulation	Tablet, capsule or liquid
Meal type	Volume, temperature, moisture
Drug dose	Amount and concentration
Route of administration	By mouth, by gastric tube, etc.
Order of administration	Pre- vs. postprandial
Time interval between food and drug administration	Phase of digestion
Owner/patient compliance	Mixing drugs and food for ease of administration

lipid, carbohydrate, protein, vitamin and mineral levels can effect changes in xenobiotic-metabolizing enzymes, resulting in altered clearance, circulating concentrations and resultant therapeutic efficacy. Naturally occurring non-nutrient food ingredients and added synthetic preservatives may similarly alter pharmacokinetics and apparent drug effectiveness.

Indirect Physiologic Effects (Pages 923-924)

The rate and extent of drug elimination are also affected by changes in blood flow and drug delivery to the principal organs of metabolism for that particular agent. Thus, postprandial alterations in blood flow to the liver may affect drug clearance from the portal and systemic circulation, and altered blood flow to the kidneys may change the rate of urinary elimination. Likewise, changes in functional morphology or pathology due to specific nutrient deficiencies or excesses can affect drug clearance.

GENERAL EFFECTS ON FOOD AND DRUG INTERACTIONS (Pages 924-925)

Ameliorating Potential Adverse Effects (Page 924)

The composition and volume of food consumed can modify the degree of GI irritation caused by concurrently administered oral drugs. Enteral and parenteral fluid intake can augment drug absorption and distribution and protect against renal damage induced by nephrotoxic agents. Supplementation with specific nutrients may prevent deficiencies secondary to adverse drug effects on nutrient absorption and metabolism.

Potentiating Drug Action (Page 924)

Specific nutrients can increase drug effects by facilitating GI absorption, improving drug distribution or decreasing drug metabolism and excretion. Furthermore, some nutrients may be necessary for optimal drug effects (e.g., arginine for nitric oxide production, cysteine for nitroglycerin action and carnitine for optimal activity of cardiac glycosides).

Impaired Drug Action (Page 924)

These adverse effects may result from inadequate amounts of the drug reaching the site of action, or nutrient interference with the drug's action. Drug action may be impaired for variable periods of time after food composition and feeding behavior are altered, if target cell receptor numbers or affinity are suppressed or long-lived biotransforming enzyme systems have been induced.

Adverse Side Effects (Pages 924-925)

Pathologic reactions to nutrient-induced changes in drug distribution and metabolism can be of greater immediate consequence than loss of disease control following impaired drug action. Drug metabolism and excretion routes may be altered, resulting in accumulation of toxic quantities of the agent itself or the products of its biotransformation. This phenomenon is similar in principle to adverse interactions between concurrently administered drugs, but may be more difficult to identify because mental recollection and written records of nutrient intake are not usually as complete as for administration of pharmaceutical agents.

EFFECTS OF NUTRIENTS ON DRUG ABSORPTION
(Pages 925-928)

General Observations (Page 925)

The absorption of orally administered drugs may be: 1) decreased, 2) delayed, 3) unaffected or 4) enhanced by the concomitant consumption of food (**Tables 27-2** to **27-4**). This interaction depends on the physical and chemical nature of the food and the drug, including such things as meal size and type, the formulation in which the drug is administered, the order in which the food and drug are ingested and the interval between their consumption.

Any food can reduce drug absorption by creating a barrier that prevents dispersion of the drug and dissolution of the active agent in GI luminal contents. Drugs are better absorbed in dilute vs. concentrated solution because of greater dissolution and more rapid gastric emptying. In people, absorption of erythromycin stearate is reduced by approximately one-half when taken with food or a small fluid volume, as com-

DIET/DRUG

Table 27-3. Examples of drugs whose absorption may be affected by concomitant food consumption.

Enhanced absorption	Reduced absorption	Delayed absorption
α-tocopherol	Amoxicillin	Acetaminophen
Carbamazepine	Ampicillin	Aspirin
Chlorothiazide	Antipyrine	Atenolol
Diazepam	Astemizole	Cefaclor
Dicumarol	Captopril	Cephalexin
Erythromycin ethylsuccinate	Cephalexin	Cephradine
Griseofulvin	Chlorpromazine	Cimetidine
Hydrochlorothiazide	Erythromycin stearate	Digoxin
Labetalol	Isoniazid	Furosemide
Lithium	Ketoconazole	Glipizide
Mebendazole	Levodopa	Hydralazine
Metoprolol	Lincomycin	Ibuprofen
Nitrofurantoin	Methacycline	Metronidazole
Phenytoin	Methyldopa	Piroxicam
Propoxyphene	Nafcillin	Quinidine
Propranolol	Penicillamine	Sulfonamides
Riboflavin	Penicillins	Theophylline
Spironolactone	Phenobarbital	
Sulfamethoxydiazine	Propantheline	
	Quinolones	
	Rifampin	
	Tetracyclines	
	Warfarin	

pared with a large fluid volume. Absorption of acetylsalicylic acid (aspirin), cephalexin, metronidazole, digoxin, hydralazine and cimetidine are similarly affected.

In people, food enhances the absorption of certain drugs (e.g., erythromycin ethylsuccinate, nitrofurantoin, hydrochlorothiazide and diazepam). The rate and quantity of drug and nutrient delivery are important, as evidenced by studies of hydralazine absorption. A bolus of nutrients impairs hydralazine uptake whereas a constant infusion over several hours does not. Total hydralazine absorption, however, is similar in both cases, although absorption is delayed by a bolus meal.

Studies of penicillin absorption in dogs after oral administration have yielded conflicting results. In a study of greyhounds receiving 20 mg/BW_{kg} ampicillin or amoxicillin per os with no food, moist food or dry food, ampicillin absorption was impeded 60 to 80% by feeding, whereas amoxicillin absorption was unaffected. In a more recent study involving a variety of breeds, absorption was impaired about 30% when ampicillin (20 mg/BW_{kg}) was administered immediately or two hours after feeding a dry food as compared with ampicillin administered to fasting dogs or

Table 27-4. Timing of drug administration relative to feeding.

Examples of drugs that should be administered after a fast to facilitate GI absorption (About one hour before or three hours after a meal).

Antibacterials
Macrolides
Erythromycin stearate   Tylosin
Penicillins
Amoxicillin   Hetacillin
Ampicillin   Methicillin
Carbenicillin   Nafcillin
Cloxacillin   Oxacillin
Dicloxacillin   Penicillin
Sulfonamides
Sulfadiazine   Sulfabromomethazine
Sulfadimethoxine   Sulfachlorpyridazine
Sulfamerazine   Sulfamethoxypyridazine
Sulfamethazine   Sulfaquinoxaline
Sulfisoxazole   Tetroxoprim
Sulfathiazole   Trimethoprim
Tetracyclines (except doxycycline)
Chlortetracycline   Oxytetracycline
Minocycline   Tetracycline
Chelating agents   Penicillamine
Laxatives   Bisacodyl

Examples of drugs that should be administered with food to minimize gastric irritation.

Antibacterials
Isoniazid   Nitrofurantoin
Metronidazole   Sulfasalazine
Anticonvulsants   Phenytoin
Primidone   Phenobarbital
Antiinflammatory agents
Aspirin   Naproxen
Indomethacin   Phenylbutazone
Diuretics   Acetazolamide
Triamterene   Thiazides
Phenothiazines
Acepromazine   Promazine
Chlorpromazine   Triflupromazine
Prochlorperazine   Trimeprazine

Examples of drugs that should be administered with large amounts of water to facilitate their action and to prevent dehydration.

Laxatives
Aluminum, calcium and magnesium salts   Psyllium
Methylcellulose

one hour before feeding. Absorption of ampicillin (20 mg/BW_{kg}) administered immediately after feeding a moist food was similarly impaired, whereas administration two hours after feeding moist food had a lesser effect on absorption.

In a comparative study of different penicillin preparations given to dogs, peak plasma concentrations of ampicillin, amoxicillin, penicillin V, phenethicillin and cloxacillin were decreased 40 to 50% by feeding immediately before drug administration; however, time to maximal concentrations was increased 0.5 to 1.5 hours only for ampicillin and amoxicillin. From these studies, it is recommended that ampicillin be given to fasted dogs and at least one hour before feeding to ensure adequate drug absorption.

Physical Incompatibilities (Page 927)

Specific nutrients (e.g., dietary fiber) can impede drug absorption across the GI mucosa by adsorbing the agent or increasing the unstirred water layer on the mucosal surface. Some of these interactions are predictable, based on the behavior of fiber in binding substances such as bile acids or decreasing the absorptive rate of solutes such as monosaccharides. Psyllium mucilloid decreases the absorption of riboflavin, β-carotene, iron, zinc and other trace elements.

Both nutritive and non-nutritive cytoprotective agents adsorb drugs and inhibit their absorption. For example, sucralfate binds tetracycline, phenytoin, cimetidine, digoxin and levothyroxine. Antacids (e.g., aluminum hydroxide) can precipitate tetracyclines, iron salts, warfarin, digoxin, quinidine, phenothiazine, indomethacin, isoniazid, sulfadiazine, prednisone and levothyroxine. Mineral supplements, including iron salts such as ferrous sulfate can decrease the absorption of methyldopa, penicillamine, tetracycline, levothyroxine and quinolone antibiotics. Calcium salts and calcium-containing foods (e.g., milk, 1 to 2 mg calcium/ml) can precipitate insoluble tetracycline chelates. Foods of plant origin may contain phytic acid, which inhibits zinc and calcium absorption, and tannins, which inhibit iron uptake. Certain liquid drug formulations, including some expectorants, elixirs, syrups, concentrates and suspensions, may gelatinize when mixed with enteral formulas.

The potential binding of pharmaceutical agents to equipment used for administration should also be considered. Adsorption of vitamin A and drugs such as phenytoin to plastic polymers used in nasogastric, gastrostomy and enterostomy tubing has been reported. Furthermore, precipitation or gelatinization of drugs by nutrients can block feeding tubes.

Physical Factors Affecting GI Absorption of Drugs (Pages 927-928)

The cephalic phase of digestion is normally initiated by the perception, visualization, smell and taste of food. This phenomenon contributes to normal GI motility, secretion, digestion and absorption of food and pharmaceutical agents. For example, when acetaminophen is administered by nasogastric tube postoperatively, its absorption is significantly reduced compared with its absorption following oral administration. Decreased GI motility due to stress, pain, luminal obstruction and postsurgical ileus may also contribute to reduced absorption.

Commercial moist and dry foods similarly affect gastric emptying; however, solid food may decrease the emptying of liquids and liquids may decrease the rate and pattern of solid emptying. Specific dietary components that affect the rate of GI transit can also alter oral drug assimilation. In dogs, meals containing cellulose or wheat bran increase the frequency of postprandial contractions; yet, only cellulose decreases duodenojejunal flow and prolongs transit time. However, bran increases mixing and onward propulsion of ingesta. Addition of guar gum induces continuous low-amplitude contractions in dogs, increases jejunal flow, but still increases transit time because of water adsorption and luminal distention. Soluble fibers (e.g., methylcellulose) increase luminal viscosity, resulting in delayed gastric emptying and increased thickness of the unstirred water layer. Thus, both delivery of drug to the intestine and contact with the mucosal surface are impeded.

Addition of fat to a meal changes intragastric distribution of solid material, induces segmental changes in antral and pyloric motility and retards gastric emptying. Intraduodenal instillation of dilute glucose solutions at a rate in excess of approximately 2 kcal (8.4 kJ)/min., regardless of tonicity, stimulates both phasic and tonic pyloric contractions, thereby inhibiting gastric emptying and delaying oral drug absorption. Propranolol and metoprolol are affected in this manner. Enterohepatic cycling of drugs (e.g., doxycycline) may be affected by rate of passage and by portal blood flow and hepatic metabolism.

Chemical Factors Affecting GI Absorption of Drugs (Page 928)

Beyond the effects of drug binding or precipitation, specific nutrients may compete for absorption by the intestinal mucosa. For instance, phenytoin absorption is impaired by concurrent administration of the B vitamins folic acid and pyridoxine. Concurrent food intake and particular food ingredients can alter gastric or intestinal pH, thereby altering drug dissolution, ionization and absorption. In addition to the effect of milk calcium content on tetracycline absorption, milk can increase gastric pH, inducing premature dissolution of enteric-coated tablets, resulting in gastric irritation, altered absorption or both.

Gastric acid secretion associated with food ingestion can assist in the dissolution and ionization of alkaline drugs. Gastric acid secretion, however, limits the rate of absorption of alkaline drugs, while promoting the

DIET/DRUG

absorption of dissolved, un-ionized acidic drugs. The subsequent release of bicarbonate-rich pancreatic secretions promotes ionization of acidic drugs, but facilitates absorption of dissolved, un-ionized alkaline drugs. Release of hydrochloric acid in the stomach typically leads to alkalinization of the blood and the postprandial "alkaline tide," establishing an ionization gradient that can affect diffusion of ionizable compounds across the GI mucosa.

By affecting the food's acidification potential, dietary cation-anion balance can alter mineral absorption and drug availability through changes in ionization. Concurrent consumption of fats can affect drug absorption, depending on the polarity and lipid solubility of the individual agent. For example, it has been well documented that lipid-soluble vitamins and the antifungal agent griseofulvin are better absorbed when taken with whole milk or a fatty meal. High-fat foods may promote the absorption of nitrofurantoin, chlorothiazide and riboflavin by delaying gastric emptying, which facilitates dissolution in the stomach before passage into the small intestine for uptake.

TRANSPORT FROM THE GI TRACT TO THE SITE OF ACTION OR METABOLISM (Pages 928-929)

Dietary factors that affect blood flow will alter the rate of delivery of absorbed drugs to their site of action or metabolism. Dehydration not only may reduce GI blood flow and absorption, but may also reduce the absorbed drug's subsequent delivery to or removal from particular tissues. Hypovolemia and reduced tissue perfusion may result in target tissue doses below the effective concentration. Decreased blood flow may reduce hepatic extraction for metabolism and excretion. Decreased urine formation may increase drug accumulation and toxicity in various organs; aminoglycoside accumulation in the renal proximal tubules is a common example. Other dietary ingredients may affect cardiac output (methylxanthines), renal blood flow (protein) or intestinal reperfusion following ischemia (antioxidants), thereby altering drug distribution.

Like many metabolites and hormones, drugs may be transported in the blood in the free form or bound to plasma proteins. Thus, changes in nutritional status that affect plasma protein synthesis will likely affect drug binding and distribution. For example, hypoalbuminemia due to low dietary protein quantity or quality can affect the distribution of antibiotics, barbiturates, cardiac glycosides and analgesics. Drugs and nutrients may influence one another's disposition because binding to plasma proteins is competitive. High postprandial free fatty acid levels can displace anionic compounds from cationic binding sites on plasma proteins. Drugs

Table 27-5. Dietary factors that may affect drug metabolism and excretion (principally through induction of phase 1 biotransformation).

Macronutrients	Micronutrients	Non-nutrients
Protein	Vitamins	Antioxidants (BHA, BHT)
Carbohydrate	Minerals	Coumarins
Fat	Essential fatty acids	Flavonoids
Fiber		Indoles
		Methylxanthines
		Organonitriles
		Phenols
		Pyrolysis byproducts
		Terpenoids

Examples of drugs whose metabolism and excretion is altered by these dietary factors

Acetaminophen	Estradiol	Pentobarbital
Allopurinol	Hexobarbital	Phenobarbital
Aminophylline	Isoniazid	Phenytoin
Cefoxitin	Meperidine	Prednisolone
Chloramphenicol	Morphine	Propranolol
Chloroquine	Oxazepam	Theophylline
Diazepam	Penicillin	Zoxazolamine

Key: BHA = butylated hydroxyanisole, BHT = butylated hydroxytoluene.

and nutrients that are competitively transported into erythrocytes may be similarly affected. This effect has been documented for the interaction between folic acid and the loop diuretics furosemide and ethacrynic acid. Dietary factors that influence acid-base metabolism can alter blood pH and intraerythrocytic pH, thereby affecting drug ionization, protein binding and cell uptake.

DIETARY EFFECTS ON DRUG METABOLISM (Pages 929-931)

The clearance of many drugs from the circulation depends on their biotransformation in the liver, kidneys and other organs with xenobiotic-metabolizing enzymes (**Table 27-5**). For drugs that are metabolized rapidly, extraction is determined principally by organ blood flow. For example, the rate-limiting step for clearance of indocyanine green and sulfobromophthalein sodium is hepatic blood flow. For drugs that are metabolized relatively slowly, clearance from the circulation is determined primarily by the quantity and affinity of enzymes responsible for their metabolism.

Hepatic drug metabolism occurs through two predominant biotransformation pathways: 1) phase I (oxidation, reduction and hydrolysis) and 2) phase II (glutathione or glucuronide conjugation, acetylation and sulfation). Phase I reactions are catalyzed principally by a family of cytochrome P450 enzymes in the microsomal mixed function oxidase system. Phase I reactions alter the functional groups of a compound. Phase II reactions are catalyzed by families of glutathione-S-transferase, glucuronyl transferase and N-acetyltransferase isoenzymes. Phase II reactions result in conjugation and altered water solubility. The outcome of phase I and II reactions is reduced pharmacologic activity and enhanced drug excretion. In some cases, phase I reactions may increase the activity or toxicity of drugs, whereas phase II reactions may alter the tissue distribution and subsequent target organs for toxicity or mutagenicity of the drug's metabolites.

Macronutrient Effects on Drug Metabolism (Pages 930-931)

Inappetence due to disease is a common cause of decreased macronutrient intake that can affect drug action. Furthermore, changes in the macronutrient composition of the diet can significantly alter hepatic drug metabolism.

Dietary Protein Intake (Page 930)

In experimental studies in rats, low dietary protein intake reduced the metabolism and increased the toxicity of pentobarbital, strychnine and zoxazolamine. The activities of the mixed function oxidase enzymes flavoprotein reductase and cytochrome b_5 are decreased by dietary protein restriction. Inducibility of cytochrome P450 by phenobarbital in rats is also decreased by feeding less dietary protein.

High-protein (e.g., 44 vs. 10% of kcal, as fed), low-carbohydrate foods (e.g., 35 vs. 70% of kcal, as fed) enhance the hepatic metabolism and excretion of many different drugs in people including acetaminophen, oxazepam, theophylline, propranolol and estradiol. Conversely, consumption of protein-restricted foods for as few as 10 days significantly decreases elimination of these drugs.

Certain essential amino acids may stimulate hepatic protein synthesis and thereby induce the hepatic mixed function oxidase system. Sulfurcontaining amino acids can promote hepatic drug metabolism by increasing glutathione synthesis and subsequent conjugation reactions. Starvation can reduce the activity of glutathione-S-transferase and the synthesis of glutathione for conjugation; events that also participate in the development of fasting hyperbilirubinemia.

Dietary protein-related changes in renal blood flow and renal tubular transport can simultaneously affect the clearance of drugs eliminated in urine. Increased dietary protein intake in dogs increases the elimination of gentamicin and reduces the potential for nephrotoxicity, presumably by stimulating renal blood flow.

Dietary Carbohydrate Intake *(Page 930)*

High carbohydrate intake in laboratory animals depresses oxidative drug metabolism. High dietary fructose, glucose and sucrose levels increase barbiturate-sleeping time and decrease in vitro metabolism of barbiturates in mice. Parenteral glucose has the same effect in dogs and cats; thus, high dietary intake of these carbohydrates would likely modify barbiturate responses in these species as well. Supplemental carbohydrate administration in rats increases liver weight, hepatic fat and glycogen deposition, but decreases hepatic mixed function oxidase activities. Carbohydrate feeding in rats can similarly decrease the microsomal activation of carcinogens such as benzo(a)pyrene and aflatoxin B_1.

In people, long-term consumption of high-carbohydrate diets depresses antipyrine and theophylline clearance. The proposed mechanism involves inhibition of the synthesis of d-aminolevulinic acid synthetase, a key enzyme in the synthesis of heme for cytochrome P450. However, carbohydrate is also required for UDP-glucuronyl transferase activity for glucuronidation of oxidized drug metabolites; short-term deprivation of carbohydrates can decrease rates of conjugation. This, too, contributes to the hyperbilirubinemia of fasting.

Dietary Fat Intake *(Page 930)*

In addition to the effects of dietary fat intake on drug absorption and plasma protein binding, lipid intake can affect hepatic xenobiotic-metabolizing enzyme activities. Foods deficient in essential fatty acids result in decreased rates of drug metabolism. Dietary lipids have been reported to be essential for optimal induction of P450 enzymes by phenobarbital. Rats fed a 20% corn oil diet for four days had twofold increases in the activities of several hepatic P450 isoenzymes (P450 2, 2A1, 2B1, 2C11, 2E1 and 3A) as compared with enzyme activities in rats fed a fat-free diet. However, there is an inverse relationship between lung P450 2B1 activity and dietary fat intake. In one study in which rats were fed 6% dietary lipid for 40 days as coconut, peanut, corn or fish oil, cytochrome P450 and epoxide hydrolase activities were highest in the fish-oil group. In this same study, UDP-glucuronyl transferase type I activity was increased by fish oil or corn oil supplementation, but reduced by coconut oil.

In another study, rats fed 10% dietary lipid for two weeks as soybean oil, lard or fish oil were exposed to pentachlorobenzene (PECB). Blood concentrations of the metabolite pentachlorophenol were highest and tissue concentrations of PECB were lowest after feeding fish oil.

Fish oils are high in polyunsaturated fatty acids, particularly of the n-3 family (eicosapentaenoic and docosahexaenoic acids), but contain relatively less n-6 fatty acids than other sources. Effects of fish oil supplementation may be due to: 1) altered cell and organelle membrane fluidity, 2) increased propensity towards oxidative damage and/or 3) specific induction of enzyme synthesis. In people, the degree of dietary fatty acid saturation has had little effect on oxidation of antipyrine or theophylline; however, the principal cytochrome P450 isoenzyme, 3A4, is sensitive to microsomal membrane characteristics. A dietary deficiency of labile methyl donors (e.g., choline or methionine) increases spontaneous and chemically induced hepatocarcinogenesis in rats because of decreased microsomal enzyme activity. Lipotrope deficiency also impairs methylation of DNA and RNA; however, a considerable portion of microsomal lipid can be removed in vitro without adversely affecting P450 activity.

Effects of Feeding Route *(Pages 930-931)*

The route of nutrient administration may also affect hepatic drug metabolism. Decreased hepatic clearance of indocyanine green in pigs fasted for 12 days is returned to normal after enteral feeding for 12 days. However, intravenous feeding with an identical formula did not improve hepatic clearance despite similar weight gains. Hepatic hydroxylation of pentobarbital and demethylation of meperidine by rats are significantly impaired following seven days of parenteral feeding with a formula that otherwise maintains hepatic drug clearance when administered enterally. Lipid-free total parenteral nutrition depresses hepatic phase I and II conjugative drug metabolism. Parenteral lipid-free nutrition for 10 days in rats decreased the hepatic activities of cytochrome P450 oxidase, p-nitroanisole demethylase and p-nitrophenol glutathione-S-transferase by one-half. Thus, the intake of macronutrients, composition of the food and route of nutritional support interact to modify drug metabolism.

Micronutrient Effects on Drug Metabolism (Page 931)

Dietary Vitamin Intake *(Page 931)*

The hepatic mixed function oxidase system requires several vitamins. Niacin and riboflavin participate directly as the principal components of the electron carriers $NADP^+$, NAD^+, FAD and FMN, which are coenzymes for cytochrome P450 reductase, DT-diaphorase and NADH-cytochrome b_5 reductase. Dietary deficiency can lead to a generalized decrease in total P450 and associated monooxygenase activities.

Folate deficiency blocks the induction of cytochrome P450 by phenobarbital, and pyridoxine (vitamin B_6) deficiency may alter cysteine conjugate b-lyase activity. Excessive dietary folate can antagonize methotrexate activity, whereas increased pyridoxine intake can increase the metabolism of levodopa, thereby reducing its effectiveness. Thiamin deficiency increases the levels of cytochrome P450 2E1, NADH-P450 reductase and cytochrome b5, but decreases the oxidation of N-nitrosodimethylamine, acetaminophen, aminopyrine, ethylmorphine, zoxazolamine and benzo(a)pyrene.

The antioxidant vitamins (A, C and E) are required for normal membrane synthesis and stability. Vitamin A deficiency decreases hepatic mixed function oxidase system activity and depresses oxidation of aminopyrine, ethylmorphine, aniline, benzo(a)pyrene and 7-ethoxycoumarin. Vitamin C deficiency decreases NADPH-P450 reductase activity and prolongs the half-life of antipyrine, acetaminophen and salicylamide. Vitamin E deficiency decreases microsomal metabolism of ethylmorphine, codeine and benzo(a) pyrene. Effects of vitamin E deficiency occur without decreases in cytochrome P450 activity, and probably relate to the antioxidant properties of tocopherol, which may prevent oxidative damage to membrane lipids. Vitamins A and D are substrates for cytochrome P450 and can competitively block the metabolism of other P450 substrates.

Dietary Mineral Intake (Page 931)

Many minerals modulate hepatic drug metabolism. Iron is required for heme synthesis in cytochromes and for metal ion-catalyzed oxidative reactions. Iron deficiency results in decreased metabolism of hexobarbital and aminopyrine. Selenium is a cofactor for glutathione peroxidase; selenium deficiency may promote oxidative damage to the microsomal system. Hypothyroidism resulting from iodide deficiency increases flavoprotein synthesis and cytochrome P450 oxidative activity. Deficiencies of zinc, magnesium and potassium decrease drug metabolism, whereas high concentrations of heavy metals (e.g., cobalt and cadmium) may block heme synthesis and thereby lower cytochrome P450 levels.

Non-Nutrient Effects on Drug Metabolism (Page 931)

Non-nutrient dietary factors can profoundly influence drug metabolism by inducing the activity of many hepatic biotransformation enzymes. Phenols (e.g., hydroxycinnamic, dihydroxycinnamic and ferulic acids) are antioxidants that block chemical carcinogenesis. Methylxanthines, including caffeine, theobromine and theophylline, competitively bind to cytochrome P450 to block oxidation of other compounds. Coumarin derivatives in vegetables and fruits induce glutathione-S-transferase activity. Organonitriles (1-cyano-2-hydroxy-3-butene, 1-cyano-3,4-epithiobutane) and indole derivatives (indole-3-carbinol, 3,3'-diindolmethane, indole-3-acetonitrile) in cruciferous plants (e.g., broccoli, cauliflower and cabbage) increase hepatic and renal glutathione concentrations and induce hepatic and renal glutathione-S-transferase activities.

Excessive organonitrile exposure induces hepatic and renal toxicity, which may impair drug metabolism, whereas small amounts may have anticarcinogenic properties. Flavonoids and terpenoids from citrus fruits can either induce or block cytochrome P450-related oxidative reactions, and can exert mutagenic or antitumorigenic effects, depending on the dose administered. Flavonoids in grapefruit juice can significantly prolong the half-life of dihydropyridine calcium channel blockers (e.g., nifedipine, felodipine and nisoldipine).

DIETARY EFFECTS ON DRUG EXCRETION (Pages 931-932)

Following P450 hydroxylation, heterocyclic amines may subsequently undergo N-acetylation, the metabolic phenotype and activity of which affects the organ and route of excretion of the metabolite. If there is "slow" N-acetyltransferase activity, most of the hydroxylated amine undergoes hepatic glucuronidation and is returned to the blood for excretion in the urine. In people, so called "slow acetylators" are predisposed to urinary bladder cancer. Those individuals with "fast" N-acetyltransferase activity appear to be predisposed to colorectal cancer, presumably through preferential colonocytic metabolism to mutagenic arylamides and acetoxyarylamines. Thus, metabolic phenotype as determined by genetics, or from enzyme induction due to dietary effects, can influence the site, route and rate of drug excretion.

Because many of the drugs excreted by the kidneys undergo active transport by anion- or cation-specific mechanisms in the renal tubular epithelium, their elimination can be altered through competitive inhibition by other charged solutes. Pharmacologically, this effect has been purposely employed by the co-administration of probenecid with penicillins to block elimination by the anion-specific renal tubular transport mechanism. Nutritionally, this effect may result from consumption of divalent cations (e.g., calcium and magnesium), which decrease renal tubular transport and accumulation of aminoglycosides such as gentamicin. As a result, urinary elimination of the antibiotic is increased and nephrotoxicity thereby reduced.

Furthermore, dietary alterations in urinary pH can affect the ionization and trapping of drugs secreted into the tubular lumina. The relatively common practice of formulating commercial feline foods to promote urinary acidification for the prevention and treatment of lower urinary tract diseases (e.g., struvite crystalluria and urolithiasis) may also affect the elimination of pharmaceutical agents excreted in the urine.

Food ingredients that stimulate bile, fecal or urine flow may affect the excretion of drugs by these routes. For example, dietary fats with choleretic properties will enhance the excretion of drug metabolites in the bile and the return of enterohepatically recycled drugs such as doxycycline. Salts of divalent cations (e.g., magnesium oxide and magnesium hydroxide) can exert a laxative effect that may increase fecal elimination of poor-

Table 27-6. Examples of the effects of nutrients on drug action.

Beneficial effects	Examples
Enhanced GI drug absorption	Fatty foods enhance absorption of griseofulvin
Prevention of undesirable drug effects	Foods minimize nausea induced by metronidazole
Enhancement of desirable drug effects	Water enhances laxative effects of psyllium
Improved drug metabolism	Enteral feeding supports metabolism of cefoxitin
Altered drug excretion	Protein promotes renal excretion of gentamicin
Detrimental effects	**Examples**
Impaired GI drug absorption	Food interferes with absorption of ampicillin
Antagonism of desirable drug effects	Folate opposes chemotherapeutic effects of methotrexate
Potentiation of undesirable drug effects	Potassium increases potential toxicity of captopril
Impaired drug metabolism	Fish oil enhances hepatic oxidation of phenobarbital
Altered drug excretion	Calcium increases urinary excretion of gentamicin

ly absorbed oral drugs and enterohepatically recycled drugs. High dietary salt content and other naturally occurring diuretics, including active loop diuretics, can enhance the excretion of drugs and their metabolites in urine.

BENEFICIAL EFFECTS OF NUTRIENTS ON DRUG ACTION (Pages 932-933)

The presence of food need not impair drug absorption, and within limits may be indicated to facilitate safe GI uptake of drugs (**Table 27-4**). Food may prevent GI irritation, modify drug-induced nausea or delay drug uptake, increasing the ultimate amount of drug absorbed (**Table 27-6**). For example, food can promote gastric acid secretion to enhance the uptake of an acidic drug such as aspirin, while simultaneously protecting the mucosa from irritation by the drug.

Consumption of food can minimize nausea induced by the concurrent administration of hypertonic salt and carbohydrate solutions. In people, micronized preparations of phenytoin are actually better absorbed in the fed rather than the fasted state. In other cases, dietary supplementation with a specific nutrient may be indicated to counteract adverse drug side effects, to prevent drug-induced nutrient imbalances or to potentiate therapeutic effects.

Provision of Nutrients to Prevent Drug-Induced Imbalances
(Page 932)

Additional energy and protein may be indicated to combat alterations in drug metabolism associated with prolonged decreases in food intake. A critical example would be the provision of nutrients during enteral- or parenteral-assisted feeding of patients incapable of voluntary food consumption. Studies of prolonged starvation and of kwashiorkor in people have demonstrated significant reductions in the metabolism of numerous drugs by both phase I and phase II hepatic biotransformation systems. These drugs include chloroquine, isoniazid, penicillin, chloramphenicol, tobramycin and cefoxitin. In addition, hypoalbuminemia-related decreases in drug binding alter the clearance of cloxacillin, streptomycin, sulfamethoxazole, sulfadiazine, digoxin, thiopentone and phenylbutazone. Malnutrition-related decreases in renal blood flow and glomerular filtration rate have caused gentamicin toxicity.

Most commercial pet foods are adequately fortified with micronutrients; therefore, supplementation is not necessary unless a homemade food is fed, nutrient intake is decreased or a specific medical indication for prescription of a nutrient as a "nutraceutical" exists. Vitamin supplementation may be indicated to counteract the effects of drugs that specifically antagonize vitamin absorption or function. These include: 1) the use of folacin to manage deficiency induced by folic acid antagonists such as methotrexate, 2) vitamin K vs. antagonists in the coumarin family, 3) tocopherol, retinol and/or ascorbic acid to counter losses due to oxidative drug damage, 4) cholecalciferol for deficiency induced by anticonvulsants such as phenytoin, 5) thiamin to replace that lost to thiaminase activity in raw fish and 6) B vitamins to replace those lost following antibiotic-induced alterations in the GI microflora.

Specific minerals may also become deficient because of binding or precipitation in the GI tract, or following enhanced fecal losses due to laxatives or urinary loss due to diuretics. Urinary electrolyte losses due to loop diuretics can lead to significant physiologic abnormalities. Trace elements such as zinc may bind to fiber or be precipitated by phytates. Oral calcium supplements may block iron absorption. Excessive use of antacids, laxatives and binding resins can result in macroelement deficiencies.

Glutathione precursors (e.g., cysteine or N-acetylcysteine) may be indicated to counter the oxidative damage induced by pharmaceutical agents such as the: 1) analgesic acetaminophen, 2) urinary antiseptic methylene blue, 3) injectable anesthetic propofol and 4) antitumor agent doxorubicin. Oxidative damage resulting from administration of oxidized lipid supplements or excessive use of n-3 fatty acid sources may also necessitate treatment with glutathione precursors or antioxidant vitamins.

Provision of additional water may be indicated for the prevention or treatment of renal damage resulting from nephrotoxic drug administration. Examples of drugs whose administration should routinely be coupled to increased water intake include cisplatin, aminoglycosides, nonsteroidal antiinflammatory drugs, analgesics and diuretics.

Provision of Nutrients to Enhance Drug Effects (Pages 932-933)

Certain nutrients may be prescribed to facilitate a drug's intended effect or to synergistically promote the target physiologic functions. Additional energy or protein can generally facilitate therapeutic drug effects by promoting optimal distribution and hepatic biotransformation activities. These additions will tend to normalize pharmacokinetics to ensure the individual patient's dose response may more closely approximate the anticipated response.

Providing adequate energy and protein to patients receiving exogenous thyroid hormones plays an integral role in the physiologic response to that supplementation. Undernutrition may result in reduced synthesis of thyroid-binding plasma proteins and subsequent changes in thyroid pharmacokinetics. Reductions in energy or protein intake suppress target tissue monodeiodination of thyroxine to the physiologically active triiodothyronine. Triiodothyronine levels decrease within 24 hours of fasting or caloric restriction, and may decline by 40 to 50% within three days. Should fasting induce increased adrenal glucocorticoid secretion, depressed target tissue triiodothyronine receptor levels may also be observed. Although these reductions in target-cell responsiveness to thyroid hormones represent an appropriate adaptation to conserve energy during starvation, the effect on exogenous thyroid hormone pharmacotherapy may be undesirable.

It is important to maintain a regular feeding schedule and consistent food for animals with diabetes mellitus to stabilize intermediary metabolism. The administration of exogenous insulin to insulin-dependent diabetics and the administration of oral hypoglycemic agents to non-insulin-dependent diabetics should be timed relative to feeding. For both forms of diabetes mellitus, specific dietary formulations are indicated to: 1) modulate GI carbohydrate uptake, 2) meet protein requirements without adversely affecting renal function and 3) moderate overall lipid metabolism to prevent ketoacidosis. (See Chapter 24.)

Dietary intake of specific minerals that modulate hormonal axes should be considered, including calcium and phosphorus intake when cholecalciferol is administered for chronic renal failure, and sodium and potassium when mineralocorticoids are replaced in hypoadrenocorticism. The trace minerals chromium and vanadium may improve glucose tolerance and facilitate management of diabetics with insulin or oral hypoglycemic agents. Specific n-3 fatty acid therapy may be used to potentiate the effects of antiinflammatory drugs, anticoagulants and antineoplastic agents. Arginine may be provided to improve nitric oxide production and to enhance immune function, glutamine to promote enterocyte metabolism, cysteine to enhance glutathione synthesis, carnitine to improve digoxin responsiveness in congestive heart failure and antioxidant vitamins to protect against oxidative damage.

Dietary fiber may be indicated along with drug therapy for a number of diseases. Increased dietary fiber intake has proved beneficial in the treatment of insulin-dependent and non-insulin-dependent diabetes mellitus by moderating glucose absorption from the GI tract. Fermentable dietary fiber increases colonic short-chain fatty acid concentrations and decreases luminal pH. As a result, these fibers may be used as the primary treatment for canine and feline colitis or as an ancillary therapy to sulfasalazine or metronidazole treatment. Soluble fibers (e.g., psyllium mucilloid) may act in this way in conjunction with other antidiarrheal treatments, or as stool softeners for use with laxatives to treat constipation. Hepatic cytochrome P450 concentrations and UDP-glucuronyl transferase activities appear to be altered by the type and quantity of fiber in the food.

Dietary buffers may be indicated in conjunction with other therapies for chronic renal failure to correct metabolic acidosis or to facilitate activity of replacement pancreatic enzymes in exocrine pancreatic disease. They may be used to enhance alkaline drug absorption from the GI tract and to promote acidic drug excretion in the urine. Alkalinization of the urine has been used clinically to reduce the ionization, renal accumulation and toxicity of aminoglycosides. Finally, buffers (e.g., sodium bicarbonate, aluminum hydroxide) can be used with H_2-receptor antagonists (e.g., cimetidine, ranitidine) or as laxatives (e.g., magnesium hydroxide, magnesium oxide).

ADVERSE EFFECTS OF NUTRIENTS ON DRUG ACTION
(Pages 933-934)

In addition to ameliorating undesirable effects on drug absorption or metabolism, specific nutrients may antagonize desired drug effects (**Table 27-6**). Excess caloric intake will complicate weight management in obese patients. Excess protein intake can adversely affect renal handling of drugs by increasing renal blood flow and drug excretion, or by promoting intraglomerular hypertension and reducing glomerular filtration in chronic renal failure. High protein intake can increase the hepatic metabolism of drugs such as the methylxanthines, resulting in reduced therapeutic efficacy.

High mineral intake can complicate drug therapy of specific disorders: 1) sodium and hypertension, 2) potassium and hypoadrenocorticism, 3) magnesium and feline lower urinary tract disease and 4) phosphorus and chronic renal failure.

Excessive dietary intake of iodine can lead to paradoxical "iodine toxicosis goiter" through what is referred to as the "Wolff-Chaikoff effect." As iodide accumulation by the thyroid gland increases, so does iodination of tyrosyl residues of thyroglobulin. However, very high iodine levels appear to cause auto-inhibition of iodide organification and thyroglobulin proteolysis, leading to thyroid hormone deficiency. This phenomenon has been observed in foals born to mares that received excessive iodine supplementation during gestation, as well as in other species.

Naturally occurring non-nutritive dietary factors that may influence drug responses include methylxanthines, which may complicate aminophylline therapy, histamine in certain types of fish, which may interfere with antihistamine treatment and tyramine in chicken livers and aged cheeses, which confound the action of monoamine oxidase inhibitors.

■ EFFECTS OF OBESITY ON DRUG METABOLISM (Page 934)

Although complicated by the metabolic effects of overnutrition during weight gain or restricted food intake during weight loss, numerous studies have documented a significant effect of obesity on drug metabolism in people and other animals. Changes in the apparent volume of distribution have been observed because of alterations in the quantity of body fat. Obesity increases the volume distribution of lipophilic drugs such as alprazolam, carbamazepine, diazepam, methotrexate, oxazepam, sufentanil and vancomycin. Obesity decreases the volume distribution of polar compounds including acetaminophen, ciprofloxacin, furosemide, gentamicin, isoniazid, sulfisoxazole and tolbutamide.

Drug clearance may be affected following changes in hepatic microsomal enzyme induction, as well as alterations in the predominant pathways used for phase II conjugation reactions. Several investigations have demonstrated enhanced biotransformation of volatile anesthetics in obese patients, resulting in increased production of the reactive intermediates typically responsible for organ toxicity. Enhanced hepatic oxidative metabolism of halothane in obese people has resulted in increased serum levels of fluoride and bromide ions; the former is associated with increased hepatotoxicity. Half-life elimination of triazolam is prolonged in obese subjects, and clearance following oral administration is reduced, presumably because of decreases in first-pass hepatic extraction.

Drug toxicity may be enhanced when the dose administered is based on total body mass, but distribution is restricted to lean body mass, resulting in higher plasma drug concentrations and greater exposure to susceptible organs. Obese rats appear to be at increased risk for gentamicin and furosemide nephrotoxicity by this mechanism. Susceptibility to the toxic effects of these drugs remains even when the dose is decreased to reflect lean body mass and to equalize drug exposure. Studies of acetaminophen toxicity in rats have shown that when obese animals are dosed according to fat-free mass, toxicity is increased because of a metabolic shift toward less sulfation and more glucuronidation. Obesity likewise appears to increase drug glucuronidation in people. Furthermore, target organs may be predisposed to drug toxicity by pre-existing obesity-related lesions such as hepatic lipidosis.

Obesity increases steroid hormone clearance in people because of enhanced aromatization and interconversion of androgens to estrogens by adipose tissue. Prednisolone and methylprednisolone succinate clearance in obese people is also increased, although potential contributions by increased cardiac output, hepatic blood flow, liver size and hydrolysis by extrahepatic carboxylesterases have not been resolved. On the other hand, methylprednisolone clearance appears to be decreased, suggesting that obesity may affect specific oxidative pathways very differently.

Although similar studies have not been conducted in companion animals, certain generalizations can be made. Obesity changes the effective dose administered for drugs given according to total body weight, whether it is increased because of poor lipid solubility or decreased because of lipophilicity. Drug dose or dosing interval may need to be adjusted to maintain therapeutic effect and protect against toxicity. Alterations in body composition concurrent with drug administration may have significant effects on clinical efficacy and margin of safety, and must be considered whenever a patient's body weight changes markedly.

SUMMARY (Pages 934-935)

Because few studies to determine the effects of food on drug metabolism have been conducted in dogs and cats, it is difficult to delineate specific feeding recommendations for drugs commonly used in veterinary practice. However, it is important to consider potential nutrient-drug interactions whenever the expected action of a prescribed drug is not seen in an individual patient. One may alter the dosing schedule relative to meals or adjust the dietary composition or dose to correct overt nutrient imbalances. Alternatively, one may determine circulating drug concentrations to detect changes in pharmacokinetics and to establish the need for a change in drug type or dose. It is clear that a standardized food, consistent feeding schedule and balanced nutrient intake are prerequisites to successful pharmacologic management of disease.

Clinical cases that illustrate and reinforce the nutritional concepts presented in this chapter can be found in Small Animal Clinical Nutrition, 4th ed., pp 936-939.

Appendix Contents

Appendices from Small Animal Clinical Nutrition, 4th ed.

[*]Appendices in bold are partially or completely included in the *Pocket Companion to
Small Animal Clinical Nutrition, 4th ed.*

Orphaned Puppy and Kitten Care

INTRODUCTION

Puppies and kittens may be considered orphaned if they lack sufficient maternal care for survival from birth to weaning. Several physiologic needs normally provided by the mother must be met to ensure survival of neonates: heat, humidity, nutrition, immunity, elimination, sanitation, security and social stimulation. A foster bitch, foster queen or the caregiver must meet these needs for orphaned puppies and kittens. Most orphans can be raised successfully with proper care and nutrition.

PHYSICAL AND SOCIAL ENVIRONMENTS

Puppies and kittens should be housed in warm draft-free enclosures. Incubators are ideal, particularly for newborns. Pet carriers, shoeboxes or cardboard boxes are suitable substitutes. The bedding should be soft, absorbent and warm. Thread-free cloth, fleece and shavings are appropriate materials and help puppies and kittens feel secure as they snuggle into them.

Neonates demonstrate a certain degree of poikilothermy and are unable to regulate body temperature well during the first four weeks of life. Puppies and kittens huddle together close to their mother, which generates an optimal microclimate, protects them against changes in environmental temperature and decreases the rate of heat loss. Orphans cannot seek protection near their mother and are more sensitive to suboptimal environmental conditions.

Without the mother, neonates can quickly become hypothermic, which leads to circulatory failure and death. Artificial heat should provide age-optimal environmental temperatures (**Tables 1** and **2**). It is best to set the heating source to establish a gradation of heat in the nest box. A gradation of environmental temperatures allows neonates to move toward or away from the heat source as needed to avoid hyperthermia, which can be as detrimental as hypothermia. Puppies and kittens can rapidly become dehydrated secondary to overheating. Maintaining humidity near 50% helps reduce water loss and maintains the moisture and health of mucous membranes.

To fulfill non-nutritive nursing needs, hand-reared puppies and kittens often nurse other littermates in the nest box. To avoid skin trauma related to excessive nursing, puppies and kittens can be housed individually

Table 1. Optimal environmental temperature for puppies.

Age	°C	°F
Immediate environment/incubator for orphans		
Week 1	29-32	84-90
Week 2	26-29	79-84
Week 3	23-26	73.4-79
Weeks 4-12	23	73.4
Environment around litter		
Week 1	24-27	75-81

Table 2. Optimal environmental temperature for kittens.

Age	°C	°F
Immediate environment/incubator for orphans		
Week 1	32-34	89.5-93
Week 2	27-29	81-84
Week 3	24-27	75-81
Weeks 4-12	24	75
Environment around litter		
Week 1	24-27	75-81

or separated by dividers. Although beneficial for alleviating problems due to non-nutritive nursing, separation of the litter reduces temperature and humidity in the immediate environment and social stimulation by littermates. Brief, but regular handling, can provide social stimulation. The stress associated with regular handling increases neural development and improves weight gain in kittens. Kittens raised without social stimulation develop abnormal behavior patterns (i.e., kittens have reduced normal exploratory behavior and become more suspicious and aggressive as adults). Similar changes may occur in puppies. Peer contact can compensate for maternal deprivation. Therefore, benefits of separating neonates must be weighed against the potential for development of abnormal behavior and increased risk for hypothermia. Puppies and kittens should interact with littermates as much as possible until weaning.

GENERAL HUSBANDRY

Puppies and kittens obtain passive systemic immunity from colostrum and passive local immunity from continued ingestion of bitch's/queen's milk. If possible, neonates should receive colostrum or bitch's/queen's milk within the first 12 to 16 hours of birth. This is particularly critical for puppies and kittens fed only milk replacers because they lack systemic and local immune protection.

Normally the mother will sever the umbilical cord. If not, it should be cut to 1.5 in. (3.5 to 4 cm) and an appropriate topical antiseptic applied. Orphaned puppies and kittens are at greater risk for infectious disease; thus, sanitary husbandry practices are important. To reduce risk for diseases, puppies and kittens should not be exposed to older animals or grouped within multiple litters. Feeding equipment and bedding should be kept clean and sanitized frequently. Caretakers should wash their hands before handling neonates and after stimulating elimination.

Puppies and kittens cannot voluntarily urinate or defecate until about three weeks of age. Until that time, they rely on the mother to stimulate the urogenital reflex to initiate elimination. Caretakers should stimulate puppies and kittens after feeding by gently swabbing the perineal region with a warm moistened cotton ball or cloth.

Often, puppies or kittens within a litter look similar; therefore, it may be difficult to tell them apart when hand rearing, especially in large litters. Different colored nail polish can be applied to the claws to help differentiate individuals; paint a different paw for each separate animal (e.g., blue front left paw, blue right rear paw, pink right front paw, etc.).

ASSESSMENT

Orphaned puppies and kittens should be thoroughly evaluated when first seen. A careful physical examination of neonate(s) and mother, if available, should be performed to detect the potential cause for abandonment. Particular attention should be given to detect common problems such as hypothermia, hypoglycemia, dehydration and congenital defects. The current nutritional and hydration status should also be noted. Puppies or kittens fostered onto another mother should be supervised initially to detect any behavioral problems between the foster parent, its young and the orphan(s). Puppies or kittens should be accepted immediately and allowed to nurse. Watch for signs of rejection or impending cannibalism by the mother.

Key Nutritional Factors
Energy
Table 3 summarizes the estimated energy requirements of neonatal puppies and kittens. A very common mistake is to underestimate the energy requirements of neonates. In the beginning, however, it is better not to overfeed to avoid diarrhea. In most cases, it is best to follow label recommendations on commercial products or feed based on caloric calculations.

Newborn puppies and kittens require about 24 kcal (100 kJ) metabolizable energy (ME)/100 g body weight. (See **Table 3** for specific energy

Table 3. Recommendations for energy intake of orphaned kittens and puppies.

Ages	kcal ME/100 g BW	kJ ME/100 g BW
Week 1	13-15	55-60
Week 2	15-20	65-85
Week 3	20	85
Week 4	$\geq$20	$\geq$85

Key: BW = body weight.

requirements based on age.) In general, kittens less than one week old will eat a volume equal to 10 to 15% of their body weight as milk or milk replacer and a volume equal to 20 to 25% of their body weight between Weeks 1 to 4. This is a reasonable target if the caloric content of the food is unknown.

Water

Water is one of the most important nutrients in orphan feeding. A normal puppy needs about 60 to 100 ml of fluid/lb body weight per day (130 to 220 ml/kg body weight). On average, orphaned puppies (and probably kittens) should receive about 180 ml/kg body weight to make orphan feeding successful. Water should be given until 180 ml/kg body weight is reached if the milk replacer doesn't provide this much water at the recommended dilution.

Digestibility

Dry matter digestibility of bitch's and queen's milk is very high (>95%). Digestibility of milk replacer formulas should be high (>90%) to allow for smaller quantities to be fed and avoid diarrhea.

Other Nutritional Factors

The preferred nutrient profile of milk, commercial milk replacers and homemade replacer formulas for nursing puppies and kittens should be similar to that of milk from bitches and queens. (See Appendix K, Small Animal Clinical Nutrition, 4th ed., pp 1064-1072.) Values recommended by the Association of American Feed Control Officials (AAFCO) for growth should be followed when the concentration of nutrients in maternal milk is unknown. (See Appendix J, Small Animal Clinical Nutrition, 4th ed., pp 1048-1063.)

It is essential that commercial milk replacers and homemade replacer formulas have adequate protein and essential amino acid content. The arginine and histidine levels in the formula are particularly important. Deficiency of these amino acids can cause cataract development in neonates and contribute to anorexia and poor growth.

Milk replacers are often fortified with iron at concentrations higher than those found in bitch's or queen's milk. Orphaned puppies and kittens, especially low birth-weight neonates born with low iron reserves, may benefit from iron intakes higher than those normally found in milk. The additional iron supports hematopoiesis and helps avoid anemia sometimes observed in three- to four-week-old neonates.

High osmolality should be avoided because it may cause hyperosmolar diarrhea and potentiate dehydration. High osmolarity may delay gastric emptying and predispose to regurgitation, vomiting and aspiration during the next meal, if the stomach is not completely empty. The osmolality of bitch's and queen's milk is approximately 569 mOsm/kg and 329 mOsm/kg, respectively.

Food Form
Foods should be liquid until puppies and kittens are three to four weeks of age when semisolid to solid foods may be introduced.

FEEDING PLAN

Success of orphan rearing depends on how well the caregiver fulfills the daily routine of hygienic measures, strict feeding schedules and all aspects of care normally provided by the bitch or queen. These measures are vital for survival of puppies and kittens early in life.

Hygiene
Hygienic measures must be more stringent for orphaned puppies and kittens because they may have received less colostrum and be more susceptible to infections than other neonates.
* Feeding materials (e.g., bottles and nipples) should be cleaned thoroughly and boiled in water between uses.
* Ingredients for homemade milk replacers should be fresh and refrigerated until used.
* Never prepare more milk replacer than can be used in 24 hours and refrigerate unused portions.
* Formulas should be discarded after one hour at room temperature.
* At least twice a week, orphans should be washed gently with a soft moistened cloth to simulate cleaning by the dam's tongue.

Select a Milk or Milk Replacer
Foods used to feed orphans may consist of bitch's/queen's milk, commercial milk replacer or homemade replacer formulas. Bitch's/queen's milk is the food of choice and will provide nutrients in the proper levels for nursing puppies and kittens. Bitch's or queen's milk is rarely available

in sufficient quantities to hand raise orphans. Commercial milk replacers are generally preferred although several homemade formulas have proved sufficient. (See Appendix K, Small Animal Clinical Nutrition, 4th ed., pp 1064-1072.) In one study, kittens fed either a commercial or homemade milk replacer had higher weight gains than kittens fed by the queen.

Commercial and homemade milk replacers should closely mimic the profile of bitch's or queen's milk. Unsupplemented ruminant milk may be used as a base for homemade formulas but does not meet the nutritional needs of puppies or kittens. Goat's milk provides no nutritional benefit over cow's milk. Appendix K (Small Animal Clinical Nutrition, 4th ed., pp 1064-1072) compares the nutrient profile of bitch's and queen's milk with that of cow's milk, various milk replacers, homemade replacer formulas and milk from other species.

Determine a Feeding Method
Fostering
The optimal means of feeding orphaned or rejected puppies or kittens is to foster them to another lactating bitch or queen. Fostering is the least labor intensive, provides optimal nutrition, reduces mortality, improves immune status, usually provides an optimal physical environment and promotes normal social development of the puppy or kitten. Unlike large animals, bitches and queens readily accept additional puppies and kittens during lactation. If several foster mothers are available, it is best to place orphans in litters with fewer than 14 days age difference. Larger puppies and kittens often crowd out smaller individuals if the age discrepancy is too large. This situation can be managed by supervised feeding until the orphans can fend for themselves. Unfortunately, foster mothers are not normally available and alternative techniques must be used.

Partial Orphan Rearing
Puppies and kittens that cannot be successfully raised by the bitch or queen for reasons of health, poor lactation performance or too large of a litter, may be left with the mother but given supplemental feeding to support nutritional needs. Supplemental food may be given by hand feedings or timed feedings using a surrogate bitch/queen. Puppies/kittens may also be reared in a communal situation. Partial orphan rearing can be accomplished by dividing the litter into two groups of equal number and size. One group remains with the mother while the other is removed and fed milk replacer. The groups are exchanged three to four times daily. It is important to feed the separated group before it is returned to the mother. As a result, the group just placed with the dam will be less inclined to nurse immediately. It is better to supplement all the puppies or kittens in the litter rather than just a few. The advantages of partial orphan rearing

are similar to those of fostering. In addition, continued access to the mother can help stimulate milk production and mothering behaviors. When using foster or surrogate mothers, it is important to monitor for signs of rejection and cannibalism. Partial orphan rearing may be necessary to assist the efforts of foster mothers. Unfortunately, foster and surrogate mothers are rarely available.

Hand Feeding

The most common method of raising orphaned puppies and kittens is hand feeding. Eyedroppers, syringes, bottles and stomach tubes are typically used to feed orphans.

BOTTLE FEEDING Bottle feeding is the preferred method for vigorous puppies and kittens with good nursing reflexes (**Figures 1** and **2**). Bottle feeding has the advantage that neonates will nurse until they are satiated and reject the milk or formula when full. However, bottle feeding can be time consuming, especially with large litters.

Most puppies and kittens will readily nurse small pet nursers available in pet stores (**Figure 3**). Feeding bottles for dolls or bottles with nipples for premature human infants are alternatives. The nipple opening should only allow one drop at a time to fall from the nipple when the bottle is inverted. A horizontal slit made with a razor blade instead of a round hole may make it easier for neonates to obtain milk or formula. Milk should be sucked—never squeezed—from the bottle. A rapid flow rate may lead to aspiration of milk and pneumonia and/or death.

Puppies and kittens should normally be held horizontally with the head in a natural position (**Figure 1**). This position reduces the risk of aspiration. Although some puppies may prefer a different position during feeding (**Figure 2**), careful observation is necessary because the risk of aspiration is increased.

TUBE FEEDING Puppies and kittens that are weak or suckle poorly may need to be tube fed. Tube feeding is quicker than bottle feeding and is often used when several orphans must be cared for by the same person(s). Bottle feeding allows puppies and kittens to control the amount of food intake, whereas tube feeding bypasses this control mechanism. Infant feeding tubes (5 to 8 Fr.) or soft urethral or intravenous catheters may be used (**Figure 3**).

The tube should be lubricated and placed in the lower esophagus, which is approximately 75% of the distance from the nose to the last rib (**Figure 4**). It is best to measure and mark the tube with an indelible marker or a piece of tape before insertion. Recheck measurements every few days to account for growth of the puppy or kitten. The orphan should normally be placed horizontally in the palm of the hand with its head in a natural position (**Figure 5**).

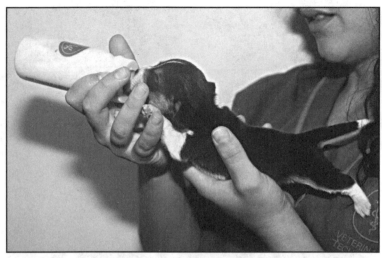

Figure 1. This is the preferred position for bottle feeding kittens and puppies. This position mimics the normal nursing position and decreases the likelihood of aspiration.

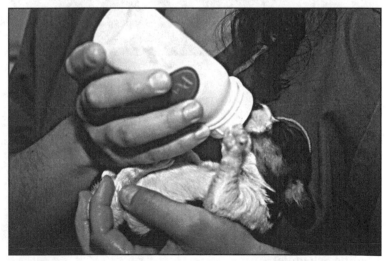

Figure 2. Some neonates prefer different positions for bottle feeding. This puppy prefers nursing in dorsal recumbency. Close observation is required because this position may predispose to aspiration.

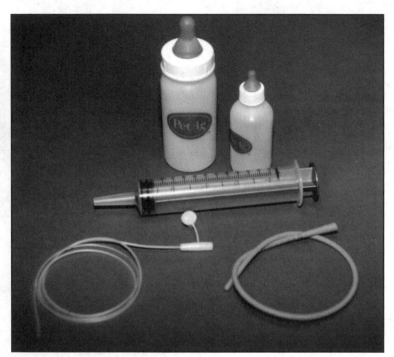

Figure 3. Various bottles and feeding tubes can be used for hand feeding orphaned puppies and kittens.

The mouth can be opened using the same hand that steadies the head. Gently advance the tube to the premeasured mark. If resistance is encountered or the animal suddenly struggles, the tube may be in the trachea. It should be removed and repositioned into the esophagus. Do not feed until proper placement is assured. After the tube is placed, attach the feeding syringe and slowly administer the warmed formula (over about one to two minutes) (**Figure 6**). The stomach may be palpated to determine the degree of distention. Administration should be stopped if the stomach becomes taut or resists formula flow. Continuation of feeding may result in overdistention and regurgitation. If regurgitation occurs, withdraw the tube and discontinue feeding until the next meal.

Feeding Schedule

Orphans should be fed at least four times daily. Very young neonates and weak individuals should preferably be fed every two to four hours. Older

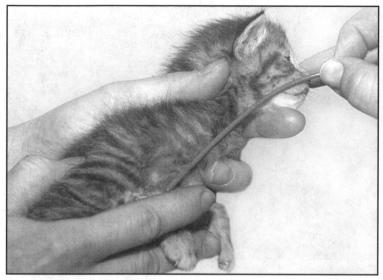

Figure 4. Feeding tubes should be premeasured and marked at a spot approximately 75% of the distance from the nose to the last rib. This placement will ensure the tube tip is in the distal esophagus.

puppies and kittens should be fed every four to six hours. Normally, one- to two-week-old puppies and kittens will obtain more than 90% of their normal daily intake in four to five meals. Milk replacer should be warmed to 38°C (100°F) and delivered slowly.

Cold foods, rapid feeding rates and overfeeding may result in regurgitation, aspiration, bloating and diarrhea. Review and correct the feeding methods if untoward signs develop. If diarrhea is observed, food volume should be reduced or diluted with water, then gradually returned to levels to meet caloric requirements over successive feedings. It is better to underfeed than overfeed neonatal puppies and kittens.

Strict hygiene as described above is especially important with hand feeding.

REASSESSMENT

Orphaned puppies or kittens should be evaluated daily for the first two weeks of life. They should remain normally hydrated, sleep quietly between feedings and gain weight at a rate similar to bitch- or queen-raised neonates.

Figure 5. Puppies and kittens should be held horizontally in the palm of the hand for tube feeding.

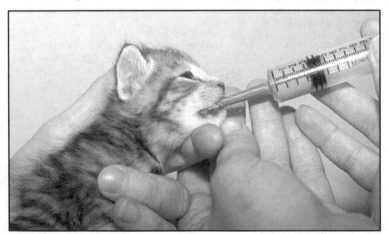

Figure 6. A lubricated tube is gently advanced to the premeasured mark and warm formula is administered over several minutes. The tube should be withdrawn and repositioned if resistance or struggling is encountered.

Alertness, eagerness to suckle, general behavior, body temperature (i.e., temperature of skin and lower limbs), body weight and stool character should be recorded daily or more often if neonates appear weak or listless.

Orphan rearing permits precise measurement of food intake. Nursing puppies should gain from 1 g body weight/2 to 5 g of milk intake during the first weeks of life. It is realistic to expect orphaned puppies to gain about the same. Kittens should grow about 100 g/week. If puppies or kittens do not thrive when fed a commercial milk replacer or homemade replacer, the nutrient content should be compared with mother's milk. (See Appendix K, Small Animal Clinical Nutrition, 4th ed., pp 1064-1072.) The dilution recommended by the manufacturer should also be checked. In some cases, it may be necessary to switch to another formula.

Puppies or kittens with rectal temperatures less than 35°C (95°F) should not be fed milk formula. At this temperature, the sucking reflex is usually absent and normal gut motility has ceased. Neonates should first be warmed slowly after receiving a warm solution of 2.5% glucose by subcutaneous injection (1 ml/30 g body weight).

ORPHANS

Nutrient Profiles of Commercial Dog and Cat Foods

Table 1. Nutrient content of moist grocery brand cat foods. This table lists products with a large market share and for which information is available. Grocery brands are those pet foods traditionally sold in grocery outlets and in some pet superstores.[*]

Foods	H_2O	kcal/g
Associate Wholesale Grocers Best Choice Tuna for Cats**	75.8	3.62
Friskies Alpo Cat Food with Chicken & Rice***	76.9	4.90
Friskies Alpo Tuna Treat**	71.1	3.90
Friskies Beef & Liver Dinner***	75.5	4.85
Friskies Turkey & Giblets Dinner***	75.7	4.91
Friskies Fancy Feast Chunky Chicken Feast***	75.6	4.88
Friskies Fancy Feast Flaked Tuna Feast***	73.9	3.83
Friskies Fancy Feast Savory Salmon Feast***	75.8	4.18
Friskies Fancy Feast Tender Liver & Chicken Feast**	76.4	5.07
Friskies Fancy Feast Turkey & Giblets Feast***	76.0	4.73
Friskies Kitten Formula, Cat Food Turkey Formula***	75.2	4.81
Friskies Kitten Formula, Ocean Whitefish Formula***	77.0	3.89
Friskies Senior Canned Cat Food Turkey & Giblets in Gravy***	76.6	4.68
Friskies Special Diet, Canned Turkey & Giblets Dinner***	77.6	4.85
Heinz 9 Lives Plus Salmon in Gravy**	77.0	3.81
Heinz 9 Lives Plus Turkey & Giblets Dinner**	76.8	4.32
Heinz Amoré Tuna Favorites, Tuna or Shrimp Entree**	74.4	3.73
Heinz Kozy Kitten Fish Dinner**	74.3	4.33
Kal Kan Optimum with Ocean Whitefish**	77.4	5.08
Kal Kan Sheba with Savory Duck in Meaty Juices**	82.1	4.23
Kal Kan Sheba with Turkey in Meaty Juices**	82.6	4.29
Kal Kan Whiskas Choice Cuts in Sauce w/Beef & Chicken**	81.6	4.15
Kal Kan Whiskas Kitty Stew**	79.3	4.86
Pedigree Katkins†	84.0	3.88
Pedigree Katkins Chunks†	78.0	4.05
Pedigree Kitekat†	85.0	3.80
Pedigree Kitekat Chunks†	79.0	3.76
Pedigree Sheba†	82.0	3.78
Pedigree Whiskas Fine Cuts***	82.0	4.11
Pedigree Whiskas Kitten***	81.0	4.21
Pedigree Whiskas Select Cuts***	80.0	3.70
Pedigree Whiskas Supermeat***	83.0	3.94
Ralston Purina Tender Beef Dinner**	74.0	4.64
Ralston Purina Tuna Dinner**	75.8	4.18
Average nutrient content of moist grocery brand cat foods	**77.9**	**4.29**

Key: H_2O = water, Prot = protein, NFE = nitrogen-free extract, Fiber = crude fiber, Ca = calcium, P = phosphorus, K = potassium, Na = sodium, Mg = magnesium, na = not available.
[*]Nutrients, except for moisture, are expressed as % dry matter. Energy density is expressed in kcal metabolizable energy/g dry matter. Kilocalories of metabolizable energy are either declared by the manufacturer or calculated based on modified Atwater values: protein = 3.5 kcal/g, fat = 8.5 kcal/g, NFE = 3.5 kcal/g. To convert to kJ, multiply kcal by 4.184.

Prot	Fat	NFE	Fiber	Ca	P	K	Na	Mg
77.2	9.3	3.6	0.4	1.32	0.95	1.07	1.03	0.13
45.0	38.5	1.3	2.34	1.52	1.26	1.07	0.83	0.08
76.9	14.7	0.0	0.4	1.66	1.56	1.14	0.69	0.10
46.5	36.5	3.5	2.23	2.14	1.38	0.97	1.17	0.07
47.0	36.7	4.0	1.35	1.82	1.41	0.92	0.91	0.07
50.8	34.0	5.3	0.4	2.54	1.76	0.79	0.93	0.07
61.3	14.9	11.5	1.9	1.76	1.23	na	0.81	0.10
56.8	24.5	3.4	1.7	2.47	1.99	1.27	0.50	0.11
55.0	36.9	0.3	1.27	0.76	1.14	1.06	0.47	0.07
55.9	32.0	1.5	1.7	0.96	1.31	1.14	0.44	0.08
51.5	34.8	1.4	1.7	1.81	1.65	1.04	0.50	0.08
65.2	17.8	2.6	0.9	3.00	2.52	1.67	0.81	0.14
48.8	32.0	7.4	1.7	1.08	1.25	1.90	0.39	0.08
48.6	34.4	6.2	1.4	0.89	0.95	1.17	0.98	0.08
62.5	12.9	15.2	0.9	0.65	0.78	0.48	1.00	0.08
42.0	28.9	11.3	0.9	2.67	2.20	0.99	2.07	0.12
75.5	11.2	3.6	0.8	1.44	0.94	1.21	0.82	0.16
38.0	24.6	26.0	3.5	1.55	1.20	1.09	0.30	0.19
37.4	41.3	7.3	1.8	2.74	2.25	0.75	0.84	0.13
63.1	23.2	1.3	2.2	2.18	1.40	1.17	1.12	0.09
60.3	25.3	0.7	2.3	2.18	1.32	1.26	0.97	0.09
44.8	22.2	20.0	1.1	2.18	1.90	0.65	1.47	0.09
51.4	33.3	6.7	1.0	1.06	1.54	0.96	0.77	0.08
36.3	30.6	13.9	2.0	2.60	1.70	na	0.60	0.09
30.5	27.0	33.2	1.4	1.60	1.50	na	1.00	0.06
37.3	29.3	14.0	2.0	2.20	1.50	na	0.7	0.10
34.2	22.4	31.8	1.4	2.10	2.00	na	1.2	0.07
62.2	18.2	6.1	1.7	1.30	1.30	na	1.1	0.07
47.8	30.7	6.2	1.7	1.90	1.90	na	1.30	0.10
50.5	29.7	6.2	2.1	1.70	1.50	na	0.80	0.06
38.5	20.1	30.2	1.5	1.40	1.60	na	1.2	0.07
48.2	27.1	8.3	2.4	1.40	1.60	na	0.90	0.08
43.9	29.3	17.4	0.8	1.77	1.42	0.96	0.58	0.09
50.4	20.7	18.9	0.4	1.69	1.41	1.12	0.62	0.10
51.2	**26.6**	**9.7**	**1.5**	**1.77**	**1.51**	**1.08**	**0.88**	**0.09**

**Analyses conducted in December 1995 and June 1996 by an independent laboratory (Woodson-Tenent Laboratories, Inc., Des Moines, IA, USA).

***Manufacturers' data.

†Data from Henderson AJ, ed. The Henston Small Animal Veterinary Vade Mecum, 15th ed. Peterborough, UK: Veterinary Business Development Ltd, 1996; 101-104.

PET FOODS

Table 2. Nutrient content of dry grocery brand cat foods. This table lists products with a large market share and for which information is available. Grocery brands are those pet foods traditionally sold in grocery outlets and in some pet superstores.*

Foods	H$_2$O	kcal/g
Friskies Alpo Dry Cat Food Gourmet Dinner**	8.3	3.70
Friskies Dry Cat Food Ocean Fish Flavor***	8.3	3.70
Friskies Fancy Feast Savory Salmon Flavor***	7.1	4.17
Friskies Fancy Feast Turkey & Giblets Flavor***	6.6	4.16
Friskies Kitten Formula Dry Cat Food***	8.3	3.70
Friskies Premium Chef's Blend 4 Delicious Flavors Dry Cat Food**	5.6	3.75
Friskies Premium Chef's Blend Dry Cat Food***	8.3	3.63
Friskies Senior Dry Cat Food***	8.5	3.74
Friskies Senior Dry Cat Food with Lamb Meal, Rice and Barley***	8.3	3.65
Friskies Special Diet Dry Cat Food***	8.3	4.01
Heinz 9 Lives Plus Tuna & Egg**	6.7	3.72
Heinz Kozy Kitten Dry Gulf Fish & Shrimp Flavor**	7.2	3.59
Kal Kan Whiskas Original Crave Recipe**	7.1	3.66
Pedigree Brekkies†	7.5	3.59
Pedigree Whiskas Cocktail†	7.5	3.66
Pedigree Whiskas Crunch†	7.5	3.58
Ralston Purina Alley Cat Poultry & Seafood Flavors**	6.1	3.82
Ralston Purina Cat Chow***	8.5	3.70
Ralston Purina Cat Chow Mature***	8.5	3.54
Ralston Purina Cat Chow Special Care**,***	7.5	4.36
Ralston Purina Deli Cat**	8.4	3.86
Ralston Purina Kit 'N Kaboodle**	6.8	3.81
Ralston Purina Kitten Chow***	8.5	3.72
Ralston Purina Meow Mix Chicken! Turkey! Salmon! Flavors**	6.3	3.80
Ralston Purina O.N.E. Chicken and Rice Formula for Cats***	8.5	4.59
Ralston Purina O.N.E. Chicken and Rice Formula for Kittens***	8.5	4.73
Average nutrient content of dry grocery brand cat foods	**7.6**	**3.84**

Key: H$_2$O = water, Prot = protein, NFE = nitrogen-free extract, Fiber = crude fiber, Ca = calcium, P = phosphorus, K = potassium, Na = sodium, Mg = magnesium, na = not available.
*Nutrients, except for moisture, are expressed as % dry matter. Energy density is expressed in kcal metabolizable energy/g dry matter. Kilocalories of metabolizable energy are either declared by the manufacturer or calculated based on modified Atwater values: protein = 3.5 kcal/g, fat = 8.5 kcal/g, NFE = 3.5 kcal/g. To convert to kJ, multiply kcal by 4.184.

Prot	Fat	NFE	Fiber	Ca	P	K	Na	Mg
33.3	10.7	46.3	2.3	1.04	1.16	0.60	0.60	0.12
35.7	11.2	42.8	2.2	1.31	1.32	0.62	0.61	0.13
34.9	19.3	37.4	2.2	1.39	1.13	0.62	0.21	0.07
34.9	19.1	37.6	2.1	1.34	1.07	0.62	0.21	0.07
39.0	11.5	38.7	2.1	1.78	1.45	0.70	0.46	0.13
34.0	11.9	44.2	2.6	1.27	1.47	0.54	0.49	0.12
36.2	10.0	43.3	2.3	1.34	1.35	0.59	0.62	0.14
32.5	10.4	49.1	2.8	1.20	0.93	0.59	0.26	0.11
31.3	9.3	50.4	2.7	1.26	1.19	0.60	0.30	0.14
33.4	16.5	41.2	2.0	1.18	1.07	0.76	0.31	0.09
33.2	10.8	46.7	1.8	1.33	1.33	0.66	0.34	0.12
31.8	8.9	49.0	3.5	1.35	1.05	0.79	0.29	0.18
32.3	10.4	46.9	2.6	1.31	1.25	0.73	0.62	0.14
32.2	8.5	51.4	2.7	1.20	1.10	na	0.80	0.10
41.7	8.3	43.1	1.6	1.00	1.10	na	0.90	0.07
20.4	5.8	67.7	2.7	1.10	0.90	na	0.50	0.05
36.8	14.6	36.9	4.1	1.43	1.18	0.85	0.44	0.18
37.1	13.2	40.3	1.6	1.72	1.34	0.78	0.32	0.13
36.8	9.7	44.5	1.8	1.38	1.09	0.82	0.35	0.12
34.1	14.6	44.9	1.0	0.99	0.94	0.84	0.30	0.08
37.3	13.0	41.5	1.4	1.20	1.08	0.93	0.37	0.14
35.1	12.6	43.1	2.3	1.34	1.08	0.92	0.40	0.14
41.0	13.8	35.5	1.8	1.54	1.46	0.80	0.33	0.14
37.0	12.4	41.4	1.5	1.37	1.22	0.95	0.52	0.13
34.4	15.5	42.6	1.4	1.15	0.96	0.85	0.22	0.07
39.1	18.4	33.8	1.4	1.21	1.09	0.86	0.33	0.11
34.8	**12.3**	**43.9**	**2.17**	**1.30**	**1.17**	**0.74**	**0.43**	**0.12**

**Analyses conducted in December 1995 and June 1996 by an independent laboratory (Woodson-Tenent Laboratories, Inc., Des Moines, IA, USA).
***Manufacturers' data.
†Data from Henderson AJ, ed. The Henston Small Animal Veterinary Vade Mecum, 15th ed. Peterborough, UK: Veterinary Business Development Ltd, 1996; 101-104.

PET FOODS

Table 3. Nutrient content of moist specialty brand cat foods. This table lists products with a large market share and for which information is available. Specialty brands are premium and super premium foods traditionally sold in pet stores, pet superstores or veterinary hospitals.*

Foods	H$_2$O	kcal/g
Eagle Pet Products Eagle Pack Beef & Liver***	76.5	5.02
Eagle Pet Products Eagle Pack Chicken, Rice & Lamb***	73.1	4.90
Eagle Pet Products Eagle Pack Poultry & White Fish***	77.4	5.22
Hill's Prescription Diet Feline g/d***	75.8	4.36
Hill's Prescription Diet Feline p/d***	71.1	4.85
Hill's Science Diet Feline Growth Tuna***	74.0	4.63
Hill's Science Diet Feline Growth***	69.6	5.43
Hill's Science Diet Feline Maintenance Beef***	75.6	4.50
Hill's Science Diet Feline Maintenance Light***	75.6	3.80
Hill's Science Diet Feline Maintenance Liver & Chicken***	75.6	4.39
Hill's Science Diet Feline Maintenance Seafood***	75.6	4.38
Hill's Science Diet Feline Maintenance Tuna***	73.4	4.47
Hill's Science Diet Feline Maintenance Turkey***	75.6	4.58
Hill's Science Diet Feline Senior Beef***	75.0	4.16
Hill's Science Diet Feline Senior Tuna***	73.0	4.27
Hill's Science Diet Feline Senior Turkey***	75.0	4.16
Iams Beef Formula***	76.1	5.75
Iams Catfish Formula***	75.6	5.70
Iams Chicken Formula***	75.5	5.77
Iams Lamb & Rice Formula***	74.1	5.63
Iams Less Active for Cats Chicken & Rice Formula***	75.6	4.93
Iams Less Active for Cats Fish & Rice Formula***	75.8	4.44
Iams Ocean Fish Formula***	76.7	5.86
Iams Turkey Formula***	74.9	5.48
Nature's Recipe Select Balance Feline Adult Formula***	71.0	4.48
Nature's Recipe Select Balance Feline Reduced Activity Formula***	75.0	4.15
Nature's Recipe Select Balance Feline Kitten Formula***	71.0	4.51
Nutro California Chicken Supreme**	80.6	4.50
Nutro Lamb & Turkey**	76.5	4.64
Nutro Salmon/Whitefish**	77.4	4.51
Nutro Veal Pate**	78.9	4.31
Pedigree Veterinary Plan 65/38***	85.4	4.45
Pedigree Veterinary Plan 75/34***	83.8	4.63
Pedigree Veterinary Plan 80/39***	81.4	4.30
Pedigree Veterinary Plan 100/40***	78.1	4.57
Average nutrient content of moist specialty brand cat foods	**75.9**	**4.74**

Key: H$_2$O = water, Prot = protein, NFE = nitrogen-free extract, Fiber = crude fiber, Ca = calcium, P = phosphorus, K = potassium, Na = sodium, Mg = magnesium, na = not available.
*Nutrients, except for moisture, are expressed as % dry matter. Energy density is expressed in kcal metabolizable energy/g dry matter. Kilocalories of metabolizable energy are either declared by the manufacturer or calculated based on modified Atwater values: protein = 3.5 kcal/g, fat = 8.5 kcal/g, NFE = 3.5 kcal/g. To convert to kJ, multiply kcal by 4.184.

Prot	Fat	NFE	Fiber	Ca	P	K	Na	Mg
46.8	36.2	8.51	1.7	1.11	1.02	0.72	0.60	0.09
39.4	33.1	19.7	1.1	1.08	0.86	1.04	0.82	0.09
45.6	38.9	8.41	0.9	1.19	1.02	0.66	0.49	0.10
35.1	19.4	37.2	3.3	0.66	0.54	0.74	0.29	0.08
48.8	31.5	11.2	0.5	1.10	0.90	0.90	0.51	0.09
44.2	28.8	18.1	1.5	1.00	0.85	1.31	0.50	0.08
49.0	36.2	6.9	0.6	1.09	0.95	0.76	0.56	0.09
45.2	25.4	19.9	2.9	0.94	0.78	0.86	0.33	0.05
45.1	14.8	27.3	7.0	0.70	0.57	0.78	0.33	0.06
45.2	25.0	20.7	2.3	0.94	0.82	0.82	0.38	0.04
45.2	25.4	20.0	2.6	0.93	0.70	0.86	0.42	0.08
43.3	23.8	26.7	1.5	0.80	0.68	1.40	0.50	0.07
45.2	25.4	20.7	2.5	0.82	0.82	0.82	0.30	0.07
40.8	20.4	28.4	5.2	0.64	0.68	0.72	0.32	0.06
40.0	20.0	33.3	1.4	0.80	0.63	1.40	0.40	0.07
40.8	20.4	27.6	5.2	0.68	0.68	0.76	0.32	0.06
47.2	31.7	14.8	0.8	1.17	1.04	0.88	0.29	0.06
49.4	30.3	13.4	1.3	1.15	0.98	0.90	0.29	0.07
46.9	31.8	14.1	1.2	1.10	1.02	0.81	0.29	0.07
45.6	32.0	14.1	2.2	1.20	1.08	0.85	0.39	0.05
44.3	19.7	27.9	1.2	1.15	1.11	1.64	0.45	0.10
44.2	17.8	30.6	0.8	1.40	1.12	1.40	0.45	0.09
46.1	32.2	13.3	2.2	1.16	1.03	1.07	0.43	0.10
48.1	33.4	11.8	1.1	0.99	0.88	0.84	0.28	0.07
42.4	27.6	23.5	1.4	0.69	0.69	na	0.34	0.10
48.0	18.0	26.4	2.3	0.68	0.68	na	0.52	0.08
45.5	29.3	15.9	1.7	1.45	1.24	na	0.41	0.10
58.1	26.6	5.99	0.5	1.08	0.98	0.83	1.34	0.06
55.8	29.9	3.92	0.9	1.96	1.36	0.64	0.90	0.07
55.8	26.9	7.92	0.4	1.42	1.06	0.84	1.06	0.07
59.2	26.2	0.38	1.0	3.47	2.18	0.57	0.71	0.09
42.5	35.6	9.6	1.4	1.85	1.78	1.10	0.98	0.10
39.5	40.1	8.6	1.2	1.05	0.99	1.05	0.80	0.10
43.0	33.3	10.8	1.6	1.08	1.08	0.86	0.91	0.16
46.1	37.4	3.7	1.4	1.42	1.37	0.87	0.68	0.08
45.9	**28.2**	**16.9**	**1.9**	**1.14**	**0.98**	**0.93**	**0.52**	**0.10**

**Analyses conducted in December 1995 and June 1996 by an independent laboratory (Woodson-Tenent Laboratories, Inc., Des Moines, IA, USA).
***Manufacturers' data.

PET FOODS

Table 4. Nutrient content of dry specialty brand cat foods. This table lists products with a large market share and for which information is available. Specialty brands are premium and super premium foods traditionally sold in pet stores, pet superstores or veterinary hospitals.*

Foods	H$_2$O	kcal/g
Diamond Maintenance***	8.4	4.25
Diamond Professional***	8.5	4.67
Eagle Pet Products Eagle Pack Cat/Kitten Pack***	6.9	4.37
Eagle Pet Products Eagle Pack Maintenance Pack***	8.0	3.88
Hill's Prescription Diet Feline g/d***	7.5	4.23
Hill's Prescription Diet Feline p/d***	7.5	4.44
Hill's Science Diet Feline Growth***	7.5	4.95
Hill's Science Diet Feline Maintenance***	7.5	4.76
Hill's Science Diet Feline Maintenance Light***	9.0	3.60
Hill's Science Diet Feline Senior***	8.0	4.04
Iams Cat Food***	7.0	4.82
Iams Eukanuba Cat Food Chicken & Rice***	7.5	5.10
Iams Eukanuba Cat Food Lamb & Rice***	7.5	5.07
Iams Kitten Food***	7.3	4.94
Iams Less Active for Cats***	7.7	4.28
Iams Natural Lamb & Rice for Cats***	7.0	4.84
Iams Ocean Fish & Rice for Cats***	6.9	4.46
Nature's Recipe Select Balance Feline Adult Formula***	8.0	4.05
Nature's Recipe Select Balance Feline Reduced Activity Formula***	8.0	3.68
Nature's Recipe Select Balance Kitten Formula***	8.0	4.15
Nutro Max Cat Adult***	8.5	4.29
Nutro Max Cat Lite***	10.0	3.70
Nutro Max Kitten***	8.9	4.33
Nutro's Natural Choice Cat***	8.2	4.05
Nutro's Natural Choice Kitten***	9.0	4.15
Pedigree Whiskas Advance Formula 8-Plus[†]	8.0	4.46
Pedigree Whiskas Advance Formula Adult[†]	8.0	4.29
Pedigree Whiskas Advance Formula Kitten/Growth[†]	8.0	4.51
Pedigree Whiskas Advance Formula Less Active[†]	8.0	3.91
Ralston Purina ProPlan Cat Adult***	7.5	4.71
Ralston Purina ProPlan Cat Growth***	7.5	5.00
Ralston Purina ProPlan Cat Lite Formula***	7.5	4.12
Ralston Purina ProPlan Turkey & Barley Formula for Cats***	8.0	4.44
Royal Canin Felinotechnique Fit32***	8.0	4.19
Royal Canin Felinotechnique Kitten34***	8.0	4.57
Royal Canin Felinotechnique Senior28***	8.0	4.78
Royal Canin Felinotechnique Sensible33***	8.0	4.95
Royal Canin Felinotechnique Slim37***	8.0	3.80
Average nutrient content of dry specialty brand cat foods	**8.1**	**4.28**

Key: H$_2$O = water, Prot = protein, NFE = nitrogen-free extract, Fiber = crude fiber, Ca = calcium, P = phosphorus, K = potassium, Na = sodium, Mg = magnesium, na = not available.
*Nutrients, except for moisture, are expressed as % dry matter. Energy density is expressed in kcal metabolizable energy/g dry matter. Kilocalories of metabolizable energy are either declared by the manufacturer or calculated based on modified Atwater values: protein = 3.5 kcal/g, fat = 8.5 kcal/g, NFE = 3.5 kcal/g. To convert to kJ, multiply kcal by 4.184.

Prot	Fat	NFE	Fiber	Ca	P	K	Na	Mg
33.2	16.8	41.2	3.0	0.88	0.76	0.65	0.27	0.11
37.4	23.2	29.8	2.6	1.13	0.99	0.66	0.38	0.10
34.8	22.6	35.1	1.8	1.29	0.75	0.67	0.43	0.06
34.6	13.2	44.2	2.2	1.20	0.87	0.65	0.43	0.09
33.4	18.9	41.9	1.4	0.51	0.55	0.75	0.34	0.05
36.1	23.7	32.8	1.1	1.24	0.94	0.69	0.43	0.08
37.1	26.8	29.1	1.2	1.30	0.96	0.64	0.35	0.11
33.8	23.0	36.9	1.0	0.85	0.74	0.67	0.32	0.07
40.7	9.0	38.0	6.8	0.96	0.79	0.68	0.30	0.06
33.7	16.3	41.6	2.5	0.87	0.68	0.88	0.29	0.08
35.5	24.0	32.4	1.8	1.14	1.01	0.75	0.31	0.09
37.7	23.7	29.8	2.1	1.03	0.92	0.94	0.55	0.10
37.7	23.7	29.7	2.1	1.03	0.92	0.80	0.39	0.10
37.7	24.9	29.9	1.5	1.28	1.02	0.97	0.65	0.10
32.4	16.6	43.3	2.0	1.09	0.92	0.76	0.32	0.09
35.4	24.3	31.4	1.8	1.12	0.99	0.97	0.47	0.10
36.4	24.8	30.6	1.6	1.15	1.04	0.86	0.30	0.10
34.8	20.1	38.0	1.6	1.02	0.83	na	0.41	0.09
34.5	9.78	46.7	3.3	0.91	0.86	na	0.38	0.09
38.0	22.3	31.5	1.6	1.08	1.03	na	0.51	0.10
36.1	21.9	33.3	2.2	1.09	1.04	0.50	0.61	0.08
37.2	11.9	39.6	4.4	1.11	1.07	0.50	0.37	0.08
41.2	23.6	25.3	2.5	1.32	1.21	0.66	0.64	0.11
33.6	16.9	41.2	1.4	1.20	0.80	0.04	0.58	0.10
35.7	19.5	35.4	1.8	1.37	1.10	0.68	0.66	0.10
35.9	25.0	31.0	1.6	1.10	0.98	na	0.98	0.09
34.8	22.8	32.6	2.2	1.10	0.98	na	0.87	0.09
35.9	26.1	29.9	1.6	1.10	0.91	na	0.87	0.09
30.4	15.2	44.6	2.2	1.10	0.98	na	0.87	0.09
34.2	16.6	42.2	1.4	0.97	0.94	0.82	0.25	0.07
36.5	22.7	33.3	1.2	1.08	1.06	0.86	0.39	0.10
34.8	9.4	47.6	2.5	1.06	0.91	0.86	0.26	0.07
35.8	18.9	36.0	2.4	1.25	1.09	0.88	0.33	0.11
34.8	14.1	41.3	2.7	1.14	1.03	0.65	0.38	0.09
37.0	21.7	31.5	2.7	1.30	1.09	0.65	0.38	0.09
30.4	25.0	36.2	2.7	0.98	0.65	0.87	0.38	0.09
35.9	23.9	30.4	2.7	1.09	1.03	0.65	0.38	0.09
40.2	10.9	37.2	4.4	1.30	1.09	0.65	0.43	0.09
35.3	**18.5**	**37.4**	**2.4**	**1.10**	**0.95**	**0.70**	**0.46**	**0.09**

**Analyses conducted in December 1995 and June 1996 by an independent laboratory (Woodson-Tenent Laboratories, Inc., Des Moines, IA, USA).
***Manufacturers' data.
†Data from Henderson AJ, ed. The Henston Small Animal Veterinary Vade Mecum, 15th ed. Peterborough, UK: Veterinary Business Development Ltd, 1996; 101-104.

PET FOODS

Table 5. Nutrient content of moist grocery brand dog foods. This table lists products with a large market share and for which information is available. Grocery brands are those pet foods traditionally sold in grocery outlets and in some pet superstores.*

Foods	H$_2$O	kcal/g
Friskies Alpo Chopped with Chicken***	76.6	4.69
Friskies Alpo Chunky with Beef**	75.6	4.84
Friskies Alpo Chunky with Liver**	76.8	4.36
Friskies Alpo Prime Cuts Gourmet Dinner**	79.1	4.10
Friskies Gourmet Cuts with Beef & Liver***	77.3	4.19
Friskies Mighty Dog Beef***	76.2	5.03
Friskies Mighty Dog Gourmet Dinner**	73.7	5.37
Friskies Mighty Dog Senior Turkey & Rice Dinner***	76.8	4.60
Heinz Cycle Adult***	81.5	4.37
Heinz Cycle Lite***	79.2	4.08
Heinz Cycle Puppy***	77.9	4.43
Heinz Cycle Senior***	81.4	5.05
Heinz Ken-L Ration Original Chunky Beef**	74.3	3.98
Heinz King Kuts Chicken & Beef**	77.6	4.12
Heinz Skippy Chunks in Gravy with Beef**	79.9	3.98
Kal Kan Pedigree Beef Choice**	81.5	4.18
Kal Kan Pedigree Choice Cuts with Beef & Liver**	79.3	4.17
Kal Kan Pedigree Chopped Combo**	79.5	4.78
Kal Kan Pedigree Select Dinners (Cesar w/Beef in Aspic)**	82.0	4.28
Kal Kan Pedigree Select Dinners (Cesar w/Chicken & Liver in Aspic)**	81.2	4.56
Kal Kan Pedigree with Chunky Chicken**	78.7	5.21
Pedigree Bounce†	81.0	4.53
Pedigree Bounce Super Chunks†	78.0	4.41
Pedigree Cesar†	84.0	4.62
Pedigree Chappie†	77.0	3.52
Pedigree Chum†	81.0	4.68
Pedigree Chum Junior†	80.0	5.15
Pedigree Chum Puppy Food†	78.0	5.18
Pedigree Chum Tender Bites†	80.0	4.70
Pedigree Pal†	80.0	4.35
Average nutrient content of moist grocery brand dog foods	**78.8**	**4.50**

Key: H$_2$O = water, Prot = protein, NFE = nitrogen-free extract, Fiber = crude fiber, Ca = calcium, P = phosphorus, K = potassium, Na = sodium, Mg = magnesium, na = not available.
*Nutrients, except for moisture, are expressed as % dry matter. Energy density is expressed in kcal metabolizable energy/g dry matter. Kilocalories of metabolizable energy are either declared by the manufacturer or calculated based on modified Atwater values: protein = 3.5 kcal/g, fat = 8.5 kcal/g, NFE = 3.5 kcal/g. To convert to kJ, multiply kcal by 4.184.

Prot	Fat	NFE	Fiber	Ca	P	K	Na	Mg
42.8	33.0	11.0	2.9	2.67	1.64	1.47	0.39	0.18
46.4	33.8	9.84	1.6	1.39	1.03	1.23	0.53	0.15
49.9	25.5	12.7	1.7	2.15	1.51	1.29	0.65	0.16
44.2	19.5	25.5	0.9	1.40	1.07	0.94	1.21	0.07
46.3	20.3	24.2	0.4	1.28	0.93	1.76	0.97	0.20
48.5	39.2	0.14	2.08	1.74	1.48	1.15	0.74	0.09
40.1	45.2	3.57	0.4	2.16	1.75	0.95	0.80	0.09
36.9	30.1	21.3	1.7	2.97	1.87	1.08	0.27	0.12
36.6	17.2	34.8	1.8	1.16	0.94	0.97	1.23	0.11
26.7	13.0	43.3	7.5	1.04	0.94	1.08	0.79	0.16
40.3	23.1	24.4	2.1	1.30	1.06	1.33	0.82	0.12
26.9	17.2	45.6	2.0	1.01	0.82	1.00	0.78	0.10
46.2	19.5	20.1	3.1	1.44	1.32	1.71	1.17	0.21
46.4	20.7	21.2	0.9	1.43	1.20	1.11	1.25	0.09
44.0	17.7	26.8	3.5	1.05	0.90	1.29	0.60	0.20
40.5	22.5	24.2	1.1	1.84	1.51	0.65	2.22	0.09
40.6	23.3	21.9	1.0	2.27	1.79	0.68	2.22	0.10
40.2	37.9	4.23	1.5	4.09	2.43	1.07	0.83	0.10
59.6	24.2	4.00	2.2	1.39	1.39	1.22	1.28	0.08
57.6	29.1	2.02	2.1	1.17	1.44	1.06	1.38	0.07
37.6	42.7	7.56	1.4	2.7	1.69	0.80	0.66	0.09
36.3	26.8	27.2	1.1	1.3	1.1	na	0.2	0.09
29.0	26.6	32.6	1.8	2.1	1.8	na	1.0	0.07
53.0	30.0	5.6	1.3	1.5	1.7	na	0.9	0.09
25.7	6.1	59.1	0.1	1.9	1.3	na	0.2	0.10
33.2	32.6	22.1	2.1	2.2	1.2	na	0.8	0.10
34.5	39.5	17.0	2.0	0.9	1.0	na	0.6	0.07
38.8	39.1	14.7	1.8	1.2	1.1	na	0.4	0.06
51.0	32.0	4.8	2.1	1.3	1.5	na	0.9	0.09
35.7	26.0	25.0	1.0	1.7	1.4	na	0.2	0.09
41.2	**27.1**	**19.9**	**1.84**	**1.73**	**1.36**	**1.14**	**0.87**	**0.11**

**Analyses conducted in December 1995 and June 1996 by an independent laboratory (Woodson-Tenent Laboratories, Inc., Des Moines, IA, USA).
***Manufacturers' data.
[†]Data from Henderson AJ, ed. The Henston Small Animal Veterinary Vade Mecum, 15th ed. Peterborough, UK: Veterinary Business Development Ltd, 1996; 90-95.

PET FOODS

Table 6. Nutrient content of dry grocery brand dog foods. This table lists products with a large market share and for which information is available. Grocery brands are those pet foods traditionally sold in grocery outlets and in some pet superstores.*

Foods	H₂O	kcal/g
Friskies Alpo Beefy Dinner***	8.3	3.64
Friskies Alpo Dry Dog Food with Lamb Meal, Rice & Barley***	6.5	3.95
Friskies Come 'N Get It Dry Dog Food***	8.3	3.70
Friskies Field Trial Bite Size***	7.0	3.73
Heinz Cycle Adult***	8.9	3.91
Heinz Cycle Lite***	9.2	3.40
Heinz Cycle Puppy***	7.0	4.24
Heinz Cycle Senior***	9.6	3.67
Heinz Ken-L Ration Gravy Train Beef Flavor**	8.5	3.76
Kal Kan Pedigree Mealtime Lamb & Rice**	8.8	3.96
Kal Kan Pedigree Mealtime Rice & Vegetables**	8.4	3.74
Kal Kan Pedigree Mealtime Small Bites**	8.4	3.74
Pedigree Chum Complete†	8.0	3.78
Pedigree Chum Complete Junior†	6.0	3.87
Pedigree Chum Complete Puppy†	7.5	4.10
Ralston Purina Chuck Wagon Stampede**	6.8	3.82
Ralston Purina Dog Chow***	10.1	4.04
Ralston Purina Fit & Trim**,***	8.5	3.10
Ralston Purina Grrravy**	6.8	3.76
Ralston Purina Hi Pro**,***	8.5	4.08
Ralston Purina Kibbles and Chunks**	11.4	3.80
Ralston Purina Mainstay**	6.9	3.52
Ralston Purina Nutrient Management Dog Food***	8.5	4.22
Ralston Purina Nutrient Management Puppy Food***	8.5	4.32
Ralston Purina O.N.E. Adult Formula**,***	8.0	4.49
Ralston Purina O.N.E. Lamb & Rice Formula***	7.5	4.55
Ralston Purina O.N.E. Lamb & Rice Formula for Puppies***	7.5	4.70
Ralston Purina O.N.E. Lite Formula***	8.5	3.54
Ralston Purina Puppy Chow***	8.5	4.03
Ralston Purina Senior***	8.0	3.50
Wafcol 20†	6.5	3.67
Wafcol Energy Plus†	6.5	3.82
Average nutrient content of dry grocery brand dog foods	**8.0**	**3.88**

Key: H₂O = water, Prot = protein, NFE = nitrogen-free extract, Fiber = crude fiber, Ca = calcium, P = phosphorus, K = potassium, Na = sodium, Mg = magnesium, na = not available.
*Nutrients, except for moisture, are expressed as % dry matter. Energy density is expressed in kcal metabolizable energy/g dry matter. Kilocalories of metabolizable energy are either declared by the manufacturer or calculated based on modified Atwater values: protein = 3.5 kcal/g, fat = 8.5 kcal/g, NFE = 3.5 kcal/g. To convert to kJ, multiply kcal by 4.184.

Prot	Fat	NFE	Fiber	Ca	P	K	Na	Mg
24.1	10.9	53.5	2.9	1.92	1.33	0.65	0.59	0.21
26.7	15.5	48.5	2.3	1.49	1.33	0.60	0.58	0.17
23.9	11.9	52.8	2.9	1.90	1.31	0.65	0.59	0.21
23.8	12.9	51.6	3.9	1.49	1.15	0.90	0.29	0.27
26.5	13.1	52.2	2.9	1.00	0.77	0.79	0.33	0.14
21.0	7.8	55.5	9.9	1.14	0.84	0.94	0.12	0.13
31.0	19.4	39.4	2.0	1.74	1.26	1.08	0.39	0.11
21.6	9.4	62.1	2.9	0.80	0.62	0.63	0.09	0.13
23.7	10.5	58.1	2.0	0.95	0.79	0.81	0.33	0.15
24.6	13.9	54.8	1.8	0.86	0.81	0.53	0.52	0.11
23.2	11.5	55.7	2.0	1.79	1.20	0.45	0.51	0.13
22.9	11.2	56.7	2.2	1.50	1.11	0.49	0.51	0.13
25.5	11.7	54.0	1.6	1.4	1.3	na	0.53	0.10
27.6	13.1	51.3	2.2	0.90	0.90	na	0.90	0.09
31.7	13.3	39.9	2.2	1.40	1.10	na	0.38	0.10
26.4	12.2	53.0	1.7	1.24	1.00	0.61	0.49	0.12
23.5	13.5	53.4	2.2	1.1	0.89	0.66	0.33	0.15
15.3	7.4	60.4	10.7	1.03	0.71	0.78	0.23	0.17
25.7	12.9	50.5	3.3	1.60	1.08	0.87	0.48	0.19
30.2	11.8	49.7	1.9	1.31	0.93	0.52	0.42	0.14
26.4	12.3	52.5	2.0	1.41	1.11	0.54	0.38	0.12
19.3	11.0	54.5	7.0	1.43	1.05	0.61	0.37	0.17
28.0	11.4	52.3	1.5	1.16	0.93	0.60	0.37	na
31.5	12.6	47.0	1.5	1.31	1.04	0.66	0.35	na
29.9	18.7	42.6	1.6	1.48	1.07	0.61	0.50	0.09
29.1	17.9	43.5	1.6	1.19	0.97	0.58	0.52	na
31.1	18.9	40.0	1.7	1.30	1.05	0.54	0.45	na
18.2	8.9	63.7	3.1	1.36	0.98	0.61	0.38	na
29.5	11.5	50.6	1.8	1.31	0.94	0.60	0.40	na
18.2	8.5	61.5	5.9	1.20	0.78	0.71	0.29	na
21.4	6.4	61.0	4.3	1.4	0.96	na	0.47	0.03
26.7	10.7	49.2	4.3	2.35	1.6	na	0.51	0.05
25.3	**12.3**	**52.2**	**3.1**	**1.36**	**1.03**	**0.67**	**0.43**	**0.14**

**Analyses conducted in December 1995 and June 1996 by an independent laboratory (Woodson-Tenent Laboratories, Inc., Des Moines, IA, USA).
***Manufacturers' data.
[†]Data from Henderson AJ, ed. The Henston Small Animal Veterinary Vade Mecum, 15th ed. Peterborough, UK: Veterinary Business Development Ltd, 1996; 90-95.

PET FOODS

Table 7. Nutrient content of moist specialty brand dog foods. This table lists products with a large market share and for which information is available. Specialty brands are premium foods traditionally sold in pet stores, pet superstores or veterinary hospitals.*

Foods	H₂O	kcal/g
AVCA Pet Nutrition Select Balance Canine Adult Formula***	72.0	4.37
Eagle Pet Products Eagle Pack Beef & Rice***	72.7	4.77
Eagle Pet Products Eagle Pack Chicken & Rice***	71.0	4.80
Eagle Pet Products Eagle Pack Lamb & Rice***	73.6	4.85
Eagle Pet Products Eagle Pack Liver & Rice***	72.8	4.71
Eagle Pet Products Eagle Pack Puppy Dinner***	73.8	4.69
Hill's Prescription Diet Canine g/d***	73.0	4.12
Hill's Prescription Diet Canine p/d***	70.0	4.72
Hill's Science Diet Canine Active Formula/Performance***	72.1	4.97
Hill's Science Diet Canine Growth***	70.0	4.14
Hill's Science Diet Canine Growth Beef & Rice Formula***	76.4	4.39
Hill's Science Diet Canine Growth Beef & Vegetable Formula***	76.3	4.36
Hill's Science Diet Canine Light***	74.0	3.28
Hill's Science Diet Canine Maintenance Beef***	76.4	4.33
Hill's Science Diet Canine Maintenance Beef & Chicken***	75.8	4.03
Hill's Science Diet Canine Maintenance Beef & Rice Formula***	79.3	4.07
Hill's Science Diet Canine Maintenance Beef & Vegetable***	79.8	4.07
Hill's Science Diet Canine Maintenance Chicken Formula***	75.7	4.01
Hill's Science Diet Canine Maintenance Turkey***	75.4	4.23
Hill's Science Diet Canine Senior Beef***	75.8	4.11
Hill's Science Diet Canine Senior Chicken Formula***	75.4	3.93
Hill's Science Diet Canine Senior Turkey***	76.0	4.15
Iams Beef Formula***	75.2	4.98
Iams Chicken Formula***	75.5	5.23
Iams Less Active Beef & Liver Formula***	76.7	4.95
Iams Less Active Chicken & Rice Formula***	76.6	4.79
Iams Puppy Formula***	68.1	4.59
Iams Turkey Formula***	75.2	5.01
Nature's Recipe Select Balance Canine Lamb & Rice Formula***	72.5	4.70
Nature's Recipe Select Balance Canine Puppy***	72.0	4.38
Nature's Recipe Select Balance Canine Reduced Activity Formula***	75.1	3.43
Pedigree Veterinary Plan 80/26***	81.5	4.32
Pedigree Veterinary Plan 90/21***	78.6	4.21
Pedigree Veterinary Plan 100/29***	77.0	4.35
Pedigree Veterinary Plan 130/33***	71.5	4.56
Average nutrient content of moist specialty brand dog foods	**75.1**	**4.42**

Key: H₂O = water, Prot = protein, NFE = nitrogen-free extract, Fiber = crude fiber, Ca = calcium, P = phosphorus, K = potassium, Na = sodium, Mg = magnesium, na = not available
*Nutrients, except for moisture, are expressed as % dry matter. Energy density is expressed in kcal metabolizable energy/g dry matter. Kilocalories of metabolizable energy are either declared by the manufacturer or calculated based on modified Atwater values: protein = 3.5 kcal/g, fat = 8.5 kcal/g, NFE = 3.5 kcal/g. To convert to kcal, multiply kcal by 4.184.

Prot	Fat	NFE	Fiber	Ca	P	K	Na	Mg
30.4	23.2	36.7	3.1	1.13	0.95	na	0.24	na
44.3	30.0	19.1	0.7	1.36	1.14	0.88	0.73	na
36.6	32.1	22.8	1.0	1.59	1.14	0.76	0.62	na
32.6	34.1	23.1	1.5	1.89	1.25	0.98	0.87	na
46.7	29.0	17.3	1.1	1.32	1.18	0.85	0.70	na
40.1	28.6	24.4	0.8	1.56	1.07	0.88	0.73	na
18.9	10.7	64.1	1.9	0.63	0.41	0.78	0.22	0.07
30.0	28.0	33.3	1.0	1.30	1.13	0.60	0.47	0.13
30.1	25.8	36.9	1.1	0.93	0.72	0.65	0.47	0.08
29.3	23.0	39.0	1.3	1.33	1.00	0.83	0.40	0.13
29.2	22.4	37.7	3.2	1.10	0.97	0.93	0.38	0.11
29.1	22.4	37.8	3.1	1.06	0.97	0.84	0.38	0.11
18.1	9.6	60.0	7.7	0.58	0.50	0.62	0.31	0.12
25.4	15.8	52.8	1.0	0.59	0.51	0.94	0.25	0.10
24.8	15.7	54.1	0.8	0.61	0.62	0.66	0.29	0.08
25.1	15.0	50.7	3.6	0.63	0.53	0.87	0.24	0.10
24.8	15.3	50.6	3.7	0.64	0.50	0.79	0.25	0.10
24.7	16.5	53.4	0.9	0.64	0.66	0.62	0.23	0.10
24.4	15.9	54.4	0.8	0.65	0.65	0.61	0.26	0.10
19.4	16.1	59.2	1.3	0.61	0.54	0.70	0.17	0.08
18.6	15.9	59.4	1.6	0.65	0.57	0.65	0.24	0.11
19.1	15.8	59.7	1.2	0.63	0.51	0.62	0.25	0.08
39.5	27.4	25.1	2.4	1.01	0.85	0.85	0.44	0.12
40.6	28.7	23.2	1.8	1.72	1.31	1.06	0.57	0.08
41.2	20.2	30.5	1.3	1.12	0.94	1.42	0.43	0.09
41.9	18.0	32.1	1.3	1.50	1.20	1.54	0.43	0.13
46.9	28.5	13.5	2.5	1.51	1.32	1.19	0.60	0.09
38.8	28.5	24.6	2.6	1.29	1.09	1.13	0.64	0.08
34.6	30.9	24.0	1.5	1.89	1.05	na	0.62	na
32.1	25.0	31.1	2.9	2.12	1.36	na	0.50	na
20.0	10.0	53.2	11.0	0.76	0.68	na	0.24	na
28.1	16.8	44.3	1.1	1.46	1.14	1.73	0.54	0.08
22.0	18.7	48.1	2.8	1.26	0.98	1.64	0.42	0.08
31.7	25.2	28.7	3.5	1.43	1.30	1.30	0.65	0.14
37.5	24.2	28.8	2.8	1.58	1.30	0.77	0.42	0.08
32.1	**22.1**	**36.9**	**2.1**	**1.21**	**0.93**	**0.89**	**0.53**	**0.10**

**Analyses conducted in December 1995 and June 1996 by an independent laboratory (Woodson-Tenent Laboratories, Inc., Des Moines, IA, USA).
***Manufacturers' data.

PET FOODS

Table 8. Nutrient content of dry specialty brand dog foods. This table lists products with a large market share and for which information is available. Specialty brands are premium and super premium foods traditionally sold in pet stores, pet superstores or veterinary hospitals.*

Foods	H₂O	kcal/g
Diamond High Protein***	8.4	4.25
Diamond Lamb, Rice & Turkey Formula***	8.2	4.09
Diamond Lite Formula***	10.0	3.76
Diamond Maintenance***	8.2	4.02
Diamond Professional***	8.4	4.50
Diamond Puppy Food***	8.1	4.53
Eagle Pet Products Eagle Pack Kennel Pack***	7.4	3.94
Eagle Pet Products Eagle Pack Lite Pack***	8.6	3.39
Eagle Pet Products Eagle Pack Maintenance Pack***	8.2	3.83
Eagle Pet Products Eagle Pack Natural Pack***	7.7	3.70
Eagle Pet Products Eagle Pack Power Pack***	6.5	4.20
Eagle Pet Products Eagle Pack Premium Select***	8.4	3.88
Eagle Pet Products Eagle Pack Puppy Pack***	6.9	4.05
Hill's Prescription Diet Canine g/d***	8.0	3.95
Hill's Prescription Diet Canine p/d***	7.5	4.51
Hill's Prescription Diet Canine p/d Large Breed***	7.5	3.46
Hill's Science Diet Canine Active Formula/Performance***	7.5	4.52
Hill's Science Diet Canine Growth***	7.5	4.26
Hill's Science Diet Canine Large Breed Growth***	7.5	3.75
Hill's Science Diet Canine Light***	9.0	3.05
Hill's Science Diet Canine Maintenance***	8.0	4.22
Hill's Science Diet Canine Senior***	8.0	3.86
Hil's Science Diet Lamb Meal & Rice Canine Growth***	7.5	4.95
Hill's Science Diet Lamb Meal & Rice Canine Maintenance***	8.0	4.25
Iams Chunks***	8.0	4.17
Iams Eukanuba Adult Maintenance Formula***	8.0	4.71
Iams Eukanuba Large Breed Puppy Formula***	8.6	4.53
Iams Eukanuba Medium Breed Puppy Formula***	6.9	4.60
Iams Eukanuba Natural Lamb & Rice***	6.9	4.42
Iams Eukanuba Natural Lamb & Rice Formula for Puppies***	7.4	4.46
Iams Eukanuba Premium Performance***	7.8	4.83
Iams Eukanuba Reduced Fat***	7.5	4.19
Iams Eukanuba Senior Maintenance Formula***	7.9	4.58
Iams Eukanuba Small Breed Puppy Formula***	7.5	4.86
Iams Eukanuba Weaning Diet***	7.5	4.86
Iams Lamb Meal & Rice for Puppies***	7.7	4.44
Iams Less Active***	7.6	4.17
Iams Minichunks***	8.0	4.17
Iams Natural Lamb Meal & Rice***	8.0	3.95
Iams Puppy***	8.1	4.66
Iams Senior***	8.0	4.23
Nature's Recipe Select Balance Canine Adult Formula***	8.0	3.97
Nature's Recipe Select Balance Canine Lamb & Rice Formula***	8.0	4.30
Nature's Recipe Select Balance Canine Puppy Formula***	8.0	4.13
Nature's Recipe Select Balance Canine Reduced Activity***	8.0	3.30

Prot	Fat	NFE	Fiber	Ca	P	K	Na	Mg
28.7	20.2	39.1	3.5	1.54	1.25	0.76	0.38	0.19
25.5	15.5	46.4	3.5	1.62	1.20	0.76	0.33	0.26
17.8	8.9	62.2	4.4	0.68	0.76	0.52	0.14	0.14
23.1	13.7	50.9	3.6	1.63	1.24	0.73	0.37	0.19
33.2	22.3	33.1	2.8	1.58	1.38	0.79	0.38	0.14
34.4	22.3	32.2	2.7	1.34	1.12	0.78	0.59	0.14
28.1	16.3	44.7	3.4	1.73	1.19	0.63	0.55	na
18.8	8.9	56.7	7.6	1.64	0.88	0.68	0.39	0.15
22.0	13.3	54.5	2.9	1.42	0.98	0.66	0.58	na
25.1	13.2	48.5	3.5	1.84	1.41	0.69	0.61	0.14
32.8	21.7	34.2	3.2	2.03	1.18	0.63	0.43	na
25.0	16.5	45.5	3.4	1.64	0.98	0.68	0.51	0.12
30.7	18.6	37.4	3.3	1.93	1.29	0.67	0.61	na
18.7	10.9	66.1	1.0	0.61	0.40	0.61	0.18	0.05
31.8	22.7	34.4	2.9	1.73	1.19	0.79	0.34	0.13
29.4	9.2	47.5	8.3	0.90	0.70	0.65	0.35	0.11
30.5	27.0	35.0	2.0	0.99	0.75	0.61	0.39	0.09
29.3	19.2	41.2	2.6	1.44	1.16	0.70	0.51	0.15
29.7	10.6	51.6	2.4	1.01	0.80	0.64	0.35	0.12
18.7	9.6	50.1	16.9	0.73	0.57	0.78	0.22	0.15
25.0	15.4	53.3	1.7	0.72	0.65	0.65	0.28	0.13
18.5	10.7	63.8	3.0	0.60	0.55	0.63	0.17	0.11
29.1	19.9	41.0	2.1	1.70	1.16	0.61	0.39	0.09
22.6	15.9	53.6	2.3	1.08	0.80	0.62	0.28	0.10
29.8	17.8	42.3	2.9	1.51	1.10	0.89	0.60	0.11
28.5	18.4	45.3	2.2	1.20	0.92	0.89	0.60	0.10
29.6	17.2	44.9	2.1	0.98	0.82	0.63	0.24	0.09
32.8	20.5	37.1	2.7	1.28	1.00	0.86	0.33	0.12
26.2	16.7	48.4	2.2	1.29	1.02	0.92	0.63	0.12
29.9	17.3	43.7	2.4	1.30	1.03	0.80	0.53	0.10
34.0	23.0	35.1	1.9	1.30	1.00	0.87	0.60	0.11
21.3	10.5	61.4	2.0	0.97	0.76	0.78	0.50	0.05
29.3	12.8	48.5	2.3	1.51	0.95	0.68	0.40	0.12
36.2	23.9	31.5	1.9	1.41	1.08	0.86	0.74	0.10
36.2	23.9	31.5	1.9	1.41	1.08	0.86	0.74	0.09
29.3	16.3	40.8	3.4	1.94	1.54	0.89	0.70	0.20
22.2	12.5	56.5	2.8	1.14	0.85	0.78	0.37	0.13
29.8	17.8	42.3	2.9	1.51	1.10	0.89	0.60	0.11
25.1	14.2	46.3	4.2	1.85	1.57	1.01	0.65	0.23
32.1	19.9	38.7	2.8	1.37	1.04	0.87	0.65	0.12
27.5	12.0	51.9	2.5	1.13	0.90	0.75	0.27	0.12
25.0	14.7	51.9	3.0	0.85	0.83	na	0.34	na
29.4	17.9	40.0	4.0	1.85	1.30	na	0.58	na
30.4	19.0	40.2	2.7	1.30	1.06	na	0.42	na
19.0	5.4	57.1	13.0	0.62	0.55	na	0.33	na

(Continued on next page.)

PET FOODS

Table 8. Nutrient content of dry specialty brand dog foods. This table lists products with a large market share and for which information is available. Specialty brands are premium and super premium foods traditionally sold in pet stores, pet superstores or veterinary hospitals* (Continued.).

Foods	H2O	kcal/g
Nutro Max & Max Mini***	8.9	4.03
Nutro Max Puppy***	8.8	4.08
Nutro Max Special***	9.7	3.58
Nutro's Natural Choice Adult***	9.2	3.85
Nutro's Natural Choice Lite***	9.9	3.42
Nutro's Natural Choice Plus***	8.9	4.17
Nutro's Natural Choice Puppy***	9.0	3.89
Pedigree Chum Advance Formula Activity Plus†	8.0	4.28
Pedigree Chum Advance Formula Adult Supreme†	8.0	4.11
Pedigree Chum Advance Formula Junior Plus†	8.0	4.20
Pedigree Chum Advance Formula Light Menu†	8.0	3.57
Pedigree Chum Advance Formula Puppy Supreme†	8.0	4.34
Pedigree Veterinary Plan 330/26***	9.7	3.65
Pedigree Veterinary Plan 340/21***	10.2	3.79
Pedigree Veterinary Plan 360/29***	8.3	3.93
Pedigree Veterinary Plan 380/33***	7.5	4.11
Ralston Purina ProPlan Dog Adult***	7.5	4.41
Ralston Purina ProPlan Dog Growth***	7.5	4.54
Ralston Purina ProPlan Dog Lite***	7.5	3.63
Ralston Purina ProPlan Dog Performance***	7.5	4.71
Ralston Purina ProPlan Natural Turkey & Barley Dog***	8.5	4.39
Ralston Purina ProPlan Natural Turkey & Barley Puppy***	7.5	4.54
Royal Canin A2***	8.0	4.72
Royal Canin AD32***	8.0	4.67
Royal Canin HE30***	8.0	4.44
Royal Canin LA23***	8.0	4.00
Royal Canin Maxi Adult 1 (GR26)***	8.0	4.50
Royal Canin Maxi Adult 2 (SGR26)***	8.0	4.49
Royal Canin Maxi Junior (AGR36)***	8.0	4.30
Royal Canin MD25***	8.0	4.24
Royal Canin Medium Adult 1 (AM25)***	8.0	4.26
Royal Canin Medium Adult 2 (SM25)***	8.0	4.27
Royal Canin Medium Junior (AM32)***	8.0	4.67
Royal Canin Mini Adult 1 (PR27)***	8.0	4.48
Royal Canin Mini Adult 2 (SPR27)***	8.0	4.49
Royal Canin Mini Junior (APR33)***	8.0	4.67
Royal Canin ST35***	8.0	4.48
Sunshine Nurture Adult Light**	7.3	3.24
Sunshine Nurture Bite Sized Adult**	8.1	4.03
Sunshine Nurture Lamb Meal & Rice**	6.5	3.87
Sunshine Nurture Puppy Growth**	6.2	4.10
Superior Brands Dr. Ballard Great Performance**	6.0	4.50
Superior Brands Dr. Ballard Oven Bake Lamb & Rice**	6.0	4.22
Superior Brands Dr. Ballard Oven Bake Maintenance**	6.0	4.25
Superior Brands Dr. Ballard Oven Bake Senior**	7.8	4.07

Prot	Fat	NFE	Fiber	Ca	P	K	Na	Mg
29.1	18.8	40.5	3.4	1.62	1.31	0.55	0.38	0.11
32.6	20.3	34.6	4.3	1.67	1.38	0.71	0.55	0.13
19.4	9.8	59.0	4.7	1.27	1.04	0.48	0.30	0.11
27.4	13.8	49.0	3.0	1.54	1.38	0.55	0.39	0.10
16.8	7.8	62.0	4.4	0.83	0.67	0.48	0.32	0.12
32.2	20.0	38.4	1.8	1.64	1.36	0.69	0.55	0.12
29.0	15.6	44.3	3.5	1.56	1.43	0.71	0.55	0.11
34.8	21.7	32.1	2.7	1.20	0.90	na	0.22	0.10
28.3	17.4	44.0	2.7	1.70	1.20	na	0.22	0.09
30.4	19.6	39.1	2.7	0.98	0.82	na	0.22	0.10
19.6	6.5	64.1	2.2	1.60	1.40	na	0.22	0.09
34.8	22.8	31.0	2.7	1.20	0.87	na	0.22	0.09
23.3	7.8	57.4	3.9	1.14	1.11	0.70	0.42	0.22
19.7	10.6	60.9	3.0	0.90	0.87	0.56	0.39	0.17
28.9	13.6	49.8	1.7	1.09	0.98	0.55	0.48	0.11
33.8	17.8	39.4	1.8	1.43	1.20	0.76	0.61	0.13
29.7	19.2	43.2	1.6	1.30	1.07	0.54	0.40	0.11
31.9	19.5	40.0	2.0	1.35	1.03	0.65	0.51	0.11
15.9	9.6	65.5	2.7	1.41	0.95	0.58	0.37	0.09
34.1	21.6	36.9	1.5	1.12	0.83	0.63	0.43	0.09
27.7	18.0	45.7	2.2	1.20	0.96	0.66	0.33	0.07
30.7	19.6	40.5	2.1	1.24	1.03	0.65	0.38	na
38.0	27.2	23.9	2.7	1.74	1.09	0.65	0.38	0.11
34.8	21.7	33.2	2.7	1.47	1.09	0.71	0.38	0.11
32.6	17.4	39.1	2.7	1.41	0.98	0.65	0.38	0.11
25.0	8.7	55.4	4.4	1.30	0.65	0.65	0.38	0.11
28.3	17.4	45.1	2.7	1.20	0.76	na	na	0.11
28.3	17.4	44.8	3.3	0.87	0.65	na	na	0.11
39.1	15.2	34.2	2.7	1.63	1.25	0.71	0.38	0.11
27.2	13.0	49.5	2.7	1.20	0.92	0.65	0.38	0.11
27.2	13.0	50.2	2.7	1.20	0.92	0.71	0.38	0.11
27.2	13.0	48.9	3.3	0.87	0.65	na	na	0.11
34.8	21.7	33.2	2.7	1.47	1.09	0.71	0.38	0.11
29.4	17.4	43.5	2.7	1.30	0.98	na	na	0.11
29.4	17.4	43.7	3.3	0.87	0.65	na	na	0.11
35.9	21.7	32.1	2.7	1.47	1.09	na	na	0.11
38.0	27.2	23.9	2.7	1.74	1.09	0.65	0.38	0.11
26.0	9.6	43.3	15.2	1.07	0.80	0.58	0.40	0.15
29.0	16.0	47.1	2.6	0.89	0.71	0.66	0.34	0.12
25.9	14.5	49.3	3.3	1.41	0.95	0.47	0.34	0.17
31.9	18.0	41.5	2.5	1.28	0.81	0.62	0.36	0.11
34.1	26.0	31.3	1.2	1.84	1.19	0.79	0.36	0.09
29.1	19.3	44.6	1.3	1.10	0.89	0.78	0.40	0.11
26.9	20.1	45.7	1.2	1.22	0.90	0.79	0.38	0.09
25.7	15.9	51.9	1.6	0.72	0.66	0.77	0.37	0.08

(Continued on next page.)

Table 8. Nutrient content of dry specialty brand dog foods. This table lists products with a large market share and for which information is available. Specialty brands are premium and super premium foods traditionally sold in pet stores, pet superstores or veterinary hospitals* (Continued.).

Foods	H_2O	kcal/g
Wafcol Puppy Food[†]	6.5	3.71
Wafcol Vegetarian[†]	6.5	3.66
Wafcol Veteran[†]	6.5	3.74
Average nutrient content of dry specialty brand dog foods	**7.9**	**4.17**

Key: H_2O = water, Prot = protein, NFE = nitrogen-free extract, Fiber = crude fiber, Ca = calcium, P = phosphorus, K = potassium, Na = sodium, Mg = magnesium, na = not available
*Nutrients, except for moisture, are expressed as % dry matter. Energy density is expressed in kcal metabolizable energy/g dry matter. Kilocalories of metabolizable energy are either declared by the manufacturer or calculated based on modified Atwater values: protein = 3.5 kcal/g, fat = 8.5 kcal/g, NFE = 3.5 kcal/g. To convert to kJ, multiply kcal by 4.184.

Prot	Fat	NFE	Fiber	Ca	P	K	Na	Mg
28.5	8.6	49.2	5.3	2.35	1.6	na	0.51	0.05
21.4	8.6	55.6	8.6	1.4	0.96	na	0.47	0.03
17.1	8.6	62.0	5.3	1.6	1.5	na	0.21	0.04
28.1	**16.3**	**45.1**	**3.3**	**1.33**	**1.01**	**0.70**	**0.42**	**0.12**

**Analyses conducted in December 1995 and June 1996 by an independent laboratory (Woodson-Tenent Laboratories, Inc., Des Moines, IA, USA).
***Manufacturers' data.
†Data from Henderson AJ, ed. The Henston Small Animal Veterinary Vade Mecum, 15th ed. Peterborough, UK: Veterinary Business Development Ltd, 1996; 90-95.

Nutrient Profiles of Treats and Snacks

CLINICAL IMPORTANCE

In the United States, up to 86% of dog owners and about 68% of cat owners regularly give treats to their animals. In 1995, canine treat sales totaled $516 million in the United States. Grocery stores carry an average of 57 and mass merchandisers about 54 different items. Treats for dogs represent 9% of the total pet food market in the United States, and about 6.4% in the United Kingdom.

Giving several treats per day can markedly affect a pet's daily cumulative caloric intake and alter the nutritional adequacy of the diet. A 7-kg (15.4-lb) adult dog receiving two snacks per day may become 30% overweight in one year, unless comparable calories are withheld from the daily food intake. A growing large-breed puppy may double its calcium intake if it receives many daily snacks. Thus, a dietary history should always include specific questions about treats, including the brand, size and number of treats given daily. This information is critical when specific nutritional problems must be ruled out (e.g., skeletal problems in large-breed puppies, adverse reactions to food and obesity). When a dog or cat is affected by disease, such as urolithiasis, diabetes mellitus, heart failure or renal disease, dietary restrictions should be respected rigorously and treats should be carefully selected or even banned. Of particular concern are caloric density, protein, calcium, phosphorus, sodium, magnesium and urinary pH.

Recommendations are based on key nutritional factors established in individual chapters in this text and are listed on a dry matter basis. Although concerns may arise about one particular nutrient, the entire nutritional profile should be considered before recommending a particular treat. The ideal treat should match the nutrient profile of the food or diet recommended for the lifestage or disease of the pet. (See respective chapters.) The treat should be considered inappropriate if one of those nutrients is significantly above or below the generally recommended level. Treats may be inappropriate if all or several of the nutrients of concern are even marginally different from recommended levels. Because the veterinary health care team has no control over how many treats the owner is going to give, it is best to base recommendations on the nutrient profile given as % dry matter.

Not all treats are complete foods; therefore, the number of treats given daily and the size of the treat are variables that influence the recommendation. A treat that only weighs 0.25 g, but has a high phosphorus level (% dry matter), may not really be contraindicated in chronic renal disease, if only

Table 1. Commercial treats for cats on a nutrient content/treat basis. This list represents products with a large market share and for which information is available.

Treats	Weight (g)	kcal ME*	DM (g)	Prot (g)	Fat (g)	Ca (mg)	P (mg)	K (mg)	Na (mg)	Mg (mg)
Heinz Pounce Treats for Cats with Tuna**	1.5	3.7	1.0	0.3	0.1	1.35	11.3	16.4	9.2	1.2
Pedigree Whiskas Kitbits with Rabbit and Turkey***	0.4	1.1	0.3	0.1	0	7.8	5.2	2	5	0.24
Ralston Purina Whisker Lickin's Kluckers Chicken-Flavored**	1.1	2.9	0.7	0.3	0.1	9.0	10.7	5.9	6.2	0.7
Thomas Cork Hartz Treatsters Chicken Flavour***	0.25	0.8	0.2	0.1	0	6.1	3.8	2.1	2.1	0.38

Key: kcal = kilocalories, ME = metabolizable energy, DM = dry matter, Prot = protein, Ca = calcium, P = phosphorus, K = potassium, Na = sodium, Mg = magnesium. To convert to kJ, multiply kcal x 4.184.
*Metabolizable energy (kcal) calculated using modified Atwater values: protein = 3.5, fat = 8.5, nitrogen-free extract = 3.5 kcal/g dry matter.
**Analyses conducted in December 1995 by an independent laboratory (Woodson-Tenent Laboratories, Inc, Des Moines, IA, USA).
***Analyses conducted in June 1996 by an independent laboratory (Woodson-Tenent Laboratories, Inc, Des Moines, IA, USA).

one or two treats are given daily. However, a treat weighing 20 g with the same phosphorus level may be unacceptable for the same animal. Likewise, veterinarians should also consider the size of the dog. It may be appropriate to give two treats to a large dog, but the same number and size of treats may be unacceptable for small dogs.

Table 2. Commercial treats for dogs on a nutrient content/treat basis. This list represents products with a large market share and for which information is available.

Treats	Weight (g)	kcal ME*	DM (g)	Prot (g)	Fat (g)	Ca (mg)	P (mg)	K (mg)	Na (mg)	Mg (mg)
Friskies Chew-eez Beef Hide Treats, Original and Beef Basted Flavors†	20	60.6	17.4	14.7	1.0	116	na	na	120	na
Heinz 100% Natural Treats Ken-L Ration**	7.6	25.9	7.1	1.3	0.5	17	49	48	41	15
Heinz Meaty Bone Medium**	18.2	64.2	16.8	2.3	1.8	8	55	71	116	20
Heinz Original Jerky Treats**	6.8	21.9	5.1	2	1.3	35	35	71	140	7
Heinz Pup-Peroni Jerky Snack Sticks Ken-L Ration**	6.6	21.3	5.2	1.8	1	55	44	59	73	7
Heinz Snausages Beef Flavor Ken-L Ration**	6.6	17.4	4.7	1.5	0.6	61	46	98	44	7
Hill's Prescription Diet Treats†	4	11.8	3.7	0.67	0.3	22	17	28	4	5
Hill's Science Diet Canine Adult Treats†	5	16.9	4.6	1.1	0.5	29	29	28.5	10.5	4
Hill's Science Diet Canine Growth Treats†	5	16.8	4.6	1.2	0.5	63.5	52.5	28	15	4.4
Hill's Science Diet Canine Senior Treats†	5	16.2	4.6	0.8	0.4	30	28.5	28	7	4.5
Hill's Science Diet Canine Light Treats†	5	14.8	4.6	0.8	0.3	29	28.5	38	11	6.6
Iams Biscuits Original Large†	42.5	137	39.1	11.1	2.8	510	383	259	64	55
Iams Biscuits Original Small†	9.5	31	8.7	2.5	0.6	114	86	58	14	12
Nabisco Milk-Bone Small**	5	16.4	4.7	1.1	0.3	71	54	30	21	7
Pedigree Chum Markies***	11.8	38.2	10.6	1.5	1.3	426	148	32	79	na
Pedigree Chum Maxi Biscrok***	19.5	61.5	18.1	2.5	1.3	646	184	53	123	na
Pedigree Chum Rask Large†	92	269	76.4	29.2	2.3	na	na	na	na	na
Pedigree Chum Rask Medium†	64	187	53.1	20.3	1.6	na	na	na	na	na
Pedigree Chum Schmackos with Beef***	8.4	25.5	7.1	2.5	0.9	73	64	97	173	na

(Continued on next page.)

Table 2. Commercial treats for dogs on a nutrient content/treat basis. This list represents products with a large market share and for which information is available (Continued.).

Treats	Weight (g)	kcal ME*	DM (g)	Prot (g)	Fat (g)	Ca (mg)	P (mg)	K (mg)	Na (mg)	Mg (mg)
Pedigree Chum Tandem with Beef and Chicken***	3.9	9.7	2.8	0.7	0.3	68	16	20	36	na
Ralston Purina Beggin Strips Original Bacon Flavor**	10.3	28.8	8	1.7	0.6	44	49	33	65	11
Ralston Purina Biscuits Medium**	10.2	36.8	9.6	2.5	1.3	114	106	91	29	21
Ralston Purina Bonz Steak Bone Shaped**	20.4	65.7	18.2	3.1	1.3	241	155	86	53	24
Spillers Latz BONZO Kleine Lieblings-Knochen***	11.4	38.4	10.4	1.9	1.1	148	55	51	109	na
Stewart Fiber Formula Dog Large Biscuits†	28.4	73.6	25.5	4.1	0.8	176	105	na	20	na
Stewart Fiber Formula Dog Medium Biscuits†	10.1	26.2	9.1	1.5	0.3	63	37	na	7	na
Stewart Lambmeal & Rice Formula Dog Biscuits†	1.5	5.2	1.35	0.12	0.11	4	2.6	na	3	7.6
Veterinary Medical Diets Medi-Treats†	4.5	13.5	4.2	0.7	0.3	30	14.5	na	7.7	na

Key: kcal = kilocalories, ME = metabolizable energy, DM = dry matter, Prot = protein, Fiber = crude fiber, Ca = calcium, P = phosphorus, K = potassium, Na = sodium, Mg = magnesium, na = not available. To convert to kJ, multiply kcal × 4.184.
*Metabolizable energy (kcal) calculated using modified Atwater values: protein = 3.5, fat = 8.5, nitrogen-free extract = 3.5 kcal/g dry matter.
**Analyses conducted in December 1995 by an independent laboratory (Woodson-Tenent Laboratories, Inc, Des Moines, IA, USA).
***Analyses conducted in June 1996 by an independent laboratory (Woodson-Tenent Laboratories, Inc, Des Moines, IA, USA).
†Information published by manufacturer.

TREATS/SNACKS

Table 3. Nutrient content of human foods often used as treats for dogs and cats in the United States on a nutrient content/treat basis.*

Foods	Weight (g)	kcal ME**	DM (g)	Prot (g)	Fat (g)	Fiber (g)	Ca (mg)	P (mg)	K (mg)	Na (mg)	Mg (mg)
Cheese											
American cheese (pasteurized)	28.4	97.2	17	6.6	8.5	na	198	219	22.7	322	na
Cheddar cheese	28.4	105	17.9	7.1	9.1	na	213	136	23.2	199	12.8
Fruit											
Apples (not pared)	34.5	17.4	5.1	0.1	0.1	0.2	2.1	3.5	38	0.3	2.8
Apples (pared)	34.5	18.3	5.4	0.1	0.2	0.3	2.4	3.5	38	0.3	1.7
Raisins	0.3	0.7	0.2	0	0	0	0.2	0.3	1.9	0.1	0.1
Ice cream											
Ice cream (10% fat)	66.5	119	24.5	3	7.1	na	99.1	76.5	120	41.9	9.3
Ice cream (12% fat)	66.5	128	25.2	2.7	8.3	na	81.8	65.8	74.5	26.6	9.3
Meat											
Bologna	23	64.4	10.1	2.8	6.3	na	1.6	29.4	52.9	299	na
Frankfurter	66.5	189	29.5	8.3	18.4	na	4.7	88.4	146	731	na
Others											
Peanut butter	5.3	30.2	5.2	1.5	2.6	0.1	3.3	21.6	35.5	32.2	na
Popcorn	6	20.9	5.8	0.8	0.3	0.1	0.7	24.4	na	0.2	na
Popcorn (with fat and salt)	9	37.8	8.7	0.9	2	0.2	0.7	19.4	na	175	na
Potato chips	28.4	149	27.8	1.5	11.3	0.5	11.3	39.4	320	284	na
Pretzels	28.4	95.6	27.1	2.8	1.3	0.1	6.2	37.1	36.9	476	na

Key: kcal = kilocalories, ME = metabolizable energy, DM = dry matter, Prot = protein, Fiber = crude fiber, Ca = calcium, P = phosphorus, K = potassium, Na = sodium, Mg = magnesium, na = not available. To convert to kJ, multiply kcal x 4.184.
*Based on references 4 and 10, Small Animal Clinical Nutrition, 4th ed., p 1091.
**Metabolizable energy (kcal) calculated using modified Atwater values: protein = 3.5, fat = 8.5, nitrogen-free extract = 3.5 kcal/g dry matter.

Table 4. Nutrient content of human foods often used as treats for dogs and cats in Europe on a nutrient content/treat basis.*

Foods	Weight (g)	kcal ME**	DM (g)	Prot (g)	Fat (g)	Fiber (g)	Ca (mg)	P (mg)	K (mg)	Na (mg)	Mg (mg)
Biscuits											
Boudoir (lady finger)	6	20	5.5	0.4	0.2	0	1.4	5.3	6.3	3.0	0.2
Dry biscuit (average)	5	20	5	0.3	0.6	0	1.6	4.2	8	11.6	0.9
Petit beurre	8	30	7.9	0.5	0.8	0	na	na	na	na	na
Speculoos	8	35	7.9	0.4	1.5	0	1.3	5.3	6.4	27	1.2
Speculoos (all wheat)	8	33	7.9	0.5	1.3	0.3	2	17	18.4	22.4	6.4
Cheese											
Camembert	10	26	4.9	2.5	2.0	0	60	30	11	79	na
Gouda	10	32	5.8	2.5	2.7	0	92	52	12	60	3
Gruyère	10	40.5	6.8	2.9	3.5	0	90	60	10	50	na
Farmesan	10	38	7.4	4	2.7	0	100	90	12.5	100	na
Chocolate											
Fondant (bitter)	25	109	25	1.4	6.8	1.7	12.5	37.5	100	2.5	25
Milk chocolate (bittersweet)	25	118	25	2	7.4	na	50	50	100	25	13.8
White chocolate	25	122	25	2	7.3	na	na	na	na	na	na
Fruit											
Dates (dried)	7	18.6	5.7	0.14	0	0.5	5	4.3	42.5	0.7	3.5
Raisins (dried)	0.3	0.6	0.2	0	0	0	0.1	0.3	2	0.1	0
Meat (cold cuts)											
Pâté de foie (liver pâté)	10	33	5.1	1.2	3.3	0	2	15	8	80	1
Salami (one slice)	10	43	6.8	1.9	4.2	0	2.8	16.7	30	152	1.9
Saucisson (average) (one slice)	10	36	5.4	1.4	3.6	0	2.5	15.4	18.3	102	0.7

Key: kcal = kilocalories, ME = metabolizable energy, DM = dry matter, Prot = protein, Fiber = crude fiber, Ca = calcium, P = phosphorus, K = potassium, Na = sodium, Cl = chloride, Mg = magnesium, na = not available. To convert to kJ, multiply kcal x 4.184.
*Based on references 6 and 8, Small Animal Clinical Nutrition, 4th ed., p 1091.
**Metabolizable energy (kcal) calculated using modified Atwater values: protein = 3.5, fat = 8.5, nitrogen-free extract = 3.5 kcal/g dry matter.

Assisted-Feeding Techniques

SYRINGE FEEDING

Patients are often fed a liquid or moist homogenized food by syringe on a short-term basis. For dogs, the syringe tip is placed between the molar teeth and cheek with the head held in a normal or lowered position (**Figure 1**). For cats, the syringe tip is placed between the four canine teeth (**Figure 2**). The patient may choose to swallow the liquid or allow it to flow out of the mouth by gravity. Because some patients refuse to swallow liquids or food, force-feeding may increase the risk of aspiration. Syringe feeding should be discontinued if the patient does not swallow food voluntarily.

Figure 1. Syringe-feeding technique for administering liquids and moist homogenized foods to dogs.

Figure 2. Syringe-feeding technique for administering liquids and moist homogenized foods to cats.

NASOESOPHAGEAL TUBE PLACEMENT

Nasoesophageal tubes are generally used for three to seven days, but are occasionally used longer (weeks). Polyurethane tubes (6 to 8 Fr., 90 to 100 cm) with or without a weighted tip and silicone feeding tubes (3.5 to 10 Fr., 20 to 105 cm) may be placed in the caudal esophagus or stomach. The preferred placement of all tubes originating cranial to the stomach is in the caudal esophagus to minimize gastric reflux and subsequent esophagitis. An 8-Fr. tube will pass through the nasal cavity of most dogs; a 5-Fr. tube is more comfortable in cats.

Measure from the nasal planum along the side of the animal to the caudal margin of the last rib to determine the length of tube to be inserted (**Figure 3**). The length of tube to insert is approximately three-fourths of this distance. Wrap a piece of adhesive tape around the tube at this point or mark how far the tube should be inserted with an indelible marker. Tape also provides a tab to secure the tube. Place a few drops of topical anesthetic (2% lidocaine or 0.5% proparacaine) into the nostril and tilt the patient's head upward for a few seconds to desensitize the nose. Lubricate the tip of the tube with a water-soluble lubricant or 2 to 5% lidocaine ointment/jelly before passage.

To pass the tube, direct the tip in a caudoventral, medial direction into the ventrolateral aspect of the external nares. The head should be held in a normal static position. As soon as the tip of the catheter reaches the medial septum at the floor of the nasal cavity in dogs, push the external nares dorsally, which opens the ventral meatus, ensuring passage of the tube into the oropharynx (**Figure 4**). To aid passage, lift the proximal end of the tube as the nose is pushed upward (**Figure 4**). Because cats lack a well-developed alar fold, insert the tube initially in a ventromedial direction and continue directly into the oropharynx. Insert the tube until the adhesive tape tab or indelible mark is reached (**Figure 5**).

To evaluate proper tube placement, 3 to 15 ml of sterile water or saline solution may be injected through the tube and the animal checked for coughing (**Figure 6**). A lateral radiograph may be taken of the neck to confirm placement of the tube in the caudal esophagus (i.e., over the larynx). After confirmation of position, secure the tube with either sutures or glue. Secure the first tape tab to the skin just lateral to the external nares. Secure a second tape tab to the skin on the dorsal nasal midline, just rostral to the level of the eyes. An Elizabethan collar helps prevent inadvertent removal of the tube (**Figure 7**).

Complications of nasoesophageal intubation include epistaxis, lack of tolerance of the procedure and inadvertent removal of the tube by the animal. Incidence of tube removal by the animal has been reported to be as high as 50% even with use of collars. Nasoesophageal tubes should not be used in vomiting patients or those with respiratory disease.

ASSISTED FEEDING

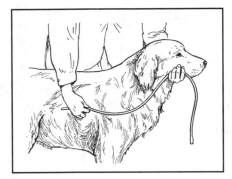

Figure 3. The length of tube to insert is determined by measuring from the nose to the last rib. Marking the tube at three-quarters of the distance between the last rib and the nose will place the end of the tube in the caudal esophagus. This location is marked with an indelible marker or a piece of adhesive tape. Tape can also serve as a suture tab to secure the tube.

Figure 4. The external nares are pushed dorsally and the proximal end of the tube is lifted to facilitate passage of the tube into the ventral nasal meatus.

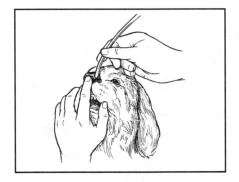

Figure 5. The tube is inserted until the indelible mark or adhesive tape tab is reached. Sutures or glue are used to secure the tape tab to the skin.

Figure 6. A test injection of sterile water or saline solution is made to ensure proper tube placement.

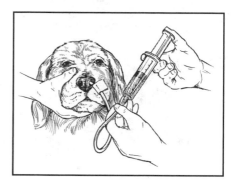

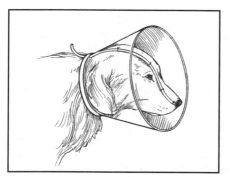

Figure 7. Securing the tube at several locations by suturing or gluing tape tabs to the skin and applying collars will help decrease inadvertent removal the tube by the animal.

PHARYNGOSTOMY TUBE PLACEMENT

In some instances, a pharyngostomy tube is used to bypass the nose and mouth of an animal requiring nutritional support (e.g., in cases of facial trauma) or when nasoesophageal tubes are not tolerated. Esophagostomy tubes or gastrostomy tubes placed percutaneously have largely replaced pharyngostomy tubes.

The animal is anesthetized, intubated and positioned in lateral recumbency. The area caudal to the mandible on either side is prepared for aseptic surgery. A 14- to 18-Fr. polyvinylchloride tube is premeasured as described in **Figure 3**, except that the tube exit site will be caudal to the mandible.

With the mouth held open with a speculum, palpate the hyoid apparatus with one finger. The tube exit site must be carefully planned to avoid interfering with the laryngeal opening and epiglottic movement. The tube should exit as far caudally and dorsally along the lateral pharyngeal wall as possible.

ASSISTED FEEDING

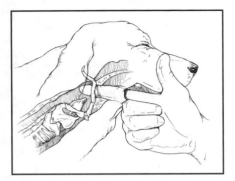

Figure 8. A finger is used to find the optimal exit site for the pharyngostomy tube. The tube should exit the pharyngeal wall as far as caudally and dorsally as possible.

Locate the hyoid apparatus with the finger inside the mouth and push outward from the pharyngeal wall at the selected exit site (**Figure 8**). Alternatively, forceps can be used to push the pharyngeal wall laterally. Locate the carotid artery with the finger placed in the oral cavity. Make a 1-cm skin incision over the bulging pharyngeal wall. Bluntly tunnel caudally through the tissues from outside to inside using long, curved forceps. Blunt dissection prevents injury to nearby nerves, and the carotid artery and jugular vein. Grasp one end of the feeding tube with forceps so it exits through the dissection site while the other end is passed down the esophagus (**Figure 9**). Secure the tube to the skin with tape and sutures.

Complications include airway obstruction, tube displacement, damage to cervical nerves and blood vessels and infection at the exit site. Placing the tube exit site caudal to the hyoid apparatus or use of large diameter tubes is much more likely to result in airway obstruction or aspiration (**Figure 10**). The animal should be observed frequently for signs of respiratory embarrassment as it recovers from anesthesia. Frequent inspection and cleansing of the tube entrance/exit site help prevent skin infection. These tubes should not be used in vomiting patients or those with respiratory disease.

Figure 9. Proper placement of a pharyngostomy tube with the tube exiting dorsal and caudal to the larynx.

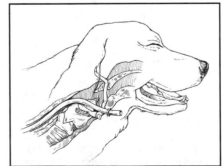

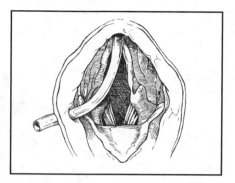

Figure 10. Inappropriate positioning of a pharyngostomy tube, as depicted here, causes the tube to course over the laryngeal opening and to interfere with movement of the epiglottis. This placement can lead to serious airway obstruction. The tube should exit the pharyngeal wall as far caudally and dorsally as possible.

ESOPHAGOSTOMY TUBE PLACEMENT

Several techniques have been described for mid-cervical placement of esophagostomy tubes in dogs and cats. A light plane of general anesthesia is used to facilitate placement of the esophagostomy tube. The entire lateral cervical region from the ventral midline to near the dorsal midline is clipped and aseptically prepared for surgery.

In one technique, appropriately sized, curved Kelly, Carmalt or similar forceps is inserted into the pharynx and then into the proximal cervical esophagus. The tip of the forceps is turned laterally and pressure is applied in an outward direction, thereby tenting up the cervical tissue so that the instrument tip can be seen and palpated externally. A small skin incision, just large enough to accommodate the feeding tube, is made over the tip of the forceps. The tip of the forceps is forced bluntly through the esophagus of small dogs and cats. A deeper incision is made to allow passage of the tip of the forceps through the esophagus of larger dogs. The tube is premeasured as described in **Figure 3** so that the distal tip resides in the mid to caudal esophagus. The distal tip of the tube is grasped with forceps, pulled into the esophagus and out through the mouth, turned around and redirected into the esophagus. The tube is then secured with tape and sutures. A light circumferential bandage containing antibiotic-impregnated gauze is then placed at the exit site.

Another technique uses a percutaneous feeding tube applicator (**Figure 11**).

Reported complications of tube esophagostomy for nutritional support include tube displacement due to vomiting or scratching by the animal and skin infection around the exit site.

ASSISTED FEEDING

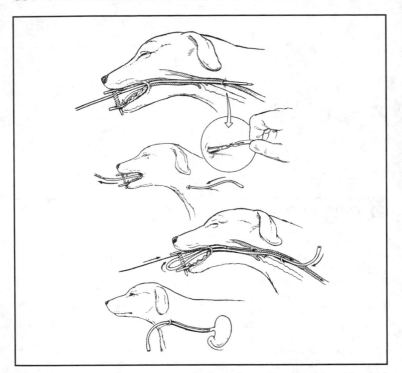

Figure 11. Insertion of a percutaneous feeding tube applicator into the midcervical esophagus. The distal tip is palpated and an incision is made through the skin and subcutaneous tissue over the tip of the applicator. The trocar is advanced through the esophageal wall and directed through the incision. The distal end of the feeding tube is secured to the eyelet of the trocar with suture material. (See also **Figure 23.**) The applicator and attached feeding tube are retracted into the esophagus and out the mouth. The feeding tube is redirected into the esophagus for final placement. A wire stylet can be inserted into the feeding tube if necessary to ease placement in the esophagus.

SURGICAL GASTROSTOMY TUBE PLACEMENT

A limited left flank celiotomy for gastrostomy tube placement provides an alternative when endoscopic or blind gastrostomy techniques are not performed. A gastrostomy tube may also be inserted when a celiotomy is performed for other reasons. General anesthesia is administered and the left flank is aseptically prepared for surgery. The prepared left paracostal area is draped and a 2- to 3-cm incision is made through the skin and subcutaneous tissue. The incision is made just caudal and parallel to the last rib, with its dorsal limit just below the ventral edge of the paravertebral epaxial musculature. The incision should be extended ventrally so that the intraperitoneal rather than the retroperitoneal space is accessed. The incision should be long enough to permit insertion of one or two fingers and a tissue forceps.

The greater curvature of the stomach is located and an Allis or Babcock tissue forceps is used to grasp and exteriorize the stomach through the incision. An assistant may pass a stomach tube and dilate the stomach with 10 to 15 ml of air/kg body weight, if difficulty is encountered locating the stomach. Exteriorizing the stomach through a small flank incision can be difficult, especially in larger, deep-chested canine breeds. The left lateral aspect of the gastric body or the caudal aspect of the fundus is selected for the ostomy site. Two pursestring sutures are placed around the selected ostomy site (**Figure 12**). A stab incision is made through the ostomy site, the tube is inserted into the stomach and the pursestring sutures are tied snugly. Tubes with 14- to 22-Fr. diameters are usually adequate.

The tube may exit the body wall through a separate stab wound or the original incision. The stomach is then fixed to the abdominal wall using a continuous suture pattern that circles the gastrostomy tube placement (**Figure 13**). After the gastropexy sutures are placed, gentle traction is applied to the external end of the tube to ensure the stomach is adjacent to the abdominal wall (**Figure 14**). The tube is secured to the skin using sutures or glue.

Potential risks include wound infection, peritonitis and dehiscence. Pressure necrosis of the stomach may also occur if excessive tension is placed on the pursestring sutures. Wrapping the intraperitoneal tube with the omentum should contain leakage to a localized site. A layer of greater omentum can also be placed over the ostomy site before the stab incision is made into the stomach.

Percutaneous gastrostomy tube placement with gastropexy using a large-bore stiff plastic stomach tube has also been described. This technique is less invasive than the technique described here and may be more convenient for some veterinary practitioners.

PERCUTANEOUS GASTROSTOMY TUBE PLACEMENT

There are two basic techniques for percutaneous placement of gastrostomy tubes. One technique uses an endoscope, whereas the other involves a "blind," nonendoscopic approach using a gastrostomy tube placement device or applicator. The advantages of percutaneous vs. surgical gastrostomy tube placement are ease and speed of placement, lower cost and less tissue trauma.

ASSISTED FEEDING

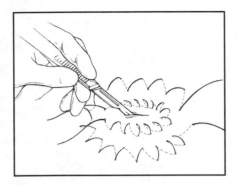

Figure 12. Two full-thickness purse-string sutures are placed concentrically around the selected gastrostomy site to help invert the stomach around the tube. A stab incision is made in the center of the suture pattern for tube placement.

Figure 13. The stomach is sutured to the abdominal wall with four preplaced mattress sutures (or a simple continuous pattern). These sutures should include the strong abdominal fascia and the gastric submucosa. Tightening the loops brings the gastric serosa and omentum snugly in contact with the peritoneum.

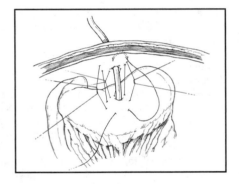

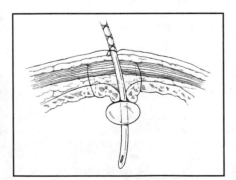

Figure 14. A mushroom-tip Pezzer catheter or one with an inflatable bulb is placed in the stomach. After the gastropexy sutures are placed, gentle traction is applied on the external end of the tube to ensure this area of the stomach is adjacent to the abdominal wall.

Percutaneous Endoscopic Gastrostomy Tubes

Percutaneous endoscopic gastrostomy (PEG) tubes are inserted with the aid of general anesthesia. The patient is placed in right lateral recumbency and an area of the left flank extending 4 to 6 inches caudal to the last rib is surgically prepared. **Figures 15** to **21** describe tube placement technique in detail. Landmarks for feeding tube placement are usually 1 to 2 cm caudal to the last rib and one-third the distance from the ventral border of the epaxial musculature to the ventral midline. Commercial 20-Fr. Pezzer catheter assembly kits are now available for small animal patients and provide cost-effective, convenient materials for PEG tube placement (**Figure 18**).

Following insertion, the tube is usually incorporated into a light bandage, with the free end brought to a convenient position for feeding. PEG tubes should be left in place for a minimum of five to seven days. Firm adhesions between the gastric serosa and the peritoneum have been reported to form within 48 to 72 hours of PEG tube placement in healthy dogs but do not reliably form in healthy cats. Adhesion formation may also be variable in undernourished animals.

The stomach should be empty when the tube is removed. Sedation or anesthesia is not generally required for tube extraction. Tubes are removed by exerting firm traction on the tube, while simultaneously applying counter-pressure around the exit site (**Figure 22**). For dogs weighing more than 10 kg, the catheter may be cut off flush with the skin, leaving the catheter tip to be passed in the feces. The resulting gastrocutaneous fistula usually heals rapidly.

Complications of PEG tube placement include vomiting, peristomal skin infection, cellulitis and pressure necrosis at the tube exit site.

Percutaneous Nonendoscopic Gastrostomy Tubes

Percutaneous gastrostomy techniques have been developed to allow convenient, cost-effective placement of feeding tubes without relying on availability of relatively expensive endoscopes. One nonendoscopic technique uses a commercial feeding tube applicator device (**Figure 23**) as previously described in the esophagostomy tube section. The other nonendoscopic technique uses a commercial gastrostomy tube placement device pressed against the stomach wall. Use of either device allows suture material to be placed through the body wall into the stomach and retrieved through the mouth, and a gastrostomy tube to be inserted as described for PEG tube placement.

ASSISTED FEEDING

Figure 15. The animal is positioned in right lateral recumbency and an endoscope is introduced. The stomach is insufflated with air so that the gastric wall comes in contact with the body wall and the spleen is displaced caudally.

Figure 16. The lighted tip of the endoscope will be seen pressing outward against the abdominal wall. A large-bore needle or an over-the-needle intravenous catheter is inserted into the stomach adjacent to the endoscope tip.

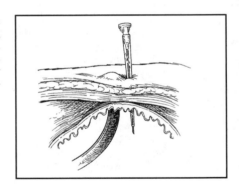

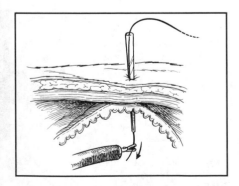

Figure 17. Nylon suture is advanced through the needle or catheter until it can be grasped with endoscopic retrieval forceps. The suture material is pulled out through the mouth as the endoscope is withdrawn.

Figure 18. Commercial 20-Fr. Pezzer catheter assembly kits provide the most convenient materials for PEG tube placement. The catheter guide is already secured to the free end of the feeding tube in commercial kits.

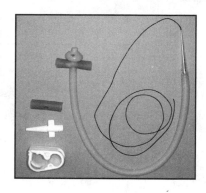

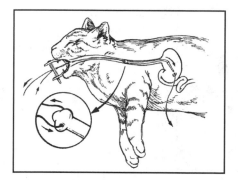

Figure 19. The lubricated catheter is drawn down the esophagus as the suture exiting the body wall is pulled. A second "safety" suture is placed through the openings in the mushroom-tip feeding tube (insert) and exits the mouth. This safety suture is used to retrieve the feeding tube from the stomach if problems occur during the placement procedure.

Figure 20. Resistance will be encountered when the catheter tip guide contacts the body wall. Steady traction and firm application of counter-pressure to the body wall will allow the guide tip to emerge through the skin (arrow). A small skin incision (2 to 3 mm) at the point of exit may help.

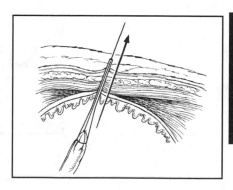

ASSISTED FEEDING

Figure 21. Gentle traction is used to bring the stomach and abdominal wall into close contact. A rubber flange is fitted down the tube and a piece of tape attached to prevent tube slippage. The tube is not usually sutured or glued to the skin. The safety suture is removed via the mouth (arrow) after the feeding tube is secured.

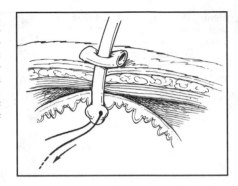

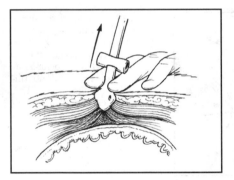

Figure 22. PEG tubes are usually removed by traction. The mushroom tip will usually collapse as it pulls through the abdominal wall. The resulting gastrocutaneous fistula usually heals rapidly.

Figure 23. A commercial gastrostomy tube applicator can be used for percutaneous nonendoscopic gastrostomy tube placement in dogs and cats. The rigid outer tube encloses a trocar that can be pushed through the stomach and abdominal wall. A suture is placed through the small hole in the trocar tip, pulled into the stomach and then pulled antegrade out through the mouth. See **Figure 11** for use of this device in esophagostomy tube placement.

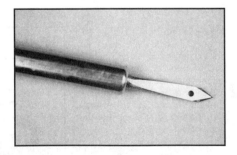

INDEX

Atlas Index